13th Biennial Conference on Classical and Quantum Relativistic Dynamics of Particles and Fields (IARD 2022)

AA007125

Journal of Physics: Conference Series Volume 2482

Prague, Czechia
5 - 9 June 2022

Editors:

Lawrence Horwitz
Petr Jizba
Gonzalo Ares de Parga
Tepper Gill
James O'Brien
Martin Land

ISBN: 978-1-7138-7453-9
ISSN: 1742-6588

Printed from e-media with permission by:

Curran Associates, Inc.
57 Morehouse Lane
Red Hook, NY 12571

proceedings.com

Some format issues inherent in the e-media version may also appear in this print version.

This work is licensed under a Creative Commons Attribution 3.0 International Licence.
Licence details: http://creativecommons.org/licenses/by/3.0/.

No changes have been made to the content of these proceedings. There may be changes to pagination and minor adjustments for aesthetics.

Printed with permission by Curran Associates, Inc. (2023)

For permission requests, please contact the Institute of Physics
at the address below.

Institute of Physics
Dirac House, Temple Back
Bristol BS1 6BE UK

Phone: 44 1 17 929 7481
Fax: 44 1 17 920 0979

techtracking@iop.org

Additional copies of this publication are available from:

Curran Associates, Inc.
57 Morehouse Lane
Red Hook, NY 12571 USA
Phone: 845-758-0400
Fax: 845-758-2633
Email: curran@proceedings.com
Web: www.proceedings.com

TABLE OF CONTENTS

Journal of Physics: Conference Series **2482** (2023) 011001 doi:10.1088/1742-6596/2482/1/011001

IARD 2022

The 13th Biennial Conference on Classical and Quantum Relativistic Dynamics of Particles and Fields

Preface

The International Association for Relativistic Dynamics was organized in February 1998 in Houston, Texas, with John R. Fanchi as president. Although the subject of relativistic dynamics has been explored, from both classical and quantum mechanical points of view, since the work of Einstein and Dirac, its most striking development has been in the framework of quantum field theory. The very accurate calculations of spectral and scattering properties, for example, of the anomalous magnetic moment of the electron and the Lamb shift in quantum electrodynamics, and many qualitative features of the strong and electroweak interactions, demonstrate the very great power of description achieved in this framework. Yet, many fundamental questions remain to be clarified, such as the structure of classical relativistic dynamical theories on the level of Hamilton and Lagrange in Minkowski space as well as on the curved manifolds of general relativity. There, moreover, remain the important questions of the covariant classical description of systems at high energy for which particle production effects are not large, such as discussed in Synge's book, The Relativistic Gas, and in Balescu's book on relativistic statistical mechanics, and the development of a consistent single and many body relativistic quantum theory. In recent years, highly accurate telescopes and advanced facilities for computation have brought a high level of interest in cosmological problems, such as the structure of galaxies (dark matter) and the apparently anomalous expansion of the universe (dark energy). Some of the papers reported here deal with these problems, as well as other fundamental related issues.

It was for this purpose, to bring together researchers from a wide variety of fields, such as particle physics, astrophysics, cosmology, foundations of relativity theory, and mathematical physics, with a common interest in relativistic dynamics, to investigate fundamental questions of this type, that this Association was founded. The second meeting took place in 2000 at Bar Ilan University in Israel, the third, in 2002, at Howard University in Washington, D.C., and the fourth, in 2004, in Saas Fee, Switzerland. Subsequent meeting took place in 2006 at the University of Connecticut Storrs, in 2008 at Aristotle University of Thessalonica, in 2010 at National Dong Hwa University, Hualien, Taiwan, in 2012 at the Galileo Galilei Institute for Theoretical Physics (GGI) in Florence, in 2014 as the University of Connecticut Storrs, Connecticut, in 2016 at Jožef Stefan Institute in Ljubljana, Slovenia, and in 2018 in Mérida, Yucatán, Mexico, under the sponsorship of the Instituto Politécnic Nacional. The 2020 meeting, planned for Czech Technical University in Prague, was successfully held online at the height of the Covid-19 pandemic, and the physical meeting in Prague was delayed to 2022.

The 2022 meeting forms the basis for the Proceedings that are recorded in this issue of the Journal of Physics: Conference Series. Along with the work of some of the founding and newer but already much engaged members of the Association, we were fortunate to have lecturers from application areas that provided strong challenges for further developments in quantum field theory, cosmological problems, and in the dynamics of systems subject to accelerations and the effects of general relativity. Topics treated in this issue include studies

Content from this work may be used under the terms of the Creative Commons Attribution 3.0 licence. Any further distribution of this work must maintain attribution to the author(s) and the title of the work, journal citation and DOI.
Published under licence by IOP Publishing Ltd

in general relativity and astrophysics, relativistic dynamics and electrodynamics, quantum theory and particles, and foundations of relativistic dynamics.

This first physical meeting of the Covid-19 era took place 6 - 9 June at Czech Technical University in Prague, as originally planned for 2020. The meeting was divided into seven plenary sessions over four days. As a result of continued travel restrictions in some areas, a small number of talks were delivered by videoconferencing. The papers presented in this volume represent extensions and refinements to the conference talks, building on feedback and discussions associated with the lectures. We once again express our gratitude to Czech Technical University, and especially the local conference chair Petr Jizba, for their generous hospitality.

We thank the **Scientific Advisory Committee** for their invaluable guidance and advice:

Stephen Adler (Institute for Advanced Study)
Itzhak Bars (University of Southern California)
Gordon Baym (University of Illinois)
Fred Cooper (Los Alamos National Laboratory)
Bei-Lok Hu (University of Maryland)
Werner Israel (University of Victoria)
E.V. Shuryak (Brookhaven National Laboratory)
L.S. Shulman (Clarkson University)
William Unruh (University of British Columbia)
Luca Lusanna (National Institute for Nuclear Physics, INFN)
Benoit Famaey (Observatoire Astronomique de Strasbourg, CNRS)

International Organizing Committee and Editorial Board of the proceedings

Lawrence Horwitz, Tel Aviv University (Israel)
larry@tauex.tau.ac.il

Petr Jizba, Czech Technical University in Prague (Czech Republic)
petr.jizba@fjfi.cvut.cz

Gonzalo Ares de Parga, Instituto Politécnico Nacional (Mexico)
gonzalo@esfm.ipn.mx

Tepper Gill, Howard University (USA)
tgill@howard.edu

James O'Brien, Springfield College (USA)
jobrien7@springfieldcollege.edu

Martin Land, Hadassah College Jerusalem (Israel)
martin@hac.ac.il

Journal of Physics: Conference Series **2482** (2023) 011001 doi:10.1088/1742-6596/2482/1/011001

Dedication

This volume is dedicated to Professor Lawrence P. Horwitz, a founder of the International Association for Relativistic Dynamics and a guiding light in relativistic classical and quantum physics for many decades. In addition to his groundbreaking work in the foundations of relativity theory, relativistic statistical mechanics, relativistic quantum mechanics, electrodynamics and gauge theories, and quantum mechanics in curved spacetime, Larry is a dedicated teacher beloved by his many students. He is generous with his time, often working side-by-side with graduate students in a spirit of collegiality and joint adventure. Continuing a long career of individual and collaborative publications, Larry graciously took time away from his latest book to review each of the papers submitted to the Proceedings of IARD 2022. He is an inspiration to younger researchers, both for the far-reaching scope of his theoretical vision and the open mindedness and seriousness he brings to discussing new ideas and innovative approaches.

In May 2022, a conference in honor of Larry's 92nd birthday was organized by Asher Yahalom at Ariel University. Larry was an ardent supporter of Ariel University since its inception, demonstrating his admirable character in the face of strong opposition. And thus, after a serving at first class academic institutions both in Israel and abroad, he continues to volunteer as a research associate at Ariel University. He has led a group of physicists and physics students to many achievements in theoretical physics, described in numerous published papers in some of the best journals. It was there that I (Asher Yahalom) was fortunate to meet Larry and start a collaboration which lasted for many years and continuous to this very day.

It was thus appropriate for Ariel University reward him by supporting a conference in honor of this 92nd birthday (originally intended for his 90th but delayed due to the Covid plague). About 90 participants were present in the conference. Many are Larry's students (and students of his students) and collaborators, joining together from first class academic institutions worldwide. The excellent lectures presented demonstrate the wide spectrum of Larry's interest, ranging from the gigantic (general relativity) to the minute (quantum mechanics and quantum field theory). It was a unique pleasure and privilege for us to be able to contribute to the organization of this conference.

On behalf of the organizers of IARD 2022 and the Ariel conference, it is an honor and pleasure to dedicate this volume to Larry Horwitz.

Martin Land
Asher Yahalom

Journal of Physics: Conference Series **2482** (2023) 011002 doi:10.1088/1742-6596/2482/1/011002

Peer Review Statement

All papers published in this volume have been reviewed through processes administered by the Editors. Reviews were conducted by expert referees to the professional and scientific standards expected of a proceedings journal published by IOP Publishing.

- **Type of peer review:** Single Anonymous
- **Conference submission management system:** Morressier
- **Number of submissions received:** 33
- **Number of submissions sent for review:** 28
- **Number of submissions accepted:** 26
- **Acceptance Rate (Submissions Accepted / Submissions Received × 100):** 78.8
- **Average number of reviews per paper:** 2
- **Total number of reviewers involved:** 7
- **Contact person for queries:**
  Name: Martin Land
  Email: martin@hac.ac.il
  Affiliation: Hadassah Academic College

Content from this work may be used under the terms of the Creative Commons Attribution 3.0 licence. Any further distribution of this work must maintain attribution to the author(s) and the title of the work, journal citation and DOI.
Published under licence by IOP Publishing Ltd

Journal of Physics: Conference Series **2482** (2023) 012001 doi:10.1088/1742-6596/2482/1/012001

$N \geq 2$ particles in the framework of *TeVeS*

L.P. Horwitz

School of Physics, Tel Aviv University, Ramat Aviv 69978, Israel
Department of Physics, Bar Ilan University, Ramat Gan 52900, Israel
Department of Physics, Ariel University, Ariel 40700, Israel

E-mail: larry@tauex.tau.ac.il

Abstract. About 40 years ago, since no viable candidate for "dark matter" was discovered, M. Milgrom and J.Bekenstein introduced a non-covariant modification of gravitational theory (MOND) to account for the anomalous rotation curves of galaxies. Bekenstein and Sanders then developed a relativistically covariant form of the theory, called *TeVeS*, involving scalar and tensor fields, which accounts well for the rotation curves as well for the observed lensing of background radiation around galaxies without the introduction of "dark matter". The dynamical behavior of a particle in such a gravitational theory has been recently discussed by Horwitz, Gershon and Schiffer. In this paper we study the dynamics of the $N \geq 2$ particle problem in the framework of the *TeVeS* theory.

1. Introduction

It has been found that galactic rotation curves are not consistent with Newtonian dynamics as they should be for the large interstellar distances in galaxies [1]. To account for this discrepancy, the idea of "dark matter" was introduced [2], which permits a reasonable fit to the data. However, no viable candidate for this matter has been found. Milgrom [3] consequently introduced the idea of modifying the gravitational force, with a construction called MOND, which was phenomenologically quite successful, even with relatively universal modification functions, without the need for "dark matter".

There was, however, no relativistically covariant formulation of the MOND theory until Bekenstein [4] and Bekenstein and Sanders [5] introduced a theory with scalar and tensor fields called *TeVeS*, successfully accounting for the galactic rotation curves and for the gravitational lensing of light by galaxies from background stars.

Horwitz, Gershon and Schiffer [6] then showed that the *TeVeS* theory can have a simple Hamiltonian formulation consistent with Einstein relativity through a conformal transformation, for which the potential in the Hamiltonian plays the role of the scalar field of *TeVeS*.

In this paper we discuss the dynamics of $N \geq 2$ particles in this framework. We argue that each particle moves in an independent MOND type gravitational field, but the particles interact in the presence of a Bekenstein-Sanders gauge field [6] which must be non-Abelian. We then show that the resulting geometry can be represented in a Kaluza-Klein structure [7].

Content from this work may be used under the terms of the Creative Commons Attribution 3.0 licence. Any further distribution of this work must maintain attribution to the author(s) and the title of the work, journal citation and DOI.
Published under licence by IOP Publishing Ltd

2. Hamiltonian for $N \geq 2$
In Stueckelberg's approach to relativistic dynamics [8], a Hamiltonian of the form ($p^\mu = \eta^{\mu\nu} p_\nu$, M a dimensional parameter)

$$K = \frac{p_\mu p^\mu}{2M} + V(x) \tag{2.1}$$

is defined, for a particle represented as an event, on the manifold x^μ, p_ν, with $V(x)$ a scalar function, invariant under the Lorentz transformations of special relativity. The (Hamilton) equations of motion are then given by

$$\dot{p}_\mu = -\frac{\partial K}{\partial x^\mu} \tag{2.2}$$

and

$$\dot{x}^\mu = \frac{\partial K}{p_\mu}. \tag{2.3}$$

The dynamics of this theory is discussed extensively in [9].

It has recently been shown that this theory can be embedded in the space-time manifold of general relativity [10].

The many body problem [11] can be treated by taking a Hamiltonian of the form

$$K = \sum_{i=1}^{N} \left\{ \frac{p_{i\mu} g^{\mu\nu} p_{i\nu}}{2M_i} + V(x_i) \right\}, \tag{2.4}$$

where we assume that each particle is affected independently by the potential energy field.

Let us first consider the one-particle case, for which

$$K = \frac{1}{2M} p_\mu g^{\mu\nu}(x) p_\nu + V(x), \tag{2.5}$$

where $V(x)$ is a world scalar field. From the canonical equations,

$$\dot{p}_\mu = -\frac{1}{2M} \frac{\partial g^{\lambda\sigma}(x)}{\partial x^\mu} p_\lambda p_\sigma - \frac{\partial V}{\partial x^\mu} \tag{2.6}$$

and

$$\dot{x}^\mu = \frac{1}{M} g^{\mu\sigma}(x) p_\sigma. \tag{2.7}$$

We then have, with the help of the identity

$$g_{\beta\sigma} \frac{\partial g^{\beta\gamma}}{\partial x^\mu} g_{\gamma\omega} = -\frac{\partial g_{\sigma\omega}}{\partial x^\mu}, \tag{2.8}$$

and its inverse, the geodesic equation, in the presence of a scalar field,

$$\ddot{x} = -\Gamma^\sigma{}_{\lambda\nu} \dot{x}^\lambda \dot{x}^\nu - \frac{1}{M} \frac{\partial V}{\partial x^\nu} g^{\sigma\nu}, \tag{2.9}$$

where

$$\Gamma^\mu_{\lambda\sigma} = \frac{1}{2} g^{\mu\nu} \left\{ \frac{\partial g_{\nu\sigma}}{\partial x^\lambda} + \frac{\partial g_{\nu\lambda}}{\partial x^\sigma} - \frac{\partial g_{\lambda\sigma}}{\partial x^\nu} \right\}$$

is the usual affine connection form [12].

For the $N \geq 2$ problem, we assume the additive form

$$K = K_1 + K_2 + \ldots K_N, \tag{2.10}$$

where

$$K_i = \frac{1}{2M_i} p_{i\mu} g^{\mu\nu}(x_i) p_{i\nu} + V(x_i). \tag{2.11}$$

We now make a conformal transformation on each of the K_i, eliminating the scalar function, by defining

$$\hat{g}^{\mu\nu}(y_i) = \frac{K_i}{K_i - V(x_i)} g^{\mu\nu}(x_i) \equiv e^{-2\phi(x_i)} g^{\mu\nu}(x_i). \tag{2.12}$$

Writing this as

$$\hat{g}^{\mu\nu}(y_i) = F(y_i) g^{\mu\nu}(x_i) \tag{2.13}$$

and setting

$$\delta {y_i}^{\mu} = F(y_i) \delta {x_i}^{\mu} \tag{2.14}$$

enables us to solve for $y_i(x)$ and the inverse [13]; however the functions $y_i(x)$ do not cover the whole space time; they are defined only in a patch in the neighborhood of each of the particles. This is sufficient to define, as for the one particle case, local Einstein geodesics on the submanifold patches $\{y_i\}$. We shall write, with this understanding, $\hat{g}^{\mu\nu}(y_i) \equiv \tilde{g}^{\mu\nu}(x_i)$ in the following.

Therefore, a many body Hamiltonian can, in the presence of a world scalar field, acquire a Milgrom type modification of Einstein gravity.

3. Many Body Dynamics in *TeVeS* and Kaluza Klein Structure

We have seen so far that the theory for many particles can treat the particles as essentially independent. However, we shall see that *TeVeS* can introduce a non-Abelian gauge field (like the Yang-Mills field [14]) which couples the dynamics of the many body system. In order to achieve a generalized *TeVeS* theory to account for lensing and gravitational redshifts, Bekenstein and Sanders [5] introduced a timelike vector field $\mathcal{U}_\mu$ satisfying $\mathcal{U}_\mu \mathcal{U}^\mu = -1$, with $\mathcal{U}^\mu = g^{\mu\nu}\mathcal{U}_\nu$. In a previous work [6] we have considered this field as a gauge field, necessarily non-Abelian to preserve the normalization condition under gauge transformations. With an additional scalar field $\Phi(x,\tau)$, we write the corresponding Hamiltonian for the many body theory

$$K = \sum_i^N K_i = \sum_i^N \left\{ \frac{1}{2M_i} (p_{i\mu} - \epsilon\mathcal{U}_\mu(x_i,\tau)) g^{\mu\nu}(x_i,\tau)(p_{i_\nu} - \epsilon\mathcal{U}_\nu(x_i,\tau)) + \Phi(x_i,\tau) \right\}, \tag{3.1}$$

defined on an N body Hilbert space with direct product basis. . Our analysis here is independent of whether the functions are symmetrized (Bose-Einstein statistics) or antisymmetrized (Fermi-Dirac statistics). The scalar field arises as a gauge field to compensate for the τ-derivative of a scalar gauge phase, depending linearly on τ, of each of the wave functions in the Stueckelberg-Schrödinger equation, in addition to the $\mathcal{U}$ fields carrying compensation for the derivative action of $p_{i\mu}$ on the non-Abelian local phase transformations of the wave functions, *i.e.*, under gauge transformations, each factor $\psi_i(x_i)$ obtains a coefficient $U(x_i,\tau) = U_0(x_i)e^{-i\varphi(x_i)\tau}$, with $U_0(x_i,\tau)$ inducing the non-Abelian gauge. It is easily seen that the $\Phi(x_i,\tau)$ additively can compensate for the Abelian part of the gauge transformation. The space-time derivative (p_μ) introduces, necessarily, a τ dependence into the gauge fields. As for the one particle case, one can eliminate the scalar potential terms by defining a modified metric. The Hamilton equations, however, assure that the Hamiltonian remains independent of τ.

For the non-Abelian part (we suppress the zero subscript),

$$\mathcal{U}'(x_i,\tau) = U(x_i)\mathcal{U}_\nu(x_i,\tau)U^{-1}(x_i,\tau) - \frac{i}{\epsilon}\frac{\partial U(x_i,\tau)}{\partial x_i{}^\mu}U^{-1}(x_i,\tau). \quad (3.2)$$

As for the Yang-Mills fields, we define the gauge and locally Lorentz covariant force field at each x,

$$f_{\mu\nu}(x,\tau) = \frac{\partial \mathcal{U}_\mu}{\partial x^\nu} - \frac{\partial \mathcal{U}_\nu}{\partial x^\mu} + i\epsilon[\mathcal{U}_\mu, \mathcal{U}_\nu]. \quad (3.3)$$

The motion of each particle gives rise to a current which, by means of the Yang-Mills propagator [14], generalized to the manifold (a procedure somewhat simplified for weak gravitational fields), contributes to this field in linear superposition. This structure will be studied in a succeeding publication.

We now define a Kaluza-Klein Hamiltonian (with $\tilde{g}^{\mu\nu}(x_i,\tau)$ of the same form as in (2.12) with $V(x_i)$ replaced by $\Phi(x_i.\tau)$)

$$K_{KK} = \sum_i \tilde{g}^{\mu\nu}(x_i,\tau)p_{i\mu}p_{i\nu} \quad (3.4)$$

with the (τ-dependent) Kaluza-Klein [6,15] metric, at each x_i

$$\tilde{g}^{\mu\nu}(x_i,\tau) = e^{-2\phi(x_i)}(g^{\mu\nu}(x_i,\tau) + \mathcal{U}^\mu(x_i,\tau)\mathcal{U}^\nu(x_i,\tau)) - e^{2\phi(x_i)}\mathcal{U}^\mu(x_i,\tau)\mathcal{U}^\nu(x_i,\tau) \quad (3.5)$$

for which the Hamiltonian has the form

$$K_{KK} = \Sigma_i e^{-2\phi(x_i)} g^{\mu\nu}(x_i,\tau)p_\mu p_\nu - 2\sinh 2\phi(x_i)(\mathcal{U}^\mu(x_i,\tau)p_\mu)^2. \quad (3.6)$$

With the definition, on $5D$,

$$g^{AB}(x_i,\tau) = \begin{pmatrix} \tilde{g}^{\mu\nu}(x_i,\tau) & \mathcal{U}^\nu(x_i,\tau) \\ \mathcal{U}^\mu(x_i,\tau) & \tilde{g}^{55}(x_i,\tau) \end{pmatrix}, \quad (3.7)$$

we have

$$g^{AB}(x_i,\tau)p_A p_B = \tilde{g}^{\mu\nu}(x_i.\tau)p_{i\mu}p_{i\nu} + 2p_{i5}(p_{i\mu}\mathcal{U}^\mu(x_i,\tau)) + p_{i5}{}^2 g^{55}(x_i,\tau). \quad (3.8)$$

Taking

$$p_{i5} = -\frac{(p_{i\mu}\mathcal{U}^\mu(x_i,\tau))}{g^{55}(x_i,\tau)}(1 \pm \sqrt{1 - 2g^{55}(x_i,\tau)\sinh 2\Phi(x_i,\tau)}, \quad (3.9)$$

the Kaluza-Klein theory provides an equivalent dynamics generated by

$$K_{KK} = \sum_i \frac{1}{2M_i} g_{AB}(x_i,\tau)p_{iA}p_{iB}. \quad (3.10)$$

4. Conclusions

In this study, we have shown that a many body dynamics can be formulated in the framework of the *TeVeS* theory of Bekenstein and Sanders. In this formulation, the Bekenstein-Sanders tensor fields as well as the scalar field arise from the requirement of gauge invariance of the Stueckelberg-Schrödinger equation embedded in the manifold of gravitation.

Acknowledgments
This paper is dedicated to the memory of J.D. Bekenstein (Z"L).

References
[1] A.Bosma, The Astrophysical Journal **86** 1825 (1981).
[2] Jan Oort, Bull. Astro. Inst. Neth. **1**,133 (1922); Fritz Zwicky, Helvetica Physica Acta **6**, 110 (1933)
[3] M. Milgrom, Astrophysical Jour. **270**, 384 (1983).
[4] J.D. Bekenstein and M. Milgrom, Astrophys. Jour.**286**, (1984).
[5] J.D. Bekenstein, Phy. Rev. D **70**,083509 (2004); R.H. Sanders, Astrophys. Jour.**480**, 492 (1997).
[6] L.P. Horwitz, A. Gershon and M.Schiffer, Found. of Physics **41**, 14(2011).
[7] A. Gershon and L.P. Horwitz, Jour. Math. Phys.**50**,102704 (2009).
[8] E.C.G. Stueckelberg, Helv. Phys. Acta,**14**, 588 (1941).
[9] Lawrence P. Horwitz, *Relativistic Quantum Mechanics*, Fundamental Theories of Physics 180, Springer, Dordrecht (2015).
[10] L.P. Horwitz, European Phys. Jour. Plus **134**, 313 (2019).
[11] Lawrence P. Horwitz and Rafael Arshansky, *Relativistic Many-Body Theory and Statistical Mechanics*, IOP Concise Physics, Morgan & Claypool, Bristol (2018).
[12] S. Weinberg, *Cosmology and Gravitation*, Wiley, New York (1972).
[13] L.P. Horwitz, A. Yahalom,J. Levitan, and M. Lewkovitz, Frontiers of Physics **12** 124501 (2017).
[14] C.N. Yang and R.L. Mills, Phys. Rev. **96**, 191 (1954).
[15] A. Gershon and L.P. Horwitz, Jour. Math. Phys. **50**, 102704 (2009).

Parametrized Relativistic Quantum Theory in Curved Spacetime

John R. Fanchi

Department of Engineering, Texas Christian University, Fort Worth, Texas, 76129 USA
j.r.fanchi@tcu.edu

Abstract. The purpose of this paper is to present a parametrized relativistic quantum theory (pRQT) in curved spacetime. The formulation of pRQT in curved spacetime is developed and applied to free particle motion in flat and curved spacetime. It provides a theory for calculating probability amplitudes in curved spacetime. Unlike other formulations of parametrized relativistic dynamics (pRD), this work assumes that the metric tensor does not depend on the invariant evolution parameter.

1. Introduction

Contemporary formulations of parametrized relativistic dynamics (pRD) in curved spacetime [1-3] assume that the metric tensor is dependent on an invariant evolution parameter. Does the metric tensor depend on an invariant evolution parameter? Here we consider the possibility that a formulation of relativistic quantum theory in curved spacetime can be developed without requiring that the metric tensor be dependent on an invariant evolution parameter.

The primary difference between this work and previous pRD publications is the assumption that the metric tensor of parametrized relativistic quantum theory (pRQT) in curved spacetime does not depend on the invariant evolution parameter. A formulation of parameter independent general relativity was developed by Fanchi [4, 5] using a variational principle that is analogous to the Hilbert-Palatini variational principle. It shows that metrics from conventional general relativity can be used in the pRQT formulation.

The formulation of pRQT in curved spacetime is presented in Section 2. It provides equations for determining probability amplitudes. The definition of expectation value associated with pRQT in curved spacetime is discussed in Section 3. The formulation is applied in Section 4 to a free scalar particle in flat spacetime. Section 4 includes a discussion of the mass of a particle and shows that both tardyons and tachyons have real mass in pRQT [6]. The pRQT formulation is then applied to a particle moving in Schwarzschild metric in Section 5. The problem posed in Section 5 is analyzed in Section 6 for a free scalar particle at a large distance from a radially symmetric mass.

2. pRQT in Curved Spacetime

A quantum theory of the behavior of a single spinless particle moving in curved spacetime is developed here using a formulation that is analogous to an extension of the four-space formulation in flat spacetime [5, 7, 8]. We begin by assuming that the behavior of the particle can be represented by a scalar conditional probability density $\rho(x|s)$. Spacetime coordinates are denoted by the spacetime contravariant four-vector x with μ^{th} component x^μ and $\mu = 0,1,2,3$. Index 0 signifies the geometric time component $x^0 = ct$ and indices 1, 2, 3 signify space components. The scalar s is an invariant evolution parameter that conditions $\rho(x|s)$. The metric tensor of the physical system is $g_{\mu\nu}$.

Content from this work may be used under the terms of the Creative Commons Attribution 3.0 licence. Any further distribution of this work must maintain attribution to the author(s) and the title of the work, journal citation and DOI.
Published under licence by IOP Publishing Ltd

According to probability theory, $\rho(x|s)$ must be positive definite and normalizable. The positive definite requirement is expressed in terms of the Born representation

$$\rho(x|s) = \Psi^*(x,s)\Psi(x,s) \geq 0 \tag{2.1}$$

where Ψ is the probability amplitude and Ψ^* is the complex conjugate of Ψ. The probability amplitude is specified to within a gauge transformation. It can be written as

$$\Psi(x,s) = \sqrt{\rho(x|s)}\, e^{i\xi(x,s)} \tag{2.2}$$

where the real scalar function ξ represents the gauge transformation.

The normalization condition is

$$\int_D \rho(x|s)\sqrt{|g(x)|}dx = 1, dx = d^4x \tag{2.3}$$

where D denotes the spacetime volume and $|g(x)|$ is the absolute value of the determinant of the metric tensor $g_{\mu\nu}$. Substituting Equation (2.1) into (2.3) gives

$$\int_D \Psi^*(x,s)\Psi(x,s)\sqrt{|g(x)|}dx = 1 \tag{2.4}$$

The probability $\Psi^*(x,s)\Psi(x,s)\sqrt{|g(x)|}dx$ is the probability that the system will be found in the volume $\sqrt{|g(x)|}dx$ in the neighborhood of spacetime point x at evolution parameter s. The normalization condition implies that the particle can be observed somewhere in space at some point in geometric time t; it does not require the particle to exist for all geometric time.

Field equations that exhibit a "minimal coupling" electromagnetic interaction are derived from the above assumptions and probability conservation. Conservation of probability in curved spacetime is expressed as an extension of the parametrized continuity equation in flat spacetime that was used to construct the Stueckelberg equation in flat spacetime [5, 7, 8]. The parametrized continuity equation in curved spacetime is

$$\frac{\partial \rho}{\partial s} + (\rho V^\mu)_{;\mu} = 0 \tag{2.5}$$

where V^μ is contravariant four-velocity, ρV^μ is probability flux, and $(\rho V^\mu)_{;\mu}$ is the divergence

$$(\rho V^\mu)_{;\mu} = \frac{\partial}{\partial x^\mu}(\rho V^\mu) + \Gamma^\mu_{\nu\mu}(\rho V^\nu) \tag{2.6}$$

The tensor $\Gamma^\mu_{\nu\mu}$ is the contracted form of the Christoffel symbol of the second kind

$$\Gamma^\sigma_{\nu\mu} = \frac{1}{2}g^{\sigma\kappa}\left(\frac{\partial g_{\kappa\nu}}{\partial x^\mu} + \frac{\partial g_{\mu\kappa}}{\partial x^\nu} - \frac{\partial\partial g_{\nu\mu}}{\partial x^\kappa}\right) \tag{2.7}$$

for metric tensor $g_{\nu\mu}$. A similar continuity equation was discussed by Pavšič [9] (Chapter 1).

The divergence $(\rho V^\mu)_{;\mu}$ of a contravariant four-vector ρV^σ is the contracted form of the covariant derivative of the contravariant four-vector

$$(\rho V^\sigma)_{;\mu} = \frac{\partial}{\partial x^\mu}(\rho V^\sigma) + \Gamma^\sigma_{\nu\mu}(\rho V^\nu) \tag{2.8}$$

The Christoffel symbol of the second kind $\Gamma^\sigma_{\nu\mu}$ is also known as the affine connection. The contracted form of $\Gamma^\sigma_{\nu\mu}$ is

$$\Gamma^\mu_{\nu\mu} = \frac{1}{2}g^{\mu\kappa}\frac{\partial g_{\mu\kappa}}{\partial x^\nu} \tag{2.9}$$

Substituting Equation (2.9) into (2.6) gives

$$(\rho V^\mu)_{;\mu} = \frac{\partial}{\partial x^\mu}(\rho V^\mu) + \left(\frac{1}{2}g^{\mu\kappa}\frac{\partial g_{\mu\kappa}}{\partial x^\nu}\right)(\rho V^\nu) \tag{2.10}$$

The four-velocity V^μ can be written as

$$V^{\mu}(x,s) = \varepsilon\left[\varepsilon_1 \frac{\partial \xi(x,s)}{\partial x_{\mu}} + \varepsilon_2 A^{\mu}(x,s)\right] \tag{2.11}$$

where $\varepsilon, \varepsilon_1, \varepsilon_2$ are real, scalar constants, ξ is a real, scalar function, and A^{μ} is a contravariant four-vector. The correspondence to physical terms is provided later. Equation (2.11) can be written in the form

$$V^{\mu} = -\frac{i\varepsilon\varepsilon_1}{2\rho}\left[\Psi^* \frac{\partial \Psi}{\partial x_{\mu}} - \Psi \frac{\partial \Psi^*}{\partial x_{\mu}}\right] + \varepsilon\varepsilon_2 A^{\mu} \tag{2.12}$$

where the Born representation of probability density in Equation (2.1) has been used to write the term $\frac{\partial \xi}{\partial x_{\mu}}$ as

$$\frac{\partial \xi}{\partial x_{\mu}} = -\frac{i}{2\rho}\left[\Psi^* \frac{\partial \Psi}{\partial x_{\mu}} - \Psi \frac{\partial \Psi^*}{\partial x_{\mu}}\right] \tag{2.13}$$

The probability flux becomes

$$\rho V^{\mu} = -\frac{i\varepsilon\varepsilon_1}{2}\left[\Psi^* \frac{\partial \Psi}{\partial x_{\mu}} - \Psi \frac{\partial \Psi^*}{\partial x_{\mu}}\right] + \varepsilon\varepsilon_2 \Psi^* \Psi A^{\mu} \tag{2.14}$$

Substituting Equations (2.1), (2.14), and (2.6) into Equation (2.5), the continuity equation, gives

$$\begin{aligned} \frac{\partial(\Psi^*\Psi)}{\partial s} &+ \frac{\partial}{\partial x^{\mu}}\left\{-\frac{i\varepsilon\varepsilon_1}{2}\left[\Psi^* \frac{\partial \Psi}{\partial x_{\mu}} - \Psi \frac{\partial \Psi^*}{\partial x_{\mu}}\right] + \varepsilon\varepsilon_2 \Psi^* \Psi A^{\mu}\right\} \\ &+ \Gamma^{\mu}_{\nu\mu}\left\{-\frac{i\varepsilon\varepsilon_1}{2}\left[\Psi^* \frac{\partial \Psi}{\partial x_{\nu}} - \Psi \frac{\partial \Psi^*}{\partial x_{\nu}}\right] + \varepsilon\varepsilon_2 \Psi^* \Psi A^{\nu}\right\} = 0 \end{aligned} \tag{2.15}$$

Expanding Equation (2.15) and using

$$\frac{\partial \Psi^*}{\partial x^{\mu}} \frac{\partial \Psi}{\partial x_{\mu}} - \frac{\partial \Psi}{\partial x^{\mu}} \frac{\partial \Psi^*}{\partial x_{\mu}} = 0 \tag{2.16}$$

gives

$$\begin{aligned} \Psi^* \frac{\partial \Psi}{\partial s} + \frac{\partial \Psi^*}{\partial s}\Psi &- \frac{i\varepsilon\varepsilon_1}{2}\left[\Psi^* \frac{\partial}{\partial x^{\mu}} \frac{\partial \Psi}{\partial x_{\mu}} - \Psi \frac{\partial}{\partial x^{\mu}} \frac{\partial \Psi^*}{\partial x_{\mu}}\right] \\ &+ \varepsilon\varepsilon_2\left[\frac{\partial \Psi^*}{\partial x^{\mu}} \Psi A^{\mu} + \Psi^* \frac{\partial \Psi}{\partial x^{\mu}} A^{\mu} + \Psi^* \Psi \frac{\partial A^{\mu}}{\partial x^{\mu}}\right] \\ &+ \Gamma^{\mu}_{\nu\mu}\left\{-\frac{i\varepsilon\varepsilon_1}{2}\left[\Psi^* \frac{\partial \Psi}{\partial x_{\nu}} - \Psi \frac{\partial \Psi^*}{\partial x_{\nu}}\right] + \varepsilon\varepsilon_2 \Psi^* \Psi A^{\nu}\right\} = 0 \end{aligned} \tag{2.17}$$

Equation (2.17) can be rewritten in the form

$$\begin{aligned} \Psi^* &\left\{\frac{\partial \Psi}{\partial s} - \frac{i\varepsilon\varepsilon_1}{2} \frac{\partial}{\partial x^{\mu}} \frac{\partial \Psi}{\partial x_{\mu}} + \varepsilon\varepsilon_2\left[A^{\mu} \frac{\partial \Psi}{\partial x^{\mu}} + \frac{1}{2}\Psi \frac{\partial A^{\mu}}{\partial x^{\mu}}\right] - \frac{i\varepsilon\varepsilon_1}{2}\Gamma^{\mu}_{\nu\mu} \frac{\partial \Psi}{\partial x_{\nu}} \right. \\ &\left. + \frac{\varepsilon\varepsilon_2}{2}\Gamma^{\mu}_{\nu\mu}\Psi A^{\nu}\right\} \\ &+ \Psi\left\{\frac{\partial \Psi^*}{\partial s} + \frac{i\varepsilon\varepsilon_1}{2} \frac{\partial}{\partial x^{\mu}} \frac{\partial \Psi^*}{\partial x_{\mu}} + \varepsilon\varepsilon_2\left[A^{\mu} \frac{\partial \Psi^*}{\partial x^{\mu}} + \frac{1}{2}\Psi^* \frac{\partial A^{\mu}}{\partial x^{\mu}}\right] \right. \\ &\left. + \frac{i\varepsilon\varepsilon_1}{2}\Gamma^{\mu}_{\nu\mu} \frac{\partial \Psi^*}{\partial x_{\nu}} + \frac{\varepsilon\varepsilon_2}{2}\Gamma^{\mu}_{\nu\mu}\Psi^* A^{\nu}\right\} = 0 \end{aligned} \tag{2.18}$$

Multiply Equation (2.18) by $i\varepsilon_1$ and rearrange:

$$\Psi^*\left\{i\varepsilon_1\frac{\partial\Psi}{\partial s}+\frac{\varepsilon\varepsilon_1^2}{2}\frac{\partial}{\partial x^\mu}\frac{\partial\Psi}{\partial x_\mu}+i\varepsilon\varepsilon_1\varepsilon_2\left[A^\mu\frac{\partial\Psi}{\partial x^\mu}+\frac{1}{2}\Psi\frac{\partial A^\mu}{\partial x^\mu}\right]\right.$$
$$\left.+\Gamma^\mu_{\nu\mu}\left[\frac{\varepsilon\varepsilon_1^2}{2}\frac{\partial\Psi}{\partial x_\nu}+\frac{i\varepsilon\varepsilon_1\varepsilon_2}{2}\Psi A^\nu\right]\right\}$$
$$=\Psi\left\{-i\varepsilon_1\frac{\partial\Psi^*}{\partial s}+\frac{\varepsilon\varepsilon_1^2}{2}\frac{\partial}{\partial x^\mu}\frac{\partial\Psi^*}{\partial x_\mu}\right.\qquad(2.19)$$
$$-i\varepsilon\varepsilon_1\varepsilon_2\left[A^\mu\frac{\partial\Psi^*}{\partial x^\mu}+\frac{1}{2}\Psi^*\frac{\partial A^\mu}{\partial x^\mu}\right]$$
$$\left.+\Gamma^\mu_{\nu\mu}\left[\frac{\varepsilon\varepsilon_1^2}{2}\frac{\partial\Psi^*}{\partial x_\nu}-\frac{i\varepsilon\varepsilon_1\varepsilon_2}{2}\Psi^*A^\nu\right]\right\}$$

Equation (2.19) can be written in the form

$$\Psi^*\{F\}=\{F^*\}\Psi=\{\Psi^*F\}^*\qquad(2.20)$$

where

$$F=i\varepsilon_1\frac{\partial\Psi}{\partial s}+\frac{\varepsilon\varepsilon_1^2}{2}\frac{\partial}{\partial x^\mu}\frac{\partial\Psi}{\partial x_\mu}+i\varepsilon\varepsilon_1\varepsilon_2\left[A^\mu\frac{\partial\Psi}{\partial x^\mu}+\frac{1}{2}\Psi\frac{\partial A^\mu}{\partial x^\mu}\right]$$
$$+\Gamma^\mu_{\nu\mu}\left[\frac{\varepsilon\varepsilon_1^2}{2}\frac{\partial\Psi}{\partial x_\nu}+\frac{i\varepsilon\varepsilon_1\varepsilon_2}{2}\Psi A^\nu\right]\qquad(2.21)$$

Equation (2.21) can be decoupled by observing that the product Ψ^*F must be real since Ψ^*F equals its complex conjugate. We can assure that Ψ^*F is real for arbitrary Ψ if we write

$$F=W\,\Psi\qquad(2.22)$$

where W is a real scalar. Combining Equations (2.21) and (2.22) gives

$$i\varepsilon_1\frac{\partial\Psi}{\partial s}+\frac{\varepsilon\varepsilon_1^2}{2}\frac{\partial}{\partial x^\mu}\frac{\partial\Psi}{\partial x_\mu}+i\frac{\varepsilon\varepsilon_1\varepsilon_2}{2}\left[A^\mu\frac{\partial\Psi}{\partial x^\mu}+\frac{\partial A^\mu\Psi}{\partial x^\mu}\right]$$
$$+\Gamma^\mu_{\nu\mu}\left[\frac{\varepsilon\varepsilon_1^2}{2}\frac{\partial\Psi}{\partial x_\nu}+\frac{i\varepsilon\varepsilon_1\varepsilon_2}{2}\Psi A^\nu\right]=W\,\Psi\qquad(2.23)$$

where

$$i\varepsilon\varepsilon_1\varepsilon_2\left[A^\mu\frac{\partial\Psi}{\partial x^\mu}+\frac{1}{2}\Psi\frac{\partial A^\mu}{\partial x^\mu}\right]=i\frac{\varepsilon\varepsilon_1\varepsilon_2}{2}\left[2A^\mu\frac{\partial\Psi}{\partial x^\mu}+\Psi\frac{\partial A^\mu}{\partial x^\mu}\right]$$
$$=i\frac{\varepsilon\varepsilon_1\varepsilon_2}{2}\left[A^\mu\frac{\partial\Psi}{\partial x^\mu}+\frac{\partial A^\mu\Psi}{\partial x^\mu}\right]\qquad(2.24)$$

Equation (2.23) can be written in a more suggestive form by writing W as the invariant

$$W=\frac{\varepsilon\varepsilon_2^2}{2}A_\mu A^\mu+V\qquad(2.25)$$

so that Equation (2.23) becomes

$$i\varepsilon_1\frac{\partial\Psi}{\partial s}+\frac{\varepsilon}{2}\left\{\varepsilon_1^2\frac{\partial}{\partial x^\mu}\frac{\partial\Psi}{\partial x_\mu}+i\varepsilon_1\varepsilon_2\left[A^\mu\frac{\partial\Psi}{\partial x^\mu}+\frac{\partial A^\mu\Psi}{\partial x^\mu}\right]-\frac{2}{\varepsilon}\frac{\varepsilon\varepsilon_2^2}{2}A_\mu A^\mu\Psi\right\}$$
$$+\Gamma^\mu_{\nu\mu}\left[\frac{\varepsilon\varepsilon_1^2}{2}\frac{\partial\Psi}{\partial x_\nu}+\frac{i\varepsilon\varepsilon_1\varepsilon_2}{2}\Psi A^\nu\right]-V\,\Psi=0\qquad(2.26)$$

Rearranging Equation (2.26) gives

$$i\varepsilon_1\frac{\partial\Psi}{\partial s}=\frac{\varepsilon}{2}\left\{-\varepsilon_1^2\frac{\partial}{\partial x^\mu}\frac{\partial\Psi}{\partial x_\mu}-i\varepsilon_1\varepsilon_2\left[A^\mu\frac{\partial\Psi}{\partial x^\mu}+\frac{\partial A^\mu\Psi}{\partial x^\mu}\right]+\varepsilon_2^2A_\mu A^\mu\Psi\right.$$
$$\left.-\Gamma^\mu_{\nu\mu}\left[\varepsilon_1^2\frac{\partial\Psi}{\partial x_\nu}+i\varepsilon_1\varepsilon_2\Psi A^\nu\right]\right\}+V\,\Psi\qquad(2.27)$$

Simplifying Equation (2.27) lets us write

$$i\varepsilon_1 \frac{\partial \Psi}{\partial s} = \frac{\varepsilon}{2}\left\{\left[\frac{\varepsilon_1}{i}\frac{\partial}{\partial x^\mu} + \varepsilon_2 A_\mu\right]\left[\frac{\varepsilon_1}{i}\frac{\partial}{\partial x_\mu} + \varepsilon_2 A^\mu\right]\Psi - \Gamma^\mu_{\nu\mu}\left[\varepsilon_1^2 \frac{\partial \Psi}{\partial x_\nu} + i\varepsilon_1\varepsilon_2 \Psi A^\nu\right]\right\} + V\Psi \tag{2.28}$$

The physical interpretation of the scalar parameters $\varepsilon, \varepsilon_1, \varepsilon_2$ in Equation (2.28) are

$$\varepsilon = \frac{1}{m}, \varepsilon_1 = \hbar, \varepsilon_2 = -\frac{e}{c} \tag{2.29}$$

The relationships in Equation (2.29) are obtained by comparing Equation (2.28) with the Klein-Gordon equation. Substituting Equation (2.29) into (2.28) gives

$$i\hbar \frac{\partial \Psi}{\partial s} = \frac{1}{2m}\left\{\left[\frac{\hbar}{i}\frac{\partial}{\partial x^\mu} - \frac{e}{c}A_\mu\right]\left[\frac{\hbar}{i}\frac{\partial}{\partial x_\mu} - \frac{e}{c}A^\mu\right]\Psi - \Gamma^\mu_{\nu\mu}\left[\hbar^2 \frac{\partial \Psi}{\partial x_\nu} - i\hbar\frac{e}{c}\Psi A^\nu\right]\right\} + V\Psi \tag{2.30}$$

Equation (2.30) is referred to here as the Stueckelberg equation in curved spacetime for a single particle.

An alternative form of Equation (2.30) is

$$i\hbar \frac{\partial \Psi}{\partial s} = \frac{1}{2m}\left\{\left[\frac{\hbar}{i}\frac{\partial}{\partial x^\mu} - \frac{e}{c}A_\mu\right]\left[\frac{\hbar}{i}\frac{\partial}{\partial x_\mu} - \frac{e}{c}A^\mu\right]\Psi - i\hbar\Gamma^\mu_{\nu\mu}\left[\frac{\hbar}{i}\frac{\partial}{\partial x_\nu} - \frac{e}{c}A^\nu\right]\Psi\right\} + V\Psi \tag{2.31}$$

Equation (2.31) can be written in the form

$$i\hbar \frac{\partial \Psi}{\partial s} = K\Psi \tag{2.32}$$

where K is the operator

$$K = \frac{1}{2m}\left\{\left[\frac{\hbar}{i}\frac{\partial}{\partial x^\mu} - \frac{e}{c}A_\mu\right]\left[\frac{\hbar}{i}\frac{\partial}{\partial x_\mu} - \frac{e}{c}A^\mu\right] - i\hbar\Gamma^\mu_{\nu\mu}\left[\frac{\hbar}{i}\frac{\partial}{\partial x_\nu} - \frac{e}{c}A^\nu\right]\right\} + V \tag{2.33}$$

A more compact form of Equation (2.33) is obtained by defining the operators

$$\pi_\sigma = \frac{\hbar}{i}\frac{\partial}{\partial x^\sigma} - \frac{e}{c}A_\sigma, \pi^\sigma = \frac{\hbar}{i}\frac{\partial}{\partial x_\sigma} - \frac{e}{c}A^\sigma \tag{2.34}$$

so that Equation (2.34) becomes

$$K = \frac{1}{2m}\{\pi_\mu \pi^\mu - i\hbar\Gamma^\mu_{\nu\mu}\pi^\nu\} + V \tag{2.35}$$

3. Expectation Values

The probabilistic formalism in Section 2 is connected to observable quantities using expectation values, a well-defined concept in probability theory. The form of the definition of expectation value that is well suited to the operators introduced in Section 2 is motivated by applying the probabilistic concept of expectation value to the calculation of two common physical quantities: four-position, and four-velocity

The expectation value of the μth component of particle four-position is

$$\langle y^\mu \rangle = \int_D \rho y^\mu \sqrt{|g(y)|}dy \tag{3.1}$$

Replacing ρ with Equation (2.1) gives the expectation value

$$\langle y^\mu \rangle = \int_D \Psi^* y^\mu \Psi \sqrt{|g(y)|}dy. \tag{3.2}$$

The rate of change of $\langle y^\mu \rangle$ with respect to evolution parameter s is obtained by differentiating Equation (3.2) with respect to s:

$$\frac{d\langle y^\mu \rangle}{ds} = \int_D y^\mu \left[\frac{\partial \rho}{\partial s}\right]\sqrt{|g(y)|}dy. \tag{3.3}$$

Journal of Physics: Conference Series **2482** (2023) 012002 doi:10.1088/1742-6596/2482/1/012002

Using the continuity equation, Equation (2.5), lets us write

$$\frac{d\langle y^\mu\rangle}{ds} = -\int_D y^\mu [(\rho V^\nu)_{;\nu}]\sqrt{|g(y)|}dy. \tag{3.4}$$

Replacing the covariant derivative $(\rho V^\nu)_{;\nu}$ with Equation (2.6) gives

$$\frac{d\langle y^\mu\rangle}{ds} = -\int_D y^\mu \left[\frac{\partial}{\partial y^\nu}(\rho V^\nu) + \Gamma^\nu_{\sigma\nu}(\rho V^\sigma)\right]\sqrt{|g(y)|}dy. \tag{3.5}$$

where ν, σ are dummy indices. Replacing $y^\mu \frac{\partial}{\partial y^\nu}(\rho V^\nu)$ with

$$y^\mu \frac{\partial}{\partial y^\nu}(\rho V^\nu) = \frac{\partial \rho V^\nu y^\mu}{\partial y^\nu} - \rho V^\nu \frac{\partial y^\mu}{\partial y^\nu} \tag{3.6}$$

lets us write

$$\frac{d\langle y^\mu\rangle}{ds} = -\left\{\int_D \left[\frac{\partial \rho V^\nu y^\mu}{\partial y^\nu} - \rho V^\nu \frac{\partial y^\mu}{\partial y^\nu}\right]\sqrt{|g(y)|}dy + \int_D y^\mu \Gamma^\nu_{\sigma\nu}(\rho V^\sigma)\sqrt{|g(y)|}dy\right\}. \tag{3.7}$$

Equation (3.7) can be written as

$$\frac{d\langle y^\mu\rangle}{ds} = -\left\{\int_D \frac{\partial \rho V^\nu y^\mu}{\partial y^\nu}\sqrt{|g(y)|}dy - \int_D \rho V^\nu \frac{\partial y^\mu}{\partial y^\nu}\sqrt{|g(y)|}dy + \int_D \Gamma^\nu_{\sigma\nu}(\rho V^\sigma y^\mu)\sqrt{|g(y)|}dy\right\}. \tag{3.8}$$

Combining the first and third terms in the curly bracketed term gives

$$\frac{d\langle y^\mu\rangle}{ds} = -\left\{\int_D \left[\frac{\partial \rho V^\nu y^\mu}{\partial y^\nu} + \Gamma^\nu_{\sigma\nu}(\rho V^\sigma y^\mu)\right]\sqrt{|g(y)|}dy - \int_D \rho V^\nu \frac{\partial y^\mu}{\partial y^\nu}\sqrt{|g(y)|}dy\right\}. \tag{3.9}$$

The bracketed term on the right-hand side of Equation (3.9) is the covariant derivative $(\rho V^\nu y^\mu)_{;\nu}$ so Equation (3.9) can be written as

$$\frac{d\langle y^\mu\rangle}{ds} = -\left\{\int_D (\rho V^\nu y^\mu)_{;\nu}\sqrt{|g(y)|}dy - \int_D \rho V^\nu \frac{\partial y^\mu}{\partial y^\nu}\sqrt{|g(y)|}dy\right\}. \tag{3.10}$$

The covariant derivative term on the right-hand side of Equation (3.10) vanishes when we use the divergence theorem

$$\int_D (\rho V^\nu y^\mu)_{;\nu}\sqrt{|g(y)|}dy = \oint_S \rho V^\nu y^\mu \sqrt{|g(y)|}dS_\nu = 0 \tag{3.11}$$

to convert the integral over the hypervolume D to an integral over the hypersurface S that encloses the hypervolume D. The hypersurface integral vanishes if we assume that probability ρ goes to zero on the hypersurface S. In this case, Equation (3.10) simplifies to the form

$$\frac{d\langle y^\mu\rangle}{ds} = \int_D \rho V^\mu \sqrt{|g(y)|}dy = \langle V^\mu\rangle \tag{3.12}$$

where we have used the Kronecker delta

$$\frac{\partial y^\mu}{\partial y^\nu} = \delta^\mu_\nu. \tag{3.13}$$

A quantum mechanical expression for $\langle V^\mu\rangle$ is obtained by writing Equation (2.14) as

$$\rho V^{\mu} = -\frac{i\varepsilon\varepsilon_1}{2}\left[\Psi^* \frac{\partial \Psi}{\partial x_\mu} - \Psi \frac{\partial \Psi^*}{\partial x_\mu}\right] + \varepsilon\varepsilon_2 \Psi^*\Psi A^{\mu}$$
$$= -\frac{i\varepsilon\varepsilon_1}{2}\left[2\Psi^* \frac{\partial \Psi}{\partial x_\mu} - \frac{\partial \Psi^*\Psi}{\partial x_\mu}\right] + \varepsilon\varepsilon_2 \Psi^*\Psi A^{\mu} \tag{3.14}$$

where

$$\Psi \frac{\partial \Psi^*}{\partial x_\mu} = \frac{\partial \Psi^*\Psi}{\partial x_\mu} - \Psi^* \frac{\partial \Psi}{\partial x_\mu} \tag{3.15}$$

has been used. Equation (3.15) can be rearranged to the form

$$\rho V^{\mu} = \Psi^* \left[-i\varepsilon\varepsilon_1 \frac{\partial}{\partial x_\mu} + \varepsilon\varepsilon_2 A^{\mu}\right]\Psi + \frac{i\varepsilon\varepsilon_1}{2}\frac{\partial \Psi^*\Psi}{\partial x_\mu} \tag{3.16}$$

Substituting Equation (3.16) into (3.12) gives

$$\langle V^{\mu}\rangle = \int_D \Psi^* \left[-i\varepsilon\varepsilon_1 \frac{\partial}{\partial x_\mu} + \varepsilon\varepsilon_2 A^{\mu}\right]\Psi\sqrt{|g(y)|}dy + \int_D \frac{i\varepsilon\varepsilon_1}{2}\frac{\partial \Psi^*\Psi}{\partial x_\mu}\sqrt{|g(y)|}dy \tag{3.17}$$

The second integral on the left-hand side of Equation (3.17) vanishes when the divergence theorem is applied and noting that ρ vanishes on the hypersurface of the hypervolume. The result is

$$\langle V^{\mu}\rangle = \int_D \Psi^* \left[-i\varepsilon\varepsilon_1 \frac{\partial}{\partial x_\mu} + \varepsilon\varepsilon_2 A^{\mu}\right]\Psi\sqrt{|g(y)|}dy. \tag{3.18}$$

Substituting the relationships in Equation (2.29) into Equation (3.18) gives

$$\langle V^{\mu}\rangle = \int_D \Psi^* \left[-i\frac{\hbar}{m}\frac{\partial}{\partial x_\mu} - \frac{e}{mc}A^{\mu}\right]\Psi\sqrt{|g(y)|}dy. \tag{3.19}$$

The bracketed term on the right-hand side of Equation (3.19) is the operator

$$\frac{\pi^{\mu}}{m} = \frac{1}{m}\left[\frac{\hbar}{i}\frac{\partial}{\partial x_\mu} - \frac{e}{c}A^{\mu}\right] \tag{3.20}$$

Substituting Equation (3.20) into (3.19) gives

$$\langle V^{\mu}\rangle = \int_D \Psi^* \left[\frac{\pi^{\mu}}{m}\Psi\right]\sqrt{|g(y)|}dy. \tag{3.21}$$

The expectation values $\langle y^{\mu}\rangle, \langle V^{\mu}\rangle$ are expressed in terms of operators in Equations (3.2) and (3.21). This suggests that a reasonable definition of the expectation value of an operator Ω should be

$$\langle \Omega\rangle = \int_D \Psi^*[\Omega\Psi]\sqrt{|g(y)|}dy. \tag{3.22}$$

where Ω operates only on Ψ. Equation (3.22) is applicable to both flat and curved spacetime.

4. Free Scalar Particle in Flat Spacetime

A free scalar particle moving in flat spacetime is modeled by Equation (2.31) with $\{V = 0, A^{\mu} = 0\}$ and the fundamental metric tensor with nonzero elements

$$g_{00} = 1 = -g_{11} = -g_{22} = -g_{33} \tag{4.1}$$

In this case Equation (2.31) becomes

$$i\hbar\frac{\partial \psi_f}{\partial s} = \frac{1}{2m}\left\{\left[\frac{\hbar}{i}\frac{\partial}{\partial x^{\mu}}\right]\left[\frac{\hbar}{i}\frac{\partial}{\partial x_{\mu}}\right] - i\hbar\Gamma^{\mu}_{\nu\mu}\left[\frac{\hbar}{i}\frac{\partial}{\partial x_{\nu}}\right]\right\}\psi_f \tag{4.2}$$

with probability amplitude ψ_f. The contracted form $\Gamma^{\mu}_{\nu\mu}$ of the Christoffel symbol of the second kind is found from Equation (2.9) to be

$$\Gamma^{\mu}_{\nu\mu} = \frac{1}{2}g^{\mu\kappa}\frac{\partial g_{\mu\kappa}}{\partial x^{\nu}} = 0 \tag{4.3}$$

for the fundamental metric tensor in Equation (4.1). Equation (4.2) simplifies to

$$i\hbar\frac{\partial\psi_f}{\partial s} = K_{flat}\psi_f \tag{4.4}$$

where K_{flat} is the operator

$$K_{flat} = \frac{1}{2m}\left[\frac{\hbar}{i}\frac{\partial}{\partial x^\mu}\right]\left[\frac{\hbar}{i}\frac{\partial}{\partial x_\mu}\right] \tag{4.5}$$

in flat spacetime. Combining Equations (4.4) and (4.5) gives the Stueckelberg equation for the free particle

$$i\hbar\frac{\partial\psi_f}{\partial s} = -\frac{\hbar^2}{2m}\frac{\partial}{\partial x^\mu}\frac{\partial}{\partial x_\mu}\psi_f = -\frac{\hbar^2}{2m}\partial_\mu\partial^\mu\psi_f \tag{4.6}$$

Equation (4.6) has the general solution

$$\Psi_f(x,s) = \int \psi_{f\kappa}(x,s)\,dk_f = \int \eta_{f\kappa}\,exp\left[i\frac{\hbar\beta_f(k_f)}{2m}s + ik_{f\mu}x^\mu\right]dk_f \tag{4.7}$$

where the integral is over energy-momentum, $\eta_{f\kappa}$ denotes normalization coefficients, and

$$\beta_f(k_f) = k_{f\mu}k_f^\mu \tag{4.8}$$

A wave packet can be constructed using Equations (4.6) to (4.8).

4.1. The Meaning of Mass

The expectation value of the four-velocity of the free particle is

$$\langle V_f^\mu\rangle = \frac{d\langle x_f^\mu\rangle}{ds} = \frac{\langle \hbar k_f^\mu\rangle}{m} = \frac{\langle p_f^\mu\rangle}{m} \tag{4.9}$$

Integrating Equation (4.9) from s to $s+\delta s$ gives

$$\delta\langle x_f^\mu\rangle = \frac{\langle p_f^\mu\rangle}{m}\delta s \tag{4.10}$$

Equation (4.10) is the expectation value of spacetime position of the free particle. The observable world-line of the free particle is given by the inner product

$$\delta\langle x_{f\mu}\rangle\delta\langle x_f^\mu\rangle = \frac{\langle p_{f\mu}\rangle\langle p_f^\mu\rangle}{m^2}(\delta s)^2 \tag{4.11}$$

Solving for m^2 gives

$$m^2 = \frac{\langle p_{f\mu}\rangle\langle p_f^\mu\rangle}{\delta\langle x_{f\mu}\rangle\delta\langle x_f^\mu\rangle}(\delta s)^2 \tag{4.12}$$

Equation (4.12) applies to both timelike (bradyonic) motion and spacelike (tachyonic) motion. We can see this by observing that in parametrized relativistic dynamics, the invariant evolution parameter increases monotonically so that $\delta s > 0$. On the other hand, the terms $\langle p_{f\mu}\rangle\langle p_f^\mu\rangle$, $\delta\langle x_{f\mu}\rangle\delta\langle x_f^\mu\rangle$ can be either timelike $\langle p_{f\mu}\rangle\langle p_f^\mu\rangle > 0$, $\delta\langle x_{f\mu}\rangle\delta\langle x_f^\mu\rangle > 0$ or spacelike $\langle p_{f\mu}\rangle\langle p_f^\mu\rangle < 0$, $\delta\langle x_{f\mu}\rangle\delta\langle x_f^\mu\rangle < 0$. Consequently, m^2 is positive for both timelike and spacelike motion because negative signs associated with spacelike motion cancel. Free tachyons in parametrized relativistic dynamics have real mass.

The conventional classical theory imposes the on-shell mass constraint

$$m_{cl}^2c^2 = p_{f\mu}p_f^\mu \tag{4.13}$$

on allowed values of energy-momentum. Subscript cl indicates that mass m_{cl} refers to the conventional classical theory. An on-shell mass relationship between mass and energy-momentum exists in pRQT in the classical limit of negligible dispersion. Energy-momentum dispersion is

$$\Delta^2 p_f \equiv \left\langle p_{f\mu} p_f^\mu \right\rangle - \langle p_{f\mu} \rangle \left\langle p_f^\mu \right\rangle \tag{4.14}$$

Substituting Equation (4.14) into (4.11) gives

$$\delta\langle x_{f\mu} \rangle \delta\left\langle x_f^\mu \right\rangle = \frac{\left\langle p_{f\mu} p_f^\mu \right\rangle - \Delta^2 p_f}{m^2} (\delta s)^2 \tag{4.15}$$

In the classical limit, dispersion is negligible and we find that m^2 is statistically related to independent variables energy and momentum by

$$m^2 c^2 = \left\{ \frac{c^2 (\delta s)^2}{\delta\langle x_{f\mu} \rangle \delta\left\langle x_f^\mu \right\rangle} \right\} \left\langle p_{f\mu} p_f^\mu \right\rangle \tag{4.16}$$

We have inserted the term c^2 on both sides of Equation (4.16) so that the bracketed term is dimensionless. In the classical limit, the magnitude of the bracketed term is one, but we need to account for the sign of $\delta\langle x_{f\mu} \rangle \delta\left\langle x_f^\mu \right\rangle$, so Equation (4.16) becomes

$$m^2 c^2 = \frac{\left\langle p_{f\mu} p_f^\mu \right\rangle}{sgn\left\{\delta\langle x_{f\mu} \rangle \delta\left\langle x_f^\mu \right\rangle\right\}} \geq 0 \tag{4.17}$$

where "sgn" says that we are accounting for the sign of the bracketed term. Equation (4.17) applies to both timelike and spacelike free particles. It says that m may be interpreted as on-shell mass for a free particle moving at any speed.

5. Free Scalar Particle in Schwarzschild Metric

The equation for a free scalar particle with mass m moving in the gravitational field of a spherically symmetric mass M can be modeled by Equation (2.31) with $\{V = 0, A^\mu = 0\}$ and the Schwarzschild metric [10] (Chapter 7), [11] (Chapter 8), [12] (Chapter 9). In this case Equation (2.31) becomes

$$i\hbar \frac{\partial \psi_{SM}}{\partial s} = K_{SM} \psi_{SM} = \frac{1}{2m} \left\{ \left[\frac{\hbar}{i} \frac{\partial}{\partial y^\mu} \right] \left[\frac{\hbar}{i} \frac{\partial}{\partial y_\mu} \right] - i\hbar \Gamma^\mu_{\nu\mu} \left[\frac{\hbar}{i} \frac{\partial}{\partial y_\nu} \right] \right\} \psi_{SM} \tag{5.1}$$

where the operator K_{SM} for the Schwarzschild metric is

$$K_{SM} = \frac{1}{2m} \left\{ \left[\frac{\hbar}{i} \frac{\partial}{\partial y^\mu} \right] \left[\frac{\hbar}{i} \frac{\partial}{\partial y_\mu} \right] - i\hbar \Gamma^\mu_{\nu\mu} \left[\frac{\hbar}{i} \frac{\partial}{\partial y_\nu} \right] \right\} \tag{5.2}$$

The contravariant four-vector for the Schwarzschild metric is $y^\mu = \{ct, r, \theta, \phi\}$ where $\{r, \theta, \phi\}$ are spherical coordinates. The invariant interval between points y^μ and $y^\mu + dy^\mu$ is

$$ds^2 = g_{\mu\nu} dy^\mu dy^\nu \tag{5.3}$$

where the nonzero elements of the Schwarzschild metric are

$$g_{00} = -A(r), g_{11} = 1/A(r), g_{22} = r^2, g_{33} = r^2 sin^2\theta, A(r) = 1 - \frac{r_S}{r} \tag{5.4}$$

The term r_S is the Schwarzschild radius

$$r_S = \frac{2MG_N}{c^2} \tag{5.5}$$

and G_N is Newton's gravitational constant. Nonzero elements of the contravariant metric tensor $g^{\mu\nu}$ of the Schwarzschild metric $g_{\mu\nu}$ in Equation (5.4) are the inverse values

$$g^{00} = -\frac{1}{A(r)}, g^{11} = A(r), g^{22} = \frac{1}{r^2}, g^{33} = \frac{1}{r^2 sin^2\theta}, A(r) = 1 - \frac{r_S}{r} \quad (5.6)$$

Combining Equations (5.4) and (5.6) shows that $g^{\mu\nu}$ is the inverse of $g_{\mu\upsilon}$:

$$g_{\mu\sigma}g^{\sigma\nu} = \delta_\mu^\nu \quad (5.7)$$

The contracted form of the Christoffel symbol of the second kind $\Gamma_{\nu\mu}^{\mu}$ in Equation (5.1) is calculated using the relationship

$$\Gamma_{\nu\mu}^{\mu} = \frac{1}{2g}\frac{\partial g}{\partial x^\nu} \quad (5.8)$$

where the determinant of the Schwarzschild metric g is

$$\mathrm{g} = -r^4 sin^2\theta \quad (5.9)$$

Combining Equations (5.8) and (5.9) gives

$$\Gamma_{0\mu}^{\mu} = 0, \Gamma_{1\mu}^{\mu} = \frac{2}{r}, \Gamma_{2\mu}^{\mu} = 2cot\theta, \Gamma_{3\mu}^{\mu} = 0 \quad (5.10)$$

Rearranging Equation (5.1) and writing the term $\Gamma_{\nu\mu}^{\mu}\left[\frac{\partial}{\partial y_\nu}\right]$ as $\Gamma_{\nu\mu}^{\mu}\left[g^{\nu\sigma}\frac{\partial}{\partial y^\sigma}\right]$ gives

$$\begin{aligned} i\hbar\frac{\partial\psi_{SM}}{\partial s} &= \frac{-\hbar^2}{2m}\left\{\frac{\partial}{\partial y^\mu}\frac{\partial}{\partial y_\mu} - \Gamma_{\nu\mu}^{\mu}\left[\frac{\partial}{\partial y_\nu}\right]\right\}\psi_{SM} \\ &= \frac{-\hbar^2}{2m}\left\{\frac{\partial}{\partial y^\mu}\frac{\partial}{\partial y_\mu} - \Gamma_{\nu\mu}^{\mu}\left[g^{\nu\sigma}\frac{\partial}{\partial y^\sigma}\right]\right\}\psi_{SM} \end{aligned} \quad (5.11)$$

Using Equations (5.6) and (5.10) in (5.11) gives

$$i\hbar\frac{\partial\psi_{SM}}{\partial s} = \frac{-\hbar^2}{2m}\left\{\frac{\partial}{\partial y^\mu}\frac{\partial}{\partial y_\mu} + \left[\frac{2}{r}A(r)\frac{\partial}{\partial r} + \frac{cot\theta}{r^2}\frac{\partial}{\partial\theta}\right]\right\}\psi_{SM} \quad (5.12)$$

The term $\frac{\partial}{\partial y^\mu}\frac{\partial}{\partial y_\mu}$ can be written in the Schwarzschild metric as

$$\begin{aligned} \frac{\partial}{\partial y^\mu}\frac{\partial}{\partial y_\mu}\psi_{SM} &= \frac{\partial}{\partial y^\mu}g^{\mu\sigma}\frac{\partial}{\partial y^\sigma}\psi_{SM} \\ &= \left\{-\frac{1}{A(r)}\frac{1}{c^2}\frac{\partial^2}{\partial t^2} + \frac{\partial}{\partial r}\left[A(r)\frac{\partial}{\partial r}\right] + \frac{1}{r^2}\frac{\partial^2}{\partial\theta^2} \right. \\ &\quad \left. + \frac{1}{r^2 sin^2\theta}\frac{\partial^2}{\partial\phi^2}\right\}\psi_{SM} \end{aligned} \quad (5.13)$$

Substituting Equation (5.13) into (5.12) gives

$$\begin{aligned} i\hbar\frac{\partial\psi_{SM}}{\partial s} = \frac{-\hbar^2}{2m}&\left\{-\frac{1}{A(r)}\frac{1}{c^2}\frac{\partial^2}{\partial t^2} + \frac{\partial}{\partial r}\left[A(r)\frac{\partial}{\partial r}\right] + \frac{1}{r^2}\frac{\partial^2}{\partial\theta^2} + \frac{1}{r^2 sin^2\theta}\frac{\partial^2}{\partial\phi^2} \right. \\ &\left. + \left[\frac{2}{r}A(r)\frac{\partial}{\partial r} + \frac{cot\theta}{r^2}\frac{\partial}{\partial\theta}\right]\right\}\psi_{SM} \end{aligned} \quad (5.14)$$

Equation (5.14) can be written in terms of Laplacian and D'Alembertian operators by rearranging terms as follows:

$$\begin{aligned} i\hbar\frac{\partial\psi_{SM}}{\partial s} = \frac{-\hbar^2}{2m}&\left\{-\frac{1}{A(r)}\frac{1}{c^2}\frac{\partial^2}{\partial t^2} + \frac{\partial}{\partial r}\left[A(r)\frac{\partial}{\partial r}\right] + \frac{2}{r}A(r)\frac{\partial}{\partial r} + \frac{1}{r^2}\frac{\partial^2}{\partial\theta^2} \right. \\ &\left. + \frac{cos\theta}{r^2 sin\theta}\frac{\partial}{\partial\theta} + \frac{1}{r^2 sin^2\theta}\frac{\partial^2}{\partial\phi^2}\right\}\psi_{SM} \end{aligned} \quad (5.15)$$

or

$$i\hbar\frac{\partial\psi_{SM}}{\partial s}=\frac{-\hbar^2}{2m}\left\{-\frac{1}{A(r)}\frac{1}{c^2}\frac{\partial^2}{\partial t^2}+\frac{1}{r^2}\frac{\partial}{\partial r}\left[r^2A(r)\frac{\partial}{\partial r}\right]+\frac{1}{r^2sin\theta}\frac{\partial}{\partial\theta}\left(sin\theta\frac{\partial}{\partial\theta}\right)+\frac{1}{r^2sin^2\theta}\frac{\partial^2}{\partial\phi^2}\right\}\psi_{SM} \tag{5.16}$$

The bracketed term in Equation (5.16) can be written in terms of the D'Alembertian operator for the Schwarzschild metric

$$\Box_{SM}=-\frac{1}{A(r)}\frac{1}{c^2}\frac{\partial^2}{\partial t^2}+\triangle_{SM} \tag{5.17}$$

where the Laplacian operator for the Schwarzschild metric is

$$\triangle_{SM}=\frac{1}{r^2}\frac{\partial}{\partial r}\left[r^2A(r)\frac{\partial}{\partial r}\right]+\frac{1}{r^2sin\theta}\frac{\partial}{\partial\theta}\left(sin\theta\frac{\partial}{\partial\theta}\right)+\frac{1}{r^2sin^2\theta}\frac{\partial^2}{\partial\phi^2} \tag{5.18}$$

Equation (5.16) becomes

$$i\hbar\frac{\partial\psi_{SM}}{\partial s}=\frac{-\hbar^2}{2m}\Box_{SM}\,\psi_{SM}=\frac{-\hbar^2}{2m}\left\{-\frac{1}{A(r)}\frac{1}{c^2}\frac{\partial^2}{\partial t^2}+\triangle_{SM}\right\}\psi_{SM} \tag{5.19}$$

In flat spacetime, $A(r)\to 1$and Equation (5.16) becomes

$$i\hbar\frac{\partial\psi_{flat}}{\partial s}=\frac{-\hbar^2}{2m}\left\{-\frac{1}{c^2}\frac{\partial^2}{\partial t^2}+\frac{1}{r^2}\frac{\partial}{\partial r}\left[r^2\frac{\partial}{\partial r}\right]+\frac{1}{r^2sin\theta}\frac{\partial}{\partial\theta}\left(sin\theta\frac{\partial}{\partial\theta}\right)+\frac{1}{r^2sin^2\theta}\frac{\partial^2}{\partial\phi^2}\right\}\psi_{flat} \tag{5.20}$$

or

$$i\hbar\frac{\partial\psi_{flat}}{\partial s}=\frac{-\hbar^2}{2m}\Box_{flat}\,\psi_{flat}=\frac{-\hbar^2}{2m}\left\{-\frac{1}{c^2}\frac{\partial^2}{\partial t^2}+\triangle_{flat}\right\}\psi_{flat} \tag{5.21}$$

with D'Alembertian $\Box_{flat}$ and Laplacian $\triangle_{flat}$ operators in flat spacetime.

6. Solution of the Free Scalar Particle in the Schwarzschild Metric

The equation for a free scalar particle with mass m moving in the gravitational field of a spherically symmetric mass M is Equation (5.16), written here in terms of the operator $L\cdot L$

$$i\hbar\frac{\partial\psi_{SM}}{\partial s}=\frac{-\hbar^2}{2m}\left\{-\frac{1}{A(r)}\frac{1}{c^2}\frac{\partial^2}{\partial t^2}+\frac{1}{r^2}\frac{\partial}{\partial r}\left[r^2A(r)\frac{\partial}{\partial r}\right]-\frac{L\cdot L}{r^2}\right\}\psi_{SM} \tag{6.1}$$

where $L\cdot L$ is the operator

$$L\cdot L=-\left[\frac{1}{sin\theta}\frac{\partial}{\partial\theta}\left(sin\theta\frac{\partial}{\partial\theta}\right)+\frac{1}{sin^2\theta}\frac{\partial^2}{\partial\phi^2}\right] \tag{6.2}$$

The nonzero elements of the Schwarzschild metric are

$$g_{00}=-A(r), g_{11}=1/A(r), g_{22}=r^2, g_{33}=r^2sin^2\theta, A(r)=1-\frac{r_S}{r} \tag{6.3}$$

where the contravariant four-vector is $y^\mu=\{ct,r,\theta,\phi\}$ and $\{r,\theta,\phi\}$ are spherical coordinates.

A trial solution for separating variables in Equation (6.1) is

$$\psi_{SM}=T(t)R(r)Y_{\ell m}(\theta,\phi)exp\left[i\frac{\hbar\beta_{SM}}{2m}s\right] \tag{6.4}$$

The term $Y_{\ell m}(\theta,\phi)$ represents orthonormal spherical harmonics and satisfies the eigenvalue equation

$$L\cdot LY_{\ell m}(\theta,\phi)=\ell(\ell+1)Y_{\ell m}(\theta,\phi)\;for\;\left\{\begin{matrix}\ell=0,1,2\ldots\\ m=0,1,\ldots,|\ell|\end{matrix}\right\} \tag{6.5}$$

Substituting Equation (6.4) into (6.1) and rearranging lets us write

$$i\hbar\frac{i\hbar\beta_{SM}}{2m}\psi_{SM}=\frac{-\hbar^2}{2m}\left\{-\frac{1}{A(r)}\frac{1}{c^2}\frac{\partial^2}{\partial t^2}+\frac{1}{r^2}\frac{\partial}{\partial r}\left[r^2A(r)\frac{\partial}{\partial r}\right]-\frac{\ell(\ell+1)}{r^2}\right\}\psi_{SM} \tag{6.6}$$

Canceling $\frac{-\hbar^2}{2m}$ gives

$$\beta_{SM}\psi_{SM}=\left\{-\frac{1}{A(r)}\frac{1}{c^2}\frac{\partial^2}{\partial t^2}+\frac{1}{r^2}\frac{\partial}{\partial r}\left[r^2A(r)\frac{\partial}{\partial r}\right]-\frac{\ell(\ell+1)}{r^2}\right\}\psi_{SM} \tag{6.7}$$

Multiplying Equation (6.7) by $1/\psi_{SM}$ and simplifying gives

$$\beta_{SM}=-\frac{1}{A(r)}\frac{1}{c^2}\frac{1}{T}\frac{\partial^2 T}{\partial t^2}+\frac{1}{R}\left\{\frac{1}{r^2}\frac{\partial}{\partial r}\left[r^2A(r)\frac{\partial}{\partial r}\right]-\frac{\ell(\ell+1)}{r^2}\right\}R \tag{6.8}$$

The $T(t)$ factor can be written as

$$T(t)=\alpha exp(\pm i\omega t) \tag{6.9}$$

where α,ω are constants. Substituting Equation (6.9) into (6.8) gives

$$\beta_{SM}=\frac{1}{A(r)}\frac{\omega^2}{c^2}+\frac{1}{R}\left\{\frac{1}{r^2}\frac{\partial}{\partial r}\left[r^2A(r)\frac{\partial}{\partial r}\right]-\frac{\ell(\ell+1)}{r^2}\right\}R \tag{6.10}$$

since

$$\frac{1}{T}\frac{\partial^2 T}{\partial t^2}=-\omega^2 \tag{6.11}$$

Multiplying Equation (6.10) by R gives

$$\beta_{SM}R=\frac{1}{A(r)}\frac{\omega^2}{c^2}R+\left\{\frac{1}{r^2}\frac{\partial}{\partial r}\left[r^2A(r)\frac{\partial}{\partial r}\right]-\frac{\ell(\ell+1)}{r^2}\right\}R \tag{6.12}$$

Equation (6.12) can be written as the radial equation

$$\left[\beta_{SM}-\frac{1}{A(r)}\frac{\omega^2}{c^2}+\frac{\ell(\ell+1)}{r^2}\right]R=\left\{\frac{1}{r^2}\frac{\partial}{\partial r}\left[r^2A(r)\frac{\partial}{\partial r}\right]\right\}R \tag{6.13}$$

upon rearrangement. Expanding the right-hand side of Equation (6.13) gives

$$\left[\beta_{SM}-\frac{1}{A(r)}\frac{\omega^2}{c^2}+\frac{\ell(\ell+1)}{r^2}\right]R=A(r)\frac{\partial^2 R}{\partial r^2}+\left[\frac{r_S}{r^2}+\frac{2A(r)}{r}\right]\frac{\partial R}{\partial r} \tag{6.14}$$

Rearranging Equation (6.14) gives the homogeneous ordinary differential equation

$$A(r)\frac{d^2R}{dr^2}+\left[\frac{r_S}{r^2}+\frac{2A(r)}{r}\right]\frac{dR}{dr}-\left[\beta_{SM}-\frac{1}{A(r)}\frac{\omega^2}{c^2}+\frac{\ell(\ell+1)}{r^2}\right]R=0 \tag{6.15}$$

Multiplying by $1/A(r)$ gives

$$\frac{d^2R}{dr^2}+p(r)\frac{dR}{dr}+q(r)R=0 \tag{6.16}$$

where

$$p(r)=\frac{1}{A(r)}\frac{r_S}{r^2}+\frac{2}{r},q(r)=\frac{1}{A(r)}\left[\frac{1}{A(r)}\frac{\omega^2}{c^2}-\beta_{SM}-\frac{\ell(\ell+1)}{r^2}\right] \tag{6.17}$$

6.1. The Large Distance Approximation

Equation (6.16) can be solved in the large distance approximation corresponding to $r \gg r_S$ and $A(r)\rightarrow 1$. In this case the particle is at a large distance from the spherically symmetric mass M. The nonzero elements of the metric become

$$g_{00}=-1,g_{11}=1,g_{22}=r^2,g_{33}=r^2sin^2\theta \tag{6.18}$$

Contrasting Equation (6.18) with (4.1) shows that the two metrics differ in spatial geometry and sign convention. The spatial geometry in Equation (4.1) is Cartesian while it is spherically symmetric in Equation (6.18).

The terms $p(r), q(r)$ in Equation (6.17) simplify to

$$p(r) \to \frac{2}{r}, q(r) \to \frac{\omega^2}{c^2} - \beta_{SM} \equiv Q \tag{6.19}$$

in the large distance approximation. The corresponding radial equation has the simplified form

$$\frac{d^2 R_{Q\ell}(r)}{dr^2} + \frac{2}{r}\frac{dR_{Q\ell}(r)}{dr} + QR_{Q\ell}(r) - \frac{\ell(\ell+1)}{r^2}R_{Q\ell}(r) = 0 \tag{6.20}$$

where the subscripts of the radial solution $R_{Q\ell}$ specify solution parameters. Multiplying by r^2 gives

$$r^2\frac{d^2 R_{Q\ell}(r)}{dr^2} + 2r\frac{dR_{Q\ell}(r)}{dr} + [Qr^2 - \ell(\ell+1)]R_{Q\ell}(r) = 0 \tag{6.21}$$

The variable transformation

$$\chi = \kappa r \tag{6.22}$$

with constant κ lets us write Equation (6.21) in the form

$$\chi^2\frac{d^2 R_{Q\ell}(\chi)}{d\chi^2} + 2\chi\frac{dR_{Q\ell}(\chi)}{d\chi} + [\chi^2 - \ell(\ell+1)]R_{Q\ell}(\chi) = 0 \tag{6.23}$$

where

$$Q = \kappa^2 \tag{6.24}$$

Solutions of Equation (6.23) that are regular at $r = 0$ are spherical Bessel functions of the first kind [13] (Chapter 9), [14] (Section 11.7), [15] (Chapter 10):

$$R_{Q\ell}(\chi) = j_\ell(\chi) = \sqrt{\frac{\pi/2}{\chi}}J_{\ell+\frac{1}{2}}(\chi) \tag{6.25}$$

The term $J_{\ell+\frac{1}{2}}(\chi)$ is the Bessel function of the first kind. As an illustration, the lowest order term is

$$R_{Q0}(\chi) = j_0(\chi) = \frac{\sin\chi}{\chi}, \chi = \kappa r = \sqrt{Q}\,r \tag{6.26}$$

7. Conclusions

A formulation of parametrized relativistic quantum theory (pRQT) in curved spacetime has been developed. The formulation is based on the simplifying assumption that the metric is independent of the evolution parameter. Given this assumption, the resulting formulation was applied to the case of a free particle in flat spacetime and the case of a particle in the Schwarzschild metric. The differential equations for the Schwarzschild metric case were presented and solved in the large distance approximation. The formulation developed here provides a basis for future applications.

References

[1] Pavšič, M. On the Stueckelberg Like Generalization of General Relativity. Journal of Physics: Conference Series, 2011. 330, 012011; doi:10.1088/1742-6596/330/1/012011
[2] Land, M. A new approach to the evolving 4+1 spacetime metric. Journal of Physics: Conference Series, 2021, Volume 1956, Number 012010. doi:10.1088/1742-6596/1956/1/012010
[3] Horwitz, L.P. Spin and entanglement in general relativity. Journal of Physics: Conference Series, 2021, Volume 1956, Number 012009 and references therein. doi:10.1088/1742-6596/1956/1/012009
[4] Fanchi, J.R. Cosmological implications of the Gibbs ensemble in parametrized relativistic classical mechanics. Phys. Rev. A 37, 1988, 3956-3962
[5] Fanchi, J.R. Parametrized Relativistic Quantum Theory, Kluwer: Dordrecht, 1993.

[6] Fanchi, J.R. Tachyon Behavior Due to Mass-State Transitions at Scattering Vertices. Physics, 2022, Volume 4, 217–228. doi.org/10.3390/physics4010016

[7] Fanchi, J.R. 4-space formulation of field equations for multicomponent eigenfunctions. J. Math. Phys. 1981, Volume 22, 794-797. doi: 10.1063/1.524985

[8] Fanchi, J.R. Manifestly Covariant Quantum Theory with Invariant Evolution Parameter in Relativistic Dynamics. Found. Phys. 2011, 41, 4-32.

[9] Pavšič, M. The Landscape of Theoretical Physics: A Global View. Kluwer: Dordrecht, 2001.

[10] Walecka, J.D. Introduction to General Relativity. World Scientific Publishing Co., Singapore, 2007.

[11] Walecka, J.D. Introduction to General Relativity. World Scientific Publishing Co., Singapore, 2007.

[12] Hartle, J. Gravity. Cambridge University Press, Cambridge, United Kingdom, 2021.

[13] Abramowitz, M. and Stegun, I.A., editors. Chapter 9. Handbook of Mathematical Functions with Formulas, Graphs, and Mathematical Tables, 1970. Original printing June 1964 by US Government Printing Office. Ninth printing, Dover Publications, New York

[14] Arfken, G. Mathematical Method for Physicists, 2nd Edition. Academic Press, New York, 1970.

[15] Olver, F. W. J., et al., editors. Chapter 10. NIST Digital Library of Mathematical Functions (DLMF), 2022. http://dlmf.nist.gov/, Release 1.1.6 of June 30, 2022; accessed July 8, 2022.

Common Features of Free Particle Wave Functions in Curved Space-times

A Moalem and A Gersten

Department of Physics, Ben Gurion University of the Negev, Beer Sheva,Israel

E-mail: Amnon.moalem@gmail.com

Abstract. Quantum equations for massless particles of any spin and for massive spin one-half particles are considered in curved space-times. It is demonstrated that in stationary axially symmetric space-times the angular wave functions up to a normalization function are the same as in a Minkowski space-time. The radial wave functions satisfy second order nonhomogeneous differential equations with three nonhomogeneous terms which depend in a unique form on the ratio of time to space curvatures. For a Dirac spin one-half particle, in addition to these three terms a fourth term which depends on the particle rest mass is added.

Introduction

In a previous contribution [1] we have demonstrated that in a global curved space-time, free massless particles satisfy the following quantum equations,

$$E_a^{\mu}\gamma^a\partial_\mu(\sqrt{e}\Phi^{(4s)}(x)) = -\frac{1}{2}\partial_\sigma E_b^{\sigma}\Phi^{(4s)}(x), \tag{1}$$

where E_a^{μ} stand for inverse vierbein fields [2], g is the determinant of a global metric tensor $g_{\mu\nu}(x)$ and $e = \sqrt{-g}$. The gamma's are $4s \times 4s$ matrices taken to be $\gamma^b = diag(\sigma^b, \ldots, \sigma^b)$; σ^b are the Pauli matrices. The wave function $\Phi^{(4s)}(x)$ and its complex conjugate $\Phi^{*(4s)}(x)$ form a basis for the representation of the Lorentz group. The basis is the sum of spin s and spin $s-1$ with altogether $(2s+1)+(2s-1) = 4s$ components. Though the expression above seems rather cumbersome to allow a general discussion of its solutions in a global metric tensor, we show below that by applying the same procedure of analysis, it is possible to spot common properties of free particle equations and their solutions in stationary axial symmetric space-times, namely, for the Minkowski, Friedmann-Robertson-Walker (FRW), Schwarzschild and Kerr space-times. All of these are exact solutions of the Einstein Equations of General Relativity [3,4]. The procedure to be applied is based on factorization of the wave function as,

$$\Phi(t, r, \vartheta, \varphi) = f(r, \vartheta)\psi(t, r, \vartheta, \varphi), \tag{2}$$

where $\psi(t, r, \vartheta, \varphi)$, here after referred to as the reduced wave function, satisfies an equation independent of the spin connection, i.e.,

$$E_a^{\mu}\gamma^a\partial_\mu\psi(t, r, \vartheta, \varphi) = 0. \tag{3}$$

Content from this work may be used under the terms of the Creative Commons Attribution 3.0 licence. Any further distribution of this work must maintain attribution to the author(s) and the title of the work, journal citation and DOI.
Published under licence by IOP Publishing Ltd

Accordingly $f(r,\vartheta)$, is a normalization function which satisfies the equation,

$$\partial_\mu \ln\left(f(r,\vartheta)\sqrt{e}\right) = -\frac{1}{2} e^b_\mu \partial_\sigma E^\sigma_b. \tag{4}$$

For stationary axially symmetric space-times $f(r,\vartheta)$ neither depends on time nor on the azimuthal coordinates and the equation above reduces to a simple integrable equation, namely,

$$\partial_\mu \ln\left(f(r,\vartheta)\sqrt{e}\right) = -\frac{1}{2} e^b_\mu \partial_\sigma E^\sigma_b, b = 1, 2. \tag{5}$$

Particularly, for the space-times mentioned above using the inverse vierbeins of Table 1 one obtains, $\partial_t \ln(f(r,\vartheta)\sqrt{e}) = 0$, $\partial_\vartheta \ln(f(r,\vartheta)\sqrt{e}) = 0$, $\partial_\varphi \ln(f(r,\vartheta)\sqrt{e}) = 0$, and,

$$\partial_\mu \ln f(r,\vartheta) = -\frac{1}{2} e^1_1 \partial_r E^1_1. \tag{6}$$

Solutions of Eq. (5) are listed in Table 1. The idea behind the factorization, Eq. (2), is to bring the reduced equation, Eq. (3), as close as possible to the corresponding equation in a Minkowski space-time. This is well accomplished for the Schwarzschild and FRW metrics. However for the Kerr metric, following this first factorization the resulting reduced equation still involves the non-diagonal E^3_0 and E^0_3 vierbeins. In order to eliminate these, we require two consecutive factorizations. As a first step let ϕ be the reduced function and solve for $f(r,\vartheta)$. With the inverse vierbein fields listed in Table 1, one finds, $e = \rho^2 \sin\vartheta$, $\partial_t \ln(f(r,\vartheta) = -\frac{1}{2} E^\lambda_b \partial_\lambda e^b_0 = 0$, $\partial_\vartheta \ln(f(r,\vartheta)\sqrt{e}) = -\frac{1}{2} E^2_2 \partial_\vartheta e^2_2 = 0$, $\partial_\varphi \ln(f(r,\vartheta)\sqrt{e}) = -\frac{1}{2} E^\lambda_b \partial_\lambda e^b_3 = 0$, and,

$$f(r,\vartheta) = \left(\frac{\sqrt{\Delta}}{\rho}\right)^{1/2} \rho \tag{7}$$

Following this step the equation for ϕ reads,

$$(E^0_0 \gamma^0 \partial_t + E^0_3 \gamma^3 \partial_t + E^1_1 \gamma^1 \partial_r + E^2_2 \gamma^2 \partial_\vartheta + E^3_0 \gamma^0 \partial_\varphi + E^3_3 \gamma^3 \partial_\varphi)\phi = 0. \tag{8}$$

In order to eliminate the two terms which involve non-diagonal vierbeins we assume that ϕ factorizes also as,

$$\phi(t, r, \vartheta, \varphi) = g(r,\vartheta)\psi(t, r, \vartheta, \varphi), \tag{9}$$

and that $\psi(t, r, \vartheta, \varphi)$ satisfies an equation which involves diagonal inverse vierbeins only, i.e.,

$$(E^0_0 \gamma^0 \partial_t + E^1_1 \gamma^1 \partial_r + E^2_2 \gamma^2 \partial_\vartheta + E^3_3 \gamma^3 \partial_\varphi)\psi = 0. \tag{10}$$

Substituting the above expression in Eq. (8) and taking $\psi \sim exp(-i\omega t)exp(im_j\varphi)$ gives the following equation for $g(r,\vartheta)$,

$$E^1_1 \gamma^1 \partial_r (\ln g) + E^2_2 \gamma^2 \partial_\vartheta (\ln g) = i[\omega E^0_3 \gamma^3 - m_j E^3_0 \gamma^0]. \tag{11}$$

To evaluate $g(r,\vartheta)$ we eliminate first the gamma matrices. To this aim we square both sides of the equation above and take their traces. Using properties of the gamma matrices (see ref [1]) one obtains,

$$[(E^1_1 \partial_r (\ln g)]^2 + [(E^2_2 \partial_\vartheta (\ln g)]^2 = -[(\omega E^0_3)^2 + (m_j E^3_0)^2]. \tag{12}$$

Finally, assuming $g(r,\vartheta) = A(r)B(\vartheta)$ and substituting E^0_3 and E^3_0 from Table 1, gives the equations,

$$\partial_r \ln A(r) = i\omega \frac{a}{\Delta} = i\omega a \frac{r - M}{(r-m)^2 + a^2 + M^2}, \tag{13}$$

and,

$$\partial_\vartheta \ln B(\vartheta) = im_j \sin\vartheta. \tag{14}$$

These two expressions give,

$$g(r,\vartheta) = exp[i(a\omega tan^{-1}(\frac{r-M}{a^2+M^2}) - m_j \cos\vartheta]. \tag{15}$$

We may now conclude that in the case of the Kerr space-time the wave function Φ is given by,

$$\Phi(t,r,\vartheta,\varphi) = g(r,\vartheta)f(r,\vartheta)\psi(t,r,\vartheta,\varphi), \tag{16}$$

with a reduced function satisfying,

$$(\frac{a^2+r^2}{\rho\sqrt{\Delta}}\gamma^0\partial_t + \frac{\sqrt{\Delta}}{\rho}\gamma^1\partial_r + \frac{1}{\rho}\gamma^2\partial_\vartheta + \frac{1}{\rho\sin\vartheta}\gamma^3\partial_\varphi)\psi(t,r,\vartheta,\varphi) = 0. \tag{17}$$

Note that in the equation above the factor $\frac{1}{\rho}$ can be eliminated and the reduced function ψ in the Kerr as well as in the other space-times, are rather similar and involves diagonal metric elements only. More over the last two terms in Eq. (17) which determine the reduced angular wave function are identically the same in the FRW, Schwarzschild and Kerr space-times. All be it, the reduced angular wave functions for these space-times are exactly the same as in the Minkowskian case. The normalization functions listed in Table 1 , ($f(r,\vartheta)$ for Minkowski, FRW and Schwarzschild space-times and $f(r,\vartheta)g(r,\vartheta)$ for the Kerr space-time), are signatures of the metrics encountered. The Minkowski space-time does not account for gravity [3] and gravity does not affect the reduced angular wave function. Therefore the impacts of gravity, whatever may be, must be due to time curvature, E_0^0 , and space curvature, E_1^1 , and these, may affect the radial wave functions only.

1. Free massless particles in Minkowski Spacetime

In the Minkowski space-time $g_{\mu\nu}(x) = diag(1,-1,-1,-1)$, $e = -detg = 1$, and taking $\Phi \sim exp(-i\omega t)exp(im_j\varphi)$ Eq.(1) reduces to,

$$\omega E_0^0\Phi = -i\gamma\cdot\nabla\Phi = H_0\Phi, \tag{18}$$

where $-i\gamma\cdot\nabla$ represents the Hamiltonian of free unperturbed states. Naturally the equation above does not depend on the spin connection. In spherical coordinates the Del operator is $\nabla = \hat{r}\partial_r + \hat{\theta}\partial_\vartheta + \hat{\phi}\partial_\varphi$ and the free Hamiltonian reads,

$$H_0 = -i\gamma\cdot\nabla = -i\gamma^r\partial_r + \gamma^\theta\frac{1}{r}\partial_\vartheta + \gamma^\phi\frac{1}{r\sin\vartheta}\partial_\varphi. \tag{19}$$

Here the matrices γ^t,γ^r, γ^θ, γ^ϕ are given via the transformation,

$$\begin{Bmatrix} \gamma^t \\ \gamma^r \\ \gamma^\theta \\ \gamma^\phi \end{Bmatrix} = \begin{bmatrix} E_0^0 & 0 & 0 & E_0^3 \\ 0 & E_1^1 & 0 & 0 \\ 0 & 0 & E_2^2 & 0 \\ E_3^0 & 0 & 0 & E_3^3 \end{bmatrix} \begin{Bmatrix} \gamma^0 \\ \gamma^1 \\ \gamma^2 \\ \gamma^3 \end{Bmatrix}. \tag{20}$$

Again the inverse vriebein fields, $E_2^2 = \frac{1}{r}$, and $E_3^3 = \frac{1}{r\sin\vartheta}$, are common to all of the space-times considered. This ascertains that the reduced angular wave function is the same as in the

Table 1: The Normalization Function

Space-times	Inverse Vierbein Fields	E^0_0/E^1_1	$N(r, \vartheta)^d$
Minkowski	$diag(1, 1, 1/r, 1/r\sin\vartheta)$	-1	$1/ar\sin\vartheta$
FRW [a]	$(1, F/a, 1/ar, 1/ar\sin\vartheta)$	a/F	$(a^3 r\sin\vartheta)^{1/2}$
Schwarzschild [b]	$(1/F^{1/2}, F^{1/2}, 1/r, 1/r\sin\vartheta)$	$1/F$	$F^{1/4}r\sin^{1/2}\vartheta$
Kerr[c]	$E^0_0 = a^2 + r^2/\rho\sqrt{\Delta}$, $E^1_1 = \Delta/\rho$, $E^2_2 = 1/\rho$	$(a^2+r^2)/\Delta$	
	$E^3_3 = 1/\rho\sin\vartheta$, $E^3_0 = a/\rho\sqrt{\Delta}$, $E^0_3 = a\sin\vartheta)$		$f(r,\vartheta)g(r,\vartheta)$

[a] $F = \sqrt{1-kr^2}$, $k = \pm 1$, a is a dimensionless parameter.
[b] $F = (1 - 2GM/r)$, G is the gravitation constant, M is the mass.
[c] $\Delta = r^2 + r^2 - 2rM$, $\rho^2 = r^2 + a^2cos^2\vartheta$, $a = J/Mc$, J is the angular momentum.
[d] $N(r,\vartheta)$ is $f(r,\vartheta)$ for the Minkowski, FRW and Schwarzschild space-times, and $f(r,\vartheta)g(r,\vartheta)$ for the Kerr space-time ;$f(r,\vartheta) = (\Delta/\rho)^{1/2}\rho\sin^{1/2}\vartheta)$; $g(r,\vartheta) = exp[ia(\omega tan^{-1}(\frac{r-M}{a^2+M^2}) - m_j\cos\vartheta)]$

Minkowskian case. This is derived rigorously below. Let us consider first the reduced angular wave function. By definition the angular momentum operator is defined as $\hat{L} = -i\hat{r}\times\nabla$. Then, $\hat{r}\times\hat{L} = i(\hat{\theta}\partial_\vartheta + \hat{\phi}\partial_\varphi)$, and,

$$\gamma\cdot\hat{r}\times\hat{L} = i\frac{1}{r}(\gamma^\theta\partial_\vartheta + \gamma^\phi\partial_\varphi) = \frac{i}{2}(E^2_2\gamma^2\partial_\vartheta + E^3_3\gamma^3\partial_\varphi) \tag{21}$$

It is straightforward to show, based on the properties of the gamma matrices, that any three vectors A and B satisfy the identity (see appendix of Ref [1])

$$(\gamma\cdot A)(\gamma\cdot B) = \gamma^0(A\cdot B) + \frac{i}{2}[\gamma\cdot A\times B]. \tag{22}$$

Then for $A = \hat{r}$ and $B = L$, $\gamma\cdot\hat{r}\times\hat{L} = i\frac{1}{r}(\gamma\cdot\hat{r})(2\gamma\cdot L)$, and we may rewrite the free particle Hamiltonian as,

$$H_0 = -i\gamma\cdot\nabla = -iE^1_1\gamma^1[\partial_r - \frac{1}{r}(2\gamma\cdot L)]. \tag{23}$$

$H_0 = -i\gamma\cdot\nabla = -iE^1_1\gamma^1[\partial_r - \frac{1}{r}(2\gamma\cdot L)]$. We identify the spin S with γ and define an angular operator K as,

$$K = (2S\cdot L + s) = J^2 - L^2 - S^2 + s, \tag{24}$$

where L, S, $J = L + S$, s are the orbital, spin, total angular momentum and the particle spin, respectively. Substituting Eq. (23) in Eq. (22) allows writing the Hamiltonian as,

$$H_0 = -iE^1_1\gamma^1[\partial_r - \frac{s}{r} - \frac{1}{r}(2S\cdot L + s)] = iE^1_1\gamma^1[\partial_r - \frac{s}{r} - \frac{1}{r}K]. \tag{25}$$

This expression of the free Hamiltonian suggests that we may construct the solution by variable separation. Indeed H_0, J^2, J_3, K and parity P form a complete set of commuting operators (see ref. [1]). The angular part of the wave function can be readily written as a spinor of spherical harmonics, i.e.,

$$\mathcal{Y}^j_{lm_j}(\vartheta,\varphi) = \Sigma_{m_j=-s,s}C(l,s,j;m_j - m_s, m_s, m_j)Y^j_{lm_j}(\vartheta,)\chi^{m_s}_s, \tag{26}$$

where $C(l, s, j; m_j - m_s, m_s, m_j)$ is a Clebsch-Gordon coefficient for combining orbital angular momentum l and spin s to a total angular momentum j, $Y^j_{lm_j}$ are spherical harmonics Eigen-functions of L^2 and L_z, $\chi^{m_s}_s$ - Eigen functions of S^2 and S_3 . For massless particles m_s can have only two values s and $-s$ [5]. The spherical harmonic spinors are orthonormal, i.e.,

$$\int \mathrm{d}\Omega(\mathcal{Y}^{j'}_{l'm'}(\Omega))^H \mathcal{Y}^j_{lm_j}(\Omega) = \delta_{jj'}\delta_{ll'}\delta_{m_j m'_j} \tag{27}$$

Furthermore, the function Φ is an Eigen function of the operators: J^2, the total angular momentum squared, J_3, the Z component of angular momentum, K the angular operator and parity P. Thus,

$$J^2\Phi^j_{m_j,\kappa_j} = j(j+1)\Phi^j_{m_j,\kappa_j} \tag{28}$$

$$J_3\Phi^j_{m_j,\kappa_j} = m_j\Phi^j_{m_j,\kappa_j} \tag{29}$$

$$K^2\Phi^j_{m_j,\kappa_j} = (\kappa_j)^2\Phi^j_{m_j,\kappa_j}. \tag{30}$$

From addition of angular momenta we may calculate the eigenvalues of the angular operator for $j = l \pm s$,

$$(\kappa_j)^2 = [s(2l+1)]^2; \kappa_j = \pm s(2l+1). \tag{31}$$

The eigenvalues of K spans all values: $-s(2l+1), -s(2l+1)+1, \cdot, \cdot, \cdot, s(2l+1)-1, s(2l+1)$. This means that for each of the cases $j = l+s$ and $j = l-s$, there exist two linearly independent solutions of the reduced equation, $R_1(r)\phi^+$ and $R_2(r)\phi^-$, where ϕ^+ and ϕ^- denote Eigen functions of the angular operator K, i.e.,

$$K\phi^+(\vartheta,\varphi) = \kappa_j\phi^+(\vartheta,\varphi), \tag{32}$$

$$K\phi^-(\vartheta,\varphi) = -\kappa_j\phi^+(\vartheta,\varphi), \tag{33}$$

Accordingly, the reduced wave function has the form,

$$\psi \sim exp(-i\omega t)\begin{Bmatrix} R_1(r)\phi^+(\vartheta,\varphi) \\ R_2(r)\phi^-(\vartheta,\varphi) \end{Bmatrix}, \tag{34}$$

where R_1 and R_2 are two independent reduced radial functions. We may turn now to consider the radial wave functions.

2. The reduced radial Wave Equations

With the free Hamiltonian, Eq. (25), the reduced function satisfies the equation,

$$iE^0_0\omega\gamma^0 = E^1_1\gamma^1[\partial_r + \frac{s}{r} - \frac{1}{r}K]\psi. \tag{35}$$

Then, by substituting $\gamma^\nu = diag(\sigma^\nu, \cdot, \cdot, \cdot, \sigma^\nu)$ Eq.(35) reduces to identical equations of bi-spinors,

$$iE^0_0\omega I^2\psi = E^1_1\sigma^1[\partial_r + \frac{s}{r} - \frac{1}{r}K]\psi. \tag{36}$$

Using Eqs. (32)-(34) the above expression gives a pair of coupled differential equations,

$$i\omega E^0_0 R_1(r)\phi^+(\theta,\phi) = E^1_1[\partial_r + \frac{s+\kappa_j}{r}]R_2(r)\phi^-(\theta,\phi), \tag{37}$$

$$i\omega E^0_0 R_2(r)\phi^-(\theta,\phi) = E^1_1[\partial_r + \frac{s-\kappa_j}{r}]R_1(r)\phi^+(\theta,\phi), \tag{38}$$

It is now straightforward to show from the expressions above that the radial equations are,

$$[d^2/dr^2 + 2\tfrac{s}{r}d/dr + \tfrac{1}{r^2}[(s(s-1) - (\kappa_j - 1)\kappa_j] + \omega^2]R_1 = [\omega^2(1 - (E^0_0/E^1_1)^2)]R_1 \\ +[dln(E^0_0/E^1_1)/dr)\tfrac{s+\kappa_j}{r}]R_1 \; + [dln(E^0_0/E^1_1)/dr]dR_1/dr \tag{39}$$

$$[d^2/dr^2 + 2\tfrac{s}{r}d/dr + \tfrac{1}{r^2}[(s(s-1) - \kappa_j(\kappa_j + 1)] + \omega^2]R_2 = [\omega^2(1 - (E^0_0/E^1_1)^2)]R_2 \\ +[dln(E^0_0/E^1_1)/dr)\tfrac{s-\kappa_j}{r}]R_2 \; + [dln(E^0_0/E^1_1)/dr]dR_2/dr \tag{40}$$

Several comments are in order. Firstly all three nonhomogeneous terms depend on the ratio of time to space curvatures, (E^0_0/E^1_1). The first of these goes with the particle energy squared, ω^2. The second and third terms involve the derivative of the logarithm of this ratio. These three terms represent the impact of gravity and the coupling ofthe gravitational field to all other fields. Secondly for a Minkowski space-time (see Table 1) $(E^0_0/E^1_1) = -1$ and the equations above become homogeneous. As suggested by the principle of equivalence [3,4] at every point of a curved space-time the metric diagonalizes to $g_{ij} = diag(1, -1, -1, -1)$ and the equations above reduce locally to be homogeneous as in the Minkowskian case. Likewise, at large distances the curvatures are expected to diminish, and for all of the space-times considered these equations become identical with the ones in a Minkowski space-time. Saying this, an arbitrary solution of Eq. (1) in axially symmetric space-times can be written as,

$$\Phi^{(4s)}(t, r, \vartheta, \varphi) = exp(-i\omega t)N(\vartheta, \varphi) \\ [A_{j,m_j}, \kappa_j(pr)R_{1,\kappa_j}(r)\phi^+(\vartheta, \varphi) + B_{j,m_j}, \kappa_j(pr)R_{1,\kappa_j}(r)\phi^-(\vartheta, \varphi)], \tag{41}$$

where the angular wave functions are those of the Minkowskian case, A_{j,m_j,κ_j} and $B_{j,m_j b\kappa_j}$ are complex coefficients depending on the particle momentum, p. A curvature is created by energy and momentum and therefore for a static and rotating space times, this ratio must be different. Indeed for the Kerr metric$(E^0_0/E^1_1 = \frac{a^2+r^2}{\sqrt{\Delta}})$ depends on rotation energy. Turning off rotation by setting the angular momentum to zero $(a = 0)$ and take $(r_s = -2GM)$ the nonhomogeneous energy term and the normalization function $N(r, \theta)$ reduce, as expected, to those of the static non-rotating Schwarschild space-time. Worth mentioning that by replacing κ_j by $-\kappa_j$ in the equation for (R_1/R_2) we obtain the equation for (R_2/R_1). Note also that for spin 1 particles, the above equations in the Minkowskian case are the familiar spherical Bessel's equations and their solutions are the spherical Bessel's functions $j_{\kappa_j}(r)$ with $\kappa_j = 1, 3, 5, \cdot, \cdot, \cdot$, and for R_1 and,$j_{-\kappa_j-1}(r)$ with $-\kappa_j - 1 = -2, -4, -6, \cdot, \cdot, \cdot$, for for R_2 .

3. Free Dirac Particles in Curved Space-time

Mass is energy and the impact of gravity on massive particles should also include a term similar to the nonhomogeneous energy term of Eqs.(39-40). To verify how the mass of a particle may modify the results presented in the previous sections, we consider the Dirac equation [6] by applying the same procedure of analysis. For free spin one-half particles, the Dirac equation in curved space-time reads,

$$H_0\Phi = -(i\gamma \cdot \nabla + m)\Phi = (E^\mu_a\gamma^a\partial_\mu + m)\Phi = 0. \tag{42}$$

Here as well we assume that Φ factorizes as in Eq.(2) and require that the reduced function satisfies the equation,

$$(E^\mu_a\gamma^a\partial_\mu + m)\psi = 0. \tag{43}$$

With this requirement, the normalization function $f(r, \vartheta)$ satisfies Eq. (4) with solutions as listed in Table 1 for massless particles. This is not surprising since the normalization function is a signature of space-time. Applying the same formalism, step by step as applied for massless particles the reduced radial wave functions are found to be,

$$[d^2/dr^2 + \tfrac{2}{r}d/dr - \tfrac{1}{r^2}(\kappa_j - 1)\kappa_j + \omega^2 - m^2]R_1 = (\omega^2 - m^2)[1 - (E_0^0/E_1^1)^2]R_1 \\ +[dln(E_0^0/E_1^1)/dr\tfrac{1+\kappa_j}{r}]R_1 + [dln(E_0^0/E_1^1)/dr]dR_1/dr \quad (44)$$

$$[d^2/dr^2 + \tfrac{2}{r}d/dr - \tfrac{1}{r^2}\kappa_j(\kappa_j + 1) + \omega^2 - m^2]R_2 = (\omega^2 - m^2)[1 - (E_0^0/E_1^1)^2])R_2 \\ +[dln(E_0^0/E_1^1)/dr\tfrac{1-\kappa_j}{r}]R_2 + [dln(E_0^0/E_1^1)/dr]dR_2/dr \quad (45)$$

Compared to Eqs.(39,40) the expressions above include a forth nonhomogeneous term which is similar to the energy term. Clearly for spin one-half particles these equations reduce, in the limit of m = 0, to Eqs.(39, 40). Likewise, at large distances and in the limit $g_{\mu\nu} \rightarrow diag(1, -1, -1, -1,)$, these equations converge to the homogeneous equations as in the Minkowskian case.

4. Summary and Conclusions

We have demonstrated that in stationary axially symmetric space-times, the wave function Φ of massless particles of any spin can be factorized into a normalization function $N(r, \vartheta)$ which encompasses the spin connection and as such is a signature of space-time, and a reduced function $\psi(t, r, \vartheta, \varphi)$ which depends on diagonal metric elements only. In particular, the elements E_2^2 and E_3^3 are restricted by symmetry and practically are the same as for the Minkowskian case. These elements determine the angular part of the wave function and therefore the reduced angular wave functions for any spin are the appropriate spherical harmonic spinors as determined for the Minkowski space-time. For free massless particles of any spin the radial equations are non-homogeneous with nonhomogeneous terms which are unique functions of the ratio of time to space curvature, (E_0^0/E_1^1). Accordingly, the reduced radial equations and their solutions are signature of a metric. Yet at large distances and locally, these equations converge to the equations of the Minkowskian case. These same nonhomogeneous terms play the same role in case of a massive free spin one-half particle, with a forth mass dependent term similar to the energy dependent term encountered for massless particles.

References

[1] A. Moalem and A. Gersten, Entropy, 23, 1205, 2021, and references mentioned therein.
[2] See for example, Jeffery Yepez, arXiv:1106.3037, (2011) 10.
[3] Robert M. Wald, General Relativity, University of Chicago Press (1084).
[4] F. D. Felice, C. J. S. Clarke, Relativity on Curved Manifolds Cambridge University Press, (1990).
[5] E. P. Wigner, On Unitary Representations of the Inhomogeneous Lorentz Group, Ann. Math. 40, 149, 1939.
[6] P. A. M. Dirac, Relativistic Wave Equations, Proc. Roy. Soc., A155, 447, 1936.

Ultra faint edge on galaxies: standard and alternative gravity explorations

James G. O'Brien[1], **William M. Kerin**[1], **Thomas L. Chiarelli**[2]

[1]Department of Mathematics, Physics and Computer Science, Springfield College, Springfield, MA 01109, USA

[2]Department of Electromechanical Engineering, Wentworth Institute of Technology, Boston, MA 02115, USA

E-mail:
jobrien7@springfieldcollege.edu, wkerin@springfieldcollege.edu, chiarellit@wit.edu

Abstract. Recently, Bizyaev et. al. (2021) conducted the first modeling of rotation curves for 153 ultra-faint, edge-on galaxies using the 3.5 m telescope at the Apache Point Observatory. These models derived high resolution rotation curves for 20 galaxies of various sizes and spiral morphologies, making new estimates of the optical scale lengths. Using the derived models, in this work we make the first fits to the rotation curves using alternative gravity, namely conformal gravity, Modified Newtonian Dynamics (MOND) and more. A robust analysis is shown including the derived rotation curve fits to the alternative gravity models, along with how the models account for empirical phenomena such as the Baryonic Tully Fisher relation and the Radial Acceleration Rule. This data is not only modern, but unique in that certain galaxies are shown to be able to be fit by standard gravity alone without dark matter.

1. Introduction

The missing matter problem in spiral galaxies has been plaguing standard gravity since the work of Rubin and Kent [1]. Although numerous alternative gravity theories have emerged in the last thirty years, none have managed to gain as much traction as the addition of dark matter as a staple of standard gravity. However, as the number of quality high resolution rotation curves has increased [2], MOND and Conformal Gravity (CG), among others (see for example Retardation Gravity [3]) have had tremendous success in fitting the expanding landscape of available data. Aside from new rotation curves fit by alternative gravity, more recent data sets of existing rotation curves [4] has helped to provide alternative gravity with a robust arsenal of examples where standard gravity with dark matter can be explained without copious amounts of dark matter. Although the rotation curve fits themselves are impressive for alternative gravity, the recent discovery of the Radial Acceleration Rule (RAR) provides another empirical test for rotation curve physics outside of the standard Baryonic Tully Fisher relation (BTF) [5], and places a bound on dark matter parameters [6]. Since the original discussion of the RAR by McGaugh Lelli and Schombert (MLS), numerous authors have expanded the number of galaxies that show the RAR trend and have posited alternative gravity as the resolution to the trend [7], attempting to answer the call of the original RAR work which stated that the empirical trend could either be explained with more fine tuned dark matter parameters or alternative gravity [6]. The masses used in the alternative theories to explain the RAR, are masses that obey the

Content from this work may be used under the terms of the Creative Commons Attribution 3.0 licence. Any further distribution of this work must maintain attribution to the author(s) and the title of the work, journal citation and DOI.
Published under licence by IOP Publishing Ltd

other , more standard, empirical test of the Baryonic Tully Fisher relation. This allows both MOND and CG to explain the missing mass in rotation curves with no free parameters outside the mass to light ratio, whereas standard dark matter requires $2n$ free parameters for a sample of n galaxies.
Expansion to the overall set of galactic rotation curves fit by alternative gravity is well enough, but this survey presents another important point of relevance. In very recent months, papers have begun to appear about galaxies that lack dark matter [8]. In the past, the shape of rotation curves could not be fit without dark matter or the modification of gravity. However, due to the newer optical scale length measurements in [9] of the galaxies in this sample, some of them can be fit without dark matter or alternative gravity. The implications of this will be discussed in the later sections.

2. The Sample

The recent 2021 survey conducted by Bizyaev et. al. was collected using the Dual Imaging Spectrograph at the Apache Point Observatory on the 3.5m telescope starting from December 2014 through August 2019 [9]. This survey consists of rotation curve models of 138 super-thin, edge-on galaxies. In order to increase the catalog of galactic rotation curves fit by alternative gravity, the authors here wished to fit all 138 galaxies, but were instead given the 20 highest resolution and most reliable data by Bizyaev et al. Even with only 20 out of the 138, the galaxies presented vary greatly in their individual input parameters and represent a diverse set of galaxies featuring distances that span a range of 9 MpC to 457 MpC. This work of Bizyaev et. al. also features new optical measurements of the galactic scale length of each of the galaxies. They compare the newly measured scale lengths to previously measured scale lengths [10], and it should be noted that systematically the new scale lengths differ by a factor of two, namely $R_0^{\text{new}} = 2R_0^{\text{old}}$ [9]. The implications of this severe of a change in the scale length will be discussed later in the analysis section.
There are several features of this survey that make it unique in terms of the galactic parameters. Outside the observation of new and much larger scale lengths, some typical rotation curve parameters were not measured by the authors such as the total luminosity and the gas masses of the galaxies. The authors instead focused only on the optical measurements to obtain the newly observed scale lengths. However, it was noted in the papers that gas and bulges are present in their models to derive the scale lengths, but no modeling parameters as such were given. A complete breakdown of the gas in edge on galaxies is troublesome to provide, so it is not surprising this information was not determined [11]. This is problematic for rotation curve fitting for any theory and thus some systematic requirements need to be established for fitting these galaxies in order to avoid bias towards any particular theory and will be discussed in the analysis section. Nonetheless once a prescription for fitting is established and followed, the 20 galaxies contained within will serve to provide some interesting insights into rotation curve physics.

3. Alternative Gravity, MOND, CG and MLS

The literature on the two main alternative gravity discussed here, MOND and CG are extensive, so here we provide a simple, short summary. The formalism of MLS being applied to individual rotation curves is not as well studied so motivation will be provided.

Unlike many alternative gravitational theories, Conformal Gravity was not conceived to solve the rotation curve problem. However, CG has enjoyed success in fitting hundreds of galactic rotation curves [12] as well as being regarded as a potential candidate for a re-normalizable, quantum metric theory of gravity. Further, the theory has extended the amount of data included in the RAR by a factor of two, posited that the RAR may be explained by CG itself [7], and has provided a first principles derivation of the BTF. Conformal gravity is derived from the fourth order conformal Weyl action instead of the Einstein Hilbert action of standard Einstein gravity. The resulting field equations can be solved for a spherical solution and the rotation curve of the galaxy ([13]) can be fit for a total number of stars $N^* = MM_\odot^{-1}$ and galactic scale length R_0 as

$$v_{CG}(R) = \sqrt{v_{NEW}^2 + \frac{N^*\gamma^* c^2 R^2}{2R_0} I_1\left(\frac{R}{2R_0}\right) K_1\left(\frac{R}{2R_0}\right) + \frac{\gamma_0 c^2 R}{2} - \kappa c^2 R}, \quad (1)$$

where $\gamma^* = 5.42 * 10^{-41}\mathrm{cm}^{-1}$, $\gamma_0 = 3.06 * 10^{-30}\mathrm{cm}^{-1}$, and $\kappa = 9.54 * 10^{-54}$ cm^{-2} are integration constants of the theory [13]. The values for the integration constants can be found in other works ([13], [14]) and are not changed for any of the galaxies that CG has fit over the last fifteen years. Writing eq. (1), the departure from the standard gravity prediction, v_{NEW} is clear where the Freeman formula is taken as the standard,

$$v_{NEW}(R) = \sqrt{\frac{N^*\beta^* c^2 R^2}{2R_0^3}\left[I_0\left(\frac{R}{2R_0}\right) K_0\left(\frac{R}{2R_0}\right) - I_1\left(\frac{R}{2R_0}\right) K_1\left(\frac{R}{2R_0}\right)\right]}, \quad (2)$$

where in both eq. (1) and eq. (2), I_n and K_n are Bessel Functions.
For almost forty years now, Modified Newtonian Dynamics still reigns as the leading alternative theory of gravity [15]. First constructed by Milgrom and later expanded by numerous authors [16], MOND was originally founded to explain the missing mass problem in spiral galaxies. Since then, MOND has helped further explore the BTF as well as the more recent RAR. Modified Newtonian Dynamics replaces dark matter using an interpolation function to modify the gravitational acceleration as $g = \frac{g_N}{\mu(g/a_0)}$ with a universal acceleration scale set by $a_0 \approx 1.21 \times 10^{-8} cm\, s^{-2}$. Although some authors have used different interpolation functions from the one previously stated and even varied the universal acceleration scale, for the work contained herein, we retain the traditional MOND formalism. The MOND prediction for the rotation curve of a spiral galaxy then takes the form,

$$v_{MOND}(R) = \sqrt{v_{NEW}^2(R) + \left(v_{NEW}^2(R)\frac{\left(\sqrt{1 + \frac{4a_0 R}{v_{NEW}^2(R)}} - 1\right)}{2}\right)}, \quad (3)$$

with v_{NEW} again as the Freeman formula of eq. (2). Thus, like CG, when the MOND prediction for rtation curves is expressed as eq. (3) the departures from standard gravity are easily resolved and still requires only one free parameter $N^* = MM_\odot^{-1}$.

In 2017, the Radial Acceleration Rule (RAR) of McGaugh, Lelli and Schombert sparked a renewed interest in rotation curves of Spiral Galaxies. Unlike most works which feature individual galactic rotation curves, the RAR showed universal trends in overall samples of galaxies, namely that the missing mass problem could be captured in a plot where the data deviated from the line of unity and thus either constrained dark matter theories or provided guidance to alternative gravity as to how to reconcile observation from theory. The fitting function highlighted in their work,

$$g_{OBS} = \frac{g_{NEW}}{[1 - \exp(-(g_{NEW}/g^\dagger)^{1/2})]}, \quad (4)$$

compared the predicted Newtonian centripetal accelerations $g_{NEW} = v_{NEW}^2/R$ to the observed centripetal accelerations $g_{OBS} = v_{OBS}^2/R$ with the inclusion of a free parameter, $g^\dagger$. For the

Table 1. Properties of the Bizyaev 2021 Galaxy Sample

Galaxy	Dist. (Mpc)	Lum. ($10^9 L_\odot$)	R_0 (kpc)	M_{gas} ($10^9 M_\odot$)	$(M_d)_C$ ($10^9 M_\odot$)	$(M/L)_C$ ($M_\odot L_\odot^{-1}$)	$(M_d)_M$ ($10^9 M_\odot$)	$((M/L))M$ ($M_\odot L_\odot^{-1}$)
0036	78	0.70	3.5	8.70	6.96	9.8	2.35	3.3
0084	114	1.18	4.4	10.69	9.53	8.1	3.16	2.7
0173	63	0.89	4.7	4.50	6.95	7.8	2.57	2.8
2339	43	0.62	3.5	7.37	5.02	8.2	1.57	2.6
2233	12	0.68	3.1	8.95	5.76	8.2	2.06	2.9
0604	95	1.44	5.9	19.97	9.14	6.5	2.92	2.1
0711	23	0.45	3.9	5.16	4.40	9.7	1.69	3.5
0847	73	1.32	8.1	48.96	40.16	30.4	17.76	13.4
1417	139	3.80	5.2	41.48	2.53	6.6	1.05	2.7
2384	103	2.30	8.5	39.70	37.50	16.3	16.20	7.2
2396	56	0.58	2.3	5.42	7.30	12.4	2.90	5.1
4247	65	1.09	3.9	13.88	6.63	6.1	1.99	1.8
4625	118	13.50	17.7	200.89	202.43	15.0	178.45	13.3
5341	109	*	1.9	7.86	88.46	*	78.07	*
7321	18	6.10	5.0	63.62	25.25	4.2	10.10	1.8
9242	21	0.36	5.6	3.20	3.47	9.4	1.78	4.9
9681	76	1.22	4.4	16.42	14.20	11.6	5.26	4.3
10043	48	5.65	8.9	161.79	76.86	13.6	42.38	7.5
11964	29	0.25	2.6	2.89	0.13	0.5	0.17	0.7
2396	56	0.58	2.3	5.42	7.30	12.4	2.90	5.1
2384	103	2.30	8.5	39.70	37.50	16.3	16.20	7.2
0035	457	*	2.5	1.65	12.12	*	5.95	*

Table 1 Entries left to right are: Galaxy name, NED distance, total luminosity, disk scale length,mass of the gas (including helium), disk mass, mass to light ratio, disk mass for MOND, mass to light ratio for MOND. Entries listed with a $*$ were not measured in the NASA Extragalactic Database.

153 galaxies from the SPARC sample, the authors showed that the best fit for the free fitting parameter took on the astonishing form of $g^\dagger = a_0$, where a_o is none other than the universal acceleration scale of MOND. This is no surprise however since in their work on the SPARC sample, the authors used MOND to derive the value of the galactic mass, $N^* = MM_\odot^{-1}$. Since their original work, O'Brien, Mannheim and Chiarelli not only extended the sample size of the RAR, but also showed that if a different theory is used to find the predicted centripetal accelerations, then $g^\dagger$ can vary [7]. This fact, makes it then instructive to rearrange and use the formula from MLS and see how based on a given mass, could make an individual rotation curve fit, and then extrapolate those over the entire sample. To this end and to avoid BIAS, we will show fits for the MLS formalism to each individual galaxy when each of CG and MOND sets the mass respectively.

4. Fits to the Bizyaev Galaxies

The observations made by Bizyaev et. al. constitute a robust survey and here we present the 20 galaxies given to us by the authors as being the highest quality contained in the survey. As previously mentioned, one of the main goals of the work of Bizyaev et. al. was to obtain more reliable measurements of the optical scale lengths of the galaxies. Compared to previous analysis of the same set of galaxies, the observers find a systematic increase by a factor of two in the

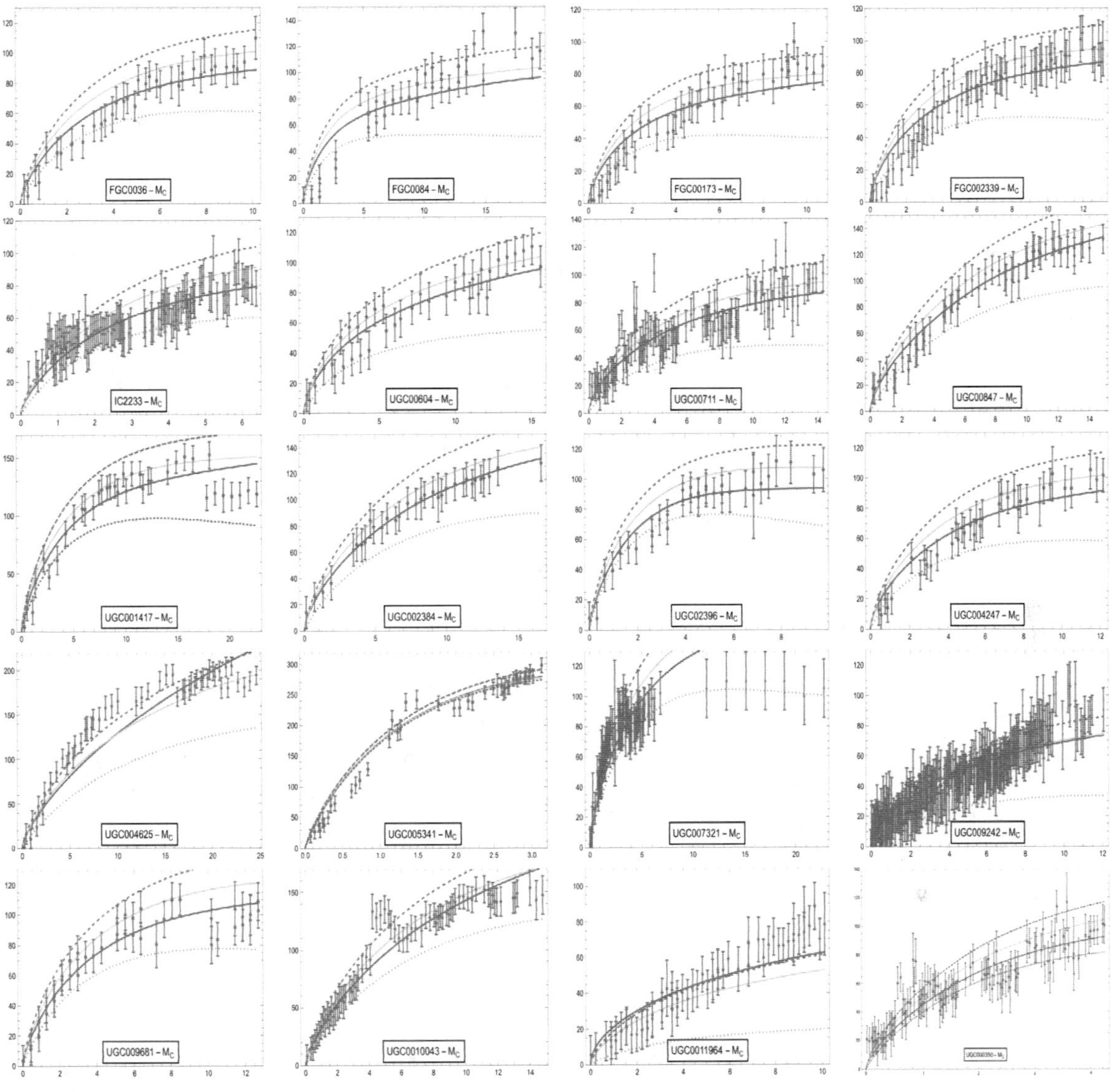

Figure 1. Fitting to the rotational velocities (km/s) of the Bizyaev sample with quoted errors as a function of radial distance (kpc) using the Conformal Gravity prediction, M_C. Blue dashed curve=Newtonian, full blue curve= conformal gravity, dashed red curve=MOND, full green curve=MLS. No dark matter is assumed.

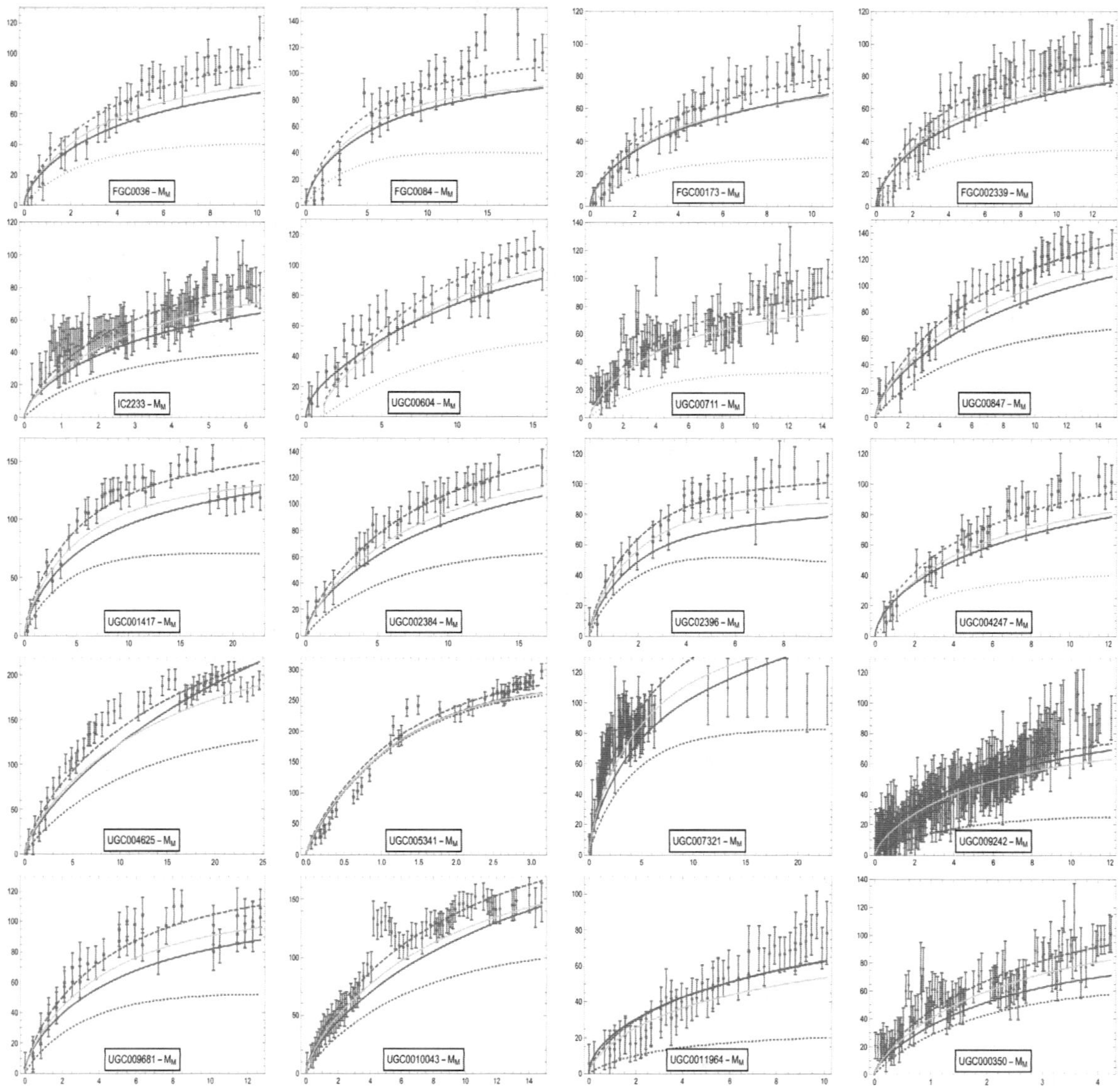

Figure 2. Fitting to the rotational velocities (km/s) of the Bizyaev sample with quoted errors as a function of radial distance (kpc) using the Modified Newtonian Dynamics prediction, M_M. Blue dashed curve=Newtonian, full blue curve= conformal gravity, dashed red curve=MOND, full green curve=MLS. No dark matter is assumed.

optical scale length as compared to the older measurements. Although Bizyaev et. al. used acceptable estimates of the distances to the galaxies, we adopt a more standardized procedure in this work. Since rotation curve fitting is extremely sensitive to distance estimates, it is of the utmost importance to use a systematic regiment for use of input parameters so as to not bias a particular theory or fit. In previous works, the authors here have established such a routine. For distances, we use the average distance as quoted by the NASA Extragalactic Database [1] (NED) and incorporate only on the distances which are from cepheids when applicable. However, for this sample, the NED contained numerous distance estimates, with almost zero of them coming from cepheids and instead from Tully Fisher methods (more on this later). To this end, we simply used the average distance stated for the galaxy. We also used the NED for the total luminosities [2] of each galaxy since no estimate was given by Bizyaev et. al.
For the modeling of the galaxies, we adopt the new quoted scale lengths. Since there was no measure or estimate of the gas mass, we adopt a standardized procedure that the gas mass is on the same order of magnitude as the disk and if a bulge is quoted, the bulge is then modeled as well. Although with no measured gas mass, one can claim that fits produced are nothing more than toy models for each galaxy, however, with the very large optical scale length of each galaxy, and the typical gas scale length being larger than the disk alone, the gas contribution will not change the shape of the fit, but only serve to lower or raise the overall mass to light ratio (more gas, less stars and vice versa). More discussion on the mass to light ratio of the individual fits will be provided in the analysis section. For each galaxy, we show the entire suite of input parameters as well as the best fit mass for the galaxies in conformal gravity (M_C) and MOND, (M_M) and the corresponding mass to light ratios in Table 1. The fits are then shown in Figures 1 and 2. In all of the fits shown, Conformal Gravity is shown in Blue, MOND is shown in Red, the MLS fit is shown in green, and standard gravity without dark matter is shown as blue dashed. The galaxy name is displayed in each fit followed by the demarcation of the mass used in that fit. Thus, there are two fits per galaxy shown in Figures 1 2 each corresponding to the theory which set the mass, conformal gravity (M_C) in Fig. 1 and MOND, (M_M) in Fig. 2 respectively. It is important to reiterate that no dark matter is assumed in any of the fits. In each of the fits, when conformal gravity is setting the mass for the plot, it clearly outperforms MOND and the converse remains true as well. Notable in each is that the MLS fit can be obtained and does not always necessarily favor the MOND fits despite the appearance of the MOND universal acceleration scale parameter. The MLS fit (green) in many of the plots can also be improved if a different value of $g^\dagger$ is used as was done in a recent work by two of the authors here [7].

5. Analysis

With both CG and MOND performing adequate fits, we can now turn to the analysis of the RAR and BTF relation for this survey of galaxies. This sample of 20 galaxies provides a total of 1421 data points. For each of these data points, we can construct the RAR plot of g_{NEW} vs g_{OBS} using g_{NEW} for both M_C and M_M. These plots are shown in Fig. 3. In each of the figures, the red line shows the line of unity (to highlight the deviation from standard gravity) and the green line represents the MLS fitting function with $g^\dagger = a_0$, and the purple line shows the best fit function if we allow $g^\dagger$ to be a free parameter in CG, thus illustrating a survey dependent departure from the original fitting parameter of MLS. As noted in the rotation curve fits alone, we see in Fig. 3(b) that although close, using $g^\dagger = a_0$ in Eq. (4) does not provide the line of best fit for this set of galaxies This survey thus adds a robust set of data points to a

[1] The NASA/IPAC Extragalactic Database (NED) is funded by the National Aeronautics and Space Administration and operated by the California Institute of Technology.

[2] The NED had no listings for luminosity for two of the galaxies and thus are not listed in Table 1.

valuable plot, and shows the same expected deviation as shown by MLS, highlighting that there is universality in the data, no matter what survey is used. Since one of the claims of the RAR was that perhaps the deviation shown in Fig. 3 could be explained by alternative gravity, we show in Fig. 4 what the expected RAR fit would look like if the x data points were replaced by the g(CG) and g(MOND) predictions instead of g(NEW) alone. It can be seen in Fig. 4 that in both cases, the line of unity is approached by both theories, with a slight favor towards CG in number of points around unity. Thus Fig. 4 highlights the fact that if we use alternative gravity, with no dark matter assumed, the RAR can be explained and fit.

Perhaps more interesting for this particular survey is the resulting fits for the baryonic Tully Fisher relation. The usual prescription of $M_T \propto v_{obs}^4$ of the BTF, is adopted here as $M_T^{\frac{1}{4}} \propto v_{obs}$ so that we can retain the natural error in the velocity. The mass used is the total baryonic mass (disk and gas) and the velocity used is that from the last observed data point. The plots for the CG and MOND prediction are shown respectively in Fig. 6. What makes Fig. 6 so interesting is that there are many outlier galaxies sitting well below the Tully Fisher expectation for both CG (Fig. 6(a)) and MOND (6(b)). When comparing these figures to many other surveys where there are a minimal amount of outliers, to have so many anomalous points in only a sample of 20 galaxies is quite unique. Further, we show in Fig. 6 the deviation prediction at high mass/high velocity for the BTF which is unique to CG [7], and we see that almost no galaxies hit this deviation. However, there are some possible explanations for what is observed in Fig. 6. First, since the scale lengths are so large in these 20 galaxies, a larger scale length tends to lower the overall mass in rotation curve fitting, which if we assume that the galaxies should be BTF compliant, would lead us to the conclusion that perhaps the previously measured, smaller scale lengths are the correct ones for these galaxies. Second, since the overall gas was not measured directly, there could be a discrepancy in the gas to disk ratio. However, here we adopted the aforementioned the equal order of magnitude of gas and disk so this would not serve to raise the mass in any of the outliers if more or less gas was actually measured since we are applying the BTF which includes the gas mass. The third possibility is a bit more complicated and should be discussed further.

As previously mentioned, galactic rotation curves are extremely sensitive to distance measurements since many of the input parameters (scale length, gas mass, luminosity etc) scale in some way with the distance [12]. Thus any small shift in a galactic distance can drastically change the output fit of a rotation curve, and thus the output mass. When selecting a distance to use for fitting, the authors here adopt the NED average distance, which is sufficient since the assumed distances of each of the 20 galaxies as stated by Bizyaev et al. are within the bounds found on the NED. However, for these 20 galaxies in particular, there are no cepheid measurements available in the NED, and instead all of the stated distance measurements are derived from Tully Fisher methods. For most galaxies in this sample, there are numerous Tully Fisher distances, and for many of them, range in a dramatic fashion, being more than factor of four in separation[3] . This is extremely problematic for fitting, but also points to a larger issue in rotation curve physics. Here, we used distances that assumed Tully Fisher, and arrived at the velocity/mass relation fits shown in Fig. 6 fits which are not behaving like Tully Fisher galaxies. This sort of circular argument of using BTF to get a distance and then use that distance to estimate a mass shows to be insufficient for this sample. A more extensive analysis of this phenomena will be explored by the authors in a future work.

If one were to increase the mass significantly, then perhaps some of the outlying galaxies in Fig. 6 would return closer to Tully Fisher. However, the general shape of rotation curves would typically prevent this by leading to an improper fit. In this sample though, for some galaxies with the largest scale lengths, allows for more mass to be added without changing the overall

[3] a simple NED search of any of the listed 20 galaxies will suffice

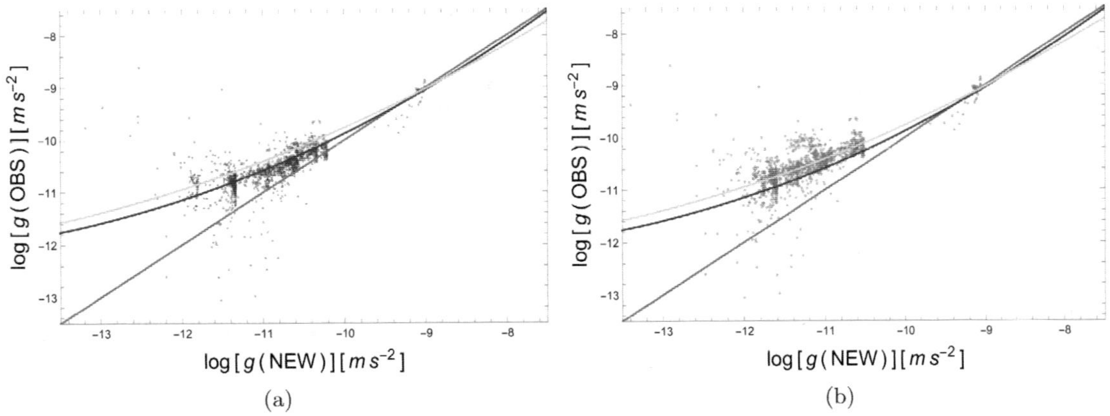

Figure 3. Fig. 3(a) shows the RAR plot using M_C, while 3(b) shows the same plot for M_M. In both figures, the green line function shows eq. 4 using $g^{\dagger} = a_0$, and the purple line shows the best fit to the data using the CG masses. In both plots the red line is the line of unity.

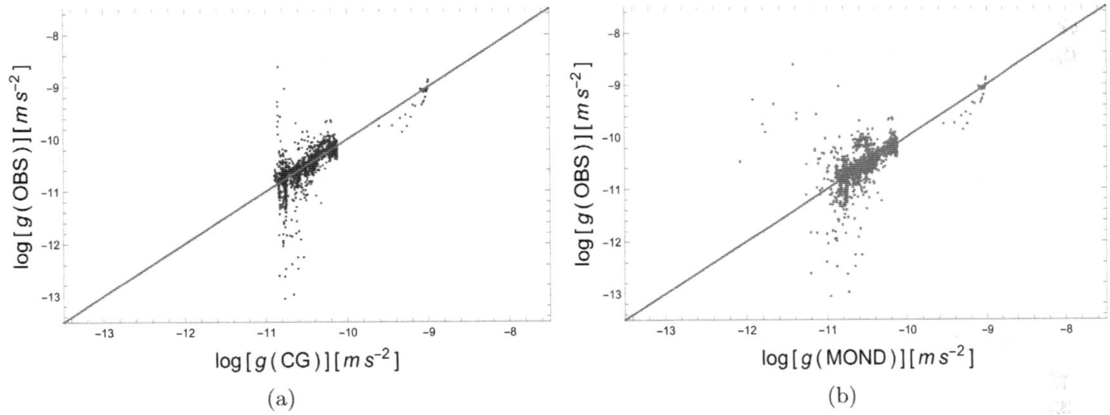

Figure 4. Fig. 4(a) shows the Modified RAR plot using g_{CG}, while 4(b) shows the same plot for g_{MOND}. In both figures the red line is the line of unity.

shape of the fit. In fact for some of these galaxies, enough disk and gas mass alone can be added to fit these galaxies with standard gravity alone without dark matter. Although this would push up the mass to light ratio's, since the total luminosities are in question anyhow, these galaxies would join a very small but growing number of galaxies that can be fit by standard gravity without dark matter. Fig. 5 shows a selection of some of the galaxies from Bizyaev et. al. as fit with a best fit mass for standard gravity alone without dark matter. It should be noted that when the authors returned the galactic scale lengths to the smaller previously measured values, the fits were no longer able to be obtained without dark matter or alternative gravity. Thus once again highlighting the importance and reliance on the validity of the newly measured scale lengths.

6. Conclusion

The 2021 Bizyaev et. al. survey provides extremely modern rotation curve data which has provided an interesting analysis for the application of standard and modified gravity. Aside

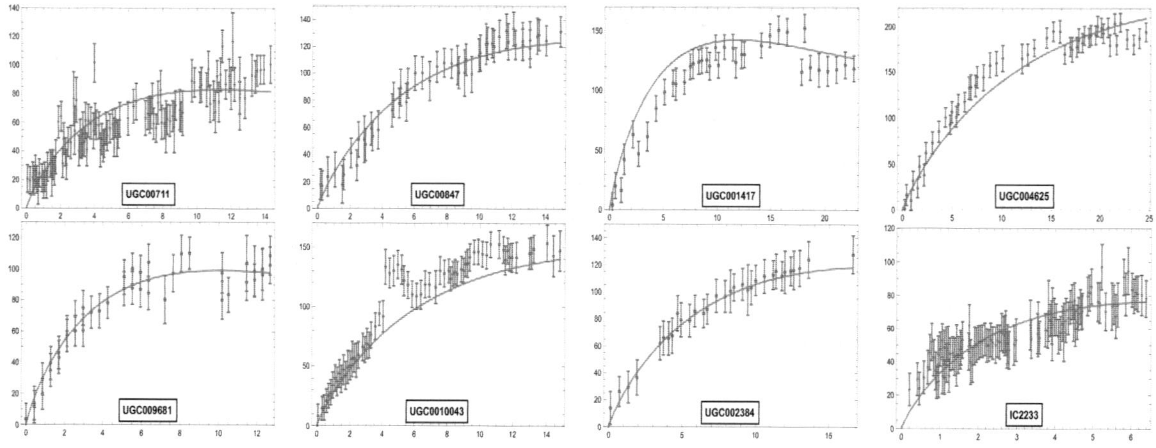

Figure 5. Fitting to the rotational velocities (km/s) of the Bizyaev sample with quoted errors as a function of radial distance (kpc) using Standard Gravity only (full orange curve). No dark matter is assumed.

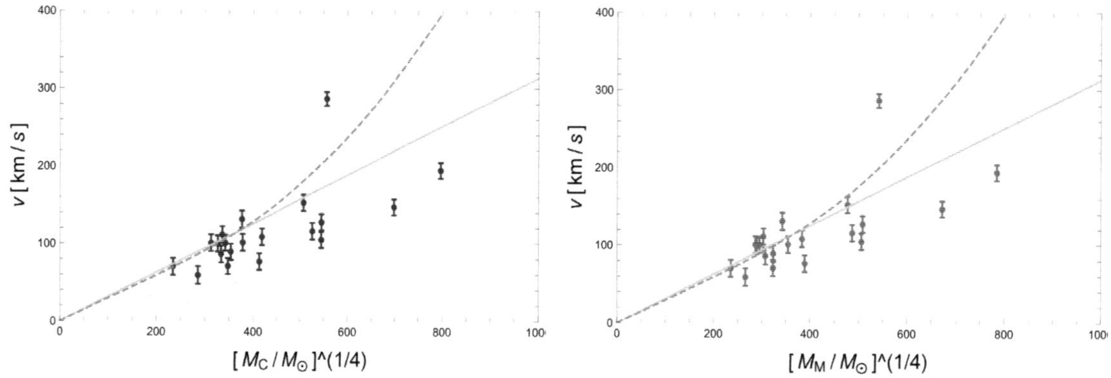

(a) (b)

Figure 6. Fig. 6(a) shows the Baryonic Tully-Fisher relation using M_C while Fig. 6(b) shows the same plot using M_M. In both figures the green line shows the pure BTF and the orange line shows the deviation of the BTF at high mass/velocities unique to CG.

from adding to the increasing pool of galaxies fit by alternative gravity, this survey does give us some more important features for discussion. This set of 1421 data points shows the trends of the RAR, while it proves to be a challenge for the BTF relation. Further, due to the large scale lengths, some of these galaxies can be modeled with no dark matter and no alternative gravity. The authors here see this set of galaxies as a survey that can be used to address the circular argument of using Tully Fisher measurements to estimate a distance and then use that distance to model a mass. Since with the current assumed set of input parameters (and the large scale lengths), this would seem to be a logical violation since using a Tully Fisher distance did not return a Tully Fisher mass. More on how this survey can be used as a test case for how to avoid this logical circle is already underway by the authors and will be featured in a future work.

Acknowledgments
J.G. O'Brien would like to thank the International Association of Relativistic Dynamics (IARD) for the opportunity to present this work at the IARD 2022 Conference in Prague. J. G. O'Brien would like to thank J. Gagnon, and F. Rueckert for their helpful discussions and support of this work. This research has made use of the NASA/IPAC Extragalactic Database (NED), which is funded by the National Aeronautics and Space Administration and operated by the California Institute of Technology.

References

[1] Rubin V C, Ford Jr W K and Thonnard N 1980 *ApJ* **238** 471

[2] Lelli F, McGaugh S S and Schombert J M 2016 *AJ* **152** 157

[3] Yahalom A 2021 *Universe* **7** ISSN 2218-1997 URL `https://www.mdpi.com/2218-1997/7/7/207`

[4] O'Brien J G, Chiarelli T L, Kerin W M and Belanger J 2021 *Journal of Physics: Conference Series* **1956** 012003 URL `https://doi.org/10.1088/1742-6596/1956/1/012003`

[5] McGaugh S S, Schombert J M, Bothun G D and de Blok W J G 2000 *ApJ* **533** L99–L102 (*Preprint* `astro-ph/0003001`)

[6] McGaugh S S, Lelli F and Schombert J M 2016 *Phys. Rev. Lett.* **117** 201101

[7] O'Brien J G, Chiarelli T L and Mannheim P D 2018 *Physics Letters B* **782** 433–439

[8] ManceraPia P E, Fraternali F, Oosterloo T, Adams E A K, Oman K A and Leisman L 2021 *Monthly Notices of the Royal Astronomical Society* **512** 3230–3242 ISSN 0035-8711 (*Preprint* `https://academic.oup.com/mnras/article-pdf/512/3/3230/43286857/stab3491.pdf`) URL `https://doi.org/10.1093/mnras/stab3491`

[9] Bizyaev D, Makarov D I, Reshetnikov V P, Mosenkov A V, Kautsch S J and Antipova A V 2021 *The Astrophysical Journal* **914** 104 URL `https://doi.org/10.3847%2F1538-4357%2Fabfb03`

[10] Bizyaev D V, Kautsch S J, Sotnikova N Y, Reshetnikov V P and Mosenkov A V 2016 *Monthly Notices of the Royal Astronomical Society* **465** 3784–3792 ISSN 0035-8711 (*Preprint* `https://academic.oup.com/mnras/article-pdf/465/4/3784/13774687/stw2972.pdf`) URL `https://doi.org/10.1093/mnras/stw2972`

[11] Kregel M and van Der Kruit P C 2004 *Monthly Notices of the Royal Astronomical Society* **352** 787–803 ISSN 0035-8711 (*Preprint* `https://academic.oup.com/mnras/article-pdf/352/3/787/18489366/352-3-787.pdf`) URL `https://doi.org/10.1111/j.1365-2966.2004.07978.x`

[12] Mannheim P D and O'Brien J G 2012 *Phys. Rev. D* **85** 124020

[13] Mannheim P D and O'Brien J G 2011 *Phys. Rev. Lett.* **106** 121101

[14] Mannheim P D 2006 *Prog. Part. Nucl. Phys.* **56** 340

[15] Milgrom M 1983 *ApJ* **270** 365

[16] Swaters R A, Sanders R H and McGaugh S S 2010 *ApJ* **718** 380–391 (*Preprint* `1005.5456`)

Applying Retardation Theory to Galaxies

Michal Wagman[1], Lawrence P. Horwitz [1,2,3] and Asher Yahalom [4,5]

[1] Department of Physics,Ariel University, Ariel 40700, Israel
[2] School of Physics, Tel Aviv University, Ramat Aviv 69978, Israel
[3] Department of Physics, Bar Ilan University, Ramat Gan 52900, Israel
[4] Faculty of Engineering,Ariel University, Ariel 40700, Israel
[5] Princeton University, Princeton, New Jersey 08543, USA

E-mail: mwgman@gmail.com, larry@tauex.tau.ac.il, asya@ariel.ac.il

Abstract. The missing mass problem has been with us since the 1970s, as Newtonian gravity using baryonic mass cannot account for various observations. We investigate the viability of retardation theory, an alternative to the Dark Matter paradigm (DM) which does not seek to modify the General Principal of Relativity but to improve solutions within it by exploring its weak field approximation to solve the said problem in a galactic context. This work presents eleven rotation curves calculated using Retardation Theory. The calculated rotation curves are compared with observed rotation curves and with those calculated using MOND. Values for the change in mass flux to mass ratio are extracted from the fitting process as a free fitting parameter. Those quantities are interpreted here and in previous works using given galactic processes. Retardation Theory was able to successfully reproduce rotation curves and a preliminary correlation with star birthrate index is seen, suggesting a possible link between galactic winds and observed rotation curves. Retardation Theory shows promising results within current observations. More research is needed to elucidate the suggested mechanism and the processes which contribute to it. Galactic mass outflows carried by galactic winds may affect rotation curves.

1. Introduction

It is well known that Newtonian Gravity can be shown to be a limiting case of General Relativity (GR). To achieve this the following conditions are imposed:

- A weak field: $g_{\mu\nu} = \eta_{\mu\nu} + h_{\mu\nu}$, where g is the metric tensor, η flat Lorentz metric and h are small deviations from the flat metric ($|h_{\mu\nu}| << 1$).
- Non relativistic motion: $\tau \cong t$, $dx^0/d\tau \cong c$, $dx^i/d\tau = v^i << c$ where τ is the proper time, t coordinate time, x spatial coordinates, c the speed of light in vacuum, v speed of the object.
- $h_{00} = 2\phi/c^2$ where ϕ is the Newtonian potential

As Newtonian Gravity has been used to explain celestial observations from its inception, primarily on the solar scale, and as GR showed Newtonian gravity can be considered a special case of GR in the weak field approximation, very little has changed when analyzing non-dense celestial objects observations such as, for example, galaxies; this is despite the fact these involve distances of many orders of magnitude beyond those for which Newtonian Gravity has been probed for experimentally.

Content from this work may be used under the terms of the Creative Commons Attribution 3.0 licence. Any further distribution of this work must maintain attribution to the author(s) and the title of the work, journal citation and DOI.
Published under licence by IOP Publishing Ltd

In order to justify discrepancies between observed phenomena and Newtonian gravity, when light is used as a tracer for mass, a new type of matter dubbed Dark Matter (DM) has been added to the Newtonian model in the 1970s [1, 2, 3]. In order to comply with said observations and structure formation simulations a list of properties has been attributed to DM [1], which will not be listed here. However, to date, 50 years since its inception, DM has not been observed nor any known particles could be identified with its properties.

The similarities between General Relativity and Electromagnetic (EM) Theory have not gone unnoticed. While exact solutions have proven challenging due to the non-linear nature of the Einstein equations, within the weak field approximation, $h_{\mu\nu}$ can be approximated to its first order, making the equations linear, therefore superposable, yielding a Retarded Potential [16, 42] similar to that of EM theory [33]:

$$\bar{h}_{\mu\nu}(\vec{x},t) = -\frac{4G}{c^2}\int \frac{T_{\mu\nu}(\vec{x}',t_{ret})}{R} d^3x', \qquad t_{ret} \equiv t - \frac{R}{c} \tag{1}$$

where $\bar{h}^{\mu\nu} = h^{\mu\nu} - 1/2\eta^{\mu\nu}h^{\lambda}_{\lambda}$, $T_{\mu\nu}$ the stress energy tensor, $t \equiv x^0/c$, $R = |\vec{R}|, \vec{R} = \vec{x} - \vec{x}'$. For non-relativistic velocities $v << c$ we have: $dv^a/dt \cong c^2/4\bar{h}^{;a}_{00}$ and the potential can be written as:

$$\phi = -\frac{c^2}{4}\bar{h}_{00} = -\frac{G}{c^2}\int \frac{T_{00}(\vec{x}',t_{ret})}{R} d^3x' = -G\int \frac{\rho(\vec{x}',t-R/c)}{R} d^3x'. \tag{2}$$

where ρ is the density function.

When focusing on the far field solutions of the Retarded Potential eq. (2), in a similar manner to EM theory, the exciting possibility of gravitational waves has risen [16, 42]. For many years, the community has focused on this possibility of Gravitational Waves, which have only recently been observed [4] with much excitement, creating new possibilities.

Since the far field approach is not suitable within galaxies, or even at their edge, as the galaxy is the source system, retardation theory [45, 46, 47, 48, 49, 50, 51, 52] focus on the near field solutions, yielding a near field regime. For a relatively small deviation in time, such as would be the case in galaxies, the Taylor expansion of the retarded potential eq. (2) yields:

$$\begin{aligned} \phi &= -G\int \frac{\rho(\vec{x}',t)}{R} d^3x' + \frac{G}{c}\int \frac{\partial \rho(\vec{x}',t)}{\partial t} d^3x' - \frac{G}{2c^2}\int R\frac{\partial^2 \rho(\vec{x}',t)}{\partial t^2} d^3x' \\ &- \int \frac{G}{R}\sum_{3}^{\infty} \frac{1}{n!}\frac{\partial^n \rho(\vec{x}',t)}{\partial t^n}\left(-\frac{R}{c}\right)^n d^3x' \end{aligned} \tag{3}$$

It is assumed terms of order $n = 3$ and higher can be neglected (we expand on this in our discussion) and as the second term does not contribute the potential can be written as

$$\phi_{ret} = -G\int \frac{\rho(\vec{x}',t)}{R} d^3x' - \frac{G}{2c^2}\int R\frac{\partial^2 \rho(\vec{x}',t)}{\partial t^2} d^3x' \tag{4}$$

This is just the Newtonian Gravity potential (1st term) with a "correction" term. When comparing with the conditions which yield Newtonian Gravity, we see that the only difference is that we calculate the potential at the retarded time t_{ret} which is expanded for small deviations from t, yielding a near field regime. Trying to apply the potential (4) to galactic rotation curves calculations, we encounter an obstacle; we do not have an explicit expression for the density time-dependent function $\rho(\vec{x},t)$. This makes calculations using the second time derivative complicated. However, we may assume that this function can be written as a combination of addition and multiplication of spatial and temporal functions:

$$\rho(\vec{r}',t) = \rho_1(\vec{r}') + m(t)n(\vec{r}') \tag{5}$$

which enables us, after some calculations, to write eq. (4) as

$$\phi_{ret} = -G\int \frac{\rho(\vec{x}',t)}{R}d^3x' - \frac{\ddot{M}}{M}\frac{G}{2c^2}\int R\rho(\vec{x}',t)d^3x' \tag{6}$$

where M is the total mass enclosed.

2. Methods

Due to the observationally short time scale (in comparison to the galactic time scale) and data extraction methods which include averaging, it is reasonable to assume $\rho(\vec{x}',t) \approx \rho(\vec{x}')$, thus temporal dependence is not explicitly expressed, for which eq. (6) becomes:

$$\phi = -G\int \frac{\rho(\vec{x}')}{R}d^3x' - \frac{\ddot{M}}{M}\frac{G}{2c^2}\int R\rho(\vec{x}')d^3x' \tag{7}$$

This work presents eleven rotation curves, for which the mid-plane circular velocity was derived from eq. (7) by:

$$v(r,0,0) = \sqrt{\Upsilon\frac{1}{r}\int \rho(\vec{x}')d^3x - \Upsilon\frac{r}{2c^2}\ddot{M}}, \qquad R \simeq r \tag{8}$$

where Υ is the mass-to-light ratio (it is assumed that the contribution of the retardation correction is only significant outside the main mass of the galaxy, for a more accurate treatment see [47]). The spatial integral was calculated numerically using the hypergeometrical method [9] according to the mass model of the galaxy; with the exception of one galaxy (NGC 3198) for which the thick disk model by Van der Kruit & Searle [37] was used (with scale height = scale length/4.8 [38]) to obtain a fit, galaxies were fitted using the double-thin mass model by Eckhardt and Pestaña [9]. The mass-to-light ratio, Υ, and the change in mass flux to mass ratio, $\ddot{M}/M$, were both used as free fitting parameters.

The data used to derive these rotation curves was taken from various literature sources [23, 27, 17, 40, 30, 43, 7, 29, 14]. The original data deprojection smoothing and processing of both the published observed photometry and the published observed rotation curves (RC) were adopted without further data processing attempts; as there are no free parameters when deriving gas velocities from the observed data, these too, when they exist, were adopted as is.

For comparison, the MONDian rotation curve for each galaxy was calculated in tandem, using the same parameters and the same mass-to-light ratio, however, unlike in various other works (for example [24]), a_0 was allowed to vary between galaxies as in its inception [26]; to date a physical justification for a universal a_0 does not exist.

3. Results

Plots are presented in Figure 1, Figure 2 and summarized in Table 1. For all plots the obtained change in mass flux (but not necessarily the mass flux) is negative, this means the flux of mass towards the center of the galaxy decreases. As can be seen from the plots, for most cases the Newtonian component contribution is dominant in the initial rise of the curve. The mass-to-light ratio diverges but values are within reason and range from 0.6 (M33) to 9 for Galaxy 3917 which is a LSB galaxy. It can be seen that the RC calculated using Retardation Theory are able to match or improve the MONDian fit. Columns 6 & 7 of Table 1 shows the values of the change in mass flux and birthrate index respectively. The data for the birthrate index was taken from Portinari et al [28] who averaged data from several sources, if data was not available for the specific galaxy type in Portinari et el. (e.g. type Sm; galaxy with a bulge) data was taken from Kennicut et. al [13]. A preliminary correlation between the mass flux change and the birthrate index can be seen, with similar values for similar galaxy types. While the value

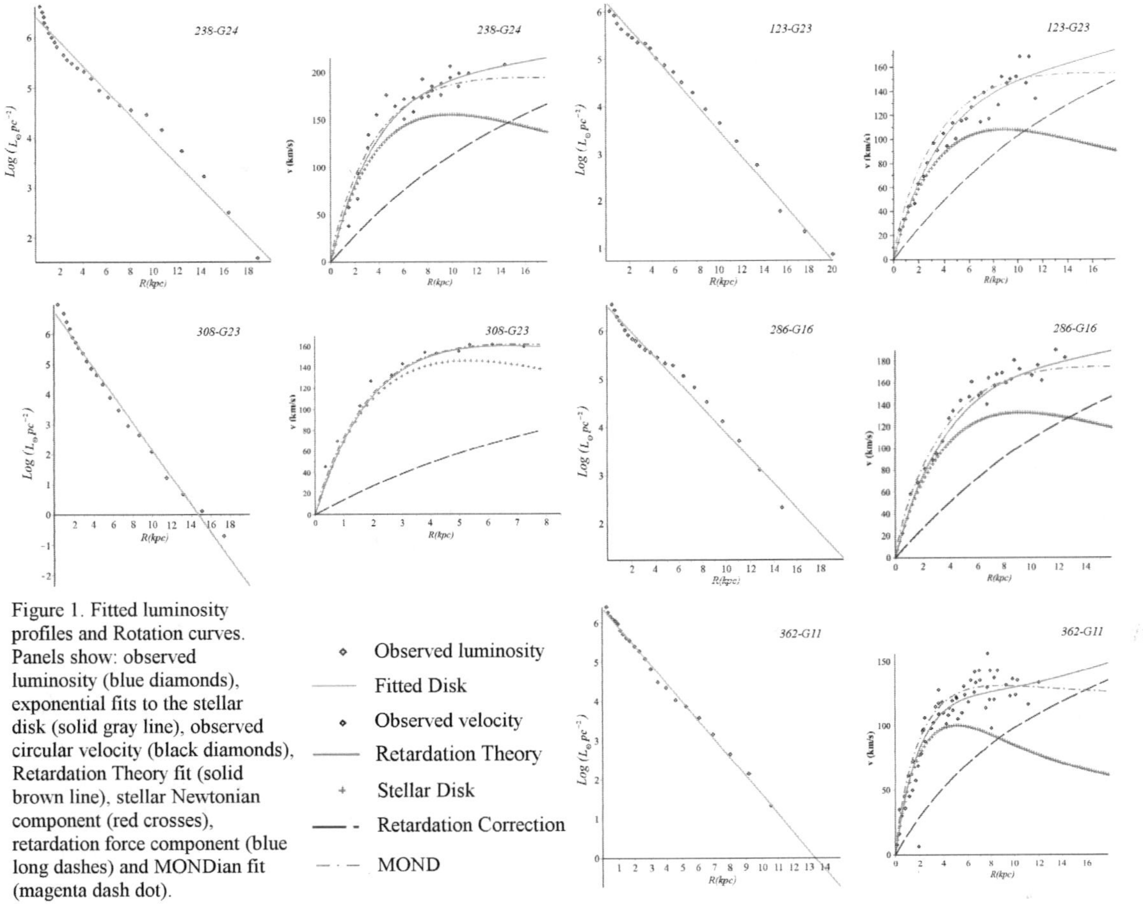

Figure 1. Fitted luminosity profiles and Rotation curves. Panels show: observed luminosity (blue diamonds), exponential fits to the stellar disk (solid gray line), observed circular velocity (black diamonds), Retardation Theory fit (solid brown line), stellar Newtonian component (red crosses), retardation force component (blue long dashes) and MONDian fit (magenta dash dot).

of $\ddot{M}/M$ suggests what seems to be quite a large decrease in flow towards the galactic center, the correlation with the birthrate index could explain this value as feedback from star formation processes has been shown to trigger mass outflows from the galaxy via thermodynamical activity and induce pressure differences [10, 21, 44].

Data for galaxies 123-G23, 238-G24, 286-G16, 308-G23 and 362-G11 was taken from Persi and Salucci [27] and Mathewson, Ford, and Buchhorn [23]. In their paper Persi and Salucci divide the data into three different quality categories, data was taken from the category deemed by the researchers as of "Excellent high quality rotation curves, suitable for individual detailed mass modelling". The original RC data did not include error bars; as we are unfamiliar with observational weighted average methods nor with the galaxies' observed data we chose to present the data as in their paper. Luminosity data was fitted using the exponential disk profile and rotation curves were calculated accordingly, both are presented in Figure 1. The mass-to-light ratios range from 0.75-1.3, which is in line with galaxy brightness. For type Sb (238-G24, 308-G23, 286-G16) the rate of change in the mass flux to mass ratio ranges between $-1.12 \cdot 10^{-24}$ to $-1.46 \cdot 10^{-24}$ and for type Sc (123-G23, 262-G11) it is $-2 \cdot 10^{-24}$ and $-3.2 \cdot 10^{-24}$ respectively. This correlates to the birthrate of 0.35 for type Sb and 0.9 for type Sc. The highest fitted mass-to-light ratio (M/L) in the sample is of 9 and is for Galaxy NGC 3197 which is a LSB, followed by NGC 3198 (M/L = 4.6) and NGC 300 (M/L = 2.5). While NGC 3198 has a small bulge, a good fit can be attained without it by subtracting the bulge data from the luminosity

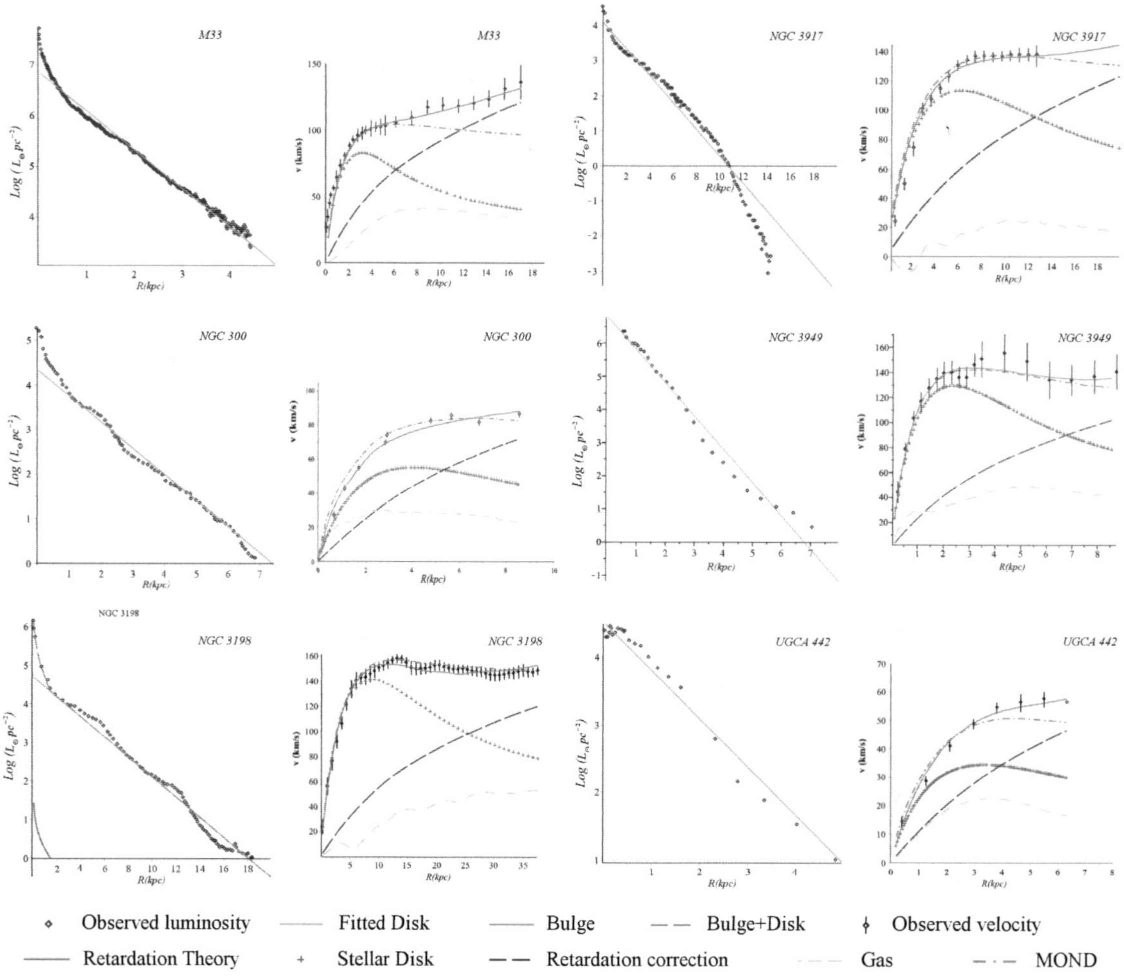

Figure 2. Fitted luminosity profiles and Rotation curves. Panels show: observed luminosity (blue diamonds), exponential fits to the stellardisk (solid gray line), bulge fit (red solid line), sum of bulge+disk (long brown dashes), observed circular velocity (black diamonds w/ error bar), Retardation Theory fit (solid brown line), stellar Newtonian component (red crosses), retardation force component (blue long dashes), gas component (short green dashes) and MONDian fit (magenta dash dot).

data, this results with brightness values similar to LSB galaxies. NGC 300 is a LSB galaxy. Therefore M/L values seem in line with the data, it is important to note that most RC fitted today use the DM paradigm which systematically underestimate the M/L ratios in order to avoid a “doughnut” shape distribution of dark matter, therefore it is expected for our M/L ratio to differ from most current literature.

The fit for galaxy NGC 3949 is not as good as others of our cases (although slightly better than MOND), as the rotation curve is not flat and exhibits a small “wave” in its shape. Richards et al. [30], similar to other literature, found that the rotation curve derived from the receding side of the HI velocity field rises faster than the approaching side, resulting in an asymmetric RC. It is possible that one side of the galaxy is more active, or its surroundings has a larger mass reservoir and therefore the change in mass flux is inconsistent throughout and too local to attain a good fit with the homogeneous mass flux change we apply; Verheijen [40] notes a faint companion 1.5 arcmin to the north that may be interacting with this galaxy.

Table 1. Rotation Curve fitting data for 11 galaxies.

Galaxy Name (1)	Galaxy Type (2)	Mass Model (3)	Scale Length (pc) (4)	Σ_0 ($M_\odot pc^{-2}$) (5)	$\ddot{M}/M$ $\left(10^{-24}/s^2\right)$ (6)	b (7)	M/L ratio ($M_\odot/L_\odot$) (8)	a_0 (MOND) (m/s^2) (9)
NGC 3198[1]	Sbc	Thick	3848	110	−0.688	0.27*	4.6	$0.5 \cdot 10^{-10}$
238 G-24[2,3]	Sb	Double	4100	602	−1.12	0.35	0.9	$1.2 \cdot 10^{-10}$
308-G23[2,3]	Sb	Double	2198	812	−1.29	0.35	1.3	$1 \cdot 10^{-10}$
286-G16[2,3]	Sb	Double	3821	665	−1.46	0.35	0.75	$1.2 \cdot 10^{-10}$
N3917[4]	Scd	Double	2600	60	−1.5	0.69	9	$0.8 \cdot 10^{-10}$
123-G23[2,3]	Sc	Double	3675	493	−2	0.9	0.7	$1.2 \cdot 10^{-10}$
362-G11[2,3]	Sc	Double	2108	572	−3.2	0.9	0.9	$1.2 \cdot 10^{-10}$
N3949[5]	Sbc	Double	995	916	−4.8	0.9	2	$1.2 \cdot 10^{-10}$
M33[6,7]	Sc	Double	1315	944	−6.23	0.9	0.6	$0.8 \cdot 10^{-10}$
NGC 300[8]	Sc	Double	1700	354	−8.17	0.9	2.5	$0.7 \cdot 10^{-10}$
UGCA 442[9]	Sm	Double	1400	92	−14.6	1.67*	1	$0.25 \cdot 10^{-10}$

Columns are: (1) Name of the galaxy. Observational data for each galaxy was taken from its respective reference (2) Galaxy type, (3) Mass model used for calculation where Double is short for double infinitesimally thin exponential disk model and Thick is short for exponential disk with a $sech^2$ distribution in the z-axis. (4) Exponential disk model scale-length given in parsecs, (5) Central mass density, (6) mass flux change-to-mass ratio, (7) star birthrate of the galaxy type from Portinari et al [28] and Kennicutt et al [13]; data retrieved from Kennicutt et al is denoted with an asterisk*. (8) mass-to-light ratio (9) MOND acceleration constant. Sources: (1) [43]; (2) [27]; (3) [23]; (4) [40]; (5) [30]; (6) [7]; (7) [29]; (8) [14]; (9) [17].

Galaxy UGCA 442 is a dwarf galaxy and its mass flux change is the highest in our sample; this result correlates well with the high star birthrate index for dwarf galaxies (1.67) as dwarf galaxies are characterized by star formation bursts and the star birthrate index relates to change in star formation rates.

The fit for galaxy M33 using retardation theory is significantly better than that of the MONDian curve. The change in mass flux is similar to other type Sc galaxies and its mass-to-light ratio is within reason (0.6) (this can be improved by using a more accurate estimation of the Newtonian and retardation contributions [47], which shows that the mass-to-light ratio in the galaxy M33 is one). This improved fit demonstrates how versatile retardation theory is and can fit various RC shapes.

4. Discussion

4.1. Origins of retardation

How can one reach an erroneous conclusion regarding the existence of "dark matter"? By ignoring retardation effects and assuming that radial velocities are a result of some mysterious substance. We obtain for a spherically symmetric distribution [5]:

$$-\frac{v_c^2}{r}\hat{r} = \vec{F}_d = -\frac{GM_d(r)}{r^2}\hat{r} \tag{9}$$

where v_c is the speed of a particle of unchanging radius r and $M_d(r)$ is the dark matter within the radius r. Comparing Equations (9) and (8), we deduce that the "dark matter" mass can be calculated as follows:

$$M_d(r) = \frac{r^2|\ddot{M}|}{2c^2} \tag{10}$$

Let us write the continuity equation:

$$\frac{\partial \rho}{\partial t} + \vec{\nabla} \cdot (\rho \vec{v}) = 0 \tag{11}$$

$\vec{\nabla}$ has its standard meaning in vector analysis. Let us now take a partial temporal derivative of equation (11) leading to:

$$\frac{\partial^2 \rho}{\partial t^2} + \vec{\nabla} \cdot (\frac{\partial \rho}{\partial t}\vec{v} + \rho \frac{\partial \vec{v}}{\partial t}) = 0. \tag{12}$$

Using equation (11) again we obtain the expression:

$$\frac{\partial^2 \rho}{\partial t^2} = \vec{\nabla} \cdot \left(\vec{\nabla} \cdot (\rho \vec{v})\vec{v} - \rho \frac{\partial \vec{v}}{\partial t}\right). \tag{13}$$

We divide the left and right hand sides of the equation by c^2 and obtain:

$$\frac{1}{c^2}\frac{\partial^2 \rho}{\partial t^2} = \vec{\nabla} \cdot \left(\frac{\vec{v}}{c}\left(\frac{\vec{v}}{c} \cdot \vec{\nabla}\rho + \rho \vec{\nabla} \cdot (\frac{\vec{v}}{c})\right) - \rho \frac{1}{c}\frac{\partial \frac{\vec{v}}{c}}{\partial t}\right). \tag{14}$$

Since $\frac{\vec{v}}{c}$ is rather small in galaxies and galaxy clusters it follows that $\frac{1}{c^2}\frac{\partial^2 \rho}{\partial t^2}$ is also small unless the density or the velocity have significant spatial derivatives. A significant acceleration $\frac{\partial \frac{\vec{v}}{c}}{\partial t}$ resulting from a considerable force can also have a decisive effect, this may be connected to star formation processes as discussed below. The depletion of available gas can also cause such gradients as we describe using a detailed model in [47]. Taking the volume integral of the left and right hand sides of equation (14) and using Gauss theorem we arrive at the following equation:

$$\frac{1}{c^2}\ddot{M} = \frac{1}{c^2}\int \frac{\partial^2 \rho}{\partial t^2} d^3x = \oint d\vec{S} \cdot \left(\frac{\vec{v}}{c}\left(\frac{\vec{v}}{c} \cdot \vec{\nabla}\rho + \rho \vec{\nabla} \cdot (\frac{\vec{v}}{c})\right) - \rho \frac{1}{c}\frac{\partial \frac{\vec{v}}{c}}{\partial t}\right). \tag{15}$$

The surface integral is taken over a surface encapsulating the galaxy or galaxy cluster. This leads according to equation (10) to a "dark matter" effect of the form:

$$M_d(r) = \frac{r^2}{2}\oint d\vec{S} \cdot \left(\frac{\vec{v}}{c}\left(\frac{\vec{v}}{c} \cdot \vec{\nabla}\rho + \rho \vec{\nabla} \cdot (\frac{\vec{v}}{c})\right) - \rho \frac{1}{c}\frac{\partial \frac{\vec{v}}{c}}{\partial t}\right). \tag{16}$$

Thus we obtain the order of magnitude estimation:

$$\left[\frac{M_d(r)}{M}\right] \approx \left(\frac{v}{c}\right)^2 \left[\frac{r}{l_\rho} + \frac{r}{l_v} + \frac{r}{l_d}\right] \tag{17}$$

In the above we define three gradient lengths:

$$l_\rho \equiv \frac{\rho}{|\vec{\nabla}\rho|}, \qquad l_v \equiv \frac{v}{|\vec{\nabla} \cdot v|}, \qquad l_d \equiv \frac{v^2}{|\partial_t v|} \tag{18}$$

We can also write:

$$\frac{1}{l_t} = \left[\frac{1}{l_\rho} + \frac{1}{l_v} + \frac{1}{l_d}\right] \tag{19}$$

in which the smallest gradient length will be the most significant one in terms of the "dark matter" phenomena. In the depletion model of [47] we assume that l_ρ associated with density gradients is the shortest length scale, this is only an assumption, other alternatives are possible as described below. For galaxies we have $(\frac{v}{c})^2 \approx 10^{-6}$, hence the factor $\frac{r}{l_t}$ should be around 10^6 to have a significant "dark matter" effect.

4.2. Correlation with star birthrate index

We have presented here a sample of galaxies for which retardation theory was able to reproduce rotation curves in good agreement with observational data, fitting different types of galaxies, with different rotation curve shapes. In the fitting process two parameters were used as free fitting parameters, the mass-to-light ratio, for which values are within reason, and the change in mass flux which is part of the retardation effect coefficient of eq. (7) (second term). As this coefficient is a function of the mass flux, the retarded effect contribution can be fine-tuned for each galaxy, resulting in a versatility which enables us to produce various RC shapes, from the prevalent flat rotation curve (308-G23, for example) to M33's rising curve. The change in the mass flux to mass ratio was considered cohesive throughout the galaxy, this is in line with a radial mass flow, or, mass flux towards the galactic center which varies with time. Radial or non-circular mass flows between the Interstellar Medium (ISM) and the Intergalactic Medium (IGM) have been recorded by various works since the 1960's [39, 32, 25]. Although this parameter was fitted for each galaxy separately, a trend in its values is seen; a preliminary correlation between the change in mass flux towards the center of the galaxy and the galaxy's star birthrate index, estimated by galaxy type, was observed.

Star formation feedback has been shown to generate mass outflow via superwinds, large-scale galactic outflows [39]. Woo et al. [44] saw that the correlation between star formation rate (SFR) and ionized gas outflow is clear, while other works [21, 22, 31, 10, 12, 34] show the velocity of these winds to be very large, and can exceed the galaxy's escape velocity. This clear correlation between SFR and mass outflows suggest that a change in the star formation rate would be expressed as a change in said mass outflow. The star birthrate index is defined by [13]:

$$b = \frac{SFR}{\langle SFR \rangle_{past}} \tag{20}$$

which relates to change in SFR. Therefore, the correlation between star birthrate index and the change in mass flux towards the galactic center may be linked via the change in mass outflows.

4.3. Truncation of the Taylor series

As the contribution of the second term of eq. (7) is large, we must justify the assumption we made yielding eq. (4), namely, neglecting terms of order $n > 2$ of equation (3); the derivatives of equation (3) are in fact of mass over time, as can be seen better in eq. (7). While one could assume depletion of Intergalactic Medium (IGM) with an exponential decay, in which case terms of order $\mathcal{O}(R/c)^3$ and higher might be shown to be neglected [48], the correlation with the birthrate index suggests that the contribution to the change in mass flux due to SFR activity to the $\ddot{M}/M$ coefficient is significant. Therefore, we delve into the processes which lead to galactic superwinds. While the field is relatively young and the mechanism is not fully understood, various observations and simulations shed light on a possible mechanism [21, 12, 39, 15]: Galactic superwinds are powered by the energy deposited in the interstellar medium by massive stars via supernovae (SNe) and stellar winds. The ejecta from hot stars and SNe in the star burst region quickly interact and mix with the surrounding gas to produce a cavity of hot material, this over-pressurized "bubbles" of hot material propagate and expand in the least resistance path, propagating laterally in the plane of the galaxy and carried vertically out of the disk. As this bubble expands it carries cold gas and dust along with it, after which it will evolve into a "blowout" phase as it accelerates outwards and fragments, propagating out into the IGM. As mentioned, the trigger mechanism of these super winds is SNe and solar winds activity, these are highly localized (~ 1 kpc) and relatively instantaneous when compared to galactic superwinds which have been observed at great distances from galaxies enriching the IGM with metals over 100 kpc away [36], with outflows spanning from 0.1 to 10 Myr [39]. As $\ddot{M}/M$ is taken to be cohesive throughout the galaxy, the change in mass flux would correlate with that of superwinds while the trigger processes are too small or too short to have a detectable effect on midplane mass flux towards the galactic center. Of high interest to this argument is Heckman et al.'s [12] work, in which they observe that for a typical outflow matter is injected at roughly zero velocity, therefore gas is "loaded" into the outflow at $v \approx v_{sys}$[1] and is accelerated up to some v_{term} which is significantly larger than v_{sys} (x4 for the largest value in their sample). As mass is loaded at "zero" and then accelerated outwards we believe it is reasonable to assume that while the mass outward acceleration in superwinds disrupt the midplane mass flux towards the galactic center, higher derivatives contribution effects which would have disrupted this pre-loaded mass are negligible in this phenomenon.

4.4. The size of the second term coefficient

The values we get for the free fitting parameter $\ddot{M}/M$ range between $-0.688 \cdot 10^{-24}s^{-2}$ to $-14.6 \cdot 10^{-24}s^{-2}$. While direct measurement of interstellar material are hard due to dust and kinematics, estimations for Galactic outflows, Galactic winds and starburst superwinds range from tens to over a thousand Solar masses per year [10]. If one assumes a galactic mass of 10 billion solar masses, this estimation would yield a ratio of mass flux to mass range of $3.17 \cdot 10^{-17}$ to $3.17 \cdot 10^{-15}$ per second (or $\dot{M}/M$). However, said outflow mass flux estimations are based on a constant average mass flow over the wind's lifetime while the observational picture does not suggest constant mass flux within the galactic disk. Mass outflows have been observed to be correlated with SFR activity; Rupke et al. [31], for example, received a linear relation between mass outflows and SFR which is larger than one ($dM/dt/SFR > 1$) while Martin [21] found that for dwarf galaxies this relation may be a few times the rate of SFR. This correlation between outflows and SFR activity means that a change in the SFR activity will be expressed as a matching or greater, change in mass outflow. More over, as detailed in the previous section, galactic super winds are drived by stellar winds and supernovae, generating blast waves [18, 41], and as Heckman et al. [12] and others (for example [34, 11, 21]) observed, the velocities of mass loaded into the super wind can be times larger than the systematic rotational velocity, it is estimated that between $10^5 - 10^8 M_\odot$ of ionized gas takes part in galactic super winds outflows [39, 12]. While direct measurements and observations of the activity in the ISM are proven very difficult due to dust, remnants and kinematics, one observation which strengthen large quantities of mass leaving the galaxies is the "missing metals problem" [8]. The ISM contains, among others, heavier than hydrogen components such as He, C, N, O, Ne, Fe and Dust (Helium, which is estimated to make up 8% of the ISM by number, contributes 40% by mass) [6]. While metals are formed exclusively in the hot cores of stars, their amounts in galaxies fall well short of the amount expected to be produced by their stars; it is estimated that galaxies retain only 20-50% of their metals while the IGM is enriched with metals to a detectable level estimated of 0.3 solar metalicity [8, 25, 12]; superwinds are thought to be the main mechanism for this enrichment, therefore carrying metals as well. Given the large mass quantities which reside in the ISM, the strong acceleration within galactic superwinds, galactic superwinds mechanism of shells removing material from the disk which only in part leaves the galaxy, and estimation of mass carried out of the galaxy of $10^5 - 10^8 M_\odot$ for gas alone, it seems that our change in the galactic midplane mass flux to mass ratio towards the galactic center range of $-0.688 \cdot 10^{-24}s^{-2}$ to $-14.6 \cdot 10^{-24}s^{-2}$ (or $\ddot{M}/M$) is within reason, though, of course, more research is needed to shed light on mechanism and quantities.

If mass were to leave the galaxy at such constant rate throughout the galaxy's life it would, of course, deplete the galaxy's entire mass making it unstable. However, its correlation with birthrate index suggests this term is not static but changes in tandem with galactic SFR processes. In eq. (7) we make the assumption that $\rho(\vec{x}', t) \approx \rho(\vec{x}')$ due to the short observational

[1] v_{sys} denotes the systematic rotational velocity, the velocity observed at the "plateau" of the galactic rotation curve.

time scale. Galactic superwinds can span from 0.1 to 10 Myr [39] therefore the size of $\ddot{M}/M$ is only static in the context of said observational time and is not static for the lifetime of the galaxy, nor for the lifetime of the galactic super wind (more details about correct and erroneous estimation of galactic mass, based on calculated second derivatives can be found in [47]).

4.5. Dwarf galaxies
Unlike most spiral galaxies, the rotation curve shape of dwarf galaxies usually does not conform to the flat rotation curve (see, for example, Swaters et. al [35] or Haghighi and Rahvar [54]) and its shape can differ from it greatly. While our sample contains only one Dwarf galaxy, the observed star formation bursts in dwarf galaxies could suggest that the retarded effect is much stronger in dwarf galaxies than in spiral galaxies; this may explain this deviation from the flat rotation curve. In addition, localized strong activity could account for non-smooth "wiggly" curves observed in some dwarf galaxies. We therefore believe that in order to further study the galactic rotation curves of dwarf galaxies, a more detailed approach, one which investigates the individual galaxy's dynamics and kinematics in tandem with the observed rotation curve should be taken; however, such research is currently beyond our scope.

The correlation with the birthrate index is an encouraging result which strengthens the theory underlying assumptions of effects due to changes in mass flux. The values used here for the galactic star birthrate index were not individually matched per galaxy but provided by works [28, 13] which try to estimate the star birthrate index, which relates to the change in SFR, by galaxy type. We therefore believe that further refining these values, investigating the mechanism and correlation between the change in mass flux and the change in SFR per galaxy may produce an additional tool for investigating galactic dynamics and kinematics, by using the relatively simple to observe rotation curve.

It is important to note, however, that the value we receive for our change in the mass flux to mass ratio coefficient is deduced by fitting and gives the net change in mass flux to mass ratio needed in order to receive good rotation curve fits. While the large mass quantities carried by galactic superwinds and their kinematics seem to be able to account for this value, more research is needed in order to elucidate the mechanism and determine accurate quantities and their exact contribution, which may differ from the required net change. Gas accretion, for example, currently estimated at $0.1 - 0.25 M_{\odot} yr^{-1}$ or even ten times as that [32], may reduce mass flux towards the galactic center by slowing down due to changing thermal and pressure differences between the ISM and IGM or due to HI depletion of the IGM. Other dynamical processes and forces may be at play as well.

5. Conclusion
It is interesting to note the similarity between Retardation Theory's potential with that of Mannheim's Conformal Gravity early linear potential [19, 20], a theory which seeks to replace GR and has been very successful reproducing rotation curves:

$$\phi^*(r) = -\frac{\beta^* c^2}{r} + \frac{\gamma^* c^2}{2} r \tag{21}$$

where $\phi^*(r)$ is the potential per unit solar mass and $\beta^* = GM_{\odot}/c^2$, to that of eq. (7). If one were to equate this with eq. (7) then γ^* would equal $\gamma^* = -G\ddot{M}/c^4$.

While other theories and paradigms seek to modify or replace current existing theories, Retardation Theory does not strive to change Newtonian theory nor the General Principal of Relativity. The theory seems elegant, with only two free parameters whose origin is clear, with good results reproducing RC for our sample and preliminary correlation of change in mass flux with star birthrate index (which relates to the change in SFR and therefore mass outflows), indicating the effect of mass outflow on rotation curves may be significant. We believe the theory has great potential for future research and are currently studying additional aspects of the theory. We also mention that using the weak gravitational field we have also obtained recently an improvement regarding the estimation of Mercury's perihelion precession [53] but this is on a completely different scale.

Acknowledgements
One of us (MW) would like to thank A. Kashi and A. M. Michaelis for useful conversations. Two of us (MW and AY) would like congratulate our senior collaborator Professor Lawrence Paul Horwitz for his ninety two birthday.

References

[1] de Swart, J. G. and Bertone, G. and van Dongen, J., How dark matter came to matter, Nature Astronomy, 2017, 1, 0059 Macmillan Publishers Limited https://doi.org/10.1038/s41550-017-0059
[2] Rubin, V.C.; Ford, W.K., Jr. Rotation of the Andromeda Nebula from a Spectroscopic Survey of Emission Regions. *Astrophys. J.* **1970**, *159*, 379.
[3] Rubin, V.C.; Ford, W.K., Jr.; Thonnard, N. Rotational Properties of 21 Sc Galaxies with a Large Range of Luminosities and Radii from NGC 4605 (R = 4kpc) to UGC 2885 (R = 122kpc). *Astrophys. J.* **1980**, *238*, 471.
[4] Abbott, B. P., Abbott, R., Abbott, T. D., et al. 2016, Phys. Rev. Lett., 116, 061102
[5] Binney, J.; & Tremaine, S. *Galactic Dynamics*; Princeton University Press: Princeton, NJ, USA, 1987.
[6] Brinks, E. 1990, in Astrophysics and Space Science Library, Vol. 161, The Interstellar Medium in Galaxies, ed. J. M. S. Harley A. Thronson Jr (Springer-Science+Business Media, B.V), 39-67
[7] Corbelli, E. 2003, Monthly Notices of the Royal Astronomical Society, 342
[8] Davé, R. & Oppenheimer, B. D. 2006, Monthly Notices of the Royal Astronomical Society, 374, 427
[9] Eckhardt, D. H. & Pestaña, J. L. G. 2002, The Astrophysical Journal Letters, 572, L135
[10] Fabian, A. 2012, Annual Review of Astronomy and Astrophysics, 50, 455
[11] Feruglio, C., Fiore, F., Carniani, S., et al. 2015, Astronomy & Astrophysics, 583, A99
[12] Heckman, T. M., Lehnert, M. D., Strickland, D. K., & Armus, L. 2000, The Astrophysical Journal Supplement Series, 129, 493
[13] Kennicutt, Jr., R. C., Tamblyn, P., & Congdon, C. E. 1994, The Astrophysical Journal, 435, 22
[14] Kent, S. M. 1987, The Astronomical Journal, 93, 816
[15] Kereŝ, D., Katz, N., Weinberg, D. H., & Davé, R. 2005, Monthly Notices of the Royal Astronomical Society, 363, 2
[16] Landau L.D., L. 1975, The classical theory of fields, 4th edn. (Pergamon)
[17] Lelli, F., McGaugh, S. S., & Schombert, J. M. 2016, The Astronomical Journal, 152, 157
[18] Mac Low, M.-M. & McCray, R. 1988, The Astrophysical Journal, 321, 776
[19] Mannheim, P. D. 1993, The Astrophysical Journal, 419, 150
[20] Mannheim, P. D. 1996, Foundations of Physics, 26, 1683
[21] Martin, C. L. 1999, The Astrophysical Journal, 513, 156
[22] Martin, C. L. 2005, The Astrophysical Journal, 621, 227
[23] Mathewson, D. S., Ford, V. L., & Buchhorn, M. 1992, The Astrophysical Journal Supplement Series, 81, 413
[24] McGaugh, S. 2008, The Astrophysical Journal, 683, 137
[25] McQuinn, M. 2016, Annual Review of Astronomy and Astrophysics, 54, 313
[26] Milgrom, M. 1983, The Astrophysical Journal, 270, 371
[27] Persic, M. & Salucci, P. 1995, Astrophysical Journal Supplement Series, 99, 501
[28] Portinari, L., Sommer-Larsen, J., & Tantalo, R. 2004, Monthly Notices of the Royal Astronomical Society, 347, 691
[29] Regan, M. W. & Vogel, S. N. 1994, The Astrophysical Journal, 434, 536
[30] Richards, E. E., van Zee, L., Barnes, K. L., et al. 2018, Monthly Notices of the Royal Astronomical Society, 476, 5127
[31] Rupke, D., Veilleux, S., & Sanders, D. 2005, The Astrophysical Journal Supplement Series, 160, 115
[32] Sancisi, R., Fraternali, F., Oosterloo, T., & van der Hulst, T. 2008, The Astronomy and Astrophysics Review, 15, 189
[33] Schwinger Julian, Lester L. DeRaad Jr, K. W. 1998, Classical Electrodynamics, Advanced Book Program (Reading, Massacusetts: Perseus Books).
[34] Strickland, D. K. & Heckman, T. M. 2009, The Astrophysical Journal, 697, 2030

[35] Swaters, R. A., Sancisi, R., van Albada, T. S., & van der Hulst, J. M. 2009, A&A, 493, 871
[36] Tumlinson, J. & Fang, T. 2005, The Astrophysical Journal, 623, L97
[37] van der Kruit, P. C. & Searle, L. 1981, Astronomy & astrophysics, 95, 105 Article number, page 7 of 8 A&A proofs: manuscript no. "AA RCGW 250521"
[38] van der Kruit, P. C. & Searle, L. 1982, Astronomy & Astrophysics, 110, 61
[39] Veilleux, S., Cecil, G., & Bland-Hawthorn, J. 2005, Annual Review of Astronomy and Astrophysics, 43, 769
[40] Verheijen, M. 1997, PhD thesis
[41] Wang, J., Nardini, E., Fabbiano, G., et al. 2014, The Astrophysical Journal, 781, 55
[42] Weinberg, S. 1972, Gravitation and Cosmology (New York, USA: John wiley and Sons, Inc), third print
[43] Wiegert, T. B. V. 2010, PhD thesis, university of Manitoba
[44] Woo, J. H., Son, D., & Rakshit, S. 2020, The Astrophysical Journal, 901, 66
[45] Yahalom, A. 2018, in Proceedings of the Material Technologies and Modeling the Tenth International Conference, Ariel University, Ariel, Israel
[46] Yahalom, A. 2019, Journal of Physics: Conference Series, 1239, 012006
[47] Asher Yahalom "Lorentz Symmetry Group, Retardation, Intergalactic Mass Depletion and Mechanisms Leading to Galactic Rotation Curves" Symmetry 2020, 12(10), 1693; https://doi.org/10.3390/sym12101693
[48] Yahalom A. Effects of Higher Order Retarded Gravity. Universe. 2021; 7(7):207.
[49] A. Yahalom "The Cosmological Decrease of Galactic Density and the Induced Retarded Gravity Effect on Rotation Curves" Proceedings of IARD 2020. 2021 J. Phys.: Conf. Ser. 1956 012002.
[50] A. Yahalom "Tully - Fisher Relations and Retardation Theory for Galaxies" International Journal of Modern Physics D, (2021), Volume No. 30, Issue No. 14, Article No. 2142008 (8 pages). ©World Scientific Publishing Company.
[51] Yahalom, A. Lensing Effects in Retarded Gravity. Symmetry 2021, 13, 1062.
[52] A. Yahalom "Lensing Effects in Galactic Retarded Gravity: Why "Dark Matter" is the Same for Both Gravitational Lensing and Rotation Curves" accepted by IJMPD, received 23 May 2022, Accepted 31 August 2022, to be published October 2022. https://doi.org/10.1142/S0218271822420184
[53] A. Yahalom "The weak field approximation of general relativity, retardation, and the problem of precession of the perihelion for mercury" Proceedings of the International Conference: COSMOLOGY ON SMALL SCALES 2022 Dark Energy and the Local Hubble Expansion Problem, Prague, September 21-24, 2022. Edited by Michal Krizek and Yuri V. Dumin, Institute of Mathematics, Czech Academy of Sciences.
[54] Zhoolideh Haghighi, M. H. & Rahvar, S. 2017, Monthly Notices of the Royal Astronomical Society, 468, 4048

A vielbein formalism for SHP general relativity

Martin Land

Department of Computer Science, Hadassah College, 37 HaNeviim Street, Jerusalem

E-mail: martin@hac.ac.il

Abstract. The 4+1 formalism in general relativity expresses the Einstein equations as a manifestly covariant initial value problem, resulting in a pair of first order evolution equations for the metric $\gamma_{\mu\nu}$ and intrinsic curvature $K_{\mu\nu}$ of *spacetime* geometry ($\mu, \nu = 0, 1, 2, 3$). This approach extends the Stueckelberg-Horwitz-Piron (SHP) framework, a covariant approach to canonical particle mechanics and field theory employing a Lorentz scalar Hamiltonian K and an external chronological parameter τ. The SHP Hamiltonian generates τ-evolution of spacetime events $x^\mu(\tau)$ or $\psi(x, \tau)$ in an *a priori* unconstrained phase space; standard relativistic dynamics can be recovered *a posteriori* by imposing symmetries that express the usual mass shell constraint for individual particles and fields as conservation laws. As a guide to posing field equations for the evolving metric, we generalize the structure of SHP electrodynamics, with particular attention to O(3,1) covariance. Thus, the 4+1 method first defines a 5D pseudo-spacetime as a direct product of spacetime geometry and chronological evolution, poses 5D field equations whose symmetry must be broken to 4D, and then implements the implied 4+1 foliation to obtain evolution equations. In this paper, we sharpen and clarify the interpretation of this decomposition by introducing a fixed orthonormal *quintrad* frame and a 5D vielbein field that by construction respects the preferred 4+1 foliation. We show that for any diagonal metric, this procedure enables the evolution equation for the metric to be replaced by an evolution equation for the vielbein field itself, simplifying calculation of the spin connection and curvature.

1. Introduction

In general relativity (GR) the problem of time [1, 2] generally refers to the conflicting roles assigned to time in relativity: on the one hand, the chronological parameter required for posing equations of motion, and on the other hand, one of four spacetime coordinates, themselves dynamical quantities to be determined by equations of motion. This conflict became apparent to Stueckelberg [3, 4] in his work on classical and quantum electrodynamics. In seeking to interpret antiparticles as particles whose trajectory reverses time direction, he saw that neither coordinate time nor the proper time of the motion could serve as the parameter of chronological evolution. He was thus led to introduce an evolution parameter τ, independent of phase space and external to the spacetime manifold.

In the Stueckelberg framework, a particle worldline is the trajectory of an event $x^\mu(\tau)$ or $\psi(x, \tau)$ generated by a Lorentz scalar Hamiltonian K, establishing a canonical system familiar from nonrelativistic physics. In a suitable potential, the velocity component $\dot{x}^0 = dx^0/d\tau$ and energy $E = M\dot{x}^0$ may change sign, producing a trajectory observed in the laboratory as a particle/antiparticle interaction. Such pair processes are classically prohibited by the phase space relation $\dot{x}^\mu \dot{x}_\mu = -c^2$, which is here demoted in status from constraint to conservation law, applicable to systems governed by a Hamiltonian satisfying $\partial_\tau K = 0$. Clearly, if $\dot{x}^2$ can

Content from this work may be used under the terms of the Creative Commons Attribution 3.0 licence. Any further distribution of this work must maintain attribution to the author(s) and the title of the work, journal citation and DOI.
Published under licence by IOP Publishing Ltd

change sign, the proper time of the motion $ds = \sqrt{-\dot{x}^2}\,d\tau$ will not be a well-behaved evolution parameter.

Stueckelberg's work was extended by Piron and Horwitz [5] who constructed a relativistic canonical many-body theory [6–10], later generalized to a gauge theory of interacting spacetime events that recovers Maxwell electrodynamics in τ-equilibrium [11–14]. By including τ in the U(1) gauge function (but not the spacetime manifold), SHP electrodynamics requires five gauge potentials, whose interaction suggests a 5D symmetry such as O(3,2) or O(4,1). But to preserve the observed Lorentz invariance of spacetime, any higher symmetry must be broken to tensor and scalar representations of O(3,1) by choosing an appropriate structure for the matter kinetic term in the action. An analogy may be seen in classical acoustics, where the pressure wave equation appears to be invariant under Lorentz-like transformations, but no relativistic effects are expected as an observer approaches the speed of sound.

Extending the Stueckelberg-Horwitz-Piron (SHP) framework to curved spacetime, one obtains a classical and quantum theory of event evolution [15, 16], leading to a scalar event density $\rho(x,\tau)$ and energy-momentum tensor $T_{\mu\nu}(x,\tau)$. Following Wheeler's summary [17] of Einstein gravity as "spacetime tells matter how to move; matter tells spacetime how to curve," the τ-evolution of the mass/energy/momentum distribution associated with these events entails the τ-evolution of spacetime curvature [18–21], described by a local metric $\gamma_{\mu\nu}(x,\tau)$. To find appropriate field equations for this metric, we generalize the 3+1 formalism in geometrodynamics [22–25] to 4+1, regarding the spacetime manifold $\mathcal{M}$ as a 4D hypersurface embedded in a 5D pseudo-spacetime $\mathcal{M}_5$. Writing 5D field equations on $\mathcal{M}_5$, the 4+1 decomposition leads to a pair of manifestly covariant first order τ-evolution equations for the local metric $\gamma_{\mu\nu}(x,\tau)$ and an extrinsic curvature $K_{\mu\nu}(x,\tau)$, along with a set of propagating constraints [19].

In [20, 21] we studied the linearized SHP theory and showed that consistency with the phenomenology of weak gravitation requires that any 5D symmetry implied in the matter terms of the field equations must be explicitly broken to tensor and scalar representations of O(3,1), as previously seen in SHP electrodynamics. This strategic symmetry breaking also insures that the standard 4D Einstein equations are recovered in τ-equilibrium.

In this paper, we present a vielbein field theory approach to the 4+1 formalism, permitting us to construct the O(3,1) symmetric SHP field equations in a systematic way. In analogy to Einstein's quatrad method [26] we define a constant orthonormal quintrad basis for the pseudo-spacetime $\mathcal{M}_5$, so the inner products of basis vectors are the components of a 5D flat Minkowski pseudo-metric. The vielbein field transforms the constant basis to a local coordinate basis whose inner products provide a 5D local metric that induces the $\gamma_{\mu\nu}(x,\tau)$ of 4D spacetime by projection. By specifying a 5D vielbein field that respects the preferred foliation of $\mathcal{M}_5$, we find that the spacetime part (a vierbein field) contains the dynamic evolution, while the normal part propagates forward, enforcing the constraints. We may thus take the 4D vierbein field as the fundamental geometrical object, from which we obtain the spin connection, covariant derivative, and curvature. This foliation of the vielbein simplifies the formulation of evolution equations whose symmetry is limited to O(3,1), and their interpretation in terms of standard approaches to GR. We show that for any diagonal metric we may replace the evolution equation for the metric with an evolution equation for the vielbein field, simplifying calculation of the spin connection and curvature.

In Section 2 we briefly review the previous work in SHP GR and the considerations associated with constructing coordinate and vielbein frames. The foliation of $\mathcal{M}_5$ and the associated decomposition of the vielbein field is described in Section 3. In Section 4 we generalize the Einstein field equations to the 5D pseudo-spacetime and obtain the symmetry broken form, exhibiting tensor and scalar representations of standard O(3,1) covariance. Section 5 reviews the intrinsic and extrinsic geometry of spacetime, posed as an embedded 4D hypersurface of $\mathcal{M}_5$, and the resulting decomposition of the Riemann tensor. These pieces are assembled to write the evolution equations for the 4D spacetime metric and extrinsic curvature. We also obtain evolution equations for the vielbein field, permitting the formulation of initial value problems in the quintrad frame. An example is given in Section 6.

2. The pseudo-spacetime $\mathcal{M}_5$

We briefly review previous work generalizing SHP electrodynamics to general relativity. For additional details, see the references mentioned above.

2.1. *SHP electrodynamics*

In flat Minkowski spacetime, with $\eta_{\mu\nu} = \text{diag}\,(-1,1,1,1)$, the free particle Hamiltonian

$$K = \frac{1}{2M} p^\mu p_\mu \tag{1}$$

leads to the action

$$S = \int d\tau \, \frac{1}{2} M \dot{x}^\mu \dot{x}_\mu \tag{2}$$

made maximally U(1) gauge invariant [11,27] by introducing five gauge fields as

$$S_{\text{SHP}} = \int d\tau \, \frac{1}{2} M \dot{x}^\mu \dot{x}_\mu + \frac{e}{c} \dot{x}^\mu a_\mu(x,\tau) + \frac{e}{c} c_5 a_5(x,\tau) \tag{3}$$

$$= \int d\tau \, \frac{1}{2} M \dot{x}^\mu \dot{x}_\mu + \frac{e}{c} \dot{x}^\beta a_\beta(x,\tau) \tag{4}$$

where we partition the Greek indices as

$$\alpha, \beta, \gamma, \delta = 0,1,2,3,5 \qquad\qquad \lambda, \mu, \nu, \rho \ldots = 0,1,2,3 \tag{5}$$

and write $x^5 = c_5\tau$ in analogy to the notation $x^0 = ct$. Because $\dot{x}^\mu \dot{x}_\mu$, $\dot{x}^\mu a_\mu$, and a_5 are O(3,1) scalars, the action is 4D Lorentz invariant as required. But we emphasize that S_{SHP} enjoys the 5D gauge invariance $a_\alpha(x,\tau) \longrightarrow a_\alpha(x,\tau) + \partial_\alpha \Lambda(x,\tau)$. For a pure gauge potential $a_\alpha = \partial_\alpha \Lambda(x,\tau)$, the interaction term is seen to be a total τ-derivative. We may regard (4) as a standard 5D action with a symmetry-breaking matter term

$$S_{5D} = \int d\tau \, \frac{1}{2} M \dot{x}^\alpha \dot{x}_\alpha + \frac{e}{c} \dot{x}^\alpha a_\alpha \xrightarrow[\dot{x}^5 \equiv c_5]{} S_{\text{SHP}} = \int d\tau \, \frac{1}{2} M \dot{x}^\mu \dot{x}_\mu + \frac{e}{c} \dot{x}^\alpha a_\alpha \tag{6}$$

associated with the constraint $\dot{x}^5 = c_5$ that restricts the phase space to $(x^\mu, \dot{x}^\mu)$. In developing the field equations for the metric we will similarly break 5D symmetry when combining geometrical terms representing 5D gauge invariance with matter terms limited to 4D Lorentz symmetry.

Variation of the SHP action with respect to x^μ provides the Lorentz force [28]

$$M\ddot{x}_\mu = \frac{e}{c}\left(\dot{x}^\nu f_{\mu\nu} + c_5 f_{\mu 5}\right) = \frac{e}{c} \dot{x}^\beta f_{\mu\beta} \tag{7}$$

$$\frac{d}{d\tau}\left(-\frac{1}{2} M \dot{x}^\mu \dot{x}_\mu\right) = c_5 \frac{e}{c} \dot{x}^\beta f_{5\beta} = c_5 \frac{e}{c} \dot{x}^\mu f_{5\mu} \tag{8}$$

where the dynamics of the field strength

$$f_{\alpha\beta} = \partial_\alpha a_\beta - \partial_\beta a_\alpha \tag{9}$$

are determined by including the kinetic term

$$S_{\text{field}} = \int d\tau d^4x f^{\alpha\beta}(x,\tau) f_{\alpha\beta}(x,\tau) \tag{10}$$

in the total action, where $f_{\mu\nu}$ is the usual second rank tensor, while $f_{5\mu}$ is a vector field strength. Therefore, SHP electrodynamics differs in significant ways from Maxwell theory in 5D. In particular equation (8) permits mass exchange between particles and fields, setting the condition for non-conservation of proper time. Nevertheless, the total mass, energy, and momentum of particles and fields are conserved [28]. Compatibility of SHP electrodynamics with Maxwell theory requires $c_5 \ll c$ and we will neglect $(c_5/c)^2$ where appropriate.

Formally raising the five-index of $f_{\alpha\beta}$ in (10) suggests a 5D flat space metric

$$\eta_{\alpha\beta} = \text{diag}(-1,1,1,1,\sigma) \tag{11}$$

where $\sigma = \pm 1$. But regarding the kinetic term as

$$f^{\alpha\beta}(x,\tau) f_{\alpha\beta}(x,\tau) = f^{\mu\nu}(x,\tau) f_{\mu\nu}(x,\tau) + 2\sigma f^{\mu}_{\ 5}(x,\tau) f_{\mu 5}(x,\tau) \tag{12}$$

we see that σ is simply the choice of sign for the vector-vector term.

2.2. *Curved spacetime*

In the parameterized SHP framework, we regard $x^\mu(\tau)$ as an irreversible physical event, occurring at time τ and spacetime coordinates x^μ. Then, $\mathcal{M}(\tau)$ represents the spacetime manifold of general relativity, a 4D block universe at τ. The scalar Hamiltonian K that generates evolution of the events of spacetime is thus said to evolve $\mathcal{M}(\tau)$ to an infinitesimally close 4D block universe $\mathcal{M}(\tau + d\tau)$. The structure of spacetime, including the past and future of coordinate time x^0, may change infinitesimally during the interval $d\tau$, and so the metric structure $\gamma_{\mu\nu}(x,\tau)$ of $\mathcal{M}(\tau)$ will be τ-dependent. A τ-independent 4D metric would thus have the character of an absolute background field, violating the goals of general relativity.

In [19] we introduced a pseudo-spacetime $\mathcal{M}_5$ as the image of an injective mapping

$$\Phi : \mathcal{M} \longrightarrow \mathcal{M}_5 = \mathcal{M} \times R \qquad\qquad X = \Phi(x,\tau) = (x, c_5\tau) \tag{13}$$

allowing us to characterize 4D spacetime $\mathcal{M}$ as a hypersurface embedded in $\mathcal{M}_5$ and borrow the mathematical tools of 3+1 geometrodynamics. The interval

$$dX = X_1 - X_2 = (x_1, c_5\tau_1) - (x_2, c_5\tau_2) \tag{14}$$

for $X \in \mathcal{M}_5$ refers to an event $x_1 \in \mathcal{M}(\tau_1)$ and an event $x_2 \in \mathcal{M}(\tau_2)$. This notion of 5D separation combines the *geometrical distance* δx^μ between arbitrary points in $\mathcal{M}(\tau)$ with the *dynamical distance* between events separated by the evolution $\mathcal{M}(\tau) \longrightarrow \mathcal{M}(\tau + \delta\tau)$. Taking the small variation $x_2 = x_1 + \delta x$ and $\tau_2 = \tau_1 + \delta\tau$, the 5D invariant interval between $X_1, X_2 \in \mathcal{M}_5$ becomes

$$\delta X^2 = \gamma_{\mu\nu}\left(\delta x^\mu + \frac{dx^\mu(\tau)}{d\tau}\delta\tau\right)\left(\delta x^\nu + \frac{dx^\nu(\tau)}{d\tau}\delta\tau\right) + \sigma c_5^2 \delta\tau^2 = g_{\alpha\beta}(x,\tau)\,\delta x^\alpha \delta x^\beta \tag{15}$$

referred to x_1 coordinates at $\tau = \tau_1$, where the spacetime metric $\gamma_{\mu\nu}$ must depend on x and τ in some manner to be determined. This interval suggests the free particle Lagrangian

$$L = \frac{1}{2} M g_{\alpha\beta}(x,\tau) \dot{x}^\alpha \dot{x}^\beta \tag{16}$$

where we must remove x^5 from the dynamical variables by asserting $\dot{x}^5 = c_5$. This constraint breaks the symmetry of the geodesic equations from

$$\ddot{x}^\gamma + \Gamma^\gamma_{\alpha\beta}\dot{x}^\alpha\dot{x}^\beta = 0 \qquad \Gamma^\gamma_{\alpha\beta} = \frac{1}{2}g^{\gamma\delta}\left(\partial_\alpha g_{\delta\beta} + \partial_\beta g_{\delta\alpha} - \partial_\delta g_{\beta\alpha}\right) \tag{17}$$

to

$$\ddot{x}^\mu + \Gamma^\mu_{\alpha\beta}\dot{x}^\alpha\dot{x}^\beta = \ddot{x}^\mu + \Gamma^\mu_{\lambda\sigma}\dot{x}^\lambda\dot{x}^\sigma + 2c_5\Gamma^\mu_{5\sigma}\dot{x}^\sigma + c_5^2\Gamma^\mu_{55} = 0 \qquad \ddot{x}^5 \equiv 0 \tag{18}$$

with $\Gamma^5_{\alpha\beta}$ playing no role in the particle dynamics.

As in 4D relativity [29] the 5D Ricci tensor is invariant under translations $x'^\alpha = x^\alpha + \Lambda^\alpha(x,\tau)$, leading to the Bianchi identity

$$\nabla_\alpha\left(R^{\alpha\beta} - \frac{1}{2}g^{\alpha\beta}R\right) = 0 \qquad \nabla_\alpha X^\beta = \partial_\alpha X^\beta + X^\gamma\Gamma^\beta_{\gamma\alpha}\,. \tag{19}$$

By constructing a description of matter satisfying $\nabla_\beta T^{\alpha\beta} = 0$, we may write Einstein field equations in 5D, although we must break their 5D symmetry to O(3,1) at the interface between field terms expressed through $R_{\alpha\beta}$ and matter terms expressed through $T^{\alpha\beta}$.

2.3. *Mass-Energy-Momentum Tensor*

We define $n(x,\tau)$ to be the number of events per spacetime volume, so that

$$j^\alpha(x,\tau) = \rho(x,\tau)\dot{x}^\alpha(\tau) = Mn(x,\tau)\dot{x}^\alpha(\tau) \tag{20}$$

is the five-component event current. The continuity equation in flat space is

$$\partial_\alpha j^\alpha = \partial_\mu j^\mu + \partial_5 j^5 = \partial_\mu j^\mu + \frac{\partial\rho}{\partial\tau} = 0 \tag{21}$$

and is generalized for a local metric to

$$\nabla_\alpha j^\alpha = 0 \tag{22}$$

where again

$$\nabla_\alpha X^\beta = \frac{\partial X^\beta}{\partial x^\alpha} + X^\gamma\Gamma^\beta_{\gamma\alpha} \tag{23}$$

is the covariant derivative of a vector. Since j^5 is a scalar (n is scalar on physical grounds) for which the covariant derivative is the partial derivative, we have

$$\nabla_5 j^5 = \frac{\partial\rho}{\partial\tau} \tag{24}$$

and the continuity equation becomes

$$\frac{\partial\rho}{\partial\tau} + \nabla_\mu j^\mu = 0. \tag{25}$$

For non-interacting particles (non-thermodynamic dust under zero pressure), we write the mass-energy-momentum tensor [30] as

$$T^{\alpha\beta} = \rho\dot{x}^\alpha\dot{x}^\beta \longrightarrow \begin{cases} T^{\mu\nu} = \rho\dot{x}^\mu\dot{x}^\nu \\ T^{5\beta} = c_5 j^\beta \end{cases} \tag{26}$$

combining the 4D components $T^{\mu\nu}$ with the current density $T^{5\beta} = \dot{x}^5\dot{x}^\beta\rho = c_5 j^\beta$. The conservation equation is

$$0 = \nabla_\beta T^{\alpha\beta} = \nabla_\beta\left(\rho\dot{x}^\alpha\dot{x}^\beta\right) = \dot{x}^\alpha\nabla_\beta\left(\rho\dot{x}^\beta\right) + \rho\dot{x}^\beta\nabla_\beta\dot{x}^\alpha = \dot{x}^\alpha\nabla_\beta j^\beta + \rho\dot{x}^\beta\nabla_\beta\dot{x}^\alpha \tag{27}$$

which vanishes by virtue of the continuity and geodesic equations

$$\nabla_\alpha j^\alpha = 0 \qquad\qquad \dot{x}^\beta\nabla_\beta\dot{x}^\alpha = \frac{D\dot{x}^\alpha}{D\tau} = 0\,. \tag{28}$$

Since the Bianchi relations are independent of dimension, the Einstein equations in 5D are

$$R_{\alpha\beta} - \frac{1}{2}g_{\alpha\beta}R = \frac{8\pi G}{c^4}T_{\alpha\beta} \tag{29}$$

with Ricci tensor $R_{\alpha\beta}$ and Ricci scalar R obtained from $g_{\alpha\beta}$. The apparent 5D symmetry of (29) must be broken to 4+1, most conveniently achieved in a quintrad frame.

2.4. *Quintrad frame*

The standard basis vectors for the tangent space $\mathcal{T}(\mathcal{M}_5)$ in a coordinate (differential) frame are

$$\mathbf{g}_\alpha = \partial_\alpha = \frac{\partial}{\partial X^\alpha} \tag{30}$$

and the basis 1-forms for the dual space $\mathcal{T}^*(\mathcal{M}_5)$ are

$$\mathbf{g}^\alpha = \mathbf{d}X^\alpha \tag{31}$$

so that

$$\mathbf{g}^\alpha\left(\mathbf{g}_\beta\right) = \mathbf{g}^\alpha\cdot\mathbf{g}_\beta = \delta^\alpha_\beta \qquad \mathbf{g}_\alpha\cdot\mathbf{g}_\beta = g_{\alpha\beta} \qquad \mathbf{g}^\alpha\cdot\mathbf{g}^\beta = g^{\alpha\beta}\,. \tag{32}$$

Extending to 5D the vierbein formalism as presented in [26] and modifying the notation slightly, we define the quintrad frames $\{\mathbf{e}_a\}$ for $\mathcal{T}(\mathcal{M}_5)$ and $\{\mathbf{e}^a\}$ for $\mathcal{T}^*(\mathcal{M}_5)$ such that

$$\mathbf{e}_a\cdot\mathbf{e}_b = \eta_{ab} \qquad \mathbf{e}^a\cdot\mathbf{e}^b = \eta^{ab} \qquad \partial_a\mathbf{e}_b = \partial_a\mathbf{e}^b = 0 \tag{33}$$

where by convention the Latin letters

$$a, b, c, \cdots = 0, 1, 2, 3, 5 \qquad\qquad \eta_{ab} = \text{diag}\left(-1, 1, 1, 1, \sigma\right) \tag{34}$$

indicate reference to the quintrad. The vielbein field provides the position-dependent components of the coordinate bases with respect to the quintrad as

$$\mathbf{g}_\alpha = E_\alpha^{\ a}(X)\,\mathbf{e}_a \qquad\qquad \mathbf{g}^\alpha = \bar{E}^\alpha_{\ a}(X)\,\mathbf{e}^a \tag{35}$$

invertible as

$$\mathbf{e}_a = e^\alpha_{\ a}(X)\,\mathbf{g}_\alpha \qquad\qquad \mathbf{e}^a = \bar{e}_\alpha^{\ a}(X)\,\mathbf{g}^\alpha \tag{36}$$

which for consistency requires

$$e^\alpha_{\ a}(X)\,E_\beta^{\ a}(X) = \delta^\alpha_\beta \qquad\qquad e^\alpha_{\ a}(X)\,E_\alpha^{\ b}(X) = \delta^b_a \tag{37}$$

with similar relations for $\bar{E}^{\alpha}{}_{a}$ and $\bar{e}_{\alpha}{}^{a}$. The duality relations impose the conditions

$$\begin{aligned}\delta^{a}_{a'} &= \mathbf{e}^{a}\cdot\mathbf{e}_{a'} = \bar{e}_{\alpha}{}^{a}\mathbf{g}^{\alpha}\cdot e^{\beta}{}_{a'}\mathbf{g}_{\beta} = \bar{e}_{\alpha}{}^{a}e^{\beta}{}_{a'}\delta^{\alpha}_{\beta} = \bar{e}_{\alpha}{}^{a}e^{\alpha}{}_{a'}\\ \delta^{\alpha}_{\alpha'} &= \mathbf{g}^{\alpha}\cdot\mathbf{g}_{\alpha'} = \bar{E}^{\alpha}{}_{a}\mathbf{e}^{a}\cdot E_{\alpha'}{}^{b}\mathbf{e}_{b} = \bar{E}^{\alpha}{}_{a}E_{\alpha'}{}^{b}\delta^{a}_{b} = \bar{E}^{\alpha}{}_{a}E_{\alpha'}{}^{a}\end{aligned} \tag{38}$$

which combined with (37) put the transformation equations into the form

$$\begin{aligned}\mathbf{g}_{\alpha} &= E_{\alpha}{}^{a}\mathbf{e}_{a} \qquad\qquad & \mathbf{e}_{a} &= e^{\alpha}{}_{a}\mathbf{g}_{\alpha}\\ \mathbf{g}^{\alpha} &= e^{\alpha}{}_{a}\mathbf{e}^{a} & \mathbf{e}^{a} &= E_{\alpha}{}^{a}\mathbf{g}^{\alpha}\end{aligned} \tag{39}$$

and so using (35) and (36) the metric is induced through

$$\begin{aligned}g_{\alpha\beta} &= \mathbf{g}_{\alpha}\cdot\mathbf{g}_{\beta} = \eta_{ab}\,E_{\alpha}{}^{a}E_{\beta}{}^{b} \qquad\qquad & g^{\alpha\beta} &= \mathbf{g}^{\alpha}\cdot\mathbf{g}^{\beta} = \eta^{ab}\,e^{\alpha}{}_{a}e^{\beta}{}_{b}\\ \eta_{ab} &= \mathbf{e}_{a}\cdot\mathbf{e}_{b} = g_{\alpha\beta}e^{\alpha}{}_{a}e^{\beta}{}_{b} & \eta^{ab} &= \mathbf{e}^{a}\cdot\mathbf{e}^{b} = g^{\alpha\beta}E_{\alpha}{}^{a}E_{\beta}{}^{b}\,.\end{aligned} \tag{40}$$

Since any vector can be written

$$\mathbf{V} = V^{\alpha}\mathbf{g}_{\alpha} = \left[V^{\alpha}E_{\alpha}{}^{a}\left(X\right)\right]\mathbf{e}_{a} = V^{a}\mathbf{e}_{a} \tag{41}$$

we may view

$$X^{a}_{b} = E_{\alpha}{}^{a}e_{a}{}^{\beta}X^{\alpha}_{\beta} \qquad\qquad X^{\alpha}_{\beta} = e^{\alpha}{}_{a}E_{\beta}{}^{b}X^{a}_{b} \tag{42}$$

as transformations between the coordinate frame and the vielbein frame.

For a tensor given with components in the vielbein frame, the covariant derivative is defined as

$$\nabla_{\alpha}X^{a}_{b} = \partial_{\alpha}X^{a}_{b} + \omega_{\alpha}{}^{a}{}_{c}X^{c}_{b} - \omega_{\alpha}{}^{c}{}_{b}X^{a}_{c} \tag{43}$$

where $\omega_{\alpha}{}^{a}{}_{c}$ is the spin connection. Writing the covariant derivative in a coordinate frame as

$$\nabla X = \left(\partial_{\alpha}X^{\beta} + \Gamma^{\beta}_{\alpha\gamma}X^{\gamma}\right)\mathbf{g}^{\alpha}\otimes\mathbf{g}_{\beta} \tag{44}$$

and transforming between frames, we are led to

$$\omega_{\alpha}{}^{b}{}_{a} = -e^{\beta}{}_{a}\left(\partial_{\alpha}E_{\beta}{}^{b}\right) + E_{\beta}{}^{b}e^{\gamma}{}_{a}\Gamma^{\beta}_{\alpha\gamma} \tag{45}$$

providing a relationship between the coordinate and spin connections. Acting with E^{a}_{δ} we find

$$\partial_{\alpha}E_{\delta}{}^{b} - \Gamma^{\gamma}_{\alpha\delta}E_{\gamma}{}^{b} + \omega_{\alpha}{}^{b}{}_{a}E_{\delta}{}^{a} = \nabla_{\alpha}E_{\delta}{}^{b} = 0 \tag{46}$$

expressing compatibility of the vielbein field. Using the known symmetries of the Christoffel connection

$$\Gamma^{\alpha}_{\gamma\beta} = \frac{1}{2}g^{\alpha\delta}\left(\partial_{\gamma}g_{\beta\delta} + \partial_{\beta}g_{\gamma\delta} - \partial_{\delta}g_{\beta\gamma}\right) = g^{\alpha\delta}\Gamma_{\gamma\beta\delta} \tag{47}$$

in (45) it is straightforward to demonstrate the antisymmetry of $(\omega_{\alpha})^{ab}$

$$(\omega_{\alpha})^{ba} = -\,(\omega_{\alpha})^{ab} \qquad \text{where} \qquad (\omega_{\alpha})^{ab} = \eta^{bb'}\omega_{\alpha}{}^{a}{}_{b'}\,. \tag{48}$$

It will prove convenient to transform the coordinate index α to a quintrad index as

$$(\omega_{c})^{ba} = e^{\alpha}{}_{c}\,(\omega_{\alpha})^{ba}\,. \tag{49}$$

Inserting the induced metric (40) into the coordinate connection (47) we are led to

$$\Gamma^{a}_{ba'} = \frac{1}{2}\left(e^{\beta}{}_{a'}\partial_b E_{\beta}{}^{a} + e^{\beta a}\partial_b E_{\beta a'} + e^{\beta}{}_{b}\partial_{a'} E_{\beta}{}^{a} + e^{\beta a}\partial_{a'} E_{\beta b} - e^{\beta}{}_{a'}\partial^{a} E_{\beta b} - e^{\beta}{}_{b}\partial^{a} E_{\beta a'}\right) \tag{50}$$

where

$$\Gamma^{a}_{ba'} = e^{\gamma}{}_{b} e^{\beta}{}_{a'} E_{\alpha}{}^{a} \Gamma^{\alpha}_{\gamma\beta} \tag{51}$$

and using (45) we express the spin connection in terms of the vielbein field as

$$(\omega_b)^{a}{}_{a'} = \frac{1}{2}\left(e^{\beta a}\partial_b E_{\beta a'} - e^{\beta}{}_{a'}\partial_b E_{\beta}{}^{a} + e^{\beta}{}_{b}\partial_{a'} E_{\beta}{}^{a} + e^{\beta a}\partial_{a'} E_{\beta b} - e^{\beta}{}_{a'}\partial^{a} E_{\beta b} - e^{\beta}{}_{b}\partial^{a} E_{\beta a'}\right) \tag{52}$$

from which we may derive the useful relation

$$(\omega_b)^{a}{}_{c} - (\omega_c)^{a}{}_{b} = e^{\beta}{}_{b}\partial_c E_{\beta}{}^{a} - e^{\beta}{}_{c}\partial_b E_{\beta}{}^{a} \,. \tag{53}$$

The 5D curvature for $\mathcal{M}_5$ in the quintrad frame is written

$$\left[\nabla_b, \nabla_a\right] X_c = X_d R^{d}_{cab} \tag{54}$$

where

$$R^{d}_{cab} = \partial_a \omega_b{}^{d}{}_{c} - \partial_b \omega_a{}^{d}{}_{c} + \omega_a{}^{d}{}_{c'}\omega_b{}^{c'}{}_{c} - \omega_b{}^{d}{}_{c'}\omega_a{}^{c'}{}_{c} \tag{55}$$

which in light of (52) contains only the vielbein field and its derivatives.

3. Foliation

3.1. Coordinate frame

By defining the scalar field $S(X) = X^5/c_5 = \tau$, the pseudo-spacetime $\mathcal{M}_5$ constructed in Section 2.2 admits the natural foliation defined by its level surfaces

$$\Sigma_{\tau_0} = \left\{X^{\alpha} \mid S(X) = X^5/c_5 = \tau_0\right\} \,. \tag{56}$$

The unit normal to the hypersurface Σ_{τ_0} was given in [19] as

$$n_{\alpha} = \sigma\frac{1}{\sqrt{|g^{55}|}}\partial_{\alpha} S(X) = \sigma\frac{1}{\sqrt{|g^{55}|}}\delta^{5}_{\alpha} \qquad\qquad g^{\alpha\beta} n_{\alpha} n_{\beta} = \sigma \tag{57}$$

where $S(X)$ = constant for $X \in \Sigma_{\tau_0}$ insures orthogonality. We similarly used this foliation to construct a coordinate frame $\{\mathbf{g}_{\alpha}\}$ for $\mathcal{T}(\mathcal{M}_5)$. For the hypersurface $\mathcal{T}(\Sigma_{\tau_0}) \subset \mathcal{T}(\mathcal{M}_5)$, we choose the four (5-component) vectors

$$(\mathbf{g}_{\mu})^{\alpha} = \partial_{\mu}\Phi^{\alpha} = \left(\frac{\partial X^{\alpha}}{\partial x^{\mu}}\right)_{\tau_0} = \delta^{\alpha}_{\mu} \,. \tag{58}$$

We may choose as the fifth basis vector for $\mathcal{T}(\mathcal{M}_5)$ the linear combination

$$\mathbf{g}_5 = N^{\mu}\mathbf{g}_{\mu} + Nn \tag{59}$$

often called the ADM parameterization [25]. Here, the 4-vector N^{μ} generalizes the shift 3-vector in 3+1 formalisms and N is the lapse function with respect to τ. Designating

$$\gamma_{\mu\nu} = g_{\mu\nu} = \mathbf{g}_{\mu}\cdot\mathbf{g}_{\nu} \tag{60}$$

we find

$$g_{5\mu} = \mathbf{g}_\mu \cdot \mathbf{g}_5 = \gamma_{\mu\mu'} N^{\mu'} = N_\mu$$
$$g_{55} = \left(N^\mu \mathbf{g}_\mu + Nn\right) \cdot \left(N^{\mu'} \mathbf{g}_{\mu'} + Nn\right) = \gamma_{\mu\mu'} N^\mu N^{\mu'} + \sigma N^2 \,. \tag{61}$$

These expressions generalize the 3+1 ADM metric decomposition to

$$g_{\alpha\beta} = \begin{bmatrix} \gamma_{\mu\nu} & N_\mu \\ N_\mu & \sigma N^2 + \gamma_{\mu\nu} N^\mu N^\nu \end{bmatrix} \qquad g^{\alpha\beta} = \begin{bmatrix} \gamma^{\mu\nu} + \sigma \frac{1}{N^2} N^\mu N^\nu & -\sigma \frac{1}{N^2} N^\mu \\ -\sigma \frac{1}{N^2} N^\mu & \sigma \frac{1}{N^2} \end{bmatrix} \tag{62}$$

and put the unit normal into the form

$$n^\alpha = \frac{1}{N} \left(-N^\mu \mathbf{g}_\mu + \mathbf{g}_5\right)^\alpha = \frac{1}{N} \left(-N^\mu \delta^\alpha_\mu + \delta^\alpha_5\right)$$
$$n_\alpha = \sigma \frac{1}{\sqrt{|g^{55}|}} \delta^5_\alpha = \sigma N \delta^5_\alpha = \sigma N \left(\mathbf{g}^5\right)_\alpha \tag{63}$$

where the second expression is implicit in parameterization (59) through $n_\alpha = g_{\alpha\beta} n^\beta$.

3.2. *Quintrad frame*

We may perform the 4+1 decomposition through the vielbein field, without direct reference to the coordinate map Φ or the scalar field $S(X)$. We partition the quintrad indices, as we did for the coordinate indices, so that the index convention is now

$$\alpha, \beta, \gamma, \delta = 0,1,2,3,5 \qquad \lambda, \mu, \nu, \rho \ldots = 0,1,2,3$$
$$a, b, c, d, = 0,1,2,3,5 \qquad k, l, m, n, \ldots = 0,1,2,3 \tag{64}$$

where the five index with respect to the quintrad frame will be denoted $\bar{5}$ when necessary to avoid confusion. In this notation we expand the frame transformations as

$$\mathbf{g}_\mu = E_\mu{}^k \mathbf{e}_k + E_\mu{}^{\bar{5}} \mathbf{e}_5 \qquad \mathbf{e}_k = e^\mu{}_k \mathbf{g}_\mu + e^5{}_k \mathbf{g}_5$$
$$\mathbf{g}_5 = E_5{}^k \mathbf{e}_k + E_5{}^{\bar{5}} \mathbf{e}_5 \qquad \mathbf{e}_5 = e^\mu{}_{\bar{5}} \mathbf{g}_\mu + e^5{}_{\bar{5}} \mathbf{g}_5 \tag{65}$$

and

$$\mathbf{g}^\mu = e^\mu{}_k \mathbf{e}^k + e^\mu{}_{\bar{5}} \mathbf{e}^5 \qquad \mathbf{e}^k = E_\mu{}^k \mathbf{g}^\mu + E_5{}^k \mathbf{g}^5$$
$$\mathbf{g}^5 = e^5{}_k \mathbf{e}^k + e^5{}_{\bar{5}} \mathbf{e}^5 \qquad \mathbf{e}^5 = E_\mu{}^{\bar{5}} \mathbf{g}^\mu + E_5{}^{\bar{5}} \mathbf{g}^5 \tag{66}$$

for the dual basis, where once again the orthogonality relations (37) are required for consistency. In the quintrad we expect the spacetime hypersurface to be spanned by $\{\mathbf{e}_k\}$ and normal to $\mathbf{e}_5$, and so assign the unit normal vector and 1-form as

$$n = \mathbf{e}_5 \qquad \bar{n} = \sigma \mathbf{e}^5 \tag{67}$$

with normalization

$$n^2 = \mathbf{e}_5 \cdot \mathbf{e}_5 = \eta_{55} = \sigma \qquad \bar{n}^2 = \sigma^2 \mathbf{e}^5 \cdot \mathbf{e}^5 = \eta^{55} = \sigma \,. \tag{68}$$

Combining this assignment with the general parameterization (59) provides

$$\mathbf{e}_5 = n = \frac{1}{N}\left(-N^\mu \mathbf{g}_\mu + \mathbf{g}_5\right) \qquad \mathbf{e}^5 = \sigma\bar{n} = \sigma^2 N\,\mathbf{g}^5 = N\,\mathbf{g}^5 \tag{69}$$

so that comparison with (65) and (66) determines four vielbein components

$$e^\mu{}_{\bar{5}} = -\frac{1}{N}N^\mu \qquad e^5{}_{\bar{5}} = \frac{1}{N} \qquad E_\mu{}^{\bar{5}} = 0 \qquad E_5{}^{\bar{5}} = N\,. \tag{70}$$

The orthogonality of the quintrad frame provides two additional conditions

$$\begin{aligned} 0 = \mathbf{e}_k \cdot \mathbf{e}_5 &= \left(e^\mu{}_k \mathbf{g}_\mu + e^5{}_k \mathbf{g}_5\right) \cdot \frac{1}{N}\left(-N^{\mu'}\mathbf{g}_{\mu'} + \mathbf{g}_5\right) \\ &= \frac{1}{N}\left(-e^\mu{}_k N_\mu + e^\mu{}_k g_{\mu 5} - e^5{}_k g_{\mu' 5} N^{\mu'} + e^5{}_k g_{55}\right) \end{aligned} \tag{71}$$

and

$$0 = \mathbf{e}^k \cdot \mathbf{e}^5 = \left(E_\mu{}^k \mathbf{g}^\mu + E_5{}^k \mathbf{g}^5\right) \cdot N\,\mathbf{g}^5 = N\left(E_\mu{}^k g^{\mu 5} + E_5{}^k g^{55}\right) \tag{72}$$

which combined with metric components from (62) provide the components

$$e^5{}_k = 0 \qquad E_5{}^k = E_\mu{}^k N^\mu\,. \tag{73}$$

Inserting $e^5{}_k = 0$ into (65) leads to

$$\mathbf{e}_k = e^\mu{}_k \mathbf{g}_\mu + e^5{}_k \mathbf{g}_5 = e^\mu{}_k \left(E_\mu{}^{k'} \mathbf{e}_{k'} + E_\mu{}^{\bar{5}} \mathbf{e}_5\right) \tag{74}$$

from which we conclude that

$$E_\mu{}^{\bar{5}} = 0 \qquad e^\mu{}_k E_\mu{}^{k'} = \delta_k^{k'}\,. \tag{75}$$

Finally, the transformations between coordinate and quintrad frames take the form

$$\begin{aligned} \mathbf{g}_\alpha &= E_\alpha{}^a \mathbf{e}_a = \delta_\alpha^\mu E_\mu{}^k \mathbf{e}_k + \delta_\alpha^5 \left(E_\mu{}^k N^\mu \mathbf{e}_k + N\mathbf{e}_5\right) \\ \mathbf{e}_a &= e^\alpha{}_a \mathbf{g}_\alpha = \delta_a^k e^\mu{}_k \mathbf{g}_\mu + \delta_a^5 \frac{1}{N}\left(-N^\mu \mathbf{g}_\mu + \mathbf{g}_5\right) \end{aligned} \tag{76}$$

and we may summarize the vielbein field as

$$\begin{aligned} E_\alpha{}^a &= \delta_\alpha^\mu \delta_k^a E_\mu{}^k + \delta_\alpha^5 \left(E_\mu{}^k N^\mu \delta_k^a + N\delta_5^a\right) \\ e^\alpha{}_a &= \delta_a^k \delta_\mu^\alpha e^\mu{}_k - \delta_a^5 \delta_\mu^\alpha \frac{1}{N} N^\mu + \delta_a^5 \delta_5^\alpha \frac{1}{N} \end{aligned} \tag{77}$$

which provides a quintrad basis for $\mathcal{T}(\mathcal{M}_5)$ with a 4+1 foliation built-in by construction. It was shown in [19] that the lapse and shift propagate with τ, enforcing the constraints on $g_{\alpha\beta}$ associated with the Bianchi relations, but are not subject to second order evolution equations. Since $e^\mu{}_k = E^\mu{}_k$ by orthogonality, the dynamical content of the Einstein equations is entirely contained in the spacetime vierbein $E_\mu{}^k$, where $\mu, k = 0, 1, 2, 3$.

3.3. *Projection onto the hypersurface*

The 4+1 decomposition of the Einstein equations is found by application of the projection operator P which acts on a vector as

$$V \in \mathcal{T}(\mathcal{M}_5) \longrightarrow V_\perp = P[V] \in \mathcal{T}(\Sigma_\tau) \subset \mathcal{T}(\mathcal{M}_5) \tag{78}$$

and in component form is

$$P[V] = \left(P^\alpha_{\alpha'} V^{\alpha'}\right) \mathbf{g}_\alpha = V^{\alpha'} \left(P^\alpha_{\alpha'} \mathbf{g}_\alpha\right) \qquad P[V] = \left(P^a_{a'} V^{a'}\right) \mathbf{e}_a = V^{a'} \left(P^a_{a'} \mathbf{e}_a\right) . \tag{79}$$

Using (63) for n_α, the components of P in the coordinate and quintrad frames are

$$\begin{aligned} P^\alpha_{\alpha'} &= \delta^\alpha_{\alpha'} - \sigma n^\alpha n_{\alpha'} = \delta^\alpha_{\alpha'} - \delta^5_{\alpha'} \left(\delta^\alpha_5 - N^\mu \delta^\alpha_\mu\right) \\ P^a_{a'} &= e_\alpha{}^a e^{\alpha'}{}_{a'} P^\alpha_{\alpha'} = \delta^a_{a'} - \delta^a_5 \delta^5_{a'} \end{aligned} \tag{80}$$

and we identify the expressions on the RHS as the component form of vectors

$$\begin{aligned} \mathbf{g}^\perp_{\alpha'} &= \mathbf{g}_{\alpha'} - \delta^5_{\alpha'} \left(\mathbf{g}_5 - N^\mu \mathbf{g}_\mu\right) = P^\alpha_{\alpha'} \mathbf{g}_\alpha \\ \mathbf{e}^\perp_{a'} &= \mathbf{e}_{a'} - \delta^5_{a'} \mathbf{e}_5 = P^a_{a'} \mathbf{e}_a . \end{aligned} \tag{81}$$

These can be rewritten

$$\mathbf{g}^\perp_\alpha = \mathbf{g}_\mu \delta^\mu_\alpha + N^\mu \mathbf{g}_\mu \delta^5_\alpha = \left(\delta^\mu_\alpha + N^\mu \delta^5_\alpha\right) \mathbf{g}_\mu \qquad \mathbf{e}^\perp_a = \delta^k_a \mathbf{e}_k \tag{82}$$

showing that $\mathbf{g}^\perp_\alpha$ and $\mathbf{e}^\perp_a$ lie in the hypersurface $\mathcal{T}(\Sigma_\tau)$, and that

$$\begin{aligned} V_\perp &= V^\alpha \mathbf{g}^\perp_\alpha & &\longrightarrow & V^\alpha_\perp &= \delta^\alpha_\mu \left(\delta^\mu_{\alpha'} + N^\mu \delta^5_{\alpha'}\right) V^{\alpha'} \\ V_\perp &= V^a \mathbf{e}^\perp_a & &\longrightarrow & V^a_\perp &= \delta^a_k \delta^k_{a'} V^{a'} . \end{aligned} \tag{83}$$

Because $V_\perp$ has four independent components we have the pull-back and push-forward relationships

$$v^\mu = \delta^\mu_\alpha V^\alpha_\perp \in \mathcal{T}(\mathcal{M}) \qquad V^\alpha_\perp = \delta^\alpha_\mu v^\mu \in \mathcal{T}(\Sigma_\tau) \tag{84}$$

where $v \in \mathcal{T}(\mathcal{M}(\tau))$ is the spacetime four-vector homeomorphic to $V_\perp \in \mathcal{T}(\Sigma_\tau)$. More generally, writing

$$\begin{aligned} v^\mu &= \delta^\mu_\alpha V^\alpha_\perp = \delta^\mu_\alpha e^\alpha{}_a V^a_\perp = \delta^\mu_\alpha e^\alpha{}_{a'} P^{a'}_a V^a = \mathcal{E}^\mu_a V^a \\ v_\mu &= \delta^\alpha_\mu V^\perp_\alpha = \delta^\alpha_\mu E_\alpha{}^{a'} V^\perp_{a'} = \delta^\alpha_\mu E_\alpha{}^{a'} P^a_{a'} V^\perp_a = \mathcal{E}^a_\mu V^\perp_a \end{aligned} \tag{85}$$

and using (80) and (77) we define the composed pull-back operators

$$\mathcal{E}^\mu_a = \delta^\mu_\alpha e^\alpha{}_{a'} P^{a'}_a = \delta^k_a e^\mu{}_k \qquad \mathcal{E}^a_\mu = \delta^\alpha_\mu E_\alpha{}^{a'} P^a_{a'} = \delta^a_k E_\mu{}^k \tag{86}$$

which map a vector in a quintrad frame for $\mathcal{T}(\mathcal{M}_5)$ to a vector in a coordinate frame for $\mathcal{T}(\mathcal{M})$, using only the spacetime part of the vielbein field. Because the projector is idempotent, we see that

$$\mathcal{E}^\mu_{a'} P^{a'}_a = \mathcal{E}^\mu_a \qquad \mathcal{E}^{a'}_\mu P^a_{a'} = \mathcal{E}^a_\mu . \tag{87}$$

Defining the normal basis components

$$\mathbf{g}^{\parallel}_\alpha = \mathbf{g}_\alpha - \mathbf{g}^\perp_\alpha = \delta^5_\alpha \left(\mathbf{g}_5 - N^\mu \mathbf{g}_\mu\right) \qquad \mathbf{e}^{\parallel}_a = \mathbf{e}_a - \mathbf{e}^\perp_a = \delta^5_a \mathbf{e}_5 \tag{88}$$

we easily verify

$$\begin{aligned}
\mathbf{g}^{\parallel}_{\alpha} \cdot \mathbf{g}^{\parallel}_{\beta} &= \sigma N^2 \delta^5_\alpha \delta^5_\beta \\
\mathbf{g}^{\perp}_{\alpha} \cdot \mathbf{g}^{\perp}_{\beta} &= \gamma_{\mu\mu'} \delta^\mu_\alpha \delta^{\mu'}_\beta + \delta^5_\alpha N_\mu \delta^\mu_\beta + \delta^5_\beta N_\mu \delta^\mu_\alpha + \gamma_{\mu\mu'} N^\mu N^{\mu'} \delta^5_\alpha \delta^5_\beta \\
\mathbf{g}^{\parallel}_{\alpha} \cdot \mathbf{g}^{\perp}_{\beta} &= 0
\end{aligned} \tag{89}$$

which decomposes the coordinate frame into tangent and normal components, and recovers the metric (62) through

$$g_{\alpha\beta} = \mathbf{g}_\alpha \cdot \mathbf{g}_\beta = \mathbf{g}^{\parallel}_{\alpha} \cdot \mathbf{g}^{\parallel}_{\beta} + \mathbf{g}^{\perp}_{\alpha} \cdot \mathbf{g}^{\perp}_{\beta} \, . \tag{90}$$

Rearranging (80) we have the completeness relations

$$\delta^\alpha_\beta = P^\alpha_\beta + \sigma n^\alpha n_\beta \qquad\qquad g_{\alpha\beta} = P_{\alpha\beta} + \sigma n_\alpha n_\beta \tag{91}$$

which are useful in the 4+1 decomposition of objects defined on $\mathcal{M}_5$. On the Σ_τ hypersurface, the projection operator $P_{\alpha\beta}$ acts as the induced metric, as can be seen through

$$\gamma_{\mu\nu} = g_{\alpha\beta} \delta^\alpha_\mu \delta^\beta_\nu = \left(P_{\alpha\beta} + \sigma n_\alpha n_\beta\right) \delta^\alpha_\mu \delta^\beta_\nu = P_{\mu\nu} \tag{92}$$

where we used $n_\alpha \delta^\alpha_\mu = \sigma N \delta^5_\mu = 0$.

4. Field equations

Decomposition of the 4D Einstein equations into evolution equations requires an expression that isolates and projects the Ricci tensor into the hypersurface. In the 3+1 formalism, one starts from the Einstein equations

$$G_{\mu\nu} = R_{\mu\nu} - \frac{1}{2} g_{\mu\nu} R = \frac{8\pi G}{c^4} T_{\mu\nu} \tag{93}$$

and takes the trace

$$g^{\mu\nu} R_{\mu\nu} - \frac{1}{2} g^{\mu\nu} g_{\mu\nu} R = \frac{8\pi G}{c^4} g^{\mu\nu} T_{\mu\nu} \tag{94}$$

using $g^{\mu\nu} g_{\mu\nu} = 4$ to obtain the trace-reversed form

$$R_{\mu\nu} = \frac{8\pi G}{c^4} \left(T_{\mu\nu} - \frac{1}{2} g_{\mu\nu} T \right) \tag{95}$$

expressing the relationship of spacetime geometry on the LHS and the distribution of matter on the RHS. But in 5D we have $g^{\alpha\beta} g_{\alpha\beta} = 5 \neq 4$ leading to the trace-reversed form

$$R_{\alpha\beta} = \frac{8\pi G}{c^4} \left(T_{\alpha\beta} + \frac{\frac{1}{2} g_{\alpha\beta}}{1 - \frac{1}{2} g^{\gamma\delta} g_{\gamma\delta}} g^{\gamma'\delta'} T_{\gamma'\delta'} \right) = \frac{8\pi G}{c^4} \left(T_{\alpha\beta} - \frac{1}{3} g_{\alpha\beta} T \right) \, . \tag{96}$$

In [21] we studied the linearized SHP theory, and showed that this issue similarly appears when solving the wave equation for a weak gravitational perturbation. Unsurprisingly, consistency with standard phenomenology requires that we replace $\eta_{\alpha\beta} \longrightarrow \widehat{\eta}_{\alpha\beta} = (-1, 1, 1, 1, 0)$ as the flat background metric. This replacement recalls the observation in Section 2.1 that $\eta_{55} = \sigma$ can be understood as the choice of sign for certain field terms, but should play no role in matter terms.

Therefore, we similarly require the substitution $g_{\alpha\beta} \longrightarrow \widehat{g}_{\alpha\beta}$ satisfying $\widehat{g}^{\alpha\beta}\,\widehat{g}_{\alpha\beta} = 4$ among the matter terms, which permits us to recover the form

$$R_{\alpha\beta} = \frac{8\pi G}{c^4}\left(T_{\alpha\beta} - \frac{1}{2}\widehat{g}_{\alpha\beta}\widehat{T}\right) \tag{97}$$

in 5D, where $\widehat{T} = \widehat{g}^{\alpha\beta}T_{\alpha\beta}$.

To obtain $\widehat{g}_{\alpha\beta}$ we write unbroken 5D Einstein equations in the quintrad frame

$$R_{ab} - \frac{1}{2}\eta_{ab}R = \frac{8\pi G}{c^4}T_{ab} \longrightarrow R_{ab} = \frac{8\pi G}{c^4}\left(T_{ab} + \frac{\frac{1}{2}\eta_{ab}}{1-\frac{1}{2}\eta^{cd}\eta_{cd}}\eta^{c'd'}T_{c'd'}\right) \tag{98}$$

where the metric η_{ab} is flat, while R_{ab} and T_{ab} are related to the coordinate frame tensors through the vielbein field. We break the 5D symmetry in the matter terms by replacing

$$\eta_{ab} \longrightarrow \widehat{\eta}_{ab} = (-1,1,1,1,0) = \delta_a^k\delta_b^l\,\eta_{kl} \tag{99}$$

on the RHS of (98), leaving the Ricci tensor R_{ab} unchanged. The SHP Einstein equations now take the form

$$R_{ab} = \frac{8\pi G}{c^4}\left(T_{ab} - \frac{1}{2}\widehat{\eta}_{ab}\widehat{T}\right) \tag{100}$$

where $\widehat{T} = \widehat{\eta}^{ab}T_{ab} = \eta^{kl}T_{kl}$. Using the vielbein field (77) we may transform (100) back to a coordinate frame, leading to

$$R_{\alpha\beta} = \frac{8\pi G}{c^4}\left(T_{\alpha\beta} - \frac{1}{2}\widehat{g}_{\alpha\beta}\widehat{T}\right) \tag{101}$$

providing the symmetry-broken local metric as

$$\widehat{g}_{\alpha\beta} = E_\alpha^{\ a}E_\beta^{\ b}\widehat{\eta}_{ab} = g_{\alpha\beta} - \delta_\alpha^5\delta_\beta^5\sigma N^2 = g_{\alpha\beta} - \sigma n_\alpha n_\beta = P_{\alpha\beta}\,. \tag{102}$$

The breaking of 5D symmetry can thus be understood as replacing the metric in the matter terms of the field equation with the projector onto the 4D hypersurface. We recall from (92) that restricted to the Σ_τ hypersurface, the projection operator $P_{\alpha\beta}$ acts as the induced metric.

Using the completeness relation (91), we decompose the mass-energy-momentum tensor into

$$T_{ab} = T_{a'b'}\left(P_a^{a'} + \sigma n^{a'}n_a\right)\left(P_b^{b'} + \sigma n^{b'}n_b\right) = S_{ab} - 2\sigma n_a p_b + n_a n_b \kappa \tag{103}$$

where

$$S_{ab} = T_{a'b'}P_a^{a'}P_b^{b'} \qquad p_b = -n^{a'}P_b^{b'}T'_{a'b'} \qquad \kappa = n^{b'}n^{a'}T_{a'b'} \tag{104}$$

representing the 4D energy momentum tensor S_{ab}, a momentum vector p_b describing the flow of mass into spacetime, and a scalar mass density κ. The trace of T_{ab} is

$$T = \eta^{ab}T_{ab} = S + \sigma\kappa \tag{105}$$

but the symmetry broken trace is

$$\widehat{T} = \widehat{g}^{\alpha\beta}T_{\alpha\beta} = P^{\alpha\beta}T_{\alpha\beta} = P^{\alpha\beta}\left(S_{\alpha\beta} + 2\sigma n_\alpha p_\beta + n_\alpha n_\beta\kappa\right) = S \tag{106}$$

where we used $P^{\alpha\beta}n_\alpha = 0$. The SHP field equations now take the form

$$R_{\alpha\beta} = \frac{8\pi G}{c^4}\left(T_{\alpha\beta} - \frac{1}{2}P_{\alpha\beta}S\right) \tag{107}$$

in a coordinate frame. In the unconstrained 8D phase space of SHP kinematics, mass is most readily understood through the independence of energy and 3-momentum in the dynamical quantity $-p^2 = -M^2 \dot{x}^\mu \dot{x}_\mu$ associated with the motion of matter events. As in SHP electrodynamics, the dynamical evolution of an event in curved space may involve a variation in particle mass, with mass transferred to the gravitational field and transferred across spacetime through field momentum p_a and mass density κ.

5. Initial value problem

5.1. Review of 4+1 decomposition

In [19] we generalized the 3+1 formalism in geometrodynamics [24,25] to 4+1 to obtain a pair of first order evolution equations for the metric $\gamma_{\mu\nu}$ and intrinsic curvature $K_{\mu\nu}$ of spacetime geometry ($\mu, \nu = 0,1,2,3$). Regarding the spacetime manifold as homeomorphic to the hypersurface Σ_τ embedded in the 5D pseudo-spacetime, the differential geometry of $\mathcal{M}$ is expressed in terms of projections of the corresponding structures for $\mathcal{M}_5$. By extension of 3+1 geometrodynamics, the initial value problem is found in the following steps:

(i) The covariant derivative D_α for $\mathcal{T}(\Sigma_\tau)$ is found by using $P_{\alpha\beta}$ to project the covariant derivative ∇_α for $\mathcal{M}_5$,

(ii) The extrinsic curvature $K_{\alpha\beta}$ is defined by projecting the covariant derivative of the unit normal n_α,

(iii) The projected curvature $\bar{R}^\delta_{\gamma\alpha\beta}$ on $\mathcal{T}(\Sigma_\tau)$ is defined through the non-commutation of projected covariant derivatives D_α and D_β,

(iv) Writing the explicit form of $P_{\alpha\beta}$ in the definition of $\bar{R}^\delta_{\gamma\alpha\beta}$ leads to the Gauss relation that decomposes the 5D curvature $R^\delta_{\gamma\alpha\beta}$ in terms of $\bar{R}^\delta_{\gamma\alpha\beta}$ and $K_{\alpha\beta}$,

(v) Projecting the 5D curvature $R^\delta_{\gamma\alpha\beta}$ on the unit normal n_α leads to the Codazzi relation providing a relationship between $K_{\alpha\beta}$ and p_α,

(vi) Lie derivatives of $P_{\alpha\beta}$ and $K_{\alpha\beta}$ along the unit normal — the direction of τ evolution — are combined with these ingredients, along with the trace-reversed field equation, to obtain τ-evolution equations for $\gamma_{\mu\nu}$ and $K_{\mu\nu}$ and a pair of constraints on the initial conditions.

5.2. Intrinsic and extrinsic curvature in the quintrad frame

In Section 2.4 we wrote the covariant derivative (43) and curvature tensor (54) for $\mathcal{M}_5$ in a quintrad frame. Here we review some details of Steps (i) to (vi) where derivation in the quintrad frame helps clarify the content of the procedure.

The 4+1 projected derivative D_a is defined as

$$(DX)_{ab_1\cdots b_n} = P_a^{a'} P_{b_1}^{b'_1} \cdots P_{b_n}^{b'_n} \left(\nabla_{a'} X_{b'_1 \cdots b'_n} \right) \tag{108}$$

and because

$$\nabla_a P_{bc} = \nabla_a \left(\eta_{bc} - \sigma n_b n_c \right) = -\sigma \left[\left(\nabla_a n_b \right) n_c + n_b \left(\nabla_a n_c \right) \right] \tag{109}$$

we have

$$(DP)_{abc} = P_a^{a'} P_b^{b'} P_c^{c'} \, \nabla_{a'} P_{b'c'} = 0 \,. \tag{110}$$

Since the $P_{aa'}$ acts as the metric on the hypersurface $\mathcal{T}(\Sigma_\tau)$, we see that D_a is the unique covariant derivative compatible with $P_{\alpha\beta}$ (and hence $\gamma_{\mu\nu}$). Pulling back the projected covariant derivative to $\mathcal{T}(\mathcal{M})$ we find

$$\begin{aligned} D_\mu v_\nu &= \mathcal{E}_\mu^a \mathcal{E}_\nu^b P_a^{a'} P_b^{b'} \left(\partial_a V_b - \omega_a{}^{a'}{}_b V_{a'} \right) = E_\mu{}^k E_\nu{}^l \left(\partial_k V_l - \omega_k{}^{l'}{}_l V_{l'} - \omega_k{}^{\bar{5}}{}_l V_{\bar{5}} \right) \\ &= \left(\partial_\mu v_\nu - \Gamma^\lambda_{\mu\nu} v_\lambda \right) - \omega_\mu{}^{\bar{5}}{}_\nu V_{\bar{5}} = \nabla^{(4)}_\mu v_\nu - \omega_\mu{}^{\bar{5}}{}_\nu V_{\bar{5}} \end{aligned} \tag{111}$$

where $\nabla^{(4)}_\mu$ contains only 4D components of the connection associated with a τ-independent spacetime. The additional $\omega_\mu{}^{\bar{5}}{}_\nu$ term suggests how D_μ retains information on the τ-evolution of the spacetime geometry. This point is sharpened by writing the extrinsic curvature

$$K_{ab} = -\left(Dn\right)_{ab} = -P^{a'}_a P^{b'}_b \nabla_{b'} n_{a'} = -P^{b'}_b \nabla_{b'} n_a \tag{112}$$

where the last equality follows from

$$0 = \nabla_{b'}\sigma = \nabla_{b'} n^2 = 2n^{a'}\left(\nabla_{b'} n_{a'}\right) . \tag{113}$$

Expanding the projector $P^{b'}_b$ we find

$$K_{ab} = -\nabla_b n_a + \sigma n_b \left(n^{b'}\nabla_{b'} n_a\right) = -\nabla_b n_a + \sigma n_b \left(-\frac{1}{N} D_a N\right) \tag{114}$$

which using $n_a = \sigma\delta^5_a$ and $n^a = \delta^a_5$ provides

$$\nabla_b n_a = \partial_b n_a - \omega_b{}^{a'}{}_a n_{a'} = -\sigma\omega_b{}^{\bar{5}}{}_a \qquad\qquad n^{b'}\nabla_{b'} n_b = -\omega_{\bar{5}}{}^{\bar{5}}{}_b \tag{115}$$

leading to

$$K_{ab} = \sigma\omega_b{}^{\bar{5}}{}_a - \sigma\delta^{\bar{5}}_b\omega_{\bar{5}}{}^{\bar{5}}{}_a = \sigma\delta^k_a\delta^l_b K_{kl} = \sigma\delta^k_a\delta^l_b\,\omega_l{}^{\bar{5}}{}_k \tag{116}$$

where $\omega_{\bar{5}}{}^{\bar{5}}{}_{\bar{5}} = 0$ by antisymmetry. Like the projected covariant derivative D_μ the extrinsic curvature $K_{\mu\nu}$ contains 5-components of the connection ω^a_{bc} not present in $\nabla^{(4)}_\mu$. Returning to the general expression (52) for ω^a_{bc} we may evaluate

$$\omega_{\bar{5}}{}^{\bar{5}}{}_b = \delta^k_b\,\omega_{\bar{5}}{}^{\bar{5}}{}_k = \delta^k_b \frac{1}{N}\partial_k N \tag{117}$$

in agreement with (114) and find the expression

$$K_{ab} = \frac{1}{2}\delta^k_a\delta^l_b\left[E_{\mu k}\left(\frac{1}{N}\partial_l N^\mu + \partial_5 e^\mu{}_l\right) + E_{\mu l}\left(\frac{1}{N}\partial_k N^\mu + \partial_5 e^\mu{}_k\right)\right] \tag{118}$$

for K_{ab} that depends on the lapse, shift, and spacetime part of the vielbein field.

The projected Riemann tensor $\bar{R}^d_{cab}$ is defined through the non-commutation of projected covariant derivatives

$$\left[D_b, D_a\right] X_c = X_d \bar{R}^d_{cab} \tag{119}$$

and describes the curvature of the hypersurface Σ_τ. Because D_μ contains projected 5-components of the connection, these components will also be present in the pull-back $\bar{R}^\rho_{\lambda\mu\nu}$ to $\mathcal{M}$ differing from the 4D curvature $R^\rho_{\lambda\mu\nu}$ associated with $\nabla^{(4)}_\mu$.

Decomposition of the 5D Riemann tensor R^d_{cab} into projected components $\bar{R}^d_{cab}$ and K_{ab} may be understood by contracting with the completeness relation (91)

$$R^c_{dab} = \left(P^{a'}_a + \sigma n_a n^{a'}\right)\left(P^{b'}_b + \sigma n_b n^{b'}\right)\left(P^c_{c'} + \sigma n^c n_{c'}\right)\left(P^{d'}_d + \sigma n_d n^{d'}\right) R^{c'}_{d'a'b'} \tag{120}$$

to obtain the sum of three projections

$$\left(P^{a'}_a P^{b'}_b P^c{}_{c'} P^{d'}_d\right) R^{c'}{}_{d'a'b'} \qquad \left(P^c_{c'} P^{a'}_a P^{b'}_b\right) n^d\, R^{c'}_{da'b'} \qquad \left(P_{aa'} P^{b'}_b\right) n^{c'} n^d\, R^{a'}_{db'c'} \tag{121}$$

where $n^{b'}n^{c'}n^d\ R^{a'}_{db'c'} = 0$ from the antisymmetry of R^d_{cab}. The first of (121) is found in Step (iv) above as the Gauss relation

$$R^{\mu}_{\nu\lambda\rho} = \bar{R}^{\mu}_{\nu\lambda\rho} - \sigma\left(K^{\mu}_{\lambda}K_{\rho\nu} - K^{\mu}_{\rho}K_{\lambda\nu}\right) . \tag{122}$$

In its original context, this shows that a curved surface embedded in a flat 3D space with $R^i_{jkl} = 0$ will derive its relevant curvature $\bar{R}^i_{jkl}$ entirely from the K_{ij} associated with the gradient of the normal to that surface.

In Step (v) above we write the definition (54) of R^c_{dab} while taking $X_d = n_d$ to obtain explicit expressions for the second and third of (121). Projecting the three remaining indices onto $\mathcal{T}(\Sigma_\tau)$ provides the Codazzi relation

$$\left(P^{a'}_a P^{b'}_b P^{c'}_c\right) n_d\ R^d_{c'a'b'} = D_b K_{ac} - D_a K_{bc} \tag{123}$$

while projecting twice onto $\mathcal{T}(\Sigma_\tau)$ and once onto $n^{c'}$ leads to

$$\left(P_{aa'}P^{b'}_b\right) n^{c'}n^d\ R^{a'}_{db'c'} = -K^c_a K_{cb} - \sigma\frac{1}{N}D_b D_a N + P^{a'}_a P^{b'}_b n^{c'}\nabla_{c'}K_{a'b'} \tag{124}$$

which we will combine with the trace-reversed field equation (107) to eliminate $R^{a'}_{db'c'}$.

5.3. Evolution equations in a coordinate frame

In Step (vi) above we construct an initial value problem for the τ-evolution of the metric and extrinsic curvature. This construction is naturally performed in a coordinate frame, because the metric in the quintrad frame is the constant η_{ab}. The τ-derivatives are found by evaluating the Lie derivatives of $\gamma_{\mu\nu}$ and $K_{\mu\nu}$ in the direction of system evolution. Defining the normal evolution vector $m_\alpha = Nn_\alpha$ we use the ADM parameterization (59) to write the Lie derivative along $\mathbf{g}_5$ as

$$\mathbf{g}_5 = N^\mu \mathbf{g}_\mu + Nn = \mathbf{N} + m \quad \longrightarrow \quad \mathcal{L}_m = \mathcal{L}_{\mathbf{g}_5} - \mathcal{L}_{\mathbf{N}} \tag{125}$$

where the Lie derivative acts as

$$\mathcal{L}_B\, A_{\alpha\beta} = B^\gamma\nabla_\gamma A_{\alpha\beta} + A_{\gamma\beta}\nabla_\alpha B^\gamma + A_{\alpha\gamma}\nabla_\beta B^\gamma . \tag{126}$$

From the definition $(\mathbf{g}_5)^\gamma = \delta^\gamma_5$ the τ-derivative is found as

$$\mathcal{L}_{\mathbf{g}_5}A_{\alpha\beta} = \delta^\gamma_5\partial_\gamma A_{\alpha\beta} + A_{\gamma\beta}\partial_\alpha\delta^\gamma_5 + A_{\alpha\gamma}\partial_\beta\delta^\gamma_5 = \partial_5 A_{\alpha\beta} = \frac{1}{c_5}\partial_\tau A_{\alpha\beta} \tag{127}$$

and the derivative along the normal evolution vector

$$\mathcal{L}_m\, A_{\alpha\beta} = m^\gamma\nabla_\gamma A_{\alpha\beta} + A_{\gamma\beta}\nabla_\alpha m^\gamma + A_{\alpha\gamma}\nabla_\beta m^\gamma \tag{128}$$

can be evaluated by rearranging (114) to write

$$\nabla_\alpha n_\beta = -K_{\alpha\beta} - n_\alpha\frac{1}{N}D_\beta N \ \longrightarrow\ \nabla_\beta m^\alpha = -NK^\alpha_\beta - n_\beta D^\alpha N + n^\alpha\nabla_\beta N . \tag{129}$$

Inserting (129) into (128) we easily evaluate

$$\mathcal{L}_m\, P_{\alpha\beta} = -2NK_{\alpha\beta} \qquad\qquad \mathcal{L}_m\, P^\alpha_\beta = 0 \tag{130}$$

where again (92) permits the substitution $\gamma_{\alpha\beta} \to P_{\alpha\beta}$. For the extrinsic curvature we find

$$\mathcal{L}_m K_{\alpha\beta} = Nn^\gamma \nabla_\gamma K_{\alpha\beta} - 2NK_{\alpha\gamma}K^\gamma{}_\beta - K_{\alpha\gamma}D^\gamma N n_\beta - K_{\beta\gamma}D^\gamma N n_\alpha \tag{131}$$

which by projecting with $P^\alpha_{\alpha'}\, P^{\beta'}{}_\beta$ and recalling (122) and (124) becomes

$$P^{\alpha'}{}_\alpha P^{\beta'}{}_\beta R_{\alpha'\beta'} = \sigma \frac{1}{N}\mathcal{L}_m K_{\alpha\beta} + \sigma \frac{1}{N} D_\alpha D_\beta N + \bar{R}_{\alpha\beta} - \sigma K K_{\alpha\beta} + 2\sigma K^\delta_\alpha K_{\beta\delta} \,. \tag{132}$$

In light of (125) we see that equations (130) and (132) provide a pair of coupled evolution equations for $\gamma_{\mu\nu}$ and $K_{\mu\nu}$, of first order in ∂_τ, in which all terms except the 5D Ricci tensor $R_{\alpha\beta}$ are derived from the projected covariant derivative. These equations become an initial value problem for general relativity by replacing $R_{\alpha\beta}$ with the Einstein equations, which by projection on (107) take the form

$$P^{\alpha'}_\alpha P^{\beta'}_\beta R_{\alpha'\beta'} = P^{\alpha'}_\alpha P^{\beta'}_\beta \left(T_{\alpha'\beta'} - \frac{1}{2} P_{\alpha'\beta'} S \right) = S_{\alpha\beta} - \frac{1}{2} P_{\alpha\beta} S \,. \tag{133}$$

Pulling back to $\mathcal{M}$ and using (125) to expand $\mathcal{L}_m$ in (130) these expressions become

$$\frac{1}{c_5}\partial_\tau \gamma_{\mu\nu} = \mathcal{L}_{\mathbf{N}}\, \gamma_{\mu\nu} - 2NK_{\mu\nu} \tag{134}$$

$$\begin{aligned} \frac{1}{c_5}\partial_\tau K_{\mu\nu} &= -D_\mu D_\nu N + \mathcal{L}_{\mathbf{N}} K_{\mu\nu} \\ &\quad + N\left\{ -\sigma \bar{R}_{\mu\nu} + K K_{\mu\nu} - 2K^\lambda_\mu K_{\nu\lambda} + \sigma \frac{8\pi G}{c^4}\left(S_{\mu\nu} - \frac{1}{2} P_{\mu\nu} S \right) \right\} \end{aligned} \tag{135}$$

which differ from the expression found in [19] using unbroken field equations by the symmetry breaking replacement $g_{\mu\nu}\,(S + \sigma\kappa) \to P_{\mu\nu} S$.

To find the constraint equations, we first contract indices in the Gauss relation

$$R - 2\sigma R_{\alpha\beta} n^\alpha n^\beta = \bar{R} - \sigma\left(K^2 - K^{\alpha\beta} K_{\alpha\beta} \right) \tag{136}$$

and use the completeness relation to evaluate the trace of $R_{\alpha\beta}$

$$R = g^{\alpha\beta} R_{\alpha\beta} = P^{\alpha\beta} R_{\alpha\beta} + \sigma R_{\alpha\beta} n^\alpha n^\beta \tag{137}$$

to obtain

$$P^{\alpha\beta} R_{\alpha\beta} - \sigma R_{\alpha\beta} n^\alpha n^\beta = \bar{R} - \sigma\left(K^2 - K^{\alpha\beta} K_{\alpha\beta} \right) \,. \tag{138}$$

Projecting the SHP field equations onto the hypersurface and the unit normal provides

$$\begin{aligned} P^{\alpha\beta} R_{\alpha\beta} &= \frac{8\pi G}{c^4} P^{\alpha\beta} \left(T_{\alpha\beta} - \frac{1}{2} P_{\alpha\beta} S \right) = -\frac{8\pi G}{c^4} S \\ R_{\alpha\beta} n^\alpha n^\beta &= \frac{8\pi G}{c^4} \left(T_{\alpha\beta} - \frac{1}{2} P_{\alpha\beta} S \right) n^\alpha n^\beta = \frac{8\pi G}{c^4} \kappa \end{aligned} \tag{139}$$

from which we obtain the Hamiltonian constraint

$$\bar{R} - \sigma\left(K^2 - K^{\mu\nu} K_{\mu\nu} \right) = -\frac{8\pi G}{c^4}\,(S + \sigma\kappa) \,. \tag{140}$$

This expression differs from the expressions found in [19] using unbroken 5D field equations by the replacement

$$-\sigma\frac{16\pi G}{c^4}\kappa \longrightarrow -\frac{8\pi G}{c^4}\left(S+\sigma\kappa\right) . \tag{141}$$

Contracting the Codazzi relation (123) to

$$P^{\beta'}_{\ \beta}n^\alpha R_{\alpha\beta'} = D_\beta K - D_\alpha K^\alpha_\beta \tag{142}$$

and evaluating the mixed projection of the field equations

$$R_{\alpha\beta}n^\alpha P^{\beta'}_{\ \beta} = \frac{8\pi G}{c^4}\left(T_{\alpha\beta'}-\frac{1}{2}P_{\alpha\beta'}S\right)n^\alpha P^{\beta'}_{\ \beta} = \frac{8\pi G}{c^4}T_{\alpha\beta'}n^\alpha P^{\beta'}_{\ \beta} = -\frac{8\pi G}{c^4}p_\beta \tag{143}$$

we find the momentum constraint

$$D_\mu K^\mu_\nu - D_\nu K = \frac{8\pi G}{c^4}p_\nu \tag{144}$$

in the same form found in [19]. Together, equations (134), (135), (140), and (144) decompose the SHP field equations into an initial value problem for the metric and extrinsic curvature.

5.4. *Evolution of the vielbein field*

In the quintrad the metric η_{ab} remains constant while the vielbein field that connects the quintrad to a coordinate frame evolves with τ. To obtain a pair of coupled evolution equations for $E_\alpha^{\ a}$ and K_{ab} we restrict our attention to line elements of the form

$$dX^2 = -g_{00}(dx^0)^2 + g_{11}(dx^1)^2 + g_{22}(dx^2)^2 + g_{33}(dx^3)^2 + \sigma g_{55}(dx^5)^2 \tag{145}$$

in rectilinear coordinates and

$$dX^2 = -g_{00}dx_0^2 + g_{11}dr^2 + g_{22}d\theta^2 + g_{33}\sin^2\theta d\phi^2 + \sigma g_{55}dx^2 \tag{146}$$

in spherical coordinates. Following the prescription described in Section (2.4) we find the coordinate frames and vielbein fields to be

$$E_\alpha^{\ a} = \text{diag}\left(A_0, A_1, A_2, A_3, A_5\right) \tag{147}$$

in the rectilinear case, and

$$E_\alpha^{\ a} = \begin{bmatrix} A_0 & 0 & 0 & 0 & 0 \\ 0 & A_1\sin\theta\cos\phi & A_1\sin\theta\sin\phi & A_1\cos\theta & 0 \\ 0 & A_2\cos\theta\cos\phi & A_2\cos\theta\sin\phi & -A_2\sin\theta & 0 \\ 0 & -A_3\sin\theta\sin\phi & A_3\sin\theta\cos\phi & 0 & 0 \\ 0 & 0 & 0 & 0 & A_5 \end{bmatrix} \tag{148}$$

in the spherical case, with $g_{\alpha\alpha} = (A_\alpha)^2$. We may write (148) as

$$E_\alpha^{\ a} = A_{(\alpha)}R_\alpha^{\ a} \tag{149}$$

where (α) indicates no summation and $R_\alpha^{\ 0} = \delta^0_\alpha$, $R_0^{\ 5} = \delta^5_\alpha$, and $R_\alpha^{\ a}$ is a standard 3D rotation matrix for $\alpha, a = 1,2,3$. We easily confirm that this vielbein also leads to the diagonal metric

$$\gamma_{\alpha\beta} = \eta_{ab}E_\alpha^{\ a}E_\beta^{\ b} = A_{(\alpha)}A_{(\beta)}\eta_{ab}R_\alpha^{\ a}R_\beta^{\ b} = A_{(\alpha)}A_{(\beta)}\eta_{\alpha\beta} = A^2_{(\alpha)}\eta_{\alpha\beta} \tag{150}$$

by virtue of its orthogonality.

For any diagonal metric we have $N^\mu = 0$, for which (118) simplifies to

$$K_{ab} = \frac{1}{2}\delta_a^k \delta_b^l \left(E_{\mu k}\partial_5 e^\mu{}_l + E_{\mu l}\partial_5 e^\mu{}_k \right) = -\frac{1}{2}\delta_a^k \delta_b^l \left(e^\mu{}_l \partial_5 E_{\mu k} + e^\mu{}_k \partial_5 E_{\mu l} \right) \tag{151}$$

where we used

$$e^\mu{}_l \partial_a E_{\mu k} + E_{\mu k}\partial_a e^\mu{}_l = \partial_a \left(e^\mu{}_l E_{\mu k} \right) = \partial_a \eta_{lk} = 0\,. \tag{152}$$

Writing the inverse vielbein field as

$$e^\alpha{}_b = \left[\left(E^{-1} \right)^T \right]^\alpha{}_b = \frac{1}{A_{(\alpha)}} R^\alpha{}_b \tag{153}$$

which we verify by

$$e^\alpha{}_b E_\alpha{}^a = \frac{1}{A_{(\alpha)}} R^\alpha{}_b A_{(\alpha)} R_\alpha{}^a = R^\alpha{}_b R_\alpha{}^a = \delta_b^a \tag{154}$$

we obtain

$$e^\alpha{}_b \partial_5 E_{\alpha a} = \eta_{aa'} e^\alpha{}_b \partial_5 E_\alpha{}^{a'} = \eta_{aa'} \frac{1}{A_{(\alpha)}} R^\alpha{}_b \partial_5 A_{(\alpha)} R_\alpha{}^{a'} = \frac{\partial_5 A_{(\alpha)}}{A_{(\alpha)}} \eta_{aa'} R^\alpha{}_b R_\alpha{}^{a'}\,. \tag{155}$$

Swapping the indices a and b we see that

$$e^\alpha{}_a \partial_5 E_{\alpha b} = \frac{\partial_5 A_{(\alpha)}}{A_{(\alpha)}} \eta_{ba'} R^\alpha{}_a R_\alpha{}^{a'} = \frac{\partial_5 A_{(\alpha)}}{A_{(\alpha)}} \eta_{ba'} R_\alpha{}^{a'} R^\alpha{}_a = \frac{\partial_5 A_{(\alpha)}}{A_{(\alpha)}} \eta_{aa''} R^\alpha{}_b R_\alpha{}^{a''} = e^\alpha{}_b \partial_5 E_{\alpha a} \tag{156}$$

from which

$$K_{kl} = -\frac{1}{2}\left(e^\mu{}_l \partial_5 E_{\mu k} + e^\mu{}_k \partial_5 E_{\mu l} \right) = -e^\mu{}_l \partial_5 E_{\mu k} \tag{157}$$

permitting us to rewrite (151) as

$$\partial_5 E_\mu{}^k = -E_\mu{}^l K_l^k\,. \tag{158}$$

Using (77) as the explicit form of the vielbein field we have

$$\partial_{\bar 5} = e^\alpha{}_5 \partial_\alpha = \left[\delta_5^k \delta_\mu^\alpha e^\mu{}_k - \delta_5^5 \frac{1}{N}\left(\delta_\mu^\alpha N^\mu - \delta_5^\alpha \right) \right] \partial_\alpha = \frac{1}{N}\partial_5 \tag{159}$$

where we used $\delta_5^k = 0$ and $N^\mu = 0$. Expression (158) now becomes the evolution equation

$$\partial_5 E_\mu{}^k = -N E_\mu{}^l K_l^k \tag{160}$$

which we compare with the evolution equation for the metric (134)

$$\left(\frac{1}{c_5}\partial_\tau - \mathcal{L}_{\mathbf{N}} \right) \gamma_{\mu\nu} = -2NK_{\mu\nu} \;\longrightarrow\; \partial_5 \gamma_{\mu\nu} = -2NK_{\mu\nu} \tag{161}$$

in the case that $N^\mu = 0$.

We now consider the coordinate evolution equation for $N^\mu = 0$

$$\partial_5 K_{\mu\nu} = -D_\mu D_\nu N + N\left\{ -\sigma \bar{R}_{\mu\nu} + KK_{\mu\nu} - 2K_\mu^\lambda K_{\nu\lambda} + \sigma \frac{8\pi G}{c^4}\left(S_{\mu\nu} - \frac{1}{2}\gamma_{\mu\nu} S \right) \right\} \tag{162}$$

found as the pullback to $\mathcal{M}$ of an expression derived on the hypersurface Σ_τ. We rewrite this as a pullback from the quintrad using

$$\mathcal{E}^a_\mu = \mathcal{E}^\alpha_\mu E_\alpha{}^a = \delta^\alpha_\mu \left[\delta^{\mu'}_\alpha \delta^a_k E_{\mu'}{}^k + \delta^5_\alpha \left(E_{\mu'}{}^k N^{\mu'} \delta^a_k + N \delta^a_5 \right) \right] = \delta^a_k E_\mu{}^k \tag{163}$$

where $N^\mu = 0$ and $\delta^5_\mu = 0$, to obtain

$$\begin{aligned} \partial_5 \left(\mathcal{E}^a_\mu \mathcal{E}^b_\nu K_{ab} \right) &= -\mathcal{E}^a_\mu D_a \mathcal{E}^b_\nu D_b N \\ &\quad + \mathcal{E}^a_\mu \mathcal{E}^b_\nu N \left\{ -\sigma \bar{R}_{ab} + K K_{ab} - 2 K^c_a K_{bc} + \sigma \frac{8\pi G}{c^4} \left(S_{ab} - \frac{1}{2} \eta_{ab} S \right) \right\} . \end{aligned} \tag{164}$$

The push forward similarly reduces to

$$V^\perp_a = e^\alpha{}_a \delta^\mu_\alpha v_\mu = \left[\delta^k_a \delta^\alpha_{\mu'} e^{\mu'}{}_k - \delta^5_a \frac{1}{N} \left(\delta^\alpha_{\mu'} N^{\mu'} - \delta^\alpha_5 \right) \right] \delta^\mu_\alpha v_\mu = \delta^k_a e^\mu{}_k v_\mu \tag{165}$$

and writing $\bar{\mathcal{E}}^\mu_a = \delta^k_a e^\mu{}_k$ we have

$$\bar{\mathcal{E}}^\mu_a \mathcal{E}^{a'}_\mu = \delta^k_a \delta^{a'}_{k'} e^\mu{}_k E_\mu{}^{k'} = \delta^k_a \delta^{a'}_k . \tag{166}$$

Applying $\bar{\mathcal{E}}^\mu_a$ to the evolution equation we are led to

$$\bar{\mathcal{E}}^\mu_k \bar{\mathcal{E}}^\nu_l \partial_5 \left(\mathcal{E}^a_\mu \mathcal{E}^b_\nu K_{ab} \right) = -D_k D_l N + N \left\{ -\sigma \bar{R}_{kl} + K K_{kl} - 2 K^j_k K_{lj} + \sigma \frac{8\pi G}{c^4} \left(S_{kl} - \frac{1}{2} \gamma_{kl} S \right) \right\} \tag{167}$$

where we used the compatibility relation (46) to write $D_a \mathcal{E}^b_\nu = 0$, and expand

$$K^c_k K_{lc} = \left(\eta^{cc'} \delta^{k'}_{c'} \delta^{l'}_k K_{k'l'} \right) \left(\delta^{k''}_l \delta^{l''}_c K_{k''l''} \right) = K^j_k K_{jl} . \tag{168}$$

On the LHS

$$\bar{\mathcal{E}}^\mu_k = \delta^{k'}_k e^\mu{}_{k'} = e^\mu{}_k \tag{169}$$

and

$$\mathcal{E}^a_\mu \mathcal{E}^b_\nu K_{ab} = \left(\delta^a_{k'} E_\mu{}^{k'} \right) \left(\delta^b_{l'} E_\nu{}^{l'} \right) \left(\delta^k_a \delta^l_b K_{kl} \right) = E_\mu{}^k E_\nu{}^l K_{kl} \tag{170}$$

so that

$$\bar{\mathcal{E}}^\mu_k \bar{\mathcal{E}}^\nu_l \partial_5 \left(\mathcal{E}^a_\mu \mathcal{E}^b_\nu K_{ab} \right) = e^\mu{}_k \left(\partial_5 E_\mu{}^{k'} \right) K_{k'l} + e^\nu{}_l \left(\partial_5 E_\nu{}^{l'} \right) K_{kl'} + \partial_5 K_{kl} \tag{171}$$

and using (160) to replace $\partial_5 E_\mu{}^k$ we find

$$\bar{\mathcal{E}}^\mu_k \bar{\mathcal{E}}^\nu_l \partial_5 \left(\mathcal{E}^a_\mu \mathcal{E}^b_\nu K_{ab} \right) = -2 N K^{k'}_k K_{k'l} + \partial_5 K_{kl} . \tag{172}$$

Inserting this into (164) leads us finally to a slightly simplified evolution equation for the extrinsic curvature

$$\partial_5 K_{kl} = -D_k D_l N + N \left\{ -\sigma \bar{R}_{kl} + K K_{kl} + \sigma \frac{8\pi G}{c^4} \left(S_{kl} - \frac{1}{2} \eta_{kl} S \right) \right\} . \tag{173}$$

We mention that the equilibrium condition for this system is found by taking $N = 1$ and $\partial_\tau E_\mu{}^k = 0$, which by (160) entails $K_{kl} = 0$. Now (173) becomes

$$0 = -\bar{R}_{kl} + \frac{8\pi G}{c^4} \left(S_{kl} - \frac{1}{2} \eta_{kl} S \right) \tag{174}$$

which we recognize as the trace-reversed form of the standard 4D Einstein field equations in the vielbein.

6. Example — evolving Schwarzschild metric

In [19] we considered an event evolving with τ-varying squared mass $-p^\mu p_\mu = -(M\dot{x})^2$, and showed, using the weak field approximation, that it induces a τ-dependent metric. This metric, in turn, induces motion in a test particle whose mass varies similarly with τ, suggesting a transfer of mass carried by the gravitational field across spacetime. We also considered a generalized Schwarzschild solution with 5D metric

$$g_{\alpha\beta} = \text{diag}\left(-B, A, r^2, r^2 \sin^2\theta, \sigma N\right) \tag{175}$$

with $B = B(r,\tau)$, $A = A(r,\tau)$, and $N = 1$. The spacetime components of the Christoffel connection and Ricci tensor are thus τ-dependent, but their functional form is unaffected, and so under boundary conditions

$$B(r,\tau) \xrightarrow[r\to\infty]{} 1 \qquad\qquad A(r,\tau) \xrightarrow[r\to\infty]{} 1 \tag{176}$$

for all τ, a solution satisfying $\bar{R}_{\mu\nu} = 0$ leads to

$$AB = 1 \qquad\qquad B = 1 + \frac{f(\tau)}{r} = 1 + \frac{2MG}{c^2 r} T(\tau) \tag{177}$$

which recovers the standard Schwarzschild solution for the function $T(\tau) = 1$. Solving the field equations to obtain $T(\tau)$ for given sources $S_{\mu\nu}$, p_μ, and κ will generally require numerical integration, but to gain a sense of the formalism we may calculate the sources for a given T. Using the unbroken 5D field equations in a coordinate frame we found

$$K_{\mu\nu} = -\frac{1}{2c_5}\partial_\tau \gamma_{\mu\nu} = -\frac{1}{c_5}\dot{T}(\tau)\,\frac{MG}{c^2 r}\,\text{diag}\left(1, \frac{1}{B^2}, 0, 0\right) \longrightarrow K = 0 \tag{178}$$

$$S_{00} = B^2 S_{11} = -\sigma\frac{c^2}{c_5^2}\frac{2M}{16\pi r}\ddot{T}(\tau) \qquad\qquad S_{22} = S_{33} = 0 \longrightarrow S = 0 \tag{179}$$

along with $\kappa \approx 0$.

In the quintrad frame, comparison of the metric (175) with (148) and (150) provides the vielbein field in the form

$$E_\alpha^{\ a} = \begin{bmatrix} \sqrt{B} & 0 & 0 & 0 & 0 \\ 0 & \sqrt{A}\sin\theta\cos\phi & \sqrt{A}\sin\theta\sin\phi & \sqrt{A}\cos\theta & 0 \\ 0 & \cos\theta\cos\phi & \cos\theta\sin\phi & -\sin\theta & 0 \\ 0 & -\sin\theta\sin\phi & \sin\theta\cos\phi & 0 & 0 \\ 0 & 0 & 0 & 0 & 1 \end{bmatrix} \tag{180}$$

where we have taken

$$A_0 = \sqrt{B} \qquad\qquad A_1 = \sqrt{A} = \frac{1}{\sqrt{B}} \qquad\qquad A_2 = A_3 = A_5 = 1\,. \tag{181}$$

leading to

$$\frac{\partial_a A_0}{A_0} = \frac{\partial_a B}{2B} \qquad\qquad \frac{\partial_a A_1}{A_1} = \frac{1}{2A}\partial_a A = -\frac{\partial_a B}{2B} \tag{182}$$

as in the standard Schwarzschild solution.

With $N = 1$, and thus $D_k D_l N = 0$, the evolution equations (160) and (173) are somewhat simplified. In (178) we found $K_{\mu\nu}$ by inverting the metric evolution equation, and here we similarly use (155) to invert the vielbein evolution equation (160) and write

$$K_{kl} = -e^{\mu}{}_k \partial_5 E_{\mu l} = -\frac{\partial_5 A_{(\mu)}}{A_{(\mu)}} \eta_{ll'} R^{\mu}{}_k R_{\mu}{}^{l'} = -\frac{\partial_5 B}{B} \left(\delta^0{}_k \eta_{l0} - \eta_{ll'} R^1{}_k R_1{}^{l'} \right) \tag{183}$$

from which we again find the vanishing trace as

$$K = \eta^{kl} K_{kl} = -\frac{\partial_5 B}{B} \eta^{kl} \left(\delta^0_0 - \delta^1_1 \right) = 0\,. \tag{184}$$

We notice that the coefficient $R^1{}_m$ is the m-component of the 3D unit vector

$$\hat{\mathbf{r}} = (\sin\theta\cos\phi, \sin\theta\sin\phi, \cos\theta) \tag{185}$$

so that

$$K_{kl} = \frac{\partial_5 B}{B} \left(\delta^0_k \delta^0_l + \hat{\mathbf{r}}_k \hat{\mathbf{r}}_l \right) \tag{186}$$

"points" in the kl-direction.

In the weak field approximation we neglected terms quadratic in $K_{\mu\nu}$ but here we readily evaluate

$$K^{kl} K_{kl} = \left(-\frac{\partial_5 A_{(\mu)}}{A_{(\mu)}} \eta^{kk'} R^{\mu}{}_{k'} R_{\mu}{}^{l} \right) \left(-\frac{\partial_5 A_{(\nu)}}{A_{(\nu)}} \eta_{ll'} R^{\nu}{}_k R_{\nu}{}^{l'} \right) = \frac{1}{2} \left(\frac{\partial_5 B}{B} \right)^2 . \tag{187}$$

Inserting (186) into the second evolution equation leads to nonlinear second order equations, as we expect to find from field equations for gravitation. Splitting (173) and the constraint equations into 3D space ($m, n = 1, 2, 3$) and 0-components, these become

$$-\partial_5 \left(\frac{\partial_5 B}{B} \right) = \sigma \frac{8\pi G}{c^4} S_{00} \qquad \hat{\mathbf{r}}_m \hat{\mathbf{r}}_n \, \partial_5 \left(\frac{\partial_5 B}{B} \right) = \sigma \frac{8\pi G}{c^4} S_{mn} \tag{188}$$

$$-\frac{1}{2} \left(\frac{\partial_5 B}{B} \right)^2 = \frac{8\pi G}{c^4} \kappa \tag{189}$$

$$-D_0 \left(\frac{\partial_5 B}{B} \right) = \frac{8\pi G}{c^4} p_0 \qquad D_m \left(\frac{\partial_5 B}{B} \hat{\mathbf{r}}^m \hat{\mathbf{r}}_n \right) = \frac{8\pi G}{c^4} p_n \tag{190}$$

from which we again find $S = \eta^{kl} S_{kl} = 0$. As for K_{kl}, we see that the spacetime energy momentum tensor S_{kl} "points" in the kl-direction. The previously neglected quadratic term (189) indicates that this solution requires a negative local mass density $\kappa < 0$. In [21] we showed that κ can be expressed as a cosmological term

$$\Lambda(x, \tau) = \sigma \frac{4\pi G}{c^4} \kappa \tag{191}$$

that is not constant, but scalar and independent of S_{kl}. For a weak field solution, it is seen that this term must be very small in magnitude.

7. Conclusion

As in SHP electrodynamics, a particle worldline is the trajectory over τ of the irreversible physical event $x^\mu(\tau)$, whose evolution is generated by a scalar Hamiltonian K. The spacetime manifold of general relativity is then $\mathcal{M}(\tau)$, a 4D block universe occurring at τ and evolving under K to an infinitesimally close 4D block universe $\mathcal{M}(\tau + d\tau)$. The local metric structure $\gamma_{\mu\nu}$ of $\mathcal{M}(\tau)$, including the coordinate past and future, must therefore be τ-dependent, and appropriate field equations must be determined to specify its evolution.

The field equations of standard general relativity (GR) express the reciprocal relationship between 4D geometry and the energy-momentum of matter present in spacetime, embodying O(3,1) covariance, general diffeomorphism invariance, and the translation invariance of the Ricci tensor that guarantees the Bianchi identity as a 4D gauge symmetry. The 3+1 formalism decomposes the field equations into a pair of coupled t-evolution equations for the geometry of 3D space, along with constraints among the initial conditions. As articulated by Wheeler [31], "A decade and more of work by Dirac, Bergmann, Schild, Pirani, Anderson, Higgs, Arnowitt, Deser, Misner, DeWitt, and others has taught us through many a hard knock that Einstein's geometrodynamics deals with the dynamics of geometry: of 3-geometry, not 4-geometry."

In SHP GR we proceed in the reverse direction, guided by SHP electrodynamics. Beginning with τ-evolution, we determine field equations possessing O(3,1) covariance and a 5D gauge symmetry associated with a Bianchi identity that obtains for the 5D Ricci tensor $R_{\alpha\beta}$ constructed on the pseudo-spacetime $\mathcal{M}_5$. This structure admits the natural foliation into 4D hyperspaces Σ_τ homeomorphic to $\mathcal{M}$, leading to the desired evolution equations for $\gamma_{\mu\nu}(x,\tau)$ and $K_{\mu\nu}(x,\tau)$ as (nearly) canonically conjugate fields [19].

The basic structure of the 4+1 formalism, generalizing the geometrical arguments of the 3+1 method to the SHP framework, was presented in [19]. In [20, 21] the linearized theory was explored, and it was shown that the evolution equations retain a 5D symmetry, possibly O(4,1) or O(3,2), that leads to deviation from standard gravitational phenomenology. As in SHP electrodynamics, the fields (geometry) may exhibit the higher symmetry in a vacuum, but the matter parts must be restricted to O(3,1). In this paper, we broke the symmetry at the interface between the field and matter parts of the field equations. To this end, we constructed the tangent space $\mathcal{T}(\mathcal{M}_5)$ and its foliation in a quintrad frame, for which the flat metric and its symmetry-broken form are easily chosen. Using the vielbein field, we transformed the symmetry-broken flat metric to its local form in a coordinate frame, leading to the desired coordinate frame field equations for the 5D metric $g_{\alpha\beta}$. Since the quintrad metric is constant and the τ-dependence is contained in the vielbein field, we obtained in an evolution equation for $E_\alpha{}^a$ that applies to any diagonal metric, and showed that it simplifies the evolution equation for $K_{\mu\nu}$. Thus, initial value problems for gravitational fields may be solved in the quintrad frame.

It follows from the 5D Bianchi identity [19–21] that the SHP field equations (107) split into

$$R_{\mu\nu} = \frac{8\pi G}{c^4}\left(T_{\mu\nu} - \frac{1}{2}P_{\mu\nu}S\right) \qquad\qquad R_{5\alpha} = \frac{8\pi G}{c^4}T_{5\alpha} \tag{192}$$

where the spacetime components are ten unconstrained field equations corresponding to the τ-evolution equations for $\gamma_{\mu\nu}$ and $K_{\mu\nu}$, and the 5-components are constraints among the initial conditions that propagate forward with τ but do not evolve under second order differential equations. This decomposition was also seen [21] in the linearized SHP theory by expanding the Ricci tensor $R_{\alpha\beta}$ for a weak perturbation to the flat metric, so that equations (107) become a 5D wave equation. This wave equation similarly splits into spacetime components from which we find the linearized evolution equations, and 5-components that become the constraint equations. In this paper we saw that the vielbein field splits into a spacetime part that contains the dynamical content, and a 5-part that embodies the constraints. The natural separation of the field equations into unconstrained evolution equations and non-evolving constraints underlies the 4+1 formalism in SHP gravitation.

References

[1] Isham C 1992 Canonical quantum gravity and the problem of time Tech. Rep. Imperial/TP/91-92/25 Blackett Laboratory, Imperial College lectures at the NATO Summer School in Salamanca URL `https://arxiv.org/abs/gr-qc/9210011`
[2] Kiefer C and Peter P 2022 *Universe* **8** 36 ISSN 2218-1997 URL `http://dx.doi.org/10.3390/universe8010036`
[3] Stueckelberg E 1941 *Helv. Phys. Acta* **14** 321–322 (In French)
[4] Stueckelberg E 1941 *Helv. Phys. Acta* **14** 588–594 (In French)
[5] Horwitz L and Piron C 1973 *Helv. Phys. Acta* **48** 316–326
[6] Horwitz L and Lavie Y 1982 *Phys. Rev. D* **26** 819–838
[7] Arshansky R and Horwitz L 1989 *J. Math. Phys.* **30** 213
[8] Arshansky R and Horwitz L 1988 *Phys. Lett. A* **131** 222–226
[9] Arshansky R and Horwitz L 1989 *J. Math. Phys.* **30** 66
[10] Arshansky R and Horwitz L 1989 *J. Math. Phys.* **30** 380
[11] Saad D, Horwitz L and Arshansky R 1989 *Found. Phys.* **19** 1125–1149
[12] Horwitz L P 2015 *Relativistic Quantum Mechanics* (Dordrecht, Netherlands: Springer)
[13] Horwitz L P and Arshansky R I 2018 *Relativistic Many-Body Theory and Statistical Mechanics* 2053-2571 (Morgan & Claypool Publishers) ISBN 978-1-6817-4948-8 URL `http://dx.doi.org/10.1088/978-1-6817-4948-8`
[14] Land M and Horwitz L P 2020 *Relativistic classical mechanics and electrodynamics* (Morgan and Claypool Publishers) URL `https://www.morganclaypoolpublishers.com/catalog_Orig/product_info.php?products_id=1489`
[15] Horwitz L P 2019 *Journal of Physics: Conference Series* **1239** 012014 URL `https://doi.org/10.1088%2F1742-6596%2F1239%2F1%2F012014`
[16] Horwitz L P 2019 *The European Physical Journal Plus* **134** 313 ISSN 2190-5444 URL `https://doi.org/10.1140/epjp/i2019-12689-7`
[17] Wheeler J A 2000 *Geons, Black Holes and Quantum Foam: A Life in Physics* (W. W. Norton & Company)
[18] Land M 2019 *Astronomische Nachrichten* **340** 983–988 URL `https://onlinelibrary.wiley.com/doi/abs/10.1002/asna.201913719`
[19] Land M 2020 *Symmetry* **12** ISSN 2073-8994 URL `https://www.mdpi.com/2073-8994/12/10/1721`
[20] Land M 2021 *Journal of Physics: Conference Series* **1956** 012010 URL `https://doi.org/10.1088/1742-6596/1956/1/012010`
[21] Land M 2022 *Universe* **8** ISSN 2218-1997 URL `https://www.mdpi.com/2218-1997/8/3/185`
[22] Gourgoulhon E 2007 3+1 formalism and bases of numerical relativity Tech. rep. Laboratoire Univers et Theories, C.N.R.S. lectures given at the General Relativity Trimester held in the Institut Henri Poincare (Paris, Sept.-Dec. 2006) and at the VII Mexican School on Gravitation and Mathematical Physics (Playa del Carmen, Mexico, 26. Nov. - 2 Dec. 2006) URL `https://arxiv.org/abs/gr-qc/0703035`
[23] Bertschinger E 2002 Hamiltonian formulation of general relativity Tech. Rep. Physics 8.962 Massachusetts Institute of Technology URL `http://web.mit.edu/edbert/GR/gr11.pdf`
[24] Blau M 2020 Lecture notes on general relativity Tech. rep. Albert Einstein Center for Fundamental Physics, Universität Bern URL `http://www.blau.itp.unibe.ch/GRLecturenotes.html`
[25] Arnowitt R L, Deser S and Misner C W 2004 *General Relativity and Gravitation* **40** 1997–2027 URL `https://arxiv.org/abs/gr-qc/0405109`
[26] Yepez J 2011 (*Preprint* 1106.2037)
[27] Land M, Shnerb N and Horwitz L 1995 *J. Math. Phys.* **36** 3263
[28] Land M and Horwitz L 1991 *Found. Phys. Lett.* **4** 61
[29] Weinberg S 1972 *Gravitation and Cosmology: Principles and Applications of the General Theory of Relativity* (New York, NY: Wiley) URL `https://cds.cern.ch/record/100595`
[30] Land M 2019 *Journal of Physics: Conference Series* **1239** 012005 URL `https://doi.org/10.1088%2F1742-6596%2F1239%2F1%2F012005`
[31] Wheeler J A 1969 *pp 615-724 of Topics in Nonlinear Physics. Zabusky, Norman J. (ed.). New York, Springer-Verlag New York, Inc., 1968.*

Pseudo-Riemann's quartics in Finsler's geometry—two-dimensional case

Yakov Itin

Mathematics Department, Jerusalem College of Technology; Jerusalem, Israel

E-mail: itin@math.huji.ac.il

Abstract. Finsler's geometry usually describes an extension of Riemmann's geometry into a direction-dependent geometric structure. Historically, the well-known Riemann's quartic length element example served as the inspiration for this construction. Surprisingly, the covariant Fresnel equation—a fundamental dispersion relation in solid-state electrodynamics—emerges as the exact same quartic expression. As a result, Riemann's quartic length expression can be regarded of as a mathematical representation of a well-known physical phenomenon. In this study, we offer numerous Riemann's quartic examples that show Finsler's geometry, even in the situation of a positive definite Euclidean signature space, is too restrictive for many applications. The strong axioms of Finsler's geometry are violated in a substantially greater number of distinctive subsets for the spaces having an indefinite (Minkowski) signature. We suggest a weaker characterization of Finsler's structure based on explicitly calculated two-dimensional examples. In tangential vector space, this concept permits singular subsets. Only open subsets of a manifold's tangent bundle are required to satisfy the strong axioms of Finsler's geometry. We demonstrate the distinctive unique subsets of the Riemann's quartic in two dimensions and briefly discuss their possible physical origins.

1. Introduction

In 1854, Riemann developed his theory of higher dimension geometry and delivered in 1854 his famous Habilitation lecture at Göttingen entitled *"Über die Hypothesen welche der Geometrie zu Grunde liegen"* (On the hypotheses which underlie geometry). His advisor and examiner, Gauss, was greatly impressed by the ideas presented in this work. The paper was only published by Dedekind in 1868, two years after Riemann's death.

Riemann wrote: *"ds is the square root of an always positive integral homogeneous function of the second order of the quantities dx."* In other words, Riemann's construction is started with the quadratic length element

$$ds = \sqrt{g_{ij}(x)dx^i dx^j}\,, \tag{1}$$

where $g_{ij}(x)$ is a symmetric positive definite tensor. Nowadays this length element serves as a well-known basis of Riemannian geometry. Its indefinite version with the tensor $g_{ij}(x)$ of the Lorentz signature provides a mathematical background of General Relativity.

In his Habilitation lecture, Riemann continued: *"...The next case in simplicity includes those manifolds in which the line-element may be expressed as the fourth root of a quartic differential expression. The investigation of this more general kind would require no really different principles, but would take considerable time and throw little new light on the theory of*

Content from this work may be used under the terms of the Creative Commons Attribution 3.0 licence. Any further distribution of this work must maintain attribution to the author(s) and the title of the work, journal citation and DOI.
Published under licence by IOP Publishing Ltd

space, especially as the results cannot be geometrically expressed, I restrict myself, therefore, to those manifolds in which the line-element is expressed as the square root of a quadratic differential expression..." We denote the quartic Riemann's expression as

$$ds = \sqrt[4]{M_{ijkl}(x)dx^i dx^j dx^k dx^l}\,. \tag{2}$$

The well-known extension of the quartic Riemann's line element is Finler's geometry that deals with a general expression for the line elements

$$ds = F(x, dx). \tag{3}$$

Here $F(x, dx)$ is an arbitrary smooth positive function that is first-order homogeneous in its second argument. After intensive study of Finsler, Cartan, Chern, and many others, this subject turned into a solid mathematical theory, see [1], [2] and the references given therein. The classical Finsler's geometry is restricted to the positive defined case, i.e., to the Euclidean signature.

Various physics applications motivate study of Finsler geometry in the *pseudo-Euclidean* domain. The well-known example from classical physics is the Randers metric that can be viewed as an attempt of geometrization of the Lagrangian of a test electric charge. Possible pseudo-Finsler modifications of gravity were discussed in [6]. Also in quantum field theories, Finsler's geometry is a rather popular construction. In particular, in [8], [9] pseudo-Finsler geometry is applied as a Lorentz symmetry violating model. In [10] Finsler-type modification of the Coulomb law and corresponding additional splitting of hydrogen energy levels was calculated. This result provides a way for an experimental verification of Finsler's structure. Another example is the covariant premetric dispersion relation in electromagnetism [5] that is presented as a general fourth order polynomial equation of the form

$$G^{ijkl}q_i q_j q_k q_l = 0, \tag{4}$$

where q_i is the wave covector, while the tensor G^{ijkl} is constructed from the characteristic parameters of the media such as *electric permittivity* and *magnetic impermeability*. Due to the well-observed birefringence effect in anisotropic optics, the covariant premetric electromagnetism must be considered as a viable theory. Remarkably, the covariant premetric dispersion relation has exactly the same form as the one exemplified by Riemann in Eq.(2).

This paper is aimed to discuss how Riemann's quadric

$$Q(x, v) = M_{ijkl}(x)v^i v^j v^k v^l \tag{5}$$

can be squeezed into the basic definitions of Finsler's metric. Our conclusion is that the for physics applications Finsler's geometry must be extended to include various singular subsets.

The paper is organized as follows: In Sect. 2, we provide the classical definition of Finsler's geometry and discuss the reasoning for its strong restriction. In Sect. 3, we apply Finsler geometry in the Euclidean domain to Riemann's quadric. We observe that even in this classical subject, the Finsler metric is degenerated on various conics of tangential vector space. So the basic definition must be modified. In Sect. 4, we discuss the extension of Finsler geometry into the non-Euclidean domain. In contrast to the familiar pure Lorentzian case, this construction is well-defined and differentiable only in a certain *conical subset of the tangent space.* Sect. 6 is devoted to a novel weakened definition of Finsler structure. We require this structure to be defined on an open subspace of the tangent bundle. We explicitly identify the singular subsets where some of the restrictions of the basic definition are broken down. In our opinion, these subsets provide important information about the physics of the corresponding processes. In the concluding section, we discuss the possible extensions and applications of the weekend Finsler's structure.

2. Finsler's construction

A generalization of Riemann's quadratic (1) and quartic (3) linear elements to a general linear element was evaluated by Finsler (1918). In this section, we briefly discuss the basic definitions of Finsler's geometry. For the details, see, e.g., [1], [2] and the references given therein.

2.1. Basic definitions

A *Finsler manifold* is a differentiable manifold M endowed with a *Finsler function* $F(x, v)$ of two different sets of variables: a point $x \in M$ and a vector $v \in T_xM$. Finsler's function is assumed to satisfy the following conditions:

(F1) *Continuity:* $F(x, v)$ is defined and continuous on the entire tangent bundle TM, i.e., for all x and v.

(F2) *Positive definiteness:* For all x and all v, Finsler's function is non-negative, $F(x, v) \geq 0$. Equality holds only at the origin $v = 0$.

(F3) *Positive homogeneity:* For all $\lambda > 0$ and all $v \neq 0$, the first order homogeneity equation holds,

$$F(x, \lambda v) = \lambda F(x, v). \tag{6}$$

(F4) *Smoothness:* For all points x and for all *nonzero* vectors v, the function $F(x, v)$ is smooth in its second argument (at least to the class C^4).

(F5) *Finsler's metric:* Finsler's metric tensor is defined as the Hessian of the Finsler function square $F(x, v)^2$

$$f_{ij}(x, v) = \frac{1}{2}\frac{\partial^2 F(x, v)^2}{\partial v^i \partial v^j}\,. \tag{7}$$

The metric is required to be non-degenerate for all points x and for all vectors $v \neq 0$.

(F6) *Euclidean signature:* The signature of the metric $f_{ij}(x, v)$ is assumed to be Euclidean. It means that at every point $x \in M$ there is a basis where the metric is equal to the unit matrix $f_{ij}(x, v) = \text{diag}(1, 1, 1, 1)$.

2.2. Motivations and comments

Some motivations for the conditions presented in the definition of the Finsler structure can be given as follows. The first group of conditions describes the norm (length) of an arbitrary vector $v \in T_xM$ at an arbitrary point $x \in M$.

(F1) Continuity condition guarantees that a line element (length of a vector) $F(x, v)$ is defined (and final) for every vector v attached at an arbitrary point x.

(F2) Positiveness of $F(x; v)$ is connected to the Euclidean signature of the space. It guarantees a positive length for an arbitrary non-zero vector v. This condition must be modified in the Minkowskian signature space, see Sect. 6. In particular, for a smooth curve $x(s) \in M$ with a tangent vector $v = \dot{x}(s)$, the Finsler function $F(x, v)$ assigns a positive length

$$\ell = \int_{s_1}^{s_2} F\big(x(s), \dot{x}(s)\big)\, ds\,. \tag{8}$$

(F3) The positive homogeneity of the degree one provides proper physical dimension of the length function. Moreover, it guarantees that the length of an arbitrary curve is invariant under positive oriented reparametrization of the curve.

The second group of conditions provides an additional structure on the tangential vector space T_xM (and on the total manifold M). The condition (F5) introduces a scalar product of two vectors u and w that depends on a chosen vector v. The corresponding metric tensor (even

Journal of Physics: Conference Series **2482** (2023) 012007 doi:10.1088/1742-6596/2482/1/012007

being dependent on v) opens a room for applying the standard tools of differential geometry—connections, curvature tensors, and so on.

(F4) The smoothness condition permits the derivatives of the function $F(x, v)$ to be singular only for $v = 0$. This restriction is necessary because already in the simplest case (1), the derivative of the square root expression is not defined for $v = 0$. One formulates the condition as differentiability of $F(x, v)$ on a subset of a bundle $TM\backslash\{0\}$, where $\{0\}$ is a null section. In fact, only the derivatives of the Lagrangian $L(x, v) = F(x, v)^2$ are in use in Eq.(15). Thus it is possible to require a bit weaker condition—smoothness of the square $F(x, v)^2$ instead of the function $F(x, v)$ itself. Such modification is useful for Riemann's quadratic expression. We will see that this condition is related to the Euclidean (elliptic) structure on the manifold. In the pseudo-Euclidean case, it must be modified to permit additional singular sets.

(F5) The Euclidean metric construction can be written in a coordinate-free form:
For each nonzero $v \in T_pM$, the Hessian of $F^2(x, v)$ is applied at two independent vectors u and w,

$$\mathbf{f}_v(u, w) := \frac{1}{2}\frac{\partial^2}{\partial s \partial t} F(x, v + su + tw)^2 \Big|_{s=t=0} \tag{9}$$

As a result, the tensor f_{ij} is symmetric. Moreover, with respect to the direction variable the metric function turns out to be a homogeneous function of the zero order.

$$f_{ij}(x, \lambda v) = f_{ij}(x, v). \tag{10}$$

Consequently, the corresponding norm $\sqrt{f_{ij} w^i w^j}$ of an arbitrary vector w^i has a proper physical dimension of length. In the simplest case of Finsler's function that is quadratic in the direction variable, Eq. (15) yields the ordinary Riemann metric. These properties (together with positive definiteness) justify the choice of the Hessian as a metric tensor.

(F6) The Euclidean signature condition require the matrix f_{ij} to be positive definite for an arbitrary vector u

$$f_{ij}(x, v) u^i u^j \geq 0. \tag{11}$$

Under a general linear transformation of the basis, the components of the matrix f_{ij} are transformed by the standard tensor rule. Thus f_{ij} is referred to as *Finsler's metric*. This metric-type tensor depends not only on the point $x \in M$ but also on the directions given by the tangent vector $v \in T_xM$. Recall that $f_{ij}(x, v)$ is not necessarily defined for $v = 0$. Using Finsler's metric, the scalar product of two arbitrary vectors u and w can be defined in the standard way

$$(u, w)|_{f(v)} = f_{ij}(x, v) u^i w^j \,. \tag{12}$$

For a given Finsler's function, this expression can be treated as a *scalar product of vectrors* u, w *with respect to the chosen vector* v. Correspondingly, a *norm of a vector* u *with respect to the chosen vector* v is defined as

$$|u|^2_{f(v)} = (u, u)|_{f(v)} = f_{ij}(x, v) u^i u^j \,. \tag{13}$$

Due to the homogeneity condition, the norm of the vector v calculated with respect to the vector v itself is equal to the square of the Finsler function

$$|v|^2_{f(v)} = (v, v)|_{f(v)} = f_{ij}(x, v) v^i v^j = F(x, v)^2 \,. \tag{14}$$

Deviation of the Finsler structure from the Riemann one is characterized by the *Cartan tensor*

$$C_{ijk} = \frac{1}{2}\frac{\partial}{\partial v^k} f_{ij}(x, v) = \frac{1}{4}\frac{\partial^3 F(x, v)^2}{\partial v^i \partial v^j \partial v^k} \,. \tag{15}$$

When $C_{ijk} = 0$ for all vectors v and for all points x, the metric f_{ij} is independent of the direction, hence the space is pure Riemannian.

Note that most of Finsler's functions, in particular those that are derived from pseudo-Riemann quartics, provide metrics that are degenerate on some subset of the vectors v. In mathematical texts, the metric condition is extended to a weaker subadditivity inequality condition. It is a generalization of the standard triangle inequality of Euclidean geometry.

(F5'–F6') *Subadditivity:*

$$F(x, v + w) \leq F(x, v) + F(x, w) \quad \text{for all} \quad x \in M \quad \text{and all} \quad v, w \in T_x M. \tag{16}$$

In this case, one does not need the metric construction at all. This extension, however, does not conform with the standard construction of Riemann's geometry (metric, connection, curvature, etc.). Moreover, it cannot be extended to the pseudo-Riemannian case.

3. Two-dimensional positive definite quartics

In order to clarify the conditions presented in the definition of Finsler's structure, we consider some explicit 2-dimensional examples. We use Beem's construction [Beem(1970)] with the second-order homogeneous Lagrangian $L(x, v)$ instead of the first-order homogeneous Finsler's function $F(x, v)$ itself. The dependence of a point x in the Lagrangian $L(x, v)$ is suppressed while the direction variable is presented in the form $v = (v^1, v^2) = (\alpha, \beta)$. Note that the 2-dimensional case has some special similarity to the 4-dimensional one. In both cases, a metrics f_{ij} of the Euclidean signature have to satisfy the condition

$$\det f_{ij} > 0. \tag{17}$$

In the 2-dimensional case, this condition is necessary and sufficient since it extracts the metrics of the signatures $(+1, +1)$ and $(-1, -1)$, so that both can be considered as Euclidean. In the 4-dimensional case, (17) is only a necessary condition since it is satisfied by the metrics of the signature $(+1, +1, +1, +1)$ and $(-1, -1, +1, +1)$. In this case, we need additional algebraic inequalities in order to extract the Euclidean signature. For the Lorentzian signature, the condition

$$\det f_{ij} < 0 \tag{18}$$

is necessary and sufficient in two-dimensional and four-dimensional cases (but not in the three-dimensional case).

3.1. Square of Euclidean quadric

The simplest positive definite quartic expression is presented as the square of the Euclidean quadratic

$$Q(v) = (\alpha^2 + \beta^2)^2 . \tag{19}$$

The unique possible Lagrangian for this expression is given by

$$L(v) = \alpha^2 + \beta^2 . \tag{20}$$

This expression is positive for all non-zero vectors v and vanishes only at the origin. Moreover, it is continuous and smooth for all $v \in V$. The corresponding Finsler metric is Riemannian

$$f_{ij} = \text{diag}(1, 1) \tag{21}$$

and well-defined for all vectors in the plan. Finsler's function

$$F(v) = \sqrt{\alpha^2 + \beta^2} \tag{22}$$

is well-defined for all v and smooth everywhere except the origin $v = (\alpha, \beta) = 0$.

3.2. One more positive defined quartic

As an example of a non-trivial Riemann's quartic of the Euclidean type, we consider the expression of the form

$$Q(v) = \alpha^4 + \beta^4. \tag{23}$$

The unique possible Lagrangian for this quartic is given by

$$L(v) = L(\alpha, \beta) = \sqrt{\alpha^4 + \beta^4}\,. \tag{24}$$

This expression is well-defined and continuous for all vectors v and has the required second order of homogeneity. Note that the expression (24) can be regarded as a special case of a quartic decomposable into a product of two positive definite quadratic factors

$$L(v) = L(\alpha, \beta) = \sqrt{(\alpha^2 - \sqrt{2}\alpha\beta + \beta^2)(\alpha^2 + \sqrt{2}\alpha\beta + \beta^2)}\,. \tag{25}$$

The vector of the partial derivatives of the Lagrangian (24)

$$L_{,i} = \frac{2}{\sqrt{\alpha^4 + \beta^4}} \begin{pmatrix} \alpha^3 \\ \beta^3 \end{pmatrix} \tag{26}$$

is continuous and non-zero for all $v \neq 0$. At the origin $v = (\alpha, \beta) = 0$, the function $L(v)$ is non-differentiable. Calculating the Hessian of $L(v)$, we obtain Finsler's metric in the form

$$f_{ij} = \frac{1}{(\alpha^4 + \beta^4)^{3/2}} \begin{pmatrix} \alpha^2(\alpha^4 + 3\beta^4) & -2\alpha^3\beta^3 \\ -2\alpha^3\beta^3 & (3\alpha^4 + \beta^4)\beta^2 \end{pmatrix}. \tag{27}$$

The determinant of this matrix

$$\det(f_{ij}) = \frac{3\alpha^2\beta^2}{\alpha^4 + \beta^4} \tag{28}$$

is non-negative for all v. It is singular only at the origin $v = (\alpha, \beta) = 0$. Moreover it vanishes on two separate axes $\alpha = 0$ and $\beta = 0$, i.e., for the vectors $v = (0, \beta)$ and $v = (\alpha, 0)$, respectively. Beyond these axes, the metric tensor is positive definite. In particular, the norm of an arbitrary nonzero vector $u = (x, y)$ with respect to the metric (27) reads

$$||u||_v^2 = \frac{1}{(\alpha^4 + \beta^4)^{3/2}} \left(\alpha^2(\alpha^4 + 3\beta^4)x^2 - 4\alpha^3\beta^3 xy + (3\alpha^4 + \beta^4)\beta^2 y^2\right), \tag{29}$$

or, equivalently,

$$||u||_v^2 = \frac{1}{(\alpha^4 + \beta^4)^{3/2}} \left((\alpha^3 x + \beta^3 y)^2 + 3\alpha^2\beta^2(\beta x - \alpha y)^2\right). \tag{30}$$

With respect to the new coordinates $(x, y) \to (\xi, \eta)$

$$\xi = \frac{\alpha^3 x + \beta^3 y}{(\alpha^4 + \beta^4)^{3/4}}, \qquad \eta = \frac{\sqrt{3}\alpha\beta(\beta x - \alpha y)}{(\alpha^4 + \beta^4)^{3/4}} \tag{31}$$

the norm expression is given in the standard Euclidean form

$$||u||_v^2 = \xi^2 + \eta^2\,. \tag{32}$$

In order to have in (31) a reversible transformation of coordinates we have to require

$$\frac{1}{(\alpha^4+\beta^4)^{3/2}}\det\begin{pmatrix}\alpha^3 & \beta^3\\ \sqrt{3}\alpha\beta^2 & -\sqrt{3}\alpha^2\beta\end{pmatrix}=-\frac{\sqrt{3}\alpha\beta}{\sqrt{\alpha^4+\beta^4}}\neq 0. \tag{33}$$

This requirement holds only for $\alpha\beta\neq 0$ that is in correspondence with the expression for the determinant (28). The expression (32) is positive beyond the axes. However on α-axis, i.e., for $\beta=0$, it takes the value $||u||_v^2 = x^2$ that is zero for nonzero vectors of the form $u=(0,y)$. A similar singular behavior emerges also in the direction based on the basis-element vectors $v=(0,\beta)$.

Due to two peculiar directions in which the geometry degenerates, Finsler's metric specified for the Lagrangian (24) cannot be regarded as pure Finslerian.

4. Two-dimensional indefinite quartics

In this section, we consider several two-dimensional examples of the indefinite quartic

$$Q(x,v)=M_{ijkl}v^iv^jv^kv^l. \tag{34}$$

In this case, the indices change in the range $i,j,\dots=1,2$. We use the presentation $v^i=(\alpha,\beta)$. The quartic $Q(x,v)$ is indefinite–it can have positive, negative, or zero value for different directions of the vector v.

In the 2-dimensional case (as well as the 4-dimensional case) the metrics f_{ij} of Lorentzian signature is distinguished by an unique condition

$$\det f_{ij}<0. \tag{35}$$

In other dimensions, we need more algebraic inequalities. Having in the mind 4-dimensional Lorentz signature of the form $(-1,+1,+1,+1)$ we refer to the vectors of positive/negative Lagrangians (squared Finsler's norms) as *spacelike* and *timelike*, respectively. The *null vectors* are vectors with a zero Lagrangian.

4.1. Square of Lorentzian quadric

We start with a simplest quartic that is presented as a square of an indefinite quadratic

$$Q(v)=(\alpha^2-\beta^2)^2\,. \tag{36}$$

This expression is positive but vanishes for non-zero vectors of the form $v=(\alpha,\pm\beta)$.

Define the Lagrangian in the standard (special relativistic) form

$$L(v)=\alpha^2-\beta^2\,. \tag{37}$$

This function is smooth for all vectors v. The Finsler metric has the standard Lorentz form

$$f_{ij}(v)=\mathrm{diag}(1,-1)\,. \tag{38}$$

Note that three characteristic sets can be determined

$$\mathcal{A} = \{v\in T_xM\big||\alpha|>|\beta|\} \Longrightarrow \text{spacelike vectors} \tag{39}$$
$$\mathcal{B} = \{v\in T_xM\big||\alpha|<|\beta|\} \Longrightarrow \text{timelike vectors} \tag{40}$$
$$\mathcal{C} = \{v\in T_xM\big||\alpha|=|\beta|\} \Longrightarrow \text{null vectors} \tag{41}$$

Consider different possible definitions of the Finsler function (norm). In fact they are not related to the specific choice of the Lagrangian.

(1) Finsler's function of the form

$$F(v) = \sqrt{\alpha^2 - \beta^2} \tag{42}$$

is defined only in the region satisfying $|\alpha| \geq |\beta|$ and differentiable in the open conic $|\alpha| > |\beta|$.

(2) Another possibility is to define the Finsler function in the absolute value form

$$F(v) = \sqrt{|\alpha^2 - \beta^2|}\,. \tag{43}$$

Recall that this construction was proposed by Beem, see (84). This function is defined for all vectors v and differentiable besides the set $\mathcal{C}$. It cannot, however, separate the time-like and space-like vectors.

(3) The third definition is based on the proposal in Sect. 5.

$$F(v) = \sqrt{|\alpha^2 - \beta^2|}\,\text{sgn}L(v) = \begin{cases} \sqrt{\alpha^2 - \beta^2} & \text{for} \quad |\alpha| \geq |\beta|; \\ -\sqrt{\beta^2 - \alpha^2} & \text{for} \quad |\alpha| < |\beta|. \end{cases} \tag{44}$$

In this case, the norm of the space-like vectors is positive and of the time-like vectors is negative. The norm of the null-vectors is zero.

4.2. Lorentzian non-birefringent space-time

We consider now a Lorentzian-type quartic of the form

$$Q(v) = \alpha^4 - \beta^4\,. \tag{45}$$

This expression is well-defined and smooth for all vectors $v = (\alpha, \beta)$. It can be expressed as a product of two quadrics–one Euclidean and one Lorentzian

$$Q(v) = (\alpha^2 + \beta^2)(\alpha^2 - \beta^2)\,. \tag{46}$$

Our first task is to construct a proper Lagrangian. Let us consider different options:

(1) Define the Lagrangian of the form

$$L(v) = L(\alpha, \beta) = \sqrt{\alpha^4 - \beta^4} = \sqrt{(\alpha^2 - \beta^2)(\alpha^2 + \beta^2)} \tag{47}$$

This function is definite and continuous only in the region $|\alpha| \geq |\beta|$ and smooth in its interior $|\alpha| > |\beta|$. Then, the characteristic sets are

$$\mathcal{A} = \{v \in T_xM \big| |\alpha| > |\beta|\} \Longrightarrow \text{spacelike vectors} \tag{48}$$

$$\mathcal{B} = \emptyset \Longrightarrow \text{timelike vectors} \tag{49}$$

$$\mathcal{C} = \{v \in T_xM \big| |\alpha| = |\beta|\} \Longrightarrow \text{null vectors} \tag{50}$$

The gradient of this function is given by

$$L_{,i} = \frac{2}{\sqrt{\alpha^4 - \beta^4}} \begin{pmatrix} \alpha^3 \\ -\beta^3 \end{pmatrix}. \tag{51}$$

This expression is continuous in the region $\mathcal{A}$. The same is true for all higher derivatives of $L(v)$. Consequently for all positive integers i, the sets of differentiability coincide with the set of positive values of $L(v)$, i.e., $\mathcal{D}_i = \mathcal{A}$. The Hessian yields Finsler's metric of the form

$$f_{ij} = \frac{1}{(\alpha^4 - \beta^4)^{3/2}} \begin{pmatrix} \alpha^2(\alpha^4 - 3\beta^4) & 2\alpha^3\beta^3 \\ 2\alpha^3\beta^3 & \beta^2(\beta^4 - 3\alpha^4) \end{pmatrix} . \tag{52}$$

The determinant of this metric,

$$f = \frac{-3\alpha^2\beta^2}{\alpha^4 - \beta^4} , \tag{53}$$

is non-positive in the set $\mathcal{A}$. It is degenerate on the strength line $\beta = 0$ (the line $\alpha = 0$ is beyond the set $\mathcal{A}$) . It means that the set of degeneration is defined as $\mathcal{E} = \{v \in T_xM | \beta = 0\}$. Besides this singular set, the metric is Lorentzian, i.e., $\mathcal{G} = \{v \in T_xM | \beta \neq 0\}$.

For the norm of an arbitrary vector $u = (x, y)$, we have

$$||u||_v^2 = \frac{1}{(\alpha^4 - \beta^4)^{3/2}} \left(\alpha^2(\alpha^4 - 3\beta^4)x^2 + 4\alpha^3\beta^3 xy + \beta^2(\beta^4 - 3\alpha^4)y^2\right) , \tag{54}$$

or, equivalently,

$$||u||_v^2 = \frac{1}{(\alpha^4 - \beta^4)^{3/2}} \left((\alpha^3 x + \beta^3 y)^2 - 3\alpha^2\beta^2(\beta x - \alpha y)^2\right) , \tag{55}$$

Let us define new coordinates $(x, y) \to (\xi, \eta)$

$$\xi = \frac{\alpha^3 x + \beta^3 y}{(\alpha^4 - \beta^4)^{3/4}}, \qquad \eta = \frac{\sqrt{3}\alpha\beta(\beta x - \alpha y)}{(\alpha^4 - \beta^4)^{3/4}} . \tag{56}$$

Then the norm of the vector u is given in the standard Lorentzian form

$$||u||_v^2 = \xi^2 - \eta^2 . \tag{57}$$

Note that the transformation of the coordinates is applicable only in the regions where it is revertible. So we have to require

$$\det \begin{pmatrix} \alpha^3 & \beta^3 \\ \sqrt{3}\alpha\beta^2 & -\sqrt{3}\alpha^2\beta \end{pmatrix} = -\sqrt{3}\alpha\beta(\alpha^4 + \beta^4) \neq 0 . \tag{58}$$

It means that the transformations are available only for $\alpha\beta \neq 0$.

(2) Let us consider a Lagrangian of the absolute value form

$$L(v) = L(\alpha, \beta) = \sqrt{|\alpha^4 - \beta^4|} = \sqrt{|\alpha^2 - \beta^2|(\alpha^2 + \beta^2)} \tag{59}$$

The function is non-negative in the whole plane and strictly positive in open conic regions beside the strength lines $|\alpha| = |\beta|$. Then, the characteristic sets are

$$\mathcal{A} = \{v \in T_xM \big| |\alpha| \neq |\beta|\} \implies \text{spacelike vectors} \tag{60}$$

$$\mathcal{B} = \emptyset \implies \text{timelike vectors} \tag{61}$$

$$\mathcal{C} = \{v \in T_xM \big| |\alpha| = |\beta|\} \implies \text{null vectors.} \tag{62}$$

Journal of Physics: Conference Series **2482** (2023) 012007 doi:10.1088/1742-6596/2482/1/012007

In the regions $|\alpha| > |\beta|$ and $|\alpha| < |\beta|$, the gradient of this function is given by

$$L_{,i} = \frac{2}{\sqrt{\alpha^4 - \beta^4}}\begin{pmatrix}\alpha^3\\ -\beta^3\end{pmatrix}, \qquad \text{and} \qquad L_{,i} = \frac{2}{\sqrt{-\alpha^4 + \beta^4}}\begin{pmatrix}-\alpha^3\\ \beta^3\end{pmatrix}, \tag{63}$$

respectively. The Hessian yields Finsler's metric of the form

$$f_{ij} = \frac{1}{|\alpha^4 - \beta^4|^{3/2}}\begin{pmatrix}\alpha^2(\alpha^4 - 3\beta^4) & 2\alpha^3\beta^3\\ 2\alpha^3\beta^3 & \beta^2(\beta^4 - 3\alpha^4)\end{pmatrix}. \tag{64}$$

The determinant of this metric reads

$$f = \frac{-3\alpha^2\beta^2}{|\alpha^4 - \beta^4|} \tag{65}$$

Consequently for $\alpha, \beta \neq 0$ the determinant is negative. In a 2-dimensional space (and in a 4-dimensional space, as well) this requirement is sufficient to deduce that the metric is Lorentzian. For this Lagrangian, the norm of an arbitrary vector can be also written in a pure Lorentzian form $||u||_v^2 = \xi^2 - \eta^2$. This form is derived by the transformations (56) with an absolute value function added into the denominator.

(3) Let us consider a Lagrangian of the form

$$L(v) = L(\alpha, \beta) = \sqrt{|\alpha^4 - \beta^4|}\,\mathrm{sgn}(Q) = \begin{cases} \sqrt{\alpha^4 - \beta^4} & \text{for} \quad |\alpha| \geq |\beta|;\\ -\sqrt{-\alpha^4 + \beta^4} & \text{for} \quad |\alpha| < |\beta|.\end{cases} \tag{66}$$

The function is definite and continuous in the whole plane. The characteristic sets are

$$\mathcal{A} = \{v \in T_xM \big| |\alpha| > |\beta|\} \implies \text{spacelike vectors} \tag{67}$$

$$\mathcal{B} = \{v \in T_xM \big| |\alpha| < |\beta|\} \implies \text{timelike vectors} \tag{68}$$

$$\mathcal{C} = \{v \in T_xM \big| |\alpha| = |\beta|\} \implies \text{null vectors.} \tag{69}$$

and strictly positive in open conic regions beside the strength lines $|\alpha| = |\beta|$. In the regions $|\alpha| > |\beta|$ and $|\alpha| < |\beta|$, the gradient of this function is given by

$$L_{,i} = \frac{2}{\sqrt{\alpha^4 - \beta^4}}\begin{pmatrix}\alpha^3\\ -\beta^3\end{pmatrix}, \qquad \text{and} \qquad L_{,i} = \frac{2}{\sqrt{-\alpha^4 + \beta^4}}\begin{pmatrix}\alpha^3\\ -\beta^3\end{pmatrix}, \tag{70}$$

respectively. The Hessian yields Finsler's metric of the form

$$f_{ij} = \frac{1}{|\alpha^4 - \beta^4|^{3/2}}\begin{pmatrix}\alpha^2(\alpha^4 - 3\beta^4) & 2\alpha^3\beta^3\\ 2\alpha^3\beta^3 & \beta^2(\beta^4 - 3\alpha^4)\end{pmatrix}\mathrm{sgn}(Q). \tag{71}$$

The determinant of this metric

$$f = \frac{-3\alpha^2\beta^2}{|\alpha^4 - \beta^4|} \tag{72}$$

is definite and negative almost everywhere. It vanishes only on the lines $|\alpha| = 0$ and $|\beta| = 0$ and degenerates on the lines $|\alpha| = |\beta|$. In this metric the squared norm of a vector can be also transformed into the standard Lorentzian form $||u||_v^2 = \xi^2 - \eta^2$.

5. Quartic as a product of two quadratics

Let us consider an example of a quartic that is expanded into a product of two factors

$$Q(v) = Q(\alpha, \beta) = \left(g_{ij}v^i v^j\right)\left(h_{ij}v^i v^j\right) \tag{73}$$

with two matrices g_{ij} and h_{ij}. Since these two matrices can be transformed simultaneously into the diagonal form, we come to expressions of three possible type

5.1. *Euclid* × *Euclid*

Let us consider a quartic expression

$$Q(v) = Q(\alpha, \beta) = (\alpha^2 + \beta^2)(\alpha^2 + k\beta^2)\,, \tag{74}$$

where k is a dimensionless *positive* numerical parameter. Thus we have a positive-definite quartic. For this quartic function, the unique possible Lagrangian is given by

$$L(v) = L(\alpha, \beta) = \sqrt{(\alpha^2 + \beta^2)(\alpha^2 + k\beta^2)}\,. \tag{75}$$

This function is definite and continuous for all vectors $v = (\alpha, \beta)$. The gradient of $L(v)$ takes the form

$$L_{,i} = \frac{1}{\sqrt{(\alpha^2 + \beta^2)(\alpha^2 + k\beta^2)}}\begin{pmatrix} 2\alpha^3 + (k+1)\alpha\beta^2 \\ (k+1)\alpha^2\beta + 2k\beta^3 \end{pmatrix}. \tag{76}$$

Thus the first order derivatives (together with the higher order ones) are defined on the whole plane except the origin. The Hessian yields Finsler's metric of the form

$$f_{ij} = \frac{1}{2\sqrt{(\alpha^2 + \beta^2)^3(\alpha^2 + k\beta^2)^3}} \cdot \begin{pmatrix} 2\alpha^6 + 3(k+1)\alpha^4\beta^2 + 6k\alpha^2\beta^4 + (k^2+k)\beta^6 & (k-1)^2\alpha^3\beta^3 \\ (k-1)^2\alpha^3\beta^3 & (k+1)\alpha^6 + 6k\alpha^4\beta^2 + 3(k^2+k)\alpha^2\beta^4 + 2k^2\beta^6 \end{pmatrix}. \tag{77}$$

We calculate the determinant of this metric by applying the Wolfram computation system. It is given by

$$\det(f_{ij}) = \frac{2(1+k)\alpha^4 - (k^2 - 10k + 1)\alpha^2\beta^2 + 2k(k+1)\beta^4}{4(\alpha^2 + \beta^2)(\alpha^2 + k\beta^2)} \tag{78}$$

Observe that for $k = 1$, we have here the standard Euclidean metric $f_{ij} = \mathrm{diag}(1,1)$, that is independent on the direction $v = (\alpha, \beta)$. For $k = -1$, we have the quartic expressions $Q(v) = \alpha^4 - \beta^4$ with the corresponding Finsler metric that we consider later.

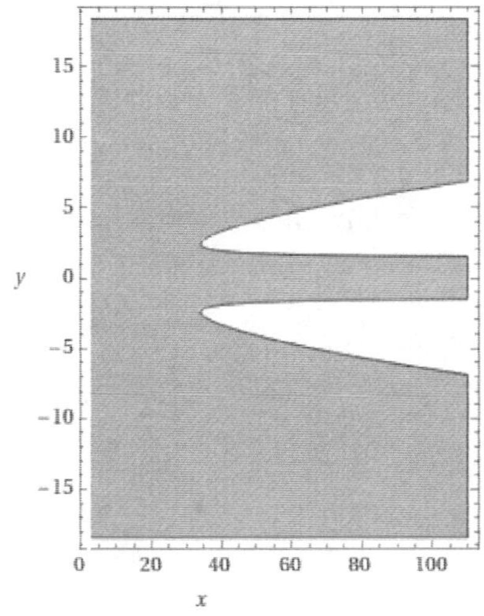

Figure 1. In the plan with the axes $x = k$ and $y = \alpha/\beta$, we depict the area (in blue) of positive determinant $\det(f_{ij}) > 0$. In this case, the metric is Euclidean. Starting with the sufficiently big values of the parameter $k \sim 30$ the determinant is negative for small angles of the vector v. In this area, the space is Lorentzian.

In Fig.(5.1), we present numerical presentation of the regions where the determinant $\det(f_{ij})$ considering as a function of two variables k and α/β is positive or negative. On the boundary curves, the metric is degenerated. The asymptotic lines $y = \alpha/\beta = \sqrt{2}$ are visual.

For large values of the parameter k, the determinant approaches the asymptotic expansion

$$\det(f_{ij}) \sim \left(\frac{2\beta^2 - \alpha^2}{\alpha^2 + \beta^2}\right)\frac{k}{4}. \tag{79}$$

It means that for the large values of k and sufficiently small tangent of the vector v, such as $\beta/\alpha < 1/\sqrt{2}$, there is a region of the negative determinant, i.e., of the Lorentzian signature. In the exterior open domain, the metric is of the Euclidean signature. These domains are bounded by the strength lines where the metric is degenerated.

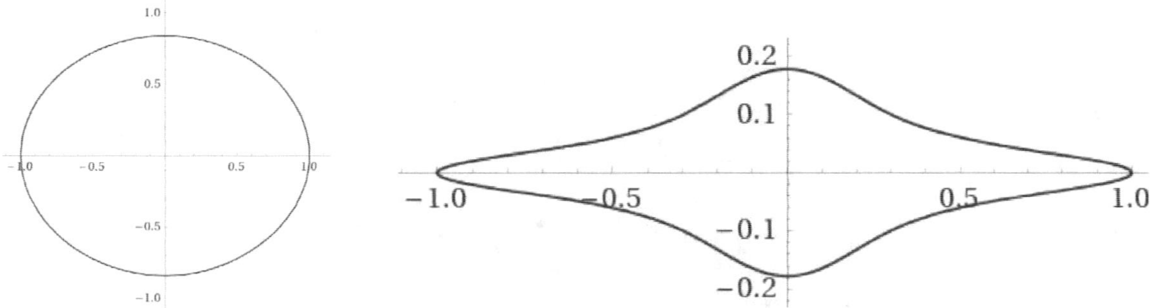

Figure 2. The graphs present the solutions of the equation $Q(v) = 1$ for different values of the parameter k. For small values of the parameter k, (the first graph) the curve is convex. For big values of the parameter k, (the second graph) the curve has convex and concave parts.

Note a non-trivial behavior of Finsler's metric in the Lorentzian regime. Since

$$f_{ij}v^iv^j = L(v) > 0, \tag{80}$$

the squared line element is positive for every $v \neq 0$.

5.2. *Euclid × Lorentz*

Let us consider a quartic expression

$$Q(v) = Q(\alpha, \beta) = (\alpha^2 + \beta^2)(\alpha^2 - k\beta^2)\,, \tag{81}$$

with a dimensionless *positive* numerical parameter k. Thus we have a product of a positive-definite quadratic and indefinite one. The Lagrangian corresponding to this quartic

$$L(v) = L(\alpha, \beta) = \sqrt{(\alpha^2 + \beta^2)|\alpha^2 - k\beta^2|}\,, \tag{82}$$

The gradient of $L(v)$ takes the form

$$L_{,i} = \frac{1}{\sqrt{(\alpha^2 + \beta^2)|\alpha^2 - k\beta^2|}} \begin{pmatrix} 2\alpha^3 + (1-k)\alpha\beta^2 \\ (1-k)\alpha^2\beta - 2k\beta^3 \end{pmatrix} \mathrm{sgn}(\alpha^2 - k\beta^2). \tag{83}$$

Thus the first order derivatives (together with the higher order ones) are defined on the whole plane except the pair of strength lines $\alpha^2 = k\beta^2$.

The Hessian yields Finsler's metric of the form

$$f_{ij} = \frac{1}{2\sqrt{(\alpha^2 + \beta^2)|\alpha^2 - k\beta^2|}|^3} \cdot \begin{pmatrix} 2\alpha^6 + 3(1-k)\alpha^4\beta^2 - 6k\alpha^2\beta^4 + (k^2 - k)\beta^6 & (k+1)^2\alpha^3\beta^3 \\ (k+1)^2\alpha^3\beta^3 & (1-k)\alpha^6 - 6k\alpha^4\beta^2 + 3(k^2 - k)\alpha^2\beta^4 + 2k^2\beta^6 \end{pmatrix}. \tag{84}$$

The determinant of this matrix takes the form

$$\det(f_{ij}) = \frac{2(1-k)\alpha^4 - (k^2 + 10k + 1)\alpha^2\beta^2 - 2k(1-k)\beta^4}{4(\alpha^2 + \beta^2)|\alpha^2 - k\beta^2|} \tag{85}$$

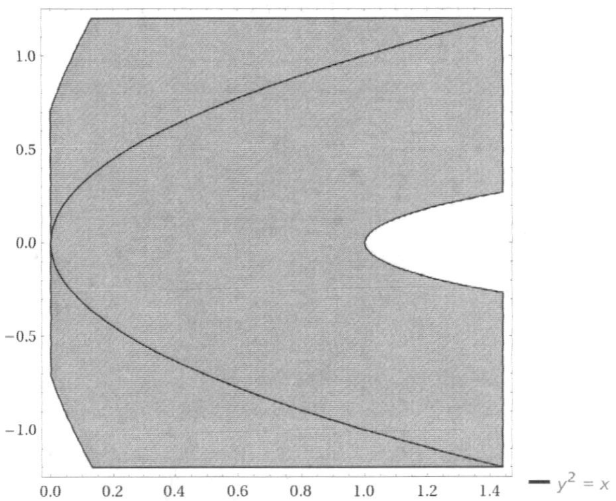

Figure 3. The solution of the inequality $\det(f_{ij}) < 0$ is presented in the blue area. The white area corresponds to the positive defined metric. The metric is degenerated at the line $\alpha^2 = k\beta^2$.

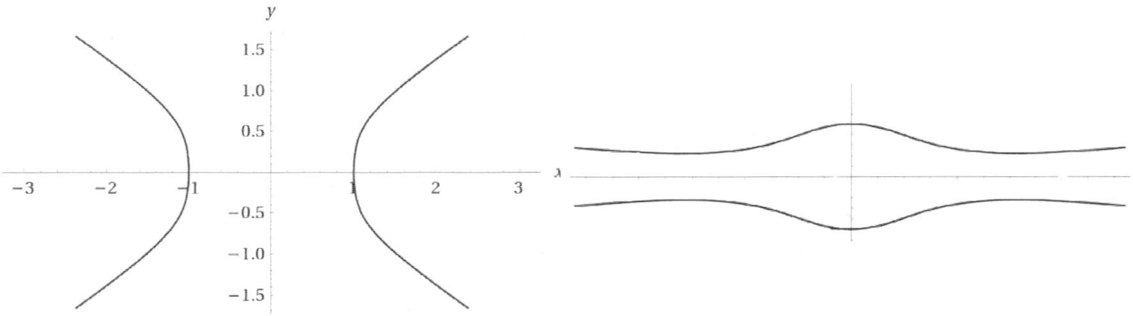

Figure 4. Unit spheres in the metric corresponding to the product quartic (81) for $k = 2$ and $k = 0.01$, respectively. The first graph is concave. On the second graph, the change of the convexity is visual.

5.3. Lorentz × Lorentz

Let us consider a quartic expression

$$Q(v) = Q(\alpha, \beta) = (\alpha^2 - \beta^2)(\alpha^2 - k\beta^2)\,, \tag{86}$$

with a dimensionless *positive* numerical parameter k. We have here a product of two indefinite factors. Consider a Lagrangian for this quadric in the form

$$L(v) = L(\alpha, \beta) = \sqrt{|(\alpha^2 - \beta^2)(\alpha^2 - k\beta^2)|}\,\mathrm{sgn}Q\,. \tag{87}$$

The gradient of the Lagrangian reads

$$L_{,i} = \frac{1}{\sqrt{|(\alpha^2 - \beta^2)(\alpha^2 - k\beta^2)|}} \begin{pmatrix} 2\alpha^3 - (1+k)\alpha\beta^2 \\ -(1+k)\alpha^2\beta + 2k\beta^3 \end{pmatrix} \mathrm{sgn}\, Q. \tag{88}$$

Finsler's metric is given by

$$f_{ij} = \frac{\mathrm{sgn}\, Q}{2\sqrt{|(\alpha^2 - \beta^2)(\alpha^2 - k\beta^2)|^3}} \cdot \begin{pmatrix} 2\alpha^6 - 3(1+k)\alpha^4\beta^2 + 6k\alpha^2\beta^4 - (k^2+k)\beta^6 & (k-1)^2\alpha^3\beta^3 \\ (k-1)^2\alpha^3\beta^3 & -(k+1)\alpha^6 + 6k\alpha^4\beta^2 - 3(k^2+k)\alpha^2\beta^4 + 2k^2\beta^6 \end{pmatrix}. \tag{89}$$

The determinant of this metric es expressed as

$$\det(f_{ij}) = -\frac{2(k+1)\alpha^4 + (k^2 - 10k + 1)\alpha^2\beta^2 + 2k(k+1)\beta^4}{4|(\alpha^2 - \beta^2)(\alpha^2 - k\beta^2)|} \tag{90}$$

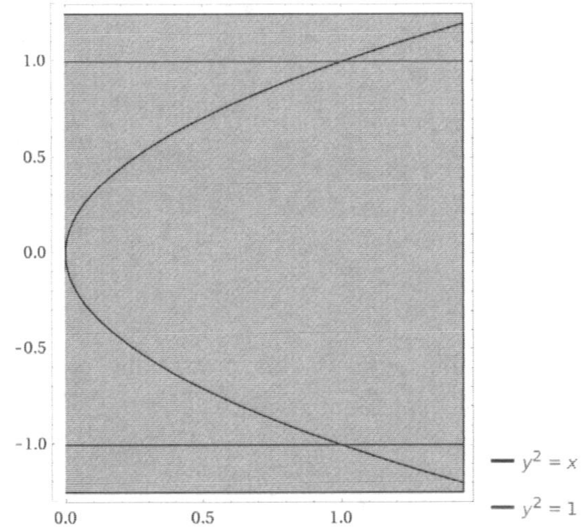

Figure 5. In the plan with the axes $x = k$ and $y = \alpha/\beta$, we depict the area (in blue) of the negative determinant $\det(f_{ij}) < 0$. In this case, the metric is Lorentzian. The metric is degenerates on the lines $\alpha^2 = \beta^2$ and $\alpha^2 = k\beta^2$

This determinant expression is strictly negative for all positive values of the parameter k and almost for all vectors v. The metric degenerates on the hypersurfaces $\alpha^2 = \beta^2$ and $\alpha^2 = k\beta^2$. It means that the space is Lorentzian beyond the degenerate surfaces.

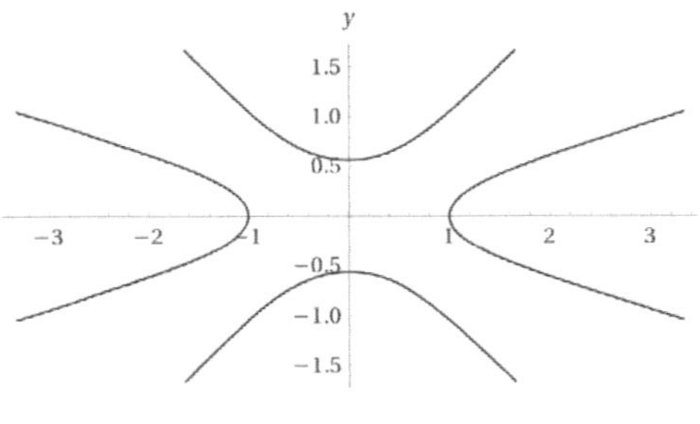

Figure 6. For all values of the parameter k, the unit spheres are hyperbolas, as in the ordinary Lorentz space.

6. Singular conics on two-dimensional space

In this section, we present the characteristic sets $\mathcal{A}, \cdots, \mathcal{H}$ for a number of the two-dimensional examples of Riemann quartic that were calculated above.

Let us look at a few general characteristics of the Finsler building made of homogeneous quartics. While the characteristic sets $\mathcal{A}, \cdots, \mathcal{H}$ are all generally distinct from one another, some of them have become the same in the case of Riemann's quartic. In particular,

- For indefinite quartics, conic singularities appear from the absolute-value definition of the Lagrangian, i.e. on the subset $Q(x, v) = 0$.
- Since Riemann's quartic is an analytic function in v, the non-differential singularities $\mathcal{D}_i$ appear only on the null-vectors conic, $Q(x, v) = 0$, then $\mathcal{C} = \mathcal{D}$. Moreover, these singularities are the same for the all order of derivation, i.e., $\mathcal{D} = \mathcal{D}_i$ for all i.

- In two-dimensional case, we cannot distinguish between the sets $\mathcal{G}$ of Lorentzian signature and the sets $\mathcal{H}$ of mixed signature.

In the following table, we denote by V the tangent vector space at a general point without the origin $V = T_xM\backslash\{0\}$. The Lagrangian is defined due to Eq.(66). The parametrization $v = (\alpha, \beta)$ is used. The areas of the positive-definite Finsler's metric are given in blue, while of the indefinite metric in yellow. Black lines describe the point where the metric is degenerated. On the red lines, the metric is not defined.

Quartic	Characteristic Sets	Diagram
$(\alpha^2+\beta^2)^2$	$\mathcal{A}=V$ $\mathcal{B}=\varnothing;$ $\mathcal{C}=\{0\}$ $\mathcal{D}=\varnothing;$ $\mathcal{E}=\varnothing;$ $\mathcal{F}=V;$ $\mathcal{G}=\varnothing$	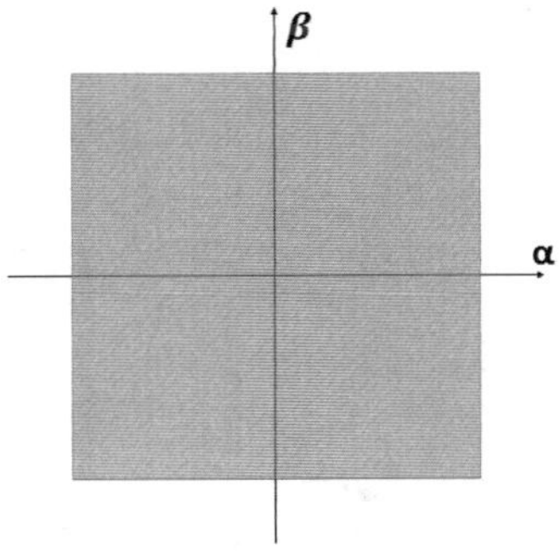
$\alpha^4+\beta^4$	$\mathcal{A}=V$ $\mathcal{B}=\varnothing;$ $\mathcal{C}=\{0\}$ $\mathcal{D}=V;$ $\mathcal{E}=\{v\in V\|\alpha\beta=0\};$ $\mathcal{F}=V\backslash\mathcal{E};$ $\mathcal{G}=\varnothing$	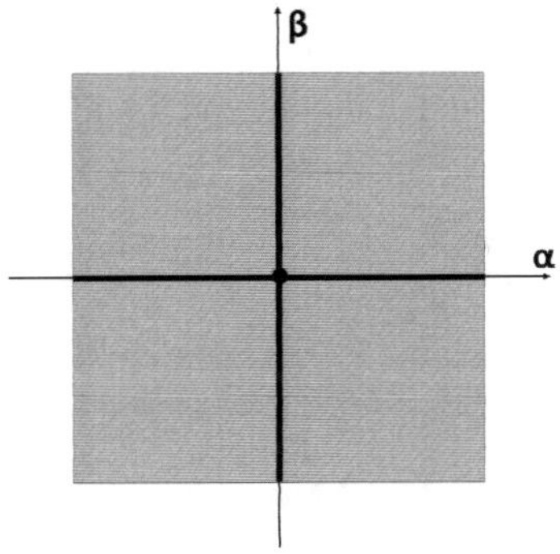
$(\alpha^2-\beta^2)^2$	$\mathcal{A}=V$ $\mathcal{B}=\varnothing;$ $\mathcal{C}=\{v\in V\|\alpha=\pm\beta\}$ $\mathcal{D}=\mathcal{C};$ $\mathcal{E}=\varnothing;$ $\mathcal{F}=\varnothing$ $\mathcal{G}=V\backslash\mathcal{C}$	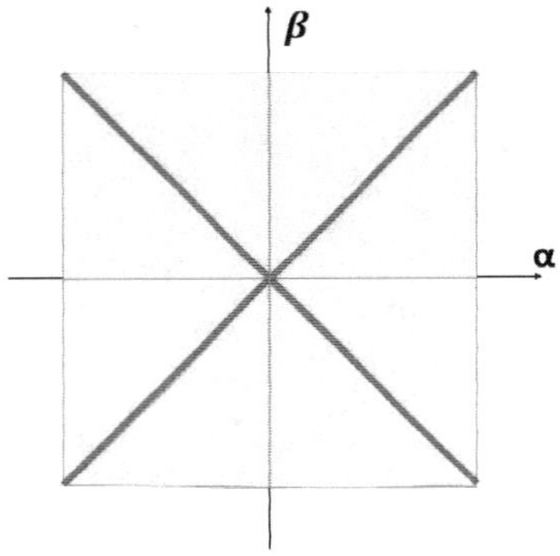

Quartic	Characteristic Sets	Diagram
$\alpha^4 - \beta^4$	$\mathcal{A} = \{v \in V \vert \alpha^2 > \beta^2\};$ $\mathcal{B} = \{v \in V \vert \alpha^2 < \beta^2\};$ $\mathcal{C} = \{v \in V \vert \alpha^2 = \beta^2\};$ $\mathcal{D} = \mathcal{C};$ $\mathcal{E} = \{v \in V \vert \alpha\beta = 0\};$ $\mathcal{F} = \varnothing$ $\mathcal{G} = V \backslash (\mathcal{C} \cup \mathcal{E})$	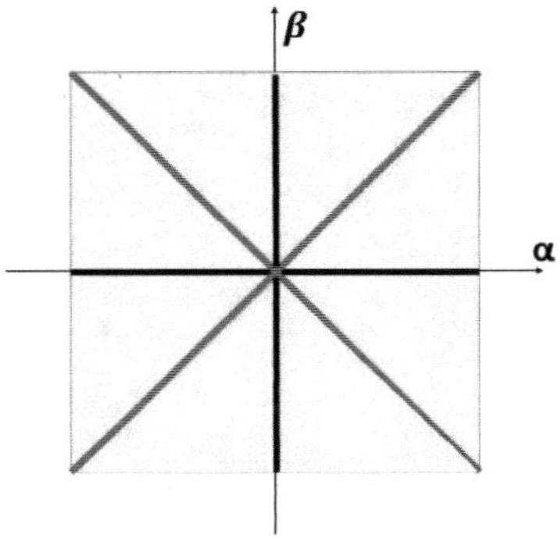
$(\alpha^2 + \beta^2)(\alpha^2 + k\beta^2)$	$\mathcal{A} = V$ $\mathcal{B} = \varnothing;$ $\mathcal{C} = \{0\}$ $\mathcal{D} = \varnothing;$ $\mathcal{E} \neq \varnothing;$ $\mathcal{F} \neq \varnothing;$ $\mathcal{G} \neq \varnothing$	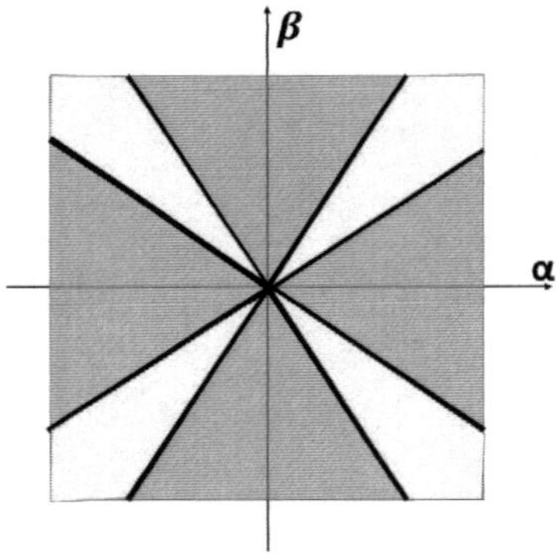
$(\alpha^2 + \beta^2)(\alpha^2 - k\beta^2)$	$\mathcal{A} = \{v \in V \vert \alpha^2 > k\beta^2\}$ $\mathcal{B} = \{v \in V \vert \alpha^2 < k\beta^2\};$ $\mathcal{C} = \{v \in V \vert \alpha^2 = k\beta^2\}$ $\mathcal{D} = \mathcal{C};$ $\mathcal{E} \neq \varnothing;$ $\mathcal{F} \neq \varnothing;$ $\mathcal{G} \neq \varnothing$	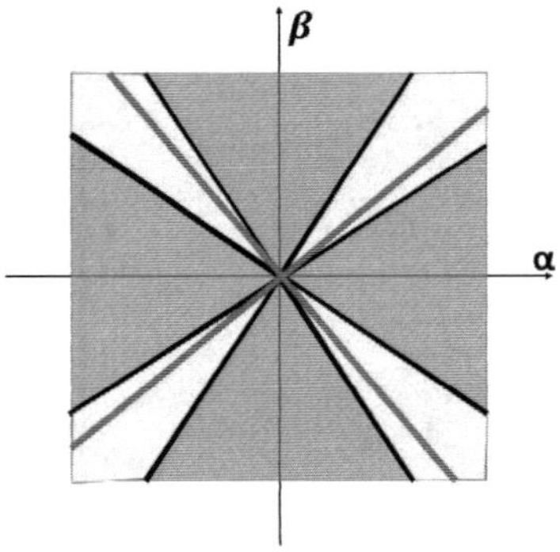
$(\alpha^2 - \beta^2)(\alpha^2 - k\beta^2)$	$\mathcal{A} = V$ $\mathcal{B} = \varnothing;$ $\mathcal{C} = \{v \in V \vert \alpha = \pm\beta\}$ $\mathcal{D} = \mathcal{C};$ $\mathcal{E} = \varnothing;$ $\mathcal{F} = \varnothing$ $\mathcal{G} = V \backslash \mathcal{C}$	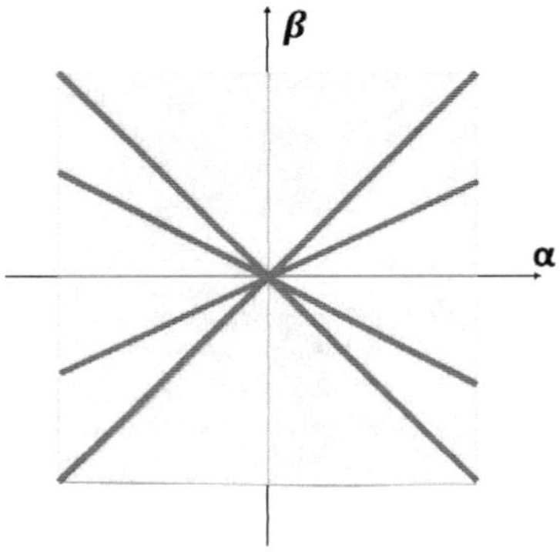

7. Conclusion

We discussed how Riemann's quartic and Finsler's geometric construction are related. We focused on Finsler's metric for various positive-definite and indefinite two-dimensional quartics. The dimensional constraint enables explicit calculations. Our results are as follows:

- For positive-definite quartics:
  - metric tensor is definite over the entire space (without the origin);
  - metric degenerates at conic subsets of the tangential space in general;
  - metrics of different signatures can coexist on the same tangential plane.
- For indefinite quartics:
  - Lagrangian is defined on the entire space;
  - there are conic subsets of the tangential space where metric tensor degenerates;
  - on some conic null space where the quartic vanishes, the metric tensor is singular;
  - signature of metric can vary over the tangential space.

The future line of consideration is to study the relations between the singular sets of Finsler's structure and observable phenomena of electromagnetic wave propagation.

References

[1] H. Rund, *The differential geometry of Finsler spaces,* Springerverlag, 1959.
[2] S.S. Chern and Z. Shen, *Riemann-Finsler Geometry* (Vol. 6). World Scientific, 2005.
[3] Chern, S. S. (1996), "Finsler geometry is just Riemannian geometry without the quadratic equation," *Notices of the American Mathematical Society,* textbf43(9), 959-963.
[4] Carnap, R. (2012). An introduction to the philosophy of science. Courier Corporation.
[5] F. W. Hehl and Yu. N. Obukhov, *Foundations of Classical Electrodynamics,* Birkhäuser: Boston, MA, 2003.
[6] C. Lämmerzahl and V. Perlick, "Finsler geometry as a model for relativistic gravity," *Int. J. Geom. Meth. Mod. Phys.* **15**, no.supp01, 1850166 (2018)
[7] Perlick, V. (2000). *Ray optics, Fermat's principle, and applications to general relativity* (Vol. 61). Springer Science & Business Media.
[8] A. Kostelecky, "Riemann-Finsler geometry and Lorentz-violating kinematics," *Phys. Lett. B* **701**, 137-143 (2011)
[9] Edwards, B. R., & Kostelecký, V. A. (2018). Riemann–Finsler geometry and Lorentz-violating scalar fields. *Physics Letters B* **786**, 319-326.
[10] Y. Itin, C. Lämmerzahl and V. Perlick, "Finsler-type modification of the Coulomb law," *Phys. Rev. D* **90**, no.12, 124057 (2014)
[11] E. Minguzzi, "Light cones in Finsler spacetime," *Commun. Math. Phys.* **334**, no.3, 1529-1551 (2015)
[Asanov(1985)] G. S. Asanov. *Finsler geometry, relativity and gauge theories,* Reidel: Dordrecht, 1985.
[Beem(1970)] J. K. Beem. "Indefinite Finsler spaces and timelike spaces," *Can. J. Math.* 1970 22:1035–1039.
[Itin(2009)] Y. Itin. On light propagation in premetric electrodynamics. *J. Phys. A* 2009, **42** 475402.
[12] Pfeifer, C., Wohlfarth, M. N. Causal structure and electrodynamics on Finsler spacetimes. *Phys. Rev. D,* 2011 **84**(4), 044039.
[Skakala(2008)] J. Skakala and M. Visser, "Pseudo-Finslerian spacetimes and multi-refringence," *Int. J. Mod. Phys. D* **19**, 1119-1146 (2010)
[13] M. Elbistan, P. M. Zhang, N. Dimakis, G. W. Gibbons and P. A. Horvathy, "Geodesic motion in Bogoslovsky-Finsler spacetimes," *Phys. Rev. D* **102**, no.2, 024014 (2020)
[14] Pfeifer, C. Finsler spacetime geometry in physics. *International Journal of Geometric Methods in Modern Physics,* **16**, (2019), 1941004.
[15] M. A. Javaloyes and M. Sánchez, "On the definition and examples of cones and Finsler spacetimes," *Ann. Sc. Norm. Sup. Pisa, Cl. Sci.* (5) XIII (2014), 813–858.
[16] Javaloyes, M. A., & Sánchez, M. (2014). Finsler metrics and relativistic spacetimes.*International Journal of Geometric Methods in Modern Physics,* 11(09), 1460032.
[17] Y. Itin, (2010). "Dispersion relation for electromagnetic waves in anisotropic media" *Physics Letters A,* 374(9), 1113-1116.
[18] L.D. Landau, E.M. Lifshitz (1984) *Electrodynamics of Continuous Media* Pergamon, Oxford

Mechanics of Spacetime with Surface Tension and Preferred Curvature

H A Perko[1]
[1]Colorado State University and Koppa Research, LLC, Office 14, 138 E. 4th Street, Loveland, CO, USA 80537

E-mail: hperko@kopparesearch.com, howie@colostate.edu

Abstract. In previous publications, it was argued that because energy and matter are aligned in the thin increment of present time for an observer, spacetime should and must have surface tension. In this paper, we review how to apply 4-dimensional continuum mechanics with imaginary time coordinates to derive a mechanical model of spacetime with surface tension. Then, to continue model development, we discuss the concept of preferred curvature from the physical chemistry of surfaces and attempt to apply those concepts to spacetime geometry. We show how the model exhibits quantum fluctuations at the Plank scale and components resembling dark matter and dark energy at the cosmic scale.

Introduction

There have been and continue to be numerous attempts to find a unified field theory in physics. The approach taken here is to find a natural mechanism that causes spacetime itself to warp, ripple, or otherwise form small particles in geometry. The proposed mechanism is surface tension [1,2,3]. From everyday experiences, we know that surface tension results in micro-phenomena including capillary waves, corpuscles, vortices, and menisci. In a similar way, it was argued in [1,2,3] that incorporating surface tension in general relativity provides a bridge to quantum mechanics and atomic particle physics.

Surface tension of spacetime is an obscure, yet present topic in modern physics. At least nineteen references can be found in quantum and gravitational literature suggesting the importance of surface tension [4-23]. A number of works discuss surface tension effects on black hole horizons, worm holes, and stellar geometries [17,18,23]. Some novel works suggest

Content from this work may be used under the terms of the Creative Commons Attribution 3.0 licence. Any further distribution of this work must maintain attribution to the author(s) and the title of the work, journal citation and DOI.
Published under licence by IOP Publishing Ltd

surface tension as an analogy for the cosmological constant [13,16,18]. Others address surface tension at the boundary of an expanding universe [4,10,21]. A few suggest the importance of surface tension in quantum field theory [11,19,22]. This work is a continuation of efforts to justify the hypothesis that spacetime itself has surface tension [1,2,3].

This paper contains a summary of the surface tension of spacetime hypothesis including the postulated stress energy tensor with negative terms. Tensors describing the curvature gradient, rate of curvature, and constitutive relation for spacetime with surface tension are presented. It is shown that components of these tensors resemble wave equations in quantum field mechanics. It is also shown that on a cosmic scale, gravity has extra terms that explain dark matter and dark energy.

Stress Energy

The case for surface tension of spacetime has been argued based on statistical thermodynamics [2,3]. Surface tension is negative stress energy in all spatial directions which thermodynamically balance the presence of nearby mass. When coordinates are rotated such that the off-diagonal time terms in the stress-energy tensor are zero, time is normal to space. This orientation is called the principal reference frame in continuum mechanics, the rest frame in modern physics, and tangent space in mathematics. Governing differential equations describing the motion of curvilinear coordinates in a stress-energy field are best derived in the principal frame. The stress-energy tensor for an infinitesimal element of spacetime with surface tension in the principal frame is given by [3],

$$T_{\nu}^{\mu} = \begin{bmatrix} dP & 0 & 0 & 0 \\ 0 & -♀ & 0 & 0 \\ 0 & 0 & -♀ & 0 \\ 0 & 0 & 0 & -♀ \end{bmatrix} \qquad (1)$$

Where dP is mass density, $-♀$ is surface energy density (a.k.a. surface tension), and $T_{\nu}^{\mu} = \Lambda^{\mu\alpha} T_{\alpha\nu}$ is the Lorentz boosted stress energy tensor from the observer's manifold to the tangent space at a point.

It has been shown using conservation of work and energy that mass and surface tension can be related thus [3],

$$\oiint_{\sigma} -♀ d\sigma = \oint_{V} dP dV \qquad (2)$$

From (2) and the divergence theorem, surface tension is related to mass by [3],

$$-\nabla \cdot ♀ = dP \qquad (3)$$

For a polar symmetric system with central mass in laboratory coordinates, (3) can be re-written as,

$$-3♀\frac{dy^0}{R} = \frac{dP}{c^2} \qquad (4)$$

where R is the spacetime separation from the center of mass to the point where surface tension is being measured, dy^0 is the basis of time at the point.

Curvature Gradient

Lorentzian geometry and continuum mechanics are very similar fields in mathematics. Both use differential geometry of manifolds to describe curved spaces with a continuous metric tensor. The mathematics of both are inherently intrinsic as neither requires a containing space. Lorentzian geometry has a pseudo-Riemannian metric and continuum mechanics has a regular (positive definite) Riemannian metric. If one makes a coordinate substitution such that time and proper time are imaginary, then it creates a complex Hilbert space where all the tools developed for continuum mechanics can be applied to evaluate spacetime geometry.

From continuum mechanics of thin shells, we know that the curvature gradient of spacetime (rotation of curvilinear coordinates in a stress-energy field) between two observations separated by $d\tau$ is given by [1,2,3],

$$\dot{u}_\nu^\mu = -\frac{1}{\text{♀}}\begin{bmatrix} \text{♀}\nabla^2 y^0 + \frac{dP}{dy^0} & 0 & 0 & 0 \\ 0 & \frac{\partial \text{♀}}{\partial y^1} & 0 & 0 \\ 0 & 0 & \frac{\partial \text{♀}}{\partial y^2} & 0 \\ 0 & 0 & 0 & \frac{\partial \text{♀}}{\partial y^3} \end{bmatrix} = \begin{bmatrix} \frac{\partial u^0}{\partial \tau} & 0 & 0 & 0 \\ 0 & \frac{\partial u^1}{\partial \tau} & 0 & 0 \\ 0 & 0 & \frac{\partial u^2}{\partial \tau} & 0 \\ 0 & 0 & 0 & \frac{\partial u^3}{\partial \tau} \end{bmatrix} \quad (5)$$

where y^ν are curvilinear coordinates at a point, u^μ is the motion of coordinates over the period of observation given by $\frac{\partial y^\mu}{\partial \tau}$.

Rate of Curvature

In this work, we use the Green-Lagrange definition of rate of curvature, which is similar to applying a small perturbation to the metric as given by,

$$\left(1 + 2D_\nu^\mu d\tau\right) ds^{\nu 2} = \widetilde{ds}^{\mu 2} \quad (6)$$

where the terms in parenthesis represent the change in the metric over the period of observation, $\dot{g}$. For small strains, $ds^\nu \approx \widetilde{ds}^\mu$, (6) is approximated by linearized engineering strain, ergo change in basis over initial basis.

$$D_\nu^\mu d\tau = \frac{\widetilde{ds}^\mu - ds^\nu}{ds^\nu} \quad (7)$$

In continuum mechanics of thin shells, the rate of curvature tensor is given by,

$$D^{\mu}_{\nu} = \begin{bmatrix} \frac{1}{2}\sum_{k=1}^{3}\left(\frac{\partial u^0}{\partial y^k}\right)^2 & \frac{1}{2}\sum_{k=1}^{3}\left(\frac{\partial u^0}{\partial y^k}\right)\left(\frac{\partial u^k}{\partial y^1}\right) & \frac{1}{2}\sum_{k=1}^{3}\left(\frac{\partial u^0}{\partial y^k}\right)\left(\frac{\partial u^k}{\partial y^2}\right) & \frac{1}{2}\sum_{k=1}^{3}\left(\frac{\partial u^0}{\partial y^k}\right)\left(\frac{\partial u^k}{\partial y^3}\right) \\ \frac{1}{2}\sum_{k=1}^{3}\left(\frac{\partial u^0}{\partial y^k}\right)\left(\frac{\partial u^k}{\partial y^1}\right) & \frac{\partial u^1}{\partial y^1}+\left(\frac{\partial u^0}{\partial y^1}\right)^2 d\tau & \frac{1}{2}\left(\frac{\partial u^1}{\partial y^2}+\frac{\partial u^2}{\partial y^1}\right)+\frac{1}{2}\left|\left(\frac{\partial u^0}{\partial y^1}\right)\left(\frac{\partial u^0}{\partial y^2}\right)\right| d\tau & \frac{1}{2}\left(\frac{\partial u^1}{\partial y^3}+\frac{\partial u^3}{\partial y^1}\right)+\frac{1}{2}\left|\left(\frac{\partial u^0}{\partial y^1}\right)\left(\frac{\partial u^0}{\partial y^3}\right)\right| d\tau \\ \frac{1}{2}\sum_{k=1}^{3}\left(\frac{\partial u^0}{\partial y^k}\right)\left(\frac{\partial u^k}{\partial y^2}\right) & \frac{1}{2}\left(\frac{\partial u^2}{\partial y^1}+\frac{\partial u^1}{\partial y^2}\right)+\frac{1}{2}\left|\left(\frac{\partial u^0}{\partial y^2}\right)\left(\frac{\partial u^0}{\partial y^1}\right)\right| d\tau & \frac{\partial u^2}{\partial y^2}+\left(\frac{\partial u^0}{\partial y^2}\right)^2 d\tau & \frac{1}{2}\left(\frac{\partial u^2}{\partial y^3}+\frac{\partial u^3}{\partial y^2}\right)+\frac{1}{2}\left|\left(\frac{\partial u^0}{\partial y^2}\right)\left(\frac{\partial u^0}{\partial y^3}\right)\right| d\tau \\ \frac{1}{2}\sum_{k=1}^{3}\left(\frac{\partial u^0}{\partial y^k}\right)\left(\frac{\partial u^k}{\partial y^3}\right) & \frac{1}{2}\left(\frac{\partial u^3}{\partial y^1}+\frac{\partial u^1}{\partial y^3}\right)+\frac{1}{2}\left|\left(\frac{\partial u^0}{\partial y^3}\right)\left(\frac{\partial u^0}{\partial y^1}\right)\right| d\tau & \frac{1}{2}\left(\frac{\partial u^3}{\partial y^2}+\frac{\partial u^2}{\partial y^3}\right)+\frac{1}{2}\left|\left(\frac{\partial u^0}{\partial y^3}\right)\left(\frac{\partial u^0}{\partial y^2}\right)\right| d\tau & \frac{\partial u^3}{\partial y^3}+\left(\frac{\partial u^0}{\partial y^3}\right)^2 d\tau \end{bmatrix} \quad (8)$$

Temporal terms above were misrepresented in earlier works [1,2,3] and have been corrected here.

The rate of curvature tensor compares closely to the Einstein tensor in general relativity. We know from mathematics of elasticity that the rate of curvature tensor for thin shells is the push forward of the Lie derivative of the metric plus the Gaussian curvature. We also know from general relativity that the Lie derivative of the metric is the 3-dimensional extrinsic curvature and that Gaussian curvature is the Ricci scalar times the metric. Thus, the rate of curvature of spacetime IS the Einstein tensor in tangent (3+1) space.

$$D^{\mu}_{\nu} = \frac{1}{2}\mathcal{L}_{\tau} g^{\mu}_{\nu} + \frac{1}{2}|\kappa_{\nu}\kappa^{\mu}| \qquad (9)$$

$$D^{\mu}_{\nu} = K_{3D} + \frac{1}{2}|\kappa_{\nu}\kappa^{\mu}| \qquad (10)$$

$$D^{\mu}_{\nu} = R^{\mu}_{\nu} + \frac{1}{2}R g^{\mu}_{\nu} \qquad (11)$$

Constitutive Relation

A covariant constitutive relationship between stress-energy and rate of deformation enforces locality, creates a map over all configurations (histories), and establishes conservation laws for energy, linear momentum, angular momentum, and other symmetries. In continuum mechanics, the general constitutive relationship is given by the tensor version of Hooke's law,

$$T^{\mu}_{\nu} = C^{\mu\alpha}_{\beta\nu} D^{\alpha}_{\beta} \qquad (12)$$

where $C^{\alpha\mu}_{\beta\nu}$ is a 4x4x4x4 matrix known as stiffness or elasticity tensor. To ensure that spacetime remain smooth and continuously differential, the elasticity tensor must be orthonormal, invertible, and non-degenerate. An anisotropic constitutive relation was proposed in [2] in an attempt to unify gravitational and quantum fields. However, [3] explained that an anisotropic elastic tensor is unnecessary for unification if the relationship between mass and stored surface energy is taken into consideration at minimum curvature.

For the current work, we shall define the constitutive relation between stress-energy and rate of deformation of spacetime as a uniform constant in accord with Einstein,

$$T^{\mu}_{\nu} = \frac{c^4}{8\pi G} D^{\mu}_{\nu} \qquad (13)$$

Journal of Physics: Conference Series **2482** (2023) 012008 doi:10.1088/1742-6596/2482/1/012008

Preferred Curvature

In the mechanics of thin shells, particularly biological systems, there is a concept known as preferred curvature wherein the lowest negative energy state (greatest surface tension) of a thin shell has a minimum preferred curvature. It is suggested here, that the minimum radius of curvature of spacetime is the Plank length, λ_p.

If we define a discrete mass, m, as the mass density integrated over the volume of three sphere with a minimum radius equal to the Plank length, then the right side of (2) becomes,

$$\oint_V dPdV = 3m \qquad (14)$$

If we integrate surface energy density (a.k.a. surface tension) over the surface of a three sphere of Plank radius, then the left side of (2) becomes,

$$\oiint_\sigma -♀ d\sigma = -12♀\pi\lambda_p^2 \qquad (15)$$

By equivalence of work and energy (2), we see,

$$-♀ = \frac{m}{4\pi\lambda_p^2} \qquad (16)$$

Quantum Mechanics Analogs

General relativity is a geometric theory. Quantum mechanics is interpreted as a probabilistic theory. Unification of general relativity and quantum mechanics requires a geometric analog for quantum mechanics. The differential geometry of continuum mechanics exists in the tangent space at all points on a manifold. If a manifold describes all possible positions, the tangent space represents all possible velocities. In an earlier section, it was explained that making a coordinate substitution such that time and proper time are imaginary, creates a positive definite complex Hilbert space. The tangent space at all points in spacetime is an orthogonal linear vector space with valid inner product consisting of state vectors $|dx_{\mu>}$ and corresponding dual vectors $< \frac{\partial}{\partial x_\mu}|$ that establish the expectation value of stress energy and curvature fields. Imposing a constitutive relation between stress energy and curvatures ensures that the space is continuously differential, separable, and complete.

Many continuum mechanics equations can be written in quantum mechanics notation. For example, (3) can be re-written as,

$$< \frac{\partial}{\partial \tau}|♀|d\tau > = dP$$

Mass density at a point is the expectation value of bras and kets formed by the spacetime separation between observations and the surface tension operator. In order to unify general relativity and quantum mechanics, it is important to recognize the mathematical similarities between differential geometry and Hilbert spaces.

In the remainder of this section, we examine other quantum mechanics analogs that arise in the geometry of spacetime with surface tension. The foregoing continuum mechanics tensors have some immediately recognizable wave equation components. The time-time component in the curvature gradient, (5), in laboratory coordinates is,

$$\frac{\partial^2 y^0}{\partial y^{1^2}} + \frac{\partial^2 y^0}{\partial y^{2^2}} + \frac{\partial^2 y^0}{\partial y^{3^2}} + \frac{1}{c^2}\frac{dP}{dy^0 ♀} = \frac{1}{c^2}\frac{\partial^2 y^0}{\partial \tau^2} \qquad (17)$$

where time and proper time have been changed to natural units by multiplying by ic. For a vacuum, (17) reduces to the wave equation,

$$\frac{\partial^2 y^0}{\partial y^{1^2}} + \frac{\partial^2 y^0}{\partial y^{2^2}} + \frac{\partial^2 y^0}{\partial y^{3^2}} = \frac{1}{c^2}\frac{\partial^2 y^0}{\partial \tau^2} \qquad (18)$$

This equation describes capillary waves, which in general relativity, represent ripples in the coordinate of time that follow the overall curvature of the metric. The waves are massless and move at the speed of light regardless of frequency and wavelength. This expression is analogous to the Klein-Gordon equation in quantum mechanics.

For the case where temporal acceleration is zero, (17) reduces to,

$$\frac{\partial^2 y^0}{\partial y^{1^2}} + \frac{\partial^2 y^0}{\partial y^{2^2}} + \frac{\partial^2 y^0}{\partial y^{3^2}} = -\frac{1}{c^2}\frac{dP}{dy^0 ♀} \qquad (19)$$

which is the classic Young-Laplace equation that describes corpuscular formation. The first three terms are the principal curvatures in the time direction. In a polar symmetric system with radius of curvature R, (19) becomes,

$$-3♀\left(\frac{dy^0}{R}\right) = \frac{dP}{c^2} \qquad (20)$$

which is the same expression intuitively derived and shown in (4).

Another quantum fluctuation is recognized in the spatial diagonal components of the deformation gradient (8). The dispersion equation is obtained by replacing the stress-energy with the corresponding rate of deformation in accord with the constitutive relation (13),

$$\frac{c^4}{8\pi G}\left(\frac{\partial^2 u^j}{\partial y^{j^2}}\right) = -♀\,\frac{\partial u^j}{\partial \tau} \qquad (21)$$

where the third order term has been omitted for simplification. For small particles with radius equal to the Plank length, λ_p, stored surface energy can be represented by a discrete mass, m, via (16). Changing to laboratory units, and noting that $\lambda_p{}^2 = \frac{\hbar G}{c^3}$ results in,

$$-\frac{\hbar}{2m}\left(\frac{\partial^2 u^j}{\partial y^{j^2}}\right) = i\,\frac{\partial u^j}{\partial \tau} \qquad (22)$$

which is analogous to the one-dimensional Schrödinger wave equation with four-velocity replacing probability density. Surface energy and the corresponding spatial strains in spacetime move in accord with a Schrödinger-like dispersion equation.

Still another quantum fluctuation follows from the fluidity of spacetime, the constitutive relationship (13), and the shear strain or off diagonal spatial terms in (8), given by,

$$\frac{\partial u^i}{\partial y^j} + \frac{\partial u^j}{\partial y^i} + \left|\left(\frac{\partial u^0}{\partial y^i}\right)\left(\frac{\partial u^0}{\partial y^j}\right)\right| d\tau = 0 \qquad (23)$$

In continuum mechanics, this is the Navier-Stokes equation for a vortex. There are three vortex solutions with half integer indices: unstable short-lived rotational vortices, stable irrotational vortices with field lines ending at a corpuscle, and a stable pair of oppositely rotating irrotational vortices entangled in time. A vortex itself does not carry mass, but does cause strain and deformation of spacetime thereby altering geodesics.

As a final comment on analogies with quantum field theory, consider the constitutive relationship itself. According to (13) and (1), spatial components of stress-energy are related to rate of deformation by,

$$-\text{♀}\ \delta^i_j = \frac{c^4}{8\pi G}\ D^i_j \qquad (24)$$

We can replace stored surface energy using the relationship (16) for discrete particles of Plank radius to obtain in laboratory units,

$$\frac{m\ \delta^i_j}{D^i_j} = -i\ \frac{\hbar}{2} \qquad (25)$$

The units of m are mass-energy. The units of D^i_j are the inverse of proper time. If we square both sides and apply the Cauchy–Schwarz inequality, then essentially,

$$m d\tau \geq \frac{\hbar}{2} \qquad (26)$$

The constitutive relationship for discrete particles of curved spacetime with surface tension at the Plank radius is analogous to the Heisenberg uncertainty principle.

Gravitational Geometry

Simply by adding and subtracting ♀ from the time-time term, the stress-energy tensor with surface tension (1) can be rewritten as,

$$T^\mu_\nu = \begin{bmatrix} dP + \text{♀} & 0 & 0 & 0 \\ 0 & 0 & 0 & 0 \\ 0 & 0 & 0 & 0 \\ 0 & 0 & 0 & 0 \end{bmatrix} - \text{♀} g^\mu_\nu \qquad (28)$$

In this form, surface tension appears as an additional mass term, $+$♀, and a negative cosmological constant, $-\text{♀} g^\mu_\nu$. In cosmology applications, the additional mass term could be interpreted as "dark matter" and the negative cosmological constant as "dark energy". Ordinary luminous matter, dP, and "dark matter", $+$♀, are interrelated by (3) and (4). A correspondence between luminous matter and dark matter is to be expected. Eloquent arguments for such a correspondence are given in [24].

In a prior work [3], it was shown that replacing surface tension in (28) with its equivalent mass relation (4), results in an equation for orbital velocity given by,

$$v = \sqrt{GM\left(\frac{1}{R} + \frac{1}{4dy^0c^2}\right)} \qquad (29)$$

where the first term represents Newton's contribution and the second term is the effect of surface tension. The time scalar, dy^0, can be set equal to unity for non-relativistic speeds.

Orbital velocities from (29) were compared with galaxy rotation curves for fifteen galaxies in [3]. Good correlation was found if the mass-to-light ratios of sample galaxies is set to values much lower than previously accepted. If the best-fit mass-to-light ratios are plotted relative to the square root of total galaxy luminosity, then an inverse correlation is found with an R^2 value of 0.96. When this relationship is re-inserted into the orbital velocity equation and the limit is taken at large values of R, the Tully-Fisher relation is derived.

Conclusions

It has been shown in this work that including surface tension in the treatment of spacetime results in quantum fluctuations that are analogous to quantum field mechanics at the minimum preferred curvature of Plank scale. It also has been shown that including surface tension in orbital dynamics at the cosmic scale can serve as an explanation for dark matter and dark energy.

References

[1] Perko, H.A. (2016) "Introducing Surface Tension to Spacetime", Journal of Physics: Conference Series, Vol. 845, 01203
[2] Perko, H.A. (2018) "Gravitation in the Surface Tension Model of Spacetime", Journal of Physics: Conference Series, Vol. 1239, 012010
[3] Perko, H.A. (2021) "Dark Matter and Dark Energy: Cosmology of Spacetime with Surface Tension", Journal of Physics: Conference Series, Vol. 1956, 012004
[4] Davidson, A. (2018) "Hydrogen-like Spectrum of Spontaneously Created Universes with deSitter Ground State", Physics Letters, B780, 29, arXiv:1708.00987
[5] Culetu, H. (2017) "Thin Shell Embedded in de Sitter Space", arXiv:1705.10381
[6] Freidel, L. and Yokokura, Y. (2015) "Non-Equilibrium Thermodynamics of Gravitational Screens", Classical and Quantum Gravity, arXiv:1405:4881
[7] Freidel, L. (2013) "Gravitational Energy, Local Holography and Non-Equilibrium Thermodynamics", arXiv:1312.1538
[8] Barut, A.O. and Pavsic, M. (1994) "Radiation Reaction and the Electromagnetic Energy-Momentum of Moving Relativistic Charged Membranes", Physics Letter B, 45-50, arXiv:1301.4837
[9] Kumar, H., Alam, S. and Ahmad, S. (2013) "Surface Tension with Normal Curvature in Curved Space-Time", General Relativistic Gravity, 45, 125-142, arXiv:1211.0128
[10] Wang, C.H., Wu, Y.H. and Hsu, S.D. (2012) "Quantum Fluctuations in Planar Domain-Wall Space-Times: A Possible Origin of Primordial Preferred Direction", Physics Letters B, 713, 6-9, arXiv:1107.1762
[11] Codello, A. and Zanusso, O. (2011) "Fluid Membranes and 2d Quantum Gravity", Physics Review D, 83, 125021, arXiv:1103.1089
[12] Culetu, H. (2010) "Boundary Stress Tensors for Spherically Symmetric Conformal Rindler Observers", Journal of Korean Physics Society, 57, (1), 17-22, arXiv:1001.4740
[13] Katti, R., Samuel, J. and Sinha, S. (2009) "The Universe in a Soap Film", Classical and Quantum Gravity, 26, 135018, arXiv:0904.1057

[14] Avelino, P.P., Martins, C.J.A.P., and Sousa, L. (2008) "Dynamics of Biased Domain Walls and the Devaluation Mechanism", Physics Review D, 78:043521, arXiv:0805.4013
[15] Cardoso, V. and Gualtieri, L. (2006) "Equilibrium Configurations of Fluids and their Stability in Higher Dimensions", Classical and Quantum Gravity, 23:7151-7198, arXiv:061004
[16] Samuel, J. and Sinha, S. (2006) "Surface Tension and the Cosmological Constant", Physics Review Letters, 97:161302, arXiv:0603804
[17] Lobo, F.S.N. (2003) "Thin Shells Around Traversable Wormholes", Proceedings of APCTP Winter School and Workshop: Quantum Gravity, Black Holes, and Wormholes, Pohang School of Environmental Engineering, South Korea, 11-14, arXiv:0401083
[18] Lobo, F.S.N. and Crawford, P. (2004) "Linearized Stability Analysis of Thin-Shell Wormholes with a Cosmological Constant", Classical and Quantum Gravity, 21, 391-404, arXiv:0311002
[19] Graham, N., et al. (2003) "Casmir Energies in Light of Quantum Field Theory", Physics Letters B, 572, 196-201, arXiv:0207205
[20] Schmidt, H.J. (1984) "Surface Layers in General Relativity and Their Relation to Surface Tensions", General Relativistic Gravity, 16:1053-1061, arXiv:0105106
[21] Guendelman, E.I. and Portnoy, J. (1999) "The Universe Out of an Elementary Particle?", Classical and Quantum Gravity, 16:3315-3320, arXiv:9901066
[22] Hajicek, P. and Bicak, J. (1997) "Gauge Invariant Hamiltonian Formalism for Spherically Symmetric Gravitating Shells", Physics Review D, 56:4706-4719, arXiv:9706022
[23] Callaway, D.J.E. (1996) "Surface Tension, Hydrophobicity, and Black Holes: The Entropic Connection", Physics Review E:53:3738-3744, arXiv:9601111
[24] Mannheim, P.D. (2005) "Alternatives to Dark Matter and Dark Energy", *The Astrophysical Journal*, 0505266v2

Time measurement with accelerating light-clocks

Uri Ben-Ya'acov

School of Engineering, Kinneret Academic College on the Sea of Galilee,
D.N. Emek Ha'Yarden 15132, Israel

E-mail: uriby@kinneret.ac.il

Abstract. The clock hypothesis in relativity states that the rate of time as measured by any clock is determined by its Minkowskian proper-time, regardless of the nature of its motion; in particular independent of its acceleration, depending only on its instantaneous velocity. However, a unique proper-time may be assigned to an accelerating clock, as to any physical system, only in the limit of being point-like. But clocks, by their very nature, must be spatially extended systems, to allow an internal periodical mechanism. Therefore the question, How does the internal structure of the clock affect the clock hypothesis?

The simplest model to examine the clock hypothesis is the so-called 'light clock', consisting of two mirrors with a light signal reflected between them. So far, such examinations were carried out mainly in the limits of point-like clocks and/or constant acceleration. Here the clock hypothesis is theoretically examined for spatially extended linearly accelerated light-clocks, parallel and vertical relative to the direction of motion, with arbitrarily varying accelerations. Using the rapidity of the clock as its evolution parameter, a Lorentz covariant analysis is neatly performed. Taking into account the spatial extension of the clock, differences between externally measured Minkowskian proper-times and the time-scale determined by the internal periodical mechanism of the clock are computed.

Although these differences are practically very minute – of order aL/c^2 for characteristic acceleration a and spatial dimension L of the clock – theoretically they cannot be ignored. They indicate inherent inconsistency between the externally measured proper-time – the physical time-line which consists of moments-of-time – and the intrinsically defined age – the internal time-line, consisting of durations, intervals-of-time.

Keywords : clock hypothesis, internal time, proper-time, relativistic age, extended relativistic systems, rapidity, relativistic rigid motion, duration

Content from this work may be used under the terms of the Creative Commons Attribution 3.0 licence. Any further distribution of this work must maintain attribution to the author(s) and the title of the work, journal citation and DOI.
Published under licence by IOP Publishing Ltd

1. Introduction

The present work was originally motivated by the question, To what extent it is possible to assign the concept of *common* or *characteristic proper-time* to composite, spatially extended systems? thus invoking the issue of (internal) *age* vs. (external) *proper-time*:

The proper-time measure of point-like bodies is well-defined in special relativity (SR) by the integral

$$\Delta\tau = \int \sqrt{dt^2 - d\vec{r}^{\,2}} = \int \sqrt{1-\vec{v}^{\,2}}dt \tag{1}$$

along their world-line $(t, \vec{r}(t))$ relative to some inertial Minkowskian reference frame. A similar concept of common, collective proper-time measure, which may also be regarded as measuring the *age* of the system, cannot in general be defined for composite, spatially extended systems because different constituents move on distinct world-lines each defining a different proper-time measure. However, while for point-like bodies the age can only be determined by the external Minkowskian measures, in composite systems there may be internal mechanisms that determine internal temporal measures, thus defining *internal age.*

Thus the question arises, What is the relation between the internal temporal measures and the external ones? Between the intrinsic time measurement and the externally measured proper-time? The present work examines this question. Towards this end it is natural to start with clocks – devices endowed with an internal periodical mechanism whose period defines the internal unit of time. Clocks are used in the following to represent any system with an internal temporal measure.

Ideal relativistic clocks are postulated to be such that their time measurement coincides with the external Minkowskian proper-time (1) associated with the clock [1, 2]. This is, essentially, the *clock hypothesis* – that the rate of time as measured by any clock coincides with its Minkowskian proper-time, regardless of the nature of its motion; in particular independent of its acceleration, depending only on its instantaneous velocity [3, 4, 5].

Ideal clocks, by their definition, must be point-like. But real clocks are composite systems consisting generically of an internal periodically moving mechanism enclosed in a rigid body or casing (for atomic clocks, the atomic structure as a whole serves as such a body). These internal mechanisms require some space, so real clocks are necessarily spatially extended; they cannot be point-like, and different parts of the clock move on different world-lines. The internal period, measured when the clock is inertially at rest, defines the fundamental time unit of the clock.

For the clock to remain reliable even when it accelerates, its internal period must remain the same as when it is at rest. Accelerations may distort the body of the clock, causing changes in the internal period. Therefore, for the clock to be reliable even when it accelerates, its body must remain rigid, at least relative to itself. In the following we thus apply Born's relativistic rigidity condition [6] which implies that at any moment there is a momentary inertial reference frame relative to which the body is at rest, and it may be satisfied with arbitrarily accelerated rectilinear rigid motion [7].

A remarkable and fundamental fact of reality which seems to be colloquially taken as obvious, but with a second thought is far from being obvious, is the fact that the ratio between the periods of different clocks which are relatively at rest remains constant, even if their mechanisms are completely different. This is recognized at least within the physical (including biological) realms (as different from, say, psychological ageing). It is this fact that allows the formation of a universal concept of time. Periodical systems that satisfy these requirements may be regarded, regardless of their particular periodical mechanism, as *universal* or *generic* clocks. A universal clock is a fundamental concept for the measurement of time.

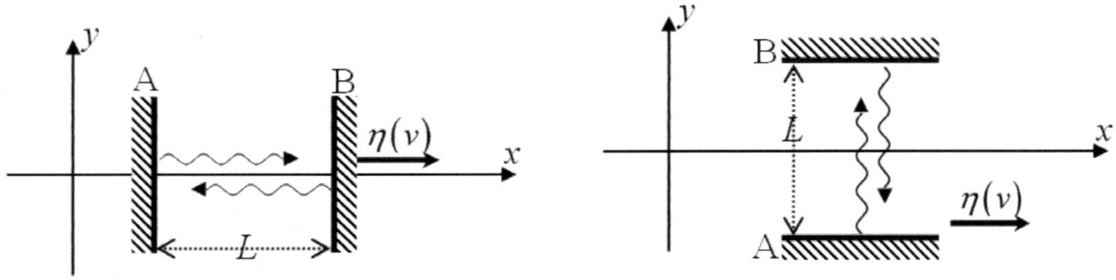

(a) parallel light-clock (b) vertical light-clock

Figure 1. The basic light-clocks.

As a consequence, we don't use in the following any particular dynamical model. The analysis is purely kinematic, and we may use the simplest model for a clock.

To examine the relation between the intrinsic time-unit of the clock and external proper-time measurement, we refer in the following to the very basic Einstein-Langevin light-clock [4, 8, 9]: A 1D device, consisting of a light source and sensor on one end and a reflecting mirror on the other end. The source sends a signal to the opposite end, the signal hits the mirror and returns to the source, and when the reflected signal hits the sensor in the emitting side then a new signal is immediately emitted. The unit of time is the time elapsed between emitting a signal and its return. If the distance between the sides is L then the unit of time is $T = 2L/c$. We seek to explore what happens to the time unit when the clock accelerates.

In this paper we consider clocks in rectilinear motion. The effect of the acceleration on the time measurement depends on the the orientation of the clock relative to its direction of motion, so we consider separately a clock oriented parallel to the direction of motion ("parallel clock" in the following) and a clock oriented perpendicular to its direction of motion ("vertical clock").

Our purpose in the following is to demonstrate, via explicit computations, that the clock hypothesis doesn't hold, in general, for real noninertial clocks: Inertial clocks, even if they are extended, share a common rest reference frame with a common time scale for all their constituents. Then, like a clock at rest, the internal time measurement coincides with the external Minkowskian proper-time measurement. But if the clock accelerates then different parts of it, being an extended system, measure the proper-times differently, and inconsistencies between the external time measurements and the internal time-scale appear.

In addition, although the time measurement by light-clocks is independent of the signals' frequencies, never-the-less these signals are Doppler-affected by the acceleration of the clock. An Appendix briefly summarizes these effects.

Notation. The convention $c = 1$ is used throughout, unless specified otherwise. Events in Minkowski space-time are $x^\mu = (x^0, x^1, x^2, x^3)$, with metric tensor $g_{\mu\nu} = \mathrm{diag}(-1, 1, 1, 1)$, $\mu, \nu = 0, 1, 2, 3$.

2. On the motion of accelerated clocks

A clock is, generically, a composite, spatially extended device, consisting of a rigid casing that contains internally moving parts. The periodical motion of the internal mechanism relative to the casing determines the fundamental time unit of the clock. (Such clocks may be regarded as autonomous. Clock types which depend on external factors, like pendulum clocks or sundials, are excluded)

A clock is inertial when there exists an inertial reference frame in which the casing is permanently at rest. For a noninertial clock to still function as a reliable time measuring device its structure must remain fixed, at least relative to itself. Otherwise, differential acceleration will cause the clock's structure to be continuously changing relative to itself, resulting in varying internal period.

Therefore, a necessary condition is that the motion of the clock as a whole be rigid, so that at any moment there is a momentary inertial reference frame relative to which the casing of the clock is at rest. This is Born's relativistic rigidity condition [6], requiring that the distance between any two point-like constituents of the casing remain unchanged throughout the motion relative to an instantaneous inertial rest frame attached to any one of these two constituents.

Born's condition may be satisfied with arbitrarily accelerated rectilinear rigid motion [7]. In this way the structure of the casing is maintained rigid relative to itself even when the clock accelerates. The rigidity requirement, that at any moment there is a momentary inertial reference frame relative to which the casing is at rest, implies that the velocities of all the casing's points are at any moment the same; different points along the system must have different accelerations which are inter-related in a very specific way, but besides this condition the acceleration may be completely arbitrary.

We start with a brief review of general rectilinear motion, arbitrarily chosen along the x-axis relative to some inertial reference frame S [10]. Consider some point A with the world-line

$$x^\mu_{\mathrm{A}} = (t, x(t), 0, 0) \tag{2}$$

Its unit 4-velocity is $u^\mu = dx^\mu_{\mathrm{A}}/d\tau = \gamma(v)\,(1, v, 0, 0)$, with $v(t) = dx/dt$ and $\gamma(v) = \left(1-v^2\right)^{-1/2} = dt/d\tau$, with the proper-time τ given by (1). The acceleration 4-vector is

$$a^\mu = \frac{du^\mu}{d\tau} = \gamma^3\,(v)\,\frac{dv}{d\tau}\,(v, 1, 0, 0) = a n^\mu\,, \tag{3}$$

with $a = \gamma^2\,(v)\,(dv/d\tau)$ the proper acceleration and $n^\mu = \gamma\,(v)\,(v, 1, 0, 0)$ the space-like unit 4-vector orthogonal to u^μ $(u \cdot n = 0)$ indicating the spatial direction of motion.

Since the velocity is common to all parts of the clock's casing, then, as long as the clock accelerates ($a \neq 0$ which is assumed in the following), it is convenient to use, as time-like evolution parameter, the rapidity [11]

$$\eta\,(v) = \tanh^{-1}(v) = \frac{1}{2}\ln\left(\frac{1+v}{1-v}\right)\,. \tag{4}$$

It is defined in this way both for single particles' velocities and for the relative velocities between inertial reference frames. It has the very useful property that under arbitrary Lorentz transformations along the x-axis,

$$x \to x' = \gamma(V)\,(x - Vt) \quad , \quad t \to t' = \gamma(V)\,(t - Vx)\,, \tag{5}$$

particles' rapidities change additively, $\eta(v) \to \eta(v') = \eta(v) - \eta(V)$, so that rapidity differences are Lorentz invariant.

In terms of the rapidity, the particle's unit 4-velocity and unit normal are

$$\begin{aligned} u^\mu(\eta) &= \frac{dx^\mu}{d\tau} = (\cosh\eta, \sinh\eta, 0, 0)\,, \\ n^\mu(\eta) &= (\sinh\eta, \cosh\eta, 0, 0) \end{aligned} \tag{6}$$

Journal of Physics: Conference Series **2482** (2023) 012009 doi:10.1088/1742-6596/2482/1/012009

From (3) then follows the basic relation between the proper acceleration, the proper-time and the rapidity

$$a = \frac{d\eta}{d\tau}, \tag{7}$$

which holds for all rectilinear motion [10, 12]. The (external) proper-time lapse between any two states of motion with rapidities η_1 and η_2, computed using (7), is then

$$\Delta\tau = \int_{\eta_1}^{\eta_2} \frac{d\eta}{a(\eta)} \tag{8}$$

Combining Eqs. (6) & (7) then follow the relations

$$\begin{aligned} dt &= \cosh\eta d\tau = \frac{\cosh\eta d\eta}{a} \\ dx &= \sinh\eta d\tau = \frac{\sinh\eta d\eta}{a} \end{aligned}, \tag{9}$$

which are integrated to yield the Minkowskian coordinates

$$t(\eta) = \int^{\eta} \frac{\cosh\eta}{a(\eta)} d\eta \quad , \quad x(\eta) = \int^{\eta} \frac{\sinh\eta}{a(\eta)} d\eta \,. \tag{10}$$

It should be noted that while the proper acceleration a is Lorentz invariant, η as a variable is defined relative to S, therefore the function $a(\eta)$ is frame-dependent.

3. Time measurement in parallel light-clocks

We start with time measurement in a parallel light-clock, in which the light signal moves forward and back parallel to the direction of motion and acceleration (Figure 1a).

3.1. Kinematics of parallel light-clocks

Suppose the clock is oriented along the x-axis which is also the direction of motion and acceleration. Let the rear end of the clock be denoted A, with the world-line (2) expressed via Eq. (10),

$$x^{\mu}_{\mathrm{A}}(\eta) = \left(\int_0^{\eta} \frac{\cosh\eta}{a(\eta)} d\eta, \int_0^{\eta} \frac{\sinh\eta}{a(\eta)} d\eta, 0, 0 \right) , \tag{11}$$

starting at rest at the origin of S, with $a(\eta)$ the proper acceleration at A. Let the front end be denoted B, and let L be the proper distance between A and B. By the rigidity assumption, this proper distance, which is measured relative to instantaneous inertial rest frames in the direction of n^{μ}, should remain unchanged. Then B is in the direction of n^{μ} relative to A, with the world-line [10]

$$x^{\mu}_{\mathrm{B}}(\eta) = x^{\mu}_{\mathrm{A}} + Ln^{\mu} = \left(\int_0^{\eta} \frac{\cosh\eta}{a(\eta)} d\eta + L\sinh\eta, \int_0^{\eta} \frac{\sinh\eta}{a(\eta)} d\eta + L\cosh\eta, 0, 0 \right) . \tag{12}$$

That both points share the same velocity is easily verified from $v_{\mathrm{B}}(\eta) = dx^1_{\mathrm{B}}/dx^o_{\mathrm{B}} = \tanh\eta = v_{\mathrm{A}}(\eta)$, asserting that they also share the same rapidity. By the rigidity condition, this is true for all the points of the clock's casing. Also, from the general analysis of rectilinear rigid motion [10] follows the proper acceleration at B

$$\frac{1}{a_{\mathrm{B}}(\eta)} = \frac{1}{a(\eta)} + L \tag{13}$$

3.2. Light signals in a parallel clock

The clock is oriented along the x-axis which is also the direction of motion and acceleration. Since we want to examine the effect of acceleration on time measurement we consider a system which is a double light-clock, so that on each side there is a light source, a mirror and a light sensor.

It is assumed throughout that the duration of emission and reception of the signals is infinitesimally smaller than the time it takes for the transmission between the mirrors.

Consider a signal emitted from A towards B. Let the emission event be at time t_1, corresponding to rapidity η_1, relative to the chosen external inertial reference frame S, arriving to B at S-time t_2 corresponding to rapidity η_2. It is important to note that the rapidity difference $\Delta\eta = \eta_2 - \eta_1$ is Lorentz invariant.

The 4-vector displacement of the signal

$$\begin{aligned}\Delta x^\mu_{\mathrm{AB}} &= x^\mu_{\mathrm{B}}(\eta_2) - x^\mu_{\mathrm{A}}(\eta_1) = \\ &= \left(\int_{\eta_1}^{\eta_2} \frac{\cosh\eta}{a(\eta)} d\eta + L\sinh\eta_2, \int_{\eta_1}^{\eta_2} \frac{\sinh\eta}{a(\eta)} d\eta + L\cosh\eta_2, 0, 0\right)\end{aligned} \tag{14}$$

satisfies the light-cone condition

$$\Delta x^0 = \Delta x^1 \quad \Rightarrow \quad \int_{\eta_1}^{\eta_2} \frac{\cosh\eta}{a(\eta)} d\eta + L\sinh\eta_2 = \int_{\eta_1}^{\eta_2} \frac{\sinh\eta}{a(\eta)} d\eta + L\cosh\eta_2 \tag{15}$$

which yields

$$L = \int_{\eta_1}^{\eta_2} \frac{e^{\eta_2-\eta}}{a(\eta)} d\eta \qquad (\mathrm{A} \to \mathrm{B}) \tag{16}$$

(16) determines η_2 in terms of L and η_1.

Similarly, for a signal emitted from B at η_1 arriving at A at η_2, the 4-vector displacement of the signal

$$\begin{aligned}\Delta x^\mu_{\mathrm{BA}} &= x^\mu_{\mathrm{A}}(\eta_2) - x^\mu_{\mathrm{B}}(\eta_1) = \\ &= \left(\int_{\eta_1}^{\eta_2} \frac{\cosh\eta}{a(\eta)} d\eta - L\sinh\eta_1, \int_{\eta_1}^{\eta_2} \frac{\sinh\eta}{a(\eta)} d\eta - L\cosh\eta_1, 0, 0\right)\end{aligned} \tag{17}$$

satisfies the light-cone condition

$$\Delta x^0 = \left|\Delta x^1\right| = -\Delta x^1 \Rightarrow \int_{\eta_1}^{\eta_2} \frac{\cosh\eta}{a(\eta)} d\eta - L\sinh\eta_1 + \int_{\eta_1}^{\eta_2} \frac{\sinh\eta}{a(\eta)} d\eta - L\cosh\eta_1 = 0 \tag{18}$$

which yields

$$L = \int_{\eta_1}^{\eta_2} \frac{e^{\eta-\eta_1}}{a(\eta)} d\eta \qquad (\mathrm{B} \to \mathrm{A}) \tag{19}$$

In either case, the proper-time lapses measured at A and B between emission and arrival are, combining (8) and (13),

$$\begin{aligned}\Delta\tau_{\mathrm{A}}(\eta_1,\eta_2) &= \int_{\eta_1}^{\eta_2}\frac{d\eta}{a(\eta)}\\ \Delta\tau_{\mathrm{B}}(\eta_1,\eta_2) &= \int_{\eta_1}^{\eta_2}\frac{d\eta}{a_{\mathrm{B}}(\eta)} = \int_{\eta_1}^{\eta_2}\left[\frac{1}{a(\eta)}+L\right]d\eta = \Delta\tau_{\mathrm{A}}(\eta_1,\eta_2)+L\Delta\eta\end{aligned} \tag{20}$$

Using relations (16) and (19) it follows (see Appendix A for details) that they satisfy the inequalities

$$e^{-\Delta\eta}L < \Delta\tau_{\mathrm{A}} < L \quad , \quad L < \Delta\tau_{\mathrm{B}} < e^{\Delta\eta}L \tag{21}$$

Clearly, different proper-times are externally measured at the two ends of the clock due to the different accelerations there.

3.3. Unit of time in a parallel clock

Consider first a signal emitted from A towards B and reflected back to A. As the clock accelerates its rapidity relative to S changes between these events. Let the emission event A_1 be at rapidity η_1 relative to the chosen external inertial reference frame S, the reflection event B_{12} be at rapidity η_{12}, and the event A_2 of arrival back to A at rapidity η_2. Applying relations (16) and (19) for the signal $A_1 \to B_{12} \to A_2$ implies the combined relations

$$\int_{\eta_1}^{\eta_{12}}\frac{e^{\eta_{12}-\eta}}{a_{\mathrm{A}}(\eta)}d\eta = \int_{\eta_{12}}^{\eta_2}\frac{e^{\eta-\eta_{12}}}{a_{\mathrm{A}}(\eta)}d\eta = L \tag{22}$$

For a signal emitted from B at $\bar{\eta}_1$ towards A, reflected back at $\bar{\eta}_{12}$ and arriving to B at $\bar{\eta}_2$, a similar relation is obtained, only with $a_{\mathrm{B}}(\eta)$ replacing $a_{\mathrm{A}}(\eta)$,

$$\int_{\bar{\eta}_1}^{\bar{\eta}_{12}}\frac{e^{\eta-\bar{\eta}_{12}}}{a_{\mathrm{B}}(\eta)}d\eta = \int_{\bar{\eta}_{12}}^{\bar{\eta}_2}\frac{e^{\bar{\eta}_{12}-\eta}}{a_{\mathrm{B}}(\eta)}d\eta = L \tag{23}$$

For the signal emitted from A and reflected at B back to A, the proper-time lapse between emission and absorption as externally measured in A, satisfies (using (22) and the fact that $e^x > 1$ for $x > 0$) the inequality

$$\Delta\tau_{\mathrm{A}} = \int_{\eta_1}^{\eta_2}\frac{d\eta}{a_{\mathrm{A}}(\eta)} < \int_{\eta_1}^{\eta_{12}}\frac{e^{\eta_{12}-\eta}}{a_{\mathrm{A}}(\eta)}d\eta + \int_{\eta_{12}}^{\eta_2}\frac{e^{\eta-\eta_{12}}}{a_{\mathrm{A}}(\eta)}d\eta = 2L \tag{24}$$

Similarly, for the signal emitted at B and reflected back at A the proper-time lapse externally measured at B between emission and absorption satisfies (using (23) and the fact that $e^x < 1$ for $x < 0$) the inequality

$$\Delta\tau_{\mathrm{B}} = \int_{\bar{\eta}_1}^{\bar{\eta}_2}\frac{d\eta}{a_{\mathrm{B}}(\eta)} > \int_{\bar{\eta}_1}^{\bar{\eta}_{12}}\frac{e^{\eta-\bar{\eta}_{12}}}{a_{\mathrm{B}}(\eta)}d\eta + \int_{\bar{\eta}_{12}}^{\bar{\eta}_2}\frac{e^{\bar{\eta}_{12}-\eta}}{a_{\mathrm{B}}(\eta)}d\eta = 2L \tag{25}$$

$T = 2L$ is the time unit relative to the clock, defining its internal time scale. Therefore, the internally measured time lapse is between the (externally measured) proper-time lapses as measured at the extreme points of the clock,

$$\Delta\tau_{\mathrm{ext,A}} < \Delta\tau_{\mathrm{int}} < \Delta\tau_{\mathrm{ext,B}}\,. \tag{26}$$

3.4. The case of hyperbolic motion
The simplest case of an accelerating clock is certainly that of hyperbolic motion with constant proper accelerations, which therefore largely abounds in the literature. It is brought now as an example for the above.

For a parallel clock with a and $a/(1+aL)$ being respectively A's and B's constant proper accelerations, it follows from both (22) and (23) that

$$e^{\eta_{12}-\eta_1} = e^{\eta_2-\eta_{12}} = 1 + aL \tag{27}$$

with same $\eta_1, \eta_{12}, \eta_2$ for signals emitted in both directions. The corresponding externally measured proper-time lapses then follow from (24) and (25),

$$\begin{aligned} \Delta\tau_{\mathrm{A}} &= \frac{2}{a}\ln(1+aL) < 2L \\ \Delta\tau_{\mathrm{B}} &= \frac{2}{a}(1+aL)\ln(1+aL) > 2L \end{aligned} \tag{28}$$

verifying the double inequality in (26). The internal period of the clock $T = 2L$ may be found to equal the externally measured proper-time lapse $\Delta\tau_\zeta = \Delta\tau_{\mathrm{A}} + \zeta\Delta\eta$ for the point of the clock which is distance ζ to A's right, with

$$\zeta = \frac{L}{\ln(1+aL)} - \frac{1}{a} \tag{29}$$

4. Time measurement in vertical light-clocks
Next we consider time measurement in vertical light-clocks, in which the source and mirrors are oriented perpendicular to the direction of motion and acceleration (Figure 1b).

4.1. Kinematics of vertical light-clocks
Unlike the parallel clock, the vertical light-clock has the virtue that all its constituents share the same externally measured proper-time.

Suppose the clock is now oriented along the y-axis, perpendicular to the direction of motion and acceleration. Both mirrors now move with the same acceleration $a(\eta)$, thus omitting the indices. With the world-line (11) for mirror A, B's world-line is now

$$x_{\mathrm{B}}^{\mu}(\eta) = \left(\int_0^{\eta} \frac{\cosh\eta}{a(\eta)}d\eta, \int_0^{\eta} \frac{\sinh\eta}{a(\eta)}d\eta, L, 0\right) \tag{30}$$

4.2. Light signals and unit of time in a vertical light-clock
Since the motion doesn't affect the apparent distance between the mirrors, signal transmission is, in principle, symmetric in both directions (for signals from A to B and from B to A), and it suffices to consider a single transmission process. Let a signal be emitted at $x_{\mathrm{A}}^{\mu}(\eta_1)$, arriving to B at $x_{\mathrm{B}}^{\mu}(\eta_2)$. Its displacement 4-vector between the mirrors is

$$\Delta x_{12}^{\mu} = x_{\mathrm{B}}^{\mu}(\eta_2) - x_{\mathrm{A}}^{\mu}(\eta_1) = \left(\int_{\eta_1}^{\eta_2} \frac{\cosh\eta}{a(\eta)}d\eta, \int_{\eta_1}^{\eta_2} \frac{\sinh\eta}{a(\eta)}d\eta, L, 0\right) \tag{31}$$

so that the light-like condition $(\Delta x_{12})^2 = 0$ transforms into

$$\left[\int_{\eta_1}^{\eta_2} \frac{\cosh\eta}{a(\eta)} d\eta\right]^2 - \left[\int_{\eta_1}^{\eta_2} \frac{\sinh\eta}{a(\eta)} d\eta\right]^2 = \left[\int_{\eta_1}^{\eta_2} \frac{e^{\eta} d\eta}{a(\eta)}\right]\left[\int_{\eta_1}^{\eta_2} \frac{e^{-\eta} d\eta}{a(\eta)}\right] = L^2 \tag{32}$$

Assuming the proper acceleration $a(\eta)$ to be positive and at least piecewise continuous, it then follows from the Cauchy-Schwarz inequality for integrals [13] that the externally measured proper-time lapse between $\eta = \eta_1$ and $\eta = \eta_2$ satisfies

$$\Delta\tau_{12} = \int_{\eta_1}^{\eta_2} \frac{d\eta}{a(\eta)} < \left[\int_{\eta_1}^{\eta_2} \frac{e^{\eta} d\eta}{a(\eta)}\right]^{1/2} \cdot \left[\int_{\eta_1}^{\eta_2} \frac{e^{-\eta} d\eta}{a(\eta)}\right]^{1/2} = L \tag{33}$$

L is the half-period or time unit relative to the clock, defining its internal time scale. Therefore, any internally measured time lapse of the clock is larger than the corresponding externally measured proper-time lapse,

$$\Delta\tau_{\text{ext}} < \Delta\tau_{\text{int}} = T\,. \tag{34}$$

The clock's time always precedes the Minkowskian time.

4.3. Geometry of signal transmission in a vertical light-clock

Let us consider now the geometry of signal transmission in the vertical clock. Let S_1 be the instantaneous co-moving inertial frame at the emission moment, moving with rapidity η_1 relative to S. Lorentz-transforming the signal's displacement vector (31) from S to S_1 it becomes

$$\begin{aligned}[\Delta x^{\mu}_{12}]_1 &= \left(\sinh\eta_1 \Delta x^1_{12} - \cosh\eta_1 \Delta x^0_{12}, \cosh\eta_1 \Delta x^1_{12} - \sinh\eta_1 \Delta x^0_{12}, L, 0\right) = \\ &= \left(\int_{\eta_1}^{\eta_2} \frac{\cosh(\eta-\eta_1)}{a(\eta)} d\eta, \int_{\eta_1}^{\eta_2} \frac{\sinh(\eta-\eta_1)}{a(\eta)} d\eta, L, 0\right)\end{aligned} \tag{35}$$

so the (x-)displacement of the mirror B relative to S_1 during the signal's travel is

$$D^{(1)}_{12} = \left[\Delta x^1_{12}\right]_1 = \int_{\eta_1}^{\eta_2} \frac{\sinh(\eta-\eta_1)}{a(\eta)} d\eta \tag{36}$$

If the motion were uniform, no acceleration, so that $\Delta x^1_{12} = v_1 \Delta x^0_{12}$, then $D^{(1)}_{12} = 0$, verifying that the transmission is perpendicular to the direction of motion. Otherwise, for the signal to reach mirror B, the emission should be inclined, relative to the y-axis as measured in S_1, with angle $\theta^{(1)}_{12} = \tan^{-1}\left(D^{(1)}_{12}/L\right)$. An alternative relation for the emission angle is obtained by applying the light-like condition to (35),

$$\begin{aligned}\left[\int_{\eta_1}^{\eta_2} \frac{\cosh(\eta-\eta_1)}{a(\eta)} d\eta\right]^2 &= L^2 + \left[\int_{\eta_1}^{\eta_2} \frac{\sinh(\eta-\eta_1)}{a(\eta)} d\eta\right]^2 = L^2 + L^2\tan^2\theta^{(1)}_{12} = \\ &= \frac{L^2}{\cos^2\theta^{(1)}_{12}}\end{aligned} \tag{37}$$

so that

$$\frac{L}{\cos\theta_{12}^{(1)}} = \int_{\eta_1}^{\eta_2} \frac{\cosh(\eta - \eta_1)}{a(\eta)} d\eta \tag{38}$$

For non-hyperbolic motion (variable $a(\eta)$), knowledge is required in advance of the plan of journey, explicitly of the acceleration, to be able to properly direct the signals between the mirrors.

As the signal arrives to B, the incidence angle relative to B is not necessarily the same as the inclination angle at emission. Let S_2 be the instantaneous co-moving inertial frame at the arrival moment, moving with rapidity η_2 relative to S. S_1 then moves with rapidity $\eta_1 - \eta_2 = -(\eta_2 - \eta_1)$ relative to S_2. Therefore, while from A the signal is sent to B inclined rightward, it is seen relative to B as arriving from the right. The displacement of mirror A relative to S_2 during the signal's travel is found in a similar way,

$$D_{12}^{(2)} = \int_{\eta_1}^{\eta_2} \frac{\sinh(\eta - \eta_2)}{a(\eta)} d\eta = -\int_{\eta_1}^{\eta_2} \frac{\sinh(\eta_2 - \eta)}{a(\eta)} d\eta\,, \tag{39}$$

and the incidence angle relative to B is $\theta_{12}^{(2)} = \tan^{-1}\left(\left|D_{12}^{(2)}\right|/L\right)$. There is also a relation similar to (38) for the incidence angle,

$$\frac{L}{\cos\theta_{12}^{(2)}} = \int_{\eta_1}^{\eta_2} \frac{\cosh(\eta_2 - \eta)}{a(\eta)} d\eta \tag{40}$$

4.4. The case of hyperbolic motion

Using again the case of hyperbolic motion as an explicit example, then for a vertical light-clock with constant acceleration a we obtain from (32)

$$L = \frac{2}{a}\sinh\left(\frac{\Delta\eta}{2}\right) > \frac{\Delta\eta}{a} = \Delta\tau \tag{41}$$

as expected from the general relation (34). The relative rapidity between the events of signal emission and arrival is independent of the initial conditions, $\Delta\eta = 2\sinh^{-1}(0.5aL)$, implying that it is unchanging for a signal bouncing forth and back between the two mirrors. The relative displacement of the mirrors is also independent of the initial conditions, the same for both mirrors,

$$D_{12} = \frac{1}{a}\int_{\eta_1}^{\eta_2} \sinh(\eta - \eta_1)\, d\eta = \frac{\cosh(\Delta\eta) - 1}{a} = \frac{aL^2}{2} \tag{42}$$

with equal emission and incidence angles, $\theta_{12} = \tan^{-1}(0.5aL)$.

Since the incidence and reflection angles are equal relative to the y-axis, the mirrors should be parallel to the x-axis, with no need to rotate them as suggested by West [8].

Journal of Physics: Conference Series **2482** (2023) 012009 doi:10.1088/1742-6596/2482/1/012009

5. Concluding remarks

In this work we compared externally measured proper-time vs. intrinsically defined age for noninertial spatially extended relativistic systems, demonstrating that they don't coincide. Discrepancies are found of order aL/c^2, so while for practical values these are very small and may be ignored, not so for matters of principle, which require absolute exactitude. Also, when we go down to Planck's scale the discrepancy is of order 1 and cannot be ignored.

In special relativity (SR) the basic grid upon which time measurements are assumed is the Minkowski space-time. The Minkowskian space-time is an assumed continuum of synchronized inertial, ideal point-like clocks. But real clocks are composite spatially extended systems with an internal periodical mechanism that defines an intrinsic measure of time. These two time-lines – the external Minkowskian proper-time and the internal, inherent, time scale of the clock – coincide only when the clock is inertial; they differ when the clock accelerates.

Real clocks are therefore not well-defined relative to Minkowski space-time, only approximately.

Clocks are used here to represent any composite system with an internal temporal measure. For such systems, the internal temporal measure, age, doesn't coincide, in general, with the external Minkowskian temporal measure. Consequently, with two different time-lines, the physical time-line appears to be not satisfactorily well-defined, only approximately; thus the Minkowski space-time is only approximately self-consistent. Something is missing in the understanding of the physical time-line.

How can we cope with this inconsistency?

The classical concept of the physical time-line is of a continuous real numbers time axis consisting of points or *moments of time*. We are used to think in terms of a succession of space-like hypersurfaces, with moment-to-moment evolution dictated by differential relations. This picture is certainly useful and sufficient for practical purposes, but it seems to be missing, not satisfactory enough, for matters of principle.

Clocks define and measure, by their periods, *intervals of time*, or *durations*; not moments of time. It is us who interpolate within these intervals to form the continuous real number picture. Indeed, the spirit of SR should conduct us to see that full covariance implies that spatial extension should be associated also with temporal extension; and therefore regard spatially extended systems also as being temporally extended. Temporal extension may then be understood to mean that systems are defined relative to intervals of time, rather than relative to space-like hypersurfaces corresponding to moments of time.

We perceive the flow of time and form the concept of time in our consciousness from and via our experiences. Our perception of time is therefore an *experiential time-line*; and our experience of time is of *durations*, in simile to the clocks' internal time-line concept. It is important to emphasize that we never experience moments of time – we always experience intervals of time, because in a moment of time, having zero duration, we cannot absorb and collect any information. We always need for that a finite, non-zero, lapse of time. This is simply demonstrated by the example of cameras taking still pictures – the camera always opens for some short interval $\Delta t > 0$, different from zero, which is necessary to collect the physical information for creating the picture. In zero time no information can be collected, and no picture can be formed in the camera. Even video cameras take in the same way discrete frames, and it is only the speed of running these frames that makes the impression of continuity and motion.

The concept of time is formed, primarily, in our consciousness, together with formation of memory, after being aware to sequential ordring of phenomena which we experience. We are then able to form a universal concept of time after realizing that the rate of time is independent of the particular phenomenon (*e.g.*, via the ratio between the periods of distinct phenomena remaining constant, even if their mechanisms are completely different).

The emphasis on the fact that we experience time in intervals, *durations* rather than moments, is central to the thesis of the Jewish-French philosopher Henri-Louis Bergson [14]. He saw time as a rolling experience, rather than a moment-by-moment experience. A century ago Einstein, who adhered to the concept of the physical time-line only, dismissed publicly Bergson's ideas. It seems now that this attitude should be seriously reviewed and revised.

Appendix A. Parallel clock inequalities

For the signal from A to B, follows directly from (16) (using the fact that $e^x > 1$ for $x > 0$) the inequality

$$L = \int_{\eta_1}^{\eta_2} \frac{e^{\eta_2-\eta}}{a(\eta)} d\eta > \int_{\eta_1}^{\eta_2} \frac{d\eta}{a(\eta)} = \Delta\tau_{\mathrm{A}} \tag{A.1}$$

Multiplying (A.1) with $\exp(-\Delta\eta)$ and using $e^x < 1$ for $x < 0$ yields a complementing inequality,

$$e^{-\Delta\eta} L = \int_{\eta_1}^{\eta_2} \frac{e^{\eta_1-\eta}}{a(\eta)} d\eta < \int_{\eta_1}^{\eta_2} \frac{d\eta}{a(\eta)} = \Delta\tau_{\mathrm{A}} \tag{A.2}$$

Replacing $a(\eta)$ with $a_{\mathrm{B}}(\eta)$ using (13) and (16) yields

$$\int_{\eta_1}^{\eta_2} \frac{e^{\eta_2-\eta}}{a_{\mathrm{B}}(\eta)} d\eta = \int_{\eta_1}^{\eta_2} e^{\eta_2-\eta} \left[\frac{1}{a(\eta)} + L\right] d\eta = L + \int_{\eta_1}^{\eta_2} e^{\eta_2-\eta} L d\eta = e^{\Delta\eta} L \tag{A.3}$$

so that, again with $e^x > 1$ for $x > 0$,

$$e^{\Delta\eta} L = \int_{\eta_1}^{\eta_2} \frac{e^{\eta_2-\eta}}{a_{\mathrm{B}}(\eta)} d\eta > \int_{\eta_1}^{\eta_2} \frac{d\eta}{a_{\mathrm{B}}(\eta)} = \Delta\tau_{\mathrm{B}} \tag{A.4}$$

Finally, multiplying (A.3) with $\exp(-\Delta\eta)$ and using $e^x < 1$ for $x < 0$ yields, in simile with (A.2), another complementing inequality,

$$L = e^{-\Delta\eta} \int_{\eta_1}^{\eta_2} \frac{e^{\eta_2-\eta}}{a_{\mathrm{B}}(\eta)} d\eta = \int_{\eta_1}^{\eta_2} \frac{e^{\eta_1-\eta}}{a_{\mathrm{B}}(\eta)} d\eta < \int_{\eta_1}^{\eta_2} \frac{d\eta}{a_{\mathrm{B}}(\eta)} = \Delta\tau_{\mathrm{B}} \tag{A.5}$$

Coupling together both (A.1) and (A.2), and similarly both (A.4) and (A.5), then yields the inequalities in (21).

Similar relations are found for signals from B to A.

Appendix B. Doppler shifts

Although the time measurement by light-clocks is independent of the signals' frequencies, nevertheless these signals are Doppler-affected by the acceleration of the clock. In this Appendix we briefly summarize these effects.

Appendix B.1. Doppler shifts of light signals in accelerated parallel light-clocks

Consider a light-clock moving in the $+x$-direction, and a signal emitted from A towards B, in the same direction, with proper frequency (relative to the source) $\omega_{\rm A}$ and wave vector $k_{\rm A}^{(+)\mu} = (\omega_{\rm A}, \omega_{\rm A}, 0, 0)$. Let the clock's velocity at the event of emission be $\vec{v}_{\rm A} = v_{\rm A}\hat{x}$ with rapidity $\eta_{\rm A} = \tanh^{-1}(v_{\rm A})$, and with velocity $\vec{v}_{\rm B} = v_{\rm B}\hat{x}$ with rapidity $\eta_{\rm B} = \tanh^{-1}(v_{\rm B})$ as the signal reaches B (all velocities and rapidities relative to some arbitrary external inertial system S). The relative velocity between the arrival and emission events is then $V_{\rm BA} = \tanh(\eta_{\rm B} - \eta_{\rm A})$; depending on the rapidity difference it is Lorentz-invariant. The wave vector relative to the arrival event is then

$$k_{\rm B}^{(+)\mu} = \sqrt{\frac{1 - V_{\rm BA}}{1 + V_{\rm BA}}}\,(\omega_{\rm A}, \omega_{\rm A}, 0, 0) = e^{\eta_{\rm A} - \eta_{\rm B}} k_{\rm A}^{(+)\mu}\,, \tag{B.1}$$

thus identifying the Doppler-shift relation as

$$\omega_{\rm B} = e^{\eta_{\rm A} - \eta_{\rm B}} \omega_{\rm A}\,. \tag{B.2}$$

For a signal in the opposite direction, $\mathrm{B} \to \mathrm{A}$, with the wave vector relative to the emission event at B, $k_{\rm B}^{(-)\mu} = (\omega_{\rm B}, -\omega_{\rm B}, 0, 0)$, and relative velocity between the arrival and emission events $V_{\rm AB} = \tanh(\eta_{\rm A} - \eta_{\rm B})$, the wave vector relative to the arrival event at A is

$$k_{\rm A}^{(-)\mu} = \sqrt{\frac{1 + V_{\rm AB}}{1 - V_{\rm AB}}}\,(\omega_{\rm B}, -\omega_{\rm B}, 0, 0) = e^{\eta_{\rm A} - \eta_{\rm B}} k_{\rm B}^{(-)\mu}\,. \tag{B.3}$$

The Doppler-shift relation is then

$$\omega_{\rm A} = e^{\eta_{\rm A} - \eta_{\rm B}} \omega_{\rm B} \tag{B.4}$$

We notice that in both cases the Doppler factor may be written as $e^{\eta_{\rm rear} - \eta_{\rm front}}$, reffering to the orientation of the clock relative to the direction of motion.

For a two-way signal, say $\mathrm{A}\,(\eta_1) \to \mathrm{B}\,(\eta_{12}) \to \mathrm{A}\,(\eta_2)$, we get, combining (B.2) and (B.4),

$$\omega_2 = e^{\eta_2 - \eta_{12}} \omega_{\rm B} = e^{\eta_2 - \eta_{12}} e^{\eta_1 - \eta_{12}} \omega_0 = e^{\eta_1 + \eta_2 - 2\eta_{12}} \omega_1 \tag{B.5}$$

In the case of hyperbolic motion $\eta_2 - \eta_{12} = \eta_{12} - \eta_1$, hence no Doppler shift between emitted and returning signals..

Appendix B.2. Doppler shifts of light signals in accelerated vertical light-clocks

Consider a signal emitted from A towards B with proper frequency (relative to the source) ω_1. Using the notation of Section 4.3, with the inclination angle $\theta_{12}^{(1)}$ the signal's wave vector relative to S_1 (the emission rest frame) is

$$k_1^\mu = \left(\omega_1, \omega_1 \sin\theta_{12}^{(1)}, \omega_1 \cos\theta_{12}^{(1)}, 0\right)\,. \tag{B.6}$$

The relative velocity between S_1 and S_2 (arrival rest frame) is $V = \tanh(\Delta\eta)$ with $\Delta\eta = \eta_2 - \eta_1$, so Lorentz transforming the wave vector (B.6) from S_1 to S_2 its components relative to the latter are

$$\begin{aligned} k_2^\mu = \Big(&\left(\cosh\Delta\eta - \sinh\Delta\eta \sin\theta_{12}^{(1)}\right)\omega_1, \\ &\left(\cosh\Delta\eta \sin\theta_{12}^{(1)} - \sinh\Delta\eta\right)\omega_1, \omega_1 \cos\theta_{12}^{(1)}, 0\Big)\,. \end{aligned} \tag{B.7}$$

The Doppler-shifted frequency is $\omega_2 = e^{-\phi_{12}}\omega_1$ with the Doppler-shift factor

$$
\begin{aligned}
e^{-\phi_{12}} &= \cosh\Delta\eta - \sinh\Delta\eta \sin\theta_{12}^{(1)} = \cos\theta_{12}^{(1)}\left[\frac{\cosh\Delta\eta}{\cos\theta_{12}^{(1)}} - \sinh\Delta\eta\tan\theta_{12}^{(1)}\right] = \\
&= \frac{1}{L}\cos\theta_{12}^{(1)}\int_{\eta_1}^{\eta_2}\frac{\cosh\Delta\eta\cosh(\eta-\eta_1) - \sinh\Delta\eta\sinh(\eta-\eta_1)}{a(\eta)}d\eta = \\
&= \frac{1}{L}\cos\theta_{12}^{(1)}\int_{\eta_1}^{\eta_2}\frac{\cosh(\eta_2-\eta)}{a(\eta)}d\eta = \frac{\cos\theta_{12}^{(1)}}{\cos\theta_{12}^{(2)}}
\end{aligned}
\tag{B.8}
$$

simply expressed in terms of the emission and incidence angles. In the case of hyperbolic motion, the equality of the angles implies that there is no Doppler shift.

References

[1] Rindler W 2006 *Relativity: Special, General, and Cosmological* (2nd ed., Oxford: Oxford University Press) p. 65

[2] Misner C W, Thorne K P and Wheeler J A 1973 *Gravitation* (San Francisco: Freeman) p. 393

[3] Mainwaring S R and Stedman G E 1993 Accelerated clock principles in special relativity *Phys. Rev. A* **47** 3611-9

[4] Kowalski F V 1996 Accelerating light clocks *Phys. Rev. A* **53** 3761-6

[5] Arthur R T W 2010 Minkowski's proper-time and the status of the clock-hypothesis, in V. Petkov (ed.) *Space, Time and Spacetime* Fundamental Theories of Physics 167 (Berlin Heidelberg: Springer-Verlag)

[6] Born M 1909 Die Theorie des starren Elektrons in der Kinematik des Relativitätsprinzips (The Theory of the Rigid Electron in the Kinematics of the Principle of Relativity) *Annalen der Physik* **335 (11)** 1-56

[7] Herglotz G 1910 Über den vom Standpunkt des Relativitätsprinzips aus als starr zu bezeichnenden Körper (On bodies that are to be designated as "rigid" from the standpoint of the relativity principle) *Annalen der Physik* **336 (2)** 393-415

Noether F 1910 Zur Kinematik des starren Körpers in der Relativtheorie *Annalen der Physik* **336 (5)** 919-44

[8] West J 2007 A light clock satisfying the clock hypothesis of special relativity *Eur. J. Phys.* **28** 693-703

[9] Kowaiski F V, Murray J and Head A C 1993 Phase measurement of light propagating in a linearly accelerating rigid-mirror system *Phys. Lett.* **A 174** 190-5

[10] Ben-Ya'acov U 2016 The 'twin paradox' in relativistic rigid motion *Eur. J. Phys.* **37** 055601

[11] Rhodes J A and Semon M D 2004 Relativistic velocity space, Wigner rotation and Thomas precession *Am. J. Phys.* **72 (7)** 943-60

[12] Minguzzi E 2005 Differential aging from acceleration: An explicit formula *Am. J. Phys.* **73 (9)** 876-80

[13] $\left[\int_a^b f^2(x)dx\right]\cdot\left[\int_a^b g^2(x)dx\right] \geq \left[\int_a^b f(x)g(x)dx\right]^2$, with equality only when $f(x) = \lambda g(x)$.
See, *e.g.*, Blinder S M Cauchy-Schwarz inequality for integrals
http://demonstrations.wolfram.com/CauchySchwarzInequalityForIntegrals/
Wolfram Demonstrations Project, published March 7 2011

[14] Bergson H L (1889) Essai sur les données immédiates de la conscience (Eng.: Time and Free Will: An Essay on the Immediate Data of Consciousness; Hebrew translation: Jerusalem: Magnes Press 2008)

Journal of Physics: Conference Series **2482** (2023) 012010 doi:10.1088/1742-6596/2482/1/012010

Comparison of Two Competing Theories of 3-Flavor Neutrino Oscillations

John R. Fanchi

Department of Engineering, Texas Christian University, Fort Worth, Texas, 76129 USA
j.r.fanchi@tcu.edu

Abstract. Neutrino oscillation observations are used to compare two competing theories of 3-flavor neutrino oscillations. The two theories considered here are the standard model of neutrino oscillations, and parametrized Relativistic Quantum Theory (pRQT). pRQT is a manifestly covariant quantum theory with invariant evolution parameter. Recent data and a neutrino mass model from each theory are used to calculate neutrino masses. The models yield significantly different predictions of neutrino masses.

1. Introduction

Experiments with solar neutrinos, atmospheric neutrinos, reactor neutrinos, and accelerator neutrinos have demonstrated that flavor mixing can occur between neutrino flavors composed of neutrino mass states. Observations of neutrino oscillations can be used to assess the validity of two theories of transitions between three neutrino flavor states $\{|v_\alpha\rangle; \alpha = e, \mu, \tau\}$ given the assumption that neutrinos are composed of up to three mass states $\{|v_j\rangle; j = 1, 2, 3\}$. Models of neutrino flavor oscillations based on the standard theory [1] and the parametrized Relativistic Quantum Theory (pRQT) show that significant differences exist between model results.

Parametrized Relativistic Quantum Theory (pRQT) is a manifestly covariant quantum theory with invariant evolution parameter. Introductions to pRQT are presented by Fanchi [2,3], Pavšič [4,5], and Horwitz [6]. A review of relativistic classical mechanics and electrodynamics in the parametrized framework is given by Land and Horwitz [7].

A model of neutrino oscillations by mass state mixing was developed within the context of pRQT by Fanchi [8-10]. The difference between the standard model of vacuum flavor mixing and the pRQT model of vacuum flavor mixing was studied by Rusov and Vlasenko [11]. Results of their model are analyzed here and updated using 2020 data [1].

Mass-state transitions in pRQT provide a mechanism for modeling neutrino oscillations. The single-body and N-body formulations of pRQT are outlined in Sections 2 and 3, respectively. An s-clock for quantifying the invariant evolution parameter s is constructed in Section 3 for use in the pRQT model of neutrino oscillations. The mass basis and flavor basis for three flavors are introduced in Section 4 and the form of the neutrino mass matrix used in the neutrino oscillation models is presented in Section 5. An algorithm for calculating neutrino masses is presented in Section 6. Results of the standard and pRQT models are updated and compared using 2020 data from the Particle Data Group [Zyla, et al., 2020] in Section 7. Conclusions are presented in Section 8.

2. Probabilistic Formulation of Parametrized Relativistic Quantum Theory

The probabilistic formulation of pRQT begins with the assumption that a physical system can be represented by a conditional probability density $\rho(x|s)$. The position four-vector x has components

Content from this work may be used under the terms of the Creative Commons Attribution 3.0 licence. Any further distribution of this work must maintain attribution to the author(s) and the title of the work, journal citation and DOI.
Published under licence by IOP Publishing Ltd

$\{x^\mu : \mu = 0,1,2,3\}$ where index 0 signifies the time component and by indices 1, 2, 3 signify three space components. The probability density $\rho(x|s)$ is conditioned by invariant evolution parameter s.

According to probability theory, the conditional probability density $\rho(x|s)$ must be positive definite and normalizable. The Born representation of the positive definite requirement is used to express the probability amplitude Ψ as

$$\rho(x|s) = \Psi^*(x,s)\Psi(x,s) \geq 0 \tag{1}$$

with normalization condition

$$\int_D \rho(x|s) d^4x = 1 \tag{2}$$

and D denotes spacetime volume. The probability amplitude Ψ is also referred to as the field or wave function of the system. It is specified to within the gauge transformation

$$\Psi(x,s) = \sqrt{\rho(x|s)}\, e^{i\xi(x,s)} \tag{3}$$

where ξ is a scalar function.

Equation (2) implies that a single particle is observable somewhere in space at some point in time. On the other hand, Equation (2) does not imply that the particle exists at all times. The conditional probability density $\rho(x|s)$ for a single particle can be expressed as the product of two conditional probabilities: $\rho(x|s) = \rho(x^1, x^2, x^3|x^0, s)\rho(x^0|s)$. The distribution $\rho(x^0|s)$ is the marginal probability density in time and is conditioned by the evolution parameter s. The particle cannot be detected anywhere in space when $\rho(x^0|s)$ is zero because the probability $\rho(x^0|s)$ of observing a particle at time x^0 given parameter s is zero. By contrast, when $\rho(x^0|s)$ is nonzero, there is a nonzero probability of observing a particle at time x^0 given parameter s [2,3,12].

The continuity equation for conservation of probability is

$$\frac{\partial \rho}{\partial s} + \frac{\partial}{\partial x_\mu}(\rho V^\mu) = 0 \tag{4}$$

Equation (4) can be combined with Equation (1) to give the probability flux

$$\rho V^\mu = -\frac{i\hbar}{2m}\left[\Psi^* \frac{\partial \Psi}{\partial x_\mu} - \Psi \frac{\partial \Psi^*}{\partial x_\mu}\right] - \frac{eA^\mu}{mc}\Psi^*\Psi \tag{5}$$

for a particle with mass m, charge e, four-velocity

$$V^\mu(x,s) = \frac{\hbar}{m}\frac{\partial \xi(x,s)}{\partial x_\mu} - \frac{e}{mc}A^\mu(x,s) \tag{6}$$

and electromagnetic four-vector potential A^μ.

The parametrized field equation

$$i\hbar\frac{\partial \Psi}{\partial s} = K\Psi \tag{7}$$

is constructed from Equations (1) to (6). The mass operator K has the form

$$K = \frac{\pi^{\mu}\pi_{\mu}}{2m} + V \tag{8}$$

with potential energy V and π^{μ} is the four-momentum operator with minimal coupling

$$\pi^{\mu} = \frac{\hbar}{i}\frac{\partial}{\partial x_{\mu}} - \frac{e}{c}A^{\mu} \tag{9}$$

The term A^{μ} is the four-vector potential. Equation (7) is called the Stueckelberg equation for a single particle. An alternative construction of an s-dependent field equation in terms of a set of fundamental postulates is known as the Stueckelberg-Horwitz-Piron (SHP) theory. The SHP theory is described by Horwitz [6].

In the pRQT formulation, the definition of the expectation value of an observable Ω is

$$\langle \Omega \rangle = \int \Psi^{*}\Omega\Psi dx \tag{10}$$

with the uncertainty principle over spacetime

$$|\Delta x_{\mu}|\,|\Delta p_{\mu}| \geq \frac{\hbar}{2} \tag{11}$$

Summation over repeated indices is not implied in Equation (11). The uncertainty principle for both energy and three-momentum is a consequence of the manifestly covariant probabilistic formulation.

3. Formulation of pRQT for an N-Body System

Extension of the probabilistic formulation for the single-particle system outlined in Section 2 to an N-body system makes it possible to combine an experimental system with a system that can function as an s-clock, that is, a system for monitoring the invariant parameter s. Field equations for the N-body system can be written as

$$i\hbar\frac{\partial \underline{\Psi}}{\partial s} = \underline{\underline{K_N}}\,\underline{\Psi} = \left\{\sum_{a=1}^{N}\frac{\pi_a^{\mu}\pi_{a\mu}}{2m_a}\underline{\underline{I}} + \underline{\underline{V}}\right\}\underline{\Psi} \tag{12}$$

where $\underline{\Psi}$ is a column vector with elements $\underline{\Psi}^{T} = [\psi_1, \psi_2, \ldots, \psi_{\Lambda}]$ and Λ labels associated internal variables. The Hermitian operator $\underline{\underline{K_N}}$ is interpreted as an N-body mass operator with a "minimal coupling" electromagnetic interaction, $\underline{\underline{I}}$ is the identity matrix, and $\underline{\underline{V}}$ represents non-electromagnetic interactions. The four-momentum operator in the bracketed term is

$$\pi_a^{\mu} = p_a^{\mu} - \frac{e}{c}A_a^{\mu} \tag{13}$$

with

$$p_a^{\mu} = \frac{\hbar}{i}\frac{\partial}{\partial x_{a\mu}} \tag{14}$$

and subscript a labels particle a for $1 \leq a \leq N$. Particle a exists at some time and some place within a $4N$ hypervolume D^{N}. The expectation value of an observable Ω in the N-body formalism is

$$\langle \Omega \rangle = \int_{D^N} \underline{\Psi}^+ \Omega \underline{\Psi} dx \tag{15}$$

where $\underline{\Psi}^+$ is the conjugate transpose of $\underline{\Psi}$ and the integral is over D^N.

The N-body formulation can be used to quantify parameter s by constructing an invariant evolution parameter s-clock. The system in Figure 1 consists of an experimental system with N_e bodies and an s-clock with N_c bodies. The experimental system should not significantly interact with the s-clock. Spacetime $\{x,t\}$ coordinates of experimental system 1 and the s-clock system 2 are used to determine parameters s_1, s_2 for system 1 and system 2, respectively. The procedure. is summarized below and discussed in more detail in the literature [3,13].

An N-body system consisting of an s-clock and a model of neutrino oscillations is constructed by considering a system with two particles. The experimental system consists of Particle 1 interacting with potential V_I. Particle 2 is a free, scalar particle that serves as a simple s-clock. Particles 1 and 2 do not interact with each other. The two-particle system is represented by the field equation

$$\left\{i\hbar\frac{\partial}{\partial s} - \left[\frac{p_1^\mu p_{1\mu}}{2m_1} + \frac{p_2^\mu p_{2\mu}}{2m_2} + V_I\right]\right\}\psi(1,2,s) = 0 \tag{16}$$

where the s-dependent eigenfunction $\psi(1,2,s)$ of the two-particle system applies to both particles 1 and 2. The eigenfunction $\psi(1,2,s)$ has a set of 4-space components y_1^μ, y_2^μ for each particle.

A solution to Equation (16) is obtained by writing $\psi(1,2,s)$ as the product of single particle eigenfunctions

$$\psi(1,2,s) = \psi(1,s)\psi(2,s) \tag{17}$$

Substituting Equation (17) into Equation (16) lets us separate Equation (16) into the two equations

$$\left\{i\hbar\frac{\partial}{\partial s} - \left[\frac{p_1^\mu p_{1\mu}}{2m_1} + V_I + \alpha_R\right]\right\}\psi(1,s) = 0 \tag{18}$$

and

$$\left\{i\hbar\frac{\partial}{\partial s} - \left[\frac{p_2^\mu p_{2\mu}}{2m_2} - \alpha_R\right]\right\}\psi(2,s) = 0 \tag{19}$$

with separation constant α_R. Particles 1 and 2 are physically independent so $\alpha_R = 0$.

Equation (19) with $\alpha_R = 0$ is the free particle equation

$$i\hbar\frac{\partial\psi_f(2,s)}{\partial s} = \left[-\frac{\hbar^2}{2m_2}\frac{\partial^2}{\partial x_2^\mu \partial x_{2\mu}}\right]\psi_f(2,s) \tag{20}$$

with the particular solution

$$\psi_f(2,s) = \eta_f^{1/2} exp\left[-\frac{i\hbar}{2m_2}(k_2^\mu k_{2\mu})s + ik_{2\mu}x_2^\mu\right] \tag{21}$$

and normalization constant η_f.

The most probable trajectory of particle 2 in the classical limit of negligible dispersion is

$$(\delta s)^2 = (s - s_0)^2 = \frac{1}{c^2}\delta\langle x_2^\mu\rangle\delta\langle x_{2\mu}\rangle \tag{22}$$

where s_0 is the invariant evolution parameter when the rest frame clocks of particles 1 and 2 are calibrated. If we synchronize the clocks at $s_0 = 0$, Equation (22) becomes

$$s^2 = \frac{1}{c^2}\delta\langle x_2^\mu\rangle\delta\langle x_{2\mu}\rangle \tag{23}$$

Now assume the motion of the free particle is linear and neglect statistical variations to obtain

$$s^2 = \delta t^2 - \frac{\delta x^2}{c^2} = \delta t^2[1 - \beta^2] \tag{24}$$

where $\{x_2^0, x_2^1\} = \{t, x\}$, and

$$\beta = \frac{v}{c} \text{ with } v \equiv \frac{\delta x}{\delta t} \tag{25}$$

The distance δx traveled by particle 2 in the interval δt can be written as L so that Equation (24) becomes

$$s = \frac{L}{v}[1 - \beta^2]^{1/2} = \frac{L}{c}\frac{[1 - \beta^2]^{1/2}}{\beta} \tag{26}$$

The invariant evolution parameter s is quantified by measuring the spacetime trajectory of particle 2 and using the resulting value of s in Equation (18) associated with particle 1. This s-clock has been used to model neutrino oscillations [8,11].

4. Mass Basis and Flavor Basis

Two theories of transitions between three neutrino flavor states $\{|\nu_\alpha\rangle; \alpha = e, \mu, \tau\}$ composed of up to three mass states $\{|\nu_j\rangle; j = 1, 2, 3\}$ are compared here. The basis of mass states $\{|\nu_j\rangle; j = 1, 2, 3\}$ is related to the basis of flavor states $\{|\nu_\alpha\rangle; \alpha = e, \mu, \tau\}$ by a unitary transformation:

$$\begin{bmatrix}|\nu_e\rangle \\ |\nu_\mu\rangle \\ |\nu_\tau\rangle\end{bmatrix} = U \begin{bmatrix}|\nu_1\rangle \\ |\nu_2\rangle \\ |\nu_3\rangle\end{bmatrix} \tag{27}$$

where the mass and flavor states are written as 3-component column vectors. The elements of the unitary matrix are written as

$$U = \begin{bmatrix} u_{e1} & u_{e2} & u_{e3} \\ u_{\mu 1} & u_{\mu 2} & u_{\mu 3} \\ u_{\tau 1} & u_{\tau 2} & u_{\tau 3}\end{bmatrix} \tag{28}$$

The inverse of the unitary matrix U is the conjugate transpose of U, thus

$$U^{-1} = (U^*)^T \tag{29}$$

with elements

$$u_{\alpha j}^{-1} = u_{j\alpha}^{*}\ ;\ j = 1,2,3 \text{ and } \alpha = e, \mu, \tau \tag{30}$$

The expanded form of the unitary transformation is

$$\begin{aligned} |\nu_e\rangle &= u_{e1}|\nu_1\rangle + u_{e2}|\nu_2\rangle + u_{e3}|\nu_3\rangle \\ |\nu_\mu\rangle &= u_{\mu 1}|\nu_1\rangle + u_{\mu 2}|\nu_2\rangle + u_{\mu 3}|\nu_3\rangle \\ |\nu_\tau\rangle &= u_{\tau 1}|\nu_1\rangle + u_{\tau 2}|\nu_2\rangle + u_{\tau 3}|\nu_3\rangle \end{aligned} \tag{31}$$

A mass basis state satisfies the temporal evolution equation

$$T|\nu_j\rangle = i\hbar\frac{\partial}{\partial\tau}|\nu_j\rangle = T_j|\nu_j\rangle \tag{32}$$

where T_j is the eigenvalue of the temporal operator $T = i\hbar\frac{\partial}{\partial\tau}$, and τ is the temporal evolution parameter. In the standard model, $(T_j)_{Std} = E_j$ where E_j is the energy of state j, and $(\tau)_{Std} = t$ with coordinate time t. In the pRQT model, $(T_j)_{pRQT} = K_j$ where K_j is the eigenvalue of state j for the mass operator K, and $(\tau)_{pRQT} = s$ with invariant evolution parameter s. Equation (32) has the formal solution

$$\begin{bmatrix} |\nu_1\rangle \\ |\nu_2\rangle \\ |\nu_3\rangle \end{bmatrix} = \begin{bmatrix} e^{-i\frac{T_1\tau}{\hbar}} & 0 & 0 \\ 0 & e^{-i\frac{T_2\tau}{\hbar}} & 0 \\ 0 & 0 & e^{-i\frac{T_3\tau}{\hbar}} \end{bmatrix} \begin{bmatrix} |\nu_1(0)\rangle \\ |\nu_2(0)\rangle \\ |\nu_3(0)\rangle \end{bmatrix} \tag{33}$$

where $\nu_j(0)$ is mass state j at $\tau = 0$.

5. Neutrino Mass Matrix Model

An analysis of the competing formalisms for vacuum-favor mixing of neutrinos within the context of the standard and pRQT models requires a basis for comparison. In this case, we seek estimates of neutrino masses based on current experimental data. To achieve this goal, we present a procedure for estimating neutrino masses from a 3×3 neutrino mass matrix M_ν using a procedure introduced by Damanik [14]. Damanik's procedure is a method for obtaining phenomenological estimates of neutrino masses from experimental estimates of mass-squared differences and mixing parameters.

Damanik [14] constructed an unperturbed neutrino mass matrix $M_{\nu 0}$

$$M_{\nu 0} = \begin{bmatrix} P & Q & Q \\ Q & P & Q \\ Q & Q & P \end{bmatrix} \tag{34}$$

that is non-singular and invariant with respect to a cyclic permutation of neutrino states, that is, $\nu_1 \rightarrow \nu_2 \rightarrow \nu_3 \rightarrow \nu_1$. Damanik used the seesaw mechanism [Gell-Mann, et al., 1979; Yanagida, et al., 1979] to help motivate the form of $M_{\nu 0}$. The presentation here focuses on mathematical arguments rather than theoretical motivations to minimize the dependence of the form of the neutrino mass matrix on a specific paradigm.

Damanik [14] determined the matrix elements P, Q by finding the eigenvalues of $M_{\nu 0}$, relating the eigenvalues to neutrino masses, and then showing that the resulting masses did not correctly predict observed mass squared differences Δm_{ij}^2 where $\Delta m_{ij}^2 = m_i^2 - m_j^2, (i,j = 1,2,3)$. To resolve these problems, Damanik introduced a parameter δ as a perturbation of diagonal elements of the neutrino mass matrix M_ν:

$$M_\nu = \begin{bmatrix} P+2\delta & Q & Q \\ Q & P-\delta & Q \\ Q & Q & P-\delta \end{bmatrix} \tag{35}$$

The form of the perturbed neutrino mass matrix M_ν includes the requirement that M_ν has the same trace as the unperturbed neutrino mass matrix $M_{\nu 0}$, thus $Tr(M_{\nu 0}) = Tr(M_\nu) = 3P$. The eigenvalues $\{\beta_1, \beta_2, \beta_3\}$ of the perturbed neutrino mass matrix M_ν are

$$\begin{aligned} \beta_1 &= P + \frac{Q}{2} + \frac{\delta}{2} - \frac{\sqrt{9\delta^2 - 6Q\delta + 9Q^2}}{2} \\ \beta_2 &= P + \frac{Q}{2} + \frac{\delta}{2} + \frac{\sqrt{9\delta^2 - 6Q\delta + 9Q^2}}{2} \\ \beta_3 &= P - Q - \delta \end{aligned} \tag{36}$$

The value of δ is obtained by finding the angle θ that diagonalizes M_ν.

The angle θ is the angle that relates the mass basis $\{|\nu_j\rangle; j = 1, 2, 3\}$ to the flavor basis $\{|\nu_\alpha\rangle; \alpha = e, \mu, \tau\}$ by the unitary transformation in Equation (27). The unitary matrix used by Damanik [14] to relate the mass basis to the flavor basis is

$$U = \begin{bmatrix} \cos\theta & -\sin\theta & 0 \\ \frac{\sin\theta}{\sqrt{2}} & \frac{\cos\theta}{\sqrt{2}} & -\frac{1}{\sqrt{2}} \\ \frac{\sin\theta}{\sqrt{2}} & \frac{\cos\theta}{\sqrt{2}} & \frac{1}{\sqrt{2}} \end{bmatrix} \tag{37}$$

with the inverse

$$U^{-1} = \begin{bmatrix} \cos\theta & \frac{\sin\theta}{\sqrt{2}} & \frac{\sin\theta}{\sqrt{2}} \\ -\sin\theta & \frac{\cos\theta}{\sqrt{2}} & \frac{\cos\theta}{\sqrt{2}} \\ 0 & -\frac{1}{\sqrt{2}} & \frac{1}{\sqrt{2}} \end{bmatrix} \tag{38}$$

The perturbation parameter δ is obtained from the relation

$$\tan^2(2\theta) = \frac{8Q^2}{(Q-3\delta)^2} \tag{39}$$

where the angle θ and the associated unitary matrix U diagonalize M_ν. The solution of Equation (39) for δ gives

$$\delta = \varepsilon Q, \varepsilon = \frac{[\tan(2\theta) - \sqrt{8}]}{3\tan(2\theta)} \tag{40}$$

The term ε is calculated given a value of θ. In this case, the perturbation parameter δ is proportional to θ.

Neutrino masses are given by the equations

$$\begin{aligned} m_1 &= P + \frac{Q}{2} + \frac{\delta}{2} - \frac{\sqrt{9\delta^2 - 6Q\delta + 9Q^2}}{2} \\ m_2 &= P + \frac{Q}{2} + \frac{\delta}{2} + \frac{\sqrt{9\delta^2 - 6Q\delta + 9Q^2}}{2} \\ m_3 &= P - Q - \delta \end{aligned} \tag{41}$$

Damanik's procedure yields a set of neutrino masses that is an inverted hierarchy, thus

$$|m_3| < |m_1| < |m_2| \tag{42}$$

6. Algorithm for Calculating Neutrino Masses

The equations in Damanik's [14] procedure presented above can be combined with experimental measurements of mass squared differences and mixing angles to provide a phenomenological estimate of the variables $\{P, Q, \delta\}$ and corresponding neutrino masses $\{m_1, m_2, m_3\}$. We begin by writing a simplified form of the neutrino mass equations. If we use the proportionality relationship between δ and θ in Equation (40), neutrino masses may be written in the form

$$\begin{aligned} m_1 &= P + a_1 Q \\ m_2 &= P + a_2 Q \\ m_3 &= P + a_3 Q \end{aligned} \tag{43}$$

where

$$\begin{aligned} a_1 Q &= \frac{Q}{2} + \frac{\delta}{2} - \frac{\sqrt{9\delta^2 - 6Q\delta + 9Q^2}}{2} \\ a_2 Q &= \frac{Q}{2} + \frac{\delta}{2} + \frac{\sqrt{9\delta^2 - 6Q\delta + 9Q^2}}{2} \\ a_3 Q &= -Q - \delta \end{aligned} \tag{44}$$

Substituting $\delta = \varepsilon Q$ into Equation (44) gives $\{a_1, a_2, a_3\}$ as functions of ε:

$$\begin{aligned} a_1 &= \frac{1}{2}\left[1 + \epsilon - \sqrt{9\varepsilon^2 - 6\varepsilon + 9}\right] \\ a_2 &= \frac{1}{2}\left[1 + \epsilon + \sqrt{9\varepsilon^2 - 6\varepsilon + 9}\right] \\ a_3 &= -(1 + \varepsilon) \end{aligned} \tag{45}$$

We use the mass squared differences Δm_{ij}^2 and the relations in Equation (43) to solve for P, Q.

The mass squared differences are

$$\begin{aligned} \Delta m_{21}^2 &= m_2^2 - m_1^2 = (P + a_2 Q)^2 - (P + a_1 Q)^2 \\ \Delta m_{32}^2 &= m_3^2 - m_2^2 = (P + a_3 Q)^2 - (P + a_2 Q)^2 \end{aligned} \tag{46}$$

with the simplified forms

$$\Delta m_{21}^2 = 2QP(a_2 - a_1) + Q^2(a_2^2 - a_1^2) \tag{47}$$

and

$$\Delta m_{32}^2 = 2QP(a_3 - a_2) + Q^2(a_3^2 - a_2^2) \tag{48}$$

The variable Q is obtained by rearranging Equations (47) and (48) so that the QP term is on the right-hand side:

$$\Delta m_{21}^2 - Q^2(a_2^2 - a_1^2) = 2QP(a_2 - a_1) \tag{49}$$

and

$$\Delta m_{32}^2 - Q^2(a_3^2 - a_2^2) = 2QP(a_3 - a_2) \tag{50}$$

Dividing Equation (49) by (50), rearranging and solving for Q^2 gives

$$Q^2 = \frac{\Delta m_{21}^2 - \frac{(a_2 - a_1)}{(a_3 - a_2)}\Delta m_{32}^2}{(a_2^2 - a_1^2) - \frac{(a_2 - a_1)}{(a_3 - a_2)}(a_3^2 - a_2^2)} \tag{51}$$

The variable P is obtained by rearranging Equations (47) and (48) so that the Q^2 term is on the right-hand side:

$$\Delta m_{21}^2 - 2QP(a_2 - a_1) = Q^2(a_2^2 - a_1^2) \tag{52}$$

and

$$\Delta m_{32}^2 - 2QP(a_3 - a_2) = Q^2(a_3^2 - a_2^2) \tag{53}$$

Dividing Equation (52) by (53), rearranging and solving for P gives

$$P = \frac{\Delta m_{21}^2 - \frac{(a_2^2 - a_1^2)}{(a_3^2 - a_2^2)}\Delta m_{32}^2}{2Q\left[(a_2 - a_1) - \frac{(a_2^2 - a_1^2)}{(a_3^2 - a_2^2)}(a_3 - a_2)\right]} \tag{54}$$

An algorithm for calculating neutrino masses using the above relationships is summarized in Table 1. It assumes that values of $\Delta m_{21}^2, \Delta m_{32}^2, \theta$ are available.

Table 1 Algorithm for Calculating Neutrino Masses

Calculate	Using	Equation
ε	θ	40
$\{a_1, a_2, a_3\}$	ε	45
Q^2	$\{a_1, a_2, a_3\}, \Delta m_{21}^2, \Delta m_{32}^2$	51
P	$Q, \{a_1, a_2, a_3\}, \Delta m_{21}^2, \Delta m_{32}^2$	54
$\{m_1, m_2, m_3\}$	$P, Q, \{a_1, a_2, a_3\}$	43

7. Comparison of Neutrino Mass Models Using 2020 Data

Table 2 validates the algorithm using the data in the Rusov-Vlasenko [11] neutrino mass calculation. Rusov and Vlasenko [11] compared the standard and pRQT models by writing

$$\alpha_{pRQT} \approx 2\alpha_{Std} \tag{55}$$

The angle θ is given by $\theta = \theta_{12}$. The masses are inverted, that is, $|m_3| < |m_1| < |m_2|$. The sum of the absolute values of the neutrino masses is less than 0.6 eV.

Table 2 Verify Algorithm with Rusov-Vlasenko [11] Results		
	Algorithm Results	
Rusov-Vlasenko [11] Data	Standard Model	pRQT Model
$\Delta m_{21}^2 \times 10^5, eV^2$	7.50	15.00
$\Delta m_{32}^2 \times 10^5, eV^2$	-2.32	-4.64
$tan^2(\theta)$	0.452	0.452
Neutrino Masses	Standard Model	pRQT Model
$\|m_1\|, eV$	0.130855	0.185056
$\|m_2\|, eV$	0.131141	0.185461
$\|m_3\|, eV$	0.121975	0.172499
$\|m_1\| + \|m_2\| + \|m_3\|, eV$	0.38397	0.54302

Table 3 repeats the Rusov-Vlasenko [11] calculations of neutrino mass using updated data from Zyla, et al. [1]. The angle θ is given by $\theta = \theta_{12}$ and the masses are inverted. The sum of the absolute values of the neutrino masses is less than 0.6 eV.

Table 3 Rusov-Vlasenko [1] Calculations Updated with Zyla, et al. [1] Data		
	Algorithm Results	
Zyla, et al. [1] Data	Standard Model	pRQT Model
$\Delta m_{21}^2 \times 10^5, eV^2$	7.53	15.06
$\Delta m_{32}^2 \times 10^5, eV^2$	-2.546	-5.092
$tan^2(\theta)$	0.443	0.443
Neutrino Masses	Standard Model	pRQT Model
$\|m_1\|, eV$	0.125714	0.177787
$\|m_2\|, eV$	0.126013	0.178210
$\|m_3\|, eV$	0.115470	0.163300
$\|m_1\| + \|m_2\| + \|m_3\|, eV$	0.36720	0.51930

The relationship in Equation (55) used by Rusov-Vlasenko [11] is not the same as the pRQT result summarized in Appendix A. The ratio

$$\frac{\alpha_{Std}}{\alpha_{pRQT}} \approx \frac{m_2^2 - m_1^2}{m_\nu(m_2 - m_1)} = \frac{m_1 + m_2}{m_\nu} \approx 2 \tag{56}$$

is simplified by assuming that $m_\nu \approx m_1 \approx m_2$ so that

$$m_\nu \approx \frac{m_1 + m_2}{2} \tag{57}$$

Equation (57) becomes

$$\alpha_{pRQT} = \frac{m_\nu(m_2 - m_1)c^4}{4\hbar E_\nu}\frac{L}{c}\frac{1}{\beta} \approx \frac{\left(\frac{m_1+m_2}{2}\right)(m_2 - m_1)c^4}{4\hbar E_\nu}\frac{L}{c}\frac{1}{\beta} = \frac{1}{2}\frac{(m_2^2 - m_1^2)c^4}{4\hbar E_\nu}\frac{L}{c}\frac{1}{\beta} \tag{58}$$

Recognizing that $\beta \approx 1$ for an ultrarelativistic neutrino and $\alpha_{Std} = \frac{(m_2^2 - m_1^2)c^4}{4\hbar E_\nu}\frac{L}{c}$ from Equation (A.8) gives

$$\alpha_{pRQT} = \frac{1}{2}\alpha_{Std} \tag{59}$$

This can be related to observations by halving the mass squared difference $\Delta m_{ji}^2 = m_j^2 - m_i^2$, thus

$$\left(\Delta m_{ji}^2\right)_{pRQT} = \frac{1}{2}\left(\Delta m_{ji}^2\right)_{Std} \tag{60}$$

By contrast, Rusov-Vlasenko [11] doubled the mass-squared difference.

Table 4 presents neutrino mass results obtained using Equation (60) and updated data from Zyla, et al. [1]. The angle θ is given by $\theta = \theta_{12}$ and the masses are inverted. The sum of the absolute values of the neutrino masses is less than 0.6 eV.

Table 4 Neutrino Masses Updated with Equation (60) and Zyla, et al. [1] Data

	Algorithm Results	
Zyla, et al. [1] Data	Standard Model	pRQT Model
$\Delta m_{21}^2 \times 10^5, eV^2$	7.53	3.77
$\Delta m_{32}^2 \times 10^5, eV^2$	-2.546	-1.27
$tan^2(\theta)$	0.443	0.443
Neutrino Masses	Standard Model	pRQT Model
$\lvert m_1 \rvert, eV$	0.125714	0.088893
$\lvert m_2 \rvert, eV$	0.126013	0.089105
$\lvert m_3 \rvert, eV$	0.115470	0.081650
$\lvert m_1 \rvert + \lvert m_2 \rvert + \lvert m_3 \rvert, eV$	0.36720	0.25965

8. Discussion

The inclusion of two temporal variables in parametrized Relativistic Quantum Theory (pRQT) yields a theory that is significantly different from theories that rely on a single temporal variable, namely the time coordinate of spacetime. The pRQT temporal variables are the time coordinate of spacetime, and an invariant evolution parameter. The difference between the temporal dependence of the standard model and the pRQT model can be directly observed in models of neutrino oscillations. This paper used 2020 data to compare neutrino mass results from the standard model of three-flavor neutrino oscillations and the pRQT model of three-flavor neutrino oscillations. Standard model neutrino masses shown in

Table 4 are in the range 0.11 eV to 0.13 eV, while pRQT neutrino masses are in the range 0.08 eV to 0.09 eV.

Neutrino mass model differences are significant for neutrino oscillation measurements. For example, the Karlsruhe Tritium Neutrino (KATRIN) Collaboration has a direct method of measuring neutrino mass based on beta decay of tritium into helium-3, an electron, and an electron antineutrino [17,18]. The KATRIN direct method has the advantage of being model independent, which makes it suitable for comparing neutrino mass predictions based on the standard model and pRQT model of three-flavor neutrino oscillations. One possible problem is that KATRIN is not currently sensitive to neutrino masses less than 0.2 eV [19,20]. The 0.2 eV limit is greater than the neutrino masses estimated here.

The neutrino mass differences shown here are dependent on the specified neutrino mass matrix and choice of variables used to conduct calculations. A more definitive comparison of competing models will depend on future studies of alternative mass matrices and associated algorithms to further incorporate all observations from all relevant experiments.

Appendix A. Outline of the Two-Flavor pRQT Model of Neutrino Oscillations

The two-flavor pRQT model of neutrino oscillations [8] is outlined here. The evolution equation in pRQT for a state may be written in terms of the evolution parameter s as

$$i\hbar \frac{\partial}{\partial s}|v_j\rangle = K_j|v_j\rangle \tag{A.1}$$

where K_j is the eigenvalue of the mass operator for mass state j. The evolution parameter dependent solution of Equation (A.1) in the mass basis for two mass states is

$$\begin{bmatrix} |v_1(s)\rangle \\ |v_2(s)\rangle \end{bmatrix} = \begin{bmatrix} e^{-iK_1 s/\hbar} & 0 \\ 0 & e^{-iK_2 s/\hbar} \end{bmatrix} \begin{bmatrix} |v_1(0)\rangle \\ |v_2(0)\rangle \end{bmatrix} \tag{A.2}$$

where

$$K_j = \hbar^2 k_j^\mu k_{j\mu}/2m_j = \hbar^2 \left[(\omega_j/c)^2 - k_j \cdot k_j\right]/2m_j \tag{A.3}$$

In pRQT, the components of the energy-momentum four-vector k_j^μ are observables and the mass m_j is a function of statistical values of k_j^μ.

In the flavor oscillation process $v_e \to v_\mu$, we begin with a pure beam of electron neutrino v_e particles and calculate the probability for forming muon neutrino v_μ particles. The pRQT result for the probability of forming the final state v_μ from the initial state v_e is

$$\begin{aligned} P_{pRQT}(v_e \to v_\mu) &= sin^2\, 2\,\theta\, sin^2 \left\{\frac{(m_2 - m_1)c^2}{4\hbar} s\right\} \\ &\equiv sin^2\, 2\,\theta\, sin^2\, \alpha_{pRQT} \end{aligned} \tag{A.4}$$

where s is temporal duration measured by an s-clock.

Flavor oscillations may be described by quantifying the behavior of two particles as outlined in Section 3. One particle propagates without interaction or oscillation from the source to the detector and serves as a "clock" for the scalar evolution parameter s. The other particle is the oscillating particle. In this application, the source and detector are separated by a distance L. The most probable trajectory of the non-interacting s-clock particle is

$$s^2 = (\delta t)^2 - \frac{(\delta x)^2}{c^2} = (\delta t)^2[1-\beta^2], \beta = \frac{v}{c}, v = \frac{\delta x}{\delta t} \tag{A.5}$$

The distance δx traveled by the s-clock particle in the interval δt is L, so we obtain

$$s = \frac{L}{c}\frac{[1-\beta^2]^{1/2}}{\beta}, \delta x = L \tag{A.6}$$

Substituting Equation (A.6) into Equation (A.4) gives

$$P_{pRQT}(v_e \to v_\mu) = sin^2 2\,\theta\, sin^2 \alpha_{pRQT}, \quad \alpha_{pRQT} = \frac{(m_2 - m_1)c^2}{4\hbar}\frac{L}{c}\frac{[1-\beta^2]^{1/2}}{\beta} \tag{A.7}$$

The result for the standard theory is

$$P_{Std}(v_e \to v_\mu) = sin^2 2\,\theta\, sin^2 \alpha_{Std}, \quad \alpha_{Std} = \frac{(m_2^2 - m_1^2)c^4}{4\hbar E_v}\frac{L}{c} \tag{A.8}$$

where E_v is the energy of the ultrarelativistic incident neutrino

$$E_v = \frac{m_v c^2}{[1-\beta^2]^{1/2}} \tag{A.9}$$

We combine Equations (A.7) and (A.9) and rearrange to simplify comparison with Equation (A.8):

$$P_{pRQT}(v_e \to v_\mu) = sin^2 2\,\theta\, sin^2 \alpha_{pRQT}, \quad \alpha_{pRQT} = \frac{(m_2 - m_1)c^2}{4\hbar}\frac{L}{c}\frac{m_v c^2}{E_v \beta} = \frac{m_v(m_2 - m_1)c^4}{4\hbar E_v}\frac{L}{c}\frac{1}{\beta} \tag{A.10}$$

The ratio of the dynamical factors $\alpha_{pRQT}, \alpha_{Std}$ is

$$\frac{\alpha_{Std}}{\alpha_{pRQT}} = \frac{m_2^2 - m_1^2}{m_v(m_2 - m_1)}\beta = \frac{m_1 + m_2}{m_v}\beta \tag{A.11}$$

The ratio of probabilities in Equations (A.8) and (A.10) is

$$\frac{P_{Std}}{P_{pRQT}} = \frac{sin^2 \alpha_{Std}}{sin^2 \alpha_{pRQT}} \tag{A.12}$$

Comparing P_{pRQT}, P_{Std} and the dynamical factors $\alpha_{pRQT}, \alpha_{Std}$ shows that the pRQT model and the conventional theory have the same dependence on the flavor mixing angle θ, but their dependence on dynamical factors differs significantly. If the mass difference between neutrino mass and flavor states is very small and the neutrinos are ultrarelativistic, then $(m_1 + m_2)/m_v \approx 2$ and $\beta \approx 1$. The ratio of dynamical factors $\alpha_{Std}/\alpha_{pRQT} \approx 2$ in this case.

References

[1] Zyla, P.A., et al. Review of Particle Physics, Particle Data Group, *Prog. Theor. Exp. Phys.* 2020, 083C01 (2020) and 2021 update; website https://pdg.lbl.gov/ accessed October 9, 2021
[2] Fanchi, J.R. *Parametrized Relativistic Quantum Theory*; Kluwer: Dordrecht, The Netherlands, 1993
[3] Fanchi, J.R. Manifestly Covariant Quantum Theory with Invariant Evolution Parameter in Relativistic Dynamics, *Found. Phys.* 2011, 41: 4–32, DOI 10.1007/s10701-009-9371-0
[4] Pavšič, M. *The Landscape of Theoretical Physics: A Global View*; Kluwer Academic Publishers: Dordrecht, 2001
[5] Pavšič, M. *Stumbling Blocks Against Unification*, World Scientific: Singapore, 2020
[6] Horwitz, L.P. *Relativistic Quantum Mechanics*, Springer: Dordrecht, 2015
[7] Land, M., Horwitz, L.P. *Relativistic Classical Mechanics and Electrodynamics*, Morgan & Claypool, 2020: Williston, Vermont; www.morganclaypool.com
[8] Fanchi, J.R. The Mass Operator and Neutrino Oscillations, *Found. Phys.* 1998, 28, 1521-1528
[9] Fanchi, J.R. Parametrized Relativistic Dynamical Framework for Neutrino Oscillations, *J. Phys.: Conf. Series* 2017, 845, 1-13, doi : 10.1088/1742-6596/845/1/012027
[10] Fanchi, J.R. Neutrino Flavor Transitions as Mass State Transitions, *Symmetry* 2019, 11(8), 948-956
[11] Rusov, V.D., Vlasenko, D.S. Quantization in relativistic classical mechanics: the Stueckelberg equation, neutrino oscillation and large-scale structure of the Universe, *J. Phys.: Conf. Series* 2012, 361, 1-15, doi:10.1088/1742-6596/361/1/012033
[12] Rusov, V.D., Vlasenko, D.S. Quantization in relativistic classical mechanics: the Stueckelberg equation, neutrino oscillation and large-scale structure of the Universe, *J. Phys.: Conf. Series* 2012, 361, 1-15, doi:10.1088/1742-6596/361/1/012033
[13] Fanchi, J.R. Parametrizing relativistic quantum mechanics, *Phys. Rev. A* 1986, 34, 1677-1681
[14] Damanik, A. Perturbed invariant under a cyclic permutation with trace of neutrino mass matrix remain constant, *Mod. Phys. Lett.* A 2011, 26 (08), 567-574
[15] Gell-Mann, M., Ramond, P., Slansky, R. Complex Spinors and Unified Theories, in *Supergravity*, P. van Nieuwenhuizen and D.Z. Freeman eds., North-Holland Publ. Co., Amsterdam, 315-321, 1979
[16] Yanagida, T. Horizontal Gauge Symmetry and Masses of Neutrinos, in *Proceedings of the Workshop on Unified Theory and the Baryon Number of the Universe*, O. Sawada and A. Sugamoto eds., KEK, Tsukuba, Japan, 95-99, 1979
[17] Brugnera, R. Homing in on the neutrino mass, *Viewpoint*, 25 November 2019, *Physics* 12, 129
[18] KATRIN Collaboration. The design, construction, and commissioning of the KATRIN experiment, *J. Inst.* 2021, 16, T08015
[19] Nucciotti, A. Still too small to be measured, *Nature Physics* Februa ry 2022, 18, 124-136
[20] KATRIN Collaborationl. Direct neutrino-mass measurement with sub-electronvolt sensitivity. *Nat ure Physics* February 2022, 18, 160–166

Clifford odd and even objects offer description of internal space of fermions and bosons, respectively, opening new insight into the second quantization of fields

Norma Susana Mankoč Borštnik

Department of Physics, University of Ljubljana
SI-1000 Ljubljana, Slovenia

E-mail: norma.mankoc@fmf.uni-lj.si

Abstract. In a long series of works the author has demonstrated that the model named the *spin-charge-family* theory offers the explanation for all in the *standard model* assumed properties of the fermion and boson fields, as well as for many of their so far observed properties if the space-time is $\geq (13+1)$ while fermions interact with gravity only. In this talk, I briefly report on the so far achievements of the theory. The main contribution demonstrates the offer of the Clifford odd and even objects for the description of the internal spaces of fermion (Clifford odd) and boson (Clifford even) fields, which is opening up a new understanding of the second quantization postulates for the fermion and boson fields: The "basis vectors" determined by the Clifford odd objects demonstrate all the properties of the internal space of fermions and transfer their anti-commutativity to their creation and annihilation operators, while the "basis vectors" determined by the Clifford even objects demonstrate all the properties of the internal space of boson fields and transfer their commutativity to their creation and annihilation operators. The toy model with $d = (5+1)$ illustrates the statements.

1. Introduction

The *standard model* (improved with neutrinos having non-zero masses) has been experimentally confirmed without raising any serious doubts so far on its assumptions, which remain unexplained, waiting for the explanation.

The *standard model* has in the literature several explanations, mostly with many new not explained assumptions. The most popular are the grand unifying theories ([1, 2, 3, 4, 5, 6] and many others.

Among the questions which must be answered are:
i. Where do fermions, quarks and leptons, with their families included, originate and why do their internal spaces differ so much from the internal spaces of boson fields?
ii. Why are charges of quarks and leptons so different, and why have the left handed family members so different charges from the right handed ones?
iii. Where do antiquarks and antileptons originate?
iv. Where do families of quarks and leptons originate? How many families exist?
v. Why do family members – quarks and leptons — manifest such different masses if they all

Content from this work may be used under the terms of the Creative Commons Attribution 3.0 licence. Any further distribution of this work must maintain attribution to the author(s) and the title of the work, journal citation and DOI.
Published under licence by IOP Publishing Ltd

start as massless?
vi. What is the origin of boson fields, of vector fields which are the gauge fields of fermions, and of the Higgs's scalars and the Yukawa couplings.
vii. How are scalar fields connected with the origin of families? How many scalar fields determine properties of the so far (and others possibly be) observed fermions and masses of weak bosons?
viii. Do possibly exist also scalar fields with the colour charges in the fundamental representation (like the Higgs scalars are doublets in the weak and hyper charge), and where, if they are, do they manifest?
ix. Can all boson fields, with the scalar fields included, have a common origin?
x. Where does the *dark matter* originate?
xi. Where does the "ordinary" matter-antimatter asymmetry originate?
xii. Where does the dark energy originate, and why is it so small?
xiii. How can we understand the second quantized fermion and boson fields?
xiv. What is the dimension of space? $(3+1)$?, $((d-1)+1)$?, ∞?
xv. Are all the fields indeed second quantized with gravity included? Furthermore, are all the systems second quantized (although we can treat them in simplified versions, like it is the first quantization and even the classical treatment), with the black holes included?
xvi. What is happening in odd dimensional spaces?
xvii. And many others.

In a long series of works ([7, 8, 9, 21, 23, 25, 26, 27, 29, 30] and the references therein), the author of this contribution has succeeded, together with collaborators, in finding the answer to the above, and also other open questions of the *standard model* and also to several open cosmological questions, with the model, named the *spin-charge-family* theory.

The theory assumes that the space has more than (3+1) dimensions; it must have $d \geq (13+1)$, so that the subgroups of the $SO(13,1)$ group, describing the internal space of fermions by the superposition of odd products of the Clifford objects γ^a's, manifest from the point of view of $d=(3+1)$-dimensional space the spins, handedness and charges assumed for massless fermions in the *standard model.* Correspondingly each irreducible representation of the $SO(13,1)$ group carrying the quantum numbers of quarks and leptons and antiquarks and antileptons, represents one of the families of fermions, the quantum numbers of which are determined by the second kind of the Clifford objects, by $\tilde{\gamma}^a$ (by $\tilde{S}^{ab}$ $(=\frac{i}{4}\{\tilde{\gamma}^a, \tilde{\gamma}^b\}_-$, indeed).

Fermions interact in $d=(13+1)$ with gravity only, with vielbeins (the gauge fields of momenta) and the two kinds of spin connection fields, the gauge fields of the two kinds of the Lorentz transformations in the internal space of fermions, of $S^{ab}=\frac{i}{4}\{\gamma^a, \gamma^b\}_-$ and of $\tilde{S}^{ab}$ $(=\frac{i}{4}\{\tilde{\gamma}^a, \tilde{\gamma}^b\}_-$.

The theory assumes a simple starting action ([10] and the references therein) for the second quantized massless fermion and antifermion fields, and the corresponding massless boson fields in $d=2(2n+1)$-dimensional space

$$\begin{aligned}
\mathcal{A} &= \int d^dx\, E\, \frac{1}{2}\,(\bar{\psi}\,\gamma^a p_{0a}\psi) + h.c. + \\
&\quad \int d^dx\, E\, (\alpha\, R + \tilde{\alpha}\, \tilde{R})\,, \\
p_{0\alpha} &= p_\alpha - \frac{1}{2}S^{ab}\omega_{ab\alpha} - \frac{1}{2}\tilde{S}^{ab}\tilde{\omega}_{ab\alpha}\,, \\
p_{0a} &= f^\alpha{}_a p_{0\alpha} + \frac{1}{2E}\,\{p_\alpha, E f^\alpha{}_a\}_-\,, \\
R &= \frac{1}{2}\,\{f^{\alpha[a}f^{\beta b]}\,(\omega_{ab\alpha,\beta} - \omega_{ca\alpha}\,\omega^c{}_{b\beta})\} + h.c.\,, \\
\tilde{R} &= \frac{1}{2}\,\{f^{\alpha[a}f^{\beta b]}\,(\tilde{\omega}_{ab\alpha,\beta} - \tilde{\omega}_{ca\alpha}\,\tilde{\omega}^c{}_{b\beta})\} + h.c.\,.
\end{aligned} \tag{1}$$

Journal of Physics: Conference Series **2482** (2023) 012011 doi:10.1088/1742-6596/2482/1/012011

Here [1] $f^{\alpha[a}f^{\beta b]} = f^{\alpha a}f^{\beta b} - f^{\alpha b}f^{\beta a}$. The vielbeins, f^{a}_{α}, and the two kinds of the spin connection fields, $\omega_{ab\alpha}$ (the gauge fields of S^{ab}) and $\tilde{\omega}_{ab\alpha}$ (the gauge fields of $\tilde{S}^{ab}$), manifest in $d=(3+1)$ as the known vector gauge fields and the scalar gauge fields taking care of masses of quarks and leptons and antiquarks and antileptons and of the weak boson fields [25] [2].

While in any even dimensional space, the superposition of odd products of γ^a's, forming the Clifford odd "basis vectors", offer the description of the internal space of fermions with the half integer spins, the superposition of even products of γ^a's, forming the Clifford even "basis vectors", offer the description of the internal space of boson fields with integer spins, manifesting as gauge fields of the corresponding Clifford odd "basis vectors".

From the point of view of $d=(3+1)$ one family of the Clifford odd "basis vectors" in $d=(13+1)$-dimensional space, with $2^{\frac{d}{2}-1}(=64)$ members manifest spins, handedness and charges of quarks and leptons and antiquarks and antileptons appearing in $2^{\frac{d}{2}-1}$ families, while their Hermitian conjugated partners appear in another group of $2^{\frac{d}{2}-1}$ members in $2^{\frac{d}{2}-1}$ families.

The Clifford even "basis vectors" appear in two groups, each with $2^{\frac{d}{2}-1}\times 2^{\frac{d}{2}-1}$ members, with the Hermitian conjugated partners within the same group and have correspondingly no families. The Clifford even "basis vectors" manifest from the point of view of $d=(3+1)$ all the properties of the vector gauge fields before the electroweak break and the scalar fields causing the electroweak break as assumed by the *standard model*.

Tensor products of either the Clifford odd or the Clifford even "basis vectors" with the basis in ordinary space form the creation and annihilation operators to which the "basis vectors" transfer either anti-commutativity (in the case of the Clifford odd "basis vectors") or commutativity (in the case of the Clifford even "basis vectors") [3]. Correspondingly the anti-commutation properties of creation and annihilation operators of fermions explain Dirac's second quantization postulates for fermion fields; the commutation properties of creation and annihilation operators for bosons explain the corresponding second quantization postulates for boson fields.

The *spin-charge-family* theory offers answers to most of the open questions presented above. The more work is put into the theory, the more answers the theory offers.

In Sect. 2 the Grassmann and the Clifford algebras are explained and creation and annihilation operators described as tensor products of the "basis vectors" offering an explanation of the internal spaces of fermion (by the Clifford odd algebra) and boson (by the Clifford even algebra) fields and the basis in ordinary space.

In Subsect. 2.1, the "basis vectors" are introduced and their properties presented.

In Subsect. 2.2, the properties of the Clifford odd and even "basis vectors" are demonstrated in the toy model in $d=(5+1)$.

In Subsect. 2.3, the properties of the creation and annihilation operators for the second quantized fields are described.

In Sect. 3 a short overview of the achievements and predictions so far of the *spin-charge-family* theory is presented.

[1] $f^{\alpha}{}_{a}$ are inverted vielbeins to $e^{a}{}_{\alpha}$ with the properties $e^{a}{}_{\alpha}f^{\alpha}{}_{b} = \delta^{a}{}_{b}$, $e^{a}{}_{\alpha}f^{\beta}{}_{a} = \delta^{\beta}_{\alpha}$, $E=\det(e^{a}{}_{\alpha})$. Latin indices $a,b,..,m,n,..,s,t,..$ denote a tangent space (a flat index), while Greek indices $\alpha,\beta,..,\mu,\nu,..\sigma,\tau,..$ denote an Einstein index (a curved index). Letters from the beginning of both the alphabets indicate a general index ($a,b,c,..$ and $\alpha,\beta,\gamma,..$), from the middle of both the alphabets the observed dimensions $0,1,2,3$ ($m,n,..$ and $\mu,\nu,..$), indexes from the bottom of the alphabets indicate the compactified dimensions ($s,t,..$ and $\sigma,\tau,..$). We assume the signature $\eta^{ab}=diag\{1,-1,-1,\cdots,-1\}$.

[2] Since the multiplication with either γ^a's or $\tilde{\gamma}^a$'s changes the Clifford odd "basis vectors" into the Clifford even objects, and even "basis vectors" commute, the action for fermions can not include an odd numbers of γ^a's or $\tilde{\gamma}^a$'s, what the simple starting action of Eq. (1) does not. In the starting action γ^a's and $\tilde{\gamma}^a$'s appear as $\gamma^0\gamma^a\hat{p}_{0a}$ or as $\gamma^0\gamma^c\, S^{ab}\omega_{abc}$ and as $\gamma^0\gamma^c\,\tilde{S}^{ab}\tilde{\omega}_{abc}$.

[3] The Clifford odd "basis vectors" transfer their anti-commutativity to creation operators and their Hermitian conjugated partners annihilation operators. The Clifford even "basis vectors" transfer their commutativity to creation operators and annihilation operators.

Sect. 4 presents what the reader could learn from the main contribution of this contribution.

2. Creation and annihilation operators for fermions and bosons

The second quantization postulates for fermions [14, 15, 16] require that the creation operators and their Hermitian conjugated partners annihilation operators, depending on a finite-dimensional basis in internal space, that is, on the space of half integer spins and on charges described by the fundamental representations of the appropriate groups, and on a continuously infinite number of momenta (or coordinates) ([10], Subsect. 3.3.1), fulfil anti-commutation relations.

The second quantization postulates for bosons [14, 15, 16] require that the creation and annihilation operators, depending on finite-dimensional basis in internal space, that is, on the space of integer spins and charges described by the adjoint representations of the same groups, and on a continuously infinite number of momenta (or coordinates) ([10], Subsect. 3.3.1), fulfil commutation relation.

The author demonstrates in this contribution that using the Clifford algebra to describe the internal space of fermions and bosons, the creation and annihilation operators which are tensor products of the internal basis and the momentum/coordinate basis, not only fulfil the appropriate anti-commutation relations (for fermions) or commutation relations (for bosons) but also have the required properties for either fermion fields (if the internal space is described with the Clifford odd products of γ^a's) or of boson fields (if the internal space is described with the Clifford even products of γ^a's). The Clifford odd and Clifford even "basic vectors" correspondingly offer the explanation for the second quantization postulates for fermions and bosons, respectively.

There are two Clifford sub-algebras, which can be used to describe the internal space of fermions and of bosons, each with 2^d members. In each of the two sub-algebras there are $2 \times 2^{\frac{d}{2}-1} \times 2^{\frac{d}{2}-1}$ Clifford odd and $2 \times 2^{\frac{d}{2}-1} \times 2^{\frac{d}{2}-1}$ Clifford even "basic vectors" which can be used to describe the internal space of fermion fields, the Clifford odd "basic vectors", and of boson fields, the Clifford even "basic vectors" in any even d. $d = (13 + 1)$ offers the explanation for all the properties of fermion fields, with families included, and boson fields which are the gauge fields of fermion fields.

In any even d, $d = 2(2n + 1)$ or $d = 4n$, any of the two Clifford sub-algebras offers twice $2^{\frac{d}{2}-1}$ irreducible representations, each with $2^{\frac{d}{2}-1}$ members, which can represent "basis vectors" and their Hermitian conjugated partners. Each irreducible representation offers in $d = (13 + 1)$ the description of the quarks and the antiquarks and the leptons and the antileptons (with the right handed neutrinos and left handed antineutrinos included in addition what is) assumed by the *standard model.*

There is obviously only one kind of fermion fields and correspondingly also of their gauge fields observed. There is correspondingly no need for two Clifford sub-algebras.

The reduction of the two sub-algebras to only one with the postulate in Eq. (7), (Ref. [10], Eq. (38)) solves this problem. At the same time, the reduction offers the quantum numbers for each of the irreducible representations of the Clifford sub-algbebra left, γ^a's when fermions are concerned ([10] Subsect. 3.2).

Boson fields have no families, as will be demonstrated.

Grassmann and Clifford algebras

The internal space of anti-commuting or commuting second quantized fields can be described by using either the Grassmann or the Clifford algebras [7, 8, 9, 29]. What follows is a short overview of Subsect.3.2 of Ref. [10] and of references cited in [10].

In Grassmann d-dimensional space there are d anti-commuting (operators) θ^a, and d anti-commuting operators which are derivatives with respect to θ^a, $\frac{\partial}{\partial \theta_a}$,

$$\begin{aligned} \{\theta^a, \theta^b\}_+ = 0\,, \qquad & \{\frac{\partial}{\partial \theta_a}, \frac{\partial}{\partial \theta_b}\}_+ = 0\,, \\ \{\theta_a, \frac{\partial}{\partial \theta_b}\}_+ &= \delta_{ab}\,, \; (a,b) = (0,1,2,3,5,\cdots,d)\,. \end{aligned} \tag{2}$$

Defining [30]

$$(\theta^a)^\dagger = \eta^{aa} \frac{\partial}{\partial \theta_a}\,, \quad \text{leads to} \quad (\frac{\partial}{\partial \theta_a})^\dagger = \eta^{aa} \theta^a\,, \tag{3}$$

with $\eta^{ab} = diag\{1, -1, -1, \cdots, -1\}$.

θ^a and $\frac{\partial}{\partial \theta_a}$ are, up to the sign, Hermitian conjugated to each other. The identity is the self adjoint member of the algebra. The choice for the following complex properties of θ^a and correspondingly of $\frac{\partial}{\partial \theta_a}$ are made

$$\begin{aligned} \{\theta^a\}^* &= (\theta^0, \theta^1, -\theta^2, \theta^3, -\theta^5, \theta^6, ..., -\theta^{d-1}, \theta^d)\,, \\ \{\frac{\partial}{\partial \theta_a}\}^* &= (\frac{\partial}{\partial \theta_0}, \frac{\partial}{\partial \theta_1}, -\frac{\partial}{\partial \theta_2}, \frac{\partial}{\partial \theta_3}, -\frac{\partial}{\partial \theta_5}, \frac{\partial}{\partial \theta_6}, ..., -\frac{\partial}{\partial \theta_{d-1}}, \frac{\partial}{\partial \theta_d})\,. \end{aligned} \tag{4}$$

There are 2^d superposition of products of θ^a, the Hermitian conjugated partners of which are the corresponding superposition of products of $\frac{\partial}{\partial \theta_a}$.

There exist two kinds of the Clifford algebra elements (operators), γ^a and $\tilde{\gamma}^a$, expressible with θ^a's and their conjugate momenta $p^{\theta a} = i\, \frac{\partial}{\partial \theta_a}$ [8], Eqs. (2, 3),

$$\begin{aligned} \gamma^a &= (\theta^a + \frac{\partial}{\partial \theta_a})\,, \quad \tilde{\gamma}^a = i\,(\theta^a - \frac{\partial}{\partial \theta_a})\,, \\ \theta^a &= \frac{1}{2}\,(\gamma^a - i\tilde{\gamma}^a)\,, \quad \frac{\partial}{\partial \theta_a} = \frac{1}{2}\,(\gamma^a + i\tilde{\gamma}^a)\,, \end{aligned} \tag{5}$$

offering together $2 \cdot 2^d$ operators: 2^d are superposition of products of γ^a and 2^d of $\tilde{\gamma}^a$. It is easy to prove if taking into account Eqs. (3, 5), that they form two anti-commuting Clifford sub-algebras, $\{\gamma^a, \tilde{\gamma}^b\}_+ = 0$, Refs. ([10] and references therein)

$$\begin{aligned} \{\gamma^a, \gamma^b\}_+ &= 2\eta^{ab} = \{\tilde{\gamma}^a, \tilde{\gamma}^b\}_+\,, \\ \{\gamma^a, \tilde{\gamma}^b\}_+ &= 0\,, \quad (a,b) = (0,1,2,3,5,\cdots,d)\,, \\ (\gamma^a)^\dagger &= \eta^{aa}\,\gamma^a\,, \quad (\tilde{\gamma}^a)^\dagger = \eta^{aa}\,\tilde{\gamma}^a\,. \end{aligned} \tag{6}$$

While the Grassmann algebra offers the description of the "anti-commuting integer spin second quantized fields" and of the "commuting integer spin second quantized fields" [33, 10], the Clifford algebras which are superposition of odd products of either γ^a's or $\tilde{\gamma}^a$'s offer the description of the second quantized half integer spin fermion fields, which from the point of the subgroups of the $SO(d-1,1)$ group manifest spins and charges of fermions and antifermions in the fundamental representations of the group and subgroups.

The superposition of even products of either γ^a's or $\tilde{\gamma}^a$'s offer the description of the commuting second quantized boson fields with integer spins (as we can see in [12] and shall see in this contribution) which from the point of the subgroups of the $SO(d-1,1)$ group manifest spins and charges in the adjoint representations of the group and subgroups.

The following *postulate*, which determines how does $\tilde{\gamma}^a$'s operate on γ^a's, reduces the two Clifford subalgebras, γ^a's and $\tilde{\gamma}^a$'s, to one, to the one described by γ^a's [20, 8, 27, 29, 30]

$$\{\tilde{\gamma}^a B = (-)^B \, i \, B\gamma^a\} \, |\psi_{oc}>\,, \tag{7}$$

with $(-)^B = -1$, if B is (a function of) an odd products of γ^a's, otherwise $(-)^B = 1$ [20], the vacuum state $|\psi_{oc}>$ is defined in Eq. (16) of Subsect. 2.1.

After the postulate of Eq. (7) it follows:
a. The Clifford sub-algebra described by $\tilde{\gamma}^a$'s looses its meaning for the description of the internal space of quantum fields.
b. The "basis vectors" which are superposition of an odd or an even products of γ^a's obey the postulates for the second quantization fields for fermions or bosons, respectively, Sect.2.1.
c. It can be proven that the relations presented in Eq. (6) remain valid also after the postulate of Eq. (7). The proof is presented in Ref. ([10], App. I, Statement 3a.
d. Each irreducible representation of the Clifford odd "basis vectors" described by γ^a's are equipped by the quantum numbers of the Cartan subalgebra members of $\tilde{S}^{ab}$, chosen in Eq. (8), as follows

$$\begin{aligned}
&\mathbf{S}^{03}, \mathbf{S}^{12}, \mathbf{S}^{56}, \cdots, \mathbf{S}^{d-1\, d}\,,\\
&S^{03}, S^{12}, S^{56}, \cdots, S^{d-1\, d}\,,\\
&\tilde{S}^{03}, \tilde{S}^{12}, \tilde{S}^{56}, \cdots, \tilde{S}^{d-1\, d}\,,\\
&\mathbf{S}^{ab} = S^{ab} + \tilde{S}^{ab} = i\,(\theta^a \frac{\partial}{\partial \theta_b} - \theta^b \frac{\partial}{\partial \theta_a})\,.
\end{aligned} \tag{8}$$

After the postulate of Eq. (7) no vector space of $\tilde{\gamma}^a$'s needs to be taken into account for the description of the internal space of either fermions or bosons, in agreement with the observed properties of fermions and bosons. Also the Grassmann algebra is reduced to only one of the Clifford sub-algebras. The operators $\tilde{\gamma}^a$'s describe from now on properties of fermion and boson "basis vectors" determined by superposition of products of odd or even numbers of γ^a's, respectively.

$\tilde{\gamma}^a$'s equip each irreducible representation of the Lorentz group (with the infinitesimal generators $S^{ab} = \frac{i}{4}\{\gamma^a, \gamma^b\}_-$) when applying on the Clifford odd "basis vectors" (which are superposition of odd products of $\gamma^{a's}$) with the family quantum numbers (determined by $\tilde{S}^{ab} = \frac{i}{4}\{\tilde{\gamma}^a, \tilde{\gamma}^b\}_-$).

Correspondingly the Clifford odd "basis vectors" (they are the superposition of odd products of γ^a's) form $2^{\frac{d}{2}-1}$ families, with the quantum number f, each family have $2^{\frac{d}{2}-1}$ members, m. They offer the description of the second quantized fermion fields.

The Clifford even "basis vectors" (they are the superposition of even products of γ^a's) have no families as we shall see in what follows, but they do carry both quantum numbers, f and m. They offer the description of the second quantized boson fields as the gauge fields of the second quantized fermion fields. The generators of the Lorentz transformations in the internal space of the Clifford even "basis vectors" are $\mathcal{S}^{ab} = S^{ab} + \tilde{S}^{ab}$.

Properties of the Clifford odd and the Clifford even"basis vectors" are discussed in the following subsection.

2.1. *"Basis vectors" of fermions and bosons*

After the reduction of the two Clifford sub-algebras to only one, Eq. (7), we only need to define "basis vectors" for the case that the internal space of second quantized fields is described by superposition of odd or even products γ^a's [4].

Let us use the technique which makes "basis vectors" products of nilpotents and projectors [8, 9, 19, 20] which are eigenvectors of the (chosen) Cartan subalgebra members, Eq. (8), of the Lorentz algebra in the space of γ^a's, either in the case of the Clifford odd or in the case of the Clifford even products of γ^a's.

There are in even dimensional spaces $\frac{d}{2}$ members of the Cartan subalgebra, Eq. (8).

One finds for any of the $\frac{d}{2}$ Cartan subalgebra member, S^{ab} or $\tilde{S}^{ab}$, both applying on a nilpotent $\stackrel{ab}{(k)}$ or on projector $\stackrel{ab}{[k]}$

$$\stackrel{ab}{(k)}:=\frac{1}{2}(\gamma^a+\frac{\eta^{aa}}{ik}\gamma^b)\,, \quad (\stackrel{ab}{(k)})^2=0\,,$$
$$\stackrel{ab}{[k]}:=\frac{1}{2}(1+\frac{i}{k}\gamma^a\gamma^b)\,, \quad (\stackrel{ab}{[k]})^2=\stackrel{ab}{[k]}$$

the relations

$$S^{ab}\stackrel{ab}{(k)}=\frac{k}{2}\stackrel{ab}{(k)}\,, \qquad \tilde{S}^{ab}\stackrel{ab}{(k)}=\frac{k}{2}\stackrel{ab}{(k)}\,,$$
$$S^{ab}\stackrel{ab}{[k]}=\frac{k}{2}\stackrel{ab}{[k]}\,, \qquad \tilde{S}^{ab}\stackrel{ab}{[k]}=-\frac{k}{2}\stackrel{ab}{[k]}\,, \tag{9}$$

with $k^2=\eta^{aa}\eta^{bb}$, demonstrating that the eigenvalues of S^{ab} on nilpotents and projectors expressed with γ^a's differ from the eigenvalues of $\tilde{S}^{ab}$ on nilpotents and projectors expressed with γ^a's, so that $\tilde{S}^{ab}$ can be used to equip each irreducible representation of S^{ab} with the "family" quantum number. [5]

We define in even d the "basis vectors" as algebraic, $*_A$, products of nilpotents and projectors so that each product is eigenvector of all $\frac{d}{2}$ Cartan subalgebra members.

We recognize in advance that the superposition of odd products of γ^a's, that is the Clifford odd "basis vectors", must include an odd number of nilpotents, at least one, while the superposition of an even products of γ^a"s, that is Clifford even "basis vectors", must include an even number of nilpotents or only projectors.

To define the Clifford odd "basis vectors", we shall see that they have properties appropriate to describe the internal space of the second quantized fermion fields, and the Clifford even "basis vectors", we shall see that they have properties appropriate to describe the internal space of the second quantized boson fields, we need to know the relations for nilpotents and projectors

$$\stackrel{ab}{(k)}: \;=\; \frac{1}{2}(\gamma^a+\frac{\eta^{aa}}{ik}\gamma^b)\,, \quad \stackrel{ab}{[k]}:=\frac{1}{2}(1+\frac{i}{k}\gamma^a\gamma^b)\,,$$
$$\stackrel{ab}{(\tilde{k})}: \;=\; \frac{1}{2}(\tilde{\gamma}^a+\frac{\eta^{aa}}{ik}\tilde{\gamma}^b)\,, \quad \stackrel{ab}{[\tilde{k}]}: \frac{1}{2}(1+\frac{i}{k}\tilde{\gamma}^a\tilde{\gamma}^b)\,, \tag{10}$$

which can be derived after taking into account Eq. (6)

$$\gamma^a\stackrel{ab}{(k)} = \eta^{aa}\stackrel{ab}{[-k]}, \quad \gamma^b\stackrel{ab}{(k)}=-ik\stackrel{ab}{[-k]}, \quad \gamma^a\stackrel{ab}{[k]}=\stackrel{ab}{(-k)}, \quad \gamma^b\stackrel{ab}{[k]}=-ik\eta^{aa}\stackrel{ab}{(-k)}\,,$$
$$\tilde{\gamma}^a\stackrel{ab}{(k)} = -i\eta^{aa}\stackrel{ab}{[k]}, \quad \tilde{\gamma}^b\stackrel{ab}{(k)}=-k\stackrel{ab}{[k]}, \quad \tilde{\gamma}^a\stackrel{ab}{[k]}=\;i\stackrel{ab}{(k)}, \quad \tilde{\gamma}^b\stackrel{ab}{[k]}=-k\eta^{aa}\stackrel{ab}{(k)}\,,$$
$$\stackrel{ab}{(k)}^{\dagger} = \eta^{aa}\stackrel{ab}{(-k)}\,, \quad (\stackrel{ab}{(k)})^2=0\,, \quad \stackrel{ab}{(k)}\stackrel{ab}{(-k)}=\eta^{aa}\stackrel{ab}{[k]}\,,$$

[4] In Ref. [10], the reader can find in Subsects. (3.2.1 and 3.2.2) definitions for the "basis vectors" for the Grassmann and the two Clifford subalgebras, which are products of nilpotents and projectors chosen to be eigenvectors of the corresponding Cartan sub-algebra members of the Lorentz algebras presented in Eq. (8).

[5] The reader can find the proof of Eq. (9) in Ref. [10], App. (I).

$$
\begin{aligned}
\overset{ab}{[k]}^{\dagger} &= \overset{ab}{[k]}\,, \quad (\overset{ab}{[k]})^2 = \overset{ab}{[k]}\,, \quad \overset{ab}{[k]}\overset{ab}{[-k]} = 0\,,\\
\overset{ab}{(k)}\overset{ab}{[k]} &= 0\,, \quad \overset{ab}{[k]}\overset{ab}{(k)} = \overset{ab}{(k)}\,, \quad \overset{ab}{(k)}\overset{ab}{[-k]} = \overset{ab}{(k)}\,, \quad \overset{ab}{[k]}\overset{ab}{(-k)} = 0\,,\\
\overset{ab}{(\tilde{k})}^{\dagger} &= \eta^{aa}\,\overset{ab}{(-\tilde{k})}\,, \quad (\overset{ab}{(\tilde{k})})^2 = 0\,, \quad \overset{ab}{(\tilde{k})}\overset{ab}{(-\tilde{k})} = \eta^{aa}\,\overset{ab}{[\tilde{k}]}\,,\\
\overset{ab}{[\tilde{k}]}^{\dagger} &= \overset{ab}{[\tilde{k}]}\,, \quad (\overset{ab}{[\tilde{k}]})^2 = \overset{ab}{[\tilde{k}]}\,, \quad \overset{ab}{[\tilde{k}]}\overset{ab}{[-\tilde{k}]} = 0\,,\\
\overset{ab}{(\tilde{k})}\overset{ab}{[\tilde{k}]} &= 0\,, \quad \overset{ab}{[\tilde{k}]}\overset{ab}{(\tilde{k})} = \overset{ab}{(\tilde{k})}\,, \quad \overset{ab}{(\tilde{k})}\overset{ab}{[-\tilde{k}]} = \overset{ab}{(\tilde{k})}\,, \quad \overset{ab}{[\tilde{k}]}\overset{ab}{(-\tilde{k})} = 0\,.
\end{aligned}
\tag{11}
$$

The relations in Eq. (11) demonstrate that the properties of the "basis vectors" which include odd number of nilpotents differ essentially from the "basis vectors" which include even number of nilpotents.

One namely recognizes:

i. Since the Hermitian conjugated partner of a nilpotent $\overset{ab}{(k)}^{\dagger}$ is $\eta^{aa}\,\overset{ab}{(-k)}$ and since neither S^{ab} nor $\tilde{S}^{ab}$ nor both can transform odd products of nilpotents to belong to one of the $2^{\frac{d}{2}-1}$ members of one of $2^{\frac{d}{2}-1}$ irreducible representations (families), the Hermitian conjugated partners of the Clifford odd "basis vectors" must belong to a different group of $2^{\frac{d}{2}-1}$ members of $2^{\frac{d}{2}-1}$ families.

Since S^{ac} transforms $\overset{ab}{(k)} *_A \overset{cd}{(k')}$ into $\overset{ab}{[-k]} *_A \overset{cd}{[-k']}$, while $\tilde{S}^{ab}$ transforms $\overset{ab}{[-k]} *_A \overset{cd}{[-k']}$ into $\overset{ab}{(-k)} *_A \overset{cd}{(-k')}$ it is obvious that the Hermitian conjugated partners of the Clifford odd "basis vectors" must belong to the same group of $2^{\frac{d}{2}-1} \times 2^{\frac{d}{2}-1}$ members. Projectors are self adjoint.

ii. Since an odd products of γ^a's anti-commute with another group of an odd product of γ^a, the Clifford odd "basis vectors" anti-commute, manifesting in a tensor product with the basis in ordinary space together with the corresponding Hermitian conjugated partners properties of the anti-commutation relations postulated by Dirac for the second quantized fermion fields.

The Clifford even "basis vectors" correspondingly fulfil the commutation relations for the second quantized boson fields.

iii. The Clifford odd "basis vectors" have all the eigenvalues of the Cartan subalgebra members equal to either $\pm\frac{1}{2}$ or to $\pm\frac{i}{2}$.

The Clifford even "basis vectors" have all the eigenvalues of the Cartan subalgebra members $\mathcal{S}^{ab}$ equal to either ± 1 and zero or to $\pm i$ and zero.

Let us define odd and even "basis vectors" as products of nilpotents and projectors in even-dimensional spaces.

a. *Clifford odd "basis vectors"*

The Clifford odd "basis vectors" must be products of an odd number of nilpotents, and the rest, up to $\frac{d}{2}$, of projectors, each nilpotent and projector must be the "eigenstate" of one of the members of the Cartan subalgebra, Eq. (8), correspondingly are the "basis vectors" eigenstates of all the members of the Lorentz algebras: S^{ab}'s determine $2^{\frac{d}{2}-1}$ members of one family, $\tilde{S}^{ab}$'s transform each member of one family to the same member of the rest of $2^{\frac{d}{2}-1}$ families.

Let us name the Clifford odd "basis vectors" $\hat{b}^{m\dagger}_f$, where m determines membership of "basis vectors" in any family, f determines a particular family. The Hermitian conjugated partner of $\hat{b}^{m\dagger}_f$ is named by $\hat{b}^{m}_f = (\hat{b}^{m\dagger}_f)^{\dagger}$.

Let us start in $d = 2(2n+1)$ with the "basis vectors" $\hat{b}^{1\dagger}_1$ which is the product of only nilpotents, all the rest members belonging to the $f = 1$ family follow by the application of S^{01},

$S^{03}, \dots, S^{0d}, S^{15}, \dots, S^{1d}, S^{5d} \dots, S^{d-2\,d}$. The algebraic product mark $*_A$ is skipped.

$$
\begin{aligned}
&d = 2(2n+1)\,,\\
&\hat{b}^{1\dagger}_1 = \overset{03}{(+i)}\overset{12}{(+)}\overset{56}{(+)}\cdots\overset{d-1\,d}{(+)}\,,\\
&\hat{b}^{2\dagger}_1 = \overset{03}{[-i]}\overset{12}{[-]}\overset{56}{(+)}\cdots\overset{d-1\,d}{(+)}\,,\\
&\cdots\\
&\hat{b}^{2^{\frac{d}{2}-1}\dagger}_1 = \overset{03}{[-i]}\overset{12}{[-]}\overset{56}{(+)}\dots\overset{d-3\,d-2}{[-]}\;\overset{d-1\,d}{[-]}\,,\\
&\cdots .
\end{aligned}
\tag{12}
$$

The Hermitian conjugated partners of the Clifford odd "basis vectors" $\hat{b}^{m\dagger}_1$, presented in Eq. (12), are

$$
\begin{aligned}
&d = 2(2n+1)\,,\\
&\hat{b}^{1}_1 = \overset{03}{(-i)}\overset{12}{(-)}\cdots\overset{d-1\,d}{(-)}\,,\\
&\hat{b}^{2}_1 = \overset{03}{[-i]}\overset{12}{[-]}\overset{56}{(-)}\cdots\overset{d-1\,d}{(-)}\,,\\
&\cdots\\
&\hat{b}^{2^{\frac{d}{2}-1}\dagger}_1 = \overset{03}{[-i]}\overset{12}{[-]}\overset{56}{(-)}\overset{78}{[-]}\dots\overset{d-3\,d-2}{[-]}\;\overset{d-1\,d}{[-]}\,,\\
&\cdots .
\end{aligned}
\tag{13}
$$

In $d = 4n$ the choice of the starting "basis vectors" with maximal number of nilpotents must have one projector

$$
\begin{aligned}
&d = 4n\,,\\
&\hat{b}^{1\dagger}_1 = \overset{03}{(+i)}\overset{12}{(+)}\cdots\overset{d-1\,d}{[+]}\,,\\
&\hat{b}^{2\dagger}_1 = \overset{03}{[-i]}\overset{12}{[-]}\overset{56}{(+)}\cdots\overset{d-1\,d}{[+]}\,,\\
&\cdots\\
&\hat{b}^{2^{\frac{d}{2}-1}\dagger}_1 = \overset{03}{[-i]}\overset{12}{[-]}\overset{56}{(+)}\dots\overset{d-3\,d-2}{[-]}\;\overset{d-1\,d}{[+]}\,,\\
&\cdots .
\end{aligned}
\tag{14}
$$

The Hermitian conjugated partners of the Clifford odd "basis vector" $\hat{b}^{m\dagger}_1$, presented in Eq. (14), follow if all nilpotents $\overset{ab}{(k)}$ are transformed into $\eta^{aa}\,\overset{ab}{(-k)}$.

For either $d = 2(2n+1)$ or for $d = 4n$ all the $2^{\frac{d}{2}-1}$ families follow by applying $\tilde{S}^{ab}$'s on all the members of the starting family. (Or one can find the starting $\hat{b}^1_f$ for all families f and then generate all the members $\hat{b}^m_f$ from $\hat{b}^1_f$ by the application of $\tilde{S}^{ab}$ on the starting member.)

It is not difficult to see that all the "basis vectors" within any family, as well as the "basis vectors" among families, are orthogonal; that is, their algebraic product is zero, and the same is true for the Hermitian conjugated partners, which can be proved by the algebraic multiplication using Eq.(11).

$$
\hat{b}^{m\dagger}_f *_A \hat{b}^{m'\dagger}_{f'} = 0\,, \quad \hat{b}^{m}_f *_A \hat{b}^{m'}_{f'} = 0\,, \quad \forall m, m', f, f'\,.
\tag{15}
$$

If we require that each family of "basis vectors" , determined by nilpotents and projectors described by γ^a's, carry the family quantum number determined by $\tilde{S}^{ab}$ and define the vacuum state on which "basis vectors" apply as

$$|\psi_{oc}>= \sum_{f=1}^{2^{\frac{d}{2}-1}} \hat{b}^{m}_{f\,*_A} \hat{b}^{m\dagger}_{f} \,|\, 1 >\,, \tag{16}$$

it follows that the Clifford odd "basis vectors" obey the relations

$$\begin{aligned}
\hat{b}^{m}_{f\,*_A}|\psi_{oc}> &= 0.\,|\psi_{oc}>\,,\\
\hat{b}^{m\dagger}_{f\,*_A}|\psi_{oc}> &= |\psi^{m}_{f}>\,,\\
\{\hat{b}^{m}_{f}, \hat{b}^{m'}_{f`}\}_{*_A +}|\psi_{oc}> &= 0.\,|\psi_{oc}>\,,\\
\{\hat{b}^{m\dagger}_{f}, \hat{b}^{m'\dagger}_{f`}\}_{*_A +}|\psi_{oc}> &= 0.\,|\psi_{oc}>\,,\\
\{\hat{b}^{m}_{f}, \hat{b}^{m'\dagger}_{f}\}_{*_A +}|\psi_{oc}> &= \delta^{mm'}\,\delta_{ff`}|\psi_{oc}>\,,
\end{aligned} \tag{17}$$

while the normalization $< \psi_{oc}|\hat{b}^{m'\dagger}_{f'} *_A \hat{b}^{m\dagger}_{f} *_A |\psi_{oc}>= \delta^{mm'}\delta_{ff'}$ is used and the anticommutation relation mean $\{\hat{b}^{m\dagger}_{f}, \hat{b}^{m'\dagger}_{f`}\}_{*_A +} = \hat{b}^{m\dagger}_{f} *_A \hat{b}^{m'\dagger}_{f`} + \hat{b}^{m'\dagger}_{f`} *_A \hat{b}^{m\dagger}_{f}$.

If we write the creation and annihilation operators as the tensor, $*_T$, products of "basis vectors" and the basis in ordinary space, the creation and annihilation operators fulfil the Dirac's anti-commutation postulates since the "basis vectors" transfer their anti-commutativity to creation and annihilation operators. It turns out that not only the Clifford odd "basis vectors" offer the description of the internal space of fermions, they offer the explanation for the second quantization postulates for fermions as well.

Table 1, presented in Subsect. 2.2, illustrates the properties of the Clifford odd "basis vectors" on the case of $d = (5+1)$.

b. *Clifford even "basis vectors"*

The Clifford even "basis vectors" must be products of an even number of nilpotents and the rest, up to $\frac{d}{2}$, of projectors, each nilpotent and projector in a product must be the "eigenstate" of one of the members of the Cartan subalgebra, Eq. (8), correspondingly are the "basis vectors" eigenstates of all the members of the Lorentz algebra: S^{ab}'s and $\tilde{S}^{ab}$'s generate from the starting "basis vector" all the $2^{\frac{d}{2}-1}\times 2^{\frac{d}{2}-1}$ members of one group, which includes as well the Hermitian conjugated partners of any member. $2^{\frac{d}{2}-1}$ members of the group are products of projectors only. They are self-adjoint.

There are two groups with $2^{\frac{d}{2}-1}\times 2^{\frac{d}{2}-1}$ members. The members of one group are not connected with the members of another group by either S^{ab}'s or $\tilde{S}^{ab}$'s or both.

Let us name the Clifford even "basis vectors" with ${}^{i}\hat{\mathcal{A}}^{m\dagger}_{f}$, where $i = (I, II)$ denotes that there are two groups of Clifford even "basis vectors", while m and f determine membership of "basis vectors" in any of the two groups I or II. Let us repeat that the Hermitian conjugated partner of any "basis vector" appears either in the case of ${}^{I}\hat{\mathcal{A}}^{m\dagger}_{f}$ or in the case of ${}^{II}\hat{\mathcal{A}}^{m\dagger}_{f}$ within the same group.

Let us write down the Clifford even "basis vectors" as a product of an even number of nilpotents and the rest of the projectors, so that the Clifford even "basis vectors" are eigenvectors of all the Cartan subalgebra members, and let us name them as follows

$$d = 2(2n+1)$$

$$\begin{aligned}
{}^{I}\hat{\mathcal{A}}^{1\dagger}_{1} &= \stackrel{03}{(+i)}\stackrel{12}{(+)}\cdots\stackrel{d-1\,d}{[+]}\,, & {}^{II}\hat{\mathcal{A}}^{1\dagger}_{1} &= \stackrel{03}{(-i)}\stackrel{12}{(+)}\cdots\stackrel{d-1\,d}{[+]}\,,\\
{}^{I}\hat{\mathcal{A}}^{2\dagger}_{1} &= \stackrel{03}{[-i]}\stackrel{12}{[-]}\stackrel{56}{(+)}\cdots\stackrel{d-1\,d}{[+]}\,, & {}^{II}\hat{\mathcal{A}}^{2\dagger}_{1} &= \stackrel{03}{[+i]}\stackrel{12}{[-]}\stackrel{56}{(+)}\cdots\stackrel{d-1\,d}{[+]}\,,
\end{aligned}$$

$$
\begin{array}{lll}
{}^{I}\hat{\mathcal{A}}^{3\dagger}_{1} = \stackrel{03}{(+i)}\stackrel{12}{(+)}\stackrel{56}{(+)} \cdots \stackrel{d-3\,d-2}{[-]}\stackrel{d-1\,d}{(-)}\,, & & {}^{II}\hat{\mathcal{A}}^{3\dagger}_{1} = \stackrel{03}{(-i)}\stackrel{12}{(+)}\stackrel{56}{(+)} \cdots \stackrel{d-3\,d-2}{[-]}\stackrel{d-1\,d}{(-)}\,, \\
\cdots & & \cdots \\
 & d = 4n & \\
{}^{I}\hat{\mathcal{A}}^{1\dagger}_{1} = \stackrel{03}{(+i)}\stackrel{12}{(+)} \cdots \stackrel{d-1\,d}{(+)}\,, & & {}^{II}\hat{\mathcal{A}}^{1\dagger}_{1} = \stackrel{03}{(-i)}\stackrel{12}{(+)} \cdots \stackrel{d-1\,d}{(+)}\,, \\
{}^{I}\hat{\mathcal{A}}^{2\dagger}_{1} = \stackrel{03}{[-i]}\stackrel{12}{[-i]}\stackrel{56}{(+)} \cdots \stackrel{d-1\,d}{(+)}\,, & & {}^{II}\hat{\mathcal{A}}^{2\dagger}_{1} = \stackrel{03}{[+i]}\stackrel{12}{[-i]}\stackrel{56}{(+)} \cdots \stackrel{d-1\,d}{(+)}\,, \\
{}^{I}\hat{\mathcal{A}}^{3\dagger}_{1} = \stackrel{03}{(+i)}\stackrel{12}{(+)}\stackrel{56}{(+)} \cdots \stackrel{d-3\,d-2}{[-]}\stackrel{d-1\,d}{[-]}\,, & & {}^{II}\hat{\mathcal{A}}^{3\dagger}_{1} = \stackrel{03}{(-i)}\stackrel{12}{(+)}\stackrel{56}{(+)} \cdots \stackrel{d-3\,d-2}{[-]}\stackrel{d-1\,d}{[-]} \\
\cdots & & \cdots
\end{array} \quad (18)
$$

There are $2^{\frac{d}{2}-1}\times 2^{\frac{d}{2}-1}$ Clifford even "basis vectors" of the kind ${}^{I}\hat{\mathcal{A}}^{m\dagger}_{f}$ and there are $2^{\frac{d}{2}-1}\times 2^{\frac{d}{2}-1}$ Clifford even "basis vectors" of the kind ${}^{II}\hat{\mathcal{A}}^{m\dagger}_{f}$.

Table 1, presented in Subsect. 2.2, illustrates properties of the Clifford odd and Clifford even "basis vectors" on the case of $d=(5+1)$. Looking at this case it is easy to evaluate properties of either even or odd "basis vectors". We shall discuss here the general results which follow after carefully inspecting properties of both kinds of "basis vectors".

The Clifford even "basis vectors" belonging to two different groups are orthogonal due to the fact that they differ in the sign of one nilpotent or one projector, or the algebraic product of a member of one group with a member of another group gives zero according to Eq. (11).

$$
{}^{I}\hat{\mathcal{A}}^{m\dagger}_{f} *_A {}^{II}\hat{\mathcal{A}}^{m\dagger}_{f} = 0 = {}^{II}\hat{\mathcal{A}}^{m\dagger}_{f} *_A {}^{I}\hat{\mathcal{A}}^{m\dagger}_{f}\,. \quad (19)
$$

The members of each of these two groups have the property

$$
{}^{I,II}\hat{\mathcal{A}}^{m\dagger}_{f} *_A {}^{I,II}\hat{\mathcal{A}}^{m'\dagger}_{f`} \rightarrow \left\{ \begin{array}{r} {}^{I,II}\hat{\mathcal{A}}^{m\dagger}_{f`}\,, \text{only one for } \forall f`\,, \\ \text{or zero}\,. \end{array} \right. \quad (20)
$$

Two "basis vectors", ${}^{I}\hat{\mathcal{A}}^{m\dagger}_{f}$ and ${}^{I}\hat{\mathcal{A}}^{m'\dagger}_{f'}$, the algebraic product, $*_A$, of which gives non zero contribution, "scatter" into the third one ${}^{I}\hat{\mathcal{A}}^{m\dagger}_{f`}$. The same is true also for the "basis vectors" belonging to the second group of the Clifford even "basis vectors ${}^{II}\hat{\mathcal{A}}^{m\dagger}_{f}$.

Let us write the commutation relations for Clifford even "basis vectors" taking into account Eq. (20).

i. Ini the case that ${}^{I}\hat{\mathcal{A}}^{m\dagger}_{f} *_A {}^{I}\hat{\mathcal{A}}^{m'\dagger}_{f`} \rightarrow {}^{I}\hat{\mathcal{A}}^{m\dagger}_{f`}$ and ${}^{I}\hat{\mathcal{A}}^{m'\dagger}_{f`} *_A {}^{I}\hat{\mathcal{A}}^{m\dagger}_{f} = 0$ it follows

$$
\{{}^{I}\hat{\mathcal{A}}^{m\dagger}_{f}, {}^{I}\hat{\mathcal{A}}^{m'\dagger}_{f`}\}_{*_A -} \rightarrow \left\{ \begin{array}{ll} {}^{I}\hat{\mathcal{A}}^{m\dagger}_{f`}\,, & (\text{if } \; {}^{I}\hat{\mathcal{A}}^{m\dagger}_{f} *_A {}^{I}\hat{\mathcal{A}}^{m'\dagger}_{f`} \rightarrow {}^{I}\hat{\mathcal{A}}^{m\dagger}_{f`} \\ & \text{and } {}^{I}\hat{\mathcal{A}}^{m'\dagger}_{f`} *_A {}^{I}\hat{\mathcal{A}}^{m\dagger}_{f} = 0)\,, \end{array} \right. \quad (21)
$$

ii. In the case that ${}^{I}\hat{\mathcal{A}}^{m\dagger}_{f} *_A {}^{I}\hat{\mathcal{A}}^{m'\dagger}_{f`} \rightarrow {}^{I}\hat{\mathcal{A}}^{m\dagger}_{f`}$ and ${}^{I}\hat{\mathcal{A}}^{m'\dagger}_{f`} *_A {}^{I}\hat{\mathcal{A}}^{m\dagger}_{f} \rightarrow {}^{I}\hat{\mathcal{A}}^{m'\dagger}_{f}$ it follows

$$
\{{}^{I}\hat{\mathcal{A}}^{m\dagger}_{f}, {}^{I}\hat{\mathcal{A}}^{m'\dagger}_{f`}\}_{*_A -} \rightarrow \left\{ \begin{array}{ll} {}^{I}\hat{\mathcal{A}}^{m\dagger}_{f`} - {}^{I}\hat{\mathcal{A}}^{m'\dagger}_{f}\,, & (\text{if } \; {}^{I}\hat{\mathcal{A}}^{m\dagger}_{f} *_A {}^{I}\hat{\mathcal{A}}^{m'\dagger}_{f`} \rightarrow {}^{I}\hat{\mathcal{A}}^{m\dagger}_{f`} \\ & \text{and } {}^{I}\hat{\mathcal{A}}^{m'\dagger}_{f`} *_A {}^{I}\hat{\mathcal{A}}^{m\dagger}_{f} \rightarrow {}^{I}\hat{\mathcal{A}}^{m'\dagger}_{f})\,, \end{array} \right. \quad (22)
$$

iii. In all other cases we have

$$
\{{}^{I}\hat{\mathcal{A}}^{m\dagger}_{f}, {}^{I}\hat{\mathcal{A}}^{m'\dagger}_{f`}\}_{*_A -} = 0\,. \quad (23)
$$

$\{{}^{I}\hat{\mathcal{A}}^{m\dagger}_{f}\,,\,{}^{I}\hat{\mathcal{A}}^{m'\dagger}_{f`}\}_{*_A -}$ means ${}^{I}\hat{\mathcal{A}}^{m\dagger}_{f} *_A {}^{I}\hat{\mathcal{A}}^{m'\dagger}_{f`} - {}^{I}\hat{\mathcal{A}}^{m'\dagger}_{f`} *_A {}^{I}\hat{\mathcal{A}}^{m\dagger}_{f}$.

It remains to evaluate the algebraic application, $*_A$, of the Clifford even "basis vectors" ${}^{I}\hat{\mathcal{A}}^{m\dagger}_{f}$ on the Clifford odd "basis vectors" $\hat{b}^{m'\dagger}_{f`}$. One finds

$$ {}^{I}\hat{\mathcal{A}}^{m\dagger}_{f`} *_A \hat{b}^{m'\dagger}_{f} \rightarrow \left\{ \begin{array}{c} \hat{b}^{m\dagger}_{f}\,, \\ \text{or zero}\,. \end{array} \right. \tag{24}$$

For each ${}^{I}\hat{\mathcal{A}}^{m\dagger}_{f}$ there are among $2^{\frac{d}{2}-1}\times 2^{\frac{d}{2}-1}$ members of the Clifford odd "basis vectors" (describing the internal space of fermion fields) $2^{\frac{d}{2}-1}$ members, $\hat{b}^{m'\dagger}_{f`}$, fulfilling the relation of Eq. (24). All the rest $(2^{\frac{d}{2}-1}\times(2^{\frac{d}{2}-1}-1))$ give zero contributions.

Eq. (24) clearly demonstrates that ${}^{I}\hat{\mathcal{A}}^{m\dagger}_{f}$ transforms the Clifford odd "basis vector" in general into another Clifford odd "basis vector", transferring to the Clifford odd "basis vector" an integer spin.

We can conclude that the Clifford even "basis vectors" offer the description of the gauge fields to the corresponding fermion fields.

While the Clifford odd "basis vectors" offer the description of the internal space of the second quantized anti-commuting fermion fields, appearing in families, the Clifford even "basis vectors" offer the description of the internal space of the second quantized commuting boson fields, having no families and manifesting as the gauge fields of the corresponding fermion fields.

2.2. Example demonstrating properties of Clifford odd and even "basis vectors" for $d = (5+1)$

Subsect. 2.2 demonstrates the properties of the Clifford odd and even "basis vectors" in the special case when $d = (5+1)$.

Table 1 presents the $64\,(=2^{d=6})$ "eigenvectors" of the Cartan subalgebra members of the Lorentz algebra S^{ab}, Eq. (8). The Clifford odd "basis vectors" — they appear in $4\,(=2^{\frac{d=6}{2}-1})$ families, each family has 4 members — are products of an odd number of nilpotents, either of three nilpotents or one nilpotent. They appear in Table 1 in the group named *odd I* $\hat{b}^{m\dagger}_{f}$. Their Hermitian conjugated partners appear in the second group named *odd II* $\hat{b}^{m}_{f}$. Within each of these two groups the members are orthogonal, Eq. (15); which means that the algebraic product of $\hat{b}^{m\dagger}_{f} *_A \hat{b}^{m'\dagger}_{f`} = 0$ for all $(m, m', f, f`)$. One can check orthogonality by using relations in Eq. (11). Equivalently, the algebraic products of their Hermitian conjugated partners are also orthogonal among themselves. The "basis vectors" and their Hermitian conjugated partners are normalized as follows,

$$ <\psi_{oc}|\hat{b}^{m}_{f} *_A \hat{b}^{m'\dagger}_{f`}|\psi_{oc}> = \delta^{mm'}\delta_{ff`}\,, \tag{25}$$

since the vacuum state $|\psi_{oc}> = \frac{1}{\sqrt{2^{\frac{d=6}{2}-1}}}\,(\stackrel{03}{[-i]}\stackrel{12}{[-]}\stackrel{56}{[-]} + \stackrel{03}{[-i]}\stackrel{12}{[+]}\stackrel{56}{[+]} + \stackrel{03}{[+i]}\stackrel{12}{[-]}\stackrel{56}{[+]} + \stackrel{03}{[+i]}\stackrel{12}{[+]}\stackrel{56}{[-]})$ is normalized to one: $<\psi_{oc}|\psi_{oc}> = 1$.

The more extended overview of the properties of the Clifford odd "basis vectors" and their Hermitian conjugated partners for the case $d = (5+1)$ can be found in Ref. [10].

The Clifford even "basis vectors" are products of an even number of nilpotents — of either two or none in this case. They are presented in Table 1 in two groups, each with $16\,(=2^{\frac{d=6}{2}-1}\times 2^{\frac{d=6}{2}-1})$ members, as *even I* $\mathcal{A}^{m\dagger}_{f}$ and *even II* $\mathcal{A}^{m\dagger}_{f}$. One can easily check, using

Table 1. $2^d = 64$ "eigenvectors" of the Cartan subalgebra of the Clifford odd and even algebras — the superposition of odd and even products of γ^a's — in $d = (5+1)$-dimensional space are presented, divided into four groups. The first group, *odd I*, is chosen to represent "basis vectors", named $\hat{b}^{m\dagger}_f$, appearing in $2^{\frac{d}{2}-1} = 4$ "families" $(f = 1,2,3,4)$, each "family" with $2^{\frac{d}{2}-1} = 4$ "family" members $(m = 1,2,3,4)$. The second group, *odd II*, contains Hermitian conjugated partners of the first group for each family separately, $\hat{b}^m_f = (\hat{b}^{m\dagger}_f)^\dagger$. Either *odd I* or *odd II* are products of an odd number of nilpotents, and of projectors. The "family" quantum numbers of $\hat{b}^{m\dagger}_f$, that is the eigenvalues of $(\tilde{S}^{03}, \tilde{S}^{12}, \tilde{S}^{56})$, are for the first *odd I* group written above each "family", the quantum numbers of the members (S^{03}, S^{12}, S^{56}) are written in the last three columns. For the Hermitian conjugated partners of *odd I*, presented in the group *odd II*, the quantum numbers (S^{03}, S^{12}, S^{56}) are presented above each group of the Hermitian conjugated partners, the last three columns tell eigenvalues of $(\tilde{S}^{03}, \tilde{S}^{12}, \tilde{S}^{56})$. The two groups with the even number of γ^a's, *even I* and *even II*, each group has their Hermitian conjugated partners within its own group, have the quantum numbers f, that is the eigenvalues of $(\tilde{S}^{03}, \tilde{S}^{12}, \tilde{S}^{56})$, written above column of four members, the quantum numbers of the members, (S^{03}, S^{12}, S^{56}), are written in the last three columns.

"basis vectors" $(\tilde{S}^{03}, \tilde{S}^{12}, \tilde{S}^{56})$	m $\rightarrow$	$f=1$ $(\frac{i}{2}, -\frac{1}{2}, -\frac{1}{2})$	$f=2$ $(-\frac{i}{2}, -\frac{1}{2}, \frac{1}{2})$	$f=3$ $(-\frac{i}{2}, \frac{1}{2}, -\frac{1}{2})$	$f=4$ $(\frac{i}{2}, \frac{1}{2}, \frac{1}{2})$	S^{03}	S^{12}	S^{56}
		03 12 56	03 12 56	03 12 56	03 12 56			
odd I $\hat{b}^{m\dagger}_f$	1	$(+i)[+][+]$	$[+i][+](+)$	$[+i](+)[+]$	$(+i)(+)(+)$	$\frac{i}{2}$	$\frac{1}{2}$	$\frac{1}{2}$
	2	$[-i](-)[+]$	$(-i)(-)(+)$	$(-i)[-][+]$	$[-i][-](+)$	$-\frac{i}{2}$	$-\frac{1}{2}$	$\frac{1}{2}$
	3	$[-i][+](-)$	$(-i)[+][-]$	$(-i)(+)(-)$	$[-i](+)[-]$	$-\frac{i}{2}$	$\frac{1}{2}$	$-\frac{1}{2}$
	4	$(+i)(-)(-)$	$[+i](-)[-]$	$[+i][-](-)$	$(+i)[-][-]$	$\frac{i}{2}$	$-\frac{1}{2}$	$-\frac{1}{2}$
(S^{03}, S^{12}, S^{56})	$\rightarrow$	$(-\frac{i}{2}, \frac{1}{2}, \frac{1}{2})$ 03 12 56	$(\frac{i}{2}, \frac{1}{2}, -\frac{1}{2})$ 03 12 56	$(\frac{i}{2}, -\frac{1}{2}, \frac{1}{2})$ 03 12 56	$(-\frac{i}{2}, -\frac{1}{2}, -\frac{1}{2})$ 03 12 56	$\tilde{S}^{03}$	$\tilde{S}^{12}$	$\tilde{S}^{56}$
odd II $\hat{b}^m_f$	1	$(-i)[+][+]$	$[+i][+](-)$	$[+i](-)[+]$	$(-i)(-)(-)$	$-\frac{i}{2}$	$-\frac{1}{2}$	$-\frac{1}{2}$
	2	$[-i](+)[+]$	$(+i)(+)(-)$	$(+i)[-][+]$	$[-i][-](-)$	$\frac{i}{2}$	$\frac{1}{2}$	$-\frac{1}{2}$
	3	$[-i][+](+)$	$(+i)[+][-]$	$(+i)(-)(+)$	$[-i](-)[-]$	$\frac{i}{2}$	$-\frac{1}{2}$	$\frac{1}{2}$
5	-1	-1						
	4	$(-i)(+)(+)$	$[+i](+)[-]$	$[+i][-](+)$	$(-i)[-][-]$	$-\frac{i}{2}$	$\frac{1}{2}$	$\frac{1}{2}$
$(\tilde{S}^{03}, \tilde{S}^{12}, \tilde{S}^{56})$	$\rightarrow$	$(-\frac{i}{2}, \frac{1}{2}, \frac{1}{2})$ 03 12 56	$(\frac{i}{2}, -\frac{1}{2}, \frac{1}{2})$ 03 12 56	$(-\frac{i}{2}, -\frac{1}{2}, -\frac{1}{2})$ 03 12 56	$(\frac{i}{2}, \frac{1}{2}, -\frac{1}{2})$ 03 12 56	S^{03}	S^{12}	S^{56}
even I ${}^{I}\mathcal{A}^m_f$	1	$[+i](+)(+)$	$(+i)[+](+)$	$[+i][+][+]$	$(+i)(+)[+]$	$\frac{i}{2}$	$\frac{1}{2}$	$\frac{1}{2}$
	2	$(-i)[-](+)$	$[-i](-)(+)$	$(-i)(-)[+]$	$[-i][-][+]$	$-\frac{i}{2}$	$-\frac{1}{2}$	$\frac{1}{2}$
	3	$(-i)(+)[-]$	$[-i][+][-]$	$(-i)[+](-)$	$[-i](+)(-)$	$-\frac{i}{2}$	$\frac{1}{2}$	$-\frac{1}{2}$
	4	$[+i][-][-]$	$(+i)(-)[-]$	$[+i](-)(-)$	$(+i)[-](-)$	$\frac{i}{2}$	$-\frac{1}{2}$	$-\frac{1}{2}$
$(\tilde{S}^{03}, \tilde{S}^{12}, \tilde{S}^{56})$	$\rightarrow$	$(\frac{i}{2}, \frac{1}{2}, \frac{1}{2})$ 03 12 56	$(-\frac{i}{2}, -\frac{1}{2}, \frac{1}{2})$ 03 12 56	$(\frac{i}{2}, -\frac{1}{2}, -\frac{1}{2})$ 03 12 56	$(-\frac{i}{2}, \frac{1}{2}, -\frac{1}{2})$ 03 12 56	S^{03}	S^{12}	S^{56}
even II ${}^{II}\mathcal{A}^m_f$	1	$[-i](+)(+)$	$(-i)[+](+)$	$[-i][+][+]$	$(-i)(+)[+]$	$-\frac{i}{2}$	$\frac{1}{2}$	$\frac{1}{2}$
	2	$(+i)[-](+)$	$[+i](-)(+)$	$(+i)(-)[+]$	$[+i][-][+]$	$\frac{i}{2}$	$-\frac{1}{2}$	$\frac{1}{2}$
	3	$(+i)(+)[-]$	$[+i][+][-]$	$(+i)[+](-)$	$[+i](+)(-)$	$\frac{i}{2}$	$\frac{1}{2}$	$-\frac{1}{2}$
	4	$[-i][-][-]$	$(-i)(-)[-]$	$[-i](-)(-)$	$(-i)[-](-)$	$-\frac{i}{2}$	$-\frac{1}{2}$	$-\frac{1}{2}$

Eq. (11), that the algebraic product ${}^{I}\mathcal{A}^{m\dagger}_f *_A {}^{II}\mathcal{A}^{m'\dagger}_{f`} = 0, \forall (m, m', f.f`)$, Eq. (19). The more extended overview of the Clifford even "basis vectors" and their Hermitian conjugated partners for the case $d = (5+1)$- can be found in Ref. [12].

While the Clifford odd "basis vectors" are (chosen to be) right handed, $\Gamma^{(5+1)} = 1$, have their Hermitian conjugated partners opposite handedness [6]

[6] The handedness $\Gamma^{(d)}$, one of the invariants of the group $SO(d)$, with the infinitesimal generators of the Lorentz group S^{ab}, is defined as $\Gamma^{(d)} = \alpha\, \varepsilon_{a_1 a_2 \ldots a_{d-1} a_d} S^{a_1 a_2} \cdot S^{a_3 a_4} \cdots S^{a_{d-1} a_d}$, with α chosen so that $\Gamma^{(d)} = \pm 1$.

While the Clifford odd "basis vectors" have half integer eigenvalues of the Cartan subalgebra members, Eq. (8), that is of S^{03}, S^{12}, S^{56} in this particular case of $d=(5+1)$, the Clifford even "basis vectors" have integer spins, obtained by $\mathcal{S}^{03}=S^{03}+\tilde{S}^{03}$, $\mathcal{S}^{12}=S^{12}+\tilde{S}^{12}$, $\mathcal{S}^{56}=S^{56}+\tilde{S}^{56}$.

Let us check what does the algebraic application, $*_A$, of ${}^{I}\hat{\mathcal{A}}^{m\dagger}_{f=3}, m=(1,2,3,4)$, presented in Table 1 in the third column of *even I*, do on the Clifford odd "basis vectors" $\hat{b}^{m=1\dagger}_{f=1}$, presented as the first *odd I* "basis vector" in Table 1. This can easily be evaluated by taking into account Eq. (10) for any m.

$$\begin{aligned}
&{}^{I}\hat{\mathcal{A}}^{m\dagger}_{3} *_A \hat{b}^{1\dagger}_{1}(\equiv\overset{03}{(+i)}\overset{12}{[+]}\overset{56}{[+]}):\\
&{}^{I}\hat{\mathcal{A}}^{1\dagger}_{3}(\equiv\overset{03}{[+i]}\overset{12}{[+]}\overset{56}{[+]}) *_A \hat{b}^{1\dagger}_{1}(\equiv\overset{03}{(+i)}\overset{12}{[+]}\overset{56}{[+]}) \rightarrow \hat{b}^{1\dagger}_{1}\,,\, {}^{I}\hat{\mathcal{A}}^{1\dagger}_{3}\,\text{is selfadjoint}\\
&{}^{I}\hat{\mathcal{A}}^{2\dagger}_{3}(\equiv\overset{03}{(-i)}\overset{12}{(-)}\overset{56}{[+]}) *_A \hat{b}^{1\dagger}_{1} \rightarrow \hat{b}^{2\dagger}_{1}(\equiv\overset{03}{[-i]}\overset{12}{(-)}\overset{56}{[+]})\,,\\
&{}^{I}\hat{\mathcal{A}}^{3\dagger}_{3}(\equiv\overset{03}{(-i)}\overset{12}{[+]}\overset{56}{(-)}) *_A \hat{b}^{1\dagger}_{1} \rightarrow \hat{b}^{3\dagger}_{1}(\equiv\overset{03}{[-i]}\overset{12}{[+]}\overset{56}{(-)})\,,\\
&{}^{I}\hat{\mathcal{A}}^{4\dagger}_{3}(\equiv\overset{03}{[+i]}\overset{12}{(-)}\overset{56}{(-)}) *_A \hat{b}^{1\dagger}_{1} \rightarrow \hat{b}^{4\dagger}_{1}(\equiv\overset{03}{(+i)}\overset{12}{(-)}\overset{56}{(-)})\,. \qquad (26)
\end{aligned}$$

The sign $\rightarrow$ means that the relation is valid up to the constant. ${}^{I}\hat{\mathcal{A}}^{1\dagger}_{3}$ is self adjoint, the Hermitian conjugated partner of ${}^{I}\hat{\mathcal{A}}^{2\dagger}_{3}$ is ${}^{I}\hat{\mathcal{A}}^{1\dagger}_{4}$, of ${}^{I}\hat{\mathcal{A}}^{3\dagger}_{3}$ is ${}^{I}\hat{\mathcal{A}}^{1\dagger}_{2}$ and of ${}^{I}\hat{\mathcal{A}}^{4\dagger}_{3}$ is ${}^{I}\hat{\mathcal{A}}^{1\dagger}_{1}$.

We can conclude that the algebraic, $*_A$, application of ${}^{I}\hat{\mathcal{A}}^{m\dagger}_{3}(\equiv\overset{03}{(-i)}\overset{12}{[+]}\overset{56}{(-)})$ on $\hat{b}^{1\dagger}_{1}$ leads to the same or another family member of the same family $f=1$, namely to $\hat{b}^{m\dagger}_{1}$, $m=(1,2,3,4)$.

Calculating the eigenvalues of the Cartan subalgebra members, Eq. (8), before and after the algebraic multiplication, $*_A$, one sees that ${}^{I}\hat{\mathcal{A}}^{m\dagger}_{3}$ carry the integer eigenvalues of the Cartan subalgebra members, namely of $\mathbf{S}^{ab}=S^{ab}+\tilde{S}^{ab}$, since they transfer when applying the Clifford odd "basis vector" to it the integer eigenvalues of the Cartan subalgebra members, changing the Clifford odd "basis vector" into another Clifford odd "basis vector".

We, therefore, find out that the algebraic application of ${}^{I}\hat{\mathcal{A}}^{m\dagger}_{3}$, $m=1,2,3,4$, on $\hat{b}^{1\dagger}_{1}$ transforms $\hat{b}^{1\dagger}_{1}$ into $\hat{b}^{m\dagger}_{1}$, $m=(1,2,3,4)$. Similarly we find that the algebraic application of ${}^{I}\hat{\mathcal{A}}^{m}_{4}$, $m=(1,2,3,4)$ on $\hat{b}^{2\dagger}_{1}$ transforms $\hat{b}^{2\dagger}_{1}$ into $\hat{b}^{m\dagger}_{1}, m=(1,2,3,4)$. The algebraic application of ${}^{I}\hat{\mathcal{A}}^{m}_{2}$, $m=(1,2,3,4)$ on $\hat{b}^{3\dagger}_{1}$ transforms $\hat{b}^{3\dagger}_{1}$ into $\hat{b}^{m\dagger}_{1}, m=(1,2,3,4)$. And the algebraic application of ${}^{I}\hat{\mathcal{A}}^{m}_{1}$, $m=(1,2,3,4)$ on $\hat{b}^{4\dagger}_{1}$ transforms $\hat{b}^{4\dagger}_{1}$ into $\hat{b}^{m\dagger}_{1}, m=(1,2,3,4)$.

The statement of Eq. (24) is therefore demonstrated on the case of $d=(5+1)$.

It remains to stress and to illustrate in the case of $d=(5+1)$ some general properties of the Clifford even "basis vectors", ${}^{I}\hat{\mathcal{A}}^{m\dagger}_{f}$ when they apply on each other. Let us denote the self-adjoint member in each group of "basis vectors" of particular f as ${}^{I}\hat{\mathcal{A}}^{m_0\dagger}_{f}$. We easily see that

$$\begin{aligned}
\{{}^{I}\hat{\mathcal{A}}^{m\dagger}_{f}, {}^{I}\hat{\mathcal{A}}^{m'\dagger}_{f}\}_- &= 0\,, \quad \text{if}\,(m,m')\neq m_0\,\text{or}\,m=m_0=m'\,,\forall f\,,\\
{}^{I}\hat{\mathcal{A}}^{m\dagger}_{f} *_A {}^{I}\hat{\mathcal{A}}^{m_0\dagger}_{f} &\rightarrow {}^{I}\hat{\mathcal{A}}^{m\dagger}_{f}\,, \quad \forall m\,,\forall f\,. \qquad (27)
\end{aligned}$$

Since a projector to the second power is the projector back, and since a projector is self-adjoint, it means that the self-adjoint Clifford even "basis vectors" are the product of projectors. In Table 1, we see that in each column of either *even* ${}^{I}\hat{\mathcal{A}}^{m\dagger}_{f}$ or of *even* ${}^{II}\hat{\mathcal{A}}^{m\dagger}_{f}$ there is one self adjoint ${}^{I,II}\hat{\mathcal{A}}^{m_0\dagger}_{f}$. We also see that two "basis vectors" ${}^{I}\hat{\mathcal{A}}^{m\dagger}_{f}$ and ${}^{I}\hat{\mathcal{A}}^{m'\dagger}_{f}$ of the same f and of $(m,m')\neq m_0$ are orthogonal. We only have to take into account Eq. (11), which tells that

$$\overset{ab}{(k)}\overset{ab}{[k]}=0\,,\quad \overset{ab}{[k]}\overset{ab}{(k)}=\overset{ab}{(k)}\,,\quad \overset{ab}{(k)}\overset{ab}{[-k]}=\overset{ab}{(k)}\,,\quad \overset{ab}{[k]}\overset{ab}{(-k)}=0.$$

These relations tell us that $= {}^{I}\hat{\mathcal{A}}^{1\dagger}_{3}$, what illustrates Eq. (23), while ${}^{I}\hat{\mathcal{A}}^{2\dagger}_{3} *_{A} {}^{I}\hat{\mathcal{A}}^{1\dagger}_{4} = {}^{I}\hat{\mathcal{A}}^{2\dagger}_{4}$ illustrating Eq. (22), while ${}^{I}\hat{\mathcal{A}}^{1\dagger}_{3} *_{A} {}^{I}\hat{\mathcal{A}}^{2\dagger}_{4} = 0$ illustrates Eq. (21).

Table 2 presents the Clifford even "basis vectors" ${}^{I}\hat{\mathcal{A}}^{m\dagger}_{f}$ for $d = (5+1)$ with the properties:

i. They are products of an even number of nilpotents, $\overset{ab}{(k)}$, with the rest up to $\frac{d}{2}$ of projectors, $\overset{ab}{[k]}$.

ii. Nilpotents and projectors are eigenvectors of the Cartan subalgebra members $\mathcal{S}^{ab} = S^{ab} + \tilde{S}^{ab}$, Eq. (8), carrying the integer eigenvalues of the Cartan subalgebra members.

iii. They have their Hetmitian conjugated partners within the same group of ${}^{I}\hat{\mathcal{A}}^{m\dagger}_{f}$ with $2^{\frac{d}{2}-1} \times 2^{\frac{d}{2}-1}$ members.

iv. They have properties of the boson gauge fields. When the Clifford even "basis vectors", ${}^{I}\hat{\mathcal{A}}^{m\dagger}_{f}$, apply the Clifford odd "basis vectors" (offering the description of the fermion fields) they transform the Clifford odd "basis vectors" into another Clifford odd "basis vectors", transferring to the Clifford odd "basis vectors" the integer spins with respect to the $SO(d-1,1)$ group, while with respect to subgroups of the $SO(d-1,1)$ group they transfer appropriate superposition of the eigenvalues (manifesting the properties of the adjoint representations of the corresponding groups).

To demonstrate that the Clifford even "basis vectors" have properties of the gauge fields of the corresponding Clifford odd "basis vectors" we study properties of the $SU(3) \times U(1)$ subgroups of the Clifford odd and Clifford even "basis vectors".

We present in Eqs. (28, 29) the superposition of members of Cartan subalgebra, Eq. (8) for S^{ab} for the Clifford odd "basis vectors", for the subgroup $SO(3,1) \times U(1)$ $(N^{3}_{\pm}, \tau)$ and for the subgroup $SU(3)$ $\times U(1)$: $(\tau', \tau^{3}, \tau^{8})$. The same relations can be used also for the corresponding operators determining the "family" quantum numbers $(\tilde{N}^{3}_{\pm}, \tilde{\tau})$ of the Clifford odd "basis vectors', if S^{ab}'s are replaced by $\tilde{S}^{ab}$'s. For the Clifford even objects $\mathbf{S}^{ab}(= S^{ab} + \tilde{S}^{ab})$ must replace S^{ab}.

$$N^{3}_{\pm}(= N^{3}_{(L,R)}) := \frac{1}{2}(S^{12} \pm i S^{03}), \quad \tau = S^{56}, \tag{28}$$

$$\begin{aligned} \tau^{3} := &\quad \frac{1}{2}(-S^{12} - iS^{03}), \qquad \tau^{8} = \frac{1}{2\sqrt{3}}(-iS^{03} + S^{12} - 2S^{56}), \\ \tau' = &\quad -\frac{1}{3}(-iS^{03} + S^{12} + S^{56}). \end{aligned} \tag{29}$$

Let us, for example, algebraically apply ${}^{I}\hat{\mathcal{A}}^{2}_{3}$ $(\equiv \overset{03}{(-i)}\overset{12}{(-)}\overset{56}{[+]})$, denoted by $\odot\odot$ in Table 2, carrying $(\tau^{3} = 0, \tau^{8} = -\frac{1}{\sqrt{3}}, \tau' = \frac{2}{3})$, represented also on Fig. 2 by $\odot\odot$, on the Clifford odd "basis vector" $\hat{b}^{1\dagger}_{4}(\equiv \overset{03}{(+i)}\overset{12}{(+)}\overset{56}{(+)})$, presented in Table 1, with $(\tau^{3} = 0, \tau^{8} = 0, \tau' = -\frac{1}{2})$, as we can calculate using Eq. (29), and which is represented in Fig. 1 by a square as a singlet. ${}^{I}\hat{\mathcal{A}}^{2}_{3}$ transforms $\hat{b}^{1\dagger}_{4}$ (by transferring to $\hat{b}^{1\dagger}_{4}$ $(\tau^{3} = 0, \tau^{8} = -\frac{1}{\sqrt{3}}, \tau' = \frac{2}{3})$) to $\hat{b}^{2\dagger}_{4}$ with $(\tau^{3} = 0, \tau^{8} = -\frac{1}{\sqrt{3}}, \tau' = \frac{1}{6})$, belonging in Fig. 1 to the triplet, denoted by $\bigcirc$. The corresponding gauge fields, presented in Fig. 2, if belonging to the sextet, would transform the triplet of quarks among themselves.

In the case of the group $SO(6)$, manifesting as $SU(3) \times U(1)$ and representing the $SU(3)$ colour group and $U(1)$ the "fermion" quantum number, embedded into $SO(13,1)$ the triplet would represent quarks and the singlet leptons. The corresponding gauge fields, presented in Fig. 2, if belonging to the sextet, would transform the triplet of quarks among themselves, changing the colour and leaving the "fermion" quantum number equal to $\frac{1}{6}$.

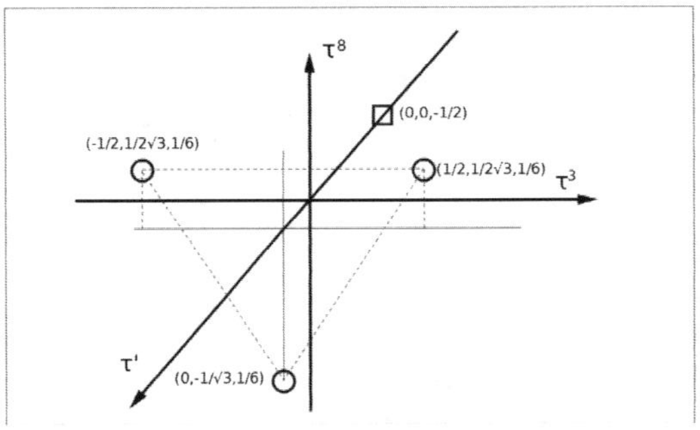

Figure 1. Representations of the subgroups $SU(3)$ and $U(1)$ of the group $SO(5,1)$, the properties of which appear in Table 1 for the Clifford odd "basis vectors", are presented. (τ^3, τ^8 and τ') can be calculated if using Eqs.(28, 29). On the abscissa axis, on the ordinate axis and on the third axis the eigenvalues of the superposition of the three Cartan subalgebra members, τ^3, τ^8, τ' are presented. One notices one triplet, denoted by ○ with the values $\tau' = \frac{1}{6}$, $(\tau^3 = -\frac{1}{2}, \tau^8 = \frac{1}{2\sqrt{3}}, \tau' = \frac{1}{6})$, $(\tau^3 = \frac{1}{2}, \tau^8 = \frac{1}{2\sqrt{3}}, \tau' = \frac{1}{6})$, $(\tau^3 = 0, \tau^8 = -\frac{1}{\sqrt{3}}, \tau' = \frac{1}{6})$, respectively, and one singlet denoted by the square. $(\tau^3 = 0, \tau^8 = 0, \tau' = -\frac{1}{2})$. The triplet and the singlet appear in four families.

We can see that ${}^{I}\hat{\mathcal{A}}^{m\dagger}_{3}$ with $(m = 2, 3, 4)$, if applied on the $SU(3)$ singlet $\hat{b}^{1\dagger}_{4}$ with $(\tau' = -\frac{1}{2}, \tau^3 = 0, \tau^8 = 0)$, transforms it to $\hat{b}^{m=2,3,4)\dagger}_{4}$, respectively, which are members of the $SU(3)$ triplet. All these Clifford even "basis vectors" have τ' equal to $\frac{2}{3}$, changing correspondingly $\tau' = -\frac{1}{2}$ into $\tau' = \frac{1}{6}$ and bringing the needed values of τ^3 and τ^8.

In Table 2 we find $(6+4)$ Clifford even "basis vectors" ${}^{I}\hat{\mathcal{A}}^{m\dagger}_{f}$ with $\tau‘ = 0$. Six of them are Hermitian conjugated to each other — the Hermitian conjugated partners are denoted by the same geometric figure on the third column. Four of them are self-adjoint and correspondingly with $(\tau' = 0, \tau^3 = 0, \tau^8 = 0)$, denoted in the third column of Table 2 by ○. The rest 6 Clifford even "basis vectors" belong to one triplet with $\tau' = \frac{2}{3}$ and (τ^3, τ^8) equal to $[(0, -\frac{1}{\sqrt{3}}), (-\frac{1}{2}, \frac{1}{2\sqrt{3}}), (\frac{1}{2}, \frac{1}{2\sqrt{3}})]$ and one antitriplet with $\tau' = -\frac{2}{3}$ and $((\tau^3, \tau^8)$ equal to $[(-\frac{1}{2}, -\frac{1}{2\sqrt{3}}), (\frac{1}{2}, -\frac{1}{2\sqrt{3}}), (0, \frac{1}{\sqrt{3}})]$.
Each triplet has Hermitian conjugated partners in antitriplet and opposite. In Table 2 the Hermitian conjugated partners of the triplet and antitriplet are denoted by the same signum: $({}^{I}\hat{\mathcal{A}}^{1\dagger}_{1}, {}^{I}\hat{\mathcal{A}}^{4\dagger}_{3})$ by ⋆⋆, $({}^{I}\hat{\mathcal{A}}^{1\dagger}_{2}, {}^{I}\hat{\mathcal{A}}^{3\dagger}_{3})$ by •, and $({}^{I}\hat{\mathcal{A}}^{2\dagger}_{3}, {}^{I}\hat{\mathcal{A}}^{1\dagger}_{4})$ by ⊙⊙.
The octet and the two triplets are presented in Fig. 2.

Fig. 2 represents the $2^{\frac{d}{2}-1} \times 2^{\frac{d}{2}-1}$ members ${}^{I}\hat{\mathcal{A}}^{m}_{f}$ of the Clifford even "basis vectors" for the case that $d = (5+1)$. The properties of ${}^{I}\hat{\mathcal{A}}^{m}_{f}$ are presented also in Table 2. There are in this case again 16 members. Manifesting the structure of subgroups $SU(3) \times U(1)$ of the group $SO(5,1)$ they are represented as eigenvectors of the superposition of the Cartan subalgebra members $(\mathcal{S}^{03}, \mathcal{S}^{12}, \mathcal{S}^{56})$, that is with $\tau^3 = \frac{1}{2}(-\mathcal{S}^{12} - i\mathcal{S}^{03})$, $\tau^8 = \frac{1}{2\sqrt{3}}(\mathcal{S}^{12} - i\mathcal{S}^{03} - 2\mathcal{S}^{56})$, and $\tau' = -\frac{1}{3}(\mathcal{S}^{12} - i\mathcal{S}^{03} + \mathcal{S}^{56})$. There are four self adjoint Clifford even "basis vectors" with $(\tau^3 = 0, \tau^8 = 0, \tau' = 0)$, one sextet of three pairs Hermitian conjugated to each other, one triplet and one antitriplet with the members of the triplet Hermitian conjugated to the corresponding members of the antitriplet and opposite. These 16 members of the Clifford even "basis vectors" ${}^{I}\hat{\mathcal{A}}^{m}_{f}$ are the boson "partners" of the Clifford odd "basis vectors" $\hat{b}^{m\dagger}_{f}$, presented in Fig. 1 for one of four families, anyone. The reader can check that the algebraic application of ${}^{I}\hat{\mathcal{A}}^{m}_{f}$, belonging to the triplet, transforms the Clifford odd singlet, denoted in Fig. 1 by a square, to one of the

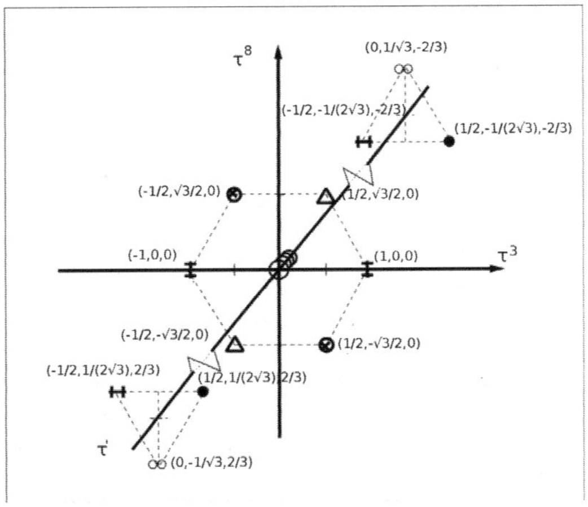

Figure 2. The Clifford even "basis vectors" ${}^{I}\hat{\mathcal{A}}^{m}_{f}$, in the case that $d = (5+1)$, are presented concerning the eigenvalues of the commuting operators of the subgroups $SU(3)$ and $U(1)$ of the group $SO(5,1)$: $\tau^3 = \frac{1}{2}\,(-\mathcal{S}^{12} - i\mathcal{S}^{03})$, $\tau^8 = \frac{1}{2\sqrt{3}}(\mathcal{S}^{12} - i\mathcal{S}^{03} - 2\mathcal{S}^{56})$, $\tau' = -\frac{1}{3}(\mathcal{S}^{12} - i\mathcal{S}^{03} + \mathcal{S}^{56})$. Their properties appear in Table 2. The abscissa axis carries the eigenvalues of τ^3, the ordinate axis of τ^8 and the third axis the eigenvalues of τ', One notices four singlets with $(\tau^3 = 0, \tau^8 = 0, \tau' = 0)$, denoted by ◯, representing four self adjoint Clifford even "basis vectors" ${}^{I}\hat{\mathcal{A}}^{m}_{f}$, one sextet of three pairs with $\tau' = 0$, Hermitian conjugated to each other, denoted by △ (with $(\tau' = 0, \tau^3 = -\frac{1}{2}, \tau^8 = -\frac{3}{2\sqrt{3}})$ and $(\tau' = 0, \tau^3 = \frac{1}{2}, \tau^8 = \frac{3}{2\sqrt{3}})$), respectively, by ‡ (with $(\tau' = 0, \tau^3 = -1, \tau^8 = 0)$ and $(\tau' = 0, \tau^3 = 1, \tau^8 = 0)$, respectively, and by ⊗ (with $(\tau' = 0, \tau^3 = \frac{1}{2}, \tau^8 = -\frac{3}{2\sqrt{3}})$ and $(\tau' = 0, \tau^3 = -\frac{1}{2}, \tau^8 = \frac{3}{2\sqrt{3}})$), respectively, and one triplet, denoted by ⋆⋆ with $(\tau' = \frac{2}{3}, \tau^3 = \frac{1}{2}, \tau^8 = \frac{1}{2\sqrt{3}})$, by • with $(\tau' = \frac{2}{3}, \tau^3 = -\frac{1}{2}, \tau^8 = \frac{1}{2\sqrt{3}})$, and by ⊙⊙ with $(\tau' = \frac{2}{3}, \tau^3 = 0, \tau^8 = -\frac{1}{\sqrt{3}})$, as well as one antitriplet, Hermitian conjugated to the triplet, denoted by ⋆⋆ with $(\tau' = -\frac{2}{3}, \tau^3 = -\frac{1}{2}, \tau^8 = -\frac{1}{2\sqrt{3}})$, by • with $(\tau' = -\frac{2}{3}, \tau^3 = \frac{1}{2}, \tau^8 = -\frac{1}{2\sqrt{3}})$, and by ⊙⊙ with $(\tau' = -\frac{2}{3}, \tau^3 = 0, \tau^8 = \frac{1}{\sqrt{3}})$.

members of the triplet, denoted in Fig. 1 by the circle ◯.

Looking at the boson fields ${}^{I}\hat{\mathcal{A}}^{m\dagger}_{f}$ from the point of view of subgroups $SU(3) \times U(1)$ of the group $SO(5+1)$ we will recognize in the part of fields forming the octet the colour gauge fields of quarks and leptons and antiquarks and antileptons.

2.3. Second quantized fermion and boson fields with internal spaces described by Clifford "basis vectors".

We learned in the previous Subsects. (2.1, 2.2) that in even dimensional spaces ($d = 2(2n+1)$ or $d = 4n$) the Clifford odd and the Clifford even "basis vectors", which are the superposition of the Clifford odd and the Clifford even products of γ^a's, respectively, offer the description of the internal spaces of fermion and boson fields.

The Clifford odd algebra offers $2^{\frac{d}{2}-1}$ "basis vectors" $\hat{b}^{m\dagger}_{f}$, appearing in $2^{\frac{d}{2}-1}$ families (with the family quantum numbers determined by $\tilde{S}^{ab} = \frac{i}{2}\{\tilde{\gamma}^a, \tilde{\gamma}^b\}_-$), which together with their $2^{\frac{d}{2}-1} \times 2^{\frac{d}{2}-1}$ Hermitian conjugated partners $\hat{b}^{m}_{f}$ fulfil the postulates for the second quantized fermion fields, Eq. (17) in this paper, Eq.(26) in Ref. [10], explaining the second quantization postulates of Dirac.

Table 2. The Clifford even "basis vectors" ${}^{I}\hat{\mathcal{A}}^{m\dagger}_{f}$, each of them is the product of projectors and an even number of nilpotents, and each is the eigenvector of all the Cartan subalgebra members, $\mathcal{S}^{03}$, $\mathcal{S}^{12}$, $\mathcal{S}^{56}$, Eq. (8), are presented for $d=(5+1)$-dimensional case. Indexes m and f determine $2^{\frac{d}{2}-1}\times 2^{\frac{d}{2}-1}$ different members ${}^{I}\hat{\mathcal{A}}^{m\dagger}_{f}$. In the third column the "basis vectors" ${}^{I}\hat{\mathcal{A}}^{m\dagger}_{f}$ which are Hermitian conjugated partners to each other (and can therefore annihilate each other) are pointed out with the same symbol. For example, with $\star\star$ are equipped the first member with $m=1$ and $f=1$ and the last member of $f=3$ with $m=4$. The sign $\bigcirc$ denotes the Clifford even "basis vectors" which are self adjoint $({}^{I}\hat{\mathcal{A}}^{m\dagger}_{f})^{\dagger}={}^{I}\hat{\mathcal{A}}^{m'\dagger}_{f'}$. It is obvious that † has no meaning, since ${}^{I}\hat{\mathcal{A}}^{m\dagger}_{f}$ are self adjoint or are Hermitian conjugated partner to another ${}^{I}\hat{\mathcal{A}}^{m'\dagger}_{f'}$. This table represents also the eigenvalues of the three commuting operators $\mathcal{N}^{3}_{L,R}$ and $\mathcal{S}^{56}$ of the subgroups $SU(2)\times SU(2)\times U(1)$ of the group $SO(5,1)$ and the eigenvalues of the three commuting operators τ^{3},τ^{8} and τ' of the subgroups $SU(3)\times U(1)$.

f	m	$*$	${}^{I}\hat{\mathcal{A}}^{m\dagger}_{f}$	$\mathcal{S}^{03}$	$\mathcal{S}^{12}$	$\mathcal{S}^{56}$	$\mathcal{N}^{3}_{L}$	$\mathcal{N}^{3}_{R}$	τ^{3}	τ^{8}	τ'
I	1	$\star\star$	$\stackrel{03}{[+i]}\stackrel{12}{(+)}\stackrel{56}{(+)}$	0	1	1.	$\frac{1}{2}$	$\frac{1}{2}$	$-\frac{1}{2}$	$-\frac{1}{2\sqrt{3}}$	$-\frac{2}{3}$
	2	$\triangle$	$\stackrel{03}{(-i)}\stackrel{12}{[-]}\stackrel{56}{(+)}$	$-i$	0	1	$\frac{1}{2}$	$-\frac{1}{2}$	$-\frac{1}{2}$	$-\frac{3}{2\sqrt{3}}$	0
	3	$\ddagger$	$\stackrel{03}{(-i)}\stackrel{12}{(+)}\stackrel{56}{[-]}$	$-i$	1	0	1	0	-1	0	0
	4	$\bigcirc$	$\stackrel{03}{[+i]}\stackrel{12}{[-]}\stackrel{56}{[-]}$	0	0	0	0	0	0	0	0
II	1	$\bullet$	$\stackrel{03}{(+i)}\stackrel{12}{[+]}\stackrel{56}{(+)}$	i	0	1	$-\frac{1}{2}$	$\frac{1}{2}$	$\frac{1}{2}$	$-\frac{1}{2\sqrt{3}}$	$-\frac{2}{3}$
	2	$\otimes$	$\stackrel{03}{[-i]}\stackrel{12}{(-)}\stackrel{56}{(+)}$	0	-1	1	$-\frac{1}{2}$	$-\frac{1}{2}$	$\frac{1}{2}$	$-\frac{3}{2\sqrt{3}}$	0
	3	$\bigcirc$	$\stackrel{03}{[-i]}\stackrel{12}{[+]}\stackrel{56}{[-]}$	0	0	0	0	0	0	0	0
	4	$\ddagger$	$\stackrel{03}{(+i)}\stackrel{12}{(-)}\stackrel{56}{[-]}$	i	-1	0	-1	0	1	0	0
III	1	$\bigcirc$	$\stackrel{03}{[+i]}\stackrel{12}{[+]}\stackrel{56}{[+]}$	0	0	0	0	0	0	0	0
	2	$\odot\odot$	$\stackrel{03}{(-i)}\stackrel{12}{(-)}\stackrel{56}{[+]}$	$-i$	-1	0	0	-1	0	$-\frac{1}{\sqrt{3}}$	$\frac{2}{3}$
	3	$\bullet$	$\stackrel{03}{(-i)}\stackrel{12}{[+]}\stackrel{56}{(-)}$	$-i$	0	-1	$\frac{1}{2}$	$-\frac{1}{2}$	$-\frac{1}{2}$	$\frac{1}{2\sqrt{3}}$	$\frac{2}{3}$
	4	$\star\star$	$\stackrel{03}{[+i]}\stackrel{12}{(-)}\stackrel{56}{(-)}$	0	-1	-1	$-\frac{1}{2}$	$-\frac{1}{2}$	$\frac{1}{2}$	$\frac{1}{2\sqrt{3}}$	$\frac{2}{3}$
IV	1	$\odot\odot$	$\stackrel{03}{(+i)}\stackrel{12}{(+)}\stackrel{56}{[+]}$	i	1	0	0	1	0	$\frac{1}{\sqrt{3}}$	$-\frac{2}{3}$
	2	$\bigcirc$	$\stackrel{03}{[-i]}\stackrel{12}{[-]}\stackrel{56}{[+]}$	0	0	0	0	0	0	0	0
	3	$\otimes$	$\stackrel{03}{[-i]}\stackrel{12}{(+)}\stackrel{56}{(-)}$	0	1	-1	$\frac{1}{2}$	$\frac{1}{2}$	$-\frac{1}{2}$	$\frac{3}{2\sqrt{3}}$	0
	4	$\triangle$	$\stackrel{03}{(+i)}\stackrel{12}{[-]}\stackrel{56}{(-)}$	i	0	-1	$-\frac{1}{2}$	$\frac{1}{2}$	$\frac{1}{2}$	$\frac{3}{2\sqrt{3}}$	0

The Clifford even algebra offers $2^{\frac{d}{2}-1}\times 2^{\frac{d}{2}-1}$ "basis vectors" of ${}^{I}\hat{\mathcal{A}}^{m\dagger}_{f}$ (and the same number of ${}^{II}\hat{\mathcal{A}}^{m\dagger}_{f}$) with the properties of the second quantized boson fields manifesting as the gauge fields of fermion fields described by the Clifford odd "basis vectors" $\hat{b}^{m\dagger}_{f}$.

The Clifford odd and the Clifford even "basis vectors" are chosen to be products of nilpotents, $\stackrel{ab}{(k)}$ (with the odd number of nilpotents if describing fermions and the even number of nilpotents if describing bosons), and projectors, $\stackrel{ab}{[k]}$. Nilpotents and projectors are (chosen to be) eigenvectors of the Cartan subalgebra members of the Lorentz algebra in the internal space of S^{ab} for the Clifford odd "basis vectors" and of $\mathcal{S}^{ab}(=S^{ab}+\tilde{S}^{ab})$ for the Clifford even "basis vectors".

To define the creation operators, either for fermions or for bosons, besides the "basis vectors" defining the internal space of fermions and bosons also the basis in ordinary space in momentum or coordinate representation is needed. Here Ref. [10], Subsect. 3.3 and App. J is overviewed.

Let us introduce the momentum part of the single particle states. The more extended version is presented in Ref. [10] in Subsect. 3.3 and in App. J.

$$\begin{aligned}
|\vec{p}> &= \hat{b}^{\dagger}_{\vec{p}}\,|\,0_p>\,, \quad <\vec{p}\,| =<0_p\,|\,\hat{b}_{\vec{p}}\,,\\
<\vec{p}\,|\,\vec{p}'> &= \delta(\vec{p}-\vec{p}') =<0_p\,|\hat{b}_{\vec{p}}\,\hat{b}^{\dagger}_{\vec{p}'}\,|\,0_p>\,,\\
&\text{leading to}\\
\hat{b}_{\vec{p'}}\,\hat{b}^{\dagger}_{\vec{p}} &= \delta(\vec{p'}-\vec{p})\,, \qquad (30)
\end{aligned}$$

with the normalization $<0_p\,|\,0_p>=1$. While the quantized operators $\hat{\vec{p}}$ and $\hat{\vec{x}}$ commute $\{\hat{p}^i\,,\hat{p}^j\}_- = 0$ and $\{\hat{x}^k\,,\hat{x}^l\}_- = 0$, it follows for $\{\hat{p}^i\,,\hat{x}^j\}_- = i\eta^{ij}$. One correspondingly finds

$$\begin{aligned}
<\vec{p}\,|\,\vec{x}> &= <0_{\vec{p}}\,|\,\hat{b}_{\vec{p}}\,\hat{b}^{\dagger}_{\vec{x}}|0_{\vec{x}}>=(<0_{\vec{x}}\,|\,\hat{b}_{\vec{x}}\,\hat{b}^{\dagger}_{\vec{p}}\,|0_{\vec{p}}>)^{\dagger}\\
\{\hat{b}^{\dagger}_{\vec{p}},\,\hat{b}^{\dagger}_{\vec{p}'}\}_- &= 0\,, \qquad \{\hat{b}_{\vec{p}},\,\hat{b}_{\vec{p}'}\}_- = 0\,, \qquad \{\hat{b}_{\vec{p}},\,\hat{b}^{\dagger}_{\vec{p}'}\}_- = 0\,,\\
\{\hat{b}^{\dagger}_{\vec{x}},\,\hat{b}^{\dagger}_{\vec{x}'}\}_- &= 0\,, \qquad \{\hat{b}_{\vec{x}},\,\hat{b}_{\vec{x}'}\}_- = 0\,, \qquad \{\hat{b}_{\vec{x}},\,\hat{b}^{\dagger}_{\vec{x}'}\}_- = 0\,,\\
\{\hat{b}_{\vec{p}},\,\hat{b}^{\dagger}_{\vec{x}}\}_- &= e^{i\vec{p}\cdot\vec{x}}\frac{1}{\sqrt{(2\pi)^{d-1}}}\,, \quad \{\hat{b}_{\vec{x}},\,\hat{b}^{\dagger}_{\vec{p}}\}_- = e^{-i\vec{p}\cdot\vec{x}}\frac{1}{\sqrt{(2\pi)^{d-1}}}\,. \qquad (31)
\end{aligned}$$

.

The internal space of either fermion or boson fields has the finite number of "basis vectors", $2^{\frac{d}{2}-1}\times 2^{\frac{d}{2}-1}$, the momentum basis is continuously infinite.

The creation operators for either fermions or bosons must be tensor products, $*_T$, of both contributions, the "basis vectors" describing the internal space of fermions or bosons and the basis in ordinary momentum or coordinate space.

The creation operators for a free massless fermion of the energy $p^0 = |\vec{p}|$, belonging to a family f and to a superposition of family members m applying on the vacuum state $|\psi_{oc}>*_T|0_{\vec{p}}>$ can be written as ([10], Subsect.3.3.2, and the references therein)

$$\hat{\mathbf{b}}^{s\dagger}_f(\vec{p}) = \sum_m c^{sm}{}_f(\vec{p})\,\hat{b}^{\dagger}_{\vec{p}}\,*_T\,\hat{b}^{m\dagger}_f\,, \qquad (32)$$

where the vacuum state for fermions $|\psi_{oc}>*_T|0_{\vec{p}}>$ includes both spaces, the internal part, Eq.(16), and the momentum part, Eq. (30) (in a tensor product for a starting single particle state with zero momentum, from which one obtains the other single fermion states of the same "basis vector" by the operator $\hat{b}^{\dagger}_{\vec{p}}$ which pushes the momentum by an amount $\vec{p}$ [7]).
The creation operators and annihilation operators for fermion fields fulfil the anti-commutation relations for the second quantized fermion fields

$$\begin{aligned}
\{\hat{\mathbf{b}}^{s'}_{f'}(\vec{p'})\,,\,\hat{\mathbf{b}}^{s\dagger}_f(\vec{p})\}_+\,|\psi_{oc}>|0_{\vec{p}}> &= \delta^{ss'}\delta_{ff'}\,\delta(\vec{p'}-\vec{p})\cdot|\psi_{oc}>|0_{\vec{p}}>\,,\\
\{\hat{\mathbf{b}}^{s'}_{f'}(\vec{p'})\,,\,\hat{\mathbf{b}}^{s}_f(\vec{p})\}_+\,|\psi_{oc}>|0_{\vec{p}}> &= 0\cdot|\psi_{oc}>|0_{\vec{p}}>\,,\\
\{\hat{\mathbf{b}}^{s'\dagger}_{f'}(\vec{p'})\,,\,\hat{\mathbf{b}}^{s\dagger}_f(\vec{p})\}_+\,|\psi_{oc}>|0_{\vec{p}}> &= 0\cdot|\psi_{oc}>|0_{\vec{p}}>\,,\\
\hat{\mathbf{b}}^{s\dagger}_f(\vec{p})\,|\psi_{oc}>|0_{\vec{p}}> &= |\psi^s_f(\vec{p})>\,,\\
\hat{\mathbf{b}}^{s}_f(\vec{p})\,|\psi_{oc}>|0_{\vec{p}}> &= 0\cdot|\psi_{oc}>|0_{\vec{p}}>\,,\\
|p^0| &= |\vec{p}|\,. \qquad (33)
\end{aligned}$$

[7] The creation operators and their Hermitian conjugated partners annihilation operators in the coordinate representation can be read in [10] and the references therein: $\hat{\mathbf{b}}^{s\dagger}_f(\vec{x},x^0) = \sum_m \hat{b}^{m\dagger}_f \int_{-\infty}^{+\infty}\frac{d^{d-1}p}{(\sqrt{2\pi})^{d-1}}\,c^{ms}{}_f\,(\vec{p})\,\hat{b}^{\dagger}_{\vec{p}}\,e^{-i(p^0x^0-\varepsilon\vec{p}\cdot\vec{x})}$ ([10], subsect. 3.3.2., Eqs. (55,57,64) and the references therein).

The creation operators $\hat{\mathbf{b}}^{s\dagger}_f(\vec{p}))$ and their Hermitian conjugated partners annihilation operators $\hat{\mathbf{b}}^{s}_f(\vec{p})$, creating and annihilating the single fermion states, respectively, fulfil when applying on the vacuum state, $|\psi_{oc}> *_T|0_{\vec{p}}>$, the anti-commutation relations for the second quantized fermions, postulated by Dirac (Ref. [10], Subsect. 3.3.1, Sect. 5). [8]

To write the creation operators for boson fields, we must take into account that boson gauge fields have the space index α, describing the α component of the boson field in the ordinary space [9]. We, therefore, add the space index α as follows

$$ {}^{\mathbf{I}}\hat{\mathcal{A}}^{\mathbf{m}\dagger}_{\mathbf{f}\alpha}(\vec{p}) = \hat{b}^{\dagger}_{\vec{p}} *_T \mathcal{C}^{m}{}_{f\alpha}\, {}^{I}\hat{\mathcal{A}}^{m\dagger}_{f} \,. \tag{34} $$

We treat free massless bosons of momentum $\vec{p}$ and energy $p^0 = |\vec{p}|$ and of particular "basis vectors" ${}^{I}\hat{\mathcal{A}}^{m\dagger}_{f}$'s which are eigenvectors of all the Cartan subalgebra members [10], $\mathcal{C}^{m}{}_{f\alpha}$ carry the space index α of the boson field. Creation operators operate on the vacuum state $|\psi_{oc_{ev}}> *_T|0_{\vec{p}}>$ with the internal space part just a constant, $|\psi_{oc_{ev}}>=|1>$, and for a starting single boson state with a zero momentum from which one obtains the other single boson states with the same "basis vector" by the operators $\hat{b}^{\dagger}_{\vec{p}}$ which push the momentum by an amount $\vec{p}$, making also $\mathcal{C}^{m}{}_{f\alpha}$ depending on $\vec{p}$.

For the creation operators for boson fields in a coordinate representation one finds using Eqs. (30, 31)

$$ {}^{\mathbf{I}}\hat{\mathcal{A}}^{\mathbf{m}\dagger}_{\mathbf{f}\alpha}(\vec{x},x^0) = \int_{-\infty}^{+\infty} \frac{d^{d-1}p}{(\sqrt{2\pi})^{d-1}}\, {}^{I}\hat{\mathcal{A}}^{m\dagger}_{f\alpha}(\vec{p})\, e^{-i(p^0x^0-\varepsilon\vec{p}\cdot\vec{x})}|_{p^0=|\vec{p}|} \,. \tag{35} $$

To understand what new does the Clifford algebra description of the internal space of fermion and boson fields, Eqs. (34, 35, 32), bring to our understanding of the second quantized fermion and boson fields and what new can we learn from this offer, we need to relate $\sum_{ab} c^{ab}\omega_{ab\alpha}$ and $\sum_{mf} {}^{I}\hat{\mathcal{A}}^{m\dagger}_{f}\mathcal{C}^{mf}_{\alpha}$, recognizing that ${}^{I}\hat{\mathcal{A}}^{m\dagger}_{f}\mathcal{C}^{mf}_{\alpha}$ are eigenstates of the Cartan subalgebra members, while $\omega_{ab\alpha}$ are not.

The gravity fields, the vielbeins and the two kinds of the spin connection fields, $f^{a}{}_{\alpha}$, $\omega_{ab\alpha}$, $\tilde{\omega}_{ab\alpha}$, respectively, are in the *spin-charge-family* theory (unifying spins, charges and families of fermions and offering not only the explanation for all the assumptions of the *standard model* but also for the increasing number of phenomena observed so far) the only boson fields in $d=(13+1)$, observed in $d=(3+1)$ besides as gravity also as all the other boson fields with the Higgs's scalars included [25].

We therefore need to relate

$$ \{\frac{1}{2}\sum_{ab} S^{ab}\,\omega_{ab\alpha}\}\sum_{m}\beta^{mf}\,\hat{\mathbf{b}}^{m\dagger}_{f}(\vec{p}) \quad \text{related to} \quad \{\sum_{m'f'} {}^{I}\hat{\mathcal{A}}^{m'\dagger}_{f'}\,\mathcal{C}^{m'f'}_{\alpha}\}\sum_{m}\beta^{mf}\,\hat{\mathbf{b}}^{m\dagger}_{f}(\vec{p}) \,, $$
$$ \forall f \,\text{and}\, \forall\, \beta^{mf} \,, $$

[8] The anti-commutation relations of Eq. (33) are valid also if we replace the vacuum state, $|\psi_{oc}>|0_{\vec{p}}>$, by the Hilbert space of the Clifford fermions generated by the tensor products multiplication, $*_{T_H}$, of any number of the Clifford odd fermion states of all possible internal quantum numbers and all possible momenta (that is of any number of $\hat{\mathbf{b}}^{s\dagger}_{f}(\vec{p})$ of any $(s,f,\vec{p})$), Ref. ([10], Sect. 5.).

[9] In the *spin-charge-family* theory the Higgs's scalars origin in the boson gauge fields with the vector index $(7,8)$, Ref. ([10], Sect. 7.4.1, and the references therein).

[10] In the general case the energy eigenstates of bosons are in superposition of ${}^{\mathbf{I}}\hat{\mathcal{A}}^{\mathbf{m}\dagger}_{\mathbf{f}}$. One example, which uses the superposition of the Cartan subalgebra eigenstates manifesting the $SU(3)\times U(1)$ subgroups of the group $SO(6)$, is presented in Fig. 2.

$$\mathcal{S}^{cd}\sum_{ab}(c^{ab}{}_{mf}\,\omega_{ab\alpha}) \quad \text{related to} \quad \mathcal{S}^{cd}\,({}^{I}\hat{\mathcal{A}}^{m\dagger}_{f}\,\mathcal{C}^{mf}_{\alpha})\,,$$
$$\forall\,(m,f),$$
$$\forall \text{ Cartan subalgebra member}\,\mathcal{S}^{cd}\,. \qquad (36)$$

Let be repeated that ${}^{I}\hat{\mathcal{A}}^{m\dagger}_{f}$ are chosen to be the eigenvectors of the Cartan subalgebra members, Eq. (8). Correspondingly we can relate a particular ${}^{I}\hat{\mathcal{A}}^{m\dagger}_{f}\mathcal{C}^{mf}_{\alpha}$ with such a superposition of $\omega_{ab\alpha}$'s which is the eigenvector with the same values of the Cartan subalgebra members as there is a particular ${}^{I}\hat{\mathcal{A}}^{m\dagger}_{f}\mathcal{C}^{mf}_{\alpha}$. We can do this in two ways:
i. Using the first relation in Eq. (36). On the left hand side of this relation S^{ab}'s apply on $\hat{b}^{m\dagger}_{f}$ part of $\mathbf{b}^{m\dagger}_{f}(\vec{p})$. On the right hand side ${}^{I}\hat{\mathcal{A}}^{m\dagger}_{f}$ apply as well on the same "basis vector" $\hat{b}^{m\dagger}_{f}$.
ii. Using the second relation, in which $\mathcal{S}^{cd}$ apply on the left hand side on $\omega_{ab\alpha}$'s,

$$\mathcal{S}^{cd}\sum_{ab} c^{ab}{}_{mf}\,\omega_{ab\alpha} \;=\; \sum_{ab} c^{ab}{}_{mf}\,i\,(\omega_{cb\alpha}\eta^{ad}-\omega_{db\alpha}\eta^{ac}+\omega_{ac\alpha}\eta^{bd}-\omega_{ad\alpha}\eta^{bc}), \qquad (37)$$

on each $\omega_{ab\alpha}$ separately; $c^{ab}{}_{mf}$ are constants to be determined from the second relation, where on the right hand side of this relation $\mathcal{S}^{cd}(=S^{cd}+\tilde{S}^{cd})$ apply on the "basis vector" ${}^{I}\hat{\mathcal{A}}^{m\dagger}_{f}$ of the corresponding gauge field.

Let us conclude this section by pointing out that either the Clifford odd "basis vectors", $\hat{b}^{m\dagger}_{f}$, or the Clifford even "basis vectors", ${}^{i}\hat{\mathcal{A}}^{m\dagger}_{f}, i=(I,II)$, have in any even d $2^{\frac{d}{2}-1}\times 2^{\frac{d}{2}-1}$ members, while $\omega_{ab\alpha}$ as well as $\tilde{\omega}_{ab\alpha}$ have each for each α $\frac{d}{2}(d-1)$ members. It is needed to find out what new this difference brings into the unifying theories of the Kaluza-Klein-like to what the *spin-charge-family* belongs.

3. Short overview and achievements of *spin-charge-family* theory

The *spin-chare-family* theory [7, 8, 21, 23, 25, 26, 27, 29, 28, 30] is a kind of the Kaluza-Klein theories [36, 37, 38, 39, 40, 41, 42, 43, 25] since it is built on the assumption that the dimension of space-time is $\geq (13+1)$ [11], and that the only interaction among fermions is the gravitational one (vielbeins, the gauge fields of momenta, and two kinds of the spin connection fields, the gauge fields of S^{ab} and of $\tilde{S}^{ab}$ [12]).

The *spin-charge-family* theory assumes (what the other Kaluza-Klein-like theories do not) that the internal spaces of fermion and boson fields are described by the Clifford odd and Clifford even algebra, respectively [13].

The theory is offering the explanation for all the assumptions of the *standard model*, unifying not only charges but also spins, charges and families, and consequently offering the explanation for the appearance of families of quarks and leptons and antiquarks and antileptons, of vector gauge fields, of Higgs's scalar field and the Yukawa couplings, for the differences in masses among quarks and leptons for the matter-antimatter asymmetry in the universe, for the dark matter, making several predictions.

[11] $d=(13+1)$ is the smallest dimension for which the subgroups of the group $SO(13,1)$ offer the description of spins and charges of fermions assumed by the *standard model* and correspondingly also of spins and charges of the corresponding boson gauge fields.
[12] If there are no fermions present, both spin connection fields are expressible with vielbeins ([10], Eq. (103)).
[13] Fermions and bosons internal spaces are assumed to be superposition of odd products of γ^a's (fermion fields) or of even products of γ^a's (boson fields) what offers the explanation for the second quantized postulates of Dirac [14]. The "basis vectors" of the internal spaces namely determine anti-commutativity or commutativity of the corresponding creation and annihilation operators.

The *spin-charge-family* theory shares with the Kaluza-Klein like theories their weak points, like: **a.** Not yet solved the quantization problem of the gravitational field [14]. **b.** The spontaneous brake of the starting symmetry which would at low energies manifest the observed almost massless fermions [37]. The spontaneously break of the starting symmetry of $SO(13+1)$ with the condensate of the two right handed neutrinos (with the family quantum numbers of the group of four families which does not include the observed three families [([17]; Table III), ([10]; Table 6)] bringing masses of the scale $\propto 10^{16}$ GeV or higher to all the vector and scalar gauge fields, which interact with the condensate [23] is promising to show the right way [32].

The scalar fields (scalar fields are the spin connection fields with the space index α higher than $(0,1,2,3)$) with the space index $(7,8)$ offer, after gaining constant non zero vacuum values, the explanation for the Higgs's scalar and the Yukawa couplings. They namely determine the mass matrices of quarks and leptons and antiquarks and antileptons. In Refs. [25, 22] it is pointed out that the spin connection gauge fields do manifest in $d=(3+1)$ as the ordinary gravity and all the observed vector and scalar gauge fields.

The *spin-charge-family* theory assumes a simple starting action for massless fermion and for the corresponding massless gauge boson fields in $d=(13+1)$-dimensional space, presented in Eq. (1).

The fermion part of the action, Eq. (1), can be rewritten in the way that it manifests in $d=(3+1)$ in the low energy regime before the electroweak break by the *standard model* postulated properties of: **i.** Quarks and leptons and antiquarks and antileptons with the required spins, handedness, charges and families [15]. (Their internal spaces are in the *spin-charge-family* theory described by the Clifford odd "basis vectors" which are eigenvectors of the Cartan subalgebra of S^{ab} and $\tilde{S}^{ab}$, Eqs. (8,29, 28)). **ii.** Couplings of fermions to the vector gauge fields which are the superposition of gauge fields $\omega^{st}{}_{\alpha}$, Sect. 6.2 in Ref. [10], with the space index $\alpha=(0,1,2,3)$ and with the charges determined by the Cartan subalgebra of S^{ab} and $\tilde{S}^{ab}$, manifesting the symmetry of space $(d-4)$, and to the scalar gauge fields [21, 24, 27, 29, 34, 35, 7, 8, 46, 47, 48, 22] with the space index $\alpha \geq 5$ and the charges determined by the Cartan subalgebra of S^{ab} and $\tilde{S}^{ab}$ (as explained in the case of the vector gauge fields), and which are superposition of either $\omega^{st}{}_{\alpha}$ or $\tilde{\omega}^{ab}{}_{\alpha}$

$$\begin{aligned}
\mathcal{L}_f &= \bar{\psi}\gamma^m(p_m - \sum_{A,i} g^{Ai}\tau^{Ai}A^{Ai}_m)\psi + \\
&\quad \{\sum_{s=7,8} \bar{\psi}\gamma^s p_{0s}\,\psi\} + \\
&\quad \{\sum_{t=5,6,9,\dots,14} \bar{\psi}\gamma^t p_{0t}\,\psi\}\,, \qquad (38)
\end{aligned}$$

with $p_{0s} = p_s - \frac{1}{2}S^{s's"}\omega_{s's"s} - \frac{1}{2}\tilde{S}^{ab}\tilde{\omega}_{abs}$, $p_{0t} = p_t - \frac{1}{2}S^{t't"}\omega_{t't"t} - \frac{1}{2}\tilde{S}^{ab}\tilde{\omega}_{abt}$, $p_{0s} = e^{\alpha}_{s}p_{0\alpha}$, $m \in (0,1,2,3)$, $s \in (7,8)$, $(s',s") \in (5,6,7,8)$, (a,b) (appearing in $\tilde{S}^{ab}$) run within either $(0,1,2,3)$ or $(5,6,7,8)$, t runs $\in (5,\dots,14)$, $(t',t")$ run either $\in (5,6,7,8)$ or $\in (9,10,\dots,14)$. The spinor function ψ represents all family members of all the $2^{\frac{7+1}{2}-1} = 8$ families.

[14] The description of the internal space of fermions and bosons as superposition of odd (for fermion fields) or even (for boson fields) products of the Clifford objects γ^a's seems very promising in looking for a new way to second quantization of all fields, with gravity included, as discussed in this contribution.

[15] In the *spin-charge-family* theory quarks and leptons shares the $SO(7,1)$ part of $SO(13,1)$ what predicts the existence of right handed neutrinos and left handed antineutrinos. The theory predicts two groups of four families, what requires the existence of the fourth family to the observed three; The stable of the upper four families contributes to the dark matter.

Journal of Physics: Conference Series **2482** (2023) 012011 doi:10.1088/1742-6596/2482/1/012011

The first line of Eq. (38) determines in $d = (3+1)$ the kinematics and dynamics of fermion fields coupled to the vector gauge fields [25, 29, 21]. The vector gauge fields are the superposition of the spin connection fields ω_{stm}, $m = (0, 1, 2, 3)$, $(s, t) = (5, 6, \cdots, 13, 14)$, and are the gauge fields of S^{st}, Subsect. (6.2.1) of Ref. [10].

The reader can find in Sect. 6 of Ref. [10] a quite detailed overview of the properties which the massless fermion and boson fields appearing in the simple starting action, Eq. (1), (bosons only as gravitational fields) manifest in $d = (3+1)$ as all the observed fermions — quarks and leptons and antiquarks, and antileptons in each family.

Table 5 and Eq. (110) of Ref. [10] explain that the scalar fields with the space index $\alpha = (7, 8)$ carry the weak charge $\tau^{13} = \pm\frac{1}{2}$ and the hyper charge $Y = \mp\frac{1}{2}$, just as assumed by the *standard model.*

Masses of families of quarks and leptons are determined by the superposition of the scalar fields, Eq. (108-120) of Ref. [10], appearing in two groups, each of them manifesting the symmetry $SU(2) \times SU(2) \times U(1)$ [16].

The scalar gauge fields with the space index $(7, 8)$ determine correspondingly the symmetry of mass matrices of quarks and leptons ([10], Eq. (111)), which appear in two groups. In Table 5 of Ref. [10]) the symmetry $SU(2) \times SU(2) \times U(1)$ for each of the two groups is presented and explained.

Although spontaneous symmetry braking of the starting symmetry has not (yet consistently enough) been studied and the coupling constants of the scalar fields among themselves and with quarks and leptons are not yet known, the known symmetry of mass matrices, presented in Eq. (111) of Ref. [10], enables to determine parameters of mass matrices from the measured data of the 3×3 sub mixing matrices and the masses of the measured three families of quarks and leptons.

Although the known 3×3 submatrix of the unitary 4×4 matrix enables to determine 4×4 matrix, the measured 3×3 mixing submatrix is even for quarks far accurately enough measured so that we only can predict the matrix elements of the 4×4 mixing matrix for quarks if assuming that masses (times c^2) of the fourth family quarks are heavy enough, that is above one TeV [44, 47]. The new measurements of the matrix elements among the observed 3 families agree better with the predictions obtained by the *sspin-charge-family* theory than the old measurements. The reader can find predictions in Refs. ([48, 49]) and the overview in Ref. ([10], Subsect. 7.3.1).

The upper group of four families offers the explanation for the Dark matter, to which the quarks and leptons from the (almost) stable of the upper four families mostly contribute. The reader can find the report on this proposal for the Dark matter origin in Ref. [47] and a short overview in Subsect. 7.3.1 of [10], where the appearance, development and properties of the Dark matter are discussed. The upper four families predict nucleons of very heavy quarks with the nuclear force among nucleons which is correspondingly very different from the known one [47, 50].

Besides the scalar fields with the space index $\alpha = (7, 8)$, which manifest in $d = (3+1)$ as scalar gauge fields with the weak and hypercharge $\pm\frac{1}{2}$ and $\mp\frac{1}{2}$, respectively, and which gaining at low energies constant values make families of quarks and leptons and the weak gauge field massive, there are in the starting action, Eqs. (1), additional scalar gauge fields with the space index $\alpha = (9, 10, 11, 12, 13, 14)$. They are with respect to the space index α, either triplets or

[16] The assumption that the symmetry $SO(13, 1)$ first breaks into $SU(3)\times U(1) \times SO(7, 1)$ makes that quarks and leptons distinguish only in the part $SU(3) \times U(1)$, while the $SO(7, 1)$ part is identical separately for quarks and leptons and separately for antiquarks and antileptons. Table 7 of Ref. [10], presenting one family which includes quarks and leptons and antiquarks and antileptons manifests these properties. The $\omega_{ab\alpha}$ with the space index $\alpha = (7, 8)$ carry with respect to the flat index ab only quantum numbers Q, Y, τ^4 (Q $(= \tau^{13} + Y)$, τ^{13} $(= \frac{1}{2}(\mathcal{S}^{56} - \mathcal{S}^{78})$, Y $(= \tau^4 + \tau^{23})$ and $\tau^4 = -\frac{1}{3}(\mathcal{S}^{9\,10} + \mathcal{S}^{11\,12} + \mathcal{S}^{13\,14})$), the flat index (ab) of $\tilde{\omega}_{ab\alpha}$, with the space index $\alpha = (7, 8)$, includes all $(0, 1, \ldots, 8)$, correspondingly forming the symmetry $SU(2)\times SU(2) \times U(1)$.

antitriplets causing transitions from antileptons into quarks and from antiquarks into quarks and back.

Their properties are presented in Ref. [23] and briefly in Table 9 and Fig. 1 of Ref. [10].

Concerning this second point we proved on the toy model of $d = (5+1)$ that the break of symmetry can lead to (almost) massless fermions [32].

In $d = (3+1)$-dimensional space — at low energies — the gauge gravitational fields manifest as the observed vector gauge fields [25], which can be quantized in the usual way. The author is in mean time trying to find out (together with the collaborators) how far can the *spin-charge-family* theory — starting in $d = (13+1)$-dimensional space with a simple and "elegant" action, Eq. (1) — reproduce in $d = (3+1)$ the observed properties of quarks and leptons [21, 23, 25, 26, 27, 29, 28, 30], the observed vector gauge fields, the scalar field and the Yukawa couplings, the appearance of the dark matter and of the matter-antimatter asymmetry, as well as the other open questions, connecting elementary fermion and boson fields and cosmology.

The work done so far on the *spin-charge-family* theory seems promising.

4. Conclusions

In the *spin-charge-family* theory [7, 8, 21, 23, 25, 26, 27, 29, 28, 30, 10] the Clifford odd algebra is used to describe the internal space of fermion fields. The Clifford odd "basis vectors" — the superposition of odd products of γ^a's — in a tensor product with the basis in ordinary space form the creation and annihilation operators, in which the anti-commutativity of the "basis vectors" is transferred to the creation and annihilation operators for fermions, offering the explanation for the second quantization postulates for fermion fields.

The Clifford odd "basis vectors" have all the properties of fermions: Half integer spins with respect to the Cartan subalgebra members of the Lorentz algebra in the internal space of fermions in even dimensional spaces ($d = 2(2n+1)$ or $d = 4n$), as discussed in Subsects. (2.1, 2.3). With respect to the subgroups of the $SO(d-1,1)$ group the Clifford odd "basis vectors" appear in the fundamental representations, as illustrated in Subsects. 2.2.

In this article it is demonstrated that the Clifford even algebra is offering the description of the internal space of boson fields. The Clifford even "basis vectors" — the superposition of even products of γ^a's — in a tensor product with the basis in ordinary space form the creation and annihilation operators which manifest the commuting properties of the second quantized boson fields, offering explanation for the second quantization postulates for boson fields [12]. The Clifford even "basis vectors" have all the properties of bosons: Integer spins with respect to the Cartan subalgebra members of the Lorentz algebra in the internal space of bosons, as discussed in Subsects. 2.1.

With respect to the subgroups of the $SO(d-1,1)$ group the Clifford even "basis vectors" manifest the adjoint representations, as illustrated in Subsect. 2.2.

There are two kinds of anti-commuting algebras [8]: The Grassmann algebra, offering in d-dimensional space $2 \,.\, 2^d$ operators (2^d θ^a's and 2^d $\frac{\partial}{\partial\theta_a}$'s, Hermitian conjugated to each other, Eq. (3)), and the two Clifford sub-algebras, each with 2^d operators named γ^a's and $\tilde{\gamma}^a$'s, respectively, [8, 19, 20], Eqs. (2-6).

The operators in each of the two Clifford sub-algebras appear in two groups of $2^{\frac{d}{2}-1} \times 2^{\frac{d}{2}-1}$ of the Clifford odd operators (the odd products of either γ^a's in one subalgebra or of $\tilde{\gamma}^a$'s in the other subalgebra), which are Hermitian conjugated to each other: In each Clifford odd group of any of the two sub-algebras, there appear $2^{\frac{d}{2}-1}$ irreducible representation each with the $2^{\frac{d}{2}-1}$ members and the group of their Hermitian conjugated partners.

There are as well the Clifford even operators (the even products of either γ^a's in one subalgebra or of $\tilde{\gamma}^a$'s in another subalgebra) which again appear in two groups of $2^{\frac{d}{2}-1}\times\, 2^{\frac{d}{2}-1}$ members each. In the case of the Clifford even objects the members of each group of $2^{\frac{d}{2}-1}\times$ $2^{\frac{d}{2}-1}$ members have the Hermitian conjugated partners within the same group, Subsect. 2.1, Table 1.

The Grassmann algebra operators are expressible with the operators of the two Clifford subalgebras and opposite, Eq. (5). The two Clifford subalgebras are independent of each other, Eq. (6), forming two independent spaces.

Either the Grassmann algebra [18, 13] or the two Clifford subalgebras can be used to describe the internal space of anticommuting objects, if the superposition of odd products of operators (θ^a's or γ^a's, or $\tilde{\gamma}^a$'s) are used to describe the internal space of these objects. The commuting objects must be superposition of even products of operators (θ^a's or γ^a's or $\tilde{\gamma}^a$'s).

No integer spin anti-commuting objects have been observed so far, and to describe the internal space of the so far observed fermions only one of the two Clifford odd subalgebras are needed.

The problem can be solved by reducing the two Clifford sub-algebras to only one, the one (chosen to be) determined by γ^{ab}'s. The decision that $\tilde{\gamma}^a$'s apply on γ^a as follows: $\{\tilde{\gamma}^a B = (-)^B\, i\, B\gamma^a\}\,|\psi_{oc}>$, Eq. (7), (with $(-)^B = -1$, if B is a function of an odd products of γ^a's, otherwise $(-)^B = 1$) enables that $2^{\frac{d}{2}-1}$ irreducible representations of $S^{ab} = \frac{i}{2}\,\{\gamma^a\,,\,\gamma^b\}_-$ (each with the $2^{\frac{d}{2}-1}$ members) obtain the family quantum numbers determined by $\tilde{S}^{ab} = \frac{i}{2}\,\{\tilde{\gamma}^a\,,\,\tilde{\gamma}^b\}_-$.

The decision to use in the *spin-charge-family* theory in $d = 2(2n+1)$, $n \geq 3$ ($d \geq (13+1)$ indeed), the superposition of the odd products of the Clifford algebra elements γ^a's to describe the internal space of fermions which interact with the gravity only (with the vielbeins, the gauge fields of momenta, and the two kinds of the spin connection fields, the gauge fields of S^{ab} and $\tilde{S}^{ab}$, respectively), Eq. (1), offers not only the explanation for all the assumed properties of fermions and bosons in the *standard model*, with the appearance of the families of quarks and leptons and antiquarks and antileptons ([10] and the references therein) and of the corresponding vector gauge fields and the Higgs's scalars included [25], but also for the appearance of the dark matter [47] in the universe, for the explanation of the matter/antimatter asymmetry in the universe [23], and for several other observed phenomena, making several predictions [35, 45, 46, 48].

Recognition that the use of the superposition of the even products of the Clifford algebra elements γ^a's to describe the internal space of boson fields, what appear to manifest all the properties of the observed boson fields, as demonstrated in this articles, makes clear that the Clifford algebra offers not only the explanation for the postulates of the second quantized anti-commuting fermion fields but also for the postulates of the second quantized boson fields.

The relations in Eq. (36)

$$
\begin{aligned}
&\{\frac{1}{2}\sum_{ab} S^{ab}\,\omega_{ab\alpha}\}\sum_{m}\beta^{mf}\,\hat{\mathbf{b}}^{m\dagger}_{f}(\vec{p}) \quad \text{to relate to} \quad \{\sum_{m'f'}{}^{I}\hat{\mathcal{A}}^{m'\dagger}_{f'}\,\mathcal{C}^{m'f'}_{\alpha}\}\sum_{m}\beta^{mf}\,\hat{\mathbf{b}}^{m\dagger}_{f}(\vec{p})\,,\\
&\qquad\qquad\qquad\qquad\qquad\qquad\qquad\qquad\quad \forall f \;\text{and}\; \forall\,\beta^{mf}\,,\\
&\mathcal{S}^{cd}\sum_{ab}(c^{ab}{}_{mf}\,\omega_{ab\alpha}) \quad \text{to relate to} \quad \mathcal{S}^{cd}\,({}^{I}\hat{\mathcal{A}}^{m\dagger}_{f}\,\mathcal{C}^{mf}_{\alpha})\,,\\
&\qquad\qquad\qquad\qquad\qquad\qquad\qquad\qquad\quad \forall\,(m,f),\\
&\qquad\qquad\qquad\qquad\qquad\qquad\qquad\qquad\quad \forall\;\text{Cartan subalgebra member}\,\mathcal{S}^{cd}\,,
\end{aligned}
$$

offers the possibility to replace the covariant derivative $p_{0\alpha}$

$$
p_{0\alpha} = p_\alpha - \frac{1}{2}S^{ab}\omega_{ab\alpha} - \frac{1}{2}\tilde{S}^{ab}\tilde{\omega}_{ab\alpha}
$$

in Eq. (1) with

$$p_{0\alpha} = p_\alpha - \sum_{mf} {}^I\hat{\mathcal{A}}^{m\dagger}_f \, {}^I\mathcal{C}^m_{f\alpha} - \sum_{mf} {}^I\stackrel{\hat{\sim}}{\mathcal{A}}{}^{m\dagger}_f \, {}^I\tilde{\mathcal{C}}^m_{f\alpha} \,,$$

where the relation among ${}^I\stackrel{\hat{\sim}}{\mathcal{A}}{}^{m\dagger}_f \, {}^I\tilde{\mathcal{C}}^m_{f\alpha}$ and ${}^{II}\stackrel{\hat{\sim}}{\mathcal{A}}{}^{m\dagger}_f \, {}^{II}\tilde{\mathcal{C}}^m_{f\alpha}$ with respect to $\omega_{ab\alpha}$ and $\tilde{\omega}_{ab\alpha}$, not discussed directly in this article, needs additional study and explanation.

Although the properties of the Clifford odd and even "basis vectors" and correspondingly of the creation and annihilation operators for fermion and boson fields are, hopefully, clearly demonstrated in this article, yet the proposed way of the second quantization of fields, the fermion and the boson ones, needs further study to find out what new can the description of the internal space of fermions and bosons brings into understanding of the second quantized fields.

It looks like that this study, showing up that the Clifford algebra can be used to describe the internal spaces of fermion and boson fields in an equivalent way offering correspondingly the explanation for the second quantization postulates for fermion and boson fields, is opening the new insight into the quantum field theory, since studies of the interaction of fermion fields with boson fields and of boson fields with boson fields so far looks very promising.

The study of properties of the second quantized boson fields, the internal space of which is described by the Clifford even algebra, has just started and needs further consideration.

Acknowledgments
The author thanks Department of Physics, FMF, University of Ljubljana, Society of Mathematicians, Physicists and Astronomers of Slovenia, for supporting the research on the *spin-charge-family* theory by offering the room and computer facilities and Matjaž Breskvar of Beyond Semiconductor for donations, in particular for the annual workshops entitled "What comes beyond the standard models".

References

[1] H. Georgi, in *Particles and Fields* (edited by C. E. Carlson), A.I.P., 1975; Google Scholar.

[2] H. Fritzsch and P. Minkowski, *Ann. Phys.* **93** (1975) 193.

[3] J. Pati and A. Salam, *Phys.Rev.* **D 8** (1973) 1240.

[4] H. Georgy and S.L. Glashow, *Phys. Rev. Lett.* **32** (1974) 438.
[5] Y. M. Cho, *J. Math. Phys.* **16** (1975) 2029.
[6] Y. M. Cho, P. G. O. Freund, *Phys. Rev.* **D 12** (1975) 1711.
[7] N. Mankoč Borštnik, "Spin connection as a superpartner of a vielbein", *Phys. Lett.* **B 292** (1992) 25-29.
[8] N. Mankoč Borštnik, "Spinor and vector representations in four dimensional Grassmann space", *J. of Math. Phys.* **34** (1993) 3731-3745.
[9] N. Mankoč Borštnik, "Unification of spin and charges in Grassmann space?", hep-th 9408002, IJS.TP.94/22, Mod. Phys. Lett.**A (10)** No.7 (1995) 587-595.
[10] N. S. Mankoč Borštnik, H. B. Nielsen, "How does Clifford algebra show the way to the second quantized fermions with unified spins, charges and families, and with vector and scalar gauge fields beyond the *standard model*", Progress in Particle and Nuclear Physics, http://doi.org/10.1016.j.ppnp.2021.103890 .
[11] N.S. Mankoč Borštnik, H.B.F. Nielsen, "Understanding the second quantization of fermions in Clifford and in Grassmann space", *New way of second quantization of fermions — Part I and Part II*, in this proceedings [arXiv:2007.03517, arXiv:2007.03516].
[12] N. S. Mankoč Borštnik, "How do Clifford algebras show the way to the second quantized fermions with unified spins, charges and families, and to the corresponding second quantized vector and scalar gauge field ", Proceedings to the 24rd Workshop "What comes beyond the standard models", 5 - 11 of July, 2021, Ed. N.S. Mankoč Borštnik, H.B. Nielsen, D. Lukman, A. Kleppe, DMFA Založništvo, Ljubljana, December 2021, [arXiv:2112.04378] .

[13] N.S. Mankoč Borštnik, H.B.F. Nielsen, "Understanding the second quantization of fermions in Clifford and in Grassmann space" *New way of second quantization of fermions — Part I and Part II*, Proceedings to the 22nd Workshop "What comes beyond the standard models", 6 - 14 of July, 2019, Ed. N.S. Mankoč Borštnik, H.B. Nielsen, D. Lukman, DMFA Založništvo, Ljubljana, December 2019, [arXiv:1802.05554v4, arXiv:1902.10628].

[14] P.A.M. Dirac *Proc. Roy. Soc. (London)*, **A 117** (1928) 610.

[15] H.A. Bethe, R.W. Jackiw, "Intermediate quantum mechanics", New York : W.A. Benjamin, 1968.

[16] S. Weinberg, "The quantum theory of fields", Cambridge, Cambridge University Press, 2015.

[17] N.S. Mankoč Borštnik, H.B.F. Nielsen, "New way of second quantized theory of fermions with either Clifford or Grassmann coordinates and *spin-charge-family* theory " [arXiv:1802.05554v4,arXiv:1902.10628].

[18] D. Lukman, N. S. Mankoč Borštnik, "Properties of fermions with integer spin described with Grassmann algebra", Proceedings to the 21st Workshop "What comes beyond the standard models", 23 of June - 1 of July, 2018, Ed. N.S. Mankoč Borštnik, H.B. Nielsen, D. Lukman, DMFA Založništvo, Ljubljana, December 2018 [arxiv:1805.06318, arXiv:1902.10628].

[19] N.S. Mankoč Borštnik, H.B.F. Nielsen, *J. of Math. Phys.* **43**, 5782 (2002) [arXiv:hep-th/0111257].

[20] N.S. Mankoč Borštnik, H.B.F. Nielsen, "How to generate families of spinors", *J. of Math. Phys.* **44** 4817 (2003) [arXiv:hep-th/0303224].

[21] N.S. Mankoč Borštnik, "Spin-charge-family theory is offering next step in understanding elementary particles and fields and correspondingly universe", Proceedings to the Conference on Cosmology, Gravitational Waves and Particles, IARD conferences, Ljubljana, 6-9 June 2016, The 10th Biennial Conference on Classical and Quantum Relativistic Dynamics of Particles and Fields, J. Phys.: Conf. Ser. 845 012017 [arXiv:1409.4981, arXiv:1607.01618v2].

[22] N.S. Mankoč Borštnik, "The attributes of the Spin-Charge-Family theory giving hope that the theory offers the next step beyond the Standard Model", Proceedings to the 12th Bienal Conference on Classical and Quantum Relativistic Dynamics of Particles and Fields IARD 2020, Prague, 1 – 4 June 2020 by ZOOM.

[23] N.S. Mankoč Borštnik, "Matter-antimatter asymmetry in the *spin-charge-family* theory", *Phys. Rev.* **D 91** (2015) 065004 [arXiv:1409.7791].

[24] N. S. Mankoč Borštnik, "How far has so far the Spin-Charge-Family theory succeeded to explain the Standard Model assumptions, the matter-antimatter asymmetry, the appearance of the Dark Matter, the second quantized fermion fields...., making several predictions", Proceedings to the 23rd Workshop "What comes beyond the standard models", 4 - 12 of July, 2020 Ed. N.S. Mankoč Borštnik, H.B. Nielsen, D. Lukman, DMFA Založništvo, Ljubljana, December 2020, [arXiv:2012.09640]

[25] N.S. Mankoč Borštnik, D. Lukman, "Vector and scalar gauge fields with respect to $d = (3+1)$ in Kaluza-Klein theories and in the *spin-charge-family theory*", *Eur. Phys. J. C* **77** (2017) 231.

[26] N.S. Mankoč Borštnik, "The *spin-charge-family* theory explains why the scalar Higgs carries the weak charge $\pm\frac{1}{2}$ and the hyper charge $\mp\frac{1}{2}$", Proceedings to the 17th Workshop "What comes beyond the standard models", Bled, 20-28 of July, 2014, Ed. N.S. Mankoč Borštnik, H.B. Nielsen, D. Lukman, DMFA Založništvo, Ljubljana December 2014, p.163-82 [arXiv:1502.06786v1] [arXiv:1409.4981].

[27] N.S. Mankoč Borštnik N S, "The spin-charge-family theory is explaining the origin of families, of the Higgs and the Yukawa couplings", *J. of Modern Phys.* **4** (2013) 823 [arXiv:1312.1542].

[28] N.S. Mankoč Borštnik, H.B.F. Nielsen, "The spin-charge-family theory offers understanding of the triangle anomalies cancellation in the standard model", *Fortschritte der Physik, Progress of Physics* (2017) 1700046.

[29] N.S. Mankoč Borštnik, "The explanation for the origin of the Higgs scalar and for the Yukawa couplings by the *spin-charge-family* theory", *J.of Mod. Physics* **6** (2015) 2244-2274, http://dx.org./10.4236/jmp.2015.615230 [arXiv:1409.4981].

[30] N.S. Mankoč Borštnik and H.B. Nielsen, "Why nature made a choice of Clifford and not Grassmann coordinates", Proceedings to the 20th Workshop "What comes beyond the standard models", Bled, 9-17 of July, 2017, Ed. N.S. Mankoč Borštnik, H.B. Nielsen, D. Lukman, DMFA Založništvo, Ljubljana, December 2017, p. 89-120 [arXiv:1802.05554v1v2].

[31] N.S. Mankoč Borštnik and H.B.F. Nielsen, "Discrete symmetries in the Kaluza-Klein theories", *JHEP* 04:165, 2014 [arXiv:1212.2362].

[32] D. Lukman, N.S. Mankoč Borštnik and H.B. Nielsen, "An effective two dimensionality cases bring a new hope to the Kaluza-Klein-like theories", *New J. Phys.* 13:103027, 2011.

[33] N. S. Mankoč Borštnik, Second quantized "anticommuting integer spin fields", sent to arXiv.

[34] A. Borštnik, N.S. Mankoč Borštnik, "Left and right handedness of fermions and bosons", J. of Phys. G: Nucl. Part. Phys.**24**(1998)963-977, hep-th/9707218.

[35] A. Borštnik Bračič, N. S. Mankoč Borštnik, "On the origin of families of fermions and their mass matrices", hep-ph/0512062, Phys Rev. **D 74** 073013-28 (2006).

[36] T. Kaluza, "On the unification problem in Physics", *Sitzungsber. d. Berl. Acad.* (1918) 204, O. Klein, "Quantum theory and five-dimensional relativity", *Zeit. Phys.* **37**(1926) 895.
[37] E. Witten, "Search for realistic Kaluza-Klein theory",*Nucl. Phys.* **B 186** (1981) 412.
[38] M. Duff, B. Nilsson, C. Pope, *Phys. Rep.* **C 130** (1984)1, M. Duff, B. Nilsson, C. Pope, N. Warner, *Phys. Lett.* **B 149** (1984) 60.
[39] T. Appelquist, H. C. Cheng, B. A. Dobrescu, *Phys. Rev.* **D 64** (2001) 035002.
[40] M. Saposhnikov, P. TinyakovP 2001 *Phys. Lett.* **B 515** (2001) 442 [arXiv:hep-th/0102161v2].
[41] C. Wetterich,*Nucl. Phys.* **B 253** (1985) 366.
[42] The authors of the works presented in *An introduction to Kaluza-Klein theories*, Ed. by H. C. Lee, World Scientific, Singapore 1983.
[43] M. Blagojević, *Gravitation and gauge symmetries*, IoP Publishing, Bristol 2002.
[44] M. Breskvar, D. Lukman, N. S. Mankoč Borštnik, "On the Origin of Families of Fermions and Their Mass Matrices — Approximate Analyses of Properties of Four Families Within Approach Unifying Spins and Charges", Proceedings to the 9th Workshop "What Comes Beyond the Standard Models", Bled, Sept. 16 - 26, 2006, Ed. by Norma Mankoč Borštnik, Holger Bech Nielsen, Colin Froggatt, Dragan Lukman, DMFA Založništvo, Ljubljana December 2006, p.25-50, hep-ph/0612250.
[45] G. Bregar, M. Breskvar, D. Lukman, N.S. Mankoč Borštnik, "Families of Quarks and Leptons and Their Mass Matrices", Proceedings to the 10th international workshop "What Comes Beyond the Standard Model", 17 -27 of July, 2007, Ed. Norma Mankoč Borštnik, Holger Bech Nielsen, Colin Froggatt, Dragan Lukman, DMFA Založništvo, Ljubljana December 2007, p.53-70, hep-ph/0711.4681.
[46] G. Bregar, M. Breskvar, D. Lukman, N.S. Mankoč Borštnik, "Predictions for four families by the Approach unifying spins and charges" *New J. of Phys.* **10** (2008) 093002, hep-ph/0606159, hep/ph-07082846.
[47] G. Bregar, N.S. Mankoč Borštnik, "Does dark matter consist of baryons of new stable family quarks?", *Phys. Rev. D* **80**, 083534 (2009), 1-16.
[48] G. Bregar, N.S. Mankoč Borštnik, "Can we predict the fourth family masses for quarks and leptons?", Proceedings (arxiv:1403.4441) to the 16 th Workshop "What comes beyond the standard models", Bled, 14-21 of July, 2013, Ed. N.S. Mankoč Borštnik, H.B. Nielsen, D. Lukman, DMFA Založništvo, Ljubljana December 2013, p. 31-51, http://arxiv.org/abs/1212.4055.
[49] G. Bregar, N.S. Mankoč Borštnik, "The new experimental data for the quarks mixing matrix are in better agreement with the *spin-charge-family* theory predictions", Proceedings to the 17th Workshop "What comes beyond the standard models", Bled, 20-28 of July, 2014, Ed. N.S. Mankoč Borštnik, H.B. Nielsen, D. Lukman, DMFA Založništvo, Ljubljana December 2014, p.20-45 [arXiv:1502.06786v1] [arxiv:1412.5866].
[50] N.S. Mankoč Borštnik, M. Rosina, "Are superheavy stable quark clusters viable candidates for the dark matter?", International Journal of Modern Physics D (IJMPD) **24** (No. 13) (2015) 1545003.

Electromagnetic properties of the quantum vacuum calculated from its structure

G. B. Mainland[1] and Bernard Mulligan[2]

[1] Department of Physics, The Ohio State University at Newark, 1179 University Dr., Newark, OH 43055, USA, [2] Department of Physics, The Ohio State University, Columbus, OH 43210, USA

E-mail: mainland.1@osu.edu, mulligan.3@osu.edu

Abstract. Maxwell's equations and quantum electrodynamics are valid for any value of the permittivity ϵ_0 of the vacuum; therefore, something additional is required to establish the value of ϵ_0. A fundamental postulate of physics is that the properties of a physical system are determined by its structure. Since ϵ_0 is an electromagnetic property of the vacuum, it should be possible to calculate ϵ_0 using Maxwell's equations and quantum electrodynamics to describe the interaction of photons with the quantum vacuum. The feature of the quantum vacuum that allows for such interactions is the presence of vacuum fluctuations that appear as particle-antiparticle pairs. To minimize the violation of energy conservation and conserve angular momentum, the pair appears with zero center-of-mass momentum in the most tightly bound state that has zero angular momentum. Because vacuum fluctuations each appear as a bound state, the permittivity of the vacuum can be calculated somewhat similarly to the way that the permittivity of a dielectric is calculated, yielding $\epsilon_0 \simeq (6\mu_o/\pi)(8e^2/\hbar)^2 = 9.10 \times 10^{-12}\text{C}/(\text{Vm})$, which is 2.8% larger than the experimental value. Formulas for the speed of light in the vacuum and the fine-structure constant follow immediately from the formula for ϵ_0.

1. Introduction

1.1. Historical Introduction

With the development of field theory it was apparent that fluctuations of particle-antiparticle pairs appear in the vacuum[1]. As early as 1934 four prominent physicists wrote about the possible effects of vacuum fluctuations: Furry & Oppenheimer[2] wrote, "Because of the polarizability of the nascent pairs, the dielectric constant of space into which no matter has been introduced differs from that of truly empty space." Also, Pauli & Weisskopf[3] wrote about the vacuum possessing magnetic and electric polarizability. Both articles point out that properties of the vacuum are affected by the presence of vacuum fluctuations. Although Furry, Oppenheimer, Pauli, and Weisskopf first discussed the fact that the vacuum possesses electric and magnetic properties 88 years ago, little progress was made in calculating properties of the quantum vacuum until two European groups discussed the problem, respectively, in 2010[4] and 2013[5]. The authors of this talk, however, were the first to use Maxwell's equations and quantum electrodynamics (as well as quantum mechanics) to describe the interaction of photons with vacuum fluctuations and calculate formulas for the permittivity ϵ_0 of the vacuum, the speed c of light in the vacuum, and the fine-structure constant α[6, 7, 8]. The values calculated for ϵ_0, c, and α are correct only to within a few percent because each quantity was calculated only to

Content from this work may be used under the terms of the Creative Commons Attribution 3.0 licence. Any further distribution of this work must maintain attribution to the author(s) and the title of the work, journal citation and DOI.
Published under licence by IOP Publishing Ltd

lowest order in the fine-structure constant.

1.2. Central Idea

Maxwell's equations and quantum electrodynamics are valid for any value of the permittivity ϵ_0 of the vacuum (or, equivalently, the speed c of light in the vacuum). In addition to Maxwell's equations and quantum electrodynamics, something additional is required to establish the value of ϵ_0. A fundamental postulate of physics is that the properties of a physical system are determined by its structure. Since ϵ_0, c, and (as will be shown) α are properties of the quantum vacuum, it should be possible to calculate ϵ_0, c, and α using Maxwell's equations, quantum electrodynamics (and quantum mechanics) to describe the interaction of photons with vacuum fluctuations.

1.3. Results

Vacuum fluctuations are polarized by photons, slowing the progress of the photons similarly to the way that an ordinary dielectric slows photons. Calculation of the polarization of each type of vacuum fluctuation by a photon and the density of each type of vacuum fluctuations that is available to interact with a photon yields the following formula:

$$\epsilon_{0\ \mathrm{theory}} = \frac{6\mu_0}{\pi}\left(\frac{8e^2}{\hbar}\right)^2 = 9.10 \times 10^{-12}\frac{\mathrm{C}}{\mathrm{Vm}}\,. \tag{1}$$

In the above formula μ_0 is the vacuum permeability, e is the magnitude of the charge on an electron, and $\hbar$ is Planck's constant divided by 2π. As will be discussed later, to nine significant figures, μ_0 is a measurement-system constant, not a property of the vacuum.

$$\epsilon_{0\ \mathrm{accepted}} = 8.85 \times 10^{-12}\frac{\mathrm{C}}{\mathrm{Vm}}\,. \tag{2}$$

$\epsilon_{0\ \mathrm{theory}}$ is 2.8% larger than the accepted value.

A theoretical formula for the speed c of light in the vacuum follows immediately from the formula $c = 1/\sqrt{\mu_0\epsilon_0}$ and the formula for $\epsilon_{0\ \mathrm{theory}}$,

$$c_{\ \mathrm{theory}} = \sqrt{\frac{\pi}{6}\frac{\hbar}{8e^2\mu_0}} = 2.96 \times 10^8\mathrm{m/s}\,. \tag{3}$$

$$c_{\ \mathrm{defined}} = 2.998 \times 10^8\mathrm{m/s}\,. \tag{4}$$

$c_{\ \mathrm{theory}}$ is 1.3% less than the defined value.

There are two postulates[9] on which special relativity depends: "**Postulate I**: The Principle of Relativity. The laws of physics are the same, or invariant, in all inertial systems; that is, the mathematical form of a physical law remains the same." "**Postulate II**: The speed of light in a vacuum is a constant, independent of the inertial system, the source, and the observer."

The above formula for the speed of light in a vacuum, which is determined by the properties of the quantum vacuum, Maxwell's equations, quantum electrodynamics, and quantum mechanics, must satisfy Postulate II. It is impossible for an observer in an inertial frame to detect relative motion of the inertial frame with respect to the quantum vacuum. As a result the observer would conclude that the inertial frame and the quantum vacuum are at rest with respect to each other. Also, the vacuum is isotropic. Therefore, the quantum vacuum is the same in all directions in every inertial frame. According to Postulate II, Maxwell's equations, quantum electrodynamics, and quantum mechanics are the same in every inertial frame. Thus the calculated value of the speed of light in the vacuum automatically satisfies Postulate II, eliminating the need for the postulate.

A formula for the fine-structure constant α follows immediately from the defining formula for α,

$$\alpha \equiv \frac{e^2}{4\pi\epsilon_0\hbar c}, \tag{5}$$

and the formulas for $\epsilon_{0\ \text{theory}}$ and c_{theory} given in (1) and (3), respectively:

$$\alpha_{\text{theory}} = \frac{1}{8^2\sqrt{3\pi/2}} = \frac{1}{138.9}\cdots, \tag{6}$$

$$\alpha_{\text{experimental}} = \frac{1}{137.0}\cdots. \tag{7}$$

α_{theory} is 1.3% smaller than $\alpha_{\text{experimental}}$.

Physicists have sought to calculate the value of the fine-structure constant for decades: in 1985 Richard Feynman wrote as follows about the fine-structure constant: "It has been a mystery ever since it was discovered more than fifty years ago, and all good theoretical physicists put this number up on their wall and worry about it.... It's one of the greatest damn mysteries of physics: a magic number that comes to us with no understanding by man. You might say the 'hand of God' wrote that number, and 'we don't know how He pushed his pencil[10].' "

Using $c = 1/\sqrt{\mu_0\epsilon_0}$ and the defining equation for α in (5), once a formula for one of the three parameters ϵ_0, c, or α has been calculated, formulas for the other two immediately follow.

2. The quantum vacuum: structure and properties

Physicists use the term "vacuum fluctuation" to describe two very different entities that here are called type 1 and type 2 vacuum fluctuations. Both type 1 and type 2 vacuum fluctuations appear out of the vacuum and then disappear back into the vacuum. From field theory it follows that vacuum energy is the energy source that creates type 1 vacuum fluctuations. Also, field theory provides a proof that type 1 vacuum fluctuations must exist for each known free field.

2.1. Structure of type 1 vacuum fluctuations

When type 1 vacuum fluctuations appear in the vacuum, conservation laws are obeyed as if the type 1 vacuum fluctuations were ordinary matter. Energy conservation is violated subject to the constraint imposed by the Heisenberg uncertainty principle. Specifically, type 1 vacuum fluctuations exist for the time Δt permitted by the Heisenberg uncertainty principle. When type 1 vacuum fluctuations appear in the vacuum, they appear as particle– anti-particle pairs. Therefore, quantum numbers such as electric charge, lepton number, and baryon number are conserved. The constituent particle and anti-particle in a type 1 vacuum fluctuation are on shell and appear as external particles in Feynman diagrams. To minimize the violation of conservation of energy and avoid violating conservation of angular momentum, if the constituent particle and antiparticle are massive, the type 1 vacuum fluctuation must appear with zero center-of-mass momentum in the least energetic bound state that has zero angular momentum. Because a type 1 vacuum fluctuation of massive particles appears in the vacuum as a bound state, calculating the permittivity of the vacuum is somewhat similar to calculating the permittivity of an ordinary dielectric consisting of atoms, molecules or both.

2.2. Interaction of type 1 vacuum fluctuations with normal matter

The feature of the quantum vacuum that permits interactions with ordinary matter is the presence of type 1 vacuum fluctuations. Type 1 vacuum fluctuations can interact with normal (meaning real, ordinary, and observable) test particles placed in the vacuum, but only in very limited ways. Specifically, the violation of energy conservation cannot exceed that allowed by the uncertainty principle. When a type 1 vacuum fluctuation vanishes, the vacuum energy required for its creation is returned to the vacuum. Type 1 vacuum fluctuations cannot do work on ordinary matter. If they could do work on ordinary matter, when a type 1 vacuum fluctuation vanished back into the vacuum, the energy associated with any work done by the type 1 vacuum fluctuation would be left behind in the physical world, violating conservation of ordinary energy.

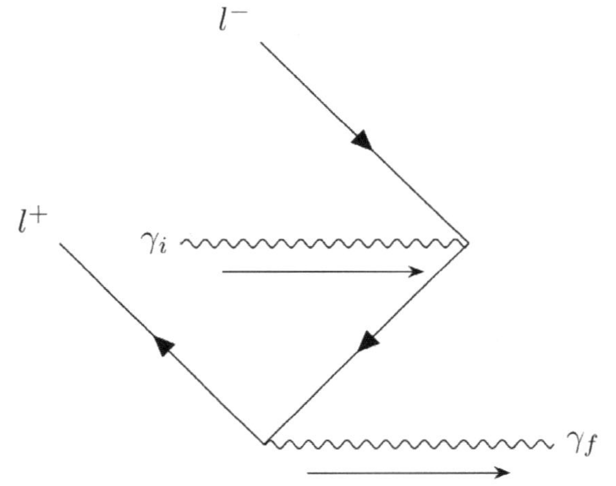

Figure 1. Feynman diagram of a normal (real, ordinary, and observable) photon (γ) interacting with a type 1 charged lepton (l^-)– charged anti-lepton (l^+) vacuum fluctuation. Note the two external photon lines.

2.3. Structure of type 2 vacuum fluctuations

Type 2 vacuum fluctuations are off shell, and are sometimes called vacuum diagrams or vacuum bubbles[1], a class of disconnected Feynman diagrams for which virtual particles appear from and then vanish back into the vacuum. Virtual particles are off shell, appear in the interior of Feynman diagrams, occur in perturbation calculations, but are not detected as normal (real, ordinary, and observable) particles. Because type 2 vacuum fluctuations are disconnected Feynman graphs, they cannot contribute to cross sections, decay rates[1], the permittivity of the vacuum, the speed of light in the vacuum, or the fine-structure constant but do contribute to vacuum energy [11].

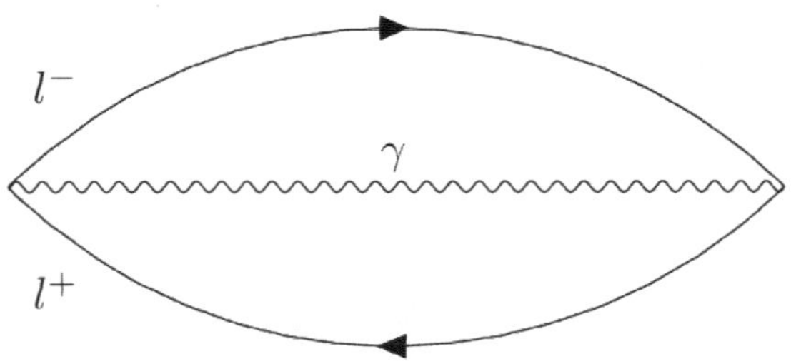

Figure 2. Feynman diagram of a type 2 charged lepton (l^-)– charged anti-lepton (l^+) – photon (γ) vacuum fluctuation. Note that the photon is represented by an interior line in the diagram.

3. Calculation of the permittivity ϵ_0 of the vacuum

As discussed previously, type 1 vacuum fluctuations consisting of a massive particle and antiparticle appear in the most tightly bound state that has zero angular momentum. Thus these type 1 vacuum fluctuations appear as transient atoms. As a consequence, the calculation of the permittivity of the vacuum has much in common with the calculation of the permittivity of a normal dielectric, which is well known and will be discussed first.

3.1. Calculation of the permittivity ϵ of a normal dielectric (ND)

From Maxwell's equation

$$\nabla \cdot \mathbf{D} = \rho \,, \tag{8}$$

where $\mathbf{D}$ is the electric displacement and ρ is the charge density, it follows that[12]

$$\text{normal dielectric: } \mathbf{D} = \epsilon_0 \mathbf{E} + \mathbf{P} \,. \tag{9}$$

In the above equation $\mathbf{E}$ is the external electric field, and $\mathbf{P}$ is the polarization density induced by the electric field. For a normal dielectric the polarization density $\mathbf{P}$ is given by

$$\text{normal dielectric: } \mathbf{P} = \sum_i N_i^{\mathrm{ND}} \mathbf{p}_i \,, \tag{10}$$

where N_i^{ND} is the number density of the ith type of atom or molecule and $\mathbf{p}_i$ is the induced electric dipole moment[13]. The induced the electric dipole moment of each type of atom or molecule in a normal dielectric is often proportional to the electric field,

$$\text{normal dielectric: } \mathbf{p}_i = K_i^{\mathrm{ND}} \mathbf{E} \,. \tag{11}$$

In the above equation K_i^{ND} is a constant that depends on the properties of the ith type of atom or molecule in the normal dielectric but is independent of the electric field $\mathbf{E}$. For the case being considered, since $\mathbf{p}_i$ is proportional to $\mathbf{E}$, it follows from (10) and (9), respectively, that $\mathbf{P}$ and $\mathbf{D}$ are proportional to $\mathbf{E}$:

$$\text{normal dielectric: } \mathbf{D} = \epsilon \mathbf{E} \,. \tag{12}$$

The proportionality constant ϵ is the permittivity of the dielectric. Using (10)-(12), (9) can be written as

$$\text{normal dielectric: } \epsilon \mathbf{E} = \epsilon_0 \mathbf{E} + \sum_i N_i^{\mathrm{ND}} K_i^{\mathrm{ND}} \mathbf{E} \,. \tag{13}$$

Cancelling $\mathbf{E}$ in the above equation yields an equation for the permittivity ϵ of the dielectric:

$$\text{normal dielectric: } \epsilon = \epsilon_0 + \sum_i N_i^{\mathrm{ND}} K_i^{\mathrm{ND}} \,. \tag{14}$$

The permittivity ϵ of a normal dielectric equals the background permittivity ϵ_0 of the vacuum plus the contribution from the polarization of the atoms and molecules in the dielectric.

3.2. Generic formula for the permittivity ϵ_0 of the vacuum

A generic formula for the permittivity ϵ_0 of the vacuum follows immediately from the above formula for the permittivity ϵ of a normal dielectric. When the vacuum *is* the dielectric, in (14) ϵ_0 disappears and $\epsilon \to \epsilon_0$. In a normal dielectric the increase in the permittivity from the value ϵ_0 in the vacuum to ϵ results entirely from the induced polarization of the atoms and molecules in the normal dielectric. Therefore, the permittivity of the vacuum must result entirely from the induced polarization of type 1 vacuum fluctuations (that appear as bound states). For the vacuum (14) becomes

$$\text{vacuum: } \epsilon_0 = \sum_i N_i^{\text{VF}} K_i^{\text{VF}} , \tag{15}$$

where N_i^{VF} is the density of vacuum fluctuations of the ith variety that are available to interact with a photon. K_i^{VF} is relatively easy to calculate from the vacuum equation corresponding to (11), namely,

$$\text{vacuum: } \mathbf{p}_i = K_i^{\text{VF}} \mathbf{E} . \tag{16}$$

For example, a type 1 electron-positron vacuum fluctuation appears as a parapositronium atom, which has zero angular momentum, and is at rest. Calculating the induced electric dipole moment of parapositronium (p-Ps) in a uniform electric field **E** is very similar to calculating the induced electric dipole moment of a normal hydrogen atom in a uniform electric field. From (16) it follows that $K_{\text{p-Ps}}^{\text{VF}}$ is just the coefficient of the external electric field **E** in the formula for the induced electric dipole moment $\mathbf{p}_{\text{p-Ps}}$.

The determination of N_i^{VF} is much more difficult. N_i^{ND} depends only on the number density of each different type of atom or molecule in the dielectric. Similarly, N_i^{VF} also depends on the number density of each different type 1 vacuum fluctuation. However, because normal dielectrics typically consist of stable atoms and molecules while type 1 vacuum fluctuations are unstable, N_i^{VF} also depends on the lifetime of the ith type 1 vacuum fluctuation and the decay rate of the ith type 1 vacuum fluctuation after it has been excited by a photon.

The type 1 vacuum fluctuations that contribute most to ϵ_0 are the electron-positron, muon – anti-muon, and tau – anti-tau fluctuations. A primary reason that quark-antiquark vacuum fluctuations do not contribute much is that the contribution to ϵ_0 is proportional to the square of the term [(energy associated with mass of a constituent particle) ÷ (binding energy of the vacuum fluctuation)]. For quarks the binding energy of the vacuum fluctuation is much larger in comparison to the energy of the constituent mass than it is for charged leptons.

4. Is the vacuum permeability μ_0 a property of the vacuum?

The value $\mu_0 \equiv 4\pi \times 10^{-7}$ N/A^2 was originally arbitrarily chosen so that the ampere in the "electromagnetic (emu)" system was equal to the rationalized meter-kilogram-second unit of current. Thus μ_0 was not a property of the vacuum but was instead a measurement-system constant. Since the value of μ_0 was chosen arbitrarily, the assigned value of μ_0 could neither be measured experimentally nor calculated theoretically. As will be discussed at the end of this section, on 20 May 2019 the definition of μ_0 was changed, changing the value of μ_0 only in the tenth significant figure.

Three equations are crucial when incorporating electromagnetism into mechanics. Each equation includes one of the constants k_1, k_2, or k_3.

1. In a vacuum the force per length F/L on one wire when two long, thin, straight, parallel wires separated by a distance r each carry a current I is

$$\frac{F}{L} = k_1 \frac{I^2}{r} . \tag{17}$$

The above formula can be derived from Ampere's Law and the formula for the magnetic force on a long, thin, straight current-carrying wire in a uniform magnetic field. In SI units, $k_1 = \mu_0/(2\pi)$.

2. The relationship between the current I and charge Q is

$$I = k_2 \frac{\text{d}Q}{\text{d}t} . \tag{18}$$

In SI units $k_2 = 1$.

3. The force F between two point-charges Q and q when they are separated by a distance r is

$$F = k_3 \frac{Qq}{r^2} \,. \tag{19}$$

In SI units k_3= the Coulomb constant$=1/(4\pi\epsilon_0)$.

The constants k_1 and k_2 were chosen to define current and charge, respectively, and the constant k_3 was determined experimentally. Using $k_1 = \mu_0/(2\pi)$ and solving for μ_0, (17), becomes

$$\mu_0 = 2\pi \left(\frac{F}{L}\right) \frac{r}{I^2} \,. \tag{20}$$

When two long, thin, straight, parallel wires in a vacuum are separated by 1m and each wire carries a 1A current, the force/length between the wires was arbitrarily chosen to be exactly 2×10^{-7}N/m so that the ampere in the "electromagnetic (emu)" system was equal to the rationalized meter-kilogram-second unit of current. Substituting $F/L = 2 \times 10^{-7}$ N/m, $r = 1$m, and $I = 1$A into (20) yields

$$\mu_0 = 2\pi(2 \times 10^{-7}\text{N/m}) \frac{1\text{m}}{(1\text{A})^2} = 4\pi \times 10^{-7}\text{N/A}^2 \,. \tag{21}$$

Of course, if the force/length had been chosen, for example, to be 4×10^{-7}N/m when $r = 1$m and $I = 1$A, which changes the definition of a 1A current, μ_0 would equal $8\pi \times 10^{-7}$N/A^2. The vacuum permeability μ_0 was not a property of the vacuum, it was a measurement-system constant that defines an ampere. As a result, the preferred name for μ_0 is now "magnetic constant".

Eq. (18) defines charge. The choice $k_2 = 1$ is made so that

$$Q = \int I \text{d}t \,. \tag{22}$$

If a constant 1 A current flows for 1 s, the total charge Q that passes through a cross section of a conductor is 1C, which defines a coulomb.

On 20 May 2019 the definition of μ_0 was changed to $\mu_0 \equiv 2\alpha h/(ce^2)$, which can be derived from the formula $c = 1/\sqrt{\mu_0\epsilon_0}$ and the defining formula for α as given in (5). The 2018 Particle Data Group[14] listed the value $\mu_0 = 4\pi \times 10^{-7}$ N/A^2. In 2020 [15] the listed value is $\mu_0 = 4\pi \times 10^{-7} \times 1.00000000055(15)$ N/A^2. The change in the definition of μ_0 affected the value of μ_0 only in the tenth significant figure.

5. Summary, discussion, and ongoing research

5.1. Summary and discussion

Formulas for the permittivity ϵ_0 of the vacuum and the speed c of light in the vacuum are calculated in terms of the magnitude of the charge e of an electron, the magnetic constant μ_0, and $\hbar$, Planck's constant divided by 2π. All of the parameters that carry units, of course, cancel in the formula for the fine-structure constant α. (As discussed this article, the value of μ_0 was originally chosen to define an Ampere; consequently, μ_0 was a measurement-system constant that could neither be measured experimentally nor calculated theoretically. On 20 May 2019 the definition of μ_0 was changed, resulting in the value of μ_0 being changed, but only in the tenth significant figure.) Because the calculations of ϵ_0, c, and α are performed only to the lowest order in the fine-structure constant, the discrepancies between the theoretical and accepted values are a few percent.

Physicist use the the term "vacuum fluctuation" to describe two structures with very different properties. Type 1 vacuum fluctuations are on shell, implying that in Feynman diagrams they appear as external particles. Vacuum energy is the energy source that creates type 1 vacuum fluctuations. The appearance of a type 1 vacuum fluctuation violates energy conservation to the extent permitted by the uncertainty principle. To conserve angular momentum and minimize the violation of conservation of energy, type 1 vacuum fluctuations appear as bound particle–antiparticle pairs in the most tightly bound state that is at rest and has zero angular momentum. While they exist, type 1 vacuum fluctuations can interact with normal matter provided there is no permanent transfer of energy between vacuum energy and normal energy.

Although type 2 vacuum fluctuations also appear from and then disappear back into the vacuum, they are a collection of virtual particles, all of which are off shell. Virtual particles are represented by internal lines in Feynman diagrams, occur in perturbation calculations, but are not detected in particle detectors. Since Type 2 vacuum fluctuations are disconnected Feynman diagrams, they do not contribute to physical quantities such as cross sections, decay rates, or formulas for ϵ_0, c, and α.

In a dielectric the increase in the permittivity from the vacuum value results entirely from the polarization of the atoms and molecules in the dielectric by an external electric field[16]. Similarly, because type 1 vacuum fluctuations appear as bound states, the permittivity ϵ_0 of the vacuum results entirely from the polarization of type 1 vacuum fluctuations by normal photons. The calculation of the permittivity of the vacuum is somewhat similar to the calculation of the permittivity of a normal dielectric. Once ϵ_0 has been calculated from the properties of the quantum vacuum, Maxwell's equations, quantum electrodynamics, and quantum mechanics, the formula for the speed of light is calculated from $c = 1/\sqrt{\mu_0\epsilon_0}$, and α is calculated using the defining formula (5) for α and the formulas (2) and (3) for ϵ_0 and c, respectively.

The calculations of ϵ_0, c, and α satisfy the following self-consistency conditions:

(i) Because of the details of the calculation, the formula for c is automatically the same in every direction in any inertial frame, eliminating the need for Postulate II of special relativity.

(ii) The formulas for ϵ_0 [which is equivalent to calculating a formula for the Coulomb coupling constant $k \equiv 1/(4\pi\epsilon_0)$], c, and α are all independent of the frequency of the photon.

5.2. Ongoing research

Since Maxwell's equations, quantum electrodynamics, quantum mechanics, and properties of the quantum vacuum have been used to predict the value of the Coulomb constant $k \equiv 1/(4\pi\epsilon_0)$, it should be possible to predict the value of the gravitational coupling constant G using general relativity, quantum gravity, quantum mechanics and properties of the quantum vacuum.

The values of ϵ_0, c, and α have only been calculated to lowest order in the fine-structure constant. If the theoretical ideas leading to formulas for these three quantities are correct, calculations of these quantities to next order in the fine-structure constant should agree with experimental values to approximately two additional significant figures.

References

[1] Bjorken J D and Drell S D 1965 *Relativistic Quantum Fields* (New York: McGraw-Hill) pp. 35-37, pp. 187-189
[2] Furry W H and Oppenheimer J R 1934 *Phys. Rev.* **45** 245 p. 261
[3] Pauli W and Weisskopf V 1994 The quantization of the scalar relativistic wave equation *Early Quantum Electrodynamics: A Source Book* ed Miller A I (Cambridge: Cambridge University Press) pp 188–205 English translation of W. Pauli and V. Weisskopf, Helvetica Physica Acta **7**, 709 (1934).
[4] Leuchs G, Villar A S and Sánchez-Soto L L 2010 *Appl. Phys. B* **100** 9–13
[5] Urban M, Couchot F, Sarazin X and Djannati-Atai A 2013 *Eur. Phys. J. D* **67** 58

[6] Mainland G B and Mulligan B 2017 *arXiv e-prints* arXiv:1705.11068 (*Preprint* 1705.11068)
[7] Mainland G B and Mulligan B 2019 *J. Phys.: Conf. Ser.* **1239** 012016
[8] Mainland G B and Mulligan B 2020 *Found of Physics* **50** 457–480
[9] Weidner R T and Sells R L 1960 *Elementary Modern Physics* (Boston: Allyn and Bacon. Inc.) pp. 44, 56
[10] Feynman R P 1985 *QED: The Strange Theory of Light and Matter* (Princeton, NJ: Princeton Univ. Press) p. 129
[11] Peskin M E and Schroeder D V 1995 *An Introduction to Quantum Field Theory* (Boulder, CO: Westview Press) p. 98
[12] Jackson J D 1999 *Classical Electrodynamics, 3rd ed.* (New Jersey: Wiley) p. 153
[13] Rossi B 1957 *Optics* (Reading, Massachusetts: Addison-Wesley) pp. 362-364
[14] Tanabashi M and others (Particle Data Group) 2018 *Phys. Rev. D* **98** 030001
[15] Zyla P A and others (Particle Data Group) 2020 *Prog. Theor. Exp. Phys.* **2020** 083C01
[16] Kramers H A 1924 *Nature* **113** 673

Entanglement, Microcausality and Gödel's Theorem

Paul O'Hara

Professor e Direttore della Cattedra Piero Pasolini: Ontologia e razionalita' scientifica, Istituto Universitario Sophia, 50064 Figline e Incisa Valdarno (FI), Italy

E-mail: paul.ohara@sophiauniversity.org

Abstract. One of the key points of Pauli's proof of the spin-statistics theorem is the *principle of microcausality*, which essentially states "that *all physical quantities at finite distance exterior to the light cone (for* $|x'_o - x''_o| < |\mathbf{x}' - \mathbf{x}''|$*) are commutable*". Indeed, Pauli was aware that if it were not valid then neither was his version of the spin-statistics theorem. In this presentation, we explore the relationship between entanglement and microcausality and point out that in the case of spin-singlet states, microcausality does not apply. Consequenly, we revise the spin statistics theorem to incorporate entanglement and to suggest some refinements to the axiomatic structure of quantum mechanics. Ironically, singlet states are $SL(2, C)$ invariant as is the Minkowski metric of special relativity, although the singlet state is often used to convey "spooky action at a distance" as if it were in violation of special relativity, which it is not. We also pose the question whether the paradoxes associated with "entanglement" can be understood as a special case of Gödel's theorem.

Key Words: entanglement, spin statistics theorem, microcausality, Gödel's theorem.

1. Introduction

More than eighty years have passed since Pauli published his version of the spin-statistics theorem[1] and in many ways the theorem has stood the test of time until now. When Pauli published his results spin entanglement was an esoteric topic. Einstein, Podolsky and Rosen [EPR] introduced the concept of a singlet state (nowadays referred to as an entangled state) to argue against the proposed axiomatic structure (especially the projection postulate) of quantum mechanics[2],[3]. In their minds, if the Copenhagen interpretation of quantum mechanics were true, it meant that either the system was incomplete or it gave way to "spooky action at a distance," which from Einstein's (deterministic) perspective meant the system permitted contradictions. Indeed, no one at the time seems to have noticed the similarities between Gödel's theorems[1] discovered in 1933 [5] and the EPR paradox. Gödel's first theorem essentially established that in an arithmetic axiomatic system questions can emerge that are not answerable

[1] Gödel's original paper titled: *On Formally Undecidable Propositions of Principia Mathematica and Related Systems* consists of thirteen propositions of which propositions IX and XI are usually designated as Gödel's first and second theorem respectively: (1) *Proposition IX: In all the formal systems referred to in Proposition VI there are undecidable problems of the restricted predicate calculus (i.e. formulae of the restricted predicate calculus for which neither universal validity nor the existence of a counter-example is provable)* In this context it should be noted that Proposition VI states *If the formal system P satisfies a certain condition of 'consistency,' then there is at least one recursive class-sign* $\mathbf{r}(\mathbf{v})$ *in P such that neither* $\mathbf{v}$ *Gen* $\mathbf{r}(\mathbf{v})$ *nor Neg* [$\mathbf{v}$ *Gen* $\mathbf{r}(\mathbf{v})$] *is 'provable' within P.* (2) *If the formal system P is 'consistent', its 'consistency' is 'unprovable' within P.*

Content from this work may be used under the terms of the Creative Commons Attribution 3.0 licence. Any further distribution of this work must maintain attribution to the author(s) and the title of the work, journal citation and DOI.
Published under licence by IOP Publishing Ltd

thus rendering it incomplete or contradictory. In other words, certain propositions or their negation can neither be proven nor disproven within the system. To avoid this paradox Gödel's theorem requires that the structure be changed by either modifying our definitions or by adding a new axiom to it, although it should not be overlooked that within this modified system further questions can emerge that would be unanswerable.

The first historical example of this (that I know of) was the discovery of irrational numbers by the Greeks more than 2000 years ago. For the Pythagoreans, all numbers were commensurate with positive integers and consequently the proposition that $\sqrt{2} = a/b$, *where a and b are positive integers* was, in the language of Gödel, a "provable formula" (p. 39). However, when they set out to write $\sqrt{2}$ as a *rational* number they obtained a mathematical contradiction, which in terms of Gödel's theorem means that the proposition was undecidable within the framework and assumptions of their geometry.[2] In other words, they had stumbled upon an undecidable formula, which meant that their assumptions about $\sqrt{2}$ led to a mathematical contradiction within the framework of their axiomatic system.

In a certain sense, the EPR paradox is analogous to the Greek discovery of irrational numbers but applied to quantum systems. Einstein et al. believed that there were hidden parameters that would explain the paradox. Transposed into the context of spin-statistics, the existence of hidden variables would also justify Pauli's interpretation of the *principle of microcausality*. However, with the work of John Bell in the 1960's [6] and the discovery of the inequality that has been called after him, followed by the subsequent experimental work of Aspect et al [7], it was clear that hidden parameter theory was being seriously challenged. The existence of entangled particles meant that "non-locality" was an issue not to be ignored. Indeed, I would say, it is analogous to the Greek discovery of irrational numbers. What for Einstein was either an incomplete or contradictory system was in fact a discovery that singlet-state events arose within the quantum system and yet could not (at least in Einstein's eyes) be explained by the system. For Einstein it meant that quantum mechanics was essentially an incomplete way to formulate the laws of (deterministic) physics. He did not realize (it seems to me) that he had actually stumbled upon the Greek analog of irrational numbers. The key was not to abolish quantum mechanics but further elaborate upon it. Indeed, since Aspect's work, Greenberger, Horne, Zeilenger [8] and others have raised the whole notion of "non-locality" to a different level in that nowadays it is an expression that is part of mainstream physics, regardless of whether one believes it or not, and it is here to stay. Transposed into spin-statistics it also means by definition that Pauli's *principle of microcausality* cannot be applied to entangled states. Consequently, as we shall see below, the spin-statistics theorem needs to be generalized to include non-locality and as a result, Fermi-Dirac statistics can be derived as a consequence of entanglement while the Bose-Einstein statistics will follow by breaking the entanglement.

In the context of quantum mechanics, the emphasis should not be on hidden parameters as such but rather on the nature of correlations.[3] Correlations can be of two kinds: those composed of distinguishable local events and related to local (hidden) variables as for example when we correlate two charged particles and those composed of indistinguishable non-local events constitutive of something new. In this latter case, the whole is more than the sum of its parts constituted by a subsistent relation between two objects[3][4]. A subsistent relation type event does not mean that cause and effect have broken down but it does mean that we are operating from a higher viewpoint, in the sense that chemistry cannot be reduced to elementary laws

[2] It should be pointed out that for the Greeks an axiomatic system was not simply an abstract set of rules but also referred to the physical world.

[3] In that regard, Alain Connes makes an interesting observation in his book [9] which reads [translated from Italian, p. 70]: *When considering two distant points, quantum causality is not completely coincidental, it is not completely independent but in the presence of "entanglement" correlations exist between one point and the other because of quantum causality.*

of physics. It means that we cannot understand the object solely by observing its parts. For example, a positronium atom in the 1-s state, created by a timelike interaction between a positron and an electron forms a spin-singlet state. The particles are isotropically spin-correlated (ISC) but separated by a spacelike difference once the singlet is formed.

In other words, the singlet cannot be reduced to a product of two one dimensional objects. It has mathematical group properties that characterize it in a unique way that are not determined by hidden parameters pertaining to each particle separately. Rather it is an emerging property that constitutes and defines the singlet state. It is composed of two indistinguishable events that unite to form a new "non-local" event defined by $SL(2, \mathcal{C})$ group properties. It seems Einstein only believed that the first type were possible, although ironically, the $SL(2, \mathcal{C})$ invariance of the spin-singlet state incorporates a higher order Lorentz invariance in keeping with special relativity. Moreover, the existence of correlations, regardless of the type means that with one measurement two pieces of information can be obtained. Before exploring the implications of subsistent relation type correlations for spin-statistics, we first need to mathematically define entanglement and specifically the form of entanglement that we refer to as isotropically spin-correlated states.

2. A definition of entanglement

Two or more states are said to be entangled if a linear combination of a tensor product of two or more such states is not factorable over the tensor product space on which it is defined. More specifically:

Definition 1 *Let $\mathcal{H} = H_1 \otimes \ldots \otimes H_n$ be a tensor product space, where each H_i is a Hilbert space. A state $|\psi\rangle \in \mathcal{H}$ is said to be entangled if it is irreducible or only partially reducible over $\mathcal{H}$, in other words if it is not completely factorable or separable.*

For example in the case of $H_1 \otimes H_2$, if $\mathbf{e}_i \in H_i$ then the state $\mathbf{e}_1 \otimes \mathbf{e}_2$ is reducible over $H_1 \otimes H_2$ while the state

$$\mathbf{e}_1 \otimes \mathbf{e}_2 - \mathbf{e}_2 \otimes \mathbf{e}_1 \tag{1}$$

is irreducible over $H_1 \otimes H_2$, provided $\mathbf{e}_1 \neq \mathbf{e}_2$. We refer to equation (1) as a singlet state. In the case of $H_1 \otimes H_2 \otimes H_3$, the state $\mathbf{e}_1 \otimes \mathbf{e}_2 \otimes \mathbf{e}_3$ is reducible (it has been expressed as a product of three factors) while a state of the form

$$\mathbf{e}_1(\mathbf{e}_2 \otimes \mathbf{e_3} + \mathbf{e}_3 \otimes \mathbf{e}_2) \tag{2}$$

is not completely factorable and consequently we would say that it is irreducible over $H_1 \otimes H_2 \otimes H_3$, provided $\mathbf{e}_2 \neq \mathbf{e}_3$.

A complete classification of all possible entangled states goes beyond the scope of this article. Other examples of entangled states are given by Greenberger et al. [10] and more recently by Baumgartner et al. who have done important work on entangled qudits [11]. Also a very interesting and recent article has appeared in [12] related to the question of separability. A complete discussion of the subject proper would be akin to the classification of polynomial fields using Galois groups. From a physics point of view, not every irreducible state is realizable, and in that context it is for the physics community to decide which states are physically meaningful. An important element when considering entangled states is the concept of indistinguishability, which refers to states that are invariant under the operations of permutations. For example, the state (1) is permutable and therefore composed of indistinguishable states, while equation (2) serves as an example of a mixed state in that it is both partially distinguishable and indistinguishable.

2.1. The Correlation Principle

At the heart of non-locality is a correlation principle that states "in the case of two correlated events one measurement (observation) simultaneously yields at least two pieces of information." For example if a coin is flipped and heads are up then necessarily tails are down. One observation has yielded two pieces of information. This can be generalized to a die in that if a die is rolled and a "six" appears up then we know a "one" is down and all of the other numbers are somewhere on the side. No hidden variable other than the correlation itself is needed to explain the (perfect) correlations and associate two pieces of information with one observation. This follows from the definition of correlation and applies both to classical and quantum states (vis-a-vis entanglement). Similarly, if one is told that a magnetic compass is pointing north then we also know that the other end of the pivot is pointing south. Moreover, if we have two or three properly made identical compasses, we know that by observing the magnetic north on one of them that all of the others will point in the same direction because they will likewise respond to the earth's magnetic field and consequently we can deduce that they are correlated and will respond identically. Here the hidden variable, which is the earth's magnetic field and the torque that it exerts on the needle allows us to predict a perfect correlation. Indeed, from one observation we instantaneously know many.

Beacause of Bell's work, it would be incorrect to associate a hidden variable comparable to the earth's magnetic field with quantum entanglement. Nevertheless, when particles are in a singlet state (equation (1)) they are correlated. This means that if we measure the spin value on one of the particles in an arbitrary direction we instantly know the spin value of the other particle in the same direction. We say that in this case the two particles are isotropically spin correlated (ISC). Another example of an ISC quantum state generated by the orthogonal vectors $\mathbf{e}_1$ and $\mathbf{e}_1^-$ is given by

$$|\psi\rangle = \mathbf{e}_1 \otimes \mathbf{e}_1 + \mathbf{e}_1^- \otimes \mathbf{e}_1^- \tag{3}$$

If a measurement were to be made on one of the particles in an arbitrary direction then by our correlation principle we would simultaneously know the value of the other. The isotropy in both of these cases can be defined in terms of rotational invariance. However, it is not immediately obvious that when such correlations exist the particles must necessarily form pairs. In that regard, it is worth noting that

$$|\psi\rangle = \mathbf{e}_1 \otimes \mathbf{e}_1 \otimes \mathbf{e}_1 + \mathbf{e}_1^- \otimes \mathbf{e}_1^- \otimes \mathbf{e}_1^- \tag{4}$$

is not ISC, although it may seem otherwise. A proof is given below.

The latter example on first appearance may appear counter intuitive seen from the perspective of hidden variables. Indeed, from a hidden parameter perspective, equation (4) might be seen as comparable to the case of the "n" compasses all aligned in the same state $\mathbf{e}_1(1) \otimes \ldots \mathbf{e}_1(i) \ldots \otimes \mathbf{e}_1(n)$ because of the earth's magnetic field, where each i represents a different compass. When we observe the direction of the needle on one of the compasses, we immediately know the direction of the needle on all of the others. Consequently, it is reasonable to ask if something similar can happen in the quantum world. The surprise is that it cannot happen since ISC states are not dependent upon hidden variables. In the case of equation (3) because of the isotropic characteristic associated with rotational invariance, it is possible to make a correlated measurement in any arbitrary direction $\theta \in [0, 2\pi]$ within the plane. This does not correspond to a polarization effect in a specific direction but rather to a polarization associated with the two particles that occurs in every direction at once. For example, it would be easy to pass photons through a polarizer device and select out those particles polarized in the x-direction. However, this is an *a posteriori* effect created by selected measurements and applies only to one direction in contrast to an ISC state.

3. Isotropically spin correlated particles

Our objective is to hold on to the notion of "non-locality" without identifying it with "action at a distance." To better grasp the nature of quantum entanglement or quantum correlations, we need to formally define the concept of isotropically spin correlated (ISC) particles [13], already referred to above. A special case of such particles would be a singlet state. ISC particles are correlated prior to any measurements and in all directions at once. In the EPR paradox, singlet state particles are introduced to convey the difficulty of simultaneously knowing two pieces of information from one observation, which led Einstein to refer to "spooky action at a distance." In effect, by measuring the spin in an arbitrary direction on one of the particles in the singlet means that the other is instantly known regardless of the reference frame. What is not clear is the uniqueness of the singlet state in that regard. If "hidden parameters" existed in Einstein's sense then it should be possible to have n ISC particles, meaning that if we measure the spin on one then we would automatically have information for the other $n-1$ particles. Moreover, even if there are no hidden parameters per se that does not guarantee that we cannot have three or more ISC particles. Therefore, it is important to prove as a first theorem that ISC particles can only occur in pairs. In a certain sense, it is a corollary to Bell's work and follows from a proper understanding of Bell's inequality. With that said, let us define a system of n ISC particles.

Definition 2 *Let $\mathcal{H}$ be as in Definition 1, with the restriction that each H_i is a two-dimensional Hilbert space. n states are said to be isotropically spin correlated if*

(i) for all θ the two state $|\psi\rangle_{ij} \in H_i \otimes H_j$ is rotationally invariant for all i, j where $i \neq j$ and $1 \leq i, j \leq n$,

(ii) for all θ and each $m \leq n$ the state $|\psi\rangle \in H_1 \otimes \ldots \otimes H_m$ can be written as

$$|\psi\rangle_m = \frac{1}{\sqrt{2}}[\mathbf{e}_1 \otimes \mathbf{e}_2 \ldots \mathbf{e}_m \pm \mathbf{e}_1^- \otimes \mathbf{e}_2^- \ldots \mathbf{e}_m^-] \tag{5}$$

where $\mathbf{e}_i$ and $\mathbf{e}_i^-$ are orthogonal.

Rotational invariance means that for $|\psi\rangle \in H_i \otimes H_j$

$$R_i(\theta) \otimes R_j(\theta) |\psi\rangle = |\psi\rangle$$

where each

$$R_i(\theta) = \begin{bmatrix} \cos(c\theta) & \sin(c\theta) \\ -\sin(c\theta) & \cos(c\theta) \end{bmatrix} \qquad c \text{ a constant}$$

It should be noticed that our focus is on spin which is considered as a two-dimensional object that can be embedded, if needed, in a higher dimensional space $L^2(\mathcal{C}) \otimes H$. Also, note the need for two terms separated by $\pm$ in equation (5). This guarantees that it is an entangled quantum state and that spin value can be known only after at least one measurement is made on a pure state. After an initial measurement, the state reduces to either $\mathbf{e}_1 \otimes \mathbf{e}_2 \ldots \mathbf{e}_m$ or $\mathbf{e}_1^- \otimes \mathbf{e}_2^- \ldots \mathbf{e}_m^-$ for each $m \leq n$, as required by Einsteinian realism.

There are parameters that do apply to elementary particle systems such as the electron. For example, when the mass or charge of an electron is accurately measured to many significant figures then automatically (given the experiment was done correctly) the value of the electron charge e or of its mass m has been determined for all electrons if they were to be measured under the same conditions. Similarly, the velocity of light in a vacuum is the same for all photons. Spin on the other hand has two permissible values associated with any direction, which we denote by either "+" or "-", analogous to an object spinning clockwise or counterclockwise on an axis and relative to a frame of reference. Oftentimes, these are referred to as "spin up" and "spin down" respectively. Measuring the spin in one direction for a particle does not necessarily allow us to predict the exact outcome for another particle. We can say that all electrons will exhibit $(+,-)$ properties of spin when measured with respect to a reference frame but we cannot say in general that if an electron is observed to be "spin up" along a particular axis that another independent electron will give the same observation. However, if the two electrons are in a spin-singlet state and we observe"+" on one of them, then we will certainly observe "-" on the other if the spins are measured in the same direction. If they are measured along two different axes making an angle of θ with respect to each other, the probability that both will exhibit the opposite spin values is given by $\cos^2(\theta/2)$ and the same spin values by $\sin^2(\theta/2)$. This is a consequence of the $SO(2)$ group properties associated with rotational invariance. Similarly, if two entangled particles were in the state given by expression (3) above then $\cos^2(\theta/2)$ will be the probability that both have the same spin values and $\sin^2(\theta/2)$ the opposite spin values. Another important point to recall is that when entangled particles are observed or measured, the entanglement is broken. This is a feature of quantum mechanics, which is different to classical physics. In the classical world, the observer is considered neutral and has no role to play. In quantum mechanics, in general, the act of observation changes the original state and one of the consequences is the following theorem:

Theorem 1 (**The Coupling Principle**) *Isotropically spin correlated particles must occur in pairs.*

Proof: *If three (or more) ISC particles existed then it would be possible to take three spin measurements in directions $\theta_1, \theta_2, \theta_3$ (cf. Figure 1) such that the set of all possible measurements would be given by*

$$S = \left\{ \begin{array}{llll} (+,+,+) & ,(+,+,-) & ,(+,-,+) & ,(-,+,+), \\ (+,-,-) & ,(-,+,-) & ,(-,-,+) & ,(-,-,-) \end{array} \right\}$$

Consider the two events

$$E_1 = \{(+,+,-),(+,-,-)\},\ E_2 = \{(+,+,-),(+,-,-),(-,+,-),(+,-,-)\}$$

Since, $E_1 \subset E_2$ then $P(E_1) \leq P(E_2)$. It follows from the $SO(2)$ properties of spin that

$$\frac{1}{2}\sin^2\frac{\theta_{31}}{2} \leq \frac{1}{2}\sin^2\frac{\theta_{23}}{2} + \frac{1}{2}\sin^2\frac{\theta_{12}}{2}$$

which was Wigner's form of Bell's inequality[13],[14]. Now take $\theta_{12} = \theta_{23} = \frac{\pi}{3}$ and $\theta_{31} = \frac{2\pi}{3}$ to obtain $\frac{1}{2} \geq \frac{1}{4}$, which is clearly a contradiction.
The theorem has been proven: *ISC particles must always occur in pairs.*

As a consequence of this theorem, states of the form (5) cannot be rotationally invariant for $m \geq 3$. At first this may seem to be almost a trivial result but in contrast to Bell's

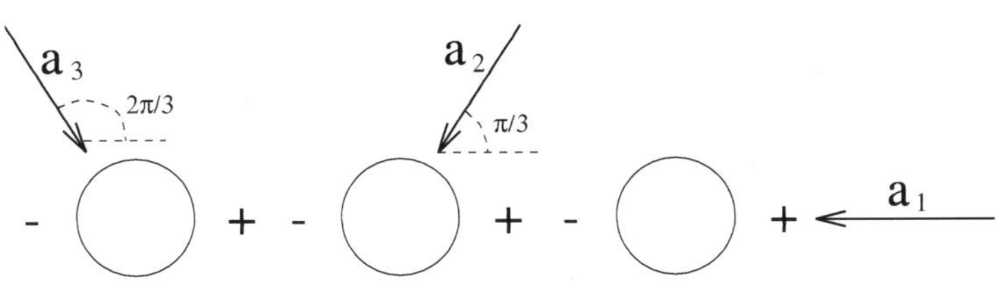

Figure 1. Three isotropically spin-correlated particles

original theorem, the specified states $(+,+,+)$ are observables and not necessarily determined by hidden parameters, which was the essence of Bell's original argument. Bell's theorem specifies that hidden variables cannot exist without giving rise to a contradiction. This theorem goes one step further and states that no three way correlation for states of the form (5) can occur without giving rise to a contradiction. For example, in the above derivation the state $(+,+,-)$ corresponds to three observations regardless of whether they are determined by hidden variables or not. In the absence of hidden variables, isotropic spin correlated particles can exist but only in pairs. The profundity of this did not go unnoticed by Einstein who referred to "spooky action at a distance," which also means (see below) that the *principle of microcausality* does not apply to ISC particles and therefore the spin-statistics theorem as proven by Pauli is no longer valid in this case. One might also object that equation (5) is a particular non-ISC state but that perhaps there are other higher order states that are ISC. In fact, Theorem 4 proves the existence of a (unique) higher dimensional ISC state that obeys Fermi-Dirac statistics for all $n \in \mathcal{N}$ but there are no hidden variables involved. Moreover, since $SO(2) \subset SL(2,\mathcal{C}) \subset SL(n,\mathcal{C})$ it also means that such higher dimensional states are pairwise rotationally invariant for any arbitrary coupling. As a result it allows us to give an alternative (physical) definition to ISC particles.

Definition 3 *Let $\mathcal{H}$ be a two dimensional Hilbert space and let $\mathbf{e}_1$ and $\mathbf{e}_1^- \in \mathcal{H}$ be unit orthoganal states. Consider the tensor product space $\mathcal{H} \otimes \mathcal{H}$ then $|\psi\rangle \in \mathcal{H} \otimes \mathcal{H}$ is an ISC state if (i) $|\psi\rangle$ is a pure state, (ii) $R(\theta)|\psi\rangle = |\psi\rangle$ and (iii) if $|\psi\rangle = |\psi_1\rangle + |\psi_2\rangle$ then $R(\theta)|\psi_1\rangle \neq |\psi_1\rangle$ and $R(\theta)|\psi_2\rangle \neq |\psi_2\rangle$.*

Two examples can be immediately be written down that fully satisfy Definition 3:

$$|\psi_1\rangle = \mathbf{a} \otimes \mathbf{b} - \mathbf{b} \otimes \mathbf{a} = (a_1 b_2 - b_1 a_2)(\mathbf{e_1} \otimes \mathbf{e_1^-} - \mathbf{e_1^-} \otimes \mathbf{e_1})$$

and

$$|\psi_2\rangle = \mathbf{a} \otimes \mathbf{b} + \mathbf{b}^- \otimes \mathbf{a}^- = (a_1 b_1 + a_2 b_2)(\mathbf{e_1} \otimes \mathbf{e_1} + \mathbf{e_1^-} \otimes \mathbf{e_1^-})$$

Usually, the expression $\mathbf{a} \otimes \mathbf{b} - \mathbf{b} \otimes \mathbf{a}$ is abbreviated as $\sqrt{2}(\mathbf{a} \wedge \mathbf{b})$ where $\sqrt{2}$ is a normalization factor associated with the Hilbert space and the symbol $\wedge$ is called the wedge-product. In other words, we define the wedge product by

$$\mathbf{a} \wedge \mathbf{b} = \frac{1}{\sqrt{2}}(\mathbf{a} \otimes \mathbf{b} - \mathbf{b} \otimes \mathbf{a}) \ . \tag{6}$$

Other vectors like $|\psi_3\rangle = |\psi_1\rangle + |\psi_2\rangle$ are also rotationally invariant but do not satisfy property (iii) of the above definition, although they do satisfy the properties of a Clifford algebra. We also point out, without going into detail, that this spin-pairing gives a natural explanation for Cooper pairs associated with superconductors [13],[15],[16].

3.1. A Clifford algebra

As mentioned, it is worth noting that the rotational invariance associated with quantum entanglement can also be associated with a Clifford algebra[17] defined by

$$\mathbf{a} \star \mathbf{b} = \frac{a.b}{\sqrt{2}}(\mathbf{e}_1 \otimes \mathbf{e}_1 + \mathbf{e}_1^- \otimes \mathbf{e}_1^-) + \mathbf{a} \wedge \mathbf{b} \tag{7}$$

if we agree to identify the scalar 1 with the isotropic vector $\frac{1}{\sqrt{2}}(\mathbf{e}_1 \otimes \mathbf{e}_1 + \mathbf{e}_1^- \otimes \mathbf{e}_1^-)$ (abbrev. $1 \leftrightarrow \frac{1}{\sqrt{2}}(\mathbf{e}_1 \otimes \mathbf{e}_1 + \mathbf{e}_1^- \otimes \mathbf{e}_1^-)$). In other words, there exists a bilinear map $\phi(\mathbf{ab}) = \phi(\mathbf{a}) \star \phi(\mathbf{b})$ such that ϕ defines an automorphism. It is a useful exercise to prove the following lemma;

Lemma 1

$$\sqrt{2}(\mathbf{a} \star \mathbf{b}) = \mathbf{a} \otimes \mathbf{b} + \mathbf{a}^- \otimes \mathbf{b}^- \tag{8}$$

The proof follows by substituting for the definitions on the right hand side and then expanding out the terms. we also note that $\mathbf{a}^- \otimes \mathbf{b}^- \neq \mathbf{b}^- \otimes \mathbf{a}^-$.

3.2. Pauli exclusion principle

In what follows, we change notation and let $\mathbf{e_1}, \mathbf{e_2}, \ldots, \mathbf{e_n}$ denote a sequence of n orthogonal unit vectors on a Hilbert space. We now derive some theorems based on the "coupling principle" which might be considered a generalization of Pauli's original exclusion principle. Indeed, within this context, if we remain within a two-dimensional framework (which can then always be embedded in higher dimensions if needed) there are essentially two ISC states in nature: $\sqrt{2}\,|\xi\rangle_1 = \mathbf{e}_1 \otimes \mathbf{e}_2 - \mathbf{e}_2 \otimes \mathbf{e}_1$ or $\sqrt{2}\,|\xi\rangle_2 = \mathbf{e}_1 \otimes \mathbf{e}_1 + \mathbf{e}_2 \otimes \mathbf{e}_2$. However, $\mathbf{e}_2$ is also what in the literature is called a spinor conjugate of $\mathbf{e}_1$ and indeed the sum $|\xi\rangle = |\xi\rangle_1 + |\xi\rangle_2$ defines, as noted above, a Clifford algebra where essentially the first term corresponds to a cross product and the second term to an inner product. Therefore, it is sufficient to work with either one, since it is easy to switch from one to another by means of the dual relationship. Moreover, if we require invariance under $SL(2,\mathcal{C})$ and not just $SO(2)$ then the singlet state is unique. Consequently, if $\mathbf{e}_1$ and $\mathbf{e}_2$ are unit vectors in $H \otimes H$ then

$$\lambda(\mathbf{e_1} \wedge \mathbf{e_2}) = \frac{\lambda}{\sqrt{2}}(\mathbf{e_1} \otimes \mathbf{e_2} - \mathbf{e_2} \otimes \mathbf{e_1}), \qquad \lambda \text{ a scalar}$$

is the unique vector that is $SL(2,\mathcal{C})$ invariant [18]. This can be extended to an L^2 space which is the ground of quantum mechanics as formulated by von Neumann [20]. In that regard, let $\mathcal{S}(q,s) \equiv L^2(q)H$ represent the space of a particle and $L^2(q_1,q_2)H_1 \otimes H_2$ the Hilbert space of two particles. Note $\mathcal{S}(q,s) \subset \mathcal{S}(q_1,s) \otimes \mathcal{S}(q_2,s) \subset L^2(q_1,q_2)H \otimes H$.

Theorem 2 *If $|\psi(q,s)\rangle \in \mathcal{S}(q,s)$ then the joint state for two indistinguishable spin entangled particles invariant under the action of $SL(2,\mathcal{C})$ but otherwise non interactive, is given by*

$$|\psi(q_1,q_2,s_1,s_2)\rangle = |\psi_1\rangle \wedge |\psi_2\rangle$$

Proof:

$$|\psi\rangle = c_1\,|\psi_1\rangle\,|\psi_2\rangle + c_2\,|\psi_2\rangle\,c_1\,|\psi_1\rangle \tag{9}$$

Since the indistinguishability has no bias, it follows that $c_1^2 = c_2^2 = \frac{1}{2}$. Therefore $c_1 = \pm c_2$. In other words, an indistinguishable unbiased two particle system must necessarily obey either a Fermi-Dirac or Bose-Einstein statistic. Of these two cases only $c_1 = -c_2$ is $SL(2,\mathcal{C})$ spin invariant. It follows that

$$|\psi(q_1,q_2,s_1,s_2)\rangle = |\psi_1\rangle \wedge |\psi_2\rangle$$

yields a form of the spin-statistics theorem.

At this stage it is a short step to derive the Pauli exclusion principle from the above lemma:

Theorem 3 *A sufficient condition for a state, representing n cyclically permutable and non-interacting particles, defined on the space $\mathcal{S}_1 \otimes \ldots \otimes \mathcal{S}_n$ to exhibit Fermi-Dirac statistics is that it contains indistinguishable ISC particles.*

Proof: *We use induction. Let $\mathcal{P}(n)$ be the proposition that $|\psi_{12\ldots n}\rangle = |\psi_1\rangle \wedge \ldots \wedge |\psi_n\rangle$. From the preceding theorem the proposition $\mathcal{P}(2)$ is true. As an induction hypothesis, we assume the theorem is true for $\mathcal{P}(n)$ then in the case of $\mathcal{P}(n+1)$*

$$|\psi_{12\ldots n+1}\rangle = \sum_{1}^{n+1} |\psi_i\rangle\,(|\psi_1\rangle \wedge \ldots \wedge |\psi_{i-1}\rangle \wedge |\psi_{i+1}\rangle \ldots \wedge |\psi_{n+1}\rangle) \tag{10}$$

$$= |\psi_1\rangle \wedge \ldots \wedge |\psi_{n+1}\rangle \tag{11}$$

which is the Fermi Dirac statistic. The Pauli exclusion principle follows by induction

[Note to facilitate reading, in the case of $n = 3$, the above can be written as [13]:

$$\begin{aligned}
|\psi(\lambda_1,\lambda_2,\lambda_3)\rangle\rangle &= \frac{1}{\sqrt{3}}\{|\psi(\lambda_1)\rangle\rangle \otimes |\psi(\lambda_2,\lambda_3)\rangle\rangle + |\psi(\lambda_2)\rangle\rangle \otimes |\psi(\lambda_3,\lambda_1)\rangle\rangle \\
&\quad + |\psi(\lambda_3)\rangle\rangle \otimes |\psi(\lambda_1,\lambda_2)\rangle\rangle\} \\
&= \frac{1}{\sqrt{3!}}\{|\psi(\lambda_1)\rangle\rangle \otimes [|\psi(\lambda_2)\rangle\rangle \otimes |\psi(\lambda_3)\rangle\rangle - |\psi(\lambda_3)\rangle\rangle \otimes |\psi(\lambda_2)\rangle\rangle] \\
&\quad + |\psi(\lambda_2)\rangle\rangle \otimes [|\psi(\lambda_3)\rangle\rangle \otimes |\psi(\lambda_1)\rangle\rangle - |\psi(\lambda_1)\rangle\rangle \otimes |\psi(\lambda_3)\rangle\rangle] \\
&\quad + |\psi(\lambda_3)\rangle\rangle \otimes [|\psi(\lambda_1)\rangle\rangle \otimes |\psi(\lambda_2)\rangle\rangle - |\psi(\lambda_2)\rangle\rangle \otimes |\psi(\lambda_1)\rangle\rangle]\} \\
&= |\psi_1(\lambda_1)\rangle\rangle \wedge |\psi_2(\lambda_2)\rangle\rangle \wedge |\psi_3(\lambda_3)\rangle\rangle\,,
\end{aligned}$$

The case of general n is similar.]

Finally, a more general theorem stating a necessary and sufficient condition is as follows [18]:

Theorem 4 *Let $V = V_1 \otimes \cdots \otimes V_n$, where each V_i is an n-dimensional vector space, and $T = T_1 \otimes \cdots \otimes T_n$ where for each i, j, $T_i = T_j$ and T_i is a linear operator on V_i. Let*

$$\begin{aligned}
v &\equiv v_1 \wedge v_2 \wedge \ldots \wedge v_n \\
&= \begin{pmatrix} v_{11} \\ \vdots \\ v_{n1} \end{pmatrix} \wedge \begin{pmatrix} v_{12} \\ \vdots \\ v_{n2} \end{pmatrix} \wedge \ldots \wedge \begin{pmatrix} v_{1n} \\ \vdots \\ v_{nn} \end{pmatrix}
\end{aligned}$$

then for $v \neq 0$

$$Tv = v \iff T \in \bigotimes_{1}^{n} SL(n,\mathcal{C}).$$

Fermi-Dirac statistics is invariant under the action of $SL(n,\mathcal{C})$.

Proof: Let $\{\mathbf{e}_1, \mathbf{e}_2 \ldots \mathbf{e}_n\}$ be an orthonormal basis of V_i, then

$$\begin{aligned}
v &= \begin{pmatrix} v_{11} \\ \vdots \\ v_{n1} \end{pmatrix} \wedge \begin{pmatrix} v_{12} \\ \vdots \\ v_{n2} \end{pmatrix} \wedge \ldots \wedge \begin{pmatrix} v_{n1} \\ \vdots \\ v_{nn} \end{pmatrix} \\
&= |v|\mathbf{e}_1 \wedge \mathbf{e}_2 \wedge \ldots \wedge \mathbf{e}_n
\end{aligned}$$

where

$$|v| = \begin{vmatrix} v_{11} & v_{12} & \cdots & v_{1n} \\ v_{21} & v_{22} & \cdots & v_{2n} \\ \vdots & \vdots & \vdots & \vdots \\ v_{n1} & v_{n2} & \cdots & v_{nn} \end{vmatrix}$$

The linearity of T gives

$$\begin{aligned}
Tv &= |v|T_1\mathbf{e}_1 \wedge T_2\mathbf{e}_2 \wedge \ldots \wedge T_n\mathbf{e}_n \\
&= |v|\begin{pmatrix} t_{11} \\ \vdots \\ t_{n1} \end{pmatrix} \wedge \begin{pmatrix} t_{12} \\ \vdots \\ t_{n2} \end{pmatrix} \wedge \ldots \wedge \begin{pmatrix} t_{n1} \\ \vdots \\ t_{nn} \end{pmatrix} \\
&= |v||T_1|\mathbf{e}_1 \wedge \mathbf{e}_2 \wedge \ldots \wedge \mathbf{e}_n, \qquad T_1 = T_2 = \ldots = T_n
\end{aligned}$$

Therefore, since $v \neq 0$ implies $|v| \neq 0$ then

$$Tv = v \Rightarrow |T_1| = 1 \text{ and } T_1 \in SL(n,\mathcal{C})$$

Conversely

$$T_1 \in SL(n, \mathcal{C}) \Rightarrow Tv = v$$

This proves the theorem. It is also shown in [18] that v is unique.

4. Entanglement and Microcausality

In the light of the above results, one might ask why they differ from Pauli's original spin-statistics theorem and if the two can be reconciled. When Pauli published his original results in 1940, the EPR paradox at the time would have been a curiosity pertaining to the axioms of quantum mechanics. The idea of non-locality ("spooky action at a distance" as Einstein would refer to it) was not taken seriously and Pauli's own formulation of microcausality reflects this [1],[19]. The position on non-locality began to change after 1964 with the publication of Bell's inequality and especially with the subsequent experiments from the early 1980's onwards, carried out by Aspect, Clauser and Zeilinger winners of the Nobel Prize in Physics 2022 "for experiments with entangled photons, establishing the violation of Bell's inequalities." At this stage "non-locality" is an established experimental fact, although that does not necessarily mean that we fully understand it. It remains an item of debate in contemporary physics with many arguing both sides of the issue, with some trying to reconcile the "non-locality" with hidden parameter theory and others believing that it is a whole new phenomenon non susceptible to classical concepts. Either way, the existence of non-locality by definition means that the *principle of microcausality* which is the basis of Pauli's results is no longer valid. To fully grasp this point, we give a short overview of Pauli's argument:

(i) He first distinguishes between vectors and spinor representations of angular momentum, which he associates with tensors of even and odd rank respectively. He then uses the second quantization procedure to obtain a set of solutions, D and D_1, of a second order wave equation.

(ii) These latter D_1 solutions, which are related to the Neumann's function and "the first Hankel cylindrical function" ([1],p.721), are then carefully eliminated as unphysical by invoking the postulate that "*all physical quantities at finite distances exterior to the light cone (for* $|x'_0 - x''_0| < |x' - x''|$*) are commutable.*" Nowadays this is referred to as microcausality.

(iii) Applying the second quantization procedure to the remaining D solutions of his equation, he distinguishes between operators obeying commutator rules and those obeying anticommutator rules which are then associated with Bose-Einstein and Fermi-Dirac statistics respectively.

(iv) He also notes ([1],p.722) that "*For integral spin the quantization according to the exclusion principle is not possible. For this result it is essential that the use of the* D_1 *function in place of the D function be, for general reasons, discarded.*"

(v) However, if we were to turn that phrase around, and if we were to accept rather than discard as a solution the non-commutable D_1 function beyond the light cone, then according to Pauli it would then be possible to quantize integral spin particles according to the exclusion principle.

(vi) In essence the D and D_1 solutions are determined by local hidden parameters. In contrast, entanglement implies that there are no hidden parameters and that entangled quantum operators separated by spacelike events can commute.

The *principle of microcausality* as interpreted by Pauli does not apply to entangled particles. There is no mention of entanglement in his paper on spin-statistics. However, he does point out that if microcausality fails then his results are no longer valid. Our approach does not depend upon the D_1 function described by Pauli nor have we explicitly referred to microcausality.

Our approach depends on the meaning of entanglement, which represents a higher viewpoint associated with the work of Gödel.

4.1. Gödel's theorem

In an article published in the IARD proceedings of 2019, Prof. Ben-Ya'accov [23] notes that "in any connected and rich enough formal structure, based on a finite number of first principles and inference rules, there will always be propositions that may be formulated within the formal system but are undecidable. Such a theory cannot be both consistent and complete." He further notes (p. 3) that "so far physics theories do not have a metaphysical view that allows formulation of metaphysical statements in ordinary physical terms; then it would have been possible to directly examine the applicability of Gödel's theorem to physics. However, we may proceed in a more subtle way, identifying the core principle of Gödel's theorem and examine in what way it may appear in physics."

In what follows, I would like to suggest that the mathematical contradiction elicited by Bell's inequality as applied to quantum entanglement in Theorem 1 (and which has as a consequence the violation of microcausality), is in accordance with Ben-Ya'accov's observation. Indeed, as he has noted "there will always be propositions that may be formulated within the formal system but are undecidable" in that "such a theory cannot be consistent or complete"[23]. In that regard it is worth noting that Einstein-Podolsky-Rosen's paper connected with quantum entanglement is titled "Can Quantum-Mechanical Description of Physical Reality Be Considered Complete?" and they conclude either that the quantum system is "not complete" or else two quantities (such as position and momentum) "cannot have simultaneous reality" (p. 778), which from the viewpoint of classical physics is absurd. Moreover, they point out that to insist on completeness "leads to a contradiction" (p. 779). Juxtaposed in this way, the EPR paradox as formulated seems to be a special case of Gödel's first theorem, a connection which we now explore more in detail both in its general form and its particular realizations.

Both of Gödel's theorems are very general in that they apply to all axiomatic systems of set theory including the metamathematical structure within which the theorems are formulated. They essentially affirm that within any mathematical structure questions will emerge (using the formal language of the structure) which cannot be decided. In the case of the first theorem, the formal language of the system will allow self-referential statements of the form "this statement is not provable" to emerge within the system without being able to decide its validity. In other words, if we were to assume it is provable then it would give rise to a mathematical contradiction within the structure in accordance with Aristotle's principle of contradiction. This would result in a vicious circle associated with self-referencing. For example, the statement that all statements are false is itself a statement, which if true is false and vice versa and thus results in a vicious circle, in that if the basic premise is true then by that very fact it is false which in turn makes in true. It becomes a never-ending story which if written as a piece of computer code would generate an endless loop unless explicitly stopped by someone (or something) outside of the system. Consequently, we say that such a proposition is undecidable in that it is incompatible with the axiomatic structure. Nevertheless, the metalanguage is able to formulate the undecidable proposition by using the human quality of self-reflection or "self-referencing"[23], since it is precisely self-referencing that allows the metastructure to emerge.[4]

Secondly, Gödel's (second) theorem affirms that "the consistency of P is unprovable in P" although it does not specify explicitly how the undecidable propositions will emerge within any specific structure. Nevertheless, all these propositions (including Gödel's original proposition) apply also to the metasystem and share one thing in common: if an attempt is made to prove

[4] Gödel's formulates a metamathematical theorem in the language of arithmetic but notes that it can be extended to all axiomatic set theoretical systems in that no mathematical system is capable of justifying itself. There will always be propositions that can be formulated within the formal system but are undecidable.

them, they all violate the principle of contradiction. To quote Gödel (p.71): "It may be noted this proof is also constructive, i.e., it permits, if a *proof* from c is produced for w, the effective derivation from c of a contradiction." Moreover, he has already earlier pointed out that this applies more broadly in that (p. 38) "there are in fact relatively simple problems in the theory of ordinary numbers which cannot be decided from thc axioms."

It is important to note that while Gödel's theorem is formulated within the metasystem it applies to any axiomatic system of set theory (footnote 62, p.69) including the metasystem itself. Moreover, once we recognize that the metalanguage is itself a mathematical language, (indeed Gödel's theorem transforms his metalanguage into the language of arithmetic to proof his theorems) then all contradictory statements emerging within any axiomatic system have the same methodological weight or value, in that they are governed by the same laws of "excluded middle" and "contradiction." It other words, reflective statements like "this statement is not provable" has the same weight as saying "$\sqrt{2}$ cannot be proven to be a rational number," in that for a formal mathematical language there is no difference between them. Broadly speaking, they are proved in the same way: each proposition is formulated within the language of the system, which when taken together with the axioms and previously derived theorems, result in a contradiction. Proofs by contradiction constitute a class of proofs into which Gödel's theorem also falls. And the second theorem is an affirmation that such undecidable questions will exist within any mathematical framework.

Mathematics per se is not self-referential. It is only within the higher context of metaphysics that self-referential statements have meaning. Once a proposition is undecidable or unprovable within the system then it generally points to an enlarged modified system in which some of the axioms are changed, substituted or modified to incorporate the new proposition. Metamathematics is itself a discovery of a higher context in which contradictory or undecidable statements within one context takes on meaning within a higher context. This is expressed very well by the Jesuit philosopher Bernard Lonergan([24], p.xxv):

> Gödel's theorem is to the effect that any set of mathematical definitions and postulates give rise to further questions that cannot be answered on the basis of definitions and postulates, say $P, Q, R, \ldots$ such that, if P is assumed, there arise questions that can be answered only by assuming Q, if Q is assumed, there arise questions that can be answered only by assuming R, and so forth. Then besides the successive lower contexts $P, Q, R, \ldots$ there also is the upper context in which Gödel's theorem is expressed. Moreover, inasmuch as the theorem is quite general, the upper context is independent of the context of any particular contexts such as, $P, Q, R, \ldots$. Finally, since there is no last, lower context that is definitive, since R will demand a context S, and S a context T, and T a context U, and so on indefinitely, the really significant context is the upper context; all lower contexts, $P, Q, R, S, T, U, \ldots$ are provisional; and they attain a definite significance only in the measure that they give access to the upper context.

For example, the discovery of $\sqrt{2}$ means the implied axiom that all numbers are commensurable with natural numbers has to be modified and extended to incorporate irrational numbers. In the context of the extended number system, $\sqrt{2}$ now has meaning. However, more undecidable propositions within the context of Gödel's theorem will emerge within the newer system. For example, the solution to the equation $x^2 + 1 = 0$ while formulated within the system of real numbers cannot be answered within that system. One is required to move into complex numbers and so on. Gödel's theorem foresees undecidable propositions arising but it does not give us an algorithm or method for discovering them. Notice that the hierarchy $\mathcal{N} \subset \mathcal{Q} \subset \mathcal{R} \subset \mathcal{C}$ is an instance of higher level viewpoints from the perspective of numbers. Similarly, in terms of infinite sets there is the theorem that the power set of any infinite set is a higher order infinity.

For example, the natural numbers $\mathcal{N}$ are a countable infinity but the power set of $\mathcal{N}$ corresponds to the real numbers.

When we move into theoretical physics, one is guided by experiment and prior mathematical models. Precisely because of our reliance on mathematics one should not be surprised that contradictions comparable to Gödel's theorem will arise if we try to proof consistency. Bell's inequality is an example of an undecidable proposition emerging within the framework of physics. All physics from an Einsteinian point of view was considered local (which would mean that position and momentum would have simultaneous reality) but instead Bell's inequality and the subsequent experimental work points towards non-local events. Some physicists take this to mean that cause and effect have broken down and consequently reject the non-local formulations of the theory (in spite of the experimental evidence) without realizing that "non-local events" are correlated and require a paradigm shift in which causality is maintained but in a highly non deterministic way that renders it "non-local." The type of contradiction associated with quantum entanglement requires that we move beyond a reductionist view of physicist.

Conceptually, quantum mechanics represents a higher viewpoint in which the higher context sublates the lower one, analogous to how the real numbers sublate the system of rational numbers. In addition, physics is also about the material world and needs not only verification of its mathematical models but also a physical domain in which higher contexts can be expressed. In that regard, it would seem that the phenomenon of quantum entanglement finds its full realization in chemistry in the sense that chemistry sublates physics. Indeed, without entanglement there would be no chemical bonding (as we know it) and without chemistry and the emerging laws, which exist because of the stability associated with entanglement, there would be no world as we know it. Chemistry, in other words, constitutes from a physics perspective a higher order viewpoint. Spin while potentially associated with individual particles exists as angular momentum only when interacting with other particles or while being measured with respect to a reference frame. Entanglement is one form of such an interaction. It cannot be deduced. It has to be first defined and then its properties can be analyzed. Moreover, this dependence upon measurable interactions or observables, places it squarely within the framework of Gödel's theorem to the extent that if we assign spin values isotropically to individual particles by means of a (hidden) parameter a contradiction arises. It can only be overcome by disregarding the (non-existent) hidden parameter and instead incorporating the measuring process (observables) into the process. Mathematically speaking quantum mechanics sublates classical mechanics and allows for new structures such as entanglement to emerge. A contradiction arises when isotropy and the probability laws associated with SO(2) relativity are included. It cannot be explained by a simple interaction and yet it can be defined by measurable observables that interact (cf. Definition 2). It is "something entirely new" to use an expression of Bohr[25].

4.2. Higher Viewpoint

Let us take a closer look at the meaning of the above in terms of commutators. First, we redefine the spin angular momentum operator by $S_i = nL_i$, where n is an integer, the subsequent Lie Algebra is defined by $[S_i, S_j] = in\epsilon_{ijk}S_k$. Note that the inclusion of the integer n permits particles of integer spin as well as particles of 1/2-integer spin to be subjected to anticommutator relations and to obey Fermi-Dirac statistics. This means that when n=1 we obtain the usual relationship for spin 1/2 particles, whereas for n=2 we obtain the usual properties for a spin 1 photon and for n=4, the properties of the spin 2 gravitons [13].

Now, consider a two particle system $|\psi(\lambda_1)\rangle \otimes |\psi(\lambda_2)\rangle \in \mathcal{S}_1 \otimes \mathcal{S}_2$ where $\mathcal{S}_1 = \mathcal{L}^2(\mathcal{R}^3) \otimes H_1$ and $\mathcal{S}_2 = \mathcal{L}^2(\mathcal{R}^3) \otimes H_2$ respectively. Note that each ket $|\psi(\lambda)\rangle \in \mathcal{S}$ can be written as $|\psi(q_1)\rangle \otimes s$, where s is a spinor. Also, let $\vec{S_1} = (S_i(q_1), S_j(q_1), S_k(q_1))$ and $\vec{S_2} = (S_i(q_2), S_j(q_2), S_k(q_2))$ be spin operators defined on the Hilbert spaces H_1 and H_2 respectively. We have already seen that

Bose-Einstein statistics follow when *no* two states are ISC. If we take this to mean that the two particles are in the triplet state then this also means that the two particles are statistically independent. Therefore, measurements on both particles separated by a spacelike distance will be represented respectively by operators of the form $S_i(q_1) \otimes I_2$ and $I_1 \otimes S_j(q_2)$. It follows, trivially, that $[S_i(q_1) \otimes I_2, I_1 \otimes S_j(q_2)] = 0$, which means that in the case of the triplet state, all measurements will commute beyond the light cone.

The singlet state defines a rotationally invariant state and obeys a Fermi-Dirac statistic. In this case, it would be inappropriate to apply the operators for independent states. Instead, we let $|\psi(\lambda_1, \lambda_2\rangle \in \mathcal{S}_1 \otimes \mathcal{S}_2$ represent the spin-singlet state of two particles. Note that the perfect correlations between them allows us to put $H_1 = H_2 = H$ and to identify $\vec{S}_1$ and $\vec{S}_2$ as follows: Let $s_1(\theta)$ and $s_2(\theta)$ represent the spin states for particles 1 and 2 respectively then for an arbitrary angle θ there exists a unit vector $\vec{n}(\theta)$ such that $\vec{S}_1.\vec{n}(\theta)(s_1(\theta)) = \pm s_1(\theta)$ if and only if $\vec{S}_2.\vec{n}(\theta)(s_2(\theta)) = \mp s_2(\theta)$. This relationship allows us to identify $s_2(\theta)$ with the orthogonal complement $s_1^{-}(\theta)$ of $s_1(\theta)$ and to put $\vec{S}_1 = \vec{S}_2$.

This identification means that $\vec{S}_1.\vec{n}(\theta) = \vec{S}_2.\vec{n}(\theta)$ (note the same θ) and therefore $[\vec{S}_1.\vec{n}(\theta), \vec{S}_2.\vec{n}(\theta)] = 0$. However, because of the nature of entanglement we also free to consider $\vec{S}_1.\vec{n}(\theta_1)$ and $\vec{S}_2.\vec{n}(\theta_2)$ where $\theta_1 \neq \theta_2$ and in this case $[\vec{S}_1.\vec{n}(\theta_1), \vec{S}_2.\vec{n}(\theta_2)] \neq 0$. In fact, $[S_i, S_j] = in\epsilon_{ijk}S_k$. Simply put, microcausality does not apply to entangled states. Entanglement permits both zero and non-zero commutators. It is an example, as noted at the beginning, of Gödel's theorem being operative at the quantum level. It does not violate quantum mechanics, on the contrary it shows the power of quantum mechanics to generate new questions within the system without it being able to offer an answer unless the axiomatic system is enlarged. It other words, the existence of entanglement means that quantum mechanics needs to be sublated into a higher viewpoint. Specifically, in this case it means that rather than asking if events are spacelike or timelike, we should first ask if they are correlated or uncorrelated and based on that answer then consider whether spacelike or timelike has any meaning.

5. Conclusion

Much has been said of both entanglement and microscausality and the main conclusion is that quantum entanglement involves a higher viewpoint that needs to be incorporated into the axioms of quantum mechanics comparable to incorporating irrational numbers into number theory or replacing Euclidean with Riemannian geometry. In an analogous way the *principle of microcausality* needs to be replaced with a correlation principle that allows for both the physics of entanglement and the physics of non-entanglement. They do not contradict each other but move us beyond a strictly reductionist approach to physics and allow for new laws and results to emerge. Already this article has indicated that Pauli's exclusion principle and Cooper pairing in superconductors can be seen from this new perspective. The laws of entanglement sublate the usual laws and principles of quantum mechanics in keeping with Bohr's observation to Einstein that the quantum world is"something entirely new" [25].

Acknowledgments

With gratitude, I dedicate this paper to Prof. Lawrence Horwitz. In 1997, I gave a presentation at the *5th Wigner Symposium* in Vienna where I presented a proof of a spin-statistics theorem in which I pointed out that Fermi-Dirac and Bose-Einstein statistics should not be distinguished by spin value but rather by the difference between entangled and non-entangled states. Prof. Horwitz who was present at the time asked me a key question in that regard: "if you are correct where was Pauli wrong?" At the time I was not able to fully answer him. Now, twenty-five years later, this paper is my official response to him.

References

[1] W. Pauli "The Connection Between Spin and Statistics",*Phys. Rev***1940**, 716-722.
[2] A. Einstein, B. Podolsky,N. Rosen"Can Quantum-Mechanical Description of Physical Reality Be Considered Complete?" *Phys. Rev.***1935**, *47*, 777-780.
[3] P. O'Hara "The Einstein-Podolsky-Rosen paradox and SU(2) relativity", *J.Phys: Conf. Ser. 1239 012021***2019**, 1-12.
[4] Aristotle,*Metaphysics*, (Everyman's Library, London, 1970),Book Z, XI, p.195; Chap VI, p 218.
[5] K. Gödel *On Formally Undecidable propositions Of Principia Mathematica and Related Systems*, 1st ed.; Dover, New York, USA, 1992; p. 57.
[6] J.S.Bell,"On The Einstein Podolsky Rosen Paradox" *Physics***1964** *1*, 195-200.
[7] A.Aspect, J. Dalibard, and G. Roger"Experimental Test of Bell's Inequalities Using Time Varying Analyzers" *Phy. Rev. Letts.***1982**, *49*, 1804-1807.
[8] D. Greenberger, M. Horne, A. Zeilinger "Going Beyond Bell's Theorem". In *Bell's Theorem, Quantum Theory, and Conceptions of the Universe*; Kafatos(ed), Menos(ed); Springer: Heidelberg, Germany; pp. 69-72.
[9] A. Connes "La Geometria e Il Mondo Dei Quanti",1st ed.; Dedalo, Bari, Italy, 2022, p.70.
[10] D.Greenberger, M.Horne, A. Shimony and A.Zeilinger "Bell's theorem without inequalities" ,*Am. J. Physics* **1990**, *58*(12), 1131-1143.
[11] B. Baumgartner, B. Hiesmayr and H. Narnhofer "A special simplex in the state space for entanled qudits ", *J. Phys. A:Math. Theor.***2007**,*40*, 7919.
[12] c. Popp and B. Hiesmeyr, arXiv:2205.11405v1[quant-ph]23 May 2022.
[13] P. O'Hara "Rotational Invariance and the Spin-Statistics Theorem",*Found. Phys.***2003**,*33*(9),1349-1368.
[14] E. P. Wigner "On hidden variables and quantum mechanical probabilities",*Am. J. Phys.***1970**, *38*, 1005.
[15] A. Kovács, G. Vassallo, P. O'Hara, F. Celani and A di Tommaso "Unified Field Theory and Occam's Razor", 1st ed., WSPC, Singapore, (2022).
[16] K. Ueda et al. "Dominant nonlocal superconducting proximity effect due to electron-electron interaction in a ballistic double nanowire",*Science Advances***2019**,*5(10)*.
[17] D. Hestenes and G. Sobczyk "Clifford Algebra and Geometric Calculus", 1st ed; Springer, 1984.
[18] P. O'Hara "A generalized spin statistics theorem", *J.Phys: Conf. Ser. 845 012030***2017**,1-12.
[19] J. Wright "Quantum Field Theory: Motivating the Axiom of Microcausality" (2012).
[20] J. von Neumann "Mathematical Foundations of Quantum Mechanics",1st ed; Princeton U, Princeton, USA,1996 (12th printing).
[21] J. Christian, arXiv:1005.4932v2[quant-ph] 8 July 2010.
[22] M. Holman,arXiv:quant-phy/0704.2038v2.
[23] U Ben-Ya'acov, "Observers, observations and referencing in physics theories," J.Phys. Conf Ser. **1956** 012019
[24] B. Lonergan ,"Insight, A Study of Human Understanding", Revised student ed., Longman and Todd, London, 1973.
[25] N.Bohr, www.marxists.org/reference/subject/philosophy/works/dk/bohr.htm

Extension of the Goldstone and the Englert-Brout-Higgs mechanisms to non-Hermitian theories

Philip D. Mannheim
Department of Physics, University of Connecticut, Storrs, CT 06269, USA
E-mail: philip.mannheim@uconn.edu

Abstract. We discuss the extension of the Goldstone and Englert-Brout-Higgs mechanisms to non-Hermitian Hamiltonians that possess an antilinear PT symmetry. We study a model due to Alexandre, Ellis, Millington and Seynaeve and show that for the spontaneous breakdown of a continuous global symmetry we obtain a massless Goldstone boson in all three of the antilinear symmetry realizations: eigenvalues real, eigenvalues in complex conjugate pairs, and eigenvalues real but eigenvectors incomplete. In this last case we show that it is possible for the Goldstone boson mode to be a zero-norm state. For the breakdown of a continuous local symmetry the gauge boson acquires a non-zero mass by the Englert-Brout-Higgs mechanism in all realizations of the antilinear symmetry, except the one where the Goldstone boson itself has zero norm, in which case, and despite the fact that the continuous local symmetry has been spontaneously broken, the gauge boson remains massless.

1. ANTILINEAR SYMMETRY

With the advent of PT theories a new chapter was opened up in quantum theory. Specifically, Bender and Boettcher [1] found that the non-Hermitian Hamiltonian $H = p^2 + ix^3$ had an energy eigenspectrum that was completely real. While a surprise, it did not actually violate any law of quantum mechanics since while Hermiticity implies reality there is no theorem that says that a non-Hermitian Hamiltonian could not have a completely real energy spectrum. Hermiticity is only **sufficient** for reality.

All the same, if a non-Hermitian Hamiltonian is nonetheless going to have a completely real eigenspectrum there would have to be some reason for this to be the case. And Bender and Boettcher identified the reason to be that this particular Hamiltonian had a particular antilinear symmetry, viz. PT (P is parity, T is time reversal), with $p \rightarrow p$, $x \rightarrow -x$ and $i \rightarrow -i$. Gradually, it was realized that antilinearity was the **necessary** condition for reality, with the **necessary** and **sufficient** condition being that the Hamiltonian have an antilinear symmetry and that its eigenstates be eigenstates of the antilinear operator (see e.g. [2] and references therein).

But there is more to PT.

Content from this work may be used under the terms of the Creative Commons Attribution 3.0 licence. Any further distribution of this work must maintain attribution to the author(s) and the title of the work, journal citation and DOI.
Published under licence by IOP Publishing Ltd

2. HOW ANTILINEAR SYMMETRY WORKS

Consider the eigenvector equation

$$i\frac{\partial}{\partial t}|\psi(t)\rangle = H|\psi(t)\rangle = E|\psi(t)\rangle. \tag{1}$$

Replace the parameter t by $-t$ and then multiply by some general antilinear operator A:

$$i\frac{\partial}{\partial t}A|\psi(-t)\rangle = AHA^{-1}A|\psi(-t)\rangle = E^*A|\psi(-t)\rangle. \tag{2}$$

If H has an antilinear symmetry so that $AHA^{-1} = H$, then

$$HA|\psi(-t)\rangle = E^*A|\psi(-t)\rangle. \tag{3}$$

(1) (Wigner): Energies can be real and have eigenfunctions that obey $A|\psi(-t)\rangle = |\psi(t)\rangle$,

(2) or energies can appear in complex conjugate pairs that have conjugate eigenfunctions ($|\psi(t)\rangle \sim \exp(-iEt)$ and $A|\psi(-t)\rangle \sim \exp(-iE^*t)$).

As to the converse, suppose we are given that the energy eigenvalues are real or appear in complex conjugate pairs. In such a case not only would E be an eigenvalue but E^* would be too. Hence, we can set $HA|\psi(-t)\rangle = E^*A|\psi(-t)\rangle$ in (2), and obtain

$$(AHA^{-1} - H)A|\psi(-t)\rangle = 0. \tag{4}$$

Then if the eigenstates of H are complete, (4) must hold for every eigenstate, to yield $AHA^{-1} = H$ as an operator identity, with H thus having an antilinear symmetry.

3. A SIMPLE EXAMPLE

Consider the 2×2 matrix

$$N = \begin{pmatrix} C+A & iB \\ iB & C-A \end{pmatrix}, \quad [PT, N] = 0, \tag{5}$$

where A, B and C are all real. The matrix N is not Hermitian but does have a PT symmetry if we set $P = \sigma_3$ and $T = K$ where K effects complex conjugation. The eigenvalues of N are given by

$$\Lambda_\pm = C \pm (A^2 - B^2)^{1/2}, \tag{6}$$

and they are **real** if $A^2 \geq B^2$ and in a **complex conjugate pair** if $A^2 < B^2$, this actually being generic to non-Hermitian but PT-symmetric systems.

In addition, if $A = B$ the matrix N only has **one** right-eigenvector and only **one** left-eigenvector, despite having two solutions to $|M - \lambda I| = 0$ (both with $\lambda = C$), viz.

$$\begin{pmatrix} C+A & iA \\ iA & C-A \end{pmatrix}\begin{pmatrix} a \\ b \end{pmatrix} = \begin{pmatrix} Ca + Aa + iAb \\ iAa + Cb - Ab \end{pmatrix} = C\begin{pmatrix} a \\ b \end{pmatrix}, \tag{7}$$

$$(e, f)\begin{pmatrix} C+A & iA \\ iA & C-A \end{pmatrix} = (eC + eA + ifA, ieA + fC - fA) = (eC, fC), \tag{8}$$

viz.

$$b = ia, \qquad \begin{pmatrix} a \\ b \end{pmatrix} = \begin{pmatrix} a \\ ia \end{pmatrix}, \qquad f = ie, \qquad (\,e, f\,) = (\,e, ie\,), \qquad (\,e, ie\,)\begin{pmatrix} a \\ ia \end{pmatrix} = 0, \tag{9}$$

and thus cannot be diagonalized by a similarity transformation. It is thus a non-diagonalizable, Jordan-block matrix.

This particular Jordan-block situation is a case where the Hamiltonian is manifestly non-diagonalizable and thus manifestly non-Hermitian and yet all eigenvalues are real. Note that the overlap of the left-eigenvector and right-eigenvector in this case is zero.

4. DIAGONALIZING ONLY IF $A \neq B$

With

$$N = \begin{pmatrix} C + A & iB \\ iB & C - A \end{pmatrix} \tag{10}$$

introduce

$$S = \frac{1}{2(A^2 - B^2)^{1/4}} \begin{pmatrix} (A+B)^{1/2} + (A-B)^{1/2} & i[(A+B)^{1/2} - (A-B)^{1/2}] \\ -i[(A+B)^{1/2} - (A-B)^{1/2}] & (A+B)^{1/2} + (A-B)^{1/2} \end{pmatrix}, \tag{11}$$

$$S^{-1} = \frac{1}{2(A^2 - B^2)^{1/4}} \begin{pmatrix} (A+B)^{1/2} + (A-B)^{1/2} & -i[(A+B)^{1/2} - (A-B)^{1/2}] \\ i[(A+B)^{1/2} - (A-B)^{1/2}] & (A+B)^{1/2} + (A-B)^{1/2} \end{pmatrix}, \tag{12}$$

$$V = \frac{1}{(A^2 - B^2)^{1/2}} \begin{pmatrix} A & iB \\ -iB & A \end{pmatrix}, \quad V^{-1} = \frac{1}{(A^2 - B^2)^{1/2}} \begin{pmatrix} A & -iB \\ iB & A \end{pmatrix}, \tag{13}$$

and they effect

$$\begin{aligned} SNS^{-1} = N' &= \begin{pmatrix} C + (A^2 - B^2)^{1/2} & 0 \\ 0 & C - (A^2 - B^2)^{1/2} \end{pmatrix}, \\ VNV^{-1} &= \begin{pmatrix} C + A & -iB \\ -iB & C - A \end{pmatrix} = N^\dagger. \end{aligned} \tag{14}$$

Here the right-eigenvectors of N that obey $NR_\pm = \Lambda_\pm R_\pm$ are given by the columns of S^{-1}, and the left-eigenvectors of N that obey $L_\pm N = \Lambda_\pm L_\pm$ are given by the rows of S. Given the right-eigenvectors one can also construct the left-eigenvectors by using the V operator when $A^2 > B^2$, with the left eigenvectors being constructed as $\langle L_\pm| = \langle R_\pm|V$. Thus as long as $A \neq B$ (S becomes undefined at $A = B$) we can diagonalize N and can construct a matrix V that effects the pseudo-Hermiticity condition $VNV^{-1} = N^\dagger$ that generalizes $H = H^\dagger$, just as is characteristic of a matrix with an antilinear symmetry ([2] and references therein). However, with $A = B$ the matrix N becomes of non-diagonalizable Jordan-block form.

5. SO WHAT HAPPENS TO THE GOLDSTONE AND ENGLERT-BROUT-HIGGS MECHANISMS?

Thus if we are going to replace Hermiticity by some antilinearity requirement, then what is going to happen to other standard quantum results that rely on Hermiticity. In [3, 2] the CPT theorem (C is charge conjugation) was derived without Hermiticity, with it being shown that with only invariance under the complex Lorentz group (the proper Lorentz group) and probability conservation the Hamiltonian would necessarily be CPT symmetric. Since CPT defaults to PT for non-relativistic systems this puts the quantum-mechanical PT program on a very secure theoretical footing.

So what about the Goldstone theorem? Does it need Hermiticity? This then was the brief of Alexandre, Ellis, Millington and Seynaeve [4] as followed up by Mannheim [5], and then by Alexandre, Ellis, Millington and Seynaeve [6] themselves. In these latter two papers the Englert-Brout-Higgs mechanism was explored in a non-Hermitian context. Subsequent follow up may be found in [7, 8, 9, 10, 11, 12, 13, 14, 15, 16, 17, 18, 19, 20, 21, 22, 23, 24, 25].

6. NON-HERMITIAN MODEL WITH A CONTINUOUS GLOBAL SYMMETRY

The model introduced by Alexandre, Ellis, Millington and Seynaeve in [4] consists of two complex (i.e. charged) scalar fields $\phi_1(x)$ and $\phi_2(x)$ with action

$$I(\phi_1,\phi_2,\phi_1^*,\phi_2^*) = \int d^4x \left[\partial_\mu\phi_1^*\partial^\mu\phi_1 + \partial_\mu\phi_2^*\partial^\mu\phi_2 + m_1^2\phi_1^*\phi_1 - m_2^2\phi_2^*\phi_2 - \mu^2(\phi_1^*\phi_2 - \phi_2^*\phi_1) - \frac{g}{4}(\phi_1^*\phi_1)^2\right], \quad (15)$$

with m_1^2, m_2^2, μ^2 and g all being real, and g being positive. Here the star symbol denotes complex conjugation, and thus Hermitian conjugation since neither of the two scalar fields possesses any internal symmetry index. Since the action is not invariant under complex conjugation, the action is not Hermitian. It is however invariant under the following CPT transformation

$$\phi_1(x_\mu) \to \phi_1^*(-x_\mu), \quad \phi_2(x_\mu) \to -\phi_2^*(-x_\mu), \quad \phi_1^*(x_\mu) \to \phi_1(-x_\mu), \quad \phi_2^*(x_\mu) \to -\phi_2(-x_\mu), \quad (16)$$

and thus has an antilinear symmetry.

As written, the action given in (15) is invariant under the electric charge transformation

$$\phi_1 \to e^{i\alpha}\phi_1, \quad \phi_1^* \to e^{-i\alpha}\phi_1^*, \quad \phi_2 \to e^{i\alpha}\phi_2, \quad \phi_2^* \to e^{-i\alpha}\phi_2, \quad (17)$$

to thus possess a standard Noether current

$$j_\mu = i(\phi_1^*\partial_\mu\phi_1 - \phi_1\partial_\mu\phi_1^*) + i(\phi_2^*\partial_\mu\phi_2 - \phi_2\partial_\mu\phi_2^*) \quad (18)$$

that is conserved in solutions to the equations of motion.

7. EQUATIONS OF MOTION

To study the dynamics associated with the action given in (15) we have found it convenient [5] to work in the component basis

$$\phi_1 = \frac{1}{\sqrt{2}}(\chi_1 + i\chi_2), \quad \phi_1^* = \frac{1}{\sqrt{2}}(\chi_1 - i\chi_2), \quad \phi_2 = \frac{1}{\sqrt{2}}(\psi_1 + i\psi_2), \quad \phi_2^* = \frac{1}{\sqrt{2}}(\psi_1 - i\psi_2), \quad (19)$$

where all four χ_1, χ_2, ψ_1, and ψ_2 are Hermitian.

In the χ_1, χ_2, ψ_1, and ψ_2 basis the action takes the form:

$$\begin{aligned} I(\chi_1,\chi_2,\psi_1,\psi_2) &= \int d^4x \Big[\frac{1}{2}\partial_\mu\chi_1\partial^\mu\chi_1 + \frac{1}{2}\partial_\mu\chi_2\partial^\mu\chi_2 + \frac{1}{2}\partial_\mu\psi_1\partial^\mu\psi_1 + \frac{1}{2}\partial_\mu\psi_2\partial^\mu\psi_2 \\ &+ \frac{1}{2}m_1^2(\chi_1^2 + \chi_2^2) - \frac{1}{2}m_2^2(\psi_1^2 + \psi_2^2) - i\mu^2(\chi_1\psi_2 - \chi_2\psi_1) - \frac{g}{16}(\chi_1^2 + \chi_2^2)^2\Big], \end{aligned} \quad (20)$$

and with the appearance of the factor i in the μ^2-dependent term, the action now has the characteristic form of the non-Hermitian but PT symmetric $p^2 + ix^3$ theory.

For this action the Euler-Lagrange equations of motion take the form

$$\begin{aligned}
-\partial_\mu\partial^\mu\chi_1 &= -m_1^2\chi_1 + i\mu^2\psi_2 + \frac{g}{4}(\chi_1^3 + \chi_1\chi_2^2),\\
-\partial_\mu\partial^\mu\chi_2 &= -m_1^2\chi_2 - i\mu^2\psi_1 + \frac{g}{4}(\chi_2^3 + \chi_2\chi_1^2),\\
-\partial_\mu\partial^\mu\psi_1 &= m_2^2\psi_1 - i\mu^2\chi_2,\\
-\partial_\mu\partial^\mu\psi_2 &= m_2^2\psi_2 + i\mu^2\chi_1.
\end{aligned} \tag{21}$$

Now we note that none of these equations is actually invariant under complex conjugation, a concern that was raised by Alexandre, Ellis, Millington and Seynaeve [4]. We shall leave resolution of this point for the moment, while noting now that its resolution will not change any of our conclusions. These equations of motion admit of a tree approximation minimum in which the scalar field expectation values obey

$$\begin{aligned}
&m_2^2\bar{\psi}_1 - i\mu^2\bar{\chi}_2 = 0, \quad m_2^2\bar{\psi}_2 + i\mu^2\bar{\chi}_1 = 0,\\
&m_1^2\bar{\chi}_1 - \frac{\mu^4}{m_2^2}\bar{\chi}_1 - \frac{g}{4}\bar{\chi}_1^3 - \frac{g}{4}\bar{\chi}_1\bar{\chi}_2^2 = 0,\\
&m_1^2\bar{\chi}_2 - \frac{\mu^4}{m_2^2}\bar{\chi}_2 - \frac{g}{4}\bar{\chi}_2^3 - \frac{g}{4}\bar{\chi}_2\bar{\chi}_1^2 = 0.
\end{aligned} \tag{22}$$

8. TREE APPROXIMATION MINIMUM

Choosing the minimum in which

$$\frac{g}{4}\bar{\chi}_1^2 = m_1^2 - \frac{\mu^4}{m_2^2}, \quad \bar{\psi}_2 = -\frac{i\mu^2\bar{\chi}_1}{m_2^2}, \quad \bar{\chi}_2 = 0, \quad \bar{\psi}_1 = 0, \tag{23}$$

and then expanding around this minimum according to

$$\chi_1 = \bar{\chi}_1 + \hat{\chi}_1, \quad \chi_2 = \hat{\chi}_2, \quad \psi_1 = \hat{\psi}_1, \quad \psi_2 = \bar{\psi}_2 + \hat{\psi}_2 \tag{24}$$

yields a first-order term in the equations of motion of the form:

$$\begin{pmatrix} -\partial_\mu\partial^\mu\hat{\chi}_1 \\ -\partial_\mu\partial^\mu\hat{\psi}_2 \\ -\partial_\mu\partial^\mu\hat{\chi}_2 \\ -\partial_\mu\partial^\mu\hat{\psi}_1 \end{pmatrix} = \begin{pmatrix} 2m_1^2 - 3\mu^4/m_2^2 & i\mu^2 & 0 & 0 \\ i\mu^2 & m_2^2 & 0 & 0 \\ 0 & 0 & -\mu^4/m_2^2 & -i\mu^2 \\ 0 & 0 & -i\mu^2 & m_2^2 \end{pmatrix} \begin{pmatrix} \hat{\chi}_1 \\ \hat{\psi}_2 \\ \hat{\chi}_2 \\ \hat{\psi}_1 \end{pmatrix} = M \begin{pmatrix} \hat{\chi}_1 \\ \hat{\psi}_2 \\ \hat{\chi}_2 \\ \hat{\psi}_1 \end{pmatrix}. \tag{25}$$

As we see, with our choice of basis, we have already block-diagonalized the mass matrix M, and so we can use our previous 2×2 matrix analysis, with the lower right 2×2 block having $C + A = -\mu^4/m_2^2$, $C - A = m_2^2$, $B = -\mu^2$. We can readily determine the mass eigenvalues, and obtain

$$|M - \lambda I| = \lambda(\lambda + \mu^4/m_2^2 - m_2^2)\left[\lambda^2 - \lambda(2m_1^2 + m_2^2 - 3\mu^4/m_2^2) + 2m_1^2m_2^2 - 2\mu^4\right]. \tag{26}$$

The mass eigenvalue solutions to $|M - \lambda I| = 0$ are thus

$$\begin{aligned}
\lambda_0 = 0, \quad \lambda_1 &= \frac{m_2^4 - \mu^4}{m_2^2}, \\
\lambda_\pm &= \frac{2m_1^2 m_2^2 + m_2^4 - 3\mu^4}{2m_2^2} \pm \frac{1}{2m_2^2}\left[(2m_1^2 m_2^2 + m_2^4 - 3\mu^4)^2 + 8\mu^4 m_2^4 - 8m_1^2 m_2^6\right]^{1/2} \\
&= \frac{2m_1^2 m_2^2 + m_2^4 - 3\mu^4}{2m_2^2} \pm \frac{1}{2m_2^2}\left[(2m_1^2 m_2^2 - m_2^4 - 3\mu^4)^2 - 4\mu^4 m_2^4\right]^{1/2}. \qquad (27)
\end{aligned}$$

We thus see that one eigenvalue is zero. This then is the Goldstone boson.

We should add that being massless does not automatically make a mode be a Goldstone boson. Thus if we set $\mu^4 = m_1^2 m_2^2$ we find that $\lambda_- = 0$, to thus give the 4×4 mass matrix a total of two zero eigenvalues, one in the upper left 2×2 sector and one in the lower right 2×2 sector. However, with this choice we find that both $\bar{\chi}_1$ and $\bar{\psi}_2$ vanish in the tree approximation minimum given in (23). But then all four of $\bar{\chi}_1$, $\bar{\psi}_2$, $\bar{\chi}_2$ and $\bar{\psi}_1$ vanish and we are in a normal, non-spontaneously broken vacuum. Since there is no symmetry that would enforce modes to be massless in the normal vacuum this is then just an artifact of the tree approximation, and neither of the two zero eigenvalues λ_0 and λ_- would remain massless in higher order. [Moreover, for the upper left 2×2 mass matrix given in (25) we have $A - B = m_1^2 - 3\mu^4/2m_2^2 - m_2^2/2 - \mu^2$, and if we set $\mu^2 = m_1 m_2$ (i.e., taking both m_1 and m_2 to be positive) this becomes $A - B = -(m_1 + m_2)^2/2$, with the upper left 2×2 matrix not being Jordan block for any values of the positive m_1 and m_2.]

However, if we do not set $\mu^4 = m_1^2 m_2^2$ then we would have a bona fide spontaneous breakdown since neither $\bar{\chi}_1$ nor $\bar{\psi}_2$ would then vanish. In that case while λ_- would become nonzero, λ_0 would stay massless, and would now indeed be a Goldstone boson, and would remain so in higher order. Now in the following we shall set $\mu^4 = m_2^4$. This would then give us a tree approximation minimum in which both λ_0 and λ_1 would vanish, and we discuss this case in detail below. And if we in addition set $m_1^2 = m_2^2$ so that $\mu^4 = m_1^2 m_2^2$ we would even have $\lambda_- = 0$, and thus with $\mu^2 = m_1^2 = m_2^2$ we have no less than three massless modes. But then we would be back to the normal vacuum and none of these modes would be a Goldstone boson and all would acquire masses in higher order. Thus in order to obtain spontaneous breakdown we must not impose $\mu^4 = m_1^2 m_2^2$. And thus if we want to impose $\mu^4 = m_2^4$ we must not also impose $m_1^2 = m_2^2$.

The author is indebted to Dr. A. Fring for raising the issue of what happens if we set λ_- to zero, and for informing him that similar concerns and an even richer structure may be obtained in a generalization of the two-field model to three or more scalar fields [7].

9. ZERO-NORM GOLDSTONE BOSON

Originally, Alexandre, Ellis, Millington and Seynaeve only showed the emergence of a Goldstone boson in the real eigenvalue realization. Here we see that it holds even if energies are in complex conjugate pairs.

And more, if $\mu_2^4 = m_2^4$ (viz. $A = B$) then λ_0 and λ_1 are both zero. But we cannot have two Goldstone bosons, so we must be missing an eigenvector. So then the Hamiltonian must not be diagonalizable, just as is the case if $A = B$. We thus extend the Goldstone theorem of Goldstone, Salam and Weinberg [26] to Jordan-block Hamiltonians.

But even more, we even find an exception [5]. Now the Goldstone boson has zero norm, carries no probability, and is not detectable. Thus spontaneous breakdown of a continuous global symmetry does not in fact require the existence of an observable massless particle, with the Goldstone theorem being evaded since the Hilbert space metric is not positive definite.

10. SPONTANEOUSLY BROKEN NON-HERMITIAN THEORY WITH A CONTINUOUS LOCAL SYMMETRY

Now that we have seen that we can consistently implement the Goldstone mechanism in a CPT-symmetric, non-Hermitian theory, it is natural to ask whether we can also implement the familiar Englert-Brout-Higgs mechanism (Englert and Brout [27], Higgs [28, 29], Guralnik, Hagen and Kibble [30]). To this end we introduce a local gauge invariance and a gauge field A_μ, and with $F_{\mu\nu} = \partial_\mu A_\nu - \partial_\nu A_\mu$ replace (15) and (17) by

$$\begin{aligned} &I(\phi_1, \phi_2, \phi_1^*, \phi_2^*, A_\mu) \\ &= \int d^4x \Big[(-i\partial_\mu + eA_\mu)\phi_1^*(i\partial^\mu + eA^\mu)\phi_1 + (-i\partial_\mu + eA_\mu)\phi_2^*(i\partial^\mu + eA^\mu)\phi_2 \\ &+m_1^2\phi_1^*\phi_1 - m_2^2\phi_2^*\phi_2 - \mu^2(\phi_1^*\phi_2 - \phi_2^*\phi_1) - \frac{g}{4}(\phi_1^*\phi_1)^2 - \frac{1}{4}F_{\mu\nu}F^{\mu\nu}\Big], \end{aligned} \tag{28}$$

and

$$\phi_1 \to e^{i\alpha(x)}\phi_1,\ \phi_1^* \to e^{-i\alpha(x)}\phi_1^*, \quad \phi_2 \to e^{i\alpha(x)}\phi_2,\ \phi_2^* \to e^{-i\alpha(x)}\phi_2,\ eA_\mu \to eA_\mu + \partial_\mu\alpha(x). \tag{29}$$

With (16), the $I(\phi_1, \phi_2, \phi_1^*, \phi_2^*, A_\mu)$ action is CPT invariant since both i and A_μ are CPT odd (spin one fields have odd CPT [31]).

We make the same decomposition of the ϕ_1 and ϕ_2 fields as in (19), and replace (20) by

$$\begin{aligned} I(\chi_1, \chi_2, \psi_1, \psi_2, A_\mu) &= \int d^4x \Big[\frac{1}{2}\partial_\mu\chi_1\partial^\mu\chi_1 + \frac{1}{2}\partial_\mu\chi_2\partial^\mu\chi_2 + \frac{1}{2}\partial_\mu\psi_1\partial^\mu\psi_1 + \frac{1}{2}\partial_\mu\psi_2\partial^\mu\psi_2 \\ &+ \frac{1}{2}m_1^2(\chi_1^2 + \chi_2^2) - \frac{1}{2}m_2^2(\psi_1^2 + \psi_2^2) - i\mu^2(\chi_1\psi_2 - \chi_2\psi_1) - \frac{g}{16}(\chi_1^2 + \chi_2^2)^2 \\ &- eA^\mu(\chi_1\partial_\mu\chi_2 - \chi_2\partial_\mu\chi_1 + \psi_1\partial_\mu\psi_2 - \psi_2\partial_\mu\psi_1) \\ &+ \frac{e^2}{2}A_\mu A^\mu\left[\chi_1^2 + \chi_2^2 + \psi_1^2 + \psi_2^2\right] - \frac{1}{4}F_{\mu\nu}F^{\mu\nu}\Big]. \end{aligned} \tag{30}$$

11. STILL MASSLESS GAUGE BOSON

In the tree approximation minimum used above in which

$$\frac{g}{4}\bar{\chi}_1^2 = m_1^2 - \frac{\mu^4}{m_2^2}, \quad \bar{\psi}_2 = -\frac{i\mu^2\bar{\chi}_1}{m_2^2}, \quad \bar{\chi}_2 = 0, \quad \bar{\psi}_1 = 0, \tag{31}$$

we induce a mass term for A_μ of the form

$$m^2(A_\mu) = e^2\left(\bar{\chi}_1^2 + \bar{\chi}_2^2 + \bar{\psi}_1^2 + \bar{\psi}_2^2\right) = e^2\bar{\chi}_1^2\left(1 - \frac{\mu^4}{m_2^4}\right) = \frac{4e^2}{g}\frac{(m_1^2m_2^2 - \mu^4)(m_2^4 - \mu^4)}{m_2^6}. \tag{32}$$

Now something quite startling happens. If we again set $\mu^4 = m_2^4$ (viz. $A = B$), we find that the gauge boson **stays massless** [5]. And yet we have spontaneously broken a continuous symmetry. So what happened to the Englert-Brout-Higgs mechanism?

What happens is that in this limit the Hamiltonian becomes Jordan-block and the Goldstone boson becomes a zero norm state. The standard proof that the gauge boson must be become massive fails because the proof was only established for a positive norm Hilbert space, where a positive norm Goldstone boson becomes the third component of a then massive gauge boson.

12. A PUZZLE, ITS RESOLUTION, AND ANOTHER PUZZLE

In the global continuous symmetry model we obtained equations of motion of the form

$$
\begin{aligned}
-\partial_\mu\partial^\mu\chi_1 &= -m_1^2\chi_1 + i\mu^2\psi_2 + \frac{g}{4}(\chi_1^3 + \chi_1\chi_2^2),\\
-\partial_\mu\partial^\mu\chi_2 &= -m_1^2\chi_2 - i\mu^2\psi_1 + \frac{g}{4}(\chi_2^3 + \chi_2\chi_1^2),\\
-\partial_\mu\partial^\mu\psi_1 &= m_2^2\psi_1 - i\mu^2\chi_2,\\
-\partial_\mu\partial^\mu\psi_2 &= m_2^2\psi_2 + i\mu^2\chi_1,
\end{aligned}
\tag{33}
$$

However, none of these equations is actually invariant under complex conjugation. Three resolutions of this issue have been presented in the literature. In their paper Alexandre, Ellis, Millington and Seynaeve [4] avoided this problem by changing the variational procedure that led to these equations of motion. Specifically, to the original action they added on a surface term, and the constraint that it put on the variational procedure then led to a set of equations of motion that then were invariant under complex conjugation. Interestingly, this led to the same mass matrix as given above, and a massless Goldstone boson state was again obtained.

While doing this is somewhat non-standard for variational procedures (but not completely so since a Hawking-Gibbons surface term plays a similar role for variation of the Einstein-Hilbert action in general relativity), it would nonetheless be more straightforward if we could avoid the introduction of such surface terms, especially since it would prevent drawing the conclusion that to implement the Goldstone theorem in the PT case one might actually have to resort to such a non-conventional variational procedure.

To this end two alternate resolutions of the problem were presented in [5]. The most straightforward is simply to reinterpret the meaning of the star symbol and identify a field such as ϕ^* as the CPT conjugate of ϕ. With this definition both the action and the equations of motion are then CPT symmetric. In fact this is in keeping with the whole PT or CPT theory program as it is PT or CPT conjugates that are the central building blocks of the program and not Hermitian conjugates, with CPT invariance being the guiding principle for constructing physical theories and not Hermiticity [2]. With this change in the meaning of the star symbol not affecting the mass matrix all the results that we found above continue to hold.

The second option identified in [5] involves making a specific similarity transformation. Specifically, with canonical conjugates $\Pi_1 = \partial_t\psi_1$ and $\Pi_2 = \partial_t\psi_1$ for ψ_1 and ψ_2 we introduce

$$
S(\psi_1) = \exp\left[\frac{\pi}{2}\int d^3x \Pi_1(\mathbf{x},t)\psi_1(\mathbf{x},t)\right], \quad S(\psi_2) = \exp\left[\frac{\pi}{2}\int d^3x \Pi_2(\mathbf{x},t)\psi_2(\mathbf{x},t)\right], \tag{34}
$$

and obtain

$$
\begin{aligned}
S(\psi_1)\psi_1 S^{-1}(\psi_1) = -i\psi_1, \quad S(\psi_1)\Pi_1 S^{-1}(\psi_1) = i\Pi_1,\\
S(\psi_2)\psi_2 S^{-1}(\psi_2) = -i\psi_2, \quad S(\psi_2)\Pi_2 S^{-1}(\psi_2) = i\Pi_2.
\end{aligned}
\tag{35}
$$

Since these transformations preserve the equal-time commutation relations $[\psi_1(\mathbf{x},t),\Pi_1(\mathbf{y},t)] = i\delta^3(\mathbf{x}-\mathbf{y})$, $[\psi_2(\mathbf{x},t),\Pi_2(\mathbf{y},t)] = i\delta^3(\mathbf{x}-\mathbf{y})$, they are fully permissible transformations that do not modify the content of the field theory. Applying (35) to $I(\chi_1,\chi_2,\psi_1,\psi_2)$ we obtain

$$S(\psi_1)S(\psi_2)I(\chi_1,\chi_2,\psi_1,\psi_2)S^{-1}(\psi_2)S^{-1}(\psi_1) = I'(\chi_1,\chi_2,\psi_1,\psi_2), \tag{36}$$

where

$$\begin{aligned} &I'(\chi_1,\chi_2,\psi_1,\psi_2) \\ &= \int d^4x\bigg[\frac{1}{2}\partial_\mu\chi_1\partial^\mu\chi_1 + \frac{1}{2}\partial_\mu\chi_2\partial^\mu\chi_2 - \frac{1}{2}\partial_\mu\psi_1\partial^\mu\psi_1 - \frac{1}{2}\partial_\mu\psi_2\partial^\mu\psi_2 + \frac{1}{2}m_1^2(\chi_1^2+\chi_2^2) \\ &+\frac{1}{2}m_2^2(\psi_1^2+\psi_2^2) - \mu^2(\chi_1\psi_2 - \chi_2\psi_1) - \frac{g}{16}(\chi_1^2+\chi_2^2)^2\bigg]. \end{aligned} \tag{37}$$

Stationary variation with respect to χ_1, χ_2, ψ_1, and ψ_2 replaces (21) by

$$\begin{aligned} -\partial_\mu\partial^\mu\chi_1 &= -m_1^2\chi_1 + \mu^2\psi_2 + \frac{g}{4}(\chi_1^3 + \chi_1\chi_2^2), \\ -\partial_\mu\partial^\mu\chi_2 &= -m_1^2\chi_2 - \mu^2\psi_1 + \frac{g}{4}(\chi_2^3 + \chi_2\chi_1^2), \\ -\partial_\mu\partial^\mu\psi_1 &= m_2^2\psi_1 + \mu^2\chi_2, \\ -\partial_\mu\partial^\mu\psi_2 &= m_2^2\psi_2 - \mu^2\chi_1, \end{aligned} \tag{38}$$

and now each one of the equations of motion is separately invariant under complex conjugation. Thus this time we do not need to reinterpret the star as a CPT transformation (though we still should of course). Finally, since all we have done is make a similarity transformation, the eigenvalues of the mass matrix do not change and the previous analysis continues to hold.

However, now we have a new problem. $I'(\chi_1,\chi_2,\psi_1,\psi_2)$ is Hermitian since all of its coefficients are real and all of its fields are Hermitian. So have we just recovered the standard Goldstone theorem after all this effort? No, since appearances can be deceptive.

13. HERMITICITY CANNOT BE DETERMINED BY INSPECTION

We know that the mass matrix has all three of the PT theory realizations, dependent on the values of its parameters. Thus we have all eigenvalues real, eigenvalues in complex conjugate pairs, and mass matrix not diagonalizable. And yet the action looks to be Hermitian in all three realizations and this simply could not be case. The conclusion: **Hermiticity cannot be determined by inspection. Hermiticity is a property of the full Hamiltonian not a property of the individual terms in it, and that depends on boundary conditions.**

The secret of PT: In $H = p^2 + ix^3$ both p and x are Hermitian when they act on their own eigenstates, i.e., on states for which you can throw away surface terms in an integration by parts. However, what matters is how p and x act on the eigenstates of H, and then you may or may not be able to throw surface terms away (and you may need to continue into the complex plane in order to be able to do so). And only if you can throw surface terms away, can you say that Hamiltonian is Hermitian. **Hermiticity is a property that is determined jointly by operators and states.**

However, you can determine antilinearity by inspection, as that does only depend on the individual terms in a Hamiltonian. Antilinearity is determined by the operators alone. Thus $H = p^2 + ix^3$ is PT symmetric regardless of boundary conditions.

14. PATH INTEGRALS AND PT SYMMETRY

As noted in [32, 2] this distinction between Hermiticity and antilinearity is seen even more sharply in the path integral formulation of quantum theory. We define a generating functional

$$W[J] = \int D[\phi] e^{i[S_{CL}(\phi)+J\phi]/\hbar}, \tag{39}$$

where $S_{CL}(\phi)$ is the classical action as evaluated on each classical path that is integrated over. You can make $W[J]$ be CPT symmetric path by path, but there is no meaning as to whether $W[J]$ might be Hermitian. Only after you have constructed the quantum Green's functions and then found the quantum Hilbert space can you even begin to ascertain whether the quantum Hamiltonian might be Hermitian, and we have seen examples in which it is not. **Antilinearity thus has primacy over Hermiticity.** Moreover, once the quantum theory is constructed the Green's functions will correspond [33, 34, 35, 32] to quantities such as $\langle \Omega_R^{CPT} | T[\phi(x)\phi(y)] | \Omega_R \rangle$ or $\langle \Omega_L | T[\phi(x)\phi(y)] | \Omega_R \rangle$, where $\langle \Omega_R^{CPT}|$ is the CPT transform of the right vacuum $|\Omega_R\rangle$ and $\langle \Omega_L|$ is the left vacuum.

Moreover, we even need CPT to correctly construct the appropriate $S_{CL}(\phi)$ before we even do the path integration. Consider a complex classical scalar field with Lagrangian $L = \partial_\mu \phi^* \partial^\mu \phi$ and now couple it to a classical electromagnetic vector potential. Which of the following do we use?

$$L_1 = [\partial_\mu - eA_\mu]\phi^*[\partial^\mu + eA^\mu]\phi, \qquad L_2 = [\partial_\mu - ieA_\mu]\phi^*[\partial^\mu + ieA^\mu]\phi. \tag{40}$$

L_1 is natural if you consider classical physics to be restricted to real numbers. (It is not, it is restricted to c-numbers, i.e., to numbers that commute and not to eigenvalues of Hermitian operators.) But only L_2 would give the correct quantum theory. So why should we use L_2? Could we argue that $i\partial_\mu$ and eA_μ are both Hermitian, and thus we should use L_2. Definitely not, as there is no meaning to Hermiticity at the classical level. But we can use CPT symmetry since at the classical level $i\partial_\mu$ and eA_μ are both CPT even. Thus even at the classical level where Hermiticity is not even definable, we still need antilinear symmetry. Thus again, **antilinearity has primacy over Hermiticity**.

15. SYMPLECTIC SYMMETRY IN CLASSICAL MECHANICS

Now it can turn out that path integral does not actually exist if the integration measure for the paths is real, but does exist if we continue the measure into the complex plane [2]. If we do have to make such a continuation in the classical theory then the resulting quantum theory is a non-Hermitian PT theory.

Thus we ask whether we can justify such a continuation of classical theories into the complex plane. With the quantum-mechanical commutation relation $[x,p] = i$ being invariant under complex transformations of the form $x \to e^{i\theta}x$, $p \to e^{-i\theta}p$, we look to see if there is an analogous complex invariance structure for Poisson brackets in classical mechanics.

Thus consider [32] a classical system with n coordinates q_i, n momenta p_i, and generic Poisson bracket

$$\{u, v\} = \sum_{i=1}^{n} \left(\frac{\partial u}{\partial q_i}\frac{\partial v}{\partial p_i} - \frac{\partial u}{\partial p_i}\frac{\partial v}{\partial q_i} \right). \tag{41}$$

If we introduce a $2n$-dimensional vector η and a $2n$-dimensional matrix J defined as

$$\eta = \begin{pmatrix} q_i \\ p_i \end{pmatrix}, \qquad J = \begin{pmatrix} 0 & I \\ -I & 0 \end{pmatrix}, \tag{42}$$

where I is an n-dimensional unit matrix, we can compactly write the generic Poisson bracket as

$$\{u,v\} = \widetilde{\frac{\partial u}{\partial \eta}} J \frac{\partial v}{\partial \eta}, \tag{43}$$

where the tilde symbol denotes transpose. If we now make a phase space transformation to a new $2n$-dimensional vector ξ according to

$$M_{ij} = \frac{\partial \xi_i}{\partial \eta_j}, \qquad \frac{\partial v}{\partial \eta} = \tilde{M}\frac{\partial v}{\partial \xi}, \qquad \widetilde{\frac{\partial u}{\partial \eta}} = \widetilde{\frac{\partial u}{\partial \xi}} M, \tag{44}$$

then in these new coordinates and momenta the Poisson bracket takes the form

$$\{u,v\} = \widetilde{\frac{\partial u}{\partial \xi}} MJ\tilde{M} \frac{\partial v}{\partial \xi}. \tag{45}$$

The Poisson bracket will thus be left invariant for any M that obeys the symplectic symmetry relation

$$MJ\tilde{M} = J. \tag{46}$$

If we introduce generators G defined according to $M = e^{i\omega G}$, such generators will obey

$$e^{i\omega G} J e^{i\omega \tilde{G}} = J, \qquad GJ + J\tilde{G} = 0. \tag{47}$$

Since the matrix J obeys $J^{-1} = \tilde{J} = -J$, the generators will obey

$$\tilde{G} = -J^{-1}GJ = JGJ. \tag{48}$$

Solutions to $\tilde{G} = JGJ$ can be broken into two classes, symmetric generators that anticommute with J, viz. those that obey

$$\tilde{G} = G, \qquad GJ + JG = 0, \tag{49}$$

and antisymmetric generators that commute with J, viz. those that obey

$$\tilde{G} = -G, \qquad GJ - JG = 0. \tag{50}$$

In $N = 2n$ dimensions there are $N(N-1)/2$ symmetric generators and N antisymmetric generators, for a total of $N(N+1)/2$ generators. These $N(N+1)/2$ generators close on the Lie algebra $Sp(N)$, the symplectic group in $2n$ dimensions. With the generic Lie algebra commutation relations being of the form $[G_i, G_j] = i\sum_k f_{ijk}G_k$ with real structure coefficients f_{ijk}, one can find representations of the symplectic algebra in which all the generators are pure imaginary. Thus if, as is standard in classical mechanics, one takes all angles ω to be real, canonical transformations effected by $e^{i\omega G}$ will transform a real η into a real ξ.

16. COMPLEX SYMPLECTIC SYMMETRY IN CLASSICAL MECHANICS

However, since the algebra of the generators makes no reference to angles, invariance of the classical Poisson brackets under canonical transformations will persist even if the ω are taken to be **complex**. The Poisson brackets of classical mechanics thus possess a broader class of invariances than those associated with real canonical transformations alone since one can transform a real η into a complex ξ and still preserve the Poisson bracket algebra.

Consequently, with both the classical Poisson brackets and the quantum commutators admitting of complex canonical transformations, for every such complex transformation we are able to construct a canonical quantization with an associated correspondence principle. Namely, for each canonically transformed Poisson bracket we associate a canonically transformed quantum commutator, with each associated set of classical coordinates being the eigenvalues of the associated transformed quantum operators.

Now while classical mechanics contains this broad class of complex symplectic transformations, they ordinarily play no role in physics since they contain no additional information that is not already contained in the real symplectic transformations alone.

However, this would not be the case if we were to encounter some form of discontinuity when we continue into the complex coordinate plane. In path integral quantization these discontinuities would occur if the path integral only existed for a domain of classical paths that were not real. In a canonical quantization these discontinuities would occur if quantum-mechanical wave functions are convergent in the domain associated with some Stokes wedges and divergent in some other Stokes wedges. (See [36] for a discussion of the role of Stokes wedges in PT theories.)

If the domain of convergence includes the real coordinate axis we are in conventional Hermitian quantum mechanics, and we can take the classical limit to be based on real numbers. However, if wave functions are only convergent in Stokes wedges that do not include the real axis, we are in a non-Hermitian realization of the theory. Now amongst such general non-Hermitian realizations there will be some that are also PT realizations. We can therefore anticipate that the ones that are PT realizations are those in which the classical domain for which the path integral exists and the quantum-mechanical domain for which wave functions exist are either the same or related in some way.

To conclude this section we note that for the simple case of a 4-dimensional phase space (viz. $n=2$) the 4-dimensional transposition matrix C that is involved in charge conjugation of Dirac spinors and effects $C^{-1}\gamma_\mu C = -\tilde{\gamma}_\mu$ can also play a role in symplectic transformations. In order to be able to use a C that is, like J, antisymmetric, orthogonal, and composed of real elements alone, we take C to be given by its representation in the Weyl basis of the Dirac gamma matrices, viz.

$$C = \begin{pmatrix} -i\sigma_2 & 0 \\ 0 & i\sigma_2 \end{pmatrix}. \tag{51}$$

Then, if we now take η to be of the form

$$\eta = \begin{pmatrix} p_1 \\ q_1 \\ q_2 \\ p_2 \end{pmatrix}, \tag{52}$$

we can write the Poisson bracket as

$$\{u, v\} = \widetilde{\frac{\partial u}{\partial \eta}} C \frac{\partial v}{\partial \eta}, \tag{53}$$

with the symplectic condition then being given by

$$MC\tilde{M} = C. \tag{54}$$

Moreover, since (54) would not be affected if we were to replace C by iC, we would then have an operator iC whose square is one, to thus be reminiscent of the PT theory C operator discussed in [36]. In such a case it would be $i\{u, v\}$ that would be defined as the Poisson bracket, and under a canonical quantization the quantum commutator $[\hat{u}, \hat{v}]$ would be identified with $\hbar$ times it.

Journal of Physics: Conference Series **2482** (2023) 012014 doi:10.1088/1742-6596/2482/1/012014

17. NOT JUST COMPLEX VARIABLES BUT COMPLEX ANALYSIS

In classical mechanics point particles move on one-dimensional real trajectories. In the classical mechanics of waves the normals to the wavefronts move on one-dimensional real trajectories. Characteristic of one-dimensional real trajectories is that the derivative at any point on the trajectory does not depend on the direction in which the derivative is taken.

So suppose we now continue trajectories into the complex plane. They will still be one-dimensional and will not become two-dimensional. But now the derivative at any point on the trajectory will depend on the direction in which the derivative is taken. However, there is a way to avoid this, namely to impose the Cauchy-Riemann equations, viz. for the complex function $f(x,y) = u(x,y) + iv(x,y)$ where $u(x,y)$ and $v(x,y)$ are both real we impose

$$\frac{\partial u}{\partial x} = \frac{\partial v}{\partial y}, \quad \frac{\partial u}{\partial y} = -\frac{\partial v}{\partial x}. \tag{55}$$

And now the derivative is uniquely defined and does not depend on the direction in which it is taken, and can thus be taken to be $df(z)/dz = \partial u/\partial x + i\partial v/\partial x$. And moreover the function $f(x,y)$ is now amenable to complex analysis.

Thus the natural generalization of real trajectories is trajectories in the complex plane that obey the Cauchy-Riemann equations. It is in this sense that the classical paths in the path integral measure are to be continued into the complex plane.

18. SUMMARY

The PT option for quantum theory is now well established. It in no way changes quantum mechanics. It simply takes advantage of an option that was there from the beginning but had been overlooked. Since having an antilinear symmetry is necessary for obtaining real eigenvalues (and also probability conservation [2]), it actually represents the most general possible option for quantum mechanics that is allowable. In fact in [2] it was noted that antilinearity should be used as the guiding principle for quantum theory rather than Hermiticity.

While antilinearity allows for real eigenvalues, it has two other realizations, realizations that are foreign to Hermitian theories. Energies could appear in complex conjugate pairs and Hamiltonians need not be diagonalizable at all. In this latter, Jordan-block, realization, the eigenstates of the Hamiltonian have a norm that is not positive but zero.

Some familiar theorems of quantum theory that are based on Hermiticity can be extended to the non-Hermitian domain, namely, the CPT theorem, the Goldstone theorem, and the Englert-Brout-Higgs mechanism. All three are found to hold in the antilinear case without needing to assume Hermiticity, and are even found to hold even if energies appear in complex conjugate pairs.

However, in the non-diagonalizable Jordan-block case something new occurs. While the spontaneous breakdown of a continuous **global** symmetry still leads to a massless Goldstone boson, in the Jordan-block case the Goldstone boson has zero norm. It is thus not observable.

And more: the spontaneous breakdown of a continuous **local** symmetry still leads to a massive gauge boson if energies are real or in complex conjugate pairs. But in the Jordan-block case the gauge boson stays massless, despite the fact that a continuous local symmetry has been spontaneously broken.

References

[1] Bender C M and Boettcher S 1998 *Phys. Rev. Lett.* **80** 5243
[2] Mannheim P D 2018 *J. Phys. A: Math. Theor.* **51** 315302
[3] Mannheim P D 2016 *Phys. Lett. B* **753** 288
[4] Alexandre J, Ellis J, Millington P and Seynaeve D 2018 *Phys. Rev. D* **98** 045001
[5] Mannheim P D 2019 *Phys. Rev. D* **99** 045006
[6] Alexandre J, Ellis J, Millington P and Seynaeve D 2019 *Phys. Rev. D* **99** 075024.
[7] Fring A and Taira T 2020 *Nucl. Phys. B* **950** 114834
[8] Alexandre J, Ellis J, Millington P and Seynaeve D 2020 *Phys. Rev. D* **101** 035008
[9] Fring A and Taira T 2020 *Phys. Rev. D* **101** 045014
[10] Mandal B P, Mourya B K and Singh A K 2020 *Eur. Phys. J. Plus* **135** 327
[11] Alexandre J, Ellis J and Millington P 2020 *Phys. Rev. D* **101** 085015
[12] Ohlsson T and Zhou S 2020 *J. Math. Phys.* **61** 052104
[13] Chernodub M N and Cortijo A 2020 *Symmetry* **12** 761
[14] Millington P 2020 *J. Phys.: Conf. Ser.* **1586** 012001
[15] Alexandre J and Mavromatos N E 2020 *Phys. Lett. B* **807** 135562
[16] Fring A and Taira T 2020 *Phys. Lett. B* **807** 135583
[17] Chernodub M N, Cortijo A and Ruggieri M 2021 *Phys. Rev. D* **104** 056023
[18] Arean D, Landsteiner K and Landea I S 2020 *SciPost Phys.* **9** 032
[19] Mavromatos N E 2021 *Jour. Phys.: Conf. Ser.* **2038** 012019
[20] Fring A and Taira T 2022 *Eur. Phys. J. Plus* **137** 716
[21] Fring A and Taira T 2020 *J. Phys. A: Math. Theor.* **53** 455701
[22] Alexandre J, Ellis J and Millington P 2020 *Phys. Rev. D* **102** 125030
[23] Fring A and Taira T 2021 *Jour. Phys.: Conf. Ser.* **2038** 012010
[24] Correa F, Fring A and Taira T 2021 *Nucl. Phys. B* **971** 115516
[25] Korchin A Y and Kovalchuk V A 2022 *Acta Phys. Polon. B* **53** 1-A2
[26] Goldstone J, Salam A and Weinberg S 1962 *Phys. Rev.* **127** 965
[27] Englert F and Brout R 1964 *Phys. Rev. Lett.* **13** 321
[28] Higgs P W 1964 *Phys. Lett.* **12** 132
[29] Higgs P W 1964 *Phys. Rev. Lett.* **13** 508
[30] Guralnik G S, Hagen C R and Kibble T W B 1964 *Phys. Rev. Lett.* **13** 585
[31] Weinberg S 1995 *The Quantum Theory of Fields: Volume I* (Cambridge: Cambridge University Press)
[32] Mannheim P D 2013 *Phil. Trans. R. Soc. A* **371** 20120060
[33] Bender C M and Mannheim P D 2008 *Phys. Rev. D* **78** 025022
[34] Mannheim P D 2009 *Ast. Soc. Pac. Conf. Ser.* **413** 279
[35] Jones H F and Rivers R J 2009 *Phys. Lett. A* **373** 3304
[36] Bender C M 2007 *Rep. Prog. Phys.* **70** 947

Foundations for QED, Feynman operator calculus, Dyson conjectures, and Einstein's dual theory

Tepper L. Gill

Department of EECS, Mathematics and Computational Physics Laboratory, Howard University, Washington DC 20059 USA, tgill@howard.edu

Gonzalo Ares de Parga

Departmento de Física, Escuela Superior de Física y Matemáticas, Instituto Politécnico Nacional COFAA
Edif 9, U. P. Adolfo Løpez Mateos, Zacatenco, Lindavista, 07738, México D.F., México
E-mail : gadpau@hotmail.com

Abstract. This paper reviews research on the foundations of quantum electrodynamics (QED). We show that there are three definitions of the proper time that follow from Einstein's theory. The first definition is used to prove that the universe has a unique clock (Newton-Horwitz-Fanchi time) available to all observers. This clock is used to briefly discuss the mathematical foundations for Feynman's time ordered operator calculus. We use this calculus to solve the first and second conjectures of Dyson for QED: that the renormalized perturbation series is asymptotic and, that the ultra-violet divergence is caused by a violation of the time-energy uncertainly relationship. The second definition gives Minkowski's version of Einstein's theory and its problems are briefly reviewed. The third definition gives the dual Newton, dual Maxwell and dual quantum theories. The theory is dual in that, for a set of n particles, every observer has two unique sets of global variables $(\mathbf{X}, t)$ and $(\mathbf{X}, \tau)$ to study the system, where $\mathbf{X}$ is the canonical center of mass. Using $(\mathbf{X}, t)$ time is relative with speed c, while in $(\mathbf{X}, \tau)$, time is unique with relative speed b. The dual Maxwell theory contains a longitudinal (dissipative) term in the $\mathbf{E}$ field wave equation, which appears instantaneously with acceleration and we predict that radiation from a cyclotron will not produce photoelectrons. It is shown that this term gives an effective mass for the photon. A major outcome is the dual unification of Newtonian mechanics and classical electrodynamics with Einstein's theory and without the need for point particles or a self-energy divergency. This means that a second quantized version will not produce a self-energy or infrared divergency. These results along with the proof of Dyson's second conjecture resolves all the problems with QED. The dual Dirac theory provides a new formula for the anomalous magnetic moment of a charged particle, which can give exact values for the electron, muon and proton g-factors.

1. Introduction

1.1. *Background and History*

In the beginning of the nineteenth century, the major problem of reconciling the transformation properties of the Newtonian and Maxwell theories was resolved in favor of Maxwell. We are now starting the twenty first century and the two theories have not been unified. Between then

Content from this work may be used under the terms of the Creative Commons Attribution 3.0 licence. Any further distribution of this work must maintain attribution to the author(s) and the title of the work, journal citation and DOI.
Published under licence by IOP Publishing Ltd

and now, many new problems have appeared and there is no clean line of deduction from a relativistic extension of Newtonian mechanics to QED in sight. Part of the reason is that the obvious danger to one's professional advancement has led each succeeding generation to take one of the following actions with respect to past problems:

(i) Ignored them (as problems).
(ii) Refer them to a covering theory, which doesn't solve them.
(iii) Deny that a problem (or problems) for one generation is a problem at all by the next generation and then claim that this change of view is a victory.

Once it was accepted that the proper Newtonian theory should be invariant under the Lorentz group, the problem was ignored until after World War Two when it was realized that quantum theory did not solve the problems left open by the classical theory. This is an example of (1), which is still with us and still ignored.

In order to by-passed the problem of infinity for a zero radius particle in the classical electrodynamics (CED) of Lorentz, Dirac replaced particles by fields. This led to an infinite field-energy at a point (self-energy, see [7]). Dirac showed that, by using both advanced and retarded fields along with a limiting procedure, he could obtained a dissipative term, which accounted for radiation reaction as an addition to the Lorentz equation, leading to the Lorentz-Dirac equation. The problems with CED (Lorentz or Dirac theory) are still with us today, and many don't know or don't consider them important anymore. In his classic on electrodynamics, Rohrlich [42] admits that the problems have not been solved at the classical level but claims that they now belong to the covering theory (QED), which is example of (2) and to some extent (3).

The recent book by Frisch [15] on classical electrodynamics gives a clear discussion of the problems, but assumes they have no solution at all and suggests that this state of affairs be accepted as a natural part of the theoretical landscape. Similar sentiments have been expressed by Schweber [46] concerning the difficulties in QED. These views represent an example that is not included in our list above, because the adoption of such a view would exclude theoretical physics from its supreme role in science, relegating it to a form of "pseudo science".

The failure to solve the classical problems forced researchers to use the Dirac theory as the basis for relativistic quantum mechanics and QED. This approach maintained the infinite self-energy divergence and introduced a few others. These problems were later by-passed by Feynman, Schwinger and Tomonaga in the late 1940's leading to the great successes of that period. At the time, there was a natural expectation that the mathematicians would eventually clean things up and find the correct way to justify the renormalization methods used in QED. However, by the early 1980's, it became clear that this would not be possible, so the next generation focused on string theory as the best way forward. Indeed, it seemed reasonable that by replacing a point by a string might eliminate both the self energy and ultraviolet divergencies. (However, this is not yet known to be true.) At present, there are no experimental confirmations or known reductions to quantum theory and/or CED. The development of the electro-weak theory and the standard model have proceeded on the assumption that the QED problems would eventually be solved. In their development, they too have introduced additional problems.

1.2. Purpose

This paper is a survey of research on the first physically consistent and mathematically correct approach to the foundations of quantum electrodynamics. A program of this scope must necessarily address the following:

(i) Resolution of the problems associated with the unification of Newtonian and Maxwell theory.

(ii) Resolution of the problems associated with radiation reaction, self-energy divergence, field-particle dichotomy and advanced interactions in CED.

(iii) Understanding of the relationship and implications of the 2.7 °K cosmic microwave background radiation (CMBR) to a preferred rest frame, a universal clock and quantum theory.

(iv) Resolution of the assumption of Minkowski's four space-time with the operational use of time as a director of motion in the Feynman formulation of QED.

(v) Understanding of the causes and provision of corrections for the physical and mathematical problems associated with the divergencies in QED.

1.3. Summary

In section two, we introduce the first definition of proper time that follows from Einstein's original approach to the special theory. We use it to prove the existence of a universal clock (which we call the Newton-Horwitz-Fanchi (NHF) clock), relate it to the 2.7 °K (CMBR), use it to provide the foundations for Feynman's time ordered operator calculus and to prove the last two remaining conjectures of Dyson concerning QED.

In section three, we discuss the second definition of proper time, its relationship to Minkowski's formulation of Einstein's theory and briefly review a few of the problems that still prevent its full implementation at the classical and quantum levels. In section four, we discuss the third definition of proper time and use it to construct the dual Newton and Maxwell theories. In section five, we introduce the dual quantum theory. We first discuss the general many-particle theory and then show that there are three possible one-particle relativistic wave equations. The last part of the section is devoted to the dual Dirac equation. We show that this equation gives a new formula for the electron g-factor, which provides the exact experimental value for the electron anomalous magnetic moment. We can also use it to obtain exact g-factors for the muon, proton and neutron.

2. Proper time I: CMBR, Feynman calculus and Dyson conjectures

Let m be the mass of a particle or the effective mass for a system of particles. We assume the particle or system is defined on phase space with variables $(t, \mathbf{x}, \mathbf{p})$ as seen by observer O, and variables $(t', \mathbf{x}', \mathbf{p}')$ as seen by observer O' (both in inertial frames). Where $\mathbf{x}, \mathbf{x}'$ is the position of the particle (or center of mass) and $(\mathbf{p}, \mathbf{p}')$ is the particle momentum (or center of mass momentum). If $\gamma^{-1}(\mathbf{v}) = \sqrt{1 - \frac{\mathbf{v}^2}{c^2}}$, $\mathbf{v} = \frac{d\mathbf{x}}{dt}$, the proper time for our two observers is defined by

$$d\tau = \gamma^{-1}(\mathbf{v})\, dt, \quad d\tau = \gamma^{-1}(\mathbf{v}')\, dt'.$$

For our first definition of proper time, we use the relations $H = \gamma(\mathbf{v})\, mc^2$ for observer O and $H' = \gamma(\mathbf{v}')\, m'c^2$ for observer O', to obtain:

$$d\tau = \frac{mc^2}{H} dt, \quad d\tau = \frac{m'c^2}{H'} dt'. \tag{2.1}$$

2.1. NHF-Time and the CMBR

In his theory, Newton assumed that an absolute time exists independent of any observer, gives a universal definition of simultaneity and ordering for events, but can only be understood mathematically. Einstein's special theory eliminated Newton's absolute time and Minkowski's additional postulate eliminated the ordering property. Feynman reintroduced the ordering property by suggesting we view a physical event as a 3-dimensional motion picture, exposing more and more of the outcome as the film unfolds.

Use of a fifth parameter to provide the missing order property of time began with Fock, then Stueckelberg, followed by Feynman and Schwinger. They considered it a global clock for Minkowski space time. Horwitz and Piron [25], and later Fanchi and Collins called it historical time. Horwitz and Fanchi were the first to realize that this clock should have physical justification. The paper by Martin Land in these proceedings provides a comprehensive account of the many contributions of L. P. Howritz to the field. Fanchi [13] is a major source of history on the subject, including many references.

In this section, we give two proofs for the existence of an absolute or NHF time. The first proof is based on the cosmological principle and equation (2.1), which was motivated by the studies by Horwitz and Fanchi (see [22]). The second proof is based on the existence of the 2.7 °K CMBR.

Theorem 2.1. *Suppose that the observable universe is homogenous, isotropic or representable in the sense that it is independent of our observed portion of the universe. Then the universe has a unique clock that is available to all observers.*

Proof. If Mc^2 is the total mass energy and H is the total energy Hamiltonian for the observable universe seen by observer O then, under the stated conditions H/Mc^2 is constant. Since this property is observer independent, every observer will obtain the same ratio. Thus, observer O', anywhere else in the universe will also have the same ratio. It now follows from (2.1) that

$$d\tau = \frac{Mc^2}{H}dt = \frac{Mc^2}{H}dt', \quad \Rightarrow t = t' \equiv \tau_N$$

and τ_N uniquely defines a clock for the universe. □

The principle of homogeneity says that the distribution of matter does not vary much with position at the large scale. A number of recent studies seem to suggest that this is not as secure a principle as believed in the past (see [32]). Likewise, the principle of isotropy has also been questioned in recent times (see [3, 45]). Thus, it is possible that the assumptions for Theorem 2.1 may not remain true if future studies support these conclusions.

2.2. 2.7 °K CMBR

In 1965, Penzias and Wilson discovered the 2.7 °K CMBR. Additional historical information and detail may be found in Peebles [34].

During the Einstein Centennial Symposium, held in 1979 at the University of Illinois at Carbondale, Dirac noted that the CMBR dipole anisotropy makes it possible to identify the relative velocity and direction of the frame where the cmbr is at rest. Measurements at the time, showed that this frame is directed towards the Hydra-Centaur with a of speed of 627km/sec. Dirac then suggested that this corresponds to the reference frame where the Big-Bang occurred, making it a privileged frame which contradicts the general theory of relativity. During his talk Wigner noted that general relativity only assures the equivalence of the laws of the nature in each reference frame and makes no claims about the initial conditions for the universe, so that a reference frame at rest with the Big-Bang does not contradict the general theory of relativity (see [9]).

We now know that this radiation represents a unique preferred frame of rest, which exists throughout the universe and is available to all observers (see [38]). This radiation is highly isotropic with anisotropy limits set at 0.001%. Furthermore, direct measurements have been made of the velocity of our Solar System and Galaxy (370 and 600 km/sec respectively). de Parga et.al have shown that, if a detector is put in the direction of Hydra-Centaur with a speed of 627km/sec, the particle number density will be independent of the angle, so that the dipolar anisotropy, which is a directional temperature will disappear (see [2]).

The following consequence of the CMBR is just as important to our understanding of the universe as it was for the resolution of the big bang debate.

Theorem 2.2. *If all observers choose a frame at rest with respect to the 2.7 °K cosmic microwave background radiation, then their clocks will be NHF and the laws of physics will be the same everywhere.*

Proof. First, the CMBR fills all space, so it's available to all observers. Second, any observer can detect their motion relative to the CMBR. Thus, each observer can transform to the CMBR frame. Since this frame is the same at every point in the universe, all clocks will give the same measurement of time. □

Remark 2.3. *Another important aspect of the CMBR is its blackbody spectrum. It was the understanding of this Planck spectrum that led to the discovery of quantum mechanics. This in addition to the experimental setup, which revealed the blackbody spectrum leads us to two conclusions:*

(i) The universe may be expanding, but it also must be closed.

(ii) Quantum mechanics is operative at the cosmic scale.

2.3. Feynman Operator Calculus and the Dyson Conjectures

The purpose of this section is to provide an intuitive based introduction to the Feynman operator calculus. Those interested in more physical motivation, mathematical detail with rigorous proofs of all results along with applications to the path integral and the Feynman formulation of quantum mechanics, are referred to [19].

2.3.1. Feynman-Dyson Space Let $\mathcal{H}$ be a Hilbert space with orthonomal basis $\{e^i\}$ and let $J = [-T, T]$ be an interval of NHF-time. For $t \in J$, set $\mathcal{H}(t) = \mathcal{H}$, and let $\mathcal{H}_\otimes = \hat{\otimes}_t \mathcal{H}(t)$ be the infinite tensor product Hilbert space over J. For each i, let $e^i_t = e^i$, $E^i = \otimes_t e^i_t$, and let $\mathcal{FD}^i \subset \mathcal{H}_\otimes$ be the smallest Hilbert space containing E^i. We call $\mathcal{FD} = \oplus_{i=1}^{\infty} \mathcal{FD}^i$ the Feynman-Dyson space over J for $\mathcal{H}$ (the film for space-time events).

2.3.2. Time Ordered Operators If $L(\mathcal{H}_\otimes)$ is a set of closed linear operators on $\mathcal{H}_\otimes$, and $\{H_I(t), t \in J\}$ is a family of generators for unitary groups, we define $L(\mathcal{H}(t)) \subset L(\mathcal{H}_\otimes)$ by:

$$L(\mathcal{H}(t)) = \left\{\mathbf{H}_I(t) \,|\mathbf{H}_I(t) = \widehat{\bigotimes_{b \geqslant s > t}} \mathrm{I}_s \otimes H_I(t) \otimes \big(\bigotimes_{t > s \geqslant -T} I_s\big)\right\},$$

where I_s is the identity operator at time s. It follows that

$$\mathbf{H}_I(t)\mathbf{H}_I(s) = \mathbf{H}_I(s)\mathbf{H}_I(t),\, t \neq s.$$

Therefore we have operators that are ordered by time, while maintaining their mathematical position (on paper).

2.3.3. Time Ordered Integrals If $\{H_I(t)|t \in J\}$ is a family of generators of strongly continuous unitary groups on $\mathcal{H}$ and $\{\mathbf{H}_I(t)|t \in J\}$ is the time ordered version then:

Theorem 2.4. *The time ordered integral* $\mathbf{Q}[t,-T] = \int_{-T}^{t} \mathbf{H}_I(s)\,ds$ *exists (a.e) and generates a strongly continuous unitary group* $\mathbf{U}[t,-T]$. *Furthermore, if* $\Psi_0 \in D$, *then*

(i)

$$\Psi(t) = \mathbf{U}[t,-T]\,\Psi_0 = \exp\left\{-\tfrac{i}{\hbar}\mathbf{Q}[t,-T]\right\}\Psi_0, \quad \textit{satisfies:}$$

(ii)

$$i\hbar\frac{\partial\Psi(t)}{\partial t} = \mathbf{H}_I(t)\,\Psi(t)\,,\ \Psi(-T) = \Psi_0.$$

2.4. Dyson's Conjectures for QED

In his analysis, simplification, and unification of the Feynman-Schwinger-Tomonaga theory, Freeman Dyson made four important conjectures concerning the foundations for QED.

(i) His first conjecture was that the divergences in QED were due to an idealized conception of measurability resulting from the infinitely precise knowledge of the space-time positions of particles implied by the Hamiltonian formulation, which is a violation of the Heisenberg uncertainty principle.

(ii) His second conjecture was that it would be more reasonable to expect the renormalized expansion to be asymptotic rather than convergent.

(iii) His third conjecture was that a certain class of Feynman graphs representing overlapping divergencies can be made to cancel out in a systematic manner (first proven by Salam [43]).

(iv) His forth conjecture was that a certain Feynman integral converges, which was necessary for a proof that the ultraviolet divergences cancel to all orders (first proven by Weinberg [48]).

Dyson's first and second conjectures depended on the provision of mathematical meaning for Feynman's operator calculus, which was used in QED to give a physical interpretation to each term in the perturbation expansion. In addition, this conjecture also required a formulation of the S-matrix which will allow one to physically explain how a violation of the Heisenberg uncertainty principle could be seen as the cause for the ultraviolet divergency. The purpose of this section is to give a physically motivated "proof" of these two conjectures (see [19] for details).

2.4.1. Dyson's second Conjecture first We prove the second conjecture first as it is essentially a mathematical problem that could not be considered without meaning for Feynman's operator calculus.

Definition 2.5. *The operator* $\mathbf{U}^w[t,a] = \exp\{w\mathbf{Q}[t,a]\}$ *is asymptotic in the sense of Poincaré if, for each* n *and each* $\Psi_a \in D\left[(\mathbf{Q}[t,a])^{n+1}\right]$ *(domain), we have*

$$\lim_{w\to 0} w^{-(n+1)}\left\{\mathbf{U}^w[t,a] - \sum_{k=1}^{n}\frac{(w\mathbf{Q}[t,a])^k}{k!}\right\}\Psi_a = \frac{\mathbf{Q}[t,a]^{n+1}}{(n+1)!}\Psi_a. \tag{2.2}$$

This is a precise definition of the operator version of an asymptotic expansion, which corresponds to the classical sense of asymptotic expansion for an infinite series of functions (as defined by Poincaré, but see [26]).

If w is a parameter (charge), and $\mathbf{Q}[t,a]$ generates a unitary group on $\mathcal{FD}$, we have:

Theorem 2.6. *For $-T < a < t \leq T$,*

(i) $\mathbf{U}^w[t,a] = exp\{w\mathbf{Q}[t,a]\}$ *is asymptotic in the sense of Poincaré.*

(ii) If $\Psi_a \in D\left[(\mathbf{Q}[t,a])^{n+1}\right]$ *and* $\Psi(t) = \mathbf{U}^w[t,a]\Psi_a$*, for each n we have:*

$$\begin{aligned}\Psi(t) &= \Psi_a + \sum_{k=1}^{n} w^k \int_a^t ds_1 \int_a^{s_1} ds_2 \cdots \int_a^{s_{k-1}} ds_k \mathbf{H}_I(s_1)\cdots\mathbf{H}_I(s_k)\Psi_a \\ &+ \int_0^w (w-\xi)^n d\xi \int_a^t ds_1 \cdots \int_a^{s_n} ds_{n+1}\mathbf{H}_I(s_1)\cdots\mathbf{H}_I(s_{n+1})\mathbf{U}^{\xi}[s_{n+1},a]\Psi_a.\end{aligned} \tag{2.3}$$

The above theorem provides a precise statement of Dyson's first conjecture ([19]). Equation (2.3) is the well-known Dyson expansion with the addition of a remainder term, which makes the expansion exact for all n. However, this does not imply that the series will always converge (see ([19] for details).

Remark 2.7. *There are also well-known special cases in which the perturbation series may actually converge to the solution. This can happen, for example, if the generator is bounded or if it is analytic in some sector. More generally, when the generator is of the form* $\mathbf{H}(t) = \mathbf{H}_0(t) + \mathbf{H}_i(t)$*, where* $\mathbf{H}_0(t)$ *is analytic and* $\mathbf{H}_i(t)$ *is some reasonable perturbation, which need not be bounded, there are conditions that allow the interaction representation to have a convergent Dyson expansion. These results can be easily formulated with easier proofs (see [20]).*

There are also cases where the renormalized series may diverge, but still respond to some summability method (usually Borel). This phenomenon is well-known in classical analysis. In field theory, things can be more complicated. A good discussion, with references, can be found in the book by Glimm and Jaffe [16].

2.4.2. Dyson's first Conjecture In explaining the basis for his first conjecture, Dyson explored the difference between the divergent Hamiltonian that one begins with and the finite S-matrix that results from renormalization. He viewed this as a contrast between a real observer and a fictitious (ideal) observer. The real observer can only determine particle positions with limited accuracy and always gets finite results.

Dyson stated that "... The ideal observer is able to disentangle a single field from its interactions with others, and measure the interaction precisely. He suggested that, from the Heisenberg's uncertainty principle, the value obtained when he measures (the interaction) is infinite." He goes on to remark that, if his analysis is correct, the problem of divergences is attributable to a (mathematically) idealized concept of measurability.

Before providing a physical basis for the first conjecture, we need the concept of an exchange operator.

Definition 2.8. *An exchange operator* $E[t,t']$ *on* $L[\mathcal{H}_\otimes]$ *is a linear map defined for pairs* t, t' *such that:*

(i) $E[t,t'] : L[\mathcal{H}(t)] \to L[\mathcal{H}(t')]$, *(isometric isomorphism),*

(ii) $E[s,t']E[t,s] = E[t,t']$,

(iii) $E[t,t']E[t',t] = \mathbf{I}_\otimes$, *(identity)*

(iv) for $s \neq t,\ t'$, $E[t,t']\mathbf{H}_I(s) = \mathbf{H}_I(s)$, for all $\mathbf{H}_I(s) \in L[\mathcal{H}(s)]$.

The above operator exchanges the time positions of a pair of operators in a more complicated expression. For later use and to make it easy to see how this operator appears, we define the experimental time ordered integral with information concentrated at time points $\{\tau_1, \tau_2, \cdots, \tau_n\}$ by:

$$\mathbf{Q}_e\left[\tau_1, \tau_2, \cdots \tau_n\right] = \sum\nolimits_{j=1}^{n} \int_{t_{j-1}}^{t_j} E\left[\tau_j, s\right] \mathbf{H}_I(s) ds$$
$$= \sum\nolimits_{j=1}^{n} \Delta t_j \left[\tfrac{1}{\Delta t_j} \int_{t_{j-1}}^{t_j} E\left[\tau_j, s\right] \mathbf{H}_I(s) ds\right].$$

In the above equation, for each interval $[t_{j-1}, t_j]$, the average is concentrated at the point τ_j. If $\Psi \in \mathcal{FD}$ and $N(t)$ is a Poisson distributed random variable with inverse time parameter λ, we define the function $\mathbf{U}[N(t), 0]\Psi$ by:

$$\mathbf{U}[N(t), 0]\Psi = \exp\left\{\mathbf{Q}_e\left[\tau_1, \tau_2 \cdots, \tau_{N(t)}\right]\right\}\Psi. \tag{2.4}$$

The function $\mathbf{U}[N(t), 0]\Psi$ is an $\mathcal{FD}$-valued random variable which represents the distribution of the number of information points that may appear on our film up to time t. In order to relate $\mathbf{U}[N(t), 0]\Psi$ to actual experimental results, we must compute its expected value.

$$\bar{\mathbf{U}}_\lambda[t, 0]\Psi = \boldsymbol{\mathcal{E}}\left\{\mathbf{U}[N(t), 0]\Psi\right\}$$
$$= \sum_{n=0}^{\infty} \boldsymbol{\mathcal{E}}\left\{\mathbf{U}[N(t), 0]\Psi \mid N(t) = n\right\} Prob[N(t) = n], \tag{2.5}$$

$$\boldsymbol{\mathcal{E}}\left\{\mathbf{U}[N(t), 0]\Psi \mid N(t) = n\right\}$$
$$= \int_0^t \tfrac{d\tau_1}{t-0} \int_{\tau_1}^t \tfrac{d\tau_2}{t-\tau_1} \cdots \int_{\tau_{n-1}}^t \tfrac{d\tau_n}{t-\tau_{n-1}} \mathbf{U}[\tau_n, \tau_{n-1} \cdots, \tau_1]\Psi = \bar{\mathbf{U}}_n[t, 0]\Psi, \tag{2.6}$$

and $Prob[N(t) = n] = \frac{(\lambda t)^n}{n!} \exp\{-\lambda t\}$. The integral distributes the time positions τ_j uniformly over the successive intervals $[t, \tau_{j-1}]$, $1 \leq j \leq n$, given that τ_{j-1} has been determined. Since interest is only in what happens when $\lambda \to \infty$ and, as the mean number of points in the film at time t is λt, we can take $\tau_j = \frac{jt}{n}$, $1 \leqslant j \leqslant n$, $\Delta t_j = \frac{t}{n}$ (for each n) and replace $\bar{\mathbf{U}}_n[t, 0]\Psi$ by $\mathbf{U}_n[t, 0]\Psi$. We continue to use τ_j, so that

$$\mathbf{U}_n[t, 0]\Psi = \exp\left\{\sum_{j=1}^{n} \int_{t_{j-1}}^{t_j} E[\tau_j, s]\mathbf{H}_I(s) ds\right\}\Psi.$$

We define our experimental evolution operator $\mathbf{U}_\lambda[t, 0]\Psi$ by

$$\mathbf{U}_\lambda[t, 0]\Psi = \sum_{n=0}^{\infty} \frac{(\lambda t)^n}{n!} \exp\{-\lambda t\}\mathbf{U}_n[t, 0]\Psi. \tag{2.7}$$

Since Borel summability is regular (i.e, $\mathbf{U}_n[t, 0]\Psi \to \mathbf{U}[t, 0]\Psi$ implies $\mathbf{U}_\lambda[t, 0]\Psi \to \mathbf{U}[t, 0]\Psi$), we have by Theorem 2.4 that:

$$\lim_{\lambda \to \infty} \mathbf{U}_\lambda[t, 0]\Psi = \mathbf{U}[t, 0]\Psi. \tag{2.8}$$

Since $\lambda \to \infty \Rightarrow \lambda^{-1} \to 0$, the average time becomes zero, providing a continuous path. This continuous path arises from averaging the sum over an infinite number of (discrete) paths. The first term corresponds to the path that created no information points (i.e., the film is blank). This event has probability $\exp\{-\lambda t\}$ (which approaches zero as $\lambda \to \infty$). The n-th term corresponds to a path that created n information points, (with probability $\frac{(\lambda t)^n}{n!}\exp\{-\lambda t\}$) etc. We now replace $\mathbf{H}_I(t)$ by $\frac{-i}{\hbar}\mathbf{H}_I(t)$, and $\mathbf{U}_\lambda[T,-T]\Psi$ by the experimental S-matrix $\mathbf{S}_\lambda[T,-T]\Psi$, so that

$$\mathbf{S}_\lambda[T,-T]\Psi = \sum_{n=0}^{\infty} \frac{(2\lambda T)^n}{n!} \exp\left[-2\lambda T\right] \mathbf{S}_n[T,-T]\Psi. \tag{2.9}$$

Combining $\exp\left[-2\lambda T\right]$ and $\mathbf{S}_n[T,-T]\Psi$, we obtain:

$$\exp\left[-2\lambda T\right] \mathbf{S}_n[T,-T]\Psi = e^{-\frac{i}{\hbar}\left\{\sum_{j=1}^{n}\int_{t_{j-1}}^{t_j}[E[\tau_j,s]\mathbf{H}_I(s)-i\lambda\hbar\mathbf{I}_\otimes]ds\right\}}\Psi, \tag{2.10}$$

$\mathbf{H}_I(t) = \int_{\mathbb{R}^3} H_I(\mathbf{x}(t),t)d\mathbf{x}(t)$ is the interaction energy, and

$$\begin{aligned} &H_I(\mathbf{x}(t),t) = -ie\mathbf{A}_\mu(\mathbf{x}(t),t)\bar{\psi}(\mathbf{x}(t),t)\gamma_\mu\psi(\mathbf{x}(t),t) \\ &-\delta mc^2\bar{\psi}(\mathbf{x}(t),t)\psi(\mathbf{x}(t),t). \end{aligned} \tag{2.11}$$

Where δmc^2 is the mass counter-term for the self-energy divergence. For a physical interpretation, we see that equation (2.9) becomes:

$$\mathbf{S}_\lambda\left[T,-T\right]\Psi = \sum_{n=0}^{\infty} \frac{(2\lambda T)^n}{n!} e^{-\frac{i}{\hbar}\left\{\sum_{j=1}^{n}\int_{t_{j-1}}^{t_j}[E[\tau_j,s]H_I(s)-i\lambda\hbar\mathbf{I}_\otimes]ds\right\}}\Psi. \tag{2.12}$$

The term $-i\lambda\hbar\mathrm{I}_\otimes$ has the physical interpretation as the absorption of photon energy of amount $\lambda\hbar$ in each subinterval $[t_{j-1}, t_j]$. Computing the limit in (2.12), we get the standard S-matrix. It follows that we must add an infinite amount of photon energy to the mathematical description of the experimental picture (at each point in time) in order to obtain the S-matrix (the ultra-violet divergence). Since the uncertainty relation is a variance, The divergence should not appear in first order perturbation but must appear in all higher-order terms. On the other hand, if the divergent terms in second order are eliminated, the same method should prevent them from appearing in all other terms of the expansion. This is precisely what happens in QED, providing a clear verification of Dyson's conjecture. Thus, the ultra-violet divergence is not a hint of some (unknown) deeper problems as many believed at the time. (It disappears if we assume a minimum time.)

Remark 2.9. *The mass counter-term δmc^2 in the interaction energy (2.11) is required to cancel the self-energy divergence. This self-energy divergence was already in Dirac's version of CED and was introduced in QED with the quantization of his theory. This plus the infrared divergency are now the only two remaining difficulties in the foundations of QED. Both of these divergences will be addressed in the dual Maxwell section.*

2.5. Conclusion I

In this section, we have:

(i) Provided two distinct proofs for the existence of a NHF-time.
(ii) Provided understanding of the relationship to and implications of the 2.7 °K CMBR relative to a preferred rest frame, the NHF-clock and quantum theory.
(iii) Provided resolution of problems with the assumption of Minkowski's four space-time and the operational use of time as a director of motion in the Feynman formulation of QED.
(iv) Provided partial understanding of the causes and provision of partial corrections for some of the physical and mathematical problems associated with the divergencies in QED.

3. Proper Time II, Minkowski's Theory

Using the same notation as in the first section, the second version of proper time is defined by:

$$\begin{aligned} d\tau &= \sqrt{1-\left(\tfrac{\mathbf{v}}{c}\right)^2}dt, \quad d\tau = \sqrt{1-\left(\frac{\mathbf{v}'}{c}\right)^2}dt' \\ d\tau^2 &= dt^2 - \tfrac{1}{c^2}d\mathbf{x}^2, \quad d\tau^2 = dt'^2 - \tfrac{1}{c^2}d\mathbf{x}'^2. \end{aligned} \tag{3.1}$$

This was first discovered by Poincaré [36] and later used by Minkowski to formulate his version of Einstein's special theory of relativity. This required the use of $d\tau$ as a metric for his four-spacetime theory. Einstein himself was the first to point out that it is not the differential of an exact one-form. In order to use his postulate, Minkowski introduced the clock of a co-moving observer as a substitute (see the notes in Sommerfeld in [35]). The new formalism became the first physical motivation for an abstract space. This made the theory attractive to mathematicians who dominated the field until the physics community was able to later join in. The excitement made Minkowski's ideas popular and helped to bring them to the attention of the masses.

In this air of euphoria no one seem to noticed that the theory did not work for two or more particles and was not an extension of Newtonian mechanics. (This is the true cause of the twin paradox.) By the time problems in attempts to merge Minkowski's theory with quantum mechanics forced researchers to take a new look at the foundations of electrodynamics, Minkowski's postulate had become sacred. When Einstein considered the extension of the special to the general theory, he was only interested in one which extended Minkowski's postulate (see Pais [33]).

Theorem 3.1. *The addition of Minkowski's postulate to the postulates of Einstein is incompatible for two particles,*

Proof. Let O and O' be two inertial observers. Without loss, we can assume both clocks begin when their origins coincide and O' is moving with uniform velocity $\mathbf{v}$ as seen by O. Let two particles, each the source of an electromagnetic field, move with velocities $\mathbf{w}_i$ $(i = 1, 2)$, as seen by O, and $\mathbf{w}'_i$ $(i = 1, 2)$, as seen by O', so that:

$$\mathbf{x}'_i = \mathbf{x}_i - \gamma(\mathbf{v})\mathbf{v}t + (\gamma(\mathbf{v}) - 1)(\mathbf{x}_i \cdot \mathbf{v}/\left\|\mathbf{v}\right\|^2)\mathbf{v} \tag{3.2}$$

and

$$\mathbf{x}_i = \mathbf{x}'_i + \gamma(\mathbf{v})\mathbf{v}t' + (\gamma(\mathbf{v}) - 1)\left(\mathbf{x}'_i \cdot \mathbf{v}/\left\|\mathbf{v}\right\|^2\right)\mathbf{v}. \tag{3.3}$$

Thus, there is clearly no problem in requiring that the positions transform as expected. However, when we try to transform the clocks, we see the problem at once since

$$t' = \gamma(\mathbf{v})\left(t - \mathbf{x}_1 \cdot \mathbf{v}/c^2\right) \quad \text{and} \quad t' = \gamma(\mathbf{v})\left(t - \mathbf{x}_2 \cdot \mathbf{v}/c^2\right). \tag{3.4}$$

This is clearly impossible unless $\mathbf{x}_1 \cdot \mathbf{v} = \mathbf{x}_2 \cdot \mathbf{v}$. This means that two observers cannot use their clocks to share information (with other observers) about two or more particles. It follows that, if we try to replace $\mathbf{x}_i$ and $\mathbf{x}'_i$ with four vectors using t and t', the first postulate fails. Thus, the three postulates are incompatible. □

3.0.1. The n-particle position problem In light of the above result, one might think that we can by-pass the problem if both observers use the center of mass and reduce it back to a one-body problem. In 1948 Pryce was the first to study the relativistic center of mass problem for 2 or more particles (see [39]). He concluded that there are three possibilities, but only one is canonical and available to all observers. His canonical representation led to the implication that the center-of-mass cannot be the three-vector part of a four-vector. This problem is almost seventy five years old and will be made explicit in the next section.

Following Pryce's investigation, in 1953 Bakamjian and Thomas were able to construct a many-particle quantum theory with a canonical center of mass, which satisfied the two postulates of Einstein, but not Minkowski's (see [5]). They conjectured that, with the addition of Minkowski's postulate, their theory would only be compatible with free particles.

In 1965, Currie, Jordan and Sudarshan [6], prove their famous "No-Interaction Theorem" for two particles, showing that Bakamjian and Thomas were correct. The theorem has been extended to an arbitrary number of particles by Leutwyler [29].

Theorem 3.2. (No-Interaction Theorem) Suppose we have a system of n particles with phase space variables $\{(\mathbf{x}_i, \mathbf{p}_i)\}_{i=1}^n$ defined on $\mathbb{R}^{3n} \times \mathbb{R}^{3n}$ with the following properties:

(i) The system has a Hamiltonian representation.
(ii) The system has a canonical representation of the Poincare group.
(iii) Each $\mathbf{x}_i$ is the vector part of a four-vector.

Then these assumptions are only compatible with free particles.

Thus, Minkowski's postulate imposes an additional condition on Einstein's theory, works for one particle, but fails for two or more particles at the classical and quantum levels.

4. Proper Time III, Einstein Dual Theory

4.0.1. One-Particle Clock For the third definition of proper time, we assume a classical interacting system of n-particles observed by O, who is able to identify each particle and attach a vector $\mathbf{x}_i$ to the ith particle, denoting its spacial distance to the origin. If $\mathbf{w}_i$ is the velocity of particle i as seen by O, let $\gamma^{-1}(\mathbf{w}_i) = \sqrt{1 - \mathbf{w}_i^2/c^2}$. The ith particle proper time is defined as before by:

$$d\tau_i = \gamma^{-1}(\mathbf{w}_i)dt, \quad \mathbf{w}_i = \frac{d\mathbf{x}_i}{dt}, \quad d\tau_i^2 = dt^2 - \tfrac{1}{c^2}d\mathbf{x}_i^2. \tag{4.1}$$

Rewrite the last term to get:

$$dt^2 = d\tau_i^2 + \tfrac{1}{c^2}d\mathbf{x}_i^2, \Rightarrow cdt = \left(\sqrt{\mathbf{u}_i^2 + c^2}\right) d\tau_i, \quad \mathbf{u}_i = \frac{d\mathbf{x}_i}{d\tau_i} = \gamma(\mathbf{w}_i)\mathbf{w}_i. \tag{4.2}$$

If we let $b_i = \sqrt{\mathbf{u}_i^2 + c^2}$, the second term in equation (4.2) becomes $cdt = b_i d\tau_i$. This leads to our first identity and our **third definition of proper time**:

$$\frac{1}{c}\frac{d}{dt} \equiv \frac{1}{b_i}\frac{d}{d\tau_i} \tag{4.3}$$

This identity provides the correct way to define the relationship between the proper time and observer time for the ith particle. If we apply the identity to $\mathbf{x}_i$, we obtain our second identity, which shows that the transformation leaves the configuration (or tangent) space variables invariant:

$$\frac{\mathbf{w}_i}{c} = \frac{1}{c}\frac{d\mathbf{x}_i}{dt} \equiv \frac{1}{b_i}\frac{d\mathbf{x}_i}{d\tau_i} = \frac{\mathbf{u}_i}{b_i}. \tag{4.4}$$

The new particle coordinates are $(\mathbf{x}_i, \tau_i)$. In this representation, the position $\mathbf{x}_i$ is uniquely defined relative to O, while τ_i is uniquely defined by the ith particle. The ith particle momentum can be represented as $\mathbf{p}_i = m_i\gamma(\mathbf{w}_i)\mathbf{w}_i = m_i\mathbf{u}_i$, where m_i is the particle rest mass. Thus, the phase space variables are also left invariant.

4.0.2. Many-Particle Clock To construct the many-particle clock, we suppose the n interacting particles have Hamiltonians H_i and total Hamiltonian $H = \sum_{i=1}^{n} H_i$. Define the effective mass energy Mc^2, and total momentum $\mathbf{P}$, by

$$Mc^2 = \sqrt{H^2 - c^2\mathbf{P}^2}, \quad \mathbf{P} = \sum_{i=1}^{n} \mathbf{p}_i.$$

We can now also represent the Hamiltonian by $H = \sqrt{c^2\mathbf{P}^2 + M^2c^4}$.

As noted before, Pryce, found three possible definitions for the center of mass position vector. However, only one is canonical and independent of the frame in which it is defined. This is the natural and necessary choice if we want a theory that provides the same physics for all observers and is compatible with quantum mechanics. Thus, for the O frame observer the canonical center of mass position $\mathbf{X}$ is defined by: (see [30]):

$$\mathbf{X} = \frac{1}{H}\sum_{i=1}^{n} H_i \mathbf{x}_i + \frac{c^2\left(\mathbf{S} \times \mathbf{P}\right)}{H\left(Mc^2 + H\right)}, \tag{4.5}$$

where $\mathbf{S}$ is the global spin of the system of particles relative to O. (It is clear that (4.5) cannot represent the vector part of a four-vector.) If there is no interaction, S, H and M are constant, and independent of the $\{\mathbf{x}_i,\, \mathbf{p}_i\}$ variables, so that:

$$\{X_i, X_j\} = \sum_{k=1}^{n} \frac{\partial X_i}{\partial \mathbf{p}_k} \cdot \frac{\partial X_j}{\partial \mathbf{x}_k} - \frac{\partial X_j}{\partial \mathbf{p}_k} \cdot \frac{\partial X_i}{\partial \mathbf{x}_k} \equiv 0.$$

However, when interaction is present, S, H and M may all depend on the $\{\mathbf{x}_i,\, \mathbf{p}_i\}$ variables, so that in general $\{X_i, X_j\} \neq 0$. Since $\mathbf{X}$ is the canonical conjugate of $\mathbf{P}$, it precisely what we need for a consistent merge with quantum mechanics.

Let $\mathbf{V}$ be the velocity of $\mathbf{X}$ with respect to O. It follows that H also has the representation $H = Mc^2\gamma(\mathbf{V})$, so that $\gamma(\mathbf{V})^{-1} = (Mc^2/H)$. In this representation, we see that $d\tau = \gamma(\mathbf{V})^{-1}dt = (Mc^2/H)dt$ does not depend on the number of particles in the system. It follows that, as long as Mc^2/H is fixed, τ is invariant, so that the number of particles can increase or decrease without changing τ. (This means that number n is not conserved and, in some cases of physical interest, may even be a integer-valued random variable).

From $dt^2 = d\tau^2 + d\mathbf{X}^2/c^2$, we see that ($\mathbf{U} = d\mathbf{X}/d\tau$)

$$c^2dt^2 = \left(c^2 + \mathbf{U}^2\right)d\tau^2 \quad \Rightarrow \quad cdt = \left(\sqrt{c^2 + \mathbf{U}^2}\right)d\tau.$$

It is easy to see that $\mathbf{U} = \gamma(\mathbf{V})\mathbf{V}$, so that $\mathbf{U}$ is constant. If we define $b = \sqrt{\mathbf{U}^2 + c^2}$, we can write $cdt = bd\tau$. Since b is constant we have: $ct = b\tau$. For observer O' the same system has velocity $\mathbf{V}'$ for the center of mass and, by the same calculations, we obtain $ct' = b'\tau$, where $b' = \sqrt{\mathbf{U}'^2 + c^2}$. This shows that a unique (operational) measure of time is available to all observers. Furthermore, τ differs from t (respectively t'), by a constant scale factor. Thus, all observers may uniquely define the local-time of the center of mass for the system of particles (independent of their chosen inertial frame). We also obtain our third identity:

$$\frac{1}{c}\frac{d}{dt} \equiv \frac{1}{b}\frac{d}{d\tau} \equiv \frac{1}{b_i}\frac{d}{d\tau_i} \tag{4.6}$$

Applying the above to $\mathbf{x}_i$ we see that:

$$\frac{1}{c}\frac{d\mathbf{x}_i}{dt} \equiv \frac{1}{b}\frac{d\mathbf{x}_i}{d\tau} \equiv \frac{1}{b_i}\frac{d\mathbf{x}_i}{d\tau_i}. \tag{4.7}$$

Theorem 4.1. *There are two sets of global variables available to O:* $(\mathbf{X}, t)$ *and* $(\mathbf{X}, \tau)$. *Use of* $(\mathbf{X}, t)$ *provides a relative definition of time and a constant speed of light; while use of* $(\mathbf{X}, \tau)$ *provides a unique definition of time and a relative definition of the speed of light, with no upper bound.*

Proof. The first part is clear. To prove the second statement, from above, we see that any other observer O' investigating the same system of particles also has two sets of global variables available: $(\mathbf{X}', t')$ with a constant speed of light and $(\mathbf{X}', \tau)$ with b' relative. We are done if we can show that Einstein's first postulate holds. Let $\mathbf{W}$ be the relative velocity between observer O and O'. Since τ is the same for both we only need the relationship between the two scale factors b and b' to satisfy the first postulate. Since $\mathbf{U} = (\mathbf{X}/\tau)$, we have:

$$t' = \gamma(\mathbf{W})\left[t - \tfrac{1}{c^2}(\mathbf{X}\cdot\mathbf{W})\right] \Rightarrow \tfrac{b'}{c}\tau = \gamma(\mathbf{W})\left[\tfrac{b}{c}\tau - \tfrac{1}{c^2}(\mathbf{X}\cdot\mathbf{W})\right] \Rightarrow$$
$$b' = \gamma(\mathbf{W})\left[b - \tfrac{1}{c}\left(\tfrac{\mathbf{X}}{\tau}\cdot\mathbf{W}\right)\right] = \gamma(\mathbf{W})\left[b - \tfrac{1}{c}(\mathbf{U}\cdot\mathbf{W})\right].$$

A similar calculation shows that $b = \gamma(\mathbf{W})\,(b' + \mathbf{U}'\cdot\mathbf{W}/c)$. This shows that each observer can have direct access to all information available to any other observer once they know their relative velocities. Thus the first postulate of Einstein is satisfied at the global level. □

Corollary 4.2. *The two sets of global variables produce mathematically equivalent theories, but not physically equivalent theories.*

Theorem 4.3. *If we use the proper time of each particle, then the special theory of Einstein holds for any many-particle system and is independent of Minkowski's postulate.*

Equations (4.6) and (4.7) provide the foundations for the "Einstein Dual Theory of Relativity".

Remark 4.4. *It is important to observe that equation (4.7) creates serious problems for any assumption, except by direct measurement, that the speed of light is c and or the velocity of objects are traveling with speeds less than c. For example, in all astronomical cases, we can only measure the ratio* β *and assume it equals* $\frac{v}{c}$. *However, since its possible that* $\beta = \frac{u}{b}$, *the object could also have a speed* $u > c$ *and we are looking at light traveling with speed* $b >> c$. *It follows that some serious retrospection may be required in both astronomy and cosmology.*

4.1. Dual Newtonian Theory

4.1.1. Canonical Dual Hamiltonian If we treat the system of particles as a single entity, then $(\mathbf{X}, \mathbf{P})$ are the natural phase space variables for the external system dynamics. We first need to identify the generator of τ translations. If $W(\mathbf{X}, \mathbf{P})$ is a dynamical parameter in $\mathbf{X}$ and $\mathbf{P}$, the time evolution of W is defined by:

$$\frac{dW}{dt} = \{H, W\} = \frac{\partial H}{\partial \mathbf{P}}\cdot\frac{\partial W}{\partial \mathbf{X}} - \frac{\partial H}{\partial \mathbf{X}}\cdot\frac{\partial W}{\partial \mathbf{P}}. \tag{4.8}$$

In order to represent the dynamics using the proper time of the system, we use the first representation $d\tau = (Mc^2/H)dt$, so that:

$$\frac{dW}{d\tau} = \frac{dt}{d\tau}\frac{dW}{dt} = \frac{H}{Mc^2}\{H, W\} = \left(\frac{H}{Mc^2}\frac{\partial H}{\partial \mathbf{P}}\right)\cdot\frac{\partial W}{\partial \mathbf{X}} - \left(\frac{H}{Mc^2}\frac{\partial H}{\partial \mathbf{X}}\right)\cdot\frac{\partial W}{\partial \mathbf{P}}.$$

The ratio H/Mc^2 is constant and Mc^2 is a well-defined (invariant) for the system, so we can determine the canonical Hamiltonian K, related to τ by:

$$\{K, W\} = \frac{H}{Mc^2}\{H, W\}, \quad K|_{\mathbf{P}=0} = H|_{\mathbf{P}=0} = Mc^2.$$

In this case:

$$\{K, W\} = \left[\frac{H}{Mc^2}\frac{\partial H}{\partial \mathbf{P}}\right]\frac{\partial W}{\partial \mathbf{X}} - \left[\frac{H}{Mc^2}\frac{\partial H}{\partial \mathbf{X}}\right]\frac{\partial W}{\partial \mathbf{P}}$$
$$= \frac{\partial}{\partial \mathbf{P}}\left[\frac{H^2}{2Mc^2} + a\right]\frac{\partial W}{\partial \mathbf{X}} - \frac{\partial}{\partial \mathbf{X}}\left[\frac{H^2}{2Mc^2} + a'\right]\frac{\partial W}{\partial \mathbf{P}},$$

we get that $a = a' = \frac{1}{2}Mc^2$, so that

$$K = \frac{H^2}{2Mc^2} + \frac{Mc^2}{2} = \frac{\mathbf{P}^2}{2M} + Mc^2, \quad \text{and} \quad \frac{dW}{d\tau} = \{K, W\}. \tag{4.9}$$

Since K does not depend on the center-of-mass position $\mathbf{X}$, it is easy to see that

$$\mathbf{U} = \frac{d\mathbf{X}}{d\tau} = \frac{\partial K}{\partial \mathbf{P}} = \frac{\mathbf{P}}{M} = \frac{1}{M}\sum_{i=1}^{n}\mathbf{p}_i = \frac{1}{M}\sum_{i=1}^{n} m_i\mathbf{u}_i = \frac{1}{M}\sum_{i=1}^{n} m_i\frac{d\mathbf{x}_i}{d\tau_i}. \tag{4.10}$$

For the O' observer, the same calculation leads to:

$$\mathbf{U}' = \frac{d\mathbf{X}'}{d\tau} = \frac{\partial K}{\partial \mathbf{P}'} = \frac{\mathbf{P}'}{M'} = \frac{1}{M'}\sum_{i=1}^{n}\mathbf{p}'_i = \frac{1}{M'}\sum_{i=1}^{n} m'_i\mathbf{u}'_i = \frac{1}{M'}\sum_{i=1}^{n} m'_i\frac{d\mathbf{x}'_i}{d\tau_i}. \tag{4.11}$$

We now observe that

$$dt = \frac{H_i}{m_i c^2}d\tau_i = \frac{H}{Mc^2}d\tau \quad \Rightarrow \quad \frac{m_i}{M}\frac{d}{d\tau_i} = \frac{H_i}{H}\frac{d}{d\tau}.$$

Thus, we can replace (4.10) and (4.11) by:

$$\frac{d\mathbf{X}}{d\tau} = \frac{\partial K}{\partial \mathbf{P}} = \frac{\mathbf{P}}{M} = \frac{1}{M}\sum_{i=1n}^{n}\mathbf{p}_i = \frac{1}{M}\sum_{i=1}^{n} m_i\mathbf{u}_i = \frac{1}{H}\sum_{i=1}^{n} H_i\frac{d\mathbf{x}_i}{d\tau} \tag{4.12}$$

and

$$\frac{d\mathbf{X}'}{d\tau} = \frac{\partial K}{\partial \mathbf{P}'} = \frac{\mathbf{P}'}{M'} = \frac{1}{M'}\sum_{i=1}^{n}\mathbf{p}'{}_i = \frac{1}{M'}\sum_{i=1}^{n} m'_i\mathbf{u}'_i = \frac{1}{H'}\sum_{i=1n}^{n} H'_i\frac{d\mathbf{x}'_i}{d\tau}. \tag{4.13}$$

Since the H_i (respectively H'_i) do not depend on τ, we can integrate both equations to get:

$$\mathbf{X} = \frac{1}{H}\sum_{i=1n}^{n} H_i\mathbf{x}_i + \mathbf{Y} \quad \text{and} \quad \mathbf{X}' = \frac{1}{H'}\sum_{i=1n}^{n} H'_i\mathbf{x}'_{\mathbf{i}} + \mathbf{Y}',$$

where $\mathbf{Y}$ and $\mathbf{Y}'$ are constants of integration. This shows that the canonical dual Hamiltonian determines the canonical position up to a constant.

To see directly that the clock transformation is also a canonical change of variables (time), which leaves phase space invariant, we have the following theorem.

Theorem 4.5. *There exists a function* $S = S(\mathbf{X},\ \mathbf{P},\ \tau)$ *such that*

$$\mathbf{P}\cdot d\mathbf{X} - Hdt \equiv \mathbf{P}\cdot d\mathbf{X} - Kd\tau + dS.$$

Proof. Set $S = [K - Mc^2]\tau$. An easy calculation, using the fact that both Mc^2 and K are conserved quantities, shows that $dS = [K - Mc^2]d\tau$. Since τ is an invariant, $d\tau$ is an exact one-form, so that dS is a total differential. □

We also have that

$$\sum_{i=1}^{n}\left[\mathbf{p}_i\cdot d\mathbf{x}_i - H_i dt\right] = \sum_{i=1}^{n}\mathbf{p}_i\cdot d\mathbf{x}_i - \sum_{i=1}^{n} H_i dt = \sum_{i=1}^{n}\mathbf{p}_i\cdot d\mathbf{x}_i - Hdt.$$

This result leads to:

Corollary 4.6. *There exists a function* $S = S\left(\{\mathbf{x}_i\},\ \{\mathbf{p}_i\},\ \tau\right)$ *such that*

$$\sum_{i=1}^{n}\mathbf{p}_i\cdot d\mathbf{x}_i - Hdt \equiv \sum_{i=1}^{n}\mathbf{p}_i\cdot d\mathbf{x}_i - Kd\tau + dS.$$

Definition 4.7. *A theory is said to be Einsteinian if at least one representation exists, which satisfies the two postulates of the special theory.*

Theorem 4.8. *Any closed system of interacting particles is Einsteinian and independent of the Minkowski postulate. Furthermore, there always exists two distinct sets of inertial frame coordinates for each observer, to describe each particle in the system and the system as a whole. The following holds:*

(i) In one frame, the speed of light is an invariant upper bound and time is relative, while in the other, time is invariant and the speed of light b, is relative with no upper bound.

(ii) For the whole system and for each particle, the equations of motion are mathematically equivalent.

We have already proven all but the last part of the above theorem, which will be completed when we review the dual Maxwell theory.

4.1.2. Global Interaction At the global level, it suffices to assume the interaction is via a potential $V(\mathbf{X})$. We can add V to the equation for H, to get:

$$\begin{aligned} &H = \sqrt{c^2\mathbf{P}^2 + M^2c^4} + V(\mathbf{X}) = H_0 + V(\mathbf{X}) \Rightarrow \\ &\frac{d\mathbf{X}}{dt} = \frac{c^2\mathbf{P}}{H_0} \quad \textit{and} \quad \frac{d\mathbf{P}}{dt} = -\nabla V(\mathbf{X}). \end{aligned} \tag{4.14}$$

For comparison, if we use the proper clock, we get:

$$\begin{aligned} &K = \frac{H^2}{2Mc^2} + \frac{Mc^2}{2} \Rightarrow \frac{d\mathbf{X}}{d\tau} = \frac{\partial K}{\partial \mathbf{P}} = \frac{H}{Mc^2}\frac{c^2\mathbf{P}}{H_0} = \frac{b}{c}\frac{d\mathbf{X}}{dt}, \\ &\frac{d\mathbf{P}}{d\tau} = \frac{\partial K}{\partial \mathbf{X}} = -\frac{H}{Mc^2}\nabla V\left(\mathbf{X}\right) = \frac{b}{c}\frac{d\mathbf{P}}{dt}. \end{aligned} \tag{4.15}$$

Comparison of (4.14) and (4.15) shows that the two clocks give mathematically equivalent equations of motion for the general system dynamics, which are not physically equivalent.

4.2. Dual Maxwell Theory

4.2.1. Dynamics of a Particle We now investigate the corresponding single particle dynamical theory. In this section, $b = b_i$, $\tau = \tau_i$ and $\mathbf{u} = \mathbf{u}_i$.

Since τ is invariant during interaction (minimal coupling), we make the natural assumption that the form of K also remains invariant. Thus, if $\sqrt{c^2\mathbf{p}^2 + m^2c^4} \to \sqrt{c^2\boldsymbol{\pi}^2 + m^2c^4} + V$, where $\mathbf{A}$ is the vector potential, $V = e\Phi$ is the potential energy, $\mathbf{E} = -\frac{1}{b}\left(\partial\mathbf{A}/\partial\tau\right) - \boldsymbol{\nabla}\Phi$ and $\boldsymbol{\pi} = \mathbf{p} - \frac{e}{c}\mathbf{A}$. In this case, K becomes:

$$K = \frac{H^2}{2mc^2} + \frac{mc^2}{2} = \frac{\boldsymbol{\pi}^2}{2m} + mc^2 + \frac{V^2}{2mc^2} + \frac{V\sqrt{c^2\boldsymbol{\pi}^2 + m^2c^4}}{mc^2}.$$

If we set $H_0 = \sqrt{c^2\boldsymbol{\pi}^2 + m^2c^4}$, use standard vector identities with $\boldsymbol{\nabla}\times\boldsymbol{\pi} = -\frac{e}{c}\mathbf{B}$, and compute Hamilton's equations, we get:

$$\frac{d\mathbf{x}}{d\tau} = \frac{\partial K}{\partial\mathbf{p}} = \frac{H}{mc^2}\left(\frac{c^2\boldsymbol{\pi}}{H_0}\right) = \frac{b}{c}\left(\frac{c^2\boldsymbol{\pi}}{H_0}\right) = \frac{b}{c}\frac{d\mathbf{x}}{dt} \tag{4.16}$$

and

$$\begin{aligned}\frac{d\mathbf{p}}{d\tau} &= \frac{b}{c}\frac{\left[\left(c^2\boldsymbol{\pi}\cdot\boldsymbol{\nabla}\right)\mathbf{A} + \frac{e}{b}\left(c^2\boldsymbol{\pi}\times\mathbf{B}\right)\right]}{H_0} - \frac{b}{c}\boldsymbol{\nabla}V\\ &= \frac{b}{c}\left[\left(\mathbf{u}\cdot\boldsymbol{\nabla}\right)\mathbf{A} + \tfrac{e}{b}\left(\mathbf{u}\times\mathbf{B}\right)\right] - \frac{b}{c}\boldsymbol{\nabla}V\\ &= \frac{b}{c}\left[e\mathbf{E} + \tfrac{e}{b}\left(\mathbf{u}\times\mathbf{B}\right) + \tfrac{e}{b}\frac{d\mathbf{A}}{d\tau}\right] \quad \Rightarrow\\ \frac{c}{b}\frac{d\boldsymbol{\pi}}{d\tau} &= \left[e\mathbf{E} + \tfrac{e}{b}\left(\mathbf{u}\times\mathbf{B}\right)\right] = \frac{d\boldsymbol{\pi}}{dt}.\end{aligned} \tag{4.17}$$

Equations (4.16) and (4.17) show that the standard and dual equations of motion are mathematically equivalent. Thus, our assumption that K remain invariant with minimal coupling is correct. This also completes the proof of Theorem 4.8 (mathematical equivalence).

4.2.2. Field of a Particle To study the field of a particle, we write Maxwell's equations (in c.g.s. units):

$$\begin{aligned}&\nabla\cdot\mathbf{B} = 0, \qquad \nabla\cdot\mathbf{E} = 4\pi\rho,\\ &\nabla\times\mathbf{E} = -\frac{1}{c}\frac{\partial\mathbf{B}}{\partial t}, \quad \nabla\times\mathbf{B} = \frac{1}{c}\left[\frac{\partial\mathbf{E}}{\partial t} + 4\pi\rho\mathbf{w}\right].\end{aligned} \tag{4.18}$$

Using equations (4.3) and (4.4) in (4.18), we have (*the mathematically identical representation*):

$$\begin{aligned}&\nabla\cdot\mathbf{B} = 0, \qquad \nabla\cdot\mathbf{E} = 4\pi\rho,\\ &\nabla\times\mathbf{E} = -\frac{1}{b}\frac{\partial\mathbf{B}}{\partial\tau}, \quad \nabla\times\mathbf{B} = \frac{1}{b}\left[\frac{\partial\mathbf{E}}{\partial\tau} + 4\pi\rho\mathbf{u}\right].\end{aligned} \tag{4.19}$$

Thus, we obtain a mathematically equivalent set of Maxwell's equations using the local time of the particle to describe its fields.

To derive the corresponding wave equations, we take the curl of the last two equations in (4.19), and use standard vector identities, to get:

$$\begin{aligned}&\frac{1}{b^2}\frac{\partial^2\mathbf{B}}{\partial\tau^2} - \frac{\mathbf{u}\cdot\mathbf{a}}{b^4}\left[\frac{\partial\mathbf{B}}{\partial\tau}\right] - \nabla^2\cdot\mathbf{B} = \frac{1}{b}\left[4\pi\nabla\times(\rho\mathbf{u})\right],\\ &\frac{1}{b^2}\frac{\partial^2\mathbf{E}}{\partial\tau^2} - \frac{\mathbf{u}\cdot\mathbf{a}}{b^4}\left[\frac{\partial\mathbf{E}}{\partial\tau}\right] - \nabla^2\cdot\mathbf{E} = -\nabla(4\pi\rho) - \frac{1}{b}\frac{\partial}{\partial\tau}\left[\frac{4\pi(\rho\mathbf{u})}{b}\right],\end{aligned} \tag{4.20}$$

where $\mathbf{a} = d\mathbf{u}/d\tau$ is the particle acceleration. Thus, a new term arises when the proper-time of the charge is used to describe its fields. This makes it clear that the local clock encodes information about the particle's interaction that is unavailable when the clock of the observer, co-moving observer or the proper clock of the center of mass is used to describe the fields. The new term in equation (4.20) is dissipative, acts to oppose the acceleration, is zero when $\mathbf{a} = 0$ or perpendicular to $\mathbf{u}$. It also arises instantaneously with the force. Furthermore, this term does not depend on the nature of the force. This is exactly what one expects of the back reaction caused by the inertial resistance of the particle to accelerated motion and, according to Wheeler and Feynman [47], is precisely what is meant by radiation reaction.

Remark 4.9. *It is of particular interest that this implies a charged particle can distinguish between inertial and accelerating frames. Thus, an observer in an elevator can always determine if they are in a state of accelerated motion. From this point of view, it is no surprise that the 2.7 °K CMBR represents a unique preferred frame of rest.*

In [21] the $\mathbf{E}$ and $\mathbf{B}$ fields for were computed directly (using only retarded potentials), to give:

$$\mathbf{E}(\mathbf{x},\tau) = \frac{q\mathbf{r}_u\left(1-\mathbf{u}^2/b^2\right)}{s^3} + \frac{q\left[\mathbf{r}\times(\mathbf{r}_u\times\mathbf{a})\right]}{b^2s^3} + \frac{q\left(\mathbf{u}\cdot\mathbf{a}\right)\left[\mathbf{r}\times(\mathbf{u}\times\mathbf{r})\right]}{b^4s^3}$$

and

$$\mathbf{B} = \frac{q\left(\mathbf{r}_u\times\mathbf{r}\right)\left(1-\mathbf{u}^2/b^2\right)}{rs^3} + \frac{q\mathbf{r}\times\left[\mathbf{r}\times(\mathbf{r}_u\times\mathbf{a})\right]}{rb^2s^3} + \frac{q\left(\mathbf{u}\cdot\mathbf{a}\right)\left[\mathbf{r}\times\mathbf{u}\right]}{b^4s^3}.$$

It is easy to see that $\mathbf{B}$ is orthogonal to $\mathbf{E}$. Except for τ in place of t, the first two terms in the above two equations are the same as (19-13) and (19-14) in [37] (pg. 299). The last term in each case is new and arises because of the dissipative terms in the two equations (4.20). These terms are zero if $\mathbf{a}$ is zero or orthogonal to $\mathbf{u}$. In the first case, there is no radiation and the particle moves with constant velocity so that the field is massless. The second case depends on the creation of motion which keeps $\mathbf{a}$ orthogonal to $\mathbf{u}$ (for example a cyclotron). Since $\mathbf{r}\times(\mathbf{u}\times\mathbf{r}) = r^2\mathbf{u} - (\mathbf{u}\cdot\mathbf{r})\,\mathbf{r}$, we see that there is a component along the direction of propagation (longitudinal). (Thus, the $\mathbf{E}$ field has a longitudinal part.) This confirms our claim that the new dissipative term is equivalent to an effective mass. This means that the cause for radiation reaction comes directly from the use of the local clock to formulate Maxwell's equations. Thus, there is no need to assume advanced potentials, self-interaction or mass renormalization along with the Lorentz-Dirac equation in order to account for radiation reaction as is required when the observer clock is used (Dirac theory). Furthermore, no assumptions about a point particle or structure of the source are required (i.e., Poincaré stresses).

Remark 4.10. *We conjecture that this effective mass is the actual source of the photoelectric effect and that the photon is a real particle of non-zero (dynamical) mass, which travels with the fields but is not a part of the field in the normal sense. If this conjecture is correct, radiation from a cyclotron (of any frequency) exposed to a metal like surface will not produce photo electrons.*

4.2.3. Global Internal Dynamics In this section, we look at the motion of one particle as seen from the global internal point of view. We assume that the system is closed or is in equilibrium with a heat bath. In this case, we can assume that Hamiltonian for each particle can be written as:

$$H_i = H_{i0} + V_i = \sqrt{c^2\boldsymbol{\pi}_i^2 + m_i^2c^4} + V_i,$$

where $\boldsymbol{\pi}_i = \mathbf{p}_i - \frac{e_i}{c}\mathbf{A}_i$, $\mathbf{A}_i = \sum_{j\neq i}\mathbf{A}_{ji}$ and $V_i = \sum_{j\neq i}V_{ji}$. We assume that $\mathbf{A}_{ji}, V_{ji}$ represents the retarded action of the vector potential (respectively scalar potential) of the j-th particle on the i-th particle. Since we allow for n to vary and that a charge at e_i may not equal the same value of e_j, we must allow for $e_i\mathbf{A}_{ji} \neq e_j\mathbf{A}_{ij}$ (respectively, $V_{ji} \neq V_{ij}$). We also do not include the customary factor of 1/2 in our definition of the scalar and vector potentials acting on particle i (see [40]).

Recalling that $\mathbf{w}_i = d\mathbf{x}_i/dt$ and $\mathbf{u}_i = d\mathbf{x}_i/d\tau_i$, we define $\mathbf{v}_i = d\mathbf{x}_i/d\tau$. From our identities, its easy to see that

$$\frac{\mathbf{w}_i}{c} = \frac{\mathbf{v}_i}{b} = \frac{\mathbf{u}_i}{b_i} \Rightarrow \gamma_i^{-1} = \sqrt{1-\left(\tfrac{\mathbf{w}_i}{c}\right)^2} = \sqrt{1-\left(\tfrac{\mathbf{v}_i}{b}\right)^2} = \sqrt{1-\left(\tfrac{\mathbf{u}_i}{b_i}\right)^2}.$$

The velocity $\mathbf{v}_i$ is the one our observer sees when he uses the global canonical proper-clock τ of the system to compute the particle velocity, while $\mathbf{w}_i$ is the one seen when he uses his clock to compute the particle velocity. if $\mathbf{U}$ is zero, $b = c$ and, from the global perspective, our theory looks like the conventional one. As the system is closed, $\mathbf{U}$ is constant and τ is linearly related to t. Since $\gamma_i^{-1} = \frac{1}{b}\sqrt{\mathbf{U}^2 + c^2 - \mathbf{v}_i^2}$, the physical interpretation is very different if $\mathbf{U}$ is not zero. Furthermore, it is easy to see that, even if $\mathbf{U}$ is zero in one frame, it will not be zero in any other frame which is in relative motion. Using $K = \frac{H^2}{2Mc^2} + \frac{Mc^2}{2}$, the equations of motion are:

$$\begin{gathered}\mathbf{v}_i = \frac{d\mathbf{x}_i}{d\tau} = \frac{\partial K}{\partial \mathbf{p}_i} = \frac{H}{Mc^2}\frac{c^2\boldsymbol{\pi}_i}{H_{i0}} = \frac{b}{c}\frac{c^2\boldsymbol{\pi}_i}{H_{i0}} = \frac{b}{c}\frac{d\mathbf{x}_i}{dt}, \\ \frac{d\mathbf{p}_i}{d\tau} = \frac{\partial K}{\partial \mathbf{p}_i} = \frac{H}{Mc^2}\sum_{k=1}^{n}\left[\frac{c^2\boldsymbol{\nabla}_i\boldsymbol{\pi}_k}{H_{i0}} - \boldsymbol{\nabla}_i V_k\right] = \frac{b}{c}\sum_{k=1}^{n}\left[\frac{c^2\boldsymbol{\nabla}_i\boldsymbol{\pi}_k}{H_{i0}} - \boldsymbol{\nabla}_i V_k\right] = \frac{b}{c}\frac{d\mathbf{p}_i}{dt}.\end{gathered} \tag{4.21}$$

It is clear that the equations in (4.21) are mathematically equivalent. Factoring out the $k = i$ term $\frac{e_i}{b}\left[(\mathbf{v}_i \cdot \nabla_i)\mathbf{A}_i + \mathbf{v}_i \times (\nabla_i \times \mathbf{A}_i)\right]$, we have:

$$\begin{aligned}\frac{c}{b}\frac{d\mathbf{p}_i}{d\tau} &= \tfrac{e_i}{b}\left[(\mathbf{v}_i \cdot \nabla_i)\mathbf{A}_i + \mathbf{v}_i \times (\nabla_i \times \mathbf{A}_i)\right] \\ &+ \sum_{k\neq i}^{n}\left\{\frac{e_k}{b}\left[(\mathbf{v}_k \cdot \nabla_i)\mathbf{A}_k + \mathbf{v}_k \times (\nabla_i \times \mathbf{A}_k)\right] - \nabla_i V_k\right\}.\end{aligned} \tag{4.22}$$

Using

$$(\mathbf{v}_i \cdot \nabla_i)\mathbf{A}_i = \frac{d\mathbf{A}_i}{d\tau} - \frac{\partial \mathbf{A}_i}{\partial \tau},$$

equation (4.22) becomes

$$\begin{aligned}\frac{c}{b}\frac{d\mathbf{p}_i}{d\tau} - \tfrac{e_i}{b}\frac{d\mathbf{A}_i}{d\tau} &= \tfrac{e_i}{b}\left[\mathbf{v}_i \times \mathbf{B}_i\right] - \tfrac{e_i}{b}\frac{\partial \mathbf{A}_i}{\partial \tau} - \nabla_i V_i \\ &+ \sum_{k\neq i}^{n}\left\{\frac{e_k}{b}\left[(\mathbf{v}_k \cdot \nabla_i)\mathbf{A}_k + \mathbf{v}_k \times (\nabla_i \times \mathbf{A}_k)\right] - \nabla_i V_k\right\}.\end{aligned} \tag{4.23}$$

Note that equation (4.23) can also be written as:

$$\begin{aligned}\frac{d\mathbf{p}_i}{dt} - \tfrac{e_i}{c}\frac{d\mathbf{A}_i}{dt} &= \tfrac{e_i}{c}\left[\mathbf{w}_i \times \mathbf{B}_i\right] - \tfrac{e_i}{c}\frac{\partial \mathbf{A}_i}{\partial t} - \nabla_i V_i \\ &+ \sum_{k\neq i}^{n}\left\{\frac{e_k}{c}\left[(\mathbf{w}_k \cdot \nabla_i)\mathbf{A}_k + \mathbf{w}_k \times (\nabla_i \times \mathbf{A}_k)\right] - \nabla_i V_k\right\}.\end{aligned} \tag{4.24}$$

Thus, equations (4.23) and (4.24) are mathematically equivalent. Set $V_i = e_i\Phi_i$ and $\mathbf{E}_i = -\frac{1}{b}\left(\partial \mathbf{A}_i/\partial\tau\right) - \nabla_i\Phi_i$, then we can write:

$$\mathbf{F}_i = \frac{e_i}{b}\left(\mathbf{v}_i \times \mathbf{B}_i\right) - \frac{e_i}{b}\frac{\partial \mathbf{A}_i}{\partial \tau} - \nabla_i V_i = e_i\mathbf{E}_i + \frac{e_i}{b}\left(\mathbf{v}_i \times \mathbf{B}_i\right).$$

We can then write equation (4.23) as:

$$\begin{aligned}\frac{c}{b}\frac{d\pi_i}{d\tau} &= \mathbf{F}_i \\ &- \sum_{k\neq i}^{n}\left\{\frac{e_k}{b}\left[\left(\mathbf{v}_k \cdot \nabla_k\right)\mathbf{A}_{ik} + \mathbf{v}_k \times \left(\mathbf{v}_k \times \mathbf{A}_{ik}\right)\right] - \nabla_k V_{ik}\right\}.\end{aligned}$$

If we now use

$$\begin{aligned}\left(\mathbf{v}_k \cdot \nabla_k\right)\mathbf{A}_{ik} &= \frac{d\mathbf{A}_{ik}}{d\tau} - \frac{\partial \mathbf{A}_{ik}}{\partial \tau}, \quad \mathbf{B}_{ik} = \nabla_k \times \mathbf{A}_{ik}, \\ \mathbf{E}_{ik} &= -\frac{1}{b}\frac{\partial \mathbf{A}_{ik}}{\partial \tau} - \nabla_k\Phi_{ik}, \quad \mathbf{F}_{ik} = e_k\mathbf{E}_{ik} + \frac{e_k}{b}\left(\mathbf{v}_k \times \mathbf{B}_{ik}\right),\end{aligned}$$

the above becomes:

$$\frac{c}{b}\frac{d\boldsymbol{\pi}_i}{d\tau} = \mathbf{F}_i - \sum_{k\neq i}^{n}\left[\mathbf{F}_{ik} - \frac{e_k}{b}\frac{d\mathbf{A}_{ik}}{d\tau}\right]. \tag{4.25}$$

If we simplify and put the last term on the other side, we have:

$$\frac{c}{b}\sum_{i=1}^{n}\frac{d\pi_i}{d\tau} + \sum_{i=1}^{n}\sum_{k\neq i}^{n}\frac{e_k}{b}\frac{d\mathbf{A}_{ik}}{d\tau} = \sum_{i=1}^{n}\mathbf{F}_i - \sum_{i=1}^{n}\sum_{k\neq i}^{n}\mathbf{F}_{ik}.$$

Performing the summations on both sides give us:

$$\begin{aligned}&\frac{c}{b}\sum_{i=1}^{n}\frac{d\pi_i}{d\tau} + \sum_{i=1}^{n}\frac{e_k}{b}\frac{d\mathbf{A}_{ik}}{d\tau} = 0 \Rightarrow \\ &\sum_{i=1}^{n}\frac{d\mathbf{p}_i}{d\tau} = 0 = \frac{d\mathbf{P}}{d\tau}.\end{aligned}$$

4.2.4. Discussion We want to first discuss the relationship between equation (4.17) and equation (4.25). For comparison, we first rewrite equation (4.17) with its indices:

$$\frac{c}{b_i}\frac{d\boldsymbol{\pi}_i}{d\tau_i} = \left[e_i\mathbf{E}_i + \tfrac{e_i}{b_i}\left(\mathbf{u}_i \times \mathbf{B}_i\right)\right] = \frac{d\boldsymbol{\pi}_i}{dt}. \tag{4.26}$$

If we use

$$\mathbf{F}_i = e_i\mathbf{E}_i + \frac{e_i}{b}\left(\mathbf{v}_i \times \mathbf{B}_i,\right).$$

We can write equation (4.25) as

$$\frac{d\boldsymbol{\pi}_i}{dt} = \frac{c}{b}\frac{d\boldsymbol{\pi}_i}{d\tau} = \left[e_i\mathbf{E}_i + \frac{e_i}{b}\left(\mathbf{v}_i \times \mathbf{B}_i,\right)\right] - \sum_{k\neq i}^{n}\left[\mathbf{F}_{ik} - \frac{e_k}{b}\frac{d\mathbf{A}_{ik}}{d\tau}\right]. \tag{4.27}$$

Equation (4.26) represents one particle in a field of force, as seen locally. It does not react via action-at-a-distance, its reaction shows up in its field via the additional term in its wave equation. When we look at the same particle from the center of mass frame (equation (4.27)), we see the force which acts on the particle and the delayed action-at-a-distance reaction force of the particle on each of the other particles in the system.

Thus, the extra term on the (far) right-hand side of equation (4.27) is the back reaction field of the i-th particle on each of the other particles in the system. This is the term that accounts for radiation reaction as action-at-a-distance. There is no self-energy (divergence), advanced potentials or any assumptions about the size or structure of the particles. It is important to point out that the mathematical equivalence is manifest in both cases (using t or τ for the center of mass). However, we must use τ_i at the local level (see Theorem 3.1).

It clear that equation (4.27) is consistent with conservation of global momentum. This along with conservation of total energy implies the following:

Theorem 4.11. (Wheeler-Feynman) *If we use the τ_i at the local level with either the $(\mathbf{X}, t)$ or $(\mathbf{X}, \tau)$ variables, the closed system of interacting charged particles exchange energy and momentum via fields at the local level and by action-at-a-distance as seen from the global internal view. In addition, at the global level, all emitted energy and momentum is absorbed internally.*

Thus, for any closed system of interacting particles, the absorption hypothesis of Wheeler and Feynman is automatically satisfied, and without the use of advanced potentials.

Remark 4.12. *Wheeler and Feynman [47] conjectured that action-at-a-distance and field theory represented different sides of the same theory. The above discussion explains why and how their conjecture is true. At the local level the individual particle sees itself immersed in a electromagnetic field, while an observer at the center of mass sees the particle being acted on by all the other particles and, the corresponding direct (delayed) back action by the particle on each of the others. This suggests that, since Maxwell's equations are spin* 1 *relativistic wave equations, the appearance of the $2.7\,^\circ K$ CMBR is a natural consequence of delayed interaction, allowing some photons to miss their mark. This would also help to explain why the probability interpretation in quantum mechanics works so well.*

4.3. Conclusion II

In this section, we have:

(i) Provided a relativistic unification of the Newtonian and Maxwell theories.

(ii) Resolved the problems associated with radiation reaction, self-energy divergence, field-particle dichotomy and advanced interactions in CED.

Recall that the self energy divergence in QED came from CED and that Feynman has shown that if the photon has a small mass, the infrared divergence in QED would disappear. It follows that a second quantized version of the dual theory will not have a self-energy or infrared divergence. Thus, this along withe the solution to Dyson's conjectures provides both understanding and correction for all the physical and mathematical problems associated with the divergencies in QED.

5. The Dual Quantum Theory

In the classical theory, the equations of motion were shown to be mathematically equivalent. In this section we introduce the dual quantum theory. In this case, the dual quantum theory is not mathematically equivalent.

5.1. The Clock Relationship

In order to introduce the dual quantum theory, we need a basic relationship between the global system clock and the clocks of the individual particles. To derive this relationship, return to our definition of the global Hamiltonian K and let W be a observable. In this case, we know that

$$\begin{aligned} \frac{dW}{d\tau} &= \{K, W\} = \frac{H}{Mc^2}\{H, W\} = \frac{H}{Mc^2}\sum_{i=1}^{n}\{H_i, W\} \\ &= \frac{H}{Mc^2}\sum_{i=1}^{n}\frac{m_ic^2}{H_i}\left[\frac{H_i}{m_ic^2}\{H_i, W\}\right] = \sum_{i=1}^{n}\frac{Hm_i}{MH_i}\{K_i, W\}. \end{aligned} \tag{5.1}$$

Using the (easily derived) fact that $d\tau_i/d\tau = Hm_i/MH_i = b_i/b$, we get

$$\frac{dW}{d\tau} = \sum_{i=1}^{n}\frac{d\tau_i}{d\tau}\{K_i, W\}. \tag{5.2}$$

Equation (5.2) allows us to relate the global system dynamics to the local systems dynamics. Let us combine equations (5.1) and (5.2), to get:

$$d\tau\{K, W\} \equiv \sum_{i=1}^{n} d\tau_i\{K_i, W\}. \tag{5.3}$$

5.2. Many Particle Theory

Let $\Psi(\mathbf{X}, t) = \Phi(\mathbf{x}_1, \mathbf{x}_2 \cdots \mathbf{x}_n; \tau_1, \tau_2 \cdots \tau_n)$ be a global wave function which satisfies:

$$i\hbar\frac{\partial\Psi}{\partial t} = H\Psi. \tag{5.4}$$

From the Poisson and Heisenberg bracket relationship, $\{\cdot,\cdot\} \to \frac{i}{\hbar}[\cdot,\cdot]$, and the first equation in (5.1), we also have that:

$$i\hbar\frac{\partial\Psi}{\partial\tau} = K\Psi. \tag{5.5}$$

From equation (5.2) or (5.3), for each i we also have:

$$i\hbar\frac{\partial\Psi}{\partial\tau_i} = K_i\Psi \tag{5.6}$$

5.3. Quantum Background

In this section we briefly discuss some the known, not known and misunderstood issues in the foundations of relativistic quantum theory.

5.3.1. The Klein-Gordon Equation

It is well-known that Schrödinger begin his work with the square root equation:

$$i\hbar\frac{\partial\Psi}{\partial t} = \sqrt{c^2\mathbf{p}^2 + m^2c^4}\,\Psi \tag{5.7}$$

However, no one knew how to directly relate this equation to physically important problems. Furthermore, this equation is nonlocal, meaning, in the terminology of the times (1920-1930), that it is represented by a power series in the momentum operator. Historically,

Schrödinger, Gordon, Klein and others attempted to circumvent this problem by starting with the relationship:

$$(H - V)^2 = c^2 \left[\mathbf{p}^2 - (e/c)\,\mathbf{A}\right]^2 + m^2c^4,$$

which led to the Klein-Gordon equation. At that time, the hope was to construct a relativistic quantum theory that would provide a natural extension of the classical theory. Schrödinger, the first to consider this approach, drop it and expanded the square root equation, taking the first two terms to derive the equation now named for him. His motivation was that the eigenvalue solutions did not correspond to the spectrum of Hydrogen. In addition, a particle interpretation was difficult because the conserved probability could take on negative values. The problems seemed insurmountable and the equation was dropped from serious consideration for a few years, but later regained favor when quantum field theory was developed.

5.3.2. The Dirac Equation The first successful attempt to resolve the question of how best to handle the square-root equation was made by Dirac in 1926 (see [8]). Dirac noted that the Pauli matrices could be used to write

$$c^2\mathbf{p}^2 + m^2c^4 = \left[c\boldsymbol{\alpha}\cdot\mathbf{p} + \boldsymbol{\beta}mc^2\right]^2, \quad \text{where,} \quad \boldsymbol{\beta} = \begin{bmatrix} I_2 & 0 \\ 0 & -I_2. \end{bmatrix} \tag{5.8}$$

and the matrix $\boldsymbol{\alpha}$ is defined by $\boldsymbol{\alpha} = (\alpha_1, \alpha_2, \alpha_3)$, with

$$\alpha_i = \begin{pmatrix} \mathbf{0} & \sigma_i \\ \sigma_i & \mathbf{0} \end{pmatrix}, \ \sigma_1 = \begin{pmatrix} 0 & 1 \\ 1 & \mathbf{0} \end{pmatrix}, \ \sigma_2 = \begin{pmatrix} 0 & -i \\ i & \mathbf{0} \end{pmatrix}, \ \sigma_3 = \begin{pmatrix} 1 & \mathbf{0} \\ \mathbf{0} & -1 \end{pmatrix}$$

Thus, Dirac showed that an alternative representation of equation (5.7) could be taken as

$$i\hbar\frac{\partial\Psi}{\partial t} = \left[c\boldsymbol{\alpha}\cdot\mathbf{p} + \boldsymbol{\beta}mc^2\right]\Psi \tag{5.9}$$

In this case, Ψ had to be viewed as a vector-valued function or spinor. To be more precise, $\Phi \in L^2(\mathbb{R}^3, C^4) = L^2(\mathbb{R}^3) \otimes C^4$ is a four-component column vector $\Psi = (\psi_1, \psi_2, \phi_1, \phi_2)^t$. In this approach, $\psi = (\psi_1, \psi_2)^t$ represented the particle (positive energy) component and $\phi = (\phi_1, \phi_2)^t$ represents the antiparticle (negative energy) component of the theory.

The Dirac equation was very successful, providing the known spectrum for Hydrogen and the correct spin value for the electron. The equation also had a problem: the negative energy solutions. This problem later led to Dirac's hole theory, where a hole was interpreted as an electron with negative energy and positive charge. This was later confirmed by the discovery of the positron. A very good discussion of the times and difficulties is best told by Schweber [46].

One might naturally suspect that Dirac would have also noticed that the positron could be interpreted as an electron moving backward in time after developing his hole theory. However, according to Schweber (see pg. 38-388 [46]), in 1940 Wheeler suggested to Feynman that one might interpret a positron as an electron moving backwards in proper time, with positive charge. He also points out that Wheeler was unaware of Stueckelberg's later work on the same subject in a 1941 paper (see pg. 388 in [46]).

In [24], Gill, Zachary and Alfred showed how to construct an exact analytical separation (diagonalization) of the full (minimal coupling) Dirac equation into particle and antiparticle components. The diagonalization was analytic in the sense that it was achieved without transforming the wave functions, as is done by the Foldy-Wouthuysen method, and revealed the nonlocal time behavior of the particle-antiparticle relationship. This led them to another interpretation of the zitterbewegung and the fact that the expected value of a velocity measurement of a Dirac particle at any instant of time is $\pm c$; namely that a Dirac particle jumps backward and forward in time at speed c, so that it can make an extended object in space appear as a point in the instantaneous present.

They also explicitly showed that the Pauli equation is not valid for the study of the Dirac hydrogen atom problem in s-states. It was concluded that there were some open mathematical problems with any attempt to show that the Dirac equation was insufficient to explain the full hydrogen spectrum. They further showed that if their perturbation method could be justified, their analysis shows that the use of cut-offs in QED is already justified by the eigenvalue analysis that supports it.

5.3.3. The Square Root Equation In 2005 Gill and Zachary [23] used the theory of fractional powers of linear operators to construct a general (analytic) representation for the square-root energy operator which is valid for all spin values. They focused on the spin 1/2 case along with a few simple yet solvable and physically interesting examples, in order to understand how to interpret the operator. Their general representation was shown to be uniquely determined by the Green's function of the corresponding Schrödinger equation. It was found that the operator had a representation as a nonlocal composite of at least three terms, depending on the magnetic field, which becomes singular at a point. In the simplest case there are two negative parts and one (hard core) positive part confined within a Compton wavelength and at the point of singularity they cancel each other providing a finite result. Furthermore, the operator could be treated like the identity outside a few Compton wavelengths. They concluded that the Dirac and square root operators cannot be seen as physically equivalent (even in the free case).

Remark 5.1. *We note that the above result is another case where mathematical equivalence is not the same as physical equivalence at the quantum level.*

5.4. Dual Quantum Equations

In this section we identify three possible dual relativistic quantum Hamiltonians for the dual theory. First we want prove the follow theorem, which relates solutions to the standard eigenvalue problem to those of the dual problem.

Theorem 5.2. *Suppose that λ is a eigenvalue for the equation:*

$$\lambda\psi = H\psi,$$

then, for the same wave function ψ the dual equation satisfies:

$$\left[\frac{\lambda^2}{2mc^2} + \frac{mc^2}{2}\right]\psi = K\psi.$$

Proof. The proof follows from direct computation, start with $\frac{\lambda^2}{2}\psi = H^2/2\psi$ and then add $\frac{mc^2}{2}\psi$ to both sides. □

Remark 5.3. *We note that in general, the eigenfunctions for the second equation will have better smoothness properties than those for the first.*

For an explicit theory, we need to consider how to introduce coupling at the quantum level. First, we return to equation (5.6), drop the index and write it as:

$$i\hbar\frac{\partial\Psi}{\partial\tau} = K\Psi = \left[\frac{H^2}{2mc^2} + \frac{mc^2}{2}\right]\Psi. \tag{5.10}$$

In addition to the Dirac Hamiltonian, there are two other possible Hamiltonians, depending on the way the potential can appear with the square-root operator:

$$\beta\sqrt{c^2\boldsymbol{\pi}^2 - ec\hbar\boldsymbol{\Sigma}\cdot\mathbf{B} + m^2c^4} + V \tag{5.11}$$

and

$$\beta\sqrt{c^2\boldsymbol{\pi}^2 - ec\hbar\boldsymbol{\Sigma}\cdot\mathbf{B} + (mc^2 + \beta V)^2}. \tag{5.12}$$

This gives us three possible dual relativistic particle equations for spin-$\frac{1}{2}$ particles.

(i) The dual Dirac equation:

$$\begin{aligned} i\hbar\frac{\partial\Psi}{\partial\tau} = &\left\{\frac{\boldsymbol{\pi}^2}{2m} + \beta V + mc^2 - \frac{e\hbar\boldsymbol{\Sigma}\cdot\mathbf{B}}{2mc}\right.\\ &\left.+\frac{V\boldsymbol{\alpha}\cdot\boldsymbol{\pi}}{mc} - \frac{i\hbar\boldsymbol{\alpha}\cdot\nabla V}{2mc} + \frac{V^2}{2mc^2}\right\}\Psi. \end{aligned} \tag{5.13}$$

(ii) The dual version of the square-root equation, using the first possibility:

$$\begin{aligned} i\hbar\frac{\partial\Psi}{\partial\tau} &= \left\{\frac{\boldsymbol{\pi}^2}{2m} - \frac{e\hbar\boldsymbol{\Sigma}\cdot\mathbf{B}}{2mc} + mc^2 + \frac{V^2}{2mc^2}\right\}\Psi \\ &+\frac{V\beta\sqrt{c^2\boldsymbol{\pi}^2 - ec\hbar\boldsymbol{\Sigma}\cdot\mathbf{B} + m^2c^4}}{2mc^2}\Psi + \frac{\beta\sqrt{c^2\boldsymbol{\pi}^2 - ec\hbar\boldsymbol{\Sigma}\cdot\mathbf{B} + m^2c^4}}{2mc^2}V\Psi. \end{aligned} \tag{5.14}$$

(iii) The dual version of the square-root equation, using the second possibility:

$$i\hbar\frac{\partial\Psi}{\partial\tau} = \{\frac{\boldsymbol{\pi}^2}{2m} + \beta V + mc^2 - \frac{e\hbar\boldsymbol{\Sigma}\cdot\mathbf{B}}{2mc} + \frac{V^2}{2mc^2}\}\Psi. \tag{5.15}$$

If $\mathbf{A}$ and V are zero, all equations reduce to:

$$i\hbar\frac{\partial\Psi}{\partial\tau} = \left\{\frac{\mathbf{p}^2}{2m} + mc^2\right\}\Psi, \tag{5.16}$$

which is the Schrödinger equation with an added mass term. This makes it easy to see that, in all cases, K is positive definite. In mathematical terms, the lower order terms are relatively bounded with respect to $\mathbf{p}^2/2m$. It follows that, unlike the Dirac and Klein-Gordon approach, we can interpret these equations as representations for actual particles. In the above equations, we have assumed that V is time independent. (However, since $\mathbf{A}(\mathbf{x},\tau)$ can have general time-dependence, $\sqrt{c^2\boldsymbol{\pi}^2 - ec\hbar\boldsymbol{\Sigma}\cdot\mathbf{B} + m^2c^4}$ need not be related to the Dirac operator by a Foldy-Wouthuysen type transformation.)

5.5. The Dual Dirac Theory

In this section, we restrict our investigation to the dual Dirac equation. Let $\mathbf{s}_p$ and $\boldsymbol{\mu}_p = 2\mu_p\mathbf{s}_p$ be the proton spin and magnetic moment operators respectively. Let $r_0 = e^2/mc^2$ be the classical electron radius, $\alpha = \frac{e^2}{\hbar c}$ be the fine structure constant and let $\boldsymbol{\alpha} = (\alpha_1, \alpha_2, \alpha_3)$ be the standard Dirac matrix. The potentials can be written as $V_0 = -mc^2r_0/r$, $\mathbf{A} = \boldsymbol{\mu}_p \times \mathbf{r}/r^3$, where the spin orientation is along the z-axis (i.e., $A_r = A_\theta = 0$ and $A_\phi = \frac{2\mu_p s_p \sin\theta}{r^2}$). In what follows, $\boldsymbol{\pi} = \mathbf{p} - \frac{e}{c}\mathbf{A}$ and π is the area of the unit circle.

The eigenvalue problem for the Dirac equation $\lambda\Psi = H_D\Psi$, with $\Psi = [\psi_1, \psi_2]$, can be written as:

$$\begin{aligned}(\lambda - V - mc^2)\psi_1 &= c(\sigma \cdot \boldsymbol{\pi})\psi_2 \\ (\lambda - V + mc^2)\psi_2 &= c(\sigma \cdot \boldsymbol{\pi})\psi_1.\end{aligned} \tag{5.17}$$

Solving the second equation for ψ_2 we have:

$$\psi_2 = c\left[\lambda - V_0 + mc^2\right]^{-1}(\sigma \cdot \boldsymbol{\pi})\,\psi_1 \tag{5.18}$$

5.5.1. The Dual Dirac Equation With $H_D = c\boldsymbol{\alpha} \cdot \boldsymbol{\pi} + mc^2\beta + V_0 = H_0 + V_0$, let $V = \frac{1}{2mc^2}[H_0V_0 + V_0H_0]$. Then, we can write the dual Dirac Hamiltonian as:

$$K_D = \frac{H_D^2}{2mc^2} + \frac{mc^2}{2} = \frac{\boldsymbol{\pi}^2}{2m} + V - \frac{e\hbar\Sigma \cdot \mathbf{B}}{2mc} + mc^2 + \frac{V_0^2}{2mc^2}, \tag{5.19}$$

5.5.2. The Eigenvalue Problem The general eigenvalue problem is:

$$\begin{aligned}E\Psi = &\left\{\frac{\boldsymbol{\pi}^2}{2m} + \beta V_0 + mc^2 - \frac{e\hbar\Sigma \cdot \mathbf{B}}{2mc}\right. \\ &\left.+\frac{V_0\alpha \cdot \boldsymbol{\pi}}{mc} - \frac{i\hbar\alpha \cdot \nabla V_0}{2mc} + \frac{V_0^2}{2mc^2}\right\}\Psi.\end{aligned} \tag{5.20}$$

Where, as before $\Psi = [\psi_1, \psi_2]^t$, with ψ_1, ψ_2 the upper and lower spinor components. For further analysis, it is convenient to split (5.20) into two equations:

$$\begin{aligned}E\psi_1 = &\left\{\frac{\boldsymbol{\pi}^2}{2m} + V + mc^2 - \frac{e\hbar\boldsymbol{\sigma} \cdot \mathbf{B}}{2mc} + \frac{V_0^2}{2mc^2}\right\}\psi_1 \\ &+\left\{\frac{V_0\boldsymbol{\sigma} \cdot \boldsymbol{\pi}}{mc} - \frac{i\hbar\boldsymbol{\sigma} \cdot \nabla V_0}{2mc}\right\}\psi_2 \\ E\psi_2 = &\left\{\frac{\boldsymbol{\pi}^2}{2m} - V + mc^2 - \frac{e\hbar\boldsymbol{\sigma} \cdot \mathbf{B}}{2mc} + \frac{V_0^2}{2mc^2}\right\}\psi_2 \\ &+\left\{\frac{V_0\boldsymbol{\sigma} \cdot \boldsymbol{\pi}}{mc} - \frac{i\hbar\boldsymbol{\sigma} \cdot \nabla V_0}{2mc}\right\}\psi_1.\end{aligned} \tag{5.21}$$

If we now use equation (5.18), in (5.21), we can now drop the second equation in (5.21) and convert the first to the stationary case, (using ψ) to get the eigenvalue equation:

$$\begin{aligned}E\psi = &\left\{\frac{\boldsymbol{\pi}^2}{2m} + V_0 + mc^2 - \frac{e\hbar\boldsymbol{\sigma} \cdot \mathbf{B}}{2mc} + \frac{V_0^2}{2mc^2}\right\}\psi \\ &+\left\{\frac{V_0\boldsymbol{\sigma} \cdot \boldsymbol{\pi}}{mc} - \frac{i\hbar\boldsymbol{\sigma} \cdot \nabla V_0}{2mc}\right\}\frac{c\boldsymbol{\sigma} \cdot \boldsymbol{\pi}}{(\lambda - V_0 + mc^2)}\psi.\end{aligned}$$

Expanding, we have:

$$\begin{aligned}E\psi = &\left\{\frac{\boldsymbol{\pi}^2}{2m} + V_0 + mc^2 - \frac{e\hbar\boldsymbol{\sigma} \cdot \mathbf{B}}{2mc} + \frac{V_0^2}{2mc^2}\right\}\psi - \frac{i\hbar\,(\boldsymbol{\sigma} \cdot \nabla V_0)\,(\boldsymbol{\sigma} \cdot \boldsymbol{\pi})}{2m\,(\lambda - V_0 + mc^2)}\psi \\ &+\frac{V_0\,(\boldsymbol{\sigma} \cdot \mathbf{p}V_0)\,(\boldsymbol{\sigma} \cdot \boldsymbol{\pi})}{m\,(\lambda - V_0 + mc^2)^2}\psi + \frac{V_0\,\,(\boldsymbol{\sigma} \cdot \boldsymbol{\pi})\,(\boldsymbol{\sigma} \cdot \boldsymbol{\pi})}{m\,(\lambda - V_0 + mc^2)}\psi.\end{aligned} \tag{5.22}$$

Since the binding energy in Hydrogen is 13ev and the rest mass of the electron is 5×10^5ev, the ratio is 2.6×10^{-5}. Thus, with $r_0 = \frac{e^2}{mc^2}$, there is little loss if we replace the denominator $\lambda - V + mc^2$ by $2mc^2(1 + \frac{r_0}{r})$ This allow us to by-pass the non-linear eigenvalue problem and equation (5.22) becomes:

$$E\psi = \left\{ \frac{\boldsymbol{\pi}^2}{2m} + V_0 + mc^2 - \frac{e\hbar\boldsymbol{\sigma}\cdot\mathbf{B}}{2mc} + \frac{V_0^2}{mc^2} \right\}\psi - \frac{i\hbar\left(\boldsymbol{\sigma}\cdot\nabla V_0\right)\left(\boldsymbol{\sigma}\cdot\boldsymbol{\pi}\right)}{4m^2c^2\left(1+\frac{r_0}{2r}\right)}\psi$$
$$+\frac{V_0\left(\boldsymbol{\sigma}\cdot\mathbf{p}V_0\right)\left(\boldsymbol{\sigma}\cdot\boldsymbol{\pi}\right)}{4m^3c^4\left(1+\frac{r_0}{2r}\right)^2}\psi + \frac{V_0\left(\boldsymbol{\sigma}\cdot\boldsymbol{\pi}\right)\left(\boldsymbol{\sigma}\cdot\boldsymbol{\pi}\right)}{2m^2c^2\left(1+\frac{r_0}{2r}\right)}\psi. \tag{5.23}$$

It is clear that a cut-off is required since the denominators in the last three terms are undefined at $r = 0$. The analysis of equation (5.23) can be found in [18]. In preparation for the next section we want focus on the fourth term in the parenthesis:

$$-\frac{e\hbar\boldsymbol{\sigma}\cdot\mathbf{B}}{2mc}$$

and the last term in (5.23),

$$\frac{V_0\left(\boldsymbol{\sigma}\cdot\boldsymbol{\pi}\right)\left(\boldsymbol{\sigma}\cdot\boldsymbol{\pi}\right)}{2m^2c^2\left(1+\frac{r_0}{2r}\right)}.$$

After some analysis of the above term and combining part of it with the first term we have (see [18]),

$$\frac{4er_0\hbar\sigma\cdot\mathbf{B}}{2mc\left(2r+r_0\right)} - \frac{e\hbar\sigma\cdot\mathbf{B}}{2mc} = -\left[1 - \frac{4r_0}{(2r+r_0)}\right]\frac{e\hbar\sigma\cdot\mathbf{B}}{2mc}, \tag{5.24}$$

In the remainder of the section, we focus on the implications of equation (5.24) and the anomalous magnetic moment.

5.5.3. Anomalous Magnetic Moment In this section, we investigate the equation (5.24) under the assumption that the charged, spin-1/2 particle does not possess any internal structure (a Dirac particle). In this case, the spin magnetic moment is given by:

$$\boldsymbol{\mu} = g\frac{e}{2mc}\mathbf{s} = g\mu_B\mathbf{s},$$

where $\hbar\mathbf{s} = \frac{\hbar\sigma}{2}$ is the intrinsic spin operator. We can also write equation (5.24) as

$$H_a = 2\left[1 - \frac{4r_0}{(2r+r_0)}\right]\mu_B\mathbf{s}\cdot\mathbf{B} \tag{5.25}$$

Thus, we have that:

$$g_r = 2\left[1 - \frac{4r_0}{(2r+r_0)}\right]. \tag{5.26}$$

Recall that the fractional part of equation (5.26) came from replacing $\lambda - V_0 + mc^2$ by $2mc^2 + \frac{e^2}{r}$ in equation (5.18). This replacement is not valid if $r = 0$, so a cutoff is required. If we take the cutoff at $r = \frac{r_0}{2}$, then $g = -2$, while if we take the cut off at $g = \lim_{r\to 0} g_r$, we obtain $g = -6$. Taking $r_e = 0.499857150068631 \times r_0$, we obtain the correct experimental result:

$$g = -2.00231930436256.$$

(We note that the value for r_e is very close to $\frac{r_0}{2}$.) If we treat the muon and proton phenomenologically we can also obtain their exact g-factors:

$$g_\mu^a = 2\left[1 - \frac{4r_0^\mu}{(2r_\mu + r_0^\mu)}\right]$$

$$g_p^a = -2\left[1 - \frac{4r_0^p}{(2r_p + r_0^p)}\right], \tag{5.27}$$

where $r_0^\mu = \frac{e^2}{m_\mu c^2}$ and $r_0^p = \frac{e^2}{m_p c^2}$.

The neutron also has a magnetic moment and g-factor. However, since it has zero charge, $r_0^n = 0$. If we make the additional assumption that $r_0^n = c_n r_0^p$ for some constant c_n we can obtain a phenomenologically g-factor for the neutron:

$$g_n^a = 2\left[1 - \frac{4r_0^p}{(2r_n + c_n r_0^p)}\right].$$

5.5.4. Discussion At the classical level we find that the standard and dual theories are mathematically equivalent. At the quantum level, the dual Dirac equation is not mathematically equivalent to the Dirac equation. The dual Dirac equation is strictly positive definite, so that there are no problems with using it as a particle equation. However, we must now directly face the existence of antiparticles.

In order to do this, let us first revisit our conceptual view of the real numbers and their representation. Recall that a field is a set $\mathbb{F}$ that has two binary operations $\oplus$ and $\odot$ that satisfies all our common experience with real numbers. Formally:

Definition 5.4. *The real numbers is a triplet* $(\mathbb{R}, +, \cdot)$*, which is a field, with* 0 *as the additive identity* (*i.e.*, $a + 0 = a$ for all $a \in \mathbb{R}$) *and* 1 *as the multiplicative identity* (*i.e.*, $a \cdot 1 = a$ for all $a \in \mathbb{R}$).

This structure was designed by mathematicians without regard to its possible use in physics. Santilli [43] defined the isodual number field for use in physics and that is what we need.

Definition 5.5. *The isodual real numbers* $(\hat{\mathbb{R}}, +, *)$ *is a field, with* $0 = \hat{0}$ *as the additive identity (i.e.,* $\hat{a} + \hat{0} = \hat{a}$ *for all* $-a = \hat{a} \in \hat{\mathbb{R}}$*) and* $\hat{1} = -1$ *as the multiplicative identity (i.e.,* $\hat{a} * \hat{1} = (-a)(-1)(-1) = \hat{a}$ *for all* $\hat{a} \in \hat{\mathbb{R}}$*).*

We note that we can obtain the isodual of any physical quantity $\hat{A}$ from the equation $A + \hat{A} = 0$.

In our theory, the evolution of a particle is formally defined on a Hilbert space $\mathcal{H}$ over the complex numbers $\mathbb{C} = \mathbb{R} + i\mathbb{R}$, with Hamiltonian K by the equation

$$i\hbar\frac{\partial\psi}{\partial\tau} = K\psi.$$

The conjugate equation is:

$$-i\hbar\frac{\partial\psi^*}{\partial\tau} = K\psi^*.$$

If we use $\hat{C}$ as our number field, we can write the above equation as:

$$\hat{i} * \hat{\hbar} * \frac{\partial\psi^*}{\partial\hat{\tau}} = \hat{K} * \psi^*$$

This approach allows us to naturally view anti-particles as particles with their proper time reversed (as first suggested by Wheeler) and their evolution defined on $\mathcal{H}^*$ over $\hat{\mathcal{C}}$. This does not imply that the time of the observer is reversed.

Remark 5.6. *Santilli [43] has shown that charge conjugation and isoduallity are equivalent for the particle-antiparticle symmetry operation.*

5.6. Conclusion III

In this section we have introduced the dual relativistic quantum theory corresponding to Einstein's special theory of relativity and Maxwell's field theory [18]. The dual classical theory was shown to be mathematically equivalent, however the dual quantum theory is not. We have found three distinct dual relativistic single particle wave equations for a spin-1/2 particle that reduce to the Schrödinger equation when minimal coupling is turned off. We have focused on the dual Dirac equation and used it to derive a new formula for the g-factor of a spin-1/2 particle. This allowed us to obtain the exact value for the electron g-factor. The formula can also be applied to the muon and the proton. Using the isodual numbers of Santilli [43], we have shown that our theory naturally interprets antiparticles as particles moving backwards in their proper time.

6. Conclusion

This paper has been a review of recent research on the foundations of quantum electrodynamics. In the second section we gave two proofs for the existence of a universal clock, which we called the NHF-clock. The first proof was based on the cosmological principle and the second was based on the existence of the 2.7 °K CMBR. We also showed that the CMBR implies that the universe is closed and that quantum mechanics is operative on the cosmological scale.

The NHF-time was then used to give a mathematically and physically correct representation theory for the Feynman operator calculus. Feynman's calculus was then used to prove the last two remaining conjectures of Dyson on the foundations for QED: that the ultra-violet divergence is caused by a violation of Heisenberg's uncertainty principle, and that the renormalized perturbation series is asymptotic.

The third section introduced the Einstein dual theory, which made it possible to give the first relativistic unification of the Newtonian and Maxwell theories, while also resolving long standing problems associated with radiation reaction, self-energy divergence, field-particle dichotomy and advanced interactions in CED. A major outcome is that, a second quantized version of the dual theory will not have a self-energy or infrared divergence. This along with the solution to Dyson's conjectures provides both understanding and correction for all the physical and mathematical problems associated with in QED.

The fourth section introduced the dual relativistic quantum theory, which is not mathematically equivalent. The dual theory has three distinct relativistic single particle wave equations for a spin-1/2 particle that all reduce to the Schrödinger equation when minimal coupling is turned off. We showed that the dual Dirac equation allowed us to derive a new formula for the g-factor of a spin-1/2 particle. We used it to obtain the exact value for the electron g-factor. The formula can also be used for the muon and the proton g-factors. Using the isodual numbers of Santilli, we have shown that our theory naturally interprets antiparticles as particles moving backwards in their (local) proper time.

Acknowledments

We would like to thank Professors Netsivi Ben-Amots, Alexander Gersten, Martin C. Land, Elliot Leib and Ruggero M. Santilli for their continued interest and support.

This paper is dedicated in honor of Professor Lawrence P. Horwitz (Larry) on the occasion of his ninety-second birthday. The first author would like to register his deep admiration and affection for the many kind ways the Larry has freely given support, helpful suggestions and advice over the last 26 years. He has always conducted himself in a manner that represents the best model of what a scientific professional should be.

Declaration
The authors certify that:

(i) they have no relevant financial or non-financial interests to disclose;
(ii) they have no conflicts of interest to declare that are relevant to the content of this manuscript;
(iii) they have no financial or proprietary interests in any material discussed in this manuscript; and
(iv) they have no affiliations with or involvement in any organization or entity with any financial interest or non-financial interest in the subject matter or materials discussed in this manuscript.

References

[1] A. Salam, *Overlapping divergence and the S-matrix*, Phys. Rev. **82**, (1951) 217-227.
[2] L. C. M. Benjamin, R. M. Antonio and A. P. Gonzalo, *The 2.7 °K blackbody radiation background reference frame*, Chin. Phys. B **19** No. 4 (2010) 04XXXX1-5.
[3] B. Javanmardi; C. Porciani; P. Kroupa; J. Pflamm-Altenburg *Probing the Isotropy of Cosmic Acceleration Traced By Type Ia Supernovae* The Astrophysical Journal Letters. **810** (1) (2015) 47.
[4] H.R Brown, Eur. J. Phys. **26**, S85 (2005).
[5] B. Bakamjian and L. H. Thomas, Phys. Rev. **92**, 1300 (1953).
[6] D. G. Currie, T. F. Jordan, and E. C. G. Sudarshan, Rev. Mod. Phys. **35**, 350 (1963).
[7] P. A. M. Dirac, *Classical theory of radiating electrons*, Proceedings of the Royal Soc. of London A **167** (1938) 148-168.
[8] P. A. M. Dirac, Proc. Roy. Soc (London) **A117** (1928) 610, **A118** (1928) 351.
[9] Einstein Centennial Symposium, October 1-5, 1979, University of Illinois at Carbondale, USA.
[10] F. Dyson, *The S-matrix in quantum electrodynamics*, Phys. Rev. **75** (1949), 1736-1755.
[11] A. Einstein, Ann. d. Phys. **17**, 891 (1905).
[12] A. Einstein, Jahrbuch Radioaktivitat **V**, 422 (1907) (Berichtigungen).
[13] J. R. Fanchi, *Confronting the ENIGMA of TIME*, World Scientific, Singapore, (2023).
[14] R. P. Feynman, Phys. Rev. **81**, 108 (1951).
[15] M. Frisch, *Inconsistency Asymmetry and Non-Locality*, Oxford University Press N. Y. (2005).
[16] J. Glimm and A. Jaffe, *Quantum Physics. A Functional Integral Point of View*, Springer, N. Y., (1987).
[17] T. L. Gill, and G. Ares de Parga,*The Einstein Dual Theory of Relativity*, Advanced Studies in Theoretical Physics **13**(8) (2019) 337-377.
[18] T. L. Gill, G. Ares de Parga, T. Morris and M, Wade, *Dual Relativistic Quantum Mechanics I*, Foundations of Physics (online) DOI 10.1007/s10701-022-00607-4, (2022).
[19] T. L. Gill and W. W. Zachary, *Foundations for relativistic quantum theory I: Feynman's operator calculus and the Dyson conjectures*, Journal of Mathematical Physics **43** (2002), 69-93.
[20] T. L. Gill and W. W. Zachary, *Functional Analysis and the Feynman operator Calculus*, Springer New York, (2016).
[21] T. L. Gill and W. W. Zachary, *Two Mathematically Equivalent Versions of Maxwell's Equations,*, Foundations of Physics (online) DOI 10.1007/s10701-009-9331-8, (2011).
[22] T.L. Gill, W.W. Zachary and J. Lindesay, *The Classical Electron Problem*, Found. Phys. **31**, 1299 (2001).
[23] T.L. Gill, W.W. Zachary, *Analytic representation of the square-root operator*, J. Phys. A: Math. Gen. **38** (2005) 1-18.
[24] T.L. Gill, W.W. Zachary, and M. Alfred, *Analytic representation of the Dirac equation*, J. Phys. A: Math. Gen. **38** (2005) 1-22.
[25] L. P. Horwitz and C. Piron, Helv. Phys. Acta **46**, 316 (1981).

[26] E. Hille, and R. S. Phillips, *Functional Analysis and Semigroups*, American Mathematical Society Colloquium Publication No. 31 ??American Mathematical Society, Providence, RI, (1957).
[27] H. A. Lorentz, *Archives Neerlandaises des Sciences Exactes et Naturelles*, **25**, 353 (1892).
[28] H. A. Lorentz, *The Theory of Electrons* B. G. Teubner, Leipzig, 1906; (reprinted by Dover, New York, 1952).
[29] H. Leutwyler. *A no-interaction theorem in classical relativistic hamiltonian particle mechanics*, Nuovo Cim., **37**, 556 (1965).
[30] G. Longhi, L. Lusanna, and J. M. Pons, J. Math. Phys. 30, 1893 (1989).
[31] H. Minkowski, Physikalische Zeitschrift **10**,104 (1909).
[32] Nathan J. Secrest; Sebastian von Hausegger; Mohamed Rameez; Roya Mohayaee; Subir Sarkar; Jacques Colin. *A Test of the Cosmological Principle with Quasars*, The Astrophysical Journal Letters, **908** (2) 2021): L51
[33] A. Pais, *Subtle is the lord: the science and life of Albert Einstein*, Oxford University Press (1982).
[34] P.J.E. Peebles, *Principles of Physical Cosmology*, Princeton University Press, London (1993).
[35] W. Perret and G. B. Jeffery (translators, with additional notes by A. Sommerfeld), *The Principle of Relativity* by H. A. Lorentz, A. Einstein, H. Minkowski and H. Weyl, Dover, New York, 1952.
[36] H. Poincaré, *Sur la dynamique de l'électron* , Rendiconti del Circolo matematico Rendiconti del Circolo di Palermo, **21**, 129-176 (1906).
[37] W. K. H. Panofsky and M. Phillips, *Classical Electricity and Magnetism*, (Second Edition), Addision-Wesley, Reading, MA. (1962).
[38] A. A. Penzias and R. W. Wilson, Ap. J. **142**, 419 (1965).
[39] M.H.L. Pryce, Proc. Roy. Soc. London A **195**, 400 (1948).
[40] G. Ares de Parga, T.L. Gill and W.W. Zachary, *The Thomas program and the canonical proper-time theory* J. of Comp. Methods in Sci. and Eng. **3** (2013) 117-134.
[41] M. Radovan *On the Nature of Time*, arXiv:1509.01498
[42] F. Rohrlich, *Classical Charged Particles*, Advanced Book Program, Addison-Wesley, New York (1990).
[43] R. M. Santilli, *Isonumbers and genonumbers of dimension 1, 2, 4, 8, their isoduals, and pseudoduals, and hidden numbers of dimension 3, 5, 6, 7*, Algebras, Groups and Geometries, Vol. **10**, 273?322 (1993).
[44] J. Schwinger, Found. Phys. **13**, 2573 (1998).
[45] Saadeh D, Feeney SM, Pontzen A, Peiris HV, McEwen, JD . *How Isotropic is the Universe*, Physical Review Letters. **117** (13) (2016) 131302.
[46] S. Schweber, *QED And The Men Who Made It* , Princeton University Press, London (1994).
[47] J.A. Wheeler and R.P. Feynman, *Interaction with the absorber as the mechanism of radiation* Rev. Mod. Phys. **17** (1949) 157-181.
[48] S. Weinberg, *High energy behavior in quantum field theory,* Phys. Rev. **118** (1960), 838-849.

High accuracy synchrotron radiation interferometry with relativistic electrons

P Klag[1], P Achenbach[1,2], T Akiyama[3], R Böhm[4], M O Distler[1], L Doria[1], P Eckert[1], A Esser[1], J Geratz[1], T Gogami[5], C Helmel[1], P Herrmann[1,6], M Hoek[1], M Kaneta[3], Y Konishi[3], R Kino[3], W Lauth[1], H Merkel[1], M Mizuno[3], U Müller[1], S Nagao[7], S N Nakamura[3,7], K Okuyama[3], J Pochodzalla[1], B S Schlimme[1], C Sfienti[1], T Shao[1,8], M Steinen[9], S Stengel[1], M Thiel[1], Y Toyama[10]

[1] Institute for Nuclear Physics, Johannes Gutenberg University, 55099 Mainz, Germany
[2] Thomas Jefferson National Accelerator Facility (JLab), Newport News, Virginia 23606, USA
[3] Graduate School of Science, Tohoku University, Sendai, Miyagi 980-8578, Japan
[4] FAIR, Planckstr. 1, 64291 Darmstadt, Germany
[5] Graduate School of Science, Kyoto University, Kyoto 606-8502, Japan
[6] GSI Helmholtzzentrum für Schwerionenforschung GmbH, 64291 Darmstadt, Germany
[7] Graduate School of Science, The University of Tokyo, Tokyo 113-0033, Japan
[8] Key Laboratory of Nuclear Physics and Ion-beam Application (MOE), Institute of Modern Physics, Fudan University, Shanghai 200433, China
[9] Helmholtz Institute Mainz (HIM), GSI Helmholtzzentrum für Schwerionenforschung, Johannes Gutenberg University, 55099 Mainz, Germany
[10] College of Engineering, Chubu University, Kasugai, Aichi 487-8501, Japan

E-mail: `pklag02@uni-mainz.de`, `pochodza@uni-mainz.de`

Abstract. A high-precision hypernuclear experiment has been performed at the Mainz Microtron (MAMI) to determine the hypertriton Λ binding energy via decay-pion spectroscopy. A key element of this measurement is an accurate calibration of the magnetic spectrometers with the MAMI beam. For such an absolute calibration with small statistical and systematic uncertainties the undulator light interference method will be applied. In this contribution the basic principle of this method is discussed and the analysis status of the measured synchrotron radiation spectra is presented.

1. Motivation and background

One of the most intriguing questions in physics is: what do we find in the inner core of dense stellar objects such as neutron stars? There has been a wide consensus in nearly all theoretical approaches for neutron star matter that hyperons may appear in these cores at densities of about twice the nuclear saturation density. However, the determination of the nuclear equation of state at such densities remains one of the biggest challenges in nuclear physics even after many decades of research. Hadrons with strangeness embedded in the nuclear environment, hypernuclei or strange atoms, are the only available tool to approach the many-body aspect of the three-flavor strong interaction.

Light hypernuclei are particularly interesting since not only phenomenological models but also ab initio studies based on chiral effective field theory including three-baryon forces can

Content from this work may be used under the terms of the Creative Commons Attribution 3.0 licence. Any further distribution of this work must maintain attribution to the author(s) and the title of the work, journal citation and DOI.
Published under licence by IOP Publishing Ltd

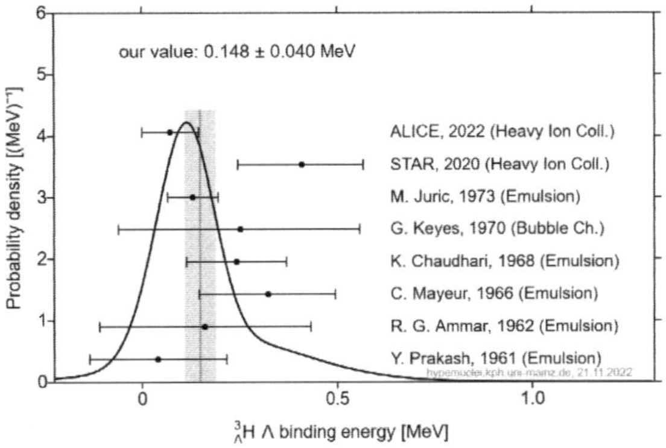

Figure 1: Ideogram of the world data set on Λ binding energy measurements for the ${}^{3}_{\Lambda}$H [1]. The error bars include statistical and systematic uncertainties.

Figure 2: Setup for the hypernuclear fragmentation experiment at MAMI. The electron beam (coming from the back) was steered through a beam transport-line magnetic chicane and was incident on the ^{9}Be target. The KAOS spectrometer (purple) was installed at 0° forward angle as a kaon tagger. The spectrometers SpekA and SpekC were used as decay-pion spectrometers. Fig. from Ref. [2].

be performed. Furthermore, lattice quantum chromodynamics calculations are within reach for such nuclei. A profound understanding of the lightest hypernuclei, the hypertriton ${}^{3}_{\Lambda}$H, is therefore a cornerstone for any strong interaction theory dealing with strange baryons.

For almost 50 years, its most precise binding energy value is given by $B_\Lambda = 130 \pm 50\,\mathrm{keV}$, averaged and compiled by Jurič et al. from emulsion experiments [3, 4]. Just recently, two new values became available, one by the STAR Collaboration [5], $B_\Lambda = 406 \pm 120\,(\mathrm{stat.}) \pm 110\,(\mathrm{syst.})\,\mathrm{keV}$, and one by the ALICE Collaboration [6], $B_\Lambda = 72 \pm 63\,(\mathrm{stat.}) \pm 36\,(\mathrm{syst.})\,\mathrm{keV}$. Remarkably, the STAR value is about 5 times larger than the one from ALICE and they differ by about two standard deviations. The STAR value also seems to be in tension with the emulsion value. This data situation along with other earlier measurements is visualized in an ideogram in Fig. 2, obtained from the Chart of Hypernuclides [1]. An average value of $\overline{B_\Lambda} = 148 \pm 40\,\mathrm{keV}$ is computed with a relative error of still more than 25 %. In the coming years, several experiments will produce more precise and accurate data for the hypertriton. The WASA-FRS Collaboration at GSI/FAIR, the E73/77 collaboration at J-PARC [7, 8], ALICE during LHC Run3 and Run4, and an experiment at ELPH at Tohoku University will improve our knowledge on the lifetime of ${}^{3}_{\Lambda}$H. New measurements of the ${}^{3}_{\Lambda}$H binding energy are expected from experiments at the Mainz Microtron (MAMI), JLab [9], from ALICE, and from the J-PARC E07 emulsion experiment. These many attempts attest the importance of the hypertriton as a benchmark for hypernuclear structure calculations.

In 2012, the first high-resolution spectroscopy of pions from decays of stopped ${}^{4}_{\Lambda}$H hypernuclei was performed by the A1 Collaboration at MAMI [10]. ${}^{4}_{\Lambda}$H hypernuclei were produced in a multi-

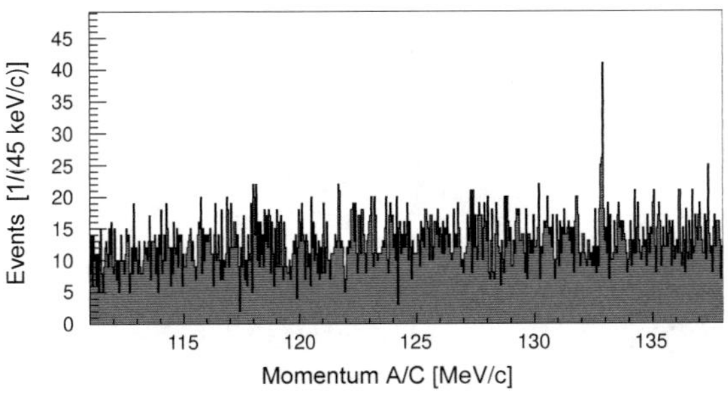

Figure 3: Momentum spectrum for strangeness tagged pions from the 2014 measurement at MAMI [11]. Mono-energetic decay-pions of ${}^{4}_{\Lambda}$H were observed at ~133 MeV/c. A signal from two-body decays of stopped ${}^{3}_{\Lambda}$H was not found at the expected momentum of ~114 MeV/c.

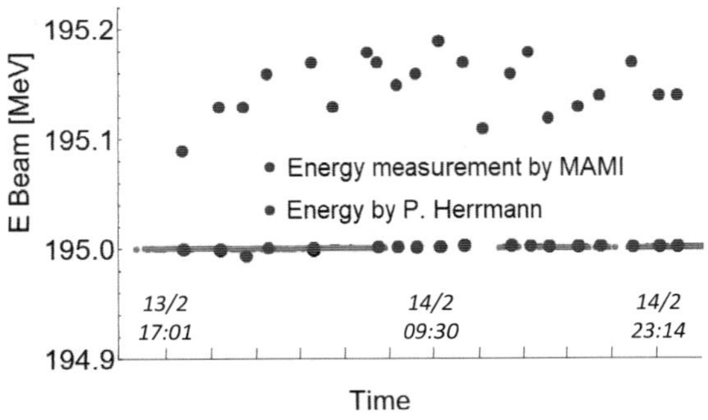

Figure 4: MAMI beam energy measured over a period of about 30 hours [12]. The blue points show MAMI measurements from the position of the beam within the Race Track Microton RTM3 [13]. The purple points are derived from a precise measurement of the deflection in a dipole magnet of the A1 beamline.

step strangeness production, nuclear fragmentation, and pionic weak decay reaction:

$$ {}^{9}\mathrm{Be}(\mathrm{e},\mathrm{e}'\mathrm{K}^{+}){}^{9}_{\Lambda}\mathrm{Li}^{*} \quad \rightarrow \quad \mathrm{X} + {}^{4}_{\Lambda}\mathrm{H}; \qquad {}^{4}_{\Lambda}\mathrm{H} \rightarrow {}^{4}\mathrm{He} + \pi^{-} \tag{1} $$

Pions were detected alternatively in one of the two spectrometers SpekA or SpekC in coincidence with positive kaons tagged in the KAOS spectrometer.

In the year 2014 the experiment at MAMI was continued with improved control of systematic uncertainties, additional background suppression, and higher luminosity. A detailed description of the 2014 experiment can be found in Refs. [2, 11]. Fig. 3 shows a representative momentum spectrum, where decay pions of ${}^{4}_{\Lambda}$H are clearly visible at around 133 MeV/c, resulting in a binding energy of $B_{\Lambda} = 2.157 \pm 0.005\,(\text{stat.}) \pm 0.077\,(\text{syst.})\,\text{MeV}$. The systematic error was strongly dominating because the spectrometer calibration was limited by the available MAMI beam energy measurement with an accuracy of 160 keV, see blue points in Fig. 4. However, the beam spread and instability are known to be much smaller. This could be verified by a precise measurement of the deflection in a dipole magnet of the A1 beamline as shown by the purple points in Fig. 4 [12]. In this study, the stability of the dipole magnet was controlled by an NMR probe whose readout (on arbitrary scale) is shown by the small red points.

As the data situation for ${}^{3}_{\Lambda}$H is clearly not satisfactory, a new high-precision experiment via decay-pion spectroscopy has been commissioned at MAMI with the goal to reach a 20 keV total error in the binding energy [14, 15]. This will be made possible by measuring the absolute value of the beam energy with a novel undulator light interference method [16].

2. Basic concept of the measurement

The method is based on the analysis of the intensity oscillation length in the synchrotron spectrum from two collinear sources, which are realized by two undulators and the electron beam of MAMI [17]. The schematic layout of the experimental setup is displayed in Fig. 5. Relativistic electrons traverse two identical undulators thus generating coherent pairs of electromagnetic wave packets as illustrated in the lower part. The time delay between the two wave packets can be controlled by changing the distance d between the two undulators.

After passing through an aperture with an opening of $a_{\mathrm{y}} \times a_{\mathrm{x}} = \pm 4\,\mathrm{mm} \times \pm 2\,\mathrm{mm}$, the radiation is dispersed according to its wavelength by a reflective grating spectrometer [18] and

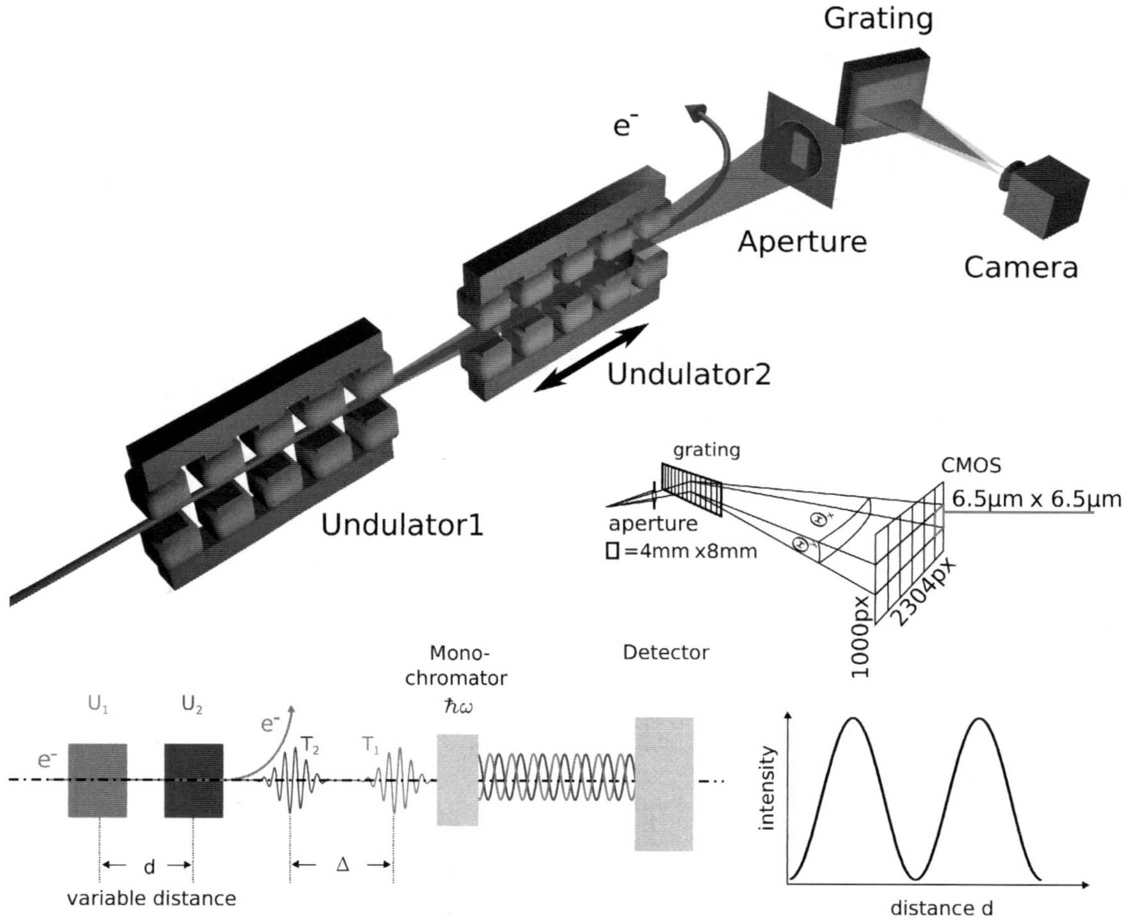

Figure 5: Schematic layout (not to scale) of the experimental setup. Relativistic electrons pass through two undulators separated by a variable distance, thus generating coherent pairs of electromagnetic wave packets as illustrated in the lower part. The insert illustrates the coordinate system of the optical system. The x-axis describes the dispersive plane and the y-axis denotes the vertical off-axis direction.

detected by a CMOS camera [19]. The CMOS chip has a pixel size of $6.5\,\mu m \times 6.5\,\mu m$ and a resolution of 2304×2304 pixels. In this setup one pixel covers a wavelength interval of $\Delta\lambda_{rad} = 0.0048\,nm$.

The intensity for a selected wavelength λ_{rad} varies periodically with the distance Δd between the two undulators as illustrated in the lower part in Fig. 5. Here, Δd is measured with respect to the closest position of the two undulators. The oscillation length λ_{osc} is directly related to the Lorentz factor of the electron beam and the wavelength of the radiation

$$\gamma^2 = \frac{1}{2}\frac{\lambda_{osc}}{\lambda_{rad}}. \tag{2}$$

Thus the energy determination is reduced to a relative distance measurement in the decimeter range and the spectroscopy in a narrow optical wavelength band.

Eq. 2 only holds in the limit of a precise alignment between the electron beam and the optical

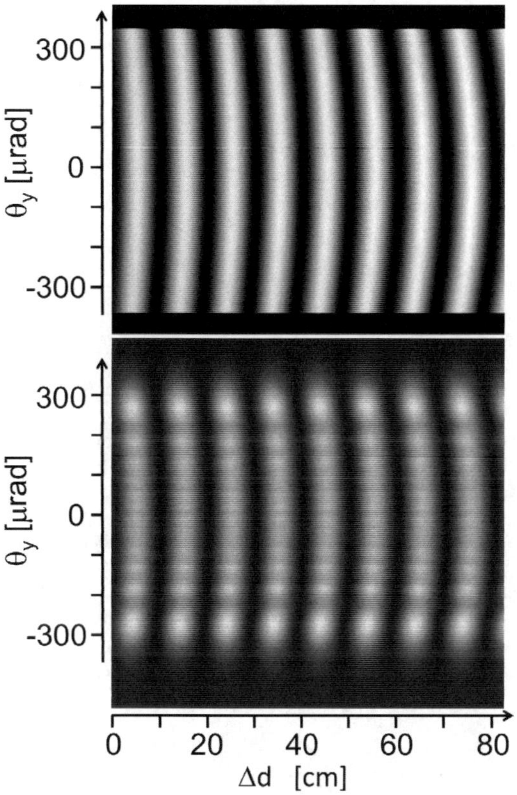

Figure 6: Top: Theoretical interference signal for a wavelength of 404.58 nm with varying distance Δd between the two undulators and its dependence on the vertical off-axis angle θ_y. The calculations do not include diffraction. Bottom: Measured interference signal for the same wavelength. Because of the finite size of the aperture $a_y = \pm 4$ mm a characteristic diffraction pattern is present.

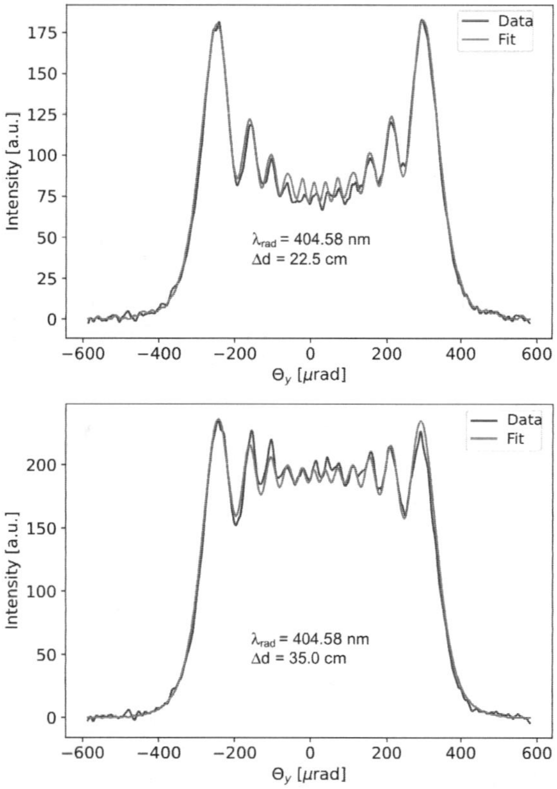

Figure 7: Two representative examples for observed radiation intensities as a function of the vertical off-axis angle θ_y for undulator distances of $\Delta d = 22.5$ cm (top) and 35.0 cm (bottom). A wavelength bin at $\lambda_{rad} = 404.58$ nm was selected. For each wavelength bin 825 spectra were recorded and analyzed.

axis [17]. For off-axis observation angles $\theta \neq 0$, Eq. 2 is modified according to

$$\gamma^2 = \frac{\lambda_{osc}}{\lambda_{osc} \cdot \theta^2 - 2\lambda_{rad}}. \quad (3)$$

This causes a bending of the interference spectrum in the non-dispersive direction.

3. Measurements and analysis of synchrotron radiation spectra

The upper part of Fig. 6 shows the theoretical variation of the interference signal with varying distance Δd between the two undulators and vertical off-axis angle θ_y, if no other limitations would apply. However, the synchrotron radiation is collimated by the finite size of the aperture with $a_y = \pm 4$ mm. As a consequence, a characteristic diffraction pattern is present in the experimentally measured spectrum as shown in the lower part of Fig. 6. This structure proves to be very useful for the determination the true $\theta_y = 0°$ direction during the fitting procedure. In the dispersive direction θ_x, the aperture is not relevant, since the grating separates wavelengths independent of the position of the incident radiation.

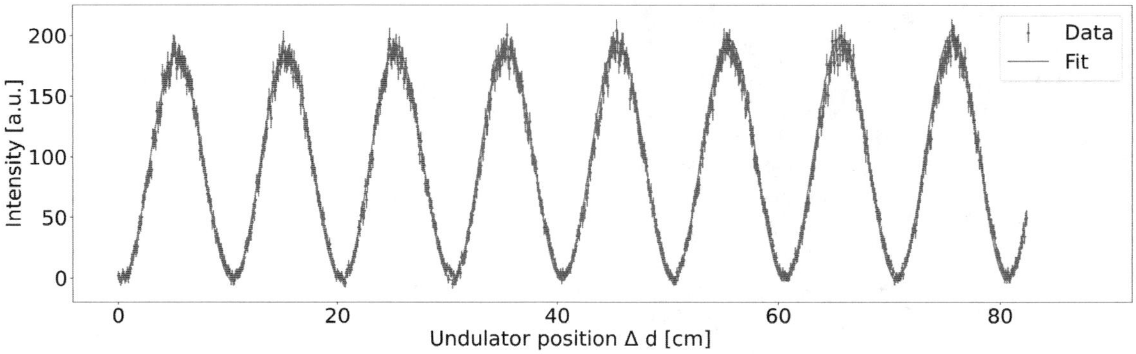

Figure 8: Intensity for $\theta_y = 0$ and $\lambda_{rad} = 404.58\,\text{nm}$ as a function of the undulator distance Δd, c.f. lower part of Fig. 5. The blue points show the measured intensity. Statistical errors are indicated in green. The red curve shows the fit discussed in the text.

Prior to the measurement, the beamline was aligned optically with a theodolite and a diode laser system at a wavelength of $\lambda_{las} = 405\,\text{nm}$. For the energy calibration of the combination of the grating spectrometer and the CMOS camera, a mercury lamp was inserted upstream of the stationary first undulator. The Hg lamp has two dominant, well known lines within the acceptable wavelength band at 404.6565 nm and 407.7837 nm [20]. In practice, the radiation from the Hg lamp, the undulators, and the laser have differing divergences: the Hg lamp emits approximately isotropically, whereas the undulator radiation is forward peaked with a typical angle of $1/\gamma \sim 0.15°$. However, both radiation fields are collimated by the rectangular aperture to about $\theta_x < 0.01°$ leading to a similar wavelength resolution. On the contrary, the divergence of the laser beam is in the range of $0.02° - 0.03°$ and thus slightly more focused. The measured width of the 407.78 nm Hg line was 4.2 pixels (FWHM), whereas the laser line was only 3.0 pixels (FWHM) wide.

In the analysis of the experimental data, 5 pixel columns of the camera were combined to a wavelength bin in order to improve the signal-to-noise ratio. Such a bin corresponds to a relative wavelength window of

$$\frac{\Delta\lambda_{rad}}{\lambda_{rad}} = \frac{0.024\,\text{nm}}{405\,\text{nm}} = 6 \cdot 10^{-5}. \tag{4}$$

This bin is sufficiently small to have no sizable effect on the determination of the MAMI beam energy after a proper calibration of the λ_{rad} scale.

Fig. 7 shows two examples of the measured interference spectra for a wavelength of $\lambda_{rad} = 404.58\,\text{nm}$ at $\Delta d = 22.5\,\text{cm}$ and $35.0\,\text{cm}$, respectively. For each selected λ_{rad} bin, 825 positions Δd of the movable undulator were analyzed. The blue data points in Fig. 8 show the observed oscillation of the intensity for $\theta_y = 0$ and $\lambda_{rad} = 404.58\,\text{nm}$ as a function of the undulator distance Δd, c.f. lower part of Fig. 5. In total, 460 different wavelength bins were measured. For every set of data, a background image has been taken which is subtracted. Even under optimal conditions offsets remain.

For each given wavelength λ_{rad}, the interference spectra for 825 measured undulator positions Δd were fitted simultaneously with only 7 parameters. The first five parameters are directly related to the experimental setup:

(i) The absolute intensity, which is, however, not relevant for the MAMI energy determination (c.f. Eq. 2).

(ii) The phase factor of the oscillation, which reflects the effective offset of Δd.

(iii) The effective distance between the aperture and the camera of approximately 5.45 m.
(iv) A displacement of the source with respect to the center of the aperture.
(v) An elevation parameter, which describes the increase of the intensity when moving the second undulator closer towards the aperture.
(vi) A parameter, which describes the residual intensity on the camera independent of the undulator radiation.
(vii) The relativistic γ factor, which determines the MAMI beam energy.

Figs. 7 and 8 include the results of the fitting procedure for $\lambda_{\rm rad} = 404.58\,\rm nm$. The Fresnel pattern in the two representative $\theta_{\rm y}$ spectra in Fig. 7 is well described. Simultaneously, the oscillating interference signal for $\theta_{\rm y} = 0°$ is correctly reproduced.

4. Summary and outlook

A precise determination of the hypertriton mass via decay-pion spectroscopy requires an accurate energy calibration of the electron beam at MAMI. For such a calibration with small statistical and systematic uncertainties the undulator light interference method has been developed. In this contribution the basic principle of this method is presented and the analysis of the measured synchrotron radiation spectra is demonstrated for a typical wavelength of $\lambda_{\rm rad} = 404.58\,\rm nm$. The interference patterns which were measured at 825 positions of the movable undulator, could be described simultaneously by a fit with only 7 parameters.

Presently, the analysis of all 460 different wavelength bins is being finalized. This will allow to determine the systematic uncertainty of the beam energy measurement with the projected high accuracy.

Acknowledgements

We would like to thank the MAMI operators, technical staff, and the accelerator group for their excellent support of the experiments. This project has received funding from the European Union's Horizon 2020 research and innovation programme under grant agreement No 824093. We acknowledge the support by Deutsche Forschungsgemeinschaft (DFG), Germany through Research Grants PO 256/7-1 and PO 256/8-1, JSPS KAKENHI grants no.18H05459 and the Graduate Program on Physics for the Universe, Tohoku University (GP-PU).

The presented data were in part collected within the framework of the PhD thesis of Pascal Klag at the Johannes Gutenberg University Mainz.

References

[1] Institut für Kernphysik Mainz 2022 Chart of Hypernuclides - Hypernuclear Structure and Decay Data visited on 2022-11-21 URL `https://hypernuclei.kph.uni-mainz.de`
[2] Schulz F 2015 *Präzisionsmessung der Λ-Separationsenergie von* $^4_\Lambda H$ *am Mainzer Mikrotron* Ph.D. thesis Johannes Gutenberg-Universität Mainz
[3] Juric M, Bohm G, Klabuhn J, Krecker U, Wysotzki F, Coremans-Bertrand G, Sacton J, Wilquet G, Cantwell T, Esmael F, Montwill A, Davis D, Kielczewska D, Pniewski T, Tymieniecka T and Zakrzewski J 1973 *Nuclear Physics B* **52** 1 – 30 ISSN 0550-3213 URL `http://www.sciencedirect.com/science/article/pii/0550321373900849`
[4] Bohm G, Klabuhn J, Krecker U, Wysotski F, Coremans G, Gajewski W, Mayeur C, Sacton J, Vilain P, Wilquet G, O'Sullivan D, Stanley D, Davis D, Fletcher E, Lovell S, Roy N, Wickens J, Filipkowski A, Garbowska-Pniewska K, Pniewski T, Skrzypczak E, Sobczak T, Allen J, Bull V, Conway A, Fishwick A and March P 1968 *Nuclear Physics B* **4** 511 – 526 ISSN 0550-3213 URL `http://www.sciencedirect.com/science/article/pii/0550321368901090`
[5] Adam J, Adamczyk L, Adams J R, Adkins J K, Agakishiev G, Aggarwal M M, Ahammed Z, Ajitanand N N, Alekseev I, Alford J, Anderson D M, Aoyama R, Aparin A, Arkhipkin D, Aschenauer E C, Ashraf M U, Attri A, Averichev G S, Bai X, Bairathi V, Barish K, Behera A, Bellwied R, Bhasin A, Bhati A K, Bhattarai P, Bielcik J, Bielcikova J, Bland L C, Bordyuzhin I G, Bouchet J, Brandenburg J D, Brandin

A V, Brown D, Bryslawskyj J, Bunzarov I, Butterworth J, Caines H, Calderón de la Barca Sánchez M, Campbell J M, Cebra D, Chakaberia I, Chaloupka P, Chang Z, Chankova-Bunzarova N, Chatterjee A, Chattopadhyay S, Chen X, Chen X, Chen J H, Cheng J, Cherney M, Christie W, Contin G, Crawford H J, Das S, Dedovich T G, Deng J, Deppner I M, Derevschikov A A, Didenko L, Dilks C, Dong X, Drachenberg J L, Draper J E, Dunlop J C, Efimov L G, Elsey N, Engelage J, Eppley G, Esha R, Esumi S, Evdokimov O, Ewigleben J, Eyser O, Fatemi R, Fazio S, Federic P, Federicova P, Fedorisin J, Feng Z, Filip P, Finch E, Fisyak Y, Flores C E, Fujita J, Fulek L, Gagliardi C A, Geurts F, Gibson A, Girard M, Grosnick D, Gunarathne D S, Guo Y, Gupta A, Guryn W, Hamad A I, Hamed A, Harlenderova A, Harris J W, He L, Heppelmann S, Heppelmann S, Herrmann N, Hirsch A, Horvat S, Huang B, Huang T, Huang X, Huang H Z, Humanic T J, Huo P, Igo G, Jacobs W W, Jentsch A, Jia J, Jiang K, Jowzaee S, Judd E G, Kabana S, Kalinkin D, Kang K, Kapukchyan D, Kauder K, Ke H W, Keane D, Kechechyan A, Khan Z, Kikoła D P, Kim C, Kisel I, Kisiel A, Kochenda L, Kocmanek M, Kollegger T, Kosarzewski L K, Kraishan A F, Krauth L, Kravtsov P, Krueger K, Kulathunga N, Kumar L, Kvapil J, Kwasizur J H, Lacey R, Landgraf J M, Landry K D, Lauret J, Lebedev A, Lednicky R, Lee J H, Li X, Li W, Li Y, Li C, Lidrych J, Lin T, Lisa M A, Liu F, Liu P, Liu Y, Liu H, Ljubicic T, Llope W J, Lomnitz M, Longacre R S, Luo X, Luo S, Ma G L, Ma L, Ma R, Ma Y G, Magdy N, Majka R, Mallick D, Margetis S, Markert C, Matis H S, Mayes D, Meehan K, Mei J C, Miller Z W, Minaev N G, Mioduszewski S, Mishra D, Mizuno S, Mohanty B, Mondal M M, Morozov D A, Mustafa M K, Nasim M, Nayak T K, Nelson J M, Nemes D B, Nie M, Nigmatkulov G, Niida T, Nogach L V, Nonaka T, Nurushev S B, Odyniec G, Ogawa A, Oh K, Okorokov V A, Olvitt D, Page B S, Pak R, Pandit Y, Panebratsev Y, Pawlik B, Pei H, Perkins C, Pluta J, Poniatowska K, Porter J, Posik M, Pruthi N K, Przybycien M, Putschke J, Quintero A, Ramachandran S, Ray R L, Reed R, Rehbein M J, Ritter H G, Roberts J B, Rogachevskiy O V, Romero J L, Roth J D, Ruan L, Rusnak J, Rusnakova O, Sahoo N R, Sahu P K, Salur S, Sandweiss J, Saur M, Schambach J, Schmah A M, Schmidke W B, Schmitz N, Schweid B R, Seger J, Sergeeva M, Seto R, Seyboth P, Shah N, Shahaliev E, Shanmuganathan P V, Shao M, Shen W Q, Shi S S, Shi Z, Shou Q Y, Sichtermann E P, Sikora R, Simko M, Singha S, Skoby M J, Smirnov N, Smirnov D, Solyst W, Sorensen P, Spinka H M, Srivastava B, Stanislaus T D S, Stewart D J, Strikhanov M, Stringfellow B, Suaide A A P, Sugiura T, Sumbera M, Summa B, Sun Y, Sun X, Sun X M, Surrow B, Svirida D N, Tang A H, Tang Z, Taranenko A, Tarnowsky T, Tawfik A, Thäder J, Thomas J H, Timmins A R, Tlusty D, Todoroki T, Tokarev M, Trentalange S, Tribble R E, Tribedy P, Tripathy S K, Trzeciak B A, Tsai O D, Ullrich T, Underwood D G, Upsal I, Van Buren G, van Nieuwenhuizen G, Vasiliev A N, Videbæk F, Vokal S, Voloshin S A, Vossen A, Wang G, Wang Y, Wang F, Wang Y, Webb G, Webb J C, Wen L, Westfall G D, Wieman H, Wissink S W, Witt R, Wu Y, Xiao Z G, Xie G, Xie W, Xu Y F, Xu J, Xu Q H, Xu N, Xu Z, Yang S, Yang Y, Yang C, Yang Q, Ye Z, Ye Z, Yi L, Yip K, Yoo I K, Yu N, Zbroszczyk H, Zha W, Zhang Z, Zhang J, Zhang S, Zhang S, Zhang J, Zhang Y, Zhang X P, Zhang J B, Zhao J, Zhong C, Zhou L, Zhou C, Zhu X, Zhu Z, Zurek M and Zyzak M (STAR) 2020 *Nature Physics* **16** 409–412 ISSN 1745-2481 URL `http://dx.doi.org/10.1038/s41567-020-0799-7`

[6] ALICE-Collaboration (ALICE) 2022 Measurement of the lifetime and Λ separation energy of $^3_\Lambda$H (*Preprint* `2209.07360`)

[7] Akaishi T, Asano H, X C, A C, Curceanu C, Del Grande R, Guaraldo C, Han C, Hashimoto T, Iliescu M, Inoue K, Ishimoto S, Itahashi K, Iwasaki M, Ma Y, Miliucci M, Noumi H, Ohnishi H, Okada S, Outa H, Piscicchia K, Sakuma F, Sato M, Scordo A, Sirghi D, Sirghi F Shirotori K, Suzuki S, Tanida K, Yamaga T, Yuan X, Zhang P, Zhang Y and Zhang H 2019 Feasibility study for $^3_\Lambda$H mesonic weak decay lifetime measurement with ^{4}He(K^-, π^0)$^4_\Lambda$H reaction URL `http://j-parc.jp/researcher/Hadron/en/pac_1901/pdf/P73_2019-06.pdf`

[8] Akaishi T, Asano H, X C, A C, Curceanu C, Del Grande R, Guaraldo C, Han C, Hashimoto T, Iliescu M, Inoue K, Ishimoto S, Itahashi K, Iwasaki M, Ma Y, Miliucci M, Noumi H, Ohnishi H, Okada S, Outa H, Piscicchia K, Sakuma F, Sato M, Scordo A, Sirghi D, Sirghi F Shirotori K, Suzuki S, Tanida K, Yamaga T, Yuan X, Zhang P, Zhang Y and Zhang H J-PARC E77 experiment: Feasibility study for $^3_\Lambda$H mesonic weak decay lifetime measurement with ^{4}He(K^-, π^0)$^4_\Lambda$H reaction URL `http://j-parc.jp/researcher/Hadron/en/pac_2001/pdf/P77_2020-01.pdf`

[9] Gogami T, Nakamura S, Garibaldi F, Markowitz P, Reinhold J, Tang L and Urciuoli G Proposal to jlab pac48, c12-19-002: High accuracy measurement of nuclear masses of λ hyperhydrogens URL `https://indico.jlab.org/event/394/contributions/6380/attachments/5334/6632/e12-19-002_JLab-PAC48_gogami_20200811.pdf`

[10] Esser A, Nagao S, Schulz F, Achenbach P, Ayerbe Gayoso C, Böhm R, Borodina O, Bosnar D, Bozkurt V, Debenjak L, Distler M O, Frišč I, Fujii Y, Gogami T, Hashimoto O, Hirose S, Kanda H, Kaneta M, Kim E, Kohl Y, Kusaka J, Margaryan A, Merkel H, Mihovilovič M, Müller U, Nakamura S N, Pochodzalla J, Rappold C, Reinhold J, Saito T R, Sanchez Lorente A, Sánchez Majos S, Schlimme B S, Schoth M, Sfienti

C, Širca S, Tang L, Thiel M, Tsukada K, Weber A and Yoshida K (A1 Collaboration) 2015 *Phys. Rev. Lett.* **114**(23) 232501 URL http://link.aps.org/doi/10.1103/PhysRevLett.114.232501

[11] Schulz F, Achenbach P, Aulenbacher S, Bericic J, Bleser S, Böhm R, Bosnar D, Correa L, Distler M, Esser A, Fonvieille H, Friŝcic I, Fujii Y, Fujita M, Gogami T, Kanda H, Kaneta M, Kegel S, Kohl Y, Kusaka W, Margaryan A, Merkel H, Mihovilovic M, Müller U, Nagao S, Nakamura S, Pochodzalla J, Lorente A S, Schlimme B, Schoth M, Sfienti C, Širca S, Steinen M, Takahashi Y, Tang L, Thiel M, Tsukada K, Tyukin A and Weber A 2016 *Nuclear Physics A* **954** 149 – 160 ISSN 0375-9474 recent Progress in Strangeness and Charm Hadronic and Nuclear Physics URL http://www.sciencedirect.com/science/article/pii/S037594741630001X

[12] Herrmann P 2021 *Vorarbeiten für genaue und präzise Messungen leichter Hyperkernmassen mit der A1-Spektrometeranlage am MAMI* Ph.D. thesis Johannes Gutenberg-Universität Mainz

[13] Kaiser K H, Aulenbacher K, Chubarov O, Dehn M, Euteneuer H, Hagenbuck F, Herr R, Jankowiak A, Jennewein P, Kreidel H J, Ludwig-Mertin U, Negrazus M, Ratschow S, Schumann S, Seidl M, Stephan G and Thomas A 2008 *Nuclear Instruments and Methods in Physics Research Section A: Accelerators, Spectrometers, Detectors and Associated Equipment* **593** 159–170 ISSN 0168-9002 URL https://www.sciencedirect.com/science/article/pii/S0168900208007341

[14] Achenbach P, Bleser S, Pochodzalla J and Steinen M 2018 *Proceedings of XVII International Conference on Hadron Spectroscopy and Structure, PoS (Hadron2017)* vol 310 p 207 URL https://pos.sissa.it/310/207/pdf

[15] Eckert P, Achenbach P, Aragonés Fontboté M, Akiyama T, Distler M O, Esser A, Geratz J, Hoek M, Itabashi K, Kaneta M, Klag P, Merkel H, Mizuno M, Müller J, Müller U, Nagao S, Nakamura S N, Nakamura Y, Okuyama K, Pochodzalla J, Sfienti C, Spreckels R, Steinen M, Thiel M, Uehara K and Toyama Y 2022 *Proceedings of Particles and Nuclei International Conference 2021 — PoS(PANIC2021)* vol 380 p 201

[16] Klag P, Achenbach P, Biroth M, Gogami T, Herrmann P, Kaneta M, Konishi Y, Lauth W, Nagao S, Nakamura S, Pochodzalla J, Roser J and Toyama Y 2018 *Nuclear Instruments and Methods in Physics Research Section A: Accelerators, Spectrometers, Detectors and Associated Equipment* **910** 147–156 ISSN 0168-9002 URL https://www.sciencedirect.com/science/article/pii/S0168900218312166

[17] Dambach S, Backe H, Doerk T, Eftekhari N, Euteneuer H, Görgen F, Hagenbuck F, Kaiser K H, Kettig O, Kube G, Lauth W, Schöpe H, Steinhof A, Tonn T and Walcher T 1998 *Phys. Rev. Lett.* **80**(25) 5473–5476 URL https://link.aps.org/doi/10.1103/PhysRevLett.80.5473

[18] HORIBA Scientific 1993 *Operators Manual: HR640* Jobin Yvon, Longjumeau, France URL http://www.horiba.com/fileadmin/uploads/Scientific/Documents/OSD/HR640.pdf

[19] Hamamatsu Photonics 2019 *Datasheet: Orca Fusion* Hamamatsu, 812 Joko-cho, Higashi-ku, Hamamatsu City, 431-3196, Japan URL https://unicam.hu/files/mikroszkopos-kamerak/hamamatsu-kamerak/cmos-kamerak/orca-fusion-c14440-20up/ORCA_Fusion_Technical_Note.pdf

[20] Kramida A, Ralchenko Y, Reader J and NIST ASD Team 2015 NIST Atomic Spectra Database(ver. 5.3), [Online]. Available: http://physics.nist.gov/asd [2016, Nov 16]. National Institute of Standards and Technology, Gaithersburg, MD.

On the electron spin and spectrum energy problems within the Fock many temporal and Feynman proper time paradigms

Anatolij K. Prykarpatski

The Department of Computer Science and Telecommunications at the Cracow University of Technology, Krakow 31-155 Poland

E-mail: pryk.anat@cybergal.com

Abstract. The quantum electron spin structure and the photon dressed electron energy spectrum problem are reanalyzed in the framework of the Fock many-temporal parameter approach. There is analyzed the spin structure and its dependence on the symmetry properties of the related representations of the basic Clifford algebra, generated by creation-annihilation operators. The self-interaction phenomenon is discussed within the quantum renormalized Lorenz constraint on a suitably reduced Fock space. The electron energy spectrum is analyzed within the many-temporal Fock and Feynman proper time paradigms.

1. Introduction

We will first discuss the electron spin and next the electron spectrum energy problems from mathematical physics point of view, using the electron Hamiltonian operator structure on the Fock space background and the related symmetry algebra properties of its representations. Subject to our physical real spacetime we assume as a fact the existence of such an elementary particle as electron, specified by its nonzero discrete charge parameter $\xi \in \mathbf{R}$, localized in the Minkowski [32, 47] spacetime $M^4 = \mathbf{R}_t \times \mathbf{R}^3_x$, where $x \in \mathbf{R}^3$ is the spatial variable, $t \in \mathbf{R}$ is the temporal variable. These coordinates are, obviously, important for specifying electron by means of external and independent observables allowed by the physical world. We naturally accept that there exist also another elementary particles like proton, photon, meson etc, considered as quants of the physical world structure, and which are indistinguishable, that is identical within each particle kind. Moreover, we accept such experimentally approved important physical phenomena as creation and annihilation of elementary quantum particles during their interaction in spacetime. At the beginning of the past 20 century both qualitative and quantitative properties of these quantum physical phenomena have been effectively enough described by classical theoretical physicists W. Pauli, E. Schrödinger, W. Heisenberg, P. Dirac and others [2, 4, 7, 11, 29, 36, 38, 39, 43, 44, 46, 47] within first quantum mechanics and next quantum field theory by means of modern mathematical tools, based on special differential-operator expressions, called *"Hamiltonians"*, jointly with the related spectral theory of special linear differential-operator Schrödinger equations in the Hilbert space $\mathcal{H}$ of complex-valued functions $f : \mathbf{R}^3 \to \mathbf{C}$, integrable with respect to the usual Lebesgue measure dx for $x \in \mathbf{R}^3$, and whose solutions possess almost all information about the evolution of quantum particle

Content from this work may be used under the terms of the Creative Commons Attribution 3.0 licence. Any further distribution of this work must maintain attribution to the author(s) and the title of the work, journal citation and DOI.
Published under licence by IOP Publishing Ltd

states in spacetime and their interaction with neighborhood. In particular, theoretical physicists soon enough understood that each quantum event in microworld, related with such elementary particles, can be well modelled by means of some special operator on the Fock space of the corresponding quantum states, and in case of electron, the mostly fundamental facts - their creations and annihilations in vacuum. These fundamental operator objects on the Fock space and the related symmetry properties of their algebraic representations proved to be governing in deepening our understanding the electron spin nature and its importance for the matter stability in the Universe.

The elementary point charged particle, like electron, its spin, energy spectrum and mass problem were inspiring many physicists [26] from the past as J. J. Thompson, G.G. Stokes, H.A. Lorentz, E. Mach, M. Abraham, P.A. M. Dirac, G.A. Schott and others. Nonetheless, their studies have not given rise up to date to a clear explanation of this phenomenon that stimulated new researchers to tackle it from different approaches based on new ideas stemming both from the classical Maxwell-Lorentz electromagnetic theory, as in [8, 20, 21, 40, 15, 16, 22, 23, 27, 28, 30], and modern quantum field theories [31, 33, 34, 35, 37, 42, 45, 48, 51] of Yang-Mills and Higgs type, as in [1, 24, 25, 50] and others, whose recent and extensive review is done in [49].

In the work we mostly concentrate on some details of quantum analysis of the energy spectrum problem for the photon dressed electron model within the Fock many-temporal approach [12, 18] and the Feynman proper time paradigm [15, 16, 13, 14] subject to deriving the quantum electromagnetic Maxwell equations. For the first time the proper time approach to classical electrodynamics and quantum mechanics was possibly suggested still in 1937 by V. Fock [17], in which, in particular, there was constructed an alternative proper time based Lagrangian description of a point charged particle under external electromagnetic field. A more detailed dynamical motivation of using the proper time approach was later presented by R. Feynman in his Lectures [15]. Based on these motivations, we have stated that the self-interacting quantum mechanism of the charged particle with its self-generated electromagnetic field consists of two physically different phenomena, whose influence on the structure of the resulting Hamilton interaction operator proved to be crucial, ensuing a modified analysis of the problem within the Lagrangian formalism.

2. Electron spin representation and related algebraic structures

Still in 1932 there were introduced the following, in some sense virtual, linear operators on a suitably chosen *"physical"* Hilbert space Φ of quantum electron states:

1) *creation operator*

$$\psi^{+}(x) : \Phi \to \Phi, x \in \mathbf{R}^3, \tag{1}$$

of quantum states, assigned to electron, localized in point $x \in \mathbf{R}^3$, and respectively,

2) *annihilation operator*

$$\psi(x) : \Phi \to \Phi, \ x \in \mathbf{R}^3, \tag{2}$$

of quantum states, assigned to electron, localized in point $x \in \mathbf{R}^3$, as the corresponding operator conjugation to the introduced above creation operator:

$$(\psi^{+}(x)h|g)_{\Phi} = (h|\psi(x))g)_{\Phi} \tag{3}$$

for all $x \in \mathbf{R}^3$ with respect to the scalar product $(\cdot|\cdot)_{\Phi}$ on Φ for all $|h), |g) \in \Phi$ for which there exists a so called *"vacuum"* vector $|\Omega_0) \in \Phi$, satisfying the determining relationship:

$$\psi(x)|\Omega_0) = 0 \tag{4}$$

for all $x \in \mathbf{R}^3$.

It is worth noting that the annihilation operator simulates the real physical process of the annihilation of an electron at spatial point $x \in \mathbf{R}^3$, sending it into the vacuum, where countless electrons exist being yet non observable. Thus, one can create new quantum states $|\Omega_n) \in \mathcal{H}^{\otimes n} := \underbrace{\mathcal{H} \otimes \mathcal{H} \otimes \cdots \otimes \mathcal{H}}$ for $n \in \mathbf{N}$ electrons, located, respectively, at points $x_j \in \mathbf{R}^3, j = \overline{1,n}$, via acting on the vacuum state $|\Omega_0)$ of the Hilbert space Φ by means of the creation operators:

$$|\Omega_n(x_1, x_2, ..., x_n) := \psi^+(x_n)\psi^+(x_{n-1})...\psi(x_1)|\Omega_0). \tag{5}$$

As a number $n \in \mathbf{N}$ of electrons is not fixed, all such quantum states (5) should be combined as the direct sum into the unique Hilbert space $\Phi \simeq \oplus_{n\in\mathbf{Z}_+}\mathcal{H}^{\otimes n}$, isometric to the Hilbert space Φ, what was done for the first time by V. Fock [17] in 1932 and called be the *Fock space* of multiparticle quantum states. Really, if to introduce on the space Φ the scalar product

$$(\tilde{f}|\tilde{g})_\Phi := \sum_{n\in\mathbf{Z}_+} \frac{1}{n!}(f_n|g_n)_{\mathcal{H}^{\otimes n}} \tag{6}$$

for any $f, g \in \Phi$, then any quantum state $f \in \Phi$ can be uniquely represented as

$$\tilde{f} = \sum_{n\in\mathbf{Z}_+} \frac{1}{\sqrt{n!}} \int_{\mathbf{R}^{3\times n}} f_n(x_1, x_2, ..., x_n)|\Omega_n(x_1, x_2, ..., x_n) \prod_{j=\overline{1,n}} dx_j \tag{7}$$

with coefficients $f_n \in \mathcal{H}^{\otimes n}, n \in \mathbf{N}$, where the basic quantum electron states $|\Omega_n) \in \Phi, n \in \mathbf{N}$, defined by the expression (5), can be taken to be orthonormal and dense in $\mathcal{H}^{\otimes n}$. Having calculated the scalar product $(\tilde{f}|\tilde{g})_\Phi$ for arbitrary vectors $\tilde{f}, \tilde{g} \in \Phi$, one easily obtains from (6) that $(\tilde{f}|\tilde{g})_\Phi = (f|g)_{\oplus_{n\in\mathbf{Z}_+}\mathcal{H}^{\otimes n}}$, meaning, evidently, that the natural mapping $\Phi \ni \tilde{f} \to f \in \oplus_{n\in\mathbf{Z}_+}\mathcal{H}^{\otimes n}$ is an isometry.

Yet now an interesting question arises: what is a nature of these multiparticle Fock subspaces $\Phi_n \simeq \mathcal{H}^{\otimes n}, n \in \mathbf{N}$, within which there can be created and annihilated many of non distinguished electrons at different points of space? Having restoring to the experimental fact [10, 11, 47] explained by W. Pauli [36] that the quantum n- electron states $|\Omega_n) \in \Phi_n$, $n \in \mathbf{N}$, persist to be invariant under spatial replacement of the electrons up to the sign, meaning nothing else than these quantum states are skew-symmetric functions of their spatial arguments with respect to the usual permutation symmetry group Σ_n, which acts on Φ_n the following way:

$$\sigma : |\Omega_n(x_1, x_2, ..., x_n) \to (-1)^\sigma |\Omega_n(x_{\sigma^{-1}(1)}, x_{\sigma^{-1}(2)}, ..., x_{\sigma^{-1}(n)}) \tag{8}$$

for any permutation $\sigma \in \Sigma_n$ with parity $(-1)^\sigma$ and $(x_1, x_2, ..., x_n) \in \mathbf{R}^{3\times n}$.

Consider now the action of the creation operator $\psi^+(x) : \Phi \to \Phi, x \in \mathbf{R}^3$, based on the definition (5):

$$\psi^+(x)|\Omega_n(x_1, x_2, ..., x_n) := |\Omega_{n+1}(x, x_1, x_2, ..., x_n), \tag{9}$$

one easily obtains that the related action

$$\psi(x)|\Omega_n(x_1, x_2, ..., x_n) = \sum_{j=\overline{1,n}} (\varepsilon)^{j-1}\delta(x - x_j)|\Omega_{n-1}(x_1, x_2, ..., \hat{x}_j, ..., x_n), \tag{10}$$

where $\varepsilon = 1$ for symmetric and $\varepsilon = -1$ for anti-symmetric Hilbert space $\mathcal{H}^{\otimes n}, n \in \mathbf{N}$. Based now on the relationships (9) and (10), it is easy to check that the following algebraic commutation expressions

$$\begin{gathered}[\psi(f_1), \psi(f_2)]_\varepsilon = 0 = [\psi^+(f_1), \psi^+(f_2)]_\varepsilon, \\ [\psi(f_1), \psi^+(f_2)]_\varepsilon = (f_1|f_2)_\mathcal{H}\end{gathered} \tag{11}$$

hold for arbitrary $f_1, f_2 \in S(\mathbf{R}^3;\mathbf{C})$, where $[\cdot,\cdot]_\varepsilon$ denotes commutator for $\varepsilon = 1$ and anticommutator for $\varepsilon = -1$, and for an operator $a(x) : \Phi \to \Phi$, $x \in \mathbf{R}^3$, the notation $a(f) := \int_{\mathbf{R}^3} f(x)a(x)dx, f \in S(\mathbf{R}^3;\mathbf{C})$, means its smearing over $\mathbf{R}^3$ with the *"density"* $f \in S(\mathbf{R}^3;\mathbf{C})$. In addition, one easily states that the following operator

$$N := \int_{\mathbf{R}^3} \psi^+(x)\psi(x)dx \tag{12}$$

on Φ is selfadjoint and integer-valued, counting amounts of particles, described by quantum states in Φ. Since for such an elementary quantum particle as electron there is realized the case $\varepsilon = 1$, that is electron is a particle, satisfying the fermi type statistics, and the basic Hilbert subspaces $\mathcal{H}^{\otimes n}, n \in \mathbf{N}$, are skew-symmetric, on which there is realized a two-valued representation of the symmetric group Σ_n. The corresponding two-valued representation Fock space $\Phi := \Phi^{(x)} \otimes \Phi_{int}$ can be, obviously, generated by the extended vacuum vector $|\Omega_0) := |\Omega_0^{(x)}) \otimes |\Omega_{int,0}) \in \Phi^{(x)} \otimes \Phi_{int}$, which is contains the isometrically imbedded Hilbert space $\Phi := \Phi^{(x)} \otimes \mathbf{C}^2 \widetilde{\to} \Phi^{(x)} \otimes \Phi_{int}$.

To understand now in more detail the structure of this Fock space Φ subject to the real electron particle evolution in the Minkowski [32] space-time $M^4 = \mathbf{R}_t \times \mathbf{R}^3_x$, we proceed to analyzing electrons states generated by the algebra of the creation-annihilation operators $\psi^+(x_j), \psi(x_k) : \mathcal{H}^{\otimes n} \to \mathcal{H}^{\otimes n}, j,k = \overline{1,n}$, satisfying the anti-commutation relationships

$$\begin{aligned} \{\psi(x_j), \psi(x_k)\} &= 0 = \{\psi^+(x_j), \psi^+(x_k)\}, \\ \{\psi(x_j), \psi^+(x_k)\} &= \delta_{jk}, \end{aligned} \tag{13}$$

where we have redefined the anti-commutator notation $[\cdot,\cdot]_{+1} := \{\cdot,\cdot\}$, and generating the so called Clifford algebra $Cl_{2n}(\psi,\psi^+)$, well known in modern mathematics. The Clifford algebra $Cl_{2n}(\psi,\psi^+), n \in \mathbf{N}$, as a mathematical physics object, is constructed by means of the usual factorization procedure the following way:

$$Cl_{2n}(\psi,\psi^+) := T_{2n}(\psi,\psi^+)/J_{2n}(\psi,\psi^+), \tag{14}$$

where $T_{2n}(\psi,\psi^+)$– the tensor algebra over $\mathbf{C}$, spanned by the operator elements $\{\psi^+(x_j), \psi(x_k) : \mathcal{H}^{\otimes n} \to \mathcal{H}^{\otimes n} : j,k = \overline{1,n}\}$, and $J_{2n}(\psi,\psi^+)$– the two-sided ideal, generated by elements $\{\psi(x_k)\psi(x_j) + \psi(x_j)\psi(x_k), \psi(x_k)\psi^+(x_j) + \psi^+(x_j)\psi(x_k) - \delta_{jk}, \psi^+(x_k)\psi^+(x_j) + \psi^+(x_j)\psi^+(x_k) : j,k = \overline{1,n}\}$, whose dimension $\dim Cl_{2n}(\psi,\psi^+) = 2^{2n}$. Moreover, as our electron Fock space Φ is also specified by the symmetry group $SU(2)$, acting on the complex space $\mathbf{C}^2 \tilde{\subset} \Phi_{int}$, imbedded into the internal Fock space Φ_{int}, whose Lie algebra $su(2)$ generators are realized by the selfadjoint Pauli spin matrices $\sigma_j \in \mathrm{End}\,\mathbf{C}^2, j = \overline{1,3}$, where

$$\sigma_1 = \begin{pmatrix} 0 & 1 \\ 1 & 0 \end{pmatrix}, \sigma_2 = \begin{pmatrix} 0 & -i \\ i & 0 \end{pmatrix}, \sigma_3 = \begin{pmatrix} 1 & 0 \\ 0 & -1 \end{pmatrix}, \tag{15}$$

satisfying the following commutator relationships:

$$[\sigma_1, \sigma_2] = \sigma_3, [\sigma_2, \sigma_3] = \sigma_1, [\sigma_3, \sigma_1] = \sigma_2, \tag{16}$$

and, respectively, forming the corresponding Clifford algebra $Cl_3(\sigma)$, reflecting the symmetry of electron quantum states subject to its internal, not depending on spatial coordinates symmetry, but related only with *"virtual"* degrees of freedom, represented by means of the *"spin"*-matrices (15) on $\mathbf{C}^2$.

Let now a matrix operator $S \in \mathbb{E}^3 \otimes \mathrm{End}(\Phi_{int})$, called a *spin operator*, be defined the following way: the projection $S|_{(p\otimes p)\otimes \mathrm{End}(\Phi_{int})} \simeq \langle \frac{\hbar p}{2|p|} | (\sigma_1, \sigma_2, \sigma_3)^\intercal \rangle \in \mathrm{End}(\mathbb{C}^2)$ on the momentum vector

$p \in \mathbb{E}^3$ of electron, where $\hat{p}\Omega_1(x) = p\Omega_1(x), \hat{p} := \frac{\hbar}{i}\nabla_x$ – the electron momentum operator at a spatial point $x \in \mathbb{R}^3$, $\Omega_1 : \mathbb{R}^3 \rightarrow \Phi_1^{(x)} \otimes \mathbb{C}^4$ is the corresponding eigenfunction and "$\hbar$" is the so called quantum Planck constant.

Concerning the Clifford algebra $Cl_3(a)$, its dimension is $\dim Cl_3(\sigma) = 8$ and describes the internal degrees of freedom of electron, whose structure is based on a weakly known physical nature of the corresponding internal quantum states, yet its representation space $\mathbf{C}^2$ makes it possible to propose some useful inferences based on their algebraic structure. There is accepted [9, 46] the following definition of the Clifford algebra representation.

Definition 1 *The left module of a Clifford algebra $Cl_m(\rho)$, - it is a linear finite-dimensional space L over $\mathbf{C}$ and a linear mapping $\rho : Cl_m(\rho) \otimes L \rightarrow L$, such that*

$$\rho(ab \otimes f) = \rho(a \otimes \rho(b \otimes f)) \tag{17}$$

for all $a, b \in Cl_m(\rho)$ and $f \in L$. In particular, a representation mapping $\rho : Cl_m(\rho) \otimes L \rightarrow L$ is a homomorphism of the corresponding algebras.

The stated above algebraic relationships (16) should be naturally realized by means of basis operators of the so called *canonical* Clifford algebra $Cl_m(\beta, \beta^+)$ of the minimal dimension $4 = \dim \mathrm{End}\, \mathbf{C}^2$, generated by some virtual creation-annihilation operators $\beta, \beta^+ : \Phi_{int} \rightarrow \Phi_{int}$ of the internal quantum states. That means that the following equivalence of the representations of $Cl_4(\beta, \beta^+)$ and $Cl_3(\sigma)$

$$\rho(Cl_m(\beta, \beta^+)) \simeq \rho(Cl_3(\sigma)) \tag{18}$$

in the internal Fock space Φ_{int} should hold, giving rise to the condition $m = 2$. As a consequence, we derive the existence of some algebraic relationships $\sigma_j = \sigma_j(\beta, \beta^+), j = \overline{1,3}$, which can be found the following way. Let us construct the following *"chirality"* operator

$$\Gamma := \exp(i\pi\beta^+\beta) : \Phi_{int} \rightarrow \Phi_{int}, \tag{19}$$

which satisfies the evident condition $\Gamma^2 = I$ and anti-commutes with operators β, β^+ :

$$\{\Gamma, \beta\} = 0 = \{\Gamma, \beta^+\}. \tag{20}$$

The latter makes it possible to determine the following projection operators:

$$P_{\pm} := (I \pm \Gamma)/2 \quad \Rightarrow \left\{ \begin{array}{c} P_+ = I - \beta^+\beta \\ P_- = \beta^+\beta \end{array} \right. , \tag{21}$$

which split the internal Fock space Φ_{int} into the direct sum: $\Phi_{int} = \Phi_{int}^{(+)} \oplus \Phi_{int}^{(-)}$, where $\Phi_{int}^{(\pm)} := P_{\pm}\Phi_{int}$. Moreover, the following condition

$$(\beta^+ + \beta)\Phi_{int}^{(+)} = \Phi_{int}^{(-)} \tag{22}$$

holds. Having now observed that $\Gamma = I - 2\beta^+\beta \in Cl_2(\beta, \beta^+)$, one easily obtains that the operator expressions

$$\sigma_1 = (\beta^+ + \beta)/2, \sigma_2 = -i(\beta^+ - \beta)/2, \sigma_3 = -\Gamma \tag{23}$$

satisfy the determining commutator relationships (16).

Remark 1 *If to consider the eight-dimensional Clifford algebra $Cl_3(\sigma)$ as a basic one, then since* $\dim Cl_3(\sigma) = 8 = \dim\left(\mathrm{End}(\mathbb{C}^2) \otimes \mathbb{C}[\varepsilon]\right)$, $\varepsilon^2 := i\sigma_1\sigma_2\sigma_3$, *where the operator* $\varepsilon = \pm\sqrt{i\sigma_1\sigma_2\sigma_3}$ *is equivalent to the "chirality" operator (19) of the electron, pointing at the existence of two non-equivalent representations of the Clifford algebra $Cl_3(\sigma)$ both on the space $\mathbb{C}^2$ by means of Pauli matrices from* $\mathrm{End}(\mathbb{C}^2) \otimes \mathbb{C}[\varepsilon = +\sqrt{i\sigma_1\sigma_2\sigma_3}]$ *and by means of the Pauli matrices from* $\mathrm{End}(\mathbb{C}^2) \otimes \mathbb{C}[\varepsilon = -\sqrt{i\sigma_1\sigma_2\sigma_3}]$, *for which the projection of the spin operator* $S \in \mathbb{E}^3 \otimes \mathrm{End}(\Phi_{int})$ *is directed along the electron momentum, or against it, respectively.*

It is easy now to observe that the basis generators of the internal symmetry group $SU(2)$, corresponding to the spin operator $S \in \mathbb{E}^3 \otimes \mathrm{End}(\Phi_{int})$, make it possible to construct an equivalent representation $\rho : Cl_3(\mu) \to \mathrm{End}(\mathbb{C}^2)$ of some one-particle four-dimensional Clifford algebra $Cl_3(\mu)$ on the space of states $\mathbb{C}^2 \simeq \Phi_{int}$ for an electron at fixed point $x \in \mathbb{R}^3$:

$$
\begin{aligned}
\rho(\mu_1(z)\,) \; &: = S_1 = \frac{\hbar}{2}\sigma_1, \rho(\mu_2(z)) := S_2 = \frac{\hbar}{2}\sigma_2, \\
\rho(\mu_3(z)) \; &: = S_3 = \frac{\hbar}{2}\sigma_3,
\end{aligned}
\tag{24}
$$

which satisfies the following commutator relationships:

$$
[S_1, S_2] = i\hbar S_3, \quad [S_2, S_3] = i\hbar S_1, \quad [S_3, S_1] = i\hbar S_2. \tag{25}
$$

The corresponding basis elements of the Clifford algebra $Cl_3(\mu)$, whose action is localized on a space Z^3_{int} of some internal parameters, satisfy the following commutator relationships:

$$
[\mu_1, \mu_2] = i\hbar\mu_3, \quad [\mu_2, \mu_3] = i\hbar\mu_1, \quad [\mu_3, \mu_1] = i\hbar\mu_2, \tag{26}
$$

which remind (yet not more!) some the real angular momentum relationships for electron localized at spatial point $x \in \mathbb{R}^3$, and namely they are called components of the electron spin operator $S \in \mathbb{E}^3 \otimes \mathrm{End}(\Phi_{int})$, since the *"vector"* $\mu := (\mu_1, \mu, \mu_3) \in \mathbb{E}^3 \otimes \mathrm{End}(\Phi_{int})$ of a selfadjoint quantum-mechanical *"angular momentum"* operator of the electron on the space $Z^3_{int} \simeq \mathbb{R}^3$ of internal parameters is formally modelled as the vector product $\mu \simeq z \times p$, $z \in Z^3_{int}$, where we took into account that the generalized internal selfadjoint momentum operator of the electron equals $p := \frac{\hbar}{i}\nabla_z$, $z \in Z^3_{int}$, imitating some observable physical quantity, corresponding to the internal spin structure of electron as a physical object of microworld.

Moreover, since a real electron exists not only in the physical spatial space $\mathbb{R}^3$, but it conserves its structure also in time, the corresponding temporal symmetry responsible for the lasting of electron states is naturally generated by means of some so called Hamiltonian operator $H_f : \Phi^{(x)} \otimes \Phi_{int} \to \Phi^{(x)} \otimes \Phi_{int}$, whose reduction on the subspace $\Phi^{(x)}_n \otimes \Phi_{int}$, $n \in \mathbb{N}$, belongs to some representation of the Clifford algebra $Cl_{2n}(\psi, \psi^+) \otimes (Cl_3(\mu) \otimes Cl_3(\mu))$, for which *a priori* there holds the following invariance relationships:

$$
S_3 : \Phi^{(x)}_n \otimes \Phi^{(\pm)}_{int} \to \Phi^{(x)}_n \otimes \Phi^{(\pm)}_{int}, \tag{27}
$$

ensuing the commutation condition $[S_3, H_f] = 0$, satisfied on the whole Fock space $\Phi = \Phi^{(x)} \otimes \Phi_{int}$. The latter is compatible with the classical Dirac [11] representation of the quantum Hamiltonian operator

$$
H_f = \int_{\mathbb{R}^3} dx \psi^+(x) \langle \alpha | \nabla_x \rangle \psi \tag{28}
$$

of electron on the Fock space $\Phi^{(x)} \otimes \Phi_{int}$, whose invariant reductions $H_{f,n} : \Phi^{(x)}_n \otimes \Phi_{int} \to \Phi^{(x)}_n \otimes \Phi_{int}$, $n \in \mathbb{N}$, belong to the Clifford algebra $Cl_{2n}(\psi, \psi^+) \otimes (Cl_3(\sigma) \otimes Cl_3(\sigma))$, $n \in \mathbb{N}$, as

$\alpha \in Cl_3(\sigma) \otimes Cl_3(\sigma) \subset \mathrm{End}(\mathbb{C}^4)$. It is worth to remark here that the Hamiltonian operator (28) of electron and the operator of the number of electrons (12) are commuting to each other, that is

$$[H_f, N] = 0 \tag{29}$$

on the whole Fock space $\Phi^{(x)} \otimes \Phi_{int}$. Really, as the temporal evolution of any operator $A : \Phi^{(x)} \otimes \Phi_{int} \to \Phi^{(x)} \otimes \Phi_{int}$ is described by means of the classical Heisenberg equation

$$\frac{\partial A}{\partial t} = \frac{i}{\hbar}[H_f, A], \tag{30}$$

where $\hbar$ denotes the before introduced physical Plank constant, the condition (29) simply means that the observable number of electrons in quantum states is conserved. Moreover, as it was shown still by P. Dirac, the formal algebraic sum of the spin $I \otimes S : \tilde{\Phi} \otimes \tilde{\Phi}_{int} \to \Phi \otimes \Phi_{int}$ and angular momentum $J \otimes I : \Phi \otimes \Phi_{int} \to \Phi \otimes \tilde{\Phi}_{int}$ operators persist to be invariant in time, that is

$$[H_f, J \otimes I + I \otimes S] = 0, \tag{31}$$

that can be easily checked by simple calculations.

Taking into account analytical properties of discussed above quantum states associated with electron and describing its evolution in the ambient Minkowski space-time, one can realize that the electron spin $S \in \mathbb{E}^3 \otimes \mathrm{End}(\Phi_{int})$ is a physical quantity, responsible for its hidden internal structure, mathematically related with the symmetry group $SU(2)$ of the electron quantum states and their representations. The latter are compatible both with the related representations of the algebra of observable operators, generated by the basic Clifford algebra $Cl_{2n}(\psi, \psi^+) \otimes (Cl_3(\sigma) \otimes Cl_3(\sigma))$ on the Fock subspaces $\Phi_n \otimes \Phi_{int}, n \in \mathbf{N}$, and with the Hamiltonian operator (28), responsible for the description of localized quantum electron energy states by means of suitably constructed antisymmetric functions from Φ_n for any $n \in \mathbf{N}$, and which can be interpreted as the mass spectrum of electron, naturally interacting with the ambient quantum electromagnetic field. The latter very interesting and important problem is analyzed below within the Feynman proper time and Fock many-temporal paradigms, applied to the physical quantum electrodynamic model, describing the quantum Dirac electron interacting with the quantized ambient Maxwell electromagnetic field.

3. The photon field dressed electron energy spectrum problem

Consider a free relativistic quantum electron *a priori* massless field described [5, 7] by means of the secondly-quantized self-adjoint Dirac-Weil type Hamiltonian operator (28)

$$H_f = \int_{\mathbf{R}^3} d^3x \psi^+ \langle \frac{ic}{\hbar} \alpha | \nabla \rangle \psi, \tag{32}$$

where $\alpha \in \mathbf{E}^3 \otimes \mathrm{End}\, M^4$ denotes the standard Dirac spin matrices vector, $c \in \mathbf{R}_+$ is the light velocity, $\langle \cdot | \cdot \rangle$ denotes the usual scalar product in the Euclidean space $\mathbf{E}^3$, $\psi : \mathbf{R}^3 \to (\mathrm{End}\, \Phi)^4$- a four-vector of the quantum annihilation operators, acting in a suitable Fock space Φ and $\psi^+ : \mathbf{R}^3 \to (\mathrm{End}\, \Phi)^4$- the corresponding adjoint co-vector of creation operators in the Fock space Φ. The following anticommuting [5, 7] operator relationships

$$\begin{aligned} \psi_j(x)\psi_l^+(y) + \psi_l^+(y)\psi_j(x) &= \delta_{jl}\delta(x-y), \\ \psi_j(x)\psi_l(y) + \psi_l(y)\psi_j(x) &= 0 = \psi_j^+(x)\psi_l^+(y) + \psi_l^+(y)\psi_j^+(x) \end{aligned} \tag{33}$$

for the Clifford algebra $Cl_{2n}(\psi, \psi^+)$ hold for any $x, y \in \mathbf{R}^3$ and $j, l \in \overline{1,4}$, being compatible with the related Heisenberg operator dynamics, generated by the fermionic Hamiltonian operator (32):

$$\partial\psi/\partial\bar{t} := \frac{i}{\hbar}[H_f, \psi], \quad \partial\psi^+/\partial\bar{t} := \frac{i}{\hbar}[H_f, \psi^+] \tag{34}$$

with respect to its own laboratory reference frame $\mathcal{K}_{\bar{t}}$, parameterized by the evolution parameter $\bar{t} \in \mathbf{R}$. The latter, in particular, means that the commutation relationships (33)

It is clear that the Hamiltonian (32) possesses no information of such an important characteristic as the electric charge $\xi \in \mathbf{R}$, which generates the own electromagnetic field interacting both with it and with other ambient charged particles. As it is usually accepted, we will model a free electromagnetic field by its bosonic self-adjoint operator four-potential $(\varphi, A) : \mathbf{R}^3 \to (\text{End } \Phi)^4$, whose evolution is generated by the self-adjoint Hamiltonian

$$H_b = 2\int_{\mathbf{R}^3} d^3k|k|^2(\langle A^+(k)|A(l)\rangle - \varphi^+(k)\varphi(k)), \tag{35}$$

acting in the before introduced common Fock space Φ and represented by means of the field operators expanded into the Fourier integrals

$$\begin{aligned} \varphi(x) \quad &: = \frac{1}{(2\pi)^{3/2}}\int_{\mathbf{R}^3} d^3k\varphi(k)\exp(i\langle k|x\rangle) + \frac{1}{(2\pi)^{3/2}}\int_{\mathbf{R}^3} d^3k\varphi^+(k)\exp(-i\langle k|x\rangle), \\ A(x) \quad &: = \frac{1}{(2\pi)^{3/2}}\int_{\mathbf{R}^3} d^3kA(k)\exp(i\langle k|x\rangle) + \frac{1}{(2\pi)^{3/2}}\int_{\mathbf{R}^3} d^3kA^+(k)\exp(-i\langle k|x\rangle)\ . \end{aligned} \tag{36}$$

The coefficients of the expansions (36) satisfy the following [7, 12, 18] commutation operator relationships:

$$\begin{aligned} [\varphi(k), \varphi^+(s)] &= -\frac{c\hbar}{2|k|}\delta(k-s), [\varphi(k), A_j(s)] = 0, \\ [\varphi(k), \varphi(s)] &= 0 = [\varphi^+(k), \varphi^+(s)], \\ [A_j(k), A_l^+(s)] &= \frac{c\hbar}{2|k|}\delta_{jl}\delta(k-s), \\ [A_j(k)\otimes, A_l(s)] &= 0 = [A_j^+(k), A_l^+(s)] \end{aligned} \tag{37}$$

for all $k, s \in \mathbf{E}^3$ and $j, l \in \overline{1,3}$, compatible with the related Heisenberg operator dynamics [7] generated by the electromagnetic field Hamiltonian (35):

$$\frac{\partial A}{\partial\tilde{t}} := \frac{i}{\hbar}[H_b, A], \frac{\partial\varphi}{\partial\tilde{t}} := \frac{i}{\hbar}[H_b, \varphi], \tag{38}$$

with respect to its own laboratory reference frame $\mathcal{K}_{\tilde{t}}$, parameterized by the temporal parameter $\tilde{t} \in \mathbf{R}$. In particular, based on the commutation relationships (37), one can check that the electric

$$E := -\nabla\varphi - \frac{1}{c}\frac{\partial A}{\partial\tilde{t}} \tag{39}$$

and magnetic

$$B := \nabla \times A \tag{40}$$

fields satisfy the vacuum operator Maxwell equations, if the following weak Lorenz type constraints

$$C_0(k)\Phi \ : \ = i(\langle k|A(k)\rangle - |k|\varphi(k))\Phi = 0, \tag{41}$$

$$C_0^+(k)\Phi \ : \ = -i(\langle k|A^+(k)\rangle - |k|\varphi^+(k))\Phi = 0,$$

hold for all $k \in \mathbf{E}^3$. As the operators $C_0(k) : \Phi \to \Phi$ and $C_0^+(k) : \Phi \to \Phi$ are commuting both to each other for all $k \in \mathbf{E}^3$ and with the Hamiltonian (35), the constraints are compatible with the evolution operator equations (38).

Remark 2 *The Hamiltonian operator (35) expression easily follows [5, 7, 18] from the well known relativistic invariant classical Fock-Podology electromagnetic Lagrangian*

$$\mathcal{L}_b \ : \ = \frac{1}{2}\int_{\mathbf{R}^3} d^3x[\langle \nabla\varphi + \frac{1}{c}\frac{\partial A}{\partial \tilde{t}}|\nabla\varphi + \frac{1}{c}\frac{\partial A}{\partial \tilde{t}}\rangle - \\ -\langle \nabla \times A|\nabla \times A\rangle - (\frac{1}{c}\frac{\partial \varphi}{\partial \tilde{t}} + \langle \nabla|A\rangle)^2] \tag{42}$$

Based on the corresponding to (42) Euler-Lagrange equations one finds that

$$\frac{1}{c^2}\frac{\partial^2 A}{\partial \tilde{t}^2} - \Delta A = 0, \qquad \frac{1}{c^2}\frac{\partial^2 \varphi}{\partial \tilde{t}^2} - \Delta\varphi = 0, \tag{43}$$

whose wave solutions allow to determine the electromagnetic fields (39) and (40) and check that the related Maxwell field equations in vacuum are satisfied if the Lorenz condition

$$C_0(\tilde{t}, x) := \frac{1}{c}\frac{\partial \varphi}{\partial \tilde{t}} + \langle \nabla|A\rangle = 0 \tag{44}$$

hold for all $(\tilde{t}, x) \in M^4$. In particular, taking into account the operator expansions (36), from (44) one obtains the corresponding operator expressions

$$C_0(k) \quad = \quad i\langle k|A^+(k)\rangle - |k|\varphi(k) \ \neq 0, \tag{45}$$

$$C_0^+(k) \quad = \quad -i\langle k|A^+(k)\rangle - |k|\varphi^+(k) \ \neq 0$$

for all $k \in \mathbf{E}^3$, contradicting the relationships, following from the evolution equations (38). As the latter relationships should be vanishing in the Fock space, it was suggested [12] to reduce the Fock space Φ to a subspace, on which there are satisfied the weak operator constraints (41).

As we are interested in describing the self-interaction of the fermionic quantum particle field $\psi : \Phi \to \Phi$ with the related self-generated bosonic electromagnetic potentials field $(\varphi, A) : \Phi \to \Phi$, we need, within the Fock many-temporal description approach [12, 18], first to consider the fermionic particle and bosonic electromagnetic fields with respect to the common reference frame $\mathcal{K}_t$ specified by the temporal parameter $t \in \mathbf{R}$. Secondly, we need to make use of the classical *"minimum interaction"* principle [6, 3] and to apply to the Hamiltonian operator expression (32):

$$H_f^{(int)} = \int_{\mathbf{R}^3} d^3x\psi^+\langle\frac{ic}{\hbar}\alpha|\nabla\rangle\psi + \int_{\mathbf{R}^3} d^3x(\xi\psi^+\psi\varphi - \psi^+\langle\frac{i\xi c}{\hbar}\alpha|A\rangle\psi), \tag{46}$$

in which the fermionic $\psi : \Phi \to \Phi$ and bosonic $(\varphi, A) : \Phi \to \Phi$ operators are commuting *a priori* to each other as quantum fields of different nature. Since the whole quantum field system consists of the fermionic particle and bosonic self-generated electromagnetic fields, its evolution is described by means of the joint Hamiltonian operator

$$H_{f-b} := H_f^{(int)} + H_b \tag{47}$$

as

$$\begin{aligned} \frac{\partial \psi}{\partial t} &:= \frac{i}{\hbar}[H_{f-b}, \psi], \quad \frac{\partial \psi^+}{\partial t} := \frac{i}{\hbar}[H_{f-b}, \psi^+], \\ \frac{\partial A}{\partial t} &:= \frac{i}{\hbar}[H_{f-b}, A], \quad \frac{\partial \varphi}{\partial t} := \frac{i}{\hbar}[H_{f-b}, \varphi] \end{aligned} \tag{48}$$

with respect to the common temporal parameter $t \in \mathbf{R}$, as in this case there is assumed that the corresponding temporal parameters $\tilde{t} \in \mathbf{R}$ and $\bar{t} \in \mathbf{R}$ coincide, that is $\tilde{t} = \bar{t} = t \in \mathbf{R}$. Simultaneously there, evidently, should be satisfied the before derived both the electromagnetic field evolution equations (38) with respect to the own reference frame $\mathcal{K}_{\tilde{t}}$ and the modified fermionic charged particle field equations

$$\frac{\partial \psi}{\partial \bar{t}} := \frac{i}{\hbar}[H_f^{(int)}, \psi], \quad \frac{\partial \psi^+}{\partial \bar{t}} := \frac{i}{\hbar}[H_f^{(int)}, \psi^+] \tag{49}$$

with respect to the own reference frame $\mathcal{K}_{\bar{t}}$.

Being mostly interested in the evolution of the quantum particle fermionic field $\psi : \Phi \to \Phi$, we can get rid of the bosonic Hamiltonian impact into (49) having applied to the Fock space Φ the unitary canonical transformation

$$\Phi \to \tilde{\Phi} := U(t)\ \Phi, \tag{50}$$

where we denoted by $U(t) : \Phi \to \Phi$ the unitary operator satisfying the determining equation $dU(t)/dt = \frac{i}{\hbar} H_f^{(int)} U(t)$ subject to the bosonic Hamiltonian operator (35) and the temporal parameter $t \in \mathbf{R}$. As a consequence of the transformation (50) we obtain the effective fermionic particle field interaction Hamiltonian operator

$$\begin{aligned} \tilde{H}_f^{(int)} &:= U(t) H_f^{(int)} U^*(t) = \\ &= \int_{\mathbf{R}^3} d^3x \psi^+ \langle \frac{ic}{\hbar} \alpha | \nabla \rangle \psi + \int_{\mathbf{R}^3} d^3x (\xi \psi^+ \psi \tilde{\varphi} - \psi^+ \langle \frac{i\xi c}{\hbar} \alpha | \tilde{A} \rangle \psi), \end{aligned} \tag{51}$$

where, by definition,

$$\tilde{A} := U(t) A U^*(t), \qquad \tilde{\varphi} := U(t) \varphi U^*(t)], \tag{52}$$

subject to which the evolution in the transformed Fock space $\tilde{\Phi}$, induced by the reduced bosonic positive definite Hamiltonian operator (35)

$$\tilde{H}_b = 2 \int_{\mathbf{R}^3} d^3k |k|^2 (\langle \tilde{A}^+(k) | \tilde{A}(l) \rangle - \tilde{\varphi}(k) \tilde{\varphi}^+(k)), \tag{53}$$

became completely eliminated. Concerning the Hamiltonian operator (53) here we need to mention that the related commutation relationships for the operators $(\tilde{\varphi}(k), \tilde{A}(k)) : \tilde{\Phi} \to \tilde{\Phi}$ and

$(\tilde{\varphi}^+(k), \tilde{A}^+(k)) : \tilde{\Phi} \to \tilde{\Phi}$ remain the same as (37), that is

$$\begin{aligned}
[\tilde{\varphi}(k), \tilde{\varphi}^+(s)] &= -\frac{c\hbar}{2|k|}\delta(k-s), \\
[\tilde{A}_j(k), \tilde{A}_l^+(s)] &= \frac{c\hbar}{2|k|}\delta_{jl}\delta(k-s), \\
[\tilde{\varphi}(k), \tilde{\varphi}(s)] &= 0 = [\tilde{\varphi}^+(k), \tilde{\varphi}^+(s)], \\
[\tilde{A}_j(k), \tilde{A}_l(s)] &= 0 = [\tilde{A}_j^+(k), \tilde{A}_l^+(s)], \quad [\tilde{\varphi}(k), \tilde{A}_j(s)] = 0,
\end{aligned} \tag{54}$$

for all $k, s \in \mathbf{E}^3$ and $j, l \in \overline{1,3}$.

Now, concerning the Hamiltonian operators (51) and (53), the following Heisenberg evolutions equations

$$\frac{\partial \psi}{\partial \bar{t}} := \frac{i}{\hbar}[\tilde{H}_f^{(int)}, \psi], \qquad \frac{\partial \psi^+}{\partial \bar{t}} := \frac{i}{\hbar}[\tilde{H}_f^{(int)}, \psi^+] \tag{55}$$

with respect to the own reference frame $\mathcal{K}_{\bar{t}}$ and the Heisenberg evolution equations

$$\frac{\partial \tilde{\varphi}}{\partial \tilde{t}} := \frac{i}{\hbar}[\tilde{H}_b, \tilde{\varphi}], \qquad \frac{\partial \tilde{A}}{\partial \tilde{t}} := \frac{i}{\hbar}[\tilde{H}_b, \tilde{A}] \tag{56}$$

with respect to the own reference frame $\mathcal{K}_{\tilde{t}}$ hold. Being further interested in the evolution equations (48), suitably rewritten in the transformed Fock space $\tilde{\Phi}$ with respect to the common temporal parameter $t \in \mathbf{R}$, we need to take into account [18] that the following functional relationships

$$\psi(t) := \psi(\bar{t}, \tilde{t})|_{\bar{t}=\tilde{t}=t}, \qquad \tilde{A}(t) := \tilde{A}(\bar{t}, \tilde{t})|_{\bar{t}=\tilde{t}=t} \tag{57}$$

hold. In particular, from (57) the evolution expressions follow

$$\begin{aligned}
\partial \psi(t)/\partial t &= \partial \psi(\bar{t}, \tilde{t})/\partial \bar{t}|_{\bar{t}=\tilde{t}=t} + \partial \psi(\bar{t}, \tilde{t})/\partial \tilde{t}|_{\bar{t}=\tilde{t}=t}, \\
\partial \tilde{A}(t)/\partial t &= \partial \tilde{A}(\bar{t}, \tilde{t})/\partial \bar{t}|_{\bar{t}=\tilde{t}=t} + \partial \tilde{A}(\bar{t}, \tilde{t})/\partial \tilde{t}|_{\bar{t}=\tilde{t}=t},
\end{aligned} \tag{58}$$

for all $t \in \mathbf{R}$. The latter will be useful when deriving the resulting quantum Maxwell electromagnetic equations.

4. The modified Lorenz constraint and quantum Maxwell equations

Before doing this, we need to take into account that the weak operator Lorenz constraints (41), rewritten in the transformed Fock space $\tilde{\Phi}$, fail to be compatible with the evolutions (56), that is

$$\begin{aligned}
[\tilde{C}_0(k), \tilde{H}_f^{(int)}] &\neq 0 \neq [\tilde{C}_0^+(k), \tilde{H}_f^{(int)}], \\
[\tilde{C}_0(k), \tilde{H}_b] &\neq 0 \neq [\tilde{C}_0^+(k), \tilde{H}_b],
\end{aligned} \tag{59}$$

as well as

$$\tilde{C}_0(k)\tilde{\Phi} \; : \; = i(\langle k, \tilde{A}(k)\rangle - |k|\tilde{\varphi}(k))\tilde{\Phi} \neq 0, \tag{60}$$

$$\tilde{C}_0^+(k)\tilde{\Phi} \; : \; = -i(\langle k, \tilde{A}^+(k)\rangle - |k|\tilde{\varphi}^+(k))\tilde{\Phi} \neq 0$$

for all $k \in \mathbf{E}^3$. Notwithstanding, it is easy enough to check that the following operators

$$\tilde{C}(k) \; : \; = \tilde{C}_0(k) + \frac{i\xi \exp(-ic|k|\bar{t})}{2|k|(2\pi)^{3/2}} \int_{\mathbf{R}^3} \exp(-i\langle k|y\rangle)\psi^+(y)\psi(y)d^3y, \tag{61}$$

$$\tilde{C}^+(k) \; : \; = \tilde{C}_0^+(k) - \frac{i\xi \exp(ic|k|\bar{t})}{2|k|(2\pi)^{3/2}} \int_{\mathbf{R}^3} \exp(i\langle k|y\rangle)\psi^+(y)\psi(y)d^3y,$$

are commuting both to each other and with the Hamiltonian operators (51) and (53):

$$[\tilde{C}(k), \tilde{C}(s)] \; = \; 0 = [\tilde{C}^+(k), \tilde{C}(s)]$$

$$[\tilde{C}(k), \tilde{H}_f^{(int)}] \; = \; 0 = [\tilde{C}^+(k), \tilde{H}_f^{(int)}], \tag{62}$$

$$[\tilde{C}(k), \tilde{H}_b] \; = \; 0 = [\tilde{C}^+(k), \tilde{H}_b]$$

for all $k, s \in \mathbf{E}^3$. Thus, the related evolution flows (55) and (56) in the transformed Fock space $\tilde{\Phi}$ should be considered under the modified constraints

$$\tilde{C}(k)\tilde{\Phi} = 0 = \tilde{C}^+(k)\tilde{\Phi} \tag{63}$$

for all $k \in \mathbf{E}^3$. Taking into account the exact expression (61), the constraints (63) can be equivalently rewritten as

$$\tilde{C}(\bar{t}; \tilde{t}, x)\tilde{\Phi} = 0, \tag{64}$$

where for all $x \in \mathbf{R}$ and the corresponding temporal parameters $\bar{t}$ and $\tilde{t} \in \mathbf{R}$

$$\begin{gathered} \tilde{C}(\bar{t}; \tilde{t}, x) := \tfrac{1}{(2\pi)^{3/2}} \int_{\mathbf{R}^3} d^3k \tilde{C}(k) \exp(i\langle k|x\rangle - i|k|\tilde{t}) + \\ + \tfrac{1}{(2\pi)^{3/2}} \int_{\mathbf{R}^3} d^3k \tilde{C}^+(k) \exp(-i\langle k|x\rangle + i|k|\tilde{t}) = \qquad , \\ \langle \nabla | \tilde{A}\rangle + \tfrac{1}{c}\tfrac{\partial \tilde{\varphi}}{\partial \tilde{t}} - \tfrac{\xi}{2\pi} \int_{\mathbf{R}^3} d^3y \Theta(c(\bar{t} - \tilde{t}), |x - y|)\psi^+(y)\psi(y) \end{gathered} \tag{65}$$

in which we put, by definition, the relativistic generalized function

$$\Theta(c(\bar{t} - \tilde{t}), |x - y|) := \frac{\delta(|x - y| + c(\bar{t} - \tilde{t})) - \delta(|x - y| + c(\bar{t} - \tilde{t}))}{2|x - y|}, \tag{66}$$

dual to the well known generalized function

$$\delta(|x - y|^2 - c^2(\bar{t} - \tilde{t})^2) = \frac{\delta(|x - y| + c(\bar{t} - \tilde{t})) + \delta(|x - y| + c(\bar{t} - \tilde{t}))}{2|x - y|}$$

solving [5] the classical relativistic wave equation.

Remark 3 *It is here worthy to mention that the above defined operator* $\tilde{C}(\bar{t}) : \tilde{\Phi} \to \tilde{\Phi}$, *depending parametrically on the bosonic temporal parameter* $\bar{t} \in \mathbf{R}$, *satisfies the relativistic wave equation*

$$\frac{1}{c^2}\frac{\partial^2 \tilde{C}}{\partial \tilde{t}^2} - \Delta \tilde{C} = 0, \tag{67}$$

that can be easily checked, if to make use of the rewritten in the Fock space wave equations (43):

$$\frac{1}{c^2}\frac{\partial^2 \tilde{A}}{\partial \tilde{t}^2} - \Delta \tilde{A} = 0, \frac{1}{c^2}\frac{\partial^2 \tilde{\varphi}}{\partial \tilde{t}^2} - \Delta \tilde{\varphi} = 0. \tag{68}$$

Now we can proceed to deriving the quantum Maxwell equations starting from the operator equations (68) and the suitably transformed to the Fock space $\tilde{\Phi}$ electromagnetic fields definitions (39) and (40):

$$(\nabla \times \tilde{B} - \frac{1}{c}\frac{\partial \tilde{E}}{\partial \tilde{t}})\tilde{\Phi} = \frac{\xi}{2\pi}\nabla \int_{\mathbf{R}^3} d^3y \Theta(c(\bar{t} - \tilde{t}), |x - y|)\psi^+(y)\psi(y)\tilde{\Phi} \tag{69}$$

and

$$\langle \nabla | \tilde{E} \rangle \tilde{\Phi} = -\frac{\xi}{2\pi}\frac{\partial}{\partial \tilde{t}} \int_{\mathbf{R}^3} d^3y \Theta(c(\bar{t} - \tilde{t}), |x - y|)\psi^+(y)\psi(y)\tilde{\Phi}, \tag{70}$$

which are considered in the weak operator sense. Taking now into account the relationships (56) and (58), one can obtain strong operator relationships for the electrical and magnetic fields

$$\tilde{E} = -\frac{1}{c}\frac{\partial \tilde{A}}{\partial \tilde{t}} - \nabla \tilde{\varphi} = -\frac{1}{c}\frac{\partial \tilde{A}}{\partial \tilde{t}} - \nabla \tilde{\varphi}, \quad \tilde{B} = \nabla \times \tilde{A}. \tag{71}$$

with respect to the common reference frame $\mathcal{K}_t$. Similarly one can easily calculate the weak operator relationship

$$(\frac{1}{c}\frac{\partial \tilde{\varphi}}{\partial t} + \langle \nabla | \tilde{A} \rangle)\tilde{\Phi} = 0, \tag{72}$$

which holds for the common temporal parameter $t \in \mathbf{R}$. Now we will calculate the weak Maxwell type operator relationships (69) and (70) with respect to the common reference frame $\mathcal{K}_t$:

$$(\nabla \times \tilde{B} - \frac{1}{c}\frac{\partial \tilde{E}}{\partial \tilde{t}})|_{\bar{t}=\tilde{t}=t}\tilde{\Phi} = \frac{\xi}{2\pi}\nabla \int_{\mathbf{R}^3} d^3y \Theta(c(\bar{t} - \tilde{t}), |x - y|)\psi^+(y)\psi(y)|_{\bar{t}=\tilde{t}=t}\tilde{\Phi} = 0 \tag{73}$$

and

$$\langle \nabla | \tilde{E} \rangle \tilde{\Phi} = -\frac{\xi}{2\pi}\frac{\partial}{\partial \tilde{t}} \int_{\mathbf{R}^3} d^3y \Theta(c(\bar{t} - \tilde{t}), |x - y|)\psi^+(y)\psi(y)|_{\bar{t}=\tilde{t}=t}\tilde{\Phi} = \xi\psi^+ \ \psi \ \tilde{\Phi}. \tag{74}$$

To calculate further the expression (73), we need to make use of the strong operator relationships (58) and find that

$$\frac{\partial \tilde{E}}{\partial \tilde{t}}|_{\bar{t}=\tilde{t}=t} = \frac{\partial \tilde{E}}{\partial t} - \frac{\partial \tilde{E}}{\partial \bar{t}}|_{\bar{t}=\tilde{t}=t} = \frac{\partial \tilde{E}}{\partial t} - \frac{i}{\hbar}[\tilde{H}_f^{(int)}, \tilde{E}] = \frac{\partial \tilde{E}}{\partial t} + \xi\psi^+ \ \alpha\psi. \tag{75}$$

Thus, from (75) and (73) one can obtain that

$$(\nabla \times \tilde{B} - \frac{1}{c}\frac{\partial \tilde{E}}{\partial t})\tilde{\Phi} = \xi\psi^+ \ \alpha\psi\tilde{\Phi} \tag{76}$$

with respect to the common reference frame $\mathcal{K}_t$. The combined together weak operator relationships (74) and (76)

$$(\nabla \times \tilde{B} - \frac{1}{c}\frac{\partial \tilde{E}}{\partial t})\tilde{\Phi} = \xi\psi^{+}\ \alpha\psi\tilde{\Phi}, \quad \langle\nabla|\tilde{E}\rangle\tilde{\Phi} = \xi\psi^{+}\ \psi\ \tilde{\Phi} \tag{77}$$

in the reduced by the weak constraint (72) Fock space $\tilde{\Phi}$ jointly with the evident strong operator relationships

$$\nabla \times \tilde{E} + \frac{1}{c}\frac{\partial \tilde{B}}{\partial \tilde{t}} = 0, \nabla \times \tilde{B} = 0 \tag{78}$$

compile the complete system of quantum Maxwell equations with respect to the common reference frame $\mathcal{K}_t$.

Really, from the Heisenberg evolution equations (55) one easily obtains the strong operator charge conservative flow relationship

$$\frac{1}{c}\frac{\partial}{\partial t}(\xi\psi^{+}\ \psi) + \langle\nabla|\xi\psi^{+}\ \alpha\psi\rangle = 0, \tag{79}$$

in which the quantity

$$\rho := \xi\psi^{+}\ \psi \tag{80}$$

is interpreted as the operator charge density and the quantity

$$J := \xi\psi^{+}\ \alpha\psi \tag{81}$$

is naturally interpreted as the operator current density in the space $\mathbf{R}^3$. Whence the weak operator equations (77) can be rewritten, taking into account the definitions (80) and (81), in the standard Maxwell equations weak form:

$$(\nabla \times \tilde{B} - \frac{1}{c}\frac{\partial \tilde{E}}{\partial t})\tilde{\Phi} = J\tilde{\Phi}, \quad \langle\nabla|\tilde{E}\rangle\tilde{\Phi} = \rho\tilde{\Phi} \tag{82}$$

under the Fock space $\tilde{\Phi}$ constraint (72). Moreover, based on the weak operator Maxwell equations (82) and the Lorenz constraint (72), one can derive easily the following weak operator wave equations

$$(\frac{1}{c^2}\frac{\partial^2\tilde{\varphi}}{\partial t^2} - \Delta\tilde{\varphi})\tilde{\Phi} = \rho\tilde{\Phi}, \quad (\frac{1}{c^2}\frac{\partial^2\tilde{A}}{\partial t^2} - \Delta\tilde{A})\tilde{\Phi} = \frac{1}{c}J\tilde{\Phi} \tag{83}$$

in respect to the common reference frame $\mathcal{K}_t$, allowing to calculate the induced by the charged electron field the quantum bosonic potentials $(\tilde{\varphi}_\xi, \tilde{A}_\xi) : \tilde{\Phi} \to \tilde{\Phi}$ in the analytical form:

$$\tilde{\varphi}_\xi = \int_{\mathbf{R}^3} \frac{\rho(t', y)d^3y}{|x-y|}, \quad \tilde{A}_\xi = \frac{1}{c}\int_{\mathbf{R}^3} \frac{J(t', y)d^3y}{|x-y|}, \tag{84}$$

where the *"retarded"* temporal parameter $t' := t - |x-y|/c \in \mathbf{R}$, making the equations (83) exactly satisfied modulo the solutions to their uniform forms.

From the analysis of the quantum electron field model, interacting with the self-generated quantum photonic field, one can infer the following important consequences:

(i) the physical effective evolution of the electron-photonic system with respect to the common reference frame $\mathcal{K}_t$ is governed by the reduced fermionic Hamiltonian operator (51), acting in the canonically transformed Fock space $\tilde{\Phi}$, reduced by the operator Lorenz constraint;

(ii) the compatibility of evolutions of the quantum electron and photonic fields with respect to the common temporal reference frame $\mathcal{K}_t$ ensues the reciprocal influence of the electron field on the photonic one and vice versa, being clearly demonstrated both by the weak field potentials operator equations (83);

(iii) subject to the basic self-interacting electron-photonic system described by the joint Hamiltonian operator (47) in the transformed Fock space $\tilde{\Phi}$, one can claim that the photonic electromagnetic impact into the quantum electron particle dynamics is decisive, as owing to it the dressed electron system can realize its self-interaction property through the physical vacuum deformation, caused by the related deformation of the weak Lorenz constraint (63) and (61).

Being experienced in the analysis of a self-interacting fermionic quantum electron field with the self-generated quantum bosonic electromagnetic field, we understand well that the influence of the electromagnetic field on the charged particle should be considered as crucial, strongly modifying the related fermionic Hamiltonian operator, describing the charged particle dynamics. As the simultaneously modified bosonic electromagnetic operator depends, owing to the self-interaction, on the electron charge and current electron field densities, the joint impact on the electron dynamics can be effectively classically modeled by means of its inertial mass parameter. In particular, within the quantum operator case the physical electron mass parameter $m_{ph} \in \mathbf{R}_+$ can be naturally defined by means of the least quantum renormalized Hamiltonian (47) energy spectrum eigenvalue

$$m_{ph} := c^{-2} \inf_{\tilde{f}\in\tilde{\Phi}|||\tilde{f}||=1} (\tilde{f}, \tilde{H}^{(int)}_{f-b}\tilde{f}), \quad \tilde{H}^{(int)}_{f-b} := \tilde{H}^{(int)}_{f} + \tilde{H}_b, \tag{85}$$

in the suitably transformed and reduced by means of the operator Lorenz type constraint (72) Fock space $\tilde{\Phi}$ with respect to the common reference frame $\mathcal{K}_t$. As the quantum spectral problem (85) is very complicated, it was recently in detail analyzed in [41] from the classical point of view. A thorough analytical study of the mass expression (85) is under elaboration.

The consequences formulated above subject to the quantum electron-photonic self-interacting phenomenon appear to be very important [41] also from classical point of view, especially for physical understanding the inertial properties of electron under action of the self-generated electromagnetic field. This and related topics are planned to be considered in detail elsewhere.

5. Classical reduction of the electron Hamiltonian and electromagnetic field evolution

Let's consider the vector position operator $\hat{x} : \tilde{\Phi} \to \tilde{\Phi}^3$ and its weak in the reduced Fock space $\tilde{\Phi}$ evolution with respect to the complete and suitably renormalized charged particle Hamiltonian operator (51). We need to take into account that the Hamiltonian operator $\tilde{H}^{(int)}_f : \tilde{\Phi} \to \tilde{\Phi}$ can be represented as

$$\tilde{H}^{(int)}_f = \int_{\mathbf{R}^3} d^3x\psi^+\langle\alpha c|\hat{p}_x\rangle\psi + \int_{\mathbf{R}^3} d^3x(\xi\psi^+\psi\tilde{\varphi}_\xi - \xi\psi^+\langle c\alpha|\tilde{A}_\xi\rangle\psi), \tag{86}$$

within which the operators $(\tilde{\varphi}_\xi, \tilde{A}_\xi) : \tilde{\Phi} \to \tilde{\Phi}^3$ are given by the nonlocal integral expressions (84) and $\hat{p}_x : \tilde{\Phi} \to \tilde{\Phi}^3$ is the locally defined charged particle ξ momentum operator $\hat{p}_x := \frac{\hbar}{i}\nabla_x$, canonically conjugated [11, 38] to the position operator $\hat{x} : \tilde{\Phi} \to \tilde{\Phi}^3$, that is

$$[\hat{p}_y, \hat{x}] = \frac{\hbar}{i}\delta(x - y) \tag{87}$$

for any $x, y \in \mathbf{R}^3$. This also, in particular, means that the position operator $\hat{x} : \tilde{\Phi} \to \tilde{\Phi}^3$ is *a priori* given in the diagonal representation: $\hat{x}\tilde{f} := x\tilde{f}$ for any vector $\tilde{f} \in \tilde{\Phi}$.

The expression (81) can be used for obtaining the classical charged particle ξ current $\bar{J}(t, x)$ expression as

$$\bar{J}(t, x) := (\Omega|\psi^+ \alpha c \psi \Omega), \tag{88}$$

where the vector $\Omega \in \tilde{\Phi}$ is the ground state of the Hamiltonian operator (86), acting in the Lorenz type reduced and suitably renormalized [7, 5, 38] Fock space $\tilde{\Phi}$. Substituting (80) into the expression (86) one obtains the classical Hamiltonian expression

$$\bar{H}_f^{(int)} = \int_{\mathbf{R}^3} d^3x \ \langle \bar{J}|\bar{p}/\bar{\rho}\rangle + \int_{\mathbf{R}^3} d^3x \xi(\bar{\rho}\bar{\varphi} - \frac{1}{c}\langle \bar{J}|\bar{A}\rangle), \tag{89}$$

within which we denoted by $\bar{p} \in T^*(\mathbf{R}^3)$ the classical electron momentum, $\bar{\rho} := (\Omega|\psi^+\psi\Omega)$, $(\bar{\varphi}, \bar{A}) := ((\Omega|\xi^{-1}\tilde{\varphi}_\xi\Omega), (\Omega|\xi^{-1}\tilde{A}_\xi\Omega))$ and took into account the previously assumed quantum massless charged particle ξ fermionic field. The expression (89) jointly with the renormalized bosonic field Hamiltonian (35) gives rise to the complete classical Hamiltonian function

$$\bar{H}_{f-b}^{(int)} = \int_{\mathbf{R}^3} d^3x[\frac{1}{2}(|\bar{E}|^2 + |\bar{B}|^2) + \langle \bar{J}|\bar{p}/\bar{\rho}\rangle + \bar{\rho}\bar{\varphi} - \frac{1}{c}\langle \bar{J}|\bar{A}\rangle], \tag{90}$$

governing the temporal evolution flow both of the charged particle flow of the density $\bar{\rho} : M^4 \to \mathbf{R}_+$ and of the electromagnetic field $(\bar{E}, \bar{B}) : M^4 \to \mathbf{E}^3 \times \mathbf{E}^3$ with respect to the laboratory reference frame $\mathcal{K}_t$. The obtained above Hamiltonian function (90) and its corresponding Lagrangian form (42) have been effectively used before in [41] for describing the classical self-interacting charged particle dynamics and its inertial properties.

6. Conclusion

In our work we reanalyzed within the framework of the Fock many-temporal parameter approach and the R. Feynman proper time paradigm the relativistic quantum electron spin and electron energy spectrum problems. We have shown that the self-interacting quantum mechanism of the charged particle with its self-generated electromagnetic field consists of two physically different phenomena, related with the representations of the Hamiltonian operator and related symmetry properties of the renormalized Lorenz constraint. Their influence on the structure of the resulting Hamilton interaction operator proves to be crucial for determining the related charged particle mass spectrum.

Acknowledgments

I am cordially appreciated to Organizers of the 13th Biennial Conference on Classical and Quantum Relativistic Dynamics of Particles and Fields (6 - 9 June 2022, Czech Technical University in Prague) for a kind invitation to present my report and for very inspiring atmosphere during the event. Especially I am thankful to Tepper L. Gill and Martin C. Land for fruitful and stimulating discussions of relativistic quantum electron spin and spectrum energy problems. I am also obliged to Taras Banakh for many useful remarks and instrumental corrections on a manuscript, to Sterna Duple, Michael Kolinsky, Vocoder Simulium, Anatolia Nicotian, Yugoslav Mukluk and Dossals Kulak for valuable comments during preparation of a manuscript. The last but not least thanks belong to the Department of Computer Science and Telecommunication at the Cracow University of Technology for support and a local travel grant.

References

[1] Annila A. The Meaning of Mass. International Journal of Theoretical and Mathematical Physics, 2(4), 2012, p. 67-78
[2] Baaquie B.E. The Theoretical Foundations of Quantum Mechanics. Springer, 2013
[3] Blackmore D., Prykarpatsky A.K. and Samoylenko V.Hr. Nonlinear dynamical systems of mathematical physics: spectral and differential-geometrical integrability analysis. World Scientific Publ., NJ, USA, 2011
[4] Blaszak M. Quantum versus Classical Mechanics and Integrability Problems. Springer, 2019
[5] Bogolubov N.N., Logunov A.A. Oksak A.I., Todorov I. T. General Principles of Quantum Field Theory. Kluwer Publ., Nertherlands, 1990
[6] Bogolubov N.N. (Jr.) and Prykarpatsky A.K. and Blackmore D. Maxwell–Lorentz Electrodynamics Revisited via the Lagrangian Formalism and Feynman Proper Time Paradigm. Mathematics 2015, 3, 190-257
[7] Bogolubov, N.N.; Shirkov, D.V. *Quantum Fields.* Addison-Wesley, 1982
[8] Brillouin L. Relativity reexamined. Academic Press Publ., New York and London, 1970
[9] Clifford W.K., Tucker, R. (ed.). Mathematical Papers. London: Macmillan, 1882.
[10] Coddens G. The geometrical meaning of spinors as a key to make sense of quantum mechanics. 2021, hal-03175981.
[11] Dirac P.A.M. The principles of quantum mechanics. Oxford, Clarendon Press, 1947
[12] Dirac P.A.M., Fock V.A., Podolsky B. On quantum electrodynamics. Sow. Physics, 2,1932, p. 468-479
[13] Dyson F.J. Feynman's proof of the Maxwell equations, Am. J. Phys. 58, 1990, p. 209-211
[14] Dyson F. J. Feynman at Cornell. Phys. Today 42 (2), 1989, p. 32-38
[15] Feynman R., Leighton R. and Sands M. The Feynman Lectures on Physics. Electrodynamics, v. 2, Addison-Wesley, Publ. Co., Massachusetts, 1964
[16] Feynman R., Leighton R. and Sands M. The Feynman lectures on physics. The modern science on the Nature. Mechanics. Space, time, motion. v. 1, Addison-Wesley, Publ. Co., Massachusetts, 1963
[17] Fock V. Die Eigenzeit in der klassischen und in der Quantenmechanik. Sow. Phys., 1937, 12, p. 404-425
[18] Fock V.A., Podolsky B. On the quantization of electromagnetic waves and the interaction of charges in Dirac's theory. Sow. Physics, 1, 1932, p. 801-817.
[19] Garling D. J. H. Clifford algebras. An introduction, London Mathematical Society Student Texts, vol. 78, Cambridge University Press, 2011.
[20] Gill T.L., Zachary W.W.Two mathematically equivalent versions of Maxwell equations. Preprint, University of Maryland, 2008
[21] Gill T.L., Zachary W.W. Two mathematically equivalent versions of Maxwell's equations. Foundations of Physics 4, 2011, 99-128
[22] Hammond R.T. Electrodynamics and Radiation reaction. Found. Physics, 43, 2013, p. 201-209
[23] Hammond R.T. Relativistic Particle Motion and Radiation Reaction. Electronic Journal of Theoretical Physics, No. 23, 2010, p. 221–258
[24] Higgs P. Broken Symmetries and the Masses of Gauge Bosons. Phys. Rev. Lett. 13 508 (1964); Spontaneous Symmetry Breakdown without Massless Bosons. Phys. Rev. 145, 1964, p. 1156
[25] t'Hooft G. Massive Yang-Mills fields. Nucl. Phys. B35, 1971, p. 167
[26] Jammer M. Concepts of Mass in Contemporary Physics and Philosophy. Princeton University Press, 2009
[27] Kosyakov B.P. Radiation in electrodynamics and in Yang-Mills theory. Sov. Phys. Usp. 35(2) 135–142 (1992)
[28] Kosyakov B.P. Introduction to the Classical theory of Particles and Fields. Springer, 2007
[29] Kumar A. Fundamental of quantum mechanics. Cambridge Univ. Press, 2018
[30] Martins Alexandre A. and Pinheiro Mario J. On the electromagnetic origin of inertia and inertial mass. Int J Theor Phys., 47, 2008. p. 2706-2715
[31] Medina R. Radiation reaction of a classical quasi-rigid extended particle. J. Phys. A: Math. Gen., 2006, p. 3801-3816
[32] Minkowski H. Raum und Zeit. Physik. Z. 10, 1909, p. 104
[33] Morozov V.B. On the question of the electromagnetic momentum of a charged body. Physics Uspekhi, 181(4), 2011, p. 389 - 392
[34] Page L., Adams N. I. (Jr.) Action and reaction between moving charges. Am. J. Phys. 13, 1945, p.141-147
[35] Pappas P.T. The original Ampere force and Biot-Savart and Lorentz force. Il Nuovo Cimento, 76B(2), 1983, p. 189-197
[36] Pauli W. Theory of Relativity. Oxford Publ., 1958
[37] Pegg D.T. Absorber theory of radiation. Rep. Prog. Phys. 1975. V. 38, p. 1339-1383
[38] Peskin M.E., Schroeder D.V. Introduction to quantum eld theory. Perseus Book, Massachusetts, USA, 1995
[39] Poincare H. Sur la Dynamique de l'Electron. C.R. Acad. Sci. (Paris) 140, 1905, p. 1504-1508
[40] Prykarpatski A.K. Classical Electromagnetic Theory Revisiting: The A.M. Ampere Law and the Vacuum Field Theory Approach. Universal Journal of Physics and Application 2(8), 2014, p. 381-413

[41] Prykarpatsky A.K., Bogolubov N.N. (Jr.). On the classical Maxwell-Lorentz electrodynamics, the electron inertia probelm, and the Feynman proper time paradigm. Ukr. J. Phys., 2016, Vol. 61, No. 3, p. 187-212
[42] Puthoff H.E. Casimir vacuum energy and the semicalssical electron. Int J Theor Phys, 46, 2007, p. 3005-3008
[43] Radovanovic V. Problem Book: Quantum Field Theory. Springer, 2006
[44] Rebenko A.I. Theory of Interacting Quantum Fields. De Gruyter, 2010
[45] Simulik V.M. The electron as a system of classical electromagnetic and scalar fields. In book: What is the electron? Edited by V.M. Simulik. Montreal: Apeiron, p.109-134.
[46] Takhtajan L.A. Lectures on Quantum Mechanics. Department of Mathematics, Stony Brook University, Stony Brook, NY 11794-3651, USA
[47] Weyl H. Group theory and quantum mechanics. 1986
[48] Wheeler J.B. and Feynman R.P. Interaction with the Absorber as the Mechanism of Radiation. Rev. Modern Phys., 17, N2-3, p. 157-181
[49] Wilczek F. Origins of Mass, MIT-CTP/4379. arXiv:1206.7114v2 [hep-ph] 22 Aug 2012
[50] Wilczek F. QCD and natural phylosophy. Ann. Henry Poincare, 4, 2003, 211-228
[51] Yaremko Y. and Tretyak V. Radiation Reaction in Classical Field Theory. LAP LAMBERT Academic Publishing, Germany, 2012

Quantum coherence of relativistic wavepackets in electromagnetic vacuum

Shih-Yuin Lin

Department of Physics, National Changhua University of Education, Changhua 500207, Taiwan

E-mail: sylin@cc.ncue.edu.tw

Abstract. We have developed a linearized effective theory with quantum mechanical Gaussian wavepacket of a charged relativistic particle coupled to quantum electromagnetic fields at a scale well below the Schwinger limit. Using this effective theory, we study the quantum decoherence of single electrons at rest due to their interactions with electromagnetic vacuum. According to the existing data of electron interference experiments in electron microscopes, the values of the parameters in our effective theory are chosen. With these values, we find that weak decoherence by vacuum fluctuations may have blurred the interference pattern in Ref. [1].

1. Introduction

In the conventional theory of quantum fields in Minkowski space, the vacuum state of quantum fields is a no-particle state with respect to laboratory observers [2]. By Lorentz symmetry, the vacuum state is also a no-particle state with respect to an observer in uniform motion. Such a "no-particle" property, anyway, does not hold for an observer in non-inertial motion even in Minkowski space [3]. In 1976, Unruh showed that an object linearly and uniformly accelerated in Minkowski vacuum of some field interacting with it will experience particles (quanta of that field) in thermal spectrum at a temperature proportional to the object's proper acceleration [4]. This is called the Unruh effect, and the temperature that the accelerated object would experience is called the Unruh temperature.

One strong evidence for Unruh's observation could be the "circular Unruh effect" pointed out by Bell and Leinaas [5], who realized that the Sokorov-Ternov effect about the spin depolarization of electrons in the storage rings of particle colliders could be interpreted as the Unruh effect. While the formulation is indeed similar [6], the centripetal acceleration of an electron in circular motion is perpendicular to the direction of its momentum, and the spectrum of quanta experienced by the electron is not strictly thermal [5]. This is different from Unruh's original setting in [4] with linear acceleration.

To get closer to Unruh's setting, Chen and Tajima [7] proposed to use an intense laser to drive electrons linearly in a very large proper acceleration (to produce a significant Unruh temperature.) They argued that the accelerated electrons may experience some thermal-like fluctuations due to the Unruh effect, and so undergo some kind of Brownian motion around the classical trajectory. The radiation originated from that Brownian motion may be observed in laboratory on top of the classical radiation. Such a quantum correction to radiation is called

Content from this work may be used under the terms of the Creative Commons Attribution 3.0 licence. Any further distribution of this work must maintain attribution to the author(s) and the title of the work, journal citation and DOI.
Published under licence by IOP Publishing Ltd

the Unruh radiation (which is *not* the black body radiation supposed to be experienced by the accelerated electrons in the effective thermal bath of field quanta.)

In a toy model of an Unruh-DeWitt(UD) harmonic-oscillator(HO) detector [4, 8] linearly and uniformly accelerated in a quantum scalar field, we found that the quantum correction to radiation corresponding to the Unruh effect does not cease at late times [9]. In Chen and Tajima's proposal, however, the driving laser field and the induced classical motion of electrons are alternating in time. In [10] the author of the present paper studied the radiation emitted by an UD HO detector in oscillatory motion. It is found that the quantum radiation originated from the nontrivial quantum fluctuations experienced by the accelerated UD HO detector during the time when acceleration is altering is much more significant than those in the period of nearly uniform acceleration. It will be interesting to see if the result is similar in a more realistic model for moving electrons.

Classical trajectories are essential in both the UD detector model and the calculations in [7] for electrons. To compare, we prefer to consider the charged particles or single electrons which can be described as wavepackets centered around their classical trajectories. We noticed that the most powerful lasers in the world to date can at most achieve an intensity of 10^{23} W/cm^2 (e.g. [11, 12]), which is well below the Schwinger limit 10^{29} W/cm^2. Thus, electrons in Chen and Tajima's proposal can be treated as quantum mechanical particles, whose wavefunctions can have single-particle interpretation safely. Just like Unruh's "particle detector" modeled in [4], we are going to consider an effective theory with a relativistic quantum mechanical charged particle coupled to quantum electromagnetic (EM) fields. The theory will be a hybrid of quantum mechanics and non-equilibrium quantum field theory.

An immediate difficulty is that the action of a relativistic particle is highly nonlinear, and the evolution of the particle wavepacket may be hard to handle. Fortunately, there exists some regime that a linearized theory with Gaussian wavepacket of particle could work well. In Ref. [13], Rosenstein and Usher have considered a fully relativistic, quantum mechanical particle in (1+1) dimensional Minkowski space, which is described by the Schödinger equation

$$i\hbar\partial_t\psi(t,x) = mc^2\sqrt{1+\left(\frac{1}{mc}\frac{\hbar}{i}\partial_x\right)^2}\,\psi(t,x), \tag{1}$$

They found a wavepacket solution whose time evolution is characterized by a parameter $c\vartheta$ (γ in [13]), which can be interpreted as the initial width of the wavepacket, and the modes of wavelength shorter than $c\vartheta$ will be suppressed in their wavepacket solution. Rosenstein and Usher further found that once the initial width $c\vartheta$ is well above the Compton wavelength of the relativistic particle, the peak value of the probability density of the wavepacket will occur around the classical trajectory of the particle.

We generalized Rosenstein and Usher's wavepacket solution to the case of a relativistic particle moving at constant speed v_0 in (1+1) dimensions in Ref. [14]. Our solution to (1) reads

$$\psi(t,x) = \sqrt{\frac{mc}{\hbar\pi\gamma_0 K_1\left(\frac{2mc^2\vartheta}{\hbar\gamma}\right)}\frac{(\vartheta+it)c}{f(t,x)}K_1\left(\frac{mc}{\hbar}f(t,x)\right)}, \tag{2}$$

where $f(t,x) \equiv \sqrt{(x-iv_0\vartheta)^2 - c^2(t-i\vartheta)^2}$, $K_1(z)$ is the modified Bessel function of the second kind, and $\gamma_0 \equiv 1/\sqrt{1-(v_0/c)^2}$. The expectation value of position with respect to the above wavepacket solution, $\langle x\rangle = \int_{-\infty}^{\infty} dx\, x|\psi|^2 = v_0 t$, is going along the classical trajectory.

In Figure 1, we show the (scaled) spatial distributions of $\rho(t,x) = |\psi(t,x)|^2$ for the wavepackets with various initial width $c\vartheta$ and their time evolution. One can see that a wavepacket of the relativistic free particle initially in a Gaussian-like shape will behave always

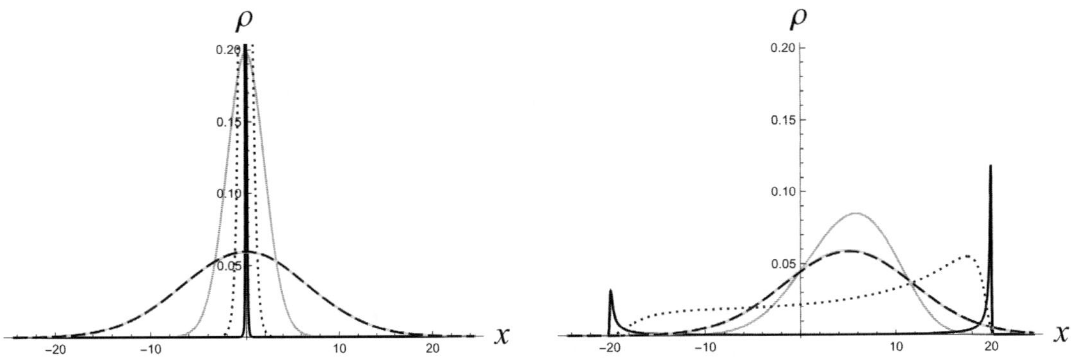

Figure 1. The (scaled) spatial distribution of $\rho(t,x) = |\psi(t,x)|^2$ at $t = 0$ (left) and $t = 20$ (right) with different initial widths $c\vartheta$. Here $c = \hbar \equiv 1$. The black solid, black dotted, gray, and black dashed curves in each plot represent ρ of the wavefunctions $\psi(t,x)$ of the relativistic free particle given in (2) with $c\vartheta = 0.1$, 1, 10, and 100 (in the unit of the Compton wavelength), respectively. All at a constant speed $v_0 = c/4$ to the right. The black solid curves have been scaled down to 1/10 of their original values. The green curves represent the Gaussian function $\rho'(t,x) = \sqrt{\alpha/\pi}e^{-\alpha(x-v_0t)^2}$, where we choose $\alpha = 0.011$ (constant in time) as a reference for the black dashed curve (ψ of $c\vartheta = 100$).

like a Gaussian wavepacket peaked around the classical trajectory if the value of parameter ϑ is sufficiently large, e.g. the dashed curve with $c\vartheta = 100\times$(Compton wavelength of the particle) in Figure 1. We have also verified that the phase of $\psi(t,x)$ evaluated along the classical trajectory $x = v_0t$ is very close to the classical action. These encourage us to construct a linearized effective theory in Section 2 for relativistic charged particles interacting with EM fields, quantize it in Section 3, and describe a relativistic charged particle as a Gaussian wavepacket centered at its classical position at time t.

In Section 4, we will consider the simplest application of our effective theory, where a charged particle is at rest in the background of zero EM fields. After working out the mode functions and two-point correlators, we will focus on quantum coherence of the particle. The values of the regulators and parameters in our effective theory can be input from the existing data of electron interference experiments in electron microscopes. We will see that the purity of the wavepacket in our theory is consistent with experiments. A short summary and outlook will be given in Section 5. An extended version of this paper is in preparation [14].

2. Relativistic particle in EM fields

Consider a relativistic particle of mass m, charge q, and worldline $z^\mu(\tau)$, interacting with electromagnetic(EM) fields A^μ, $\mu = 0,1,2,3$. The dynamics of the combined system is described by the action $S = S_z + S_I + S_F$ with

$$S_z = -mc\int d\tau\sqrt{-\frac{dz_\mu}{d\tau}\frac{dz^\mu}{d\tau}} = -mc^2\int dt\sqrt{1-\frac{1}{c^2}\frac{dz_i}{dt}\frac{dz^i}{dt}} \tag{3}$$

$$S_I = \frac{q}{c}\int d^4x\int d\tau\frac{dz^\mu}{d\tau}\delta^4\left[x - z(\tau)\right]A_\mu(x) \tag{4}$$

$$S_F = \int dtd^3x\left[-\frac{1}{4}F_{\mu\nu}F^{\mu\nu} - \frac{\alpha}{2}\left(\partial_\mu A^\mu\right)^2\right]. \tag{5}$$

Here $F_{\mu\nu} = \partial_\mu A_\nu - \partial_\nu A_\mu$, τ is the proper time of the particle, α is an arbitrary constant, and we are working in the Minkowski coordinates with $\eta_{\mu\nu} = \mathrm{diag}\{-1,1,1,1\}$. S_F is the action of free

Journal of Physics: Conference Series **2482** (2023) 012018 doi:10.1088/1742-6596/2482/1/012018

electromagnetic fields to be constrained by the Lorenz gauge $\partial_\mu A^\mu = 0$. $S_z + S_I$ is form-invariant under a reparametrization $\tau \to \tau' = \tau'(\tau)$, meaning that the number of the degrees of freedom of $z^\mu(\tau)$ is $4 - 1 = 3$. To fix the reparametrization symmetry, we choose the Minkowski-time gauge $z^0 = ct$ for the particle.

Let $z^i(t) = \bar{z}^i(t) + \tilde{z}^i(t)$, $i = 1, 2, 3$, and $A^\mu(t, \mathbf{x}) = \bar{A}^\mu(t, \mathbf{x}) + \tilde{A}^\mu(t, \mathbf{x})$, where $\bar{z}^i(t)$ and $\bar{A}^\mu(t, \mathbf{x})$ are the classical solutions to the Euler-Lagrange equations $\delta S/\delta z^i(t) = 0$ and $\delta S/\delta A^\mu_{\mathbf{x}}(t) = 0$. Assume that the deviations $\tilde{z}^i$ and $\tilde{A}^\mu$ from classcal solutions are so small that we can neglect the higher order terms and truncate S to the quadratic order,

$$\begin{aligned} S \approx\ & S[\bar{z}, \bar{A}] + \frac{1}{2}\int dt\, dt' \Bigg\{ \sum_{i,j} \tilde{z}^i(t) \left.\frac{\delta^2 S}{\delta z^i(t)\delta z^j(t')}\right|_{\bar{z},\bar{A}} \tilde{z}^j(t') + \\ & 2\sum_{i,\mu,\mathbf{x}} \tilde{z}^i(t) \left.\frac{\delta^2 S}{\delta z^i(t)\delta A^\mu_{\mathbf{x}}(t')}\right|_{\bar{z},\bar{A}} \tilde{A}^\mu_{\mathbf{x}}(t') + \sum_{\mu,\mathbf{x},\nu,\mathbf{y}} \tilde{A}^\nu_{\mathbf{y}}(t) \left.\frac{\delta^2 S}{\delta A^\nu_{\mathbf{y}}(t)\delta A^\mu_{\mathbf{x}}(t')}\right|_{\bar{z},\bar{A}} \tilde{A}^\mu_{\mathbf{x}}(t') \Bigg\} \\ =\ & S[\bar{z}, \bar{A}] + \\ & \int dt \frac{m}{2}\bar{\gamma}(t)\bar{M}_{ij}(t)\dot{\tilde{z}}^i(t)\dot{\tilde{z}}^j(t) + \int dt \sum_{\mathbf{x}} \left[-\frac{1}{4}\tilde{F}_{\mu\nu}\tilde{F}^{\mu\nu}(x) - \frac{\alpha}{2}\left(\partial_\mu \tilde{A}^\mu(x)\right)^2 \right] + \\ & q\int dt \sum_{\mathbf{x}} \left[\frac{\bar{v}^\mu}{c}\tilde{z}^i\partial_i\tilde{A}^{\mathbf{x}}_\mu + \frac{\dot{\tilde{z}}^i}{c}\tilde{A}^{\mathbf{x}}_i + \frac{\dot{\tilde{z}}^j}{c}\tilde{z}^i\partial_i\bar{A}^{\mathbf{x}}_j + \frac{\bar{v}^\mu}{2c}\tilde{z}^i\tilde{z}^j\partial_i\partial_j\bar{A}^{\mathbf{x}}_\mu \right] \delta^3(\mathbf{x} - \bar{\mathbf{z}}) \end{aligned} \quad (6)$$

with the linear terms of $\tilde{z}$ and $\tilde{A}^\mu$ vanishing after introducing the Euler-Lagrangian equations. Here we have chosen $z^0 = ct$ and so $\bar{v}^\mu \equiv (c, \bar{v}^j) = (c, \partial_t\bar{z}^j)$, $\dot{\tilde{z}}^i \equiv \partial_t\tilde{z}^i$, $\tilde{F}_{\mu\nu} \equiv \partial_\mu\tilde{A}_\nu - \partial_\nu\tilde{A}_\mu$, $\sum_{\mathbf{x}} \equiv \int d^3x$, $\bar{\gamma} \equiv 1/\sqrt{1 - \frac{\bar{v}_i\bar{v}^i}{c^2}}$, and $\bar{M}_{ij}(t) \equiv \eta_{ij} + \frac{\bar{\gamma}^2}{c^2}\bar{v}_i\bar{v}_j$.

The classical action would be the phase of the quantum state evaluated around the peak of the Gaussian wavepacket. The Euler-Lagrange equations up to the classical level read

$$\partial_t\Big[m\bar{\gamma}\bar{v}_i(t)\Big] = \frac{q}{c}\bar{F}_{i\mu}(\bar{z}(t))\bar{v}^\mu(t), \quad (7)$$

$$\partial_\nu\bar{F}^{\nu\mu}(x) + \alpha\partial^\mu\partial_\nu\bar{A}^\nu(x) = -\frac{q}{c}\bar{v}^\mu(t)\delta^3(\mathbf{x} - \bar{\mathbf{z}}(t)), \quad (8)$$

In the Lorentz gauge $\partial_\nu\bar{A}^\nu_{\mathbf{x}}(t) \equiv \partial_\nu\bar{A}^\nu(x) = 0$, Eq. (8) reduces to

$$\left(-\frac{1}{c^2}\partial_t^2 + \nabla^2\right)\bar{A}^\mu_{\mathbf{x}}(t) = -\frac{q}{c}\bar{v}^\mu(t)\delta^3(\mathbf{x} - \bar{\mathbf{z}}(t)), \quad (9)$$

where the four components of A^μ decouple. One immediately obtains a solution

$$\bar{A}^\mu_{\mathbf{x}}(t) = \bar{A}^\mu_{[0]\mathbf{x}}(t) + \bar{A}^\mu_{[1]\mathbf{x}}(t) \quad (10)$$

where $\bar{A}^\mu_{[0]\mathbf{x}}(t)$ is a homogeneous solution to $\left(-\frac{1}{c^2}\partial_t^2 + \nabla^2\right)\bar{A}^\mu_{[0]\mathbf{x}} = 0$, assumed to be regular around the worldline of the charged particle, and $\bar{A}^\mu_{[1]\mathbf{x}}(t)$ is the retarded field sourced from the charged particle, namely,

$$\bar{A}^\mu_{[1]\mathbf{x}}(t) = \int_{-\infty}^t dt' \int d^3x' G_{ret}(t, \mathbf{x}; t', \mathbf{x}')q\bar{v}^\mu(t')\delta^3(\mathbf{x}' - \bar{\mathbf{z}}(t')). \quad (11)$$

The retarded Green's function $G_{ret}(t, \mathbf{x}; t', \mathbf{x}') = \frac{1}{4\pi}\delta(\sigma)\theta(t - t')$, where $\sigma(x, x') \equiv -\frac{1}{2}(x_\mu - x'_\mu)(x^\mu - x'^\mu)$ is the Synge's world function, diverges as $x' \to x$. This implies that the self field established by the charged particle diverges around the particle as the particle size goes to zero. Following the same method in [15, 9], we regularize the retarded Green's function by a UV cutoff Λ as

$$G^{\Lambda}_{ret}(t,\mathbf{x};t',\mathbf{x}') \equiv \frac{1}{4\pi}\sqrt{\frac{8}{\pi}}\Lambda^2 e^{-2\Lambda^4\sigma^2(x,x')}\theta(t-t')\theta(\sigma), \tag{12}$$

which approaches $G_{ret}(t,\mathbf{x};t',\mathbf{x}')$ as $\Lambda \to \infty$. Replace the G_{ret} in (11) by G^{Λ}_{ret}, then insert (10) with (11) into (7), we obtain the Lorentz-Abraham-Dirac (LAD) equation [16, 17, 18],

$$\bar{m}\partial_t\left[\bar{\gamma}\bar{v}_i(t)\right] = \frac{q}{c}\bar{F}^{[0]}_{i\mu}(\bar{z}(t))\bar{v}^{\mu}(t) + \bar{\Gamma}_i(t) + O(\Lambda^{-1}) \tag{13}$$

for large Λ, where

$$\bar{m} \equiv m + \Delta_m \qquad \text{with} \qquad \Delta_m \equiv \frac{q^2 2^{\frac{3}{4}}\Gamma\left(\frac{5}{4}\right)}{4\pi c^2\sqrt{\pi}}\Lambda, \tag{14}$$

is the classical renormalized (or field-corrected) mass with contributions from the self field proportional to Λ, $\bar{F}^{[0]}_{\mu\nu} \equiv \partial_\mu\bar{A}^{[0]}_\nu - \partial_\nu\bar{A}^{[0]}_\mu$ is the external EM field strength, and the radiation reaction force of order Λ^0 (the LAD force) reads

$$\bar{\Gamma}_i = \frac{q^2}{4\pi c^3}\frac{2}{3}w^{\mu}_i\partial_t\left[\bar{\gamma}\partial_t\left(\bar{\gamma}\bar{v}_\mu\right)\right], \tag{15}$$

which contains the third derivative of the particle's position, $\partial_t\partial_t\bar{v}_i(t) = \partial^3_t\bar{z}_i(t)$. Here, $w^{\mu}_{\nu} \equiv \delta^{\mu}_{\nu} + \frac{\bar{\gamma}^2}{c^2}\bar{v}_\nu\bar{v}^\mu$.

Later we will treat the charged particle quantum-mechanically and assume that the probability of finding the charged particle in the Universe is conserved. We expect that our effective theory would break down beyond the charged particle production scale. Thus we set the UV cutoff $\Lambda = \bar{m}c/h = 1/\lambda_C$ in (12), where λ_C is the Compton wavelength of the particle. For electrons, $q^2 \approx 2.3\times10^{-28}$ N $\cdot$ m^2, $\bar{m} = 9.1\times10^{-31}$ kg, and $\lambda_C \approx 2.4\times10^{-12}$ m. Write $\bar{m} = m(1+\chi)$ in (14), one can see that the mass correction due to dressing up the self EM fields, $\chi \approx 7.9\times10^{-5} \ll 1$, is only a small portion of electron mass here.

3. Quantization

From (6), the conjugate momenta of the deviations $\tilde{z}^i$ and $\tilde{A}^{\mu}_{\mathbf{x}}$ read

$$\tilde{p}_i = \frac{\delta S}{\delta\partial_t\tilde{z}^i} = m\bar{\gamma}\bar{M}_{ij}\dot{\tilde{z}}^j + \frac{q}{c}\left(\tilde{A}^{\bar{\mathbf{z}}}_i + \tilde{z}^j\partial_j\bar{A}^{\bar{\mathbf{z}}}_i\right), \tag{16}$$

$$\tilde{\pi}^i_{\mathbf{x}} = \frac{\delta S}{\delta\partial_t\tilde{A}^{\mathbf{x}}_i} = \frac{1}{c}\tilde{F}^{i0}_{\mathbf{x}}, \qquad \tilde{\pi}^0_{\mathbf{x}} = \frac{\delta S}{\delta\partial_t\tilde{A}^{\mathbf{x}}_0} = \frac{\alpha}{c}\partial_\mu\tilde{A}^{\mu}_{\mathbf{x}}, \tag{17}$$

and the Hamiltonian of the quardratic part of S defined on the t-slice is given by

$$\begin{aligned}\tilde{H}_2 = {} & \frac{\bar{M}^{ij}}{2m\bar{\gamma}}\left[\tilde{p}_i - \frac{q}{c}\left(\tilde{A}^{\bar{\mathbf{z}}}_i + \tilde{z}^k\partial_k\bar{A}^{\bar{\mathbf{z}}}_i\right)\right]\left[\tilde{p}_j - \frac{q}{c}\left(\tilde{A}^{\bar{\mathbf{z}}}_j + \tilde{z}^l\partial_l\bar{A}^{\bar{\mathbf{z}}}_j\right)\right] \\ & + \sum_{\mathbf{x}}\left\{\frac{c^2}{2}\tilde{\pi}^{\mathbf{x}}_i\tilde{\pi}^i_{\mathbf{x}} - \frac{c^2}{2\alpha}\left(\tilde{\pi}^0_{\mathbf{x}}\right)^2 + c\tilde{\pi}^i_{\mathbf{x}}\partial_i\tilde{A}^{\mathbf{x}}_0 + c\tilde{\pi}^0_{\mathbf{x}}\partial_i\tilde{A}^i_{\mathbf{x}} + \frac{1}{4}\tilde{F}^{\mathbf{x}}_{ij}\tilde{F}^{ij}_{\mathbf{x}}\right\} \\ & - \frac{q}{c}\left(\bar{v}^\mu\tilde{z}^i\partial_i\tilde{A}^{\bar{\mathbf{z}}}_\mu + \frac{\bar{v}^\mu}{2}\tilde{z}^i\tilde{z}^j\partial_i\partial_j\bar{A}^{\bar{\mathbf{z}}}_\mu\right),\end{aligned} \tag{18}$$

where $\bar{M}^{ij}$ is the inverse matrix of $\bar{M}_{ij}$ defined by $\bar{M}^{ij}\bar{M}_{jk} = \delta^i_k$.

In our linearized theory described by the above quadratic Hamiltonian $\hat{H}_2$, the Hamilton equations (or the Euler-Lagrange equations) yield

$$\partial_t\left(m\bar{\gamma}\bar{M}_{ij}\dot{\tilde{z}}^j\right) = \frac{q}{c}\left[\tilde{F}^{\bar{\mathbf{z}}}_{i\mu}\bar{v}^\mu + \left(\tilde{z}^j\partial_j\bar{F}^{\bar{\mathbf{z}}}_{i\mu}\right)\bar{v}^\mu + \bar{F}^{\bar{\mathbf{z}}}_{ij}\dot{\tilde{z}}^j\right], \tag{19}$$

$$\partial_\mu\tilde{F}^{\mu\nu}_{\mathbf{x}} + \alpha\partial^\nu\partial_\mu\tilde{A}^\mu_{\mathbf{x}} = -\frac{q}{c}\tilde{V}^\nu\delta^3(\mathbf{x}-\bar{\mathbf{z}}), \tag{20}$$

where $\tilde{V}^\nu \equiv (-c\tilde{z}^j\partial_j,\ \dot{\tilde{z}}^i - \bar{v}^i\tilde{z}^j\partial_j)$.

We quantize our linearized theory by promoting the perturbative variables $\tilde{z}^i$ and $\tilde{A}^i_{\mathbf{x}}$ to the operators $\hat{z}^i$ and $\hat{A}^i_{\mathbf{x}}$, and introducing the equal-time commutation relations

$$[\hat{z}^i, \hat{p}_j] = i\hbar\delta^i_j, \qquad [\hat{A}^\mu_{\mathbf{x}}, \hat{\pi}^\nu_{\mathbf{y}}] = i\hbar\eta^{\mu\nu}\delta^3(\mathbf{x}-\mathbf{y}). \tag{21}$$

Assume the charged particle and the EM fields are not coupled until $t=0$, namely, set $q=0$ for $t<0$ in (16), (19) and (20). Then, for $t\le 0$, we have

$$\hat{p}_i = m\bar{\gamma}\bar{M}_{ij}\dot{\hat{z}}^j, \qquad \dot{\hat{p}}^j = 0, \tag{22}$$

$$\partial_\mu\hat{F}^{\mu\nu}_{\mathbf{x}} + \alpha\partial^\nu\partial_\mu\hat{A}^\mu_{\mathbf{x}} = 0, \tag{23}$$

where $\hat{z}^j$ and $\hat{A}^\mu_{\mathbf{x}}$ are operators. Thus, before the coupling is switched on, $\hat{p}_i$ are independent of time, and the deviation from the classical worldline of the particle,

$$\hat{z}^i(t) = \hat{z}^i(t_0) + \hat{p}_j\int_{t_0}^{t} d\tilde{t}\,\frac{\bar{M}^{ij}(\tilde{t})}{m\bar{\gamma}(\tilde{t})} \tag{24}$$

with a constant $t_0 \le 0$, is increasing in t. As for the EM fields, following the canonical quantization in the Lorentz-Feynman gauge ($\alpha=1$) [2], we write

$$\hat{A}^\mu_{[0]\mathbf{x}}(t) = \sum_{\mathbf{k}}\sum_{\lambda=0}^{3}\left[\epsilon^\mu_{(\lambda)\mathbf{k}}e^{-i\omega t+i\mathbf{k}\cdot\mathbf{x}}\,\hat{b}^{(\lambda)}_{\mathbf{k}} + \epsilon^{\mu*}_{(\lambda)\mathbf{k}}e^{i\omega t-i\mathbf{k}\cdot\mathbf{x}}\,\hat{b}^{(\lambda)\dagger}_{\mathbf{k}}\right] \tag{25}$$

for $t\le 0$, with $\omega = |\mathbf{k}|c$, $\sum_{\mathbf{k}} \equiv \int\frac{c\,d^3k}{(2\pi)^3}\sqrt{\frac{\hbar}{2\omega}}$, and

$$[\hat{b}^{(\lambda)}_{\mathbf{k}}, \hat{b}^{(\lambda')\dagger}_{\mathbf{k}'}] = (2\pi)^3\eta^{(\lambda)(\lambda')}\delta^3(\mathbf{k}-\mathbf{k}'). \tag{26}$$

Here $\epsilon^\mu_{(0)\mathbf{k}} = \delta^\mu_0$ is the unit temporal vector in the t-direction, $\epsilon^\mu_{(3)\mathbf{k}} = (0, \mathbf{k}/|\mathbf{k}|)$ is the longitudinal unit vector, and $\epsilon^\mu_{(1)\mathbf{k}}$ and $\epsilon^\mu_{(2)\mathbf{k}}$ are the two orthonormal polarization vectors perpendicular to $\epsilon^\mu_{(3)\mathbf{k}}$ and $\epsilon^\mu_{(0)\mathbf{k}}$. We will adopt the conventional physical states $|\psi_{phys}\rangle$ of EM fields in the Lorentz gauge as those satisfying $\left(\hat{b}^{(0)}_{\mathbf{k}} - \hat{b}^{(3)}_{\mathbf{k}}\right)|\psi_{phys}\rangle = 0$ [2]. Note that the Minkowski vacuum state $|0_M\rangle$ of EM fields is a physical state.

Suppose the particle-field interaction is switched on at $t=0$. For $t>0$, the evolution of the system is governed by the quadratic Hamiltonian $\hat{H}_2$ with all the dynamical variables $\tilde{\mathcal{O}}$ in (18) promoted to the operators $\hat{\mathcal{O}}$. Since our effective theory is linear, the Heisenberg equations for the operators have the same form as Eqs. (19) and (20) while the deviations there are replaced by the operators. Then one can write

$$\hat{z}^i(t) = \sum_{j=1}^{3}\left[\mathcal{Z}^i_{z^j}(t)\hat{z}^j + \mathcal{Z}^i_{p_j}(t)\hat{p}_j\right] + \sum_{\mathbf{k}}\sum_{\lambda=0}^{3}\left[\mathcal{Z}^i_{(\lambda)\mathbf{k}}(t)\hat{b}^{(\lambda)}_{\mathbf{k}} + \mathcal{Z}^{i*}_{(\lambda)\mathbf{k}}(t)\hat{b}^{(\lambda)\dagger}_{\mathbf{k}}\right], \tag{27}$$

$$\hat{A}^\mu_{\mathbf{x}}(t) = \sum_{j=1}^{3}\left[\mathcal{A}^\mu_{z^j}(t,\mathbf{x})\hat{z}^j + \mathcal{A}^\mu_{p_j}(t,\mathbf{x})\hat{p}_j\right] + \sum_{\mathbf{k}}\sum_{\lambda=0}^{3}\left[\mathcal{A}^\mu_{(\lambda)\mathbf{k}}(t,\mathbf{x})\hat{b}^{(\lambda)}_{\mathbf{k}} + \mathcal{A}^{\mu*}_{(\lambda)\mathbf{k}}(t,\mathbf{x})\hat{b}^{(\lambda)\dagger}_{\mathbf{k}}\right], \tag{28}$$

and the conjugate momenta

$$\hat{p}_i(t) = m\bar{\gamma}\bar{M}_{ij}(t)\dot{\hat{z}}^j(t) + \frac{q}{c}\left[\hat{A}_i^{\bar{\mathbf{z}}(t)}(t) + \hat{z}^j\partial_j\bar{A}_i^{\bar{\mathbf{z}}(t)}(t)\right], \tag{29}$$

$$\hat{\pi}^i_{\mathbf{x}}(t) = \frac{1}{c}\hat{F}^{i0}_{\mathbf{x}}(t), \qquad \hat{\pi}^0_{\mathbf{x}}(t) = \frac{1}{c}\partial_\mu\hat{A}^\mu_{\mathbf{x}}(t), \tag{30}$$

from (16)-(17). Here and below, $\hat{z}^j \equiv \hat{z}^j(t_0)$, $\hat{p}_l \equiv \hat{p}_l(t_0)$ and $\hat{b}^{(\lambda)}_{\mathbf{k}} \equiv \hat{b}^{(\lambda)}_{\mathbf{k}}(t_0)$ for some $t_0 \le 0$ are understood.

Compare (27) and (28) with the free operators (24) and (25), one can read off the free mode functions as

$$\mathcal{Z}^i_{[0]z^j}(t) = \delta^i_j, \qquad \mathcal{Z}^i_{[0]p_j}(t) = \int_{t_0}^t d\tilde{t}\,\frac{\bar{M}^{ij}(\tilde{t})}{m\bar{\gamma}(\tilde{t})}, \qquad \mathcal{Z}^i_{[0](\lambda)\mathbf{k}}(t) = 0, \tag{31}$$

$$\mathcal{A}^\mu_{[0]z^j} = \mathcal{A}^\mu_{[0]p_j} = 0, \qquad \mathcal{A}^\mu_{[0](\lambda)\mathbf{k}}(t,\mathbf{x}) = \epsilon^\mu_{(\lambda)\mathbf{k}}e^{-i\omega t + i\mathbf{k}\cdot\mathbf{x}}. \tag{32}$$

For $t \le 0$, one has $\mathcal{Z}^i_\Omega(t) = \mathcal{Z}^i_{[0]\Omega}(t)$ and $\mathcal{A}^\mu_\Omega(t,\mathbf{x}) = \mathcal{A}^\mu_{[0]\Omega}(t,\mathbf{x})$ with $\Omega = z^j$, p_j, and $(\lambda)\mathbf{k}$.

Applying the commutation relations (21) to the Heisenberg equations, one can see that the evolution equations for the mode functions are in the same form as the Euler-Lagrange equations of the deviations (19) and (20), namely,

$$\partial_t\left(m\bar{\gamma}\bar{M}_{ij}\dot{\mathcal{Z}}^j_\Omega\right) = \frac{q}{c}\varsigma(t)\left[\mathcal{F}^{\bar{\mathbf{z}}}_{\Omega i\mu}\bar{v}^\mu + \left(\mathcal{Z}^j_\Omega\partial_j\bar{F}^{\bar{\mathbf{z}}}_{i\mu}\right)\bar{v}^\mu + \bar{F}^{\bar{\mathbf{z}}}_{ij}\dot{\mathcal{Z}}^j_\Omega\right], \tag{33}$$

$$\partial_\mu\mathcal{F}^{\mu\nu}_\Omega(t,\mathbf{x}) + \alpha\partial^\nu\partial_\mu\mathcal{A}^\mu_\Omega(t,\mathbf{x}) = -\frac{q}{c}\mathcal{V}^\nu_\Omega\delta^3(\mathbf{x}-\bar{\mathbf{z}}), \tag{34}$$

by virtue of the linearity of our quadratic Hamiltonian $\hat{H}_2$. Here $\mathcal{F}^\Omega_{\mu\nu} = \partial_\mu\mathcal{A}^\Omega_\nu - \partial_\nu\mathcal{A}^\Omega_\nu$,

$$\mathcal{V}^0_\Omega(t) \equiv -c\varsigma(t)\mathcal{Z}^j_\Omega(t)\partial_j, \qquad \mathcal{V}^i_\Omega(t) \equiv \partial_t\left[\varsigma(t)\mathcal{Z}^i_\Omega(t)\right] - \bar{v}^i\mathcal{Z}^j_\Omega(t)\partial_j, \tag{35}$$

and we have put in by hand a switching function $\varsigma(t)$ (vanishing for $t \le 0$) for the coupling to suppress the non-adiabatic effect due to sudden switching. Note that our switching function $\varsigma(t)$ only works for quantized deviations $\mathcal{Z}^j_\Omega$ and $\mathcal{A}^\mu_\Omega$, and we assume that the particle-field coupling in the classical equations (7), (8), and (13) has been switched on at past infinity for simplicity.

In the Feynman gauge $\alpha = 1$, (34) can be written as

$$\partial_\nu\partial^\nu\mathcal{A}^\mu_\Omega(t,\mathbf{x}) = -\frac{q}{c}\mathcal{V}^\mu_\Omega(t)\delta^3(\mathbf{x}-\bar{\mathbf{z}}(t)), \tag{36}$$

where the four components $\mu = 0,1,2,3$ decouple like the classical equation (9). The above equation has a general solution similar to (10), which reads

$$\mathcal{A}^\mu_\Omega(t,\mathbf{x}) = \mathcal{A}^\mu_{[0]\Omega}(t,\mathbf{x}) + \mathcal{A}^\mu_{[1]\Omega}(t,\mathbf{x}), \tag{37}$$

where

$$\mathcal{A}^\mu_{[1]\Omega}(t,\mathbf{x}) = q\int dt'd^3x'\,G_{ret}(t,\mathbf{x};t',\mathbf{x}')\mathcal{V}^\mu_\Omega(t')\delta^3(\mathbf{x}'-\bar{\mathbf{z}}(t')). \tag{38}$$

Using the same regularization and renormalization as those for the classical theory in previous section, we obtain the quantum counterpart of the LAD equation (qLAD),

$$\begin{aligned}\partial_t\left\{\bar{\gamma}\bar{M}_{ij}\partial_t\left[\mathsf{m}(t)\mathcal{Z}^j_\Omega(t)\right]\right\} = \sum_{n=1}^{3}\frac{\partial\bar{\Gamma}_i}{\partial\left(\partial^n_t\bar{z}^j\right)}\partial^n_t[\varsigma\mathcal{Z}^j_\Omega] + \\ \frac{q}{c}\varsigma\left\{\bar{v}^\mu\mathcal{F}^{[0]\Omega}_{i\mu}(t,\bar{\mathbf{z}}(t)) + \bar{v}^\mu\mathcal{Z}^j_\Omega\partial_j\bar{F}^{[0]}_{i\mu}(t,\bar{\mathbf{z}}) + \bar{F}^{[0]}_{ij}(t,\bar{\mathbf{z}})\dot{\mathcal{Z}}^j_\Omega\right\} + O(\Lambda^{-1}),\end{aligned} \tag{39}$$

where the quantum counterpart of the LAD force is, explicitly,

$$\sum_{n=1}^{3}\frac{\partial\bar{\Gamma}_i}{\partial\left(\partial_t^n\bar{z}^j\right)}\partial_t^n[\varsigma\mathcal{Z}_\Omega^j]=\frac{q^2\bar{\gamma}^4}{4\pi c^5}\times$$
$$\left\{\frac{2c^2}{3\bar{\gamma}^2}\bar{M}_{ij}\partial_t^3[\varsigma\mathcal{Z}_\Omega^j]+2\left[\bar{\upsilon}^k\dot{\bar{\upsilon}}_k\eta_{ij}+\left(\dot{\bar{\upsilon}}_i+2\frac{\bar{\gamma}^2}{c^2}\bar{\upsilon}^k\dot{\bar{\upsilon}}_k\bar{\upsilon}_i\right)\bar{\upsilon}_j\right]\partial_t^2[\varsigma\mathcal{Z}_\Omega^j]\right.$$
$$+2\left[\frac{1}{3}\bar{\upsilon}^k\ddot{\bar{\upsilon}}_k\eta_{ij}+\frac{\bar{\gamma}^2}{c^2}\left(\bar{\upsilon}^k\dot{\bar{\upsilon}}_k\right)^2\eta_{ij}+\frac{1}{3}\bar{\upsilon}_i\ddot{\bar{\upsilon}}_j+\frac{2}{3}\ddot{\bar{\upsilon}}_i\bar{\upsilon}_j+\frac{4\bar{\gamma}^2}{3c^2}\bar{\upsilon}^k\ddot{\bar{\upsilon}}_k\bar{\upsilon}_i\bar{\upsilon}_j\right.$$
$$\left.\left.+\dot{\bar{\upsilon}}_i\dot{\bar{\upsilon}}_j+\frac{\bar{\gamma}^2}{c^2}\bar{\upsilon}^k\dot{\bar{\upsilon}}_k\left(4\dot{\bar{\upsilon}}_i\bar{\upsilon}_j+2\bar{\upsilon}_i\dot{\bar{\upsilon}}_j\right)+6\left(\frac{\bar{\gamma}^2}{c^2}\bar{\upsilon}^k\dot{\bar{\upsilon}}_k\right)^2\bar{\upsilon}_i\bar{\upsilon}_j\right]\partial_t[\varsigma\mathcal{Z}_\Omega^j]\right\},\tag{40}$$

and the renormalized mass is defined as

$$\mathsf{\bar{m}}(t)\equiv m+\Delta_m\varsigma(t)\tag{41}$$

with the same mass correction Δ_m in (14) for the classical LAD equation (13). $\mathsf{\bar{m}}(t)$ grows from m to $\bar{m}$ as t increases from 0 to ∞, showing how the bare energy of the particle gradually put on the energy of quantum fields as the particle-field coupling is switched on smoothly in time.

4. Wavepacket of charged particle at rest

Consider a charged particle situated at rest in Minkowski vacuum with zero background fields $\bar{F}^{\mu\nu}_{[0]}=0$. Let the particle's worldline be $\bar{z}^\mu(t)=(ct,0,0,0)$. Then its four velocity is $\bar{\upsilon}^\mu=(c,0,0,0)$, and Eq. (39) reads

$$\partial_t^2\left\{\mathsf{\bar{m}}(t)\mathcal{Z}_i^\Omega(t)\right\}=q\varsigma(t)\mathcal{F}_{i0}^{[0]\Omega}(t,\mathbf{0})+s\bar{m}\partial_t^3[\varsigma\mathcal{Z}_i^\Omega]+O(s\Lambda^{-1})\tag{42}$$

after the particle-field coupling is switched on at $t=0$. Here the small parameter s is defined as

$$s\equiv\frac{q^2}{6\pi c^3\bar{m}}\tag{43}$$

with unit of time. For electrons, the time scale $s\approx 5.0\times 10^{-25}$ s corresponds to a length scale $cs=r_0/(6\pi)$, which is well below the classical electron radius $r_0\equiv\frac{q^2}{\bar{m}c^2}\approx 2.8\times 10^{-15}$ m.

4.1. Mode functions

Below in this paper we set $t_0\to 0-$ for simplicity. Assume the absolute value of the counterpart of the LAD force, $|s\partial_t^3(\varsigma\mathcal{Z}_i^\Omega)|$, is always much smaller than the other terms in (42). Then we would be allowed to write the solution to (42) in a series of s as

$$\mathcal{Z}_j^\Omega(t)=\sum_{n=0}^{\infty}s^n\mathcal{Z}_{\{n\}j}^\Omega(t),\tag{44}$$

and require $\mathcal{Z}_{\{n\}j}^\Omega(0)=\dot{\mathcal{Z}}_{\{n\}j}^\Omega(0)=0$ for $n\geq 1$.

From Eq.(32), we have

$$\mathcal{F}_{(\lambda)j0}^{[0]\mathbf{k}}(t,\mathbf{0})=\mathcal{E}_{(\lambda)j}^{\mathbf{k}}e^{-i\omega t},\tag{45}$$

where

$$\mathcal{E}_{(\lambda)\mathbf{k}}^{j}=\left(-ik^j,i\frac{\omega}{c}\epsilon_{(1)\mathbf{k}}^j,i\frac{\omega}{c}\epsilon_{(2)\mathbf{k}}^j,ik^j\right),\tag{46}$$

and $\mathcal{F}^{[0]z^l}_{j0} = \mathcal{F}^{[0]p_l}_{j0} = 0$. Inserting these $\mathcal{F}^{[0]\Omega}_{j0}$ into Eq. (42), and matching the initial condition (31) at $t = 0$, we find the lowest orders of the series solution (44) for $t \geq 0$ below:

$$\mathcal{Z}^j_{z^{j'}} = \frac{m\delta^j_{j'}}{\bar{\mathsf{m}}(t)}\Big\{1 + s\partial_t\tilde{\varsigma} + s^2\partial_t\left(\tilde{\varsigma}\partial_t\tilde{\varsigma}\right) + \cdots\Big\}, \tag{47}$$

$$\mathcal{Z}^j_{p_{j'}} = \frac{\eta^{jj'}}{\bar{\mathsf{m}}(t)}\Big\{(t - t_0) + s\partial_t\Big[(t-t_0)\tilde{\varsigma}\Big] + s^2\partial_t\Big[\tilde{\varsigma}\partial_t\Big((t-t_0)\tilde{\varsigma}\Big)\Big] + \cdots\Big\}, \tag{48}$$

$$\mathcal{Z}^j_{(\lambda)\mathbf{k}}(t) = \frac{q\mathcal{E}^j_{(\lambda)\mathbf{k}}}{\bar{\mathsf{m}}(t)}\Big\{f_\omega + s\partial_t[\tilde{\varsigma}f_\omega] + s^2\partial_t[\tilde{\varsigma}\partial_t(\tilde{\varsigma}f_\omega)] + \cdots\Big\}, \tag{49}$$

where

$$\tilde{\varsigma}(t) \equiv \frac{(m + \Delta_m)\varsigma(t)}{m + \Delta_m\varsigma(t)}, \tag{50}$$

with $\tilde{\varsigma} = 0$ at $t = 0$, $\tilde{\varsigma} \to 1$ as $t \to \infty$, and $f_\omega(t)$ is defined by

$$\mathcal{Z}^{\{0\}j}_{(\lambda)\mathbf{k}}(t) = \frac{q}{\bar{\mathsf{m}}(t)}\int_{-\infty}^{t} d\tilde{t}\, K(t,\tilde{t})\varsigma(\tilde{t})e^{-i\omega\tilde{t}}\mathcal{E}^j_{(\lambda)\mathbf{k}} \equiv \frac{q}{\bar{\mathsf{m}}(t)}\mathcal{E}^j_{(\lambda)\mathbf{k}}f_\omega(t). \tag{51}$$

Here, the evolution kernel $K(t,\tilde{t}) = t - \tilde{t}$ is the solution to $\partial_t^2 K(t,\tilde{t}) = 0$ satisfying the boundary conditions $K(\tilde{t},\tilde{t}) = 0$ and $\partial_t K(t,\tilde{t}) = 1$, and the switching function has been assumed to have $\varsigma(0) = \dot{\varsigma}(0) = \ddot{\varsigma}(0) = 0$.

4.2. Particle correlators

Suppose at $t = 0$ the initial state of the combined system is $\rho^{\mathbf{I}} = \rho^{\mathbf{I}}_P \otimes \rho^{\mathbf{I}}_F$, which is a direct product of a Gaussian state $\rho^{\mathbf{I}}_P$ of the charged particle and the Minkowski vacuum state $\rho^{\mathbf{I}}_F = |0_M\rangle\langle 0_M|$ of the EM fields. By virtue of the linearity of our effective theory, the symmetrized two-point correlators of the renormalized charged particle's deviation from classical trajectory split into two parts, e.g.,

$$\langle \hat{z}^j(t), \hat{z}^{j'}(t)\rangle \equiv \mathrm{Tr}\left[\rho^{\mathbf{I}}\frac{1}{2}\left(\hat{z}^j(t)\hat{z}^{j'}(t) + \hat{z}^{j'}(t)\hat{z}^j(t)\right)\right] = \langle \hat{z}^j(t), \hat{z}^{j'}(t)\rangle_P + \langle \hat{z}^j(t), \hat{z}^{j'}(t)\rangle_F, \tag{52}$$

with the expansion of $\hat{z}^j(t)$ given in (27) [1]. Here the P-part,

$$\begin{aligned}\langle \hat{z}^j(t), \hat{z}^{j'}(t)\rangle_P &\equiv \mathrm{Tr}\left[\rho^{\mathbf{I}}_P\sum_{l,l'}\left\{\left(\mathcal{Z}^j_{z^l}(t)\hat{z}^l + \mathcal{Z}^j_{p_l}(t)\hat{p}_l\right), \left(\mathcal{Z}^{j'}_{z^{l'}}(t)\hat{z}^{l'} + \mathcal{Z}^{j'}_{p_{l'}}(t)\hat{p}_{l'}\right)\right\}\right] \\ &= \sum_{l,l'}\Big[\langle\hat{z}^l,\hat{z}^{l'}\rangle_{\mathrm{I}}\,\mathcal{Z}^j_{z^l}(t)\mathcal{Z}^{j'}_{z^{l'}}(t) + \langle\hat{p}^l,\hat{p}^{l'}\rangle_{\mathrm{I}}\,\mathcal{Z}^j_{p^l}(t)\mathcal{Z}^{j'}_{p^{l'}}(t) + \\ &\qquad \langle\hat{z}^l,\hat{p}^{l'}\rangle_{\mathrm{I}}\,\mathcal{Z}^j_{p^l}(t)\mathcal{Z}^{j'}_{z^{l'}}(t) + \langle\hat{p}^l,\hat{z}^{l'}\rangle_{\mathrm{I}}\,\mathcal{Z}^j_{z^l}(t)\mathcal{Z}^{j'}_{p^{l'}}(t)\Big]\end{aligned} \tag{53}$$

with $\langle\hat{\mathcal{O}}^l,\hat{\mathcal{O}}^{l'}\rangle_{\mathrm{I}} \equiv \mathrm{Tr}\left[\rho^{\mathbf{I}}_P\{\hat{\mathcal{O}}^l,\hat{\mathcal{O}}^{l'}\}\right]$, depends on the initial state of the particle $\rho^{\mathbf{I}}_P$ only, and the F-part,

$$\langle \hat{z}^j(t), \hat{z}^{j'}(t)\rangle_F \equiv \lim_{t'\to t}\sum_{\mathbf{k},\mathbf{k}'}\frac{1}{2}\left(\mathcal{Z}^j_{(\lambda)\mathbf{k}}(t)\mathcal{Z}^{j'*}_{(\lambda')\mathbf{k}'}(t') + \mathcal{Z}^{j'}_{(\lambda)\mathbf{k}}(t')\mathcal{Z}^{j*}_{(\lambda')\mathbf{k}'}(t)\right)\langle 0_M|\hat{b}^{(\lambda)}_{\mathbf{k}}\hat{b}^{(\lambda')\dagger}_{\mathbf{k}'}|0_M\rangle \tag{54}$$

[1] The "P-part" and "F-part" correlators are the counterparts of the "a-part" and "v-part" correlators, respectively, in [9] and our series of papers on the Unruh-DeWitt detectors afterward.

depends on the initial state of the field $|0_M\rangle$. In (54), $\sum_{\mathbf{k}}$ has been defined below (25), and $\langle 0_M|\hat{b}^{(\lambda)}_{\mathbf{k}}\hat{b}^{(\lambda')\dagger}_{\mathbf{k}'}|0_M\rangle = (2\pi)^3\eta^{(\lambda)(\lambda')}\delta^3(\mathbf{k}-\mathbf{k}')$ from (26).

Substituting the mode functions $\mathcal{Z}^j_\Omega$ obtained in Section 4.1 to the above expressions, we can calculate the two-point correlators order by order in s. Recall that we have neglected the $O(s\Lambda^{-1})$ terms in Eq. (42) for mode functions. Since $s^2 \ll s\Lambda^{-1}$ for electrons, the $O(s^2)$ terms of the mode functions are negligible. It is therefore sufficient to calculate the P-part of the two-point correlators of the particle up to s^2-order terms, and the F-part correlators up to s^3-order [2]. The s^3-order terms in the F-part correlators may become greater than the s^2-order terms at late times, as we will see in Figure 2, while the higher-order terms would be negligible for electrons.

The zeroth-order F-part of the particle-deviation correlator can be formally written as

$$\begin{aligned}\langle \hat{z}^j(t), \hat{z}^{j'}(t)\rangle_F^{\{0\}} &\equiv \lim_{t'\to t}\int \frac{cd^3k}{(2\pi)^3}\sqrt{\frac{\hbar}{2\omega}}\int\frac{cd^3k'}{(2\pi)^3}\sqrt{\frac{\hbar}{2\omega'}}\times\\ &\quad \frac{1}{2}\left(\mathcal{Z}^{\{0\}j}_{(\lambda)\mathbf{k}}(t)\mathcal{Z}^{\{0\}j'*}_{(\lambda')\mathbf{k}'}(t') + \mathcal{Z}^{\{0\}j'}_{(\lambda)\mathbf{k}}(t')\mathcal{Z}^{\{0\}j*}_{(\lambda')\mathbf{k}'}(t)\right)(2\pi)^3\eta^{(\lambda)(\lambda')}\delta^3(\mathbf{k}-\mathbf{k}')\\ &= \frac{\hbar c}{(2\pi)^3}\mathrm{Re}\lim_{t'\to t}\int_0^\infty \frac{\omega^2}{c^2}\frac{d\omega}{2\omega c}\int_0^{2\pi}d\varphi\int_0^\pi d\theta\,\sin\theta\frac{q\mathcal{E}^j_{(\lambda)\mathbf{k}}}{\bar{\mathsf{m}}(t)}f_\omega(t)\frac{q\mathcal{E}^{(\lambda)j'*}_{\mathbf{k}}}{\bar{\mathsf{m}}(t')}f^*_\omega(t') \end{aligned}\quad (55)$$

after inserting (51) into the formula. Given

$$\int_0^{2\pi}d\varphi\int_0^\pi d\theta\,\sin\theta\,\mathcal{E}^{(\lambda)j}_{\mathbf{k}}\mathcal{E}^{j'*}_{(\lambda)\mathbf{k}} = \int d\Omega\left(\frac{\omega^2}{c^2}\eta^{jj'} - k^jk^{j'}\right) = \frac{8\pi}{3}\frac{\omega^2}{c^2}\eta^{jj'}, \quad (56)$$

where $\mathbf{k} = (\frac{\omega}{c}\sin\theta\cos\varphi, \frac{\omega}{c}\sin\theta\sin\varphi, \frac{\omega}{c}\cos\theta) = \frac{\omega}{c}\vec{\epsilon}_{(3)\mathbf{k}}$, we have

$$\begin{aligned}\langle \hat{z}^j(t), \hat{z}^{j'}(t)\rangle_F^{\{0\}} &= \frac{\hbar q}{6\pi^2c^3}\frac{\eta^{jj'}}{\bar{\mathsf{m}}^2(t)}\mathrm{Re}\lim_{t'\to t}\int_0^\infty d\omega\,\omega^3e^{-\omega\epsilon}\times\\ &\quad \int_0^t d\tilde{t}K(t,\tilde{t})\varsigma(\tilde{t})e^{-i\omega\tilde{t}}\int_0^{t'}d\tilde{t}'K(t',\tilde{t}')\varsigma(\tilde{t}')e^{i\omega\tilde{t}'}, \end{aligned}\quad (57)$$

where we have introduced a factor $e^{-\omega\epsilon}$ with a regulator $\epsilon > 0$ in the ω integration to control the divergence, and the lower limits of the $\tilde{t}$ and $\tilde{t}$ integrals are zero since $\varsigma(\tau) = 0$ for $\tau \le 0$.

Let us adopt a simple switching function

$$\mathsf{S}^{[n]}(\tau) \equiv \begin{cases}\left(1 - e^{-\tau/T}\right)^n & \text{for} \quad \tau \ge 0,\\ 0 & \text{for} \quad \tau < 0,\end{cases} \quad (58)$$

where the constant T sets the time scale of switching on. The N-th derivative of $\mathsf{S}^{[n]}(0)$, $N = 0, 1, \cdots, n-1$, are vanishing at the initial moment $t = 0$. Requiring the higher-derivatives in (42) continuous at the initial moment while keeping the calculation doable, below we choose $n = 4$ such that the first, second, and third derivatives of $\varsigma(\tau)$ are all vanishing at the initial moment $\tau = 0$.

Let $\varsigma(t) = \mathsf{S}^{[4]}(t)$. To deal with the divergence in the coincidence limit $t' \to t$, we further introduce two positive regulators ϵ_0 and ϵ_1 and shift the upper and lower limits of the $\tilde{t}'$

[2] There is an extra s in the overall factor of the F-part correlators [compare (57) and (59)], which is the product of the overall factor $q/\bar{\mathsf{m}}$ of the mode functions $\mathcal{Z}^j_{(\lambda)\mathbf{k}}$ in (49).

integration in (57) to $t' = t - \epsilon_1$ and ϵ_0, namely,

$$\langle \hat{z}^j(t), \hat{z}^{j'}(t) \rangle_F^{\{0\}} = \frac{\hbar s \bar{m} \eta^{jj'}}{\pi \bar{\mathsf{m}}^2(t)} \mathrm{Re} \lim_{\epsilon,\epsilon_0,\epsilon_1 \to 0} \int_0^\infty d\omega\, \omega^3 e^{-\omega\epsilon} \times$$
$$\int_0^t d\tilde{t}(t-\tilde{t}) \left(1 - e^{-\tilde{t}/T}\right)^4 e^{-i\omega\tilde{t}} \int_{\epsilon_0}^{t-\epsilon_1} d\tilde{t}'(t - \epsilon_1 - \tilde{t}') \left(1 - e^{-(\tilde{t}'-\epsilon_0)/T}\right)^4 e^{i\omega\tilde{t}'} \tag{59}$$

with the switching function (58) inserted. The regulators ϵ_1 and ϵ_0 correspond to the resolution of the evolution time in a specific experiment. As we learned in Appendix A of Ref.[19], one should require $\epsilon \ll \epsilon_0$ and ϵ_1, though both ϵ_1 and ϵ_0 should be small. Then, with the aid of computer algebra, we find

$$\langle \hat{z}^j(t), \hat{z}^{j'}(t) \rangle_F^{\{0\}} = \frac{\hbar s \bar{m} \eta^{jj'}}{\pi \bar{\mathsf{m}}^2(t)} \Big[Q_2(t^*) t^{*2} + Q_1(t^*) t^* + 2 \ln t^*$$
$$- \left(1 - e^{-t^*}\right)^8 \left(\gamma_e + \ln \frac{\epsilon_1}{T}\right) + 2\gamma_e + Q_0(t^*) + O\left(\frac{\epsilon_0}{T}, \frac{\epsilon_1}{T}, \frac{\epsilon^2}{\epsilon_1^2}\right) \Big], \tag{60}$$

where $t^* \equiv t/T$ is dimensionless, $\gamma_e \approx 0.577216$ is the Euler's constant, and we have assumed $\epsilon \ll \epsilon_1 \approx \epsilon_0 \ll T$ while the value of t is no less than $t_0 + \epsilon_0 + \epsilon_1$ around the initial moment $t_0 = 0$, due to the uncertiainty of time tagging. The functions Q_2, Q_1, and Q_0 depend on the choice of the switching function. At late times ($t^* \gg 1$), $Q_2(t^*) \to \frac{72}{35} \ln \frac{4}{3} \approx 0.5918$, $Q_1(t^*) \to -\frac{4}{7} - \frac{19888}{3675} \ln 2 + \frac{2988}{1225} \ln 3 \approx -1.6428$, and $Q_0(t^*) \to -\frac{9}{490} + \frac{205966}{128625} \ln 2 - \frac{32041}{42875} \ln 3 \approx 0.2706$. These constants also depend on the choice of the switching function, while the values of them are roughly of the same order with different choices of n ($n \geq 2$) in (58) for our switching function ς.

Inserting the switching function $\varsigma(t) = \mathsf{S}^{[4]}(t)$ in (58) into the mode functions (47) and (48), the P-part correlators (53) up to $O(s^2)$ can be obtained straightforwardly. The closed forms of the $O(s^2)$ and $O(s^3)$ corrections to the F-part of the particle correlators $\langle \hat{z}^j(t), \hat{z}^{j'}(t) \rangle$, as well as the correlators $\langle \hat{p}^j(t), \hat{p}^{j'}(t) \rangle$ and $\langle \hat{z}^j(t), \hat{p}^{j'}(t) \rangle$ can be obtained in a similar way. While we do not put these full closed-form results in this paper because of their lengths, we do use them to obtain the numerical results in Figure 2 and the purity in next section. In Figure 2, one can see that in each correlator the correction by the qLAD force (dashed and dotted curves) is small compared with the zeroth order (solid curves), while in the left plot the contributions of order s^3 to $\langle \hat{p}^i, \hat{p}^j \rangle_F$ [dotted curve, the $s(s/\epsilon_1)^2$ terms in our result] are greater than those of order s^2 [dashed curve] at late times, because the $s(s/\epsilon_1)$ terms in the latter almost cancel when $\epsilon_1 \ll T$.

4.3. Quantum coherence of the electron wavepacket

The reduced state of the charged particle can be obtained by tracing out the field degrees of freedom in the quantum state of the combined system. In the presence of the particle-field coupling, the reduced state of the charged particle is a mixed state. For Gaussian states of a single charged particle considered in this paper, the two-point correlators of the particle can completely determine the density matrix of the reduced state [20, 21], from which the purity, von Neumann entropy, and the effective temperature of the particle can be extracted [20, 22]. The key quantity is the uncertainty function

$$\mathcal{U}(\tau) \equiv \sqrt{|\det \mathbf{C}|} \tag{61}$$

where $\mathbf{C}$ is the 6×6 covariance matrix

$$\mathbf{C} = \begin{pmatrix} \langle \hat{p}_i(\tau), \hat{p}_j(\tau) \rangle & \langle \hat{z}^i(\tau), \hat{p}_j(\tau) \rangle \\ \langle \hat{p}_i(\tau), \hat{z}^j(\tau) \rangle & \langle \hat{z}^i(\tau), \hat{z}^j(\tau) \rangle \end{pmatrix} \tag{62}$$

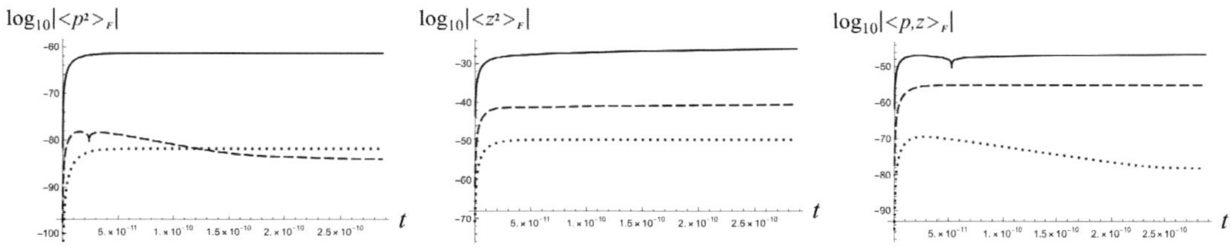

Figure 2. Time evolution of the order of magnitude of the particle correlators, with $\langle \hat{p}^j, \hat{p}^{j'} \rangle_F = \eta^{jj'} \langle p^2(t) \rangle_F$, $\langle \hat{z}^j, \hat{z}^{j'} \rangle_F = \eta^{jj'} \langle z^2(t) \rangle_F$, and $\langle \hat{p}^j, \hat{z}^{j'} \rangle_F = \eta^{jj'} \langle p(t) z(t) \rangle_F$ in the above plots. The solid, dashed, and dotted curves represent the contributions at the order of s (without taking the qLAD force (40) into account), s^2 (first order), and s^3 (second order of the qLAD correction), respectively. Here we set the time scale of the switching function $T = 10^{-11}$ s, and the time resolution of the experiment $\epsilon_0 = \epsilon_1 = 1.4 \times 10^{-14}$ s.

The purity of the reduced state of the particle then reads

$$\mathcal{P} = \frac{(\hbar/2)^3}{\mathcal{U}}. \tag{63}$$

For our charged particle at rest, $\langle z_I | \{ \hat{\mathcal{O}}_j(\tau), \hat{\mathcal{O}}'_{j'}(\tau) \} | z_I \rangle \propto \eta_{jj'}$ for $\mathcal{O}, \mathcal{O}' = z, p$, so the three degrees of freedom of the particle motion decouple, and

$$\mathcal{U} = \prod_{j=1}^{3} u_j, \tag{64}$$

where

$$u_j(t) = \sqrt{\langle \hat{p}_j(t), \hat{p}_j(t) \rangle \langle \hat{z}_j(t), \hat{z}_j(t) \rangle - \left(\langle \hat{p}_j(t), \hat{z}_j(t) \rangle \right)^2} \tag{65}$$

is the uncertainty function in the x^j direction. Suppose that when the particle-field coupling is switched on, the particle is in a nearly pure state, and so $u_j(0) \approx \hbar/2$. After the particle-field coupling is switched on, the F-part of each correlators becomes nonzero, and u_j increases from $\hbar/2$. At late times ($t \gg T$), the deviation of u_i from $\hbar/2$ would be dominated by the t^2 terms of the correlators.

In Ref. [1], Tonomura *et al.* clearly demonstrated that a train of coherent single electrons can gradually form an interference pattern after passing through a biprism. While our wavepacket for a charged particle at rest in vacuum would not form any interference fringes since our wavepacket is neither for a moving particle (longitudinal wavevector $k_3 = 0$ here) nor for a superposition of two quantum states moving in non-parallel directions (we have transverse wavevector $\pm k_1 = 0$), the coherent lengths in transverse and longitudinal directions in Ref. [1] can still help us to choose reasonable values of the regulators ϵ, ϵ_0, ϵ_1, and T, introduced earlier in this paper.

According to [1], the coherent electrons tunneling out of a field-emission electron gun (FEG) are accelerated to a speed $v \approx 50$ keV $= 1.2 \times 10^8$ m/s $\approx 0.4c$, corresponding to a de Broglie wavelength of 5.4×10^{-12} m. At this speed, $\gamma = 1/\sqrt{1 - v^2/c^2}$ is about 1.1 only, and so electron's time dilation is not significant for laboratory observers. The distance from the source to the screen in their electron microscope is about 1.5 m, and in most of the journey the single electrons are approximately in inertial motion. Since the interference pattern does form on the image plane, the electron wavepacket in [1] must have nearly full quantum coherence, namely, its purity must still be of order 1, after flying for $t_F \approx 1.2 \times 10^{-8}$ s to the image plane, and the short acceleration stage at early times as well as the passages through the electron lenses did not have too much impact on the electrons' coherence.

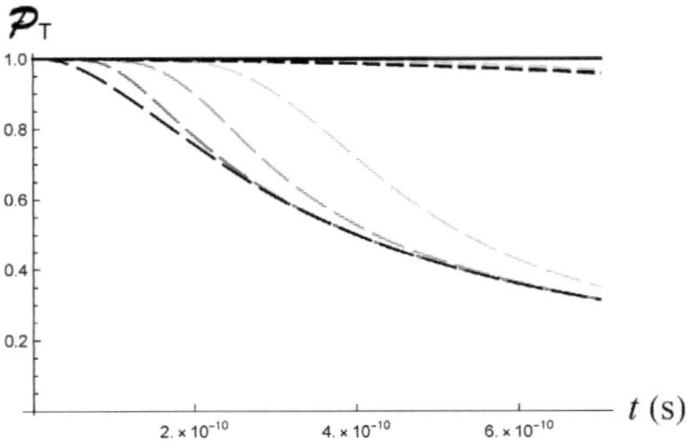

Figure 3. Early-time evolution of the transverse components of purity $\mathcal{P}_\mathsf{T} \equiv \hbar/(2u_\mathsf{T})$ after the particle-field coupling is switched on at $t = 0$. Here we set the time scales of the switching function (from dark to light curves) $T = 10^{-11}$, 5×10^{-11}, 10^{-10}, and 2×10^{-10} s, all have the time resolution $\epsilon_0 = \epsilon_1 = 1.4 \times 10^{-14}$ s. The curves represent the electron wavepackets with initial width $\sqrt{\langle \hat{z}_j^2(0)\rangle} = 5 \times 10^n$ m, $n = -11$ (long dashed), -10 (short dashed), and -9 (solid), respectively, while $\langle \hat{p}_j^2(0)\rangle = \hbar^2/(4\langle \hat{z}_j^2(0)\rangle)$ and $\langle \hat{p}_j(0), \hat{z}_j(0)\rangle = 0$ for these curves. The contributions of the qLAD force is not resolvable in the plot. One can see that the black and gray curves for each initial widths tend to converge after transient, indicating that their late-time values are not sensitive to the choice of T.

The diameter of the virtual source of a FEG applied in Ref. [1] is about 5 nm [23, 24], so we set the initial width of the wavepacket $\sqrt{\langle\{\hat{z}^\mathsf{T}(0), \hat{z}^\mathsf{T}(0)\}\rangle} = 5 \times 10^{-9}$ m in the transverse directions $\mathsf{T} = 1$ or 2, while we choose $\langle\{\hat{p}^\mathsf{T}(0), \hat{z}^\mathsf{T}(0)\}\rangle = 0$ for simplicity. These lead to a transverse spreading rate of the wavepacket $v^\mathsf{T}(0) = m^{-1}\sqrt{\langle\{\hat{p}^\mathsf{T}(0), \hat{p}^\mathsf{T}(0)\}\rangle} \approx 1.2 \times 10^4$ m/s. From (53) with (47) and (48) inserted, we find the P-part correlator $\langle\{\hat{z}^\mathsf{T}(t_F), \hat{z}^\mathsf{T}(t_F)\}\rangle_P \approx (141\ \mu\text{m})^2$ when arriving at the image plane. This is consistent with the transverse coherence length of electrons, 140 μm at the screen, given in [1].

Due to the thermal and quantum fluctuations of the cathode, electrons just emitted by a FEG operating at room temperature typically have an energy spread $\Delta E \approx 0.3$ eV [24]. This implies an uncertainty $\Delta t \sim h/\Delta E \approx 1.4 \times 10^{-14}$ s of time tagging to the history (or the characteristic longitudinal coherence time [25]) of an emitted electron. The same energy spread also corresponds to the longitudinal coherent length $\Delta z_3 \sim v\Delta t \approx 1.7 \times 10^{-6}$ m [25], which is close to the value of the longitudinal coherence length 1 μm (corresponding to $\Delta E \approx 0.5$ eV) given in Refs. [1, 23] as the longitudinal length of the electron wavepacket.

The time resolution of single electron detectors can reach the order of 0.1 ns (e.g. TimePix4 [26]), yet this is still much greater than the time scale of Δt. Considering Δt as the ultimate time resolution in this experiment, let us set $\epsilon_0 = \epsilon_1 = \Delta t = 1.4 \times 10^{-14}$ s. From our observation in Figure 1, we set $\epsilon = 100\, t_C = 8.09 \times 10^{-19}$ s, which is significantly longer than electron's Compton time t_C but still much shorter than ϵ_0 and ϵ_1. This fits our assumption learned from [19]. Moreover, in this paper we choose the time scale of the switching function $T = 10\epsilon_0$, which is basically undetectable compared with the time resolution of single electron detectors 0.1 ns and the flying proper-time $\tau_F = t_F/\gamma \approx 1.1 \times 10^{-8}$ s, as it should be.

Inserting the closed form of all the P-part and F-part correlators with the above chosen values for the parameters into (65), we obtain

$$\mathcal{P}_\mathsf{T} \equiv \frac{\hbar}{2u_\mathsf{T}} \approx 0.88 \tag{66}$$

in the transverse directions $\mathsf{T} = 1, 2$ when the wavepacket arrives at the screen [3]. In this case, the late-time correction to $u_{\mathsf{T}}^2 - (\hbar/2)^2$ is mainly contributed by $\langle\{\hat{p}_{\mathsf{T}}(t), \hat{p}_{\mathsf{T}}(t)\}\rangle_F \langle \hat{z}_{\mathsf{T}}(t), \hat{z}_{\mathsf{T}}(t)\rangle_P \approx 0.28\times (\hbar^2/4)$ (because $\langle \hat{z}_{\mathsf{T}}(t), \hat{p}_{\mathsf{T}}(t)\rangle_P \propto t^2$ can be relatively large at late times when $t \gg T$). If we choose a larger value for ϵ_1 (and ϵ_0) to get a lower time resolution, then the value of $\mathcal{P}_{\mathsf{T}}$ will be closer to 1.

In the longitudinal direction, almost all the parameters are the same except the initial width of the wavepacket is now $\sqrt{\langle\{\hat{z}_3(0), \hat{z}_3(0)\}\rangle} = \Delta z_3 = 1.7 \times 10^{-6}$ m. We get

$$\mathcal{P}_3 \equiv \frac{\bar{h}}{2u_3} \approx 0.99998. \tag{67}$$

So the purity of the single electron wavepacket around the image plane is $\mathcal{P} = \mathcal{P}_1\mathcal{P}_2\mathcal{P}_3 \approx (0.88)^2 \times 1 \approx 0.78$. Such a weak but sufficiently significant decoherence indicates that vacuum fluctuations may be a major source of blurring the interference pattern obtained in Ref. [1].

In Figure 3 we show the early-time evolution of $\mathcal{P}_j$ for various initial widths $\sqrt{\langle \hat{z}_j^2(0)\rangle}$ of the wavepacket. One can see that the smaller initial widths we choose, the faster decoherence will occur. Our results here are not sensitive to the choice of the time scale T of our artificial switching function $\varsigma(t)$. For the same choice of the value of ϵ_0 (and we choose $\epsilon_1 = \epsilon_0$), our calculation [without the restriction $T \gg \epsilon_0$ and ϵ_1 in (60)] shows that the purity is of the same order from $T \approx t_F$ all the way down to $T \approx \epsilon_1/2$.

5. Summary and Outlook

We have constructed a linearized effective theory for a relativistic charged particle interacting with EM fields. The quantized theory describes the interplay between a quantum mechanical wavepacket of the particle and quantum EM fields. The equation of motion for the mode functions of the particle is a quantum counterpart of the LAD equation, where the radiation-reaction force is derived. A Gaussian wavepacket of the relativistic particle will be always Gaussian and moving along a classical trajectory.

The values of the parameters in our effective theory can be learned from the data of electron interference experiments in electron microscopes. With those values we find that weak decoherence may have been present in the blurred interference pattern in Ref. [1].

Following the method in Ref.[9, 20], it will be straightforward to calculate the Unruh effect on electron trajectories and quantum corrections to the radiation emitted by single charged wavepackets in our effective theory. We are also extending our model to the cases with two relativistic charged particles to investigate the dynamics of quantum entanglement between them. More details and results will be presented in Ref. [14].

Acknowledgments

This work is supported by the National Science and Technology Council of Taiwan under grant No. NSTC 111-2112-M-018-007.

References

[1] Tonomura A, Endo J, Matsuda T, Kawasaki T and Ezawa H 1989 Demonstration of single-electron buildup of an interference pattern *Am. J. Phys.* **57** 117
[2] Itzykson C and Zuber J-B 1980 *Quantum Field Theory* (McGraw-Hill, New York)
[3] Davies P C W 1975 Scalar particle production in Schwarzschild and Rindler metrics *J Phys A: Math Gen* **8** 609

[3] If we insert the travel proper time $\tau_F = t_F/\gamma \approx 1.11 \times 10^{-8}$ s instead of t_F to take the time dilation into account, then $\mathcal{P}_{\mathsf{T}}$ will be 0.90, a little closer to 1. As an estimate to the order of magnitude, we will keep using t_F in this paper.

[4] Unruh W G, 1976 Notes on black hole evaporation *Phys. Rev. D* **14** 3251
[5] Bell J S and Leinaas J M 1983 Electrons as accelerated thermometers *Nucl. Phys.* **B212** 131
[6] Akhmedov E T and Singleton D 2007 On the relation between Unruh and Sokolov Ternov effects *Int. J. Mod. Phys. A* **22** 4797
[7] Chen P and Tajima T 1999 Testing Unruh radiation with untraintense lasers *Phys. Rev. Lett.* **83** 256
[8] DeWitt B S 1979 Quantum gravity: the new synthesis, in *General Relativity: an Einstein Centenary Survey*, edited by S. W. Hawking and W. Israel (Cambridge University Press, Cambridge).
[9] Lin S-Y and Hu B L 2006 Accelerated detector - quantum field correlations: From vacuum fluctuations to radiation flux *Phys. Rev. D* **73** 124018.
[10] Lin S-Y 2017 Quantum radiation by an Unruh-DeWitt detector in oscillatory motion *JHEP* **11 (2017)** 102
[11] Doria D, Cernaianu M O, Ghenuche P, Stutman D, Tanaka K A, Ticosa C and Ur C A 2020 Overview of ELI-NP status and laser commissioning experiments with 1PW and 10PW class-lasers *JINST* **15** C09053
[12] Gan Z *et al.* 2021 The Shanghai superintense ultrafast laser facility (SULF) project *Progress in Ultrafast Intense Laser Science XVI, Topics in Applied Physics 141* ed K Yamanouchi, K Midorikawa and L Roso (Berlin: Springer) p 199
[13] Rosenstein B and Usher M 1987 Explicit illustration of causality violation: Noncausal relativistic wave-packet evolution *Phys. Rev.* D **36** 2381
[14] Lin S-Y and Hu B L 2023 Relativistic particle wavepacket in quantum electromagnetic fields: quantum coherence, and the Unruh effect (in preparation)
[15] Johnson P R and Hu B L 2002 Stochastic theory of relativistic particles moving in a quantum field 2: Scalar Abraham-Lorentz-Dirac-Langevin equation, radiation reaction and vacuum fluctuations *Phys. Rev. D* **65** 065015
[16] Lorentz H A 1892 *Arch. Ner. Sci. Exact. Nat.* **XXV** 363; *The Theory of Electrons*, 2nd ed. (Dover, New York, 1952).
[17] Abraham M 1905 *Theorie der Elektrizität*, Vol. II (Leipzig: Teubner)
[18] Dirac P A M 1938 Classical Theory of Radiating Electrons *Proc. R. Soc. London A* **167** 148
[19] Lin S-Y and Hu B L 2010 Entanglement creation between two causally disconnected objects *Phys. Rev. D* **81** 045019
[20] Lin S-Y and Hu B L 2007 Backreaction and the Unruh effect: new insights from exact solutions of uniformly accelerated detectors *Phys. Rev. D* **76** 064008
[21] Lin S-Y 2012 Instantaneous spatially local projective measurements are consistent in a relativistic quantum field *Ann. Phys. (Amsterdam)* **327** 3102
[22] Doukas J, Lin S-Y, Hu B L and Mann R B 2013 Unruh effect under non-equilibrium conditions: oscillatory motion of an Unruh-DeWitt detector *JHEP* **11 (2013)** 119.
[23] Tonomura A 1998 *The Quantum World Unveiled by Electron Waves* (Singapore: World Scientific)
[24] Glossary of TEM/SEM Terms, JEOL Ltd. https://www.jeol.co.jp/en/words/emterms/index.html (Retrieved 2 October 2022)
[25] Silverman M P, Strange W, and Spence J C H 1995 The brightest beam in science: New directions in electron microscopy and interferometry *Am. J. Phys.* **63** 800
[26] Llopart X et al. 2022 Timepix4, a large area pixel detector readout chip which can be tiled on 4 sides providing sub-200 ps timestamp binning *JINST* **17** C01044

Signatures of 1.5 MeV leptons in nuclear reactions

Andras Kovacs[1], **Valery Zatelepin**[2], **Dmitry Baranov** [2]

[1]BroadBit Energy Technologies, Espoo, Finland
[2]INLEAS laboratory, Moscow, Russia

E-mail: andras.kovacs@broadbit.com

Abstract. While experimental physics progressed tremendously since the 1970s, the neutron model has remained essentially unchanged. Motivated by developments in both experiments and theory, which we briefly review in section 1, we propose that the initial neutron decay step is not the emission of an 80 GeV mass boson particle, but the emission of a much lighter lepton particle. On the basis of well-known neutron data, in section 2 we estimate that this new lepton's mass is 1.5 MeV.

Historically, investigations of deuteron photo-dissociation led nuclear scientists to assume that no electron-like particle is associated with neutron decay. We therefore re-examine these experiments in section 3. We demonstrate that deuteron photo-dissociation leads to $2p^+ + e^-$ products at high photon energies. Our calculations show why a deuteron always breaks up into $p + n$ particles at <3 MeV photon energy.

Sections 4 - 7 discuss the properties and interactions of the 1.5 MeV lepton particle. Numerous investigations, including our own experiments, demonstrate the presence of negative elementary charges within atomic nuclei. The emission or absorption of negative nuclear charges involves the emission or absorption of a new lepton particle, which always decays into an electron. Various mass measurement methods converge to the same result: the emitted or absorbed lepton is approximately three times heavier than an ordinary electron. Specifically, we measure its mass to be 1553.5 keV.

Our work demonstrates that, despite being a single particle, the neutron comprises a positive and a negative elementary charge. To make sense of the neutron structure, it is necessary to firstly understand the proton's and the newly discovered 1.5 MeV lepton's internal structures. In section 9, we apply our results to better understand the neutron's properties.

1. Motivation

1.1. A brief history of the proton model

Before the 1970s, most scientists viewed the proton as an elementary particle. Starting from the 1970s, scientists working at high energy particle colliders proposed that protons and neutrons are not elementary particles, but comprise smaller sub-particles. According to their model, a proton and a neutron both comprise three quark sub-particles. The existence of quarks has been suggested initially in the 1960s, based on the theoretical efforts by Gell-Mann to model baryons and mesons [3], which were observed in a great variety during high energy nuclear experiments. While Gell-Mann's original quark theory required the F_2 momentum distribution to peak at $x = \frac{1}{3}$, reference [6] demonstrates that this is not the case because the experimentally observed F_2 peak is at $x = \frac{1}{9}$. This deviation from the required peak at $x = \frac{1}{3}$ was explained away

Content from this work may be used under the terms of the Creative Commons Attribution 3.0 licence. Any further distribution of this work must maintain attribution to the author(s) and the title of the work, journal citation and DOI.
Published under licence by IOP Publishing Ltd

via the hypothesis that the three quarks originally thought to form the proton are the so-called "valence quarks", which are swimming in the background of "sea quarks" [4]. These so-called sea quarks are a collection of quark-antiquark pairs, radiated by the three valence quarks. However, the calculations of 1970s still showed that the valence quarks together with the sea quarks only accounted for 54% of the proton's momentum [2]. A further hypothesis was added to supplement the momentum shortfall of the quarks; chargeless particles called gluons were introduced into the proton model [5]. Since gluons have no electric charge, the thinking was that they are there, but the electrons probing the proton in deep inelastic scattering cannot see them. These hypothesized gluons were assigned the missing proton momentum, and the resulting proton model became the quark-gluon model that it is today. But even with this model of "valence quarks" swimming in the background of "sea quarks and gluons", there seemed to be an angular momentum deficit with respect to the measured angular momentum of the proton, and therefore the presence of "virtual strange quarks" was also postulated in the 1990s [7].

Although the quark-based model was inspired by the great variety of mesons, the proposed quark masses do not add up the masses of observed mesons. According to quark proponents, this is explained by a negative binding energy between quarks: any particle's valence quark masses are only a small percentage of the total particle mass, with the bulk of the particle mass coming from particles which represent the binding force: i.e. virtual quarks and gluons. Moreover, the valence quark : virtual quark : gluon mass ratio is allowed to vary from particle to particle in order to match the observed masses. Now what is the physical meaning of negative binding energy? By definition, negative binding energy means a meta-stable bound state. This model implies that individual quarks should be easily observable upon the break-up of their meta-stable binding. However, quark proponents also postulated that the meta-stable bonds between quarks can never be dissociated. It appears that the quark-based proton model has become embraced without any direct evidence, and maintaining it requires a forever expanding list of postulates.

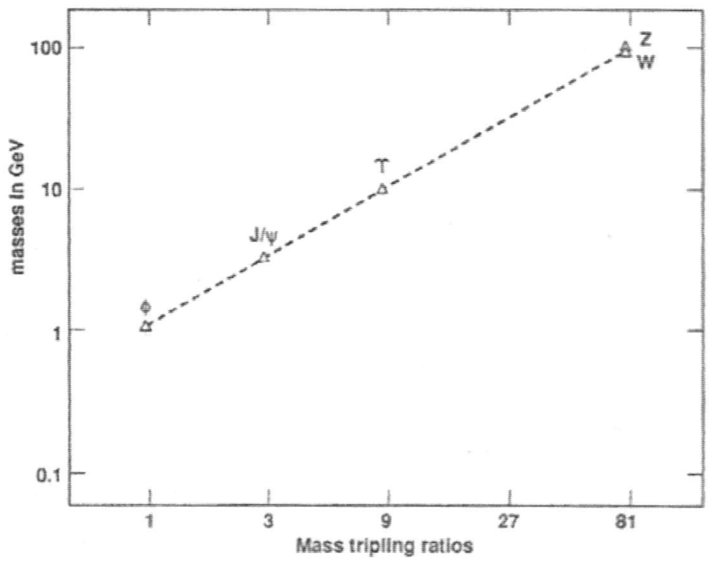

Figure 1.1. The mass tripling pattern of certain mesons, reproduced from [13].

Recently, an electromagnetic proton model has emerged [14], which restores proton's elementary particle status. This proton model employs the same methodology as the successful electron model [9, 25] which gives tangible answers to fundamental questions, such as: what is the electron made of?, what generates the electron's spin?, what is the meaning of the experimentally measured Compton radius?, and what is the meaning of the experimentally measured classical electron radius? The proton model of reference [14] gives tangible answers to questions, such as: what is the proton made of?, what generates the proton's spin?, what generates the proton's anapole magnetic moment?, and what is the meaning of the experimentally measured proton radius? While the preceding quark-based proton model violates Maxwell's equation via the so-called "renormalization" procedure and proposes complicated interactions among experimentally non-observable virtual particles, which are referred to as "sea quarks" and "gluons", the proton model of reference [14] respects Maxwell's equation and does not require the presence of energy conservation violating virtual particles. The authors of reference [14] demonstrate that their proton model matches all data in high energy electron-proton scattering experiments, without employing postulates or hidden parameters.

The proton's elementary particle status implies that the neutron should comprise positive and negative elementary charges.

Regarding mesons, an alternative modeling approach is to look for simple patterns of their mass variations. As illustrated in figure 1.1, some meson particles form a simple mass-tripling pattern [13]. Coincidentally, our present work demonstrates a similar, but electron related mass-tripling pattern.

1.2. A brief history of the neutron decay model

Regarding neutron decay, it was proposed in the 1970s that the nuclear beta-emission of electrons and antineutrinos is being mediated by 80 GeV mass W boson particles, i.e. an $X_Z^A \rightarrow X_Z^{A+1} + W^-(80GeV) \rightarrow X_Z^{A+1} + e^- + \bar{\nu}_e$ process, which temporarily violates energy conservation. The emission of this 80 GeV mass particle is supposedly allowed by Heisenberg uncertainty; i.e. nucleons are thought activate this uncertainty only above a specific neutron:proton ratio threshold, and keep it de-activated below that threshold. However, it seems to be a violation of the Heisenberg uncertainty principle to model it with an on-off switch.

Presently, claims about the structure and interactions of 80 GeV particles with 10^{-25} sec half-life are highly speculative. Claiming that the emission and absorption of neutrinos is mediated by 80 GeV virtual particles is a particularly extraordinary claim, and should require extraordinary evidence. However, the presence of a W boson was never experimentally observed during neutron decay.

The hypothesis of 80 GeV virtual particles is experimentally contradicted by the beta decay of the ^{163}Dy nucleus. It is known since 1992 that the ^{163}Dy nucleus beta decays in a fully stripped Dy^{66+} state, while the same ^{163}Dy nucleus is stable in a neutral atom [15]. Such beta decay of Dy^{66+} produces a bound state electron, which has a negative energy state with respect to a free electron. While most nuclear theorists insist that any beta decay reaction is completely independent from electromagnetism, it is clear that the electric potential is the controlling parameter of this beta decay process.

If beta decay was triggered by the formation of an 80 GeV mass W boson particle, such process would not care about the presence or absence of orbital electrons. Since the presence or absence of orbital electrons is the actual control parameter of the ^{163}Dy beta decay, it is impossible that any W boson particle is involved in this process.

A similar experimental contradiction is also presented by the beta decay of the ^{210}Bi nucleus. ^{210}Bi beta decays in its ground state, but remains stable in its 271 keV excited state, which is a long-lived excitation state. If beta decay was triggered by the formation of an 80 GeV mass W boson particle, such process would not care about the presence or absence of a nuclear excitation which has 5 orders of magnitude lower energy.

Based on these shortcomings, it is reasonable to consider an alternative model of neutron decays. A more preferable neutron decay model avoids violating any conservation laws.

2. Signatures of 1.5 MeV leptons in neutron decay

Motivated by section 1, we consider what type of particle is released during neutron decays. Since the neutron decay end products are a proton and an electron, this released particle always decays into an electron; i.e. it is some type of a lepton. We refer to this hypothetical lepton as the e_n^- particle.

In a free-particle state, the neutron half-life is 610 seconds. Let us assume that the first step during neutron decay is its transformation into bound p^+ and e_n^- particles. Energy conservation requires the following relationship between these particle masses:

$$m_n = m_p + m_{en} - E_{binding}/c^2 \quad (2.1)$$

$$m_n = m_p + \gamma_L m_{en} - E_{potential}/c^2 \quad (2.2)$$

where m_{en} is the rest mass of released e_n^- particle, and γ_L is the Lorentz boost factor. The above two equations are two equivalent ways of expressing energy conservation.

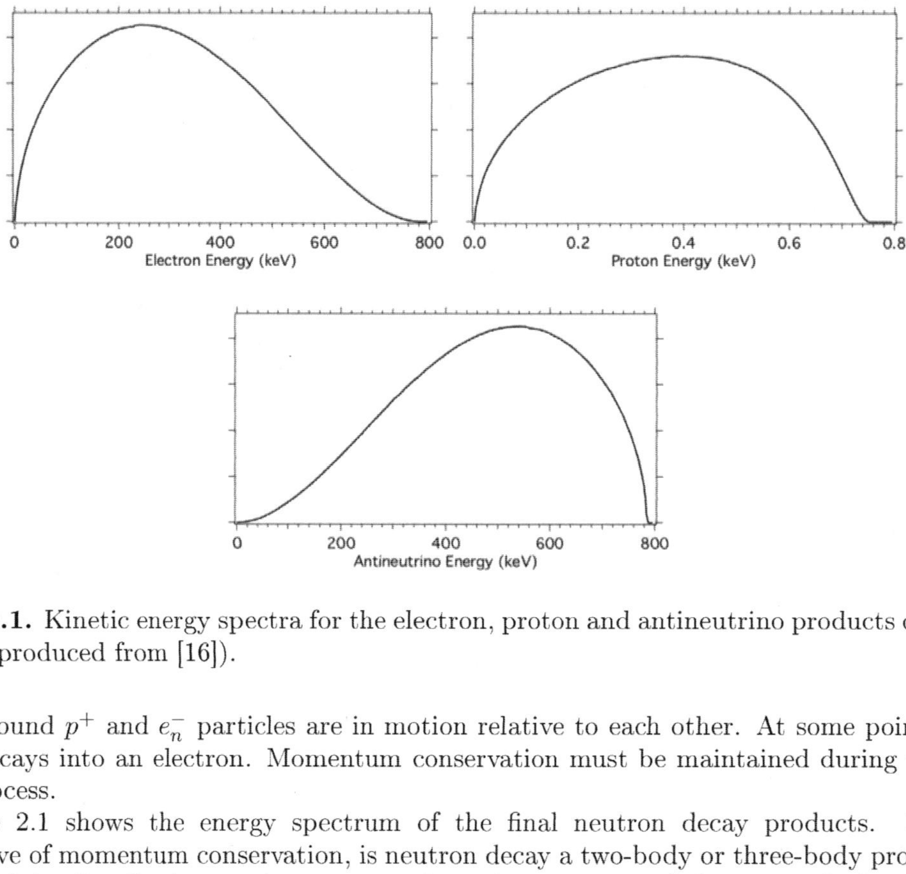

Figure 2.1. Kinetic energy spectra for the electron, proton and antineutrino products of neutron decay (reproduced from [16]).

The bound p^+ and e_n^- particles are in motion relative to each other. At some point, the e_n^- lepton decays into an electron. Momentum conservation must be maintained during the entire decay process.

Figure 2.1 shows the energy spectrum of the final neutron decay products. From the perspective of momentum conservation, is neutron decay a two-body or three-body problem? At one end of the distribution, we have zero antineutrino energy, and thus a two-body problem of momentum conservation. At the other end of the distribution, both the electron and proton have zero momentum; i.e. the neutrino field carries away all the decay energy without impacting any momentum to either the electron or proton. From the perspective of the two charged particles, momentum conservation is therefore a two-body problem at both ends of the distribution spectrum, and we assume that it remains a two-body problem in the entire spectrum.

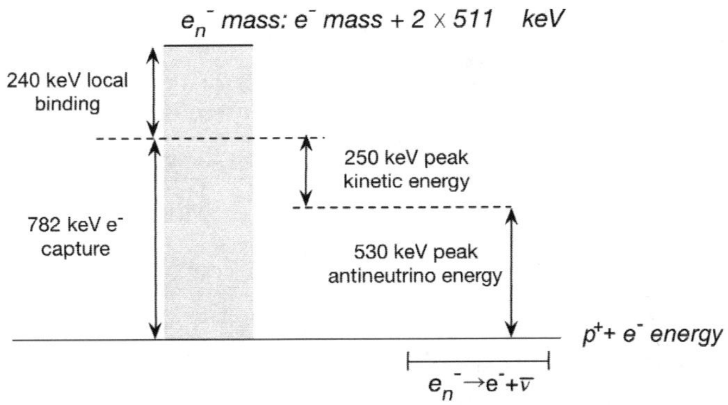

Figure 2.2. The energy diagram of the short-lived bound state between a p^+ and an e_n^- lepton

First estimation of the emitted lepton mass: As the separating p^+ and e_n^- particles must maintain zero net momentum, the ratio between their kinetic energies equals the inverse ratio of their respective masses. We use the peak energy values, which correspond to the highest probability, as the characteristic kinetic energy values. Figure 2.1 shows that the ratio of proton and electron energy peaks is $\frac{250\,keV}{0.4\,keV} = 625 \approx \frac{1836.15}{3} = \frac{m_{p^+}}{3m_{e^-}}$. This data indicates that the intermediate e_n^- particle is inherently 3 times heavier than an ordinary electron.

Based on the actual neutron mass value, about 240 keV $p^+ - e_n^-$ binding energy would be released if the starting particles are p^+ and e_n^-. Conversely, if the starting particles are an ordinary proton and an electron, the required relativistic electron mass is 240 keV less than 3 electron masses; i.e. the electron must have at least 782 keV energy for neutron formation. Figure 2.2 illustrates the resulting $p^+ - e_n^-$ energy diagram.

3. Signatures of 1.5 MeV leptons in deuteron photo-dissociation

3.1. The historic interpretation of deuteron photo-dissociation

After Chadwick's discovery of the neutron in 1932, there were a lot of discussions whether it is an elementary particle or a hydrogen-like atom formed from electron and proton [1]. For example, Heisenberg and Rutherford were among those who argued that Chadwick's particle is a small hydrogen atom. Pauli however stated that the neutron should be viewed as an elementary particle. To decide who is right, Chadwick's team irradiated deuterons by 2.62 MeV photons, and saw deuterons splitting into a proton and a neutron, without the emission of any electron-like particle. Consequently, most nuclear physicists rejected the nuclear electron concept. With the discovery of the neutron's magnetic moment in 1940, most scientists also rejected Pauli's elementary particle neutron concept, as their elementary particle model assumed the absence of internal charge distribution.

Since deuteron photo-dissociation experiments became a historic water-shed, we now re-examine the methodology and assumptions behind these experiments. The first such experiments were performed by Chadwick and Goldhaber in 1935 [26]: they passed 2.62 MeV gamma rays, emitted by a thorium source, into deuterium gas. The dissociation generated energetic protons with mean proton energy of 0.185 MeV. By conservation of momentum, the other released particle must carry away the same momentum. Since neutrons have about the same mass, a dissociation into a proton and a neutron means 0.185 MeV kinetic energy of the generated neutrons. These values imply a deuteron binding energy of $2.62 - 2 \times 0.185 = 2.25$ MeV. This value matches the experimental deuteron binding energy, measured by photon energy detection upon proton-neutron reactions. Thus the deuteron photo-dissociation into a proton and a neutron has been proven. Later work showed that the photon energy threshold for deuteron photo-dissociation is 2.26 MeV.

The observation of only proton plus neutron reaction products at this photon energy was historically thought to be a proof for the absence of electron-like particles, based on the idea that energetic photons should primarily ionize away any light particles. The implicit assumption here is that an electron-like particle would more readily interact with >2.26 MeV photons than a proton, and would thus become ionized from the nucleus. We now examine this assumption.

3.2. Deuteron photo-dissociation cross section calculation

3.2.1. The photoelectric effect on nuclear electrons During the chemical photo-dissociation of a H_2 molecule, the radiation field can either break the H_2 molecule into two H atoms [30] or it can ionize the H_2 molecule into an H_2^+ ion and an electron. Analogously, the nuclear photoelectric effect may yield either the $^2H^+ \rightarrow p^+ + n$ or the $^2H^+ \rightarrow p^+ + p^+ + e_n^-$ reaction pathway. We now determine the relative probability of these two pathways.

Playing with the idea of an emerging electron-like sub-particle, we assume that the radiation field splits the deuteron into bound $p^+ + p^+ + e_n^-$ sub-particles. We start by defining a three-body wavefunction. A reasonable wavefunction model is to consider the three-body wavefunction being a Gaussian function of the sub-particles' distance:

$$\psi = De^{-\frac{\beta}{2}(\mathbf{r}_{p1}-\mathbf{r}_{en})^2}e^{-\frac{\beta}{2}(\mathbf{r}_{p2}-\mathbf{r}_{en})^2}e^{-\frac{\beta}{2}(\mathbf{r}_{p1}-\mathbf{r}_{p2})^2}$$

where $\mathbf{r}_{p1}$ and $\mathbf{r}_{p2}$ denote the two protons' position, and $\mathbf{r}_{en}$ denotes the lepton's position. Such wavefunction maximizes the Shannon entropy of its sub-particles. D and β are two unknown parameters, which can be calculated from two boundary conditions: the deuteron's charge radius is 2.1 fm, and the wavefunction is normalized to a unit probability over the whole space. We thus get $\beta = \frac{1}{24}$ fm^{-2} and $D = 0.0512$ fm^{-3}. The Hamiltonian operator of light-matter interaction is:

$$H = -\sum_i \frac{q_i}{m_i}\hat{\mathbf{A}}(\mathbf{r}_i)\cdot\hat{\mathbf{p}}_i$$

where q_i, m_i, $\mathbf{r}_i$, and $\hat{\mathbf{p}}_i$ are the given sub-particle's charge, mass, position, and momentum operator, while the $\hat{\mathbf{A}}$ operator corresponds to the electromagnetic wave's vector potential.

Upon photo-dissociation into $2p^+ + e_n^-$ reaction products, the wavefunction of the final state comprises three harmonic functions of the three resulting particles. Using the above Hamiltonian, we calculate the cross-section corresponding to the deuteron wavefunctions' transition from its initial ground state into its final dissociated state. Using 1.5 MeV e_n^- mass, we obtain the orange curve of figure 3.2, which is far below the experimental photo-dissociation cross-section. In the 1930s, theoreticians expected this orange curve to be orders of magnitude higher than the experimental cross-section, mainly because an electron-like particle is much lighter than the other particles. In the light of actually performing the calculation, we can see that it is the low density of states associated with an electron-like particle which causes the $^2H^+ \rightarrow p^+ + n$ reaction pathway to have much higher probability than the $^2H^+ \rightarrow 2p^+ + e_n^-$ reaction pathway.

3.2.2. The Compton scattering of nuclear electrons At high photon energy, Compton scattering becomes the dominant photo-ionization process of ordinary electrons. Therefore, one cannot neglect the role of Compton scattering in deuteron photo-dissociation. Any particle's Compton scattering cross-section is given by the Klein-Nishina formula. 20-100 MeV photons have much less energy than the proton mass, and thus protons' Compton scattering is negligible in this energy range. On the other hand, this energy range is much higher energy than either the e^- or e_n^- mass. Therefore, according to the Klein-Nishina formula, the e^- and e_n^- Compton scattering

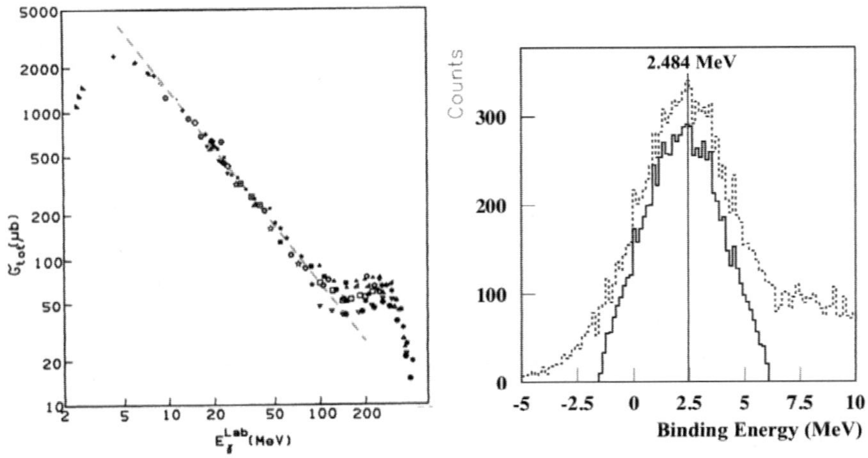

Figure 3.1. Deuteron photo-dissociation by high-energy radiation. Left: the p^+ counting based photo-dissociation cross section (microbarns), as a function of photon energy, reproduced from chapter 6 of [28]. Right: the binding energy among deuteron constituents, measured upon photo-dissociation by 30-50 MeV photons, reproduced from [29].

cross sections vary at the same rate with respect to the photon energy. Within a certain photon energy range, if deuteron photo-dissociation is dominated by the e_n^- lepton's Compton scattering process, then this Compton scattering cross section can be approximated by the total deuteron photo-dissociation cross-section.

The electron's Compton scattering cross-section values at 20 and 100 MeV photon energy are shown in table 1, based on the data of reference [27]. Regarding the e_n^- particle, its Compton scattering cross-section is identified with $\frac{1}{2}$ of the deuteron photo-dissociation cross section in the 20-100 MeV photon energy range. A $\frac{1}{2}$ multiplier must be applied to the p^+ counting based experimental cross-section measurement because the $^2H^+ \rightarrow 2p^+ + e_n^-$ Compton scattering produces two protons. The obtained values at 20 and 100 MeV photon energy are displayed in the last column of table 1. As expected, the electron's and nuclear electron's cross sections vary at a similar rate with respect to the photon energy, which is a signature of Compton scattering.

	e^- (Xe)	e^- (Pb)	e_n^- (^2H)
E_γ=20 MeV	14.9 mb	13.7 mb	0.29 mb
E_γ=100 MeV	2.02 mb	1.87 mb	0.035 mb
σ_{20}/σ_{100}	**7.4**	**7.5**	**8.3**

Table 1. A comparison between the e^- and e_n^- Compton scattering cross sections at 20 and 100 MeV photon energy. The millibarn values are the Compton scattering cross-sections of the particles identified in the first row.

When the nuclear electron is removed via Compton scattering, the deuteron breaks up via the $^2H^+ \rightarrow 2p^+ + e_n^-$ pathway. Writing this process as $^2H^+ \rightarrow p^+ + n \rightarrow 2p^+ + e_n^-$, one may see that such three-body break-up requires 240+2226 keV = 2464 keV energy input. We used here the 240 keV $p^+ - e_n^-$ binding energy estimation obtained in the previous section. As shown in the right side of figure 3.1, the binding energy which was measured upon deuteron photo-dissociation by 30-50 MeV photons indeed peaks at this energy, and not at 2.23 MeV energy. The two curves peaking at 2.484 MeV only differ in the applied signal filtering condition, which is explained in reference [29].

By comparing the $\boldsymbol{e_n^-}$ and $\boldsymbol{e^-}$ columns of table 1, we can estimate the deuteron-bound e_n^- charge radius according to the following formula:

$$r_{en} = r_e \sqrt{\frac{\sigma_{en}}{\sigma_e}}$$

The data of table 1 yields $2.82\sqrt{\frac{0.035}{1.9}}$ ≈0.39 fm charge radius for the deuteron-bound e_n^-, using the 100 MeV photon energy value. The analogous analysis can be applied also to the ^{3}He isotope: at 100 MeV photon energy its p^+ counting based photo-disintegration cross section is 0.13 mb, and thus its $p^+/3$ counting based photo-disintegration cross section is 0.043 mb. The e_n^- lepton's charge radius calculation according to the above formula yields $2.82\sqrt{\frac{0.043}{1.9}}$ ≈0.42 fm charge radius for the ^{3}He case. The very similar e_n^- charge radii in ^{2}H and ^{3}He further demonstrate that the e_n^- lepton's Compton scattering is the dominating photo-dissociation reaction at 100 MeV photon energy.

Second estimation of the emitted lepton mass: Using 1.5 MeV e_n^- mass and 0.4 fm charge radius parameters, we apply the Klein-Nishina formula, assuming deuteron break-up when the scattered photon deposits >2.484 MeV energy. We thus obtain the red curve of figure 3.2. As anticipated, the Compton scattering cross-section indeed approaches the experimental photo-dissociation cross-section at high photon energies, using $\frac{1}{2}p^+$ counting. In other words, nuclear electrons' Compton scattering dominates deuteron break-up at high photon energies. At 100 MeV photon energy, the $\frac{1}{2}p^+$ counting photo-dissociation cross-section is 0.035 mb, while the e_n^- Compton scattering cross-section evaluates to 0.031 mb. This good convergence validates our 1.5 MeV estimation for the emitted e_n^- mass.

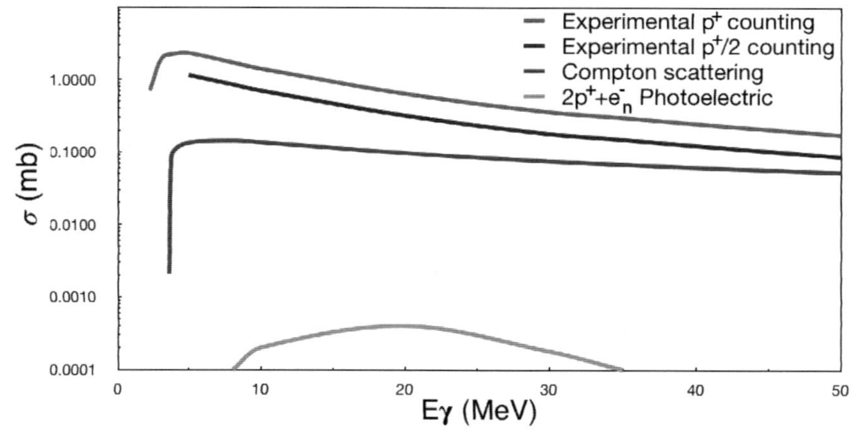

Figure 3.2. The experimental deuteron photo-dissociation cross-section, and the two calculated processes which yield $2p^+ + e_n^-$ dissociation products.

It is important to note from figure 3.2 that the Compton scattering cross section falls to zero near 3.2 MeV photon energy. Below this energy, the only remaining deuteron dissociation reaction is the $^2H^+ \rightarrow p^+ + n$ photoelectric process.

3.2.3. Deuteron photo-dissociation summary As seen in figure 3.2, the two processes which yield $2p^+ + e_n^-$ dissociation products rapidly fall to zero in the <5 MeV photon energy range.

Specifically, our calculation shows that the probability of $^2H^+ \rightarrow 2p^+ + e_n^-$ reaction is zero when the photon energy is less than 3.2 MeV. Our results clarify why the experimenters of the 1930s saw only $p + n$ photo-dissociation products.

On the other hand, e_n^- Compton scattering becomes the dominant deuteron break-up mechanism when photon energy reaches around 30 MeV. In this photon energy range, the $e^- - e^+$ pair creation process also has a high cross-section. Up to now, experimentalists assumed that any detected electrons originate from pair creation events, thereby not noticing the electrons originating from the deuteron break-up.

4. ^4He and two e_n^- particles may form a neutral composite

The emission of electrically neutral four-nucleon particles has been observed in various high-energy experiments [11]. The authors of [11] calculate 420 keV binding energy holding such particles together, and they also mention prior experimental studies that obtained the same binding energy value. The scientific literature refers to such four-nucleon objects as "tetra-neutrons", as they were thought to be a meta-stable composite of four neutrons. However, the dissociation of "tetra-neutrons" into individual neutrons has never been observed.

In this section we review experiments which demonstrate that "tetra-neutrons" are in fact short-lived $\left({}^4He + 2e_n^-\right)_{bound}$ particles. In turn, the existence of electrically neutral $\left({}^4He + 2e_n^-\right)_{bound}$ evidences the nuclear presence of this new lepton particle.

4.1. Lead photo-dissociation

The authors of [10] irradiated deuterated materials by 2 MeV photons. They observed deuterium fusion reactions, evidenced by the energetic neutrons that were absent in the hydrated materials based control experiments. The sample materials were enclosed in a lead shielded chamber. These experiments also document the effect of 2 MeV photons on the lead material, but without any interpretation of the obtained results. Here, we focus on the reactions of this irradiated lead material. After some experiment runs, the red signal of figure 4.1 shows the appearance of ^{210}Pb, which is an unstable isotope with 22 year half-life. As can be seen by comparing the red and green signals of figure 4.1, a subsequent irradiation run transmutes ^{210}Pb to ^{214}Pb. This transmutation is evidenced by: i) the disappearance of the ^{210}Pb, ii) the appearance of ^{214}Pb, and iii) the appearance of ^{214}Bi which is a decay product of the short-lived ^{214}Pb.

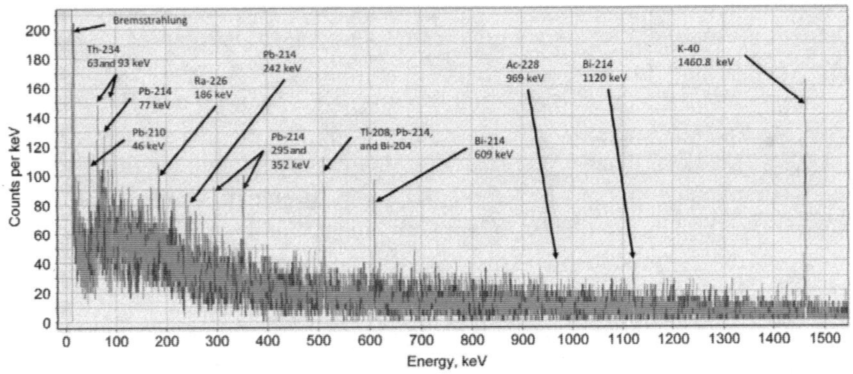

Figure 4.1. The gamma spectrum of an already irradiated lead material, before (red signal) and after (green signal) a 4-hour run of 2 MeV photon irradiation. The signatures of ^{210}Pb to ^{214}Pb transmutation are clearly seen. Reproduced from [10].

How did ^{210}Pb transmute to ^{214}Pb? The involved experiment run was performed with hydrated sample in the lead chamber, which did not emit neutrons [10]. Regarding the lead material itself, neutron photo-dissociation requires >6.7 MeV for any of its isotopes, which is far larger than the applied photon energy. Moreover, in the case of neutron-induced transmutations we would see the signatures of ^{211}Pb and ^{213}Pb isotopes as well, which have similar half-life and gamma intensity as ^{214}Pb. Therefore, experimental data contradicts individual neutron absorption, and shows that four neutral nucleons were simultaneously absorbed by ^{210}Pb. However, the simultaneous photo-dissociation of four neutrons requires much higher energy than 2 MeV.

What requires less than 2 MeV energy is to strip away an alpha particle and two electrons from any lead isotope. The remaining lead nucleus appears to have lost "four neutrons". In order to generate ^{210}Pb to ^{214}Pb transmutation, the involved alpha particle and two e_n^- particles must be emitted together, as a bound particle:

$$^{208}Pb \quad \overrightarrow{2\,MeV\,\gamma} \quad ^{204}Pb + \left(^4He + 2e_n^-\right)_{bound}$$

The above photo-dissociation scheme is the only reasonable explanation for the observed transmutations, which occur upon the subsequent absorption of $\left(^4He + 2e_n^-\right)$ particles. We note that the resulting ^{214}Pb isotope decays back to ^{210}Pb via the consecutive emission of two electrons and an alpha particle, and never decays via neutron emission.

The appearance of e_n^- leptons during lead photo-dissociation is analogous to their above-discussed appearance during deuteron photo-dissociation.

4.2. The evolution of ^{242}Pu concentration in nuclear reactors

In nuclear reactor cores, ^{239}Pu is produced via a neutron capture by ^{238}U. Further neutron captures generate heavier plutonium isotopes. Starting from ^{238}U, it takes four consecutive neutron captures to get to ^{242}Pu, and thus its concentration should be very low in the initial stages of the fuel cycle. On the other hand, if $\left(^4He + 2e_n^-\right)$ particles were produced in nuclear reactor cores, then ^{242}Pu would be generated via a single capture reaction:

$$^{238}U + \left(^4He + 2e_n^-\right)_{bound} \rightarrow {}^{242}U \rightarrow {}^{242}Np + e^- \rightarrow {}^{242}Pu + 2e^-$$

We discovered in the preceding section that the production of $\left(^4He + 2e_n^-\right)$ particles is triggered by gamma radiation. Gamma radiation is also present in nuclear reactor cores, which motivates examining the concentration of ^{242}Pu in order to determine whether the above reaction takes place. In the absence of the above reaction, the experimental ^{242}Pu concentration must evolve according to the neutron capture calculation. In the presence of the above reaction, the experimental ^{242}Pu concentration must be higher than what is predicted by the neutron capture calculation, especially in the initial stages of the fuel cycle.

A recent study [17] analyzed over 250 fuel samples from multiple reactors, at various stages of burn-up. Under a low burn-up condition of <10 GW-day per fuel ton, the author of [17] finds on average 3 times higher than predicted ^{242}Pu concentration. In the same samples, the average ^{239}Pu concentrations match predictions, which means that the neutron capture calculations are correctly done. The largest concentration excess with respect neutron capture calculations is 400% for ^{242}Pu, while only 39% for ^{239}Pu. These large discrepancies demonstrate that the capture of $\left(^4He + 2e_n^-\right)$ particles is the main source of ^{242}Pu during the initial stages of the fuel cycle.

As the fuel cycles progress, the probability of ^{242}Pu production via four consecutive neutron captures rises steeply. The discrepancy between experimental and predicted ^{242}Pu concentrations becomes gradually smaller with rising burn-up rate [17]. Approximately the same result is found in reference [18]: relative to neutron capture calculations, 15-20% higher experimental ^{242}Pu/^{239}Pu ratio is observed in highly burnt up nuclear fuel. The correctness of the authors' neutron capture calculation is validated by the matching concentration of other isotopes, such as the ^{146}Nd/^{145}Nd ratio [18]. Consecutive neutron capture thus becomes the main source of ^{242}Pu only in the final stages of the fuel cycle.

In summary, the evolution of ^{242}Pu concentration demonstrates the presence of short-lived $(^4He + 2e_n^-)$ particles in nuclear reactor cores.

5. The fission of excited ^{9}Be

The ^{9}Be nucleus comprises two alpha sub-particles, a nuclear electron, and a proton. It takes only 1665 keV to separate a neutron away from it, which is less energy than any of the ^{9}Be excitation energy levels. The energies of its lowest three excited states are shown in figure 5.1. It has been observed that any of these excitations break up the ^{9}Be nucleus with very close to 100% probability [12]. The 1684 keV and 2780 keV excitations decay by neutron emission; the remaining ^{8}Be nucleus then promptly splits into two alpha particles.

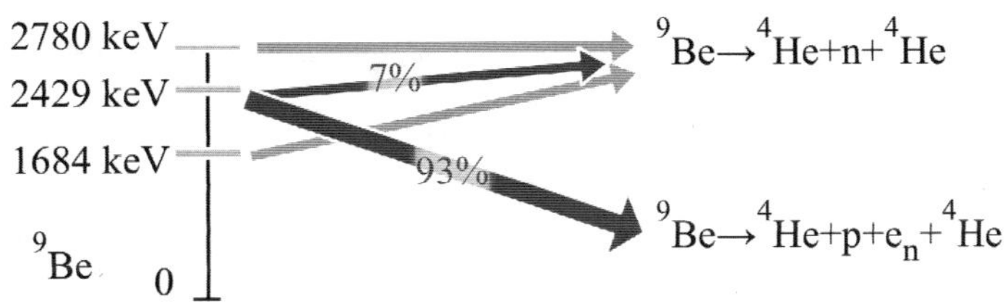

Figure 5.1. The lowest three excitation levels of ^{9}Be and their subsequent break-up reactions.

Interestingly, the 2429 keV excitation decays by neutron emission in only 7% of cases [12]. While, nuclear data tables do not say what break-up products emerge in the other 93% of cases, there is no other possibility than the $^9Be \rightarrow ^4 He + ^4He + p + e_n$ reaction shown in figure 5.1. The reasons for the impossibility of other reaction products are: i) ^{5}He and ^{5}Li do not exist, and ii) a proton separation into $^8Li + p$ would require over 16 MeV energy. Therefore, the 2429 keV excitation of ^{9}Be decays mainly by emitting a proton and an electron. Such a prompt release of an electron upon nuclear break-up demonstrates that a negative elementary charge must be present in the ^{9}Be nucleus.

6. A precise measurement of the 1.5 MeV lepton mass

In the preceding sections, we found signatures of a 1.5 MeV new lepton which emerges from certain nuclear reactions. Here, we determine its mass more precisely.

It follows from the results of the previous section that negatively charged particles are emitted during the fission of certain excited nuclei. What happens if an emitted e_n^- lepton is captured by another nucleus, prior to its decay into e^-? The nuclear capture of ordinary electrons generates neutrino emission, which is a consequence of the isospin change between the e^- and e_n^- particles. However, if the e_n^- particle's isospin does not change during nuclear capture; its capture may involve just gamma emission. In that case, the binding energy of the e_n^- capture process is emitted as gamma radiation. The measurement of gamma radiation peaks during the nuclear capture of e_n^- is therefore a direct measurement of their binding energy to a given nucleus. On this basis, we seek experimental data of gamma emission from electron capture capable nuclei.

6.1. Nuclear electron capture by ^{58}Ni

The authors of [19] applied hot hydrogen treatment to a nickel metal. They observed unexpected gamma emission from hydrated nickel, and could not explain the source of the obtained gamma

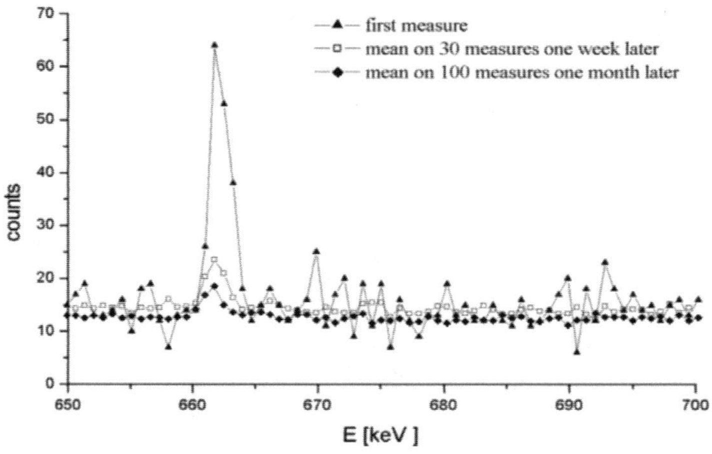

Figure 6.1. The gamma peak corresponding to a nuclear electron capture by ^{58}Ni. Measurement data from [19].

peak. As shown in figure 6.1, this gamma peak is at 661.5 keV. At the same time, they also observed a simultaneous emission of neutrons, at low intensity. Neutron emission is a signature of nuclear fission process. With reference to section 5, a nuclear fission process may involve the release of e_n^- particles as well. We therefore interpret the origin of the gamma peak shown in figure 6.1 as e_n^- capture by certain nuclei.

The ^{58}Ni isotope of nickel is capable of electron capture via the $^{58}Ni + e_n^- \rightarrow {}^{58}Co$ reaction. According to figure 6.1, the binding energy of the e_n^- capture is E_b=661.5 keV. In the case of an ordinary electron, the $^{58}Ni + e^- \rightarrow {}^{58}Co$ reaction is endothermic by E_{ec}=-381.6 keV. Because of the same reaction end product, the difference between E_b and E_{ec} must correspond to the mass difference of the incoming particles. We may therefore determine the e_n^- mass from the following equation:

$$m_{en}c^2 - m_ec^2 = E_b - E_{ec} \tag{6.1}$$

The above equation yields 1554 keV for the e_n^- mass, which matches well the preceding $m_{en} \approx 3m_e$ results.

6.2. Nuclear electron capture by ^1H

We found a second unexpected gamma emission peak during experiments on water vapor under high voltage spark discharges. The measurements were done under air atmosphere, and a shielded chamber was employed to minimize the background noise. The experimental details are given in section 10. This experiment is in fact a laboratory analogue of a natural lightning discharge. The physical processes during lightning discharges have been rather well studied, and several authors noted the production of neutrons by lightnings [20, 21, 22]. The author of [21] discusses similar neutron production by artificial lightning discharges in the laboratory. Via a detailed examination of the lightning process signatures, the authors of [22] determined that the produced neutrons originate from the $^{14}N \rightarrow {}^{13}N + n^0$ fission reaction. The resulting ^{13}N isotope then further decays to ^{13}C. The observation of such fission neutrons yet again signals the possible presence of e_n^- particles as well. We therefore interpret the origin of the gamma peak shown in figure 6.2 as e_n^- capture by some nucleus. This interpretation is further corroborated by a time correlation analysis between the spark discharges and gamma photon emissions, which shows that the peak of figure 6.2 occurs right after each spark discharge, and has microsecond scale duration. The

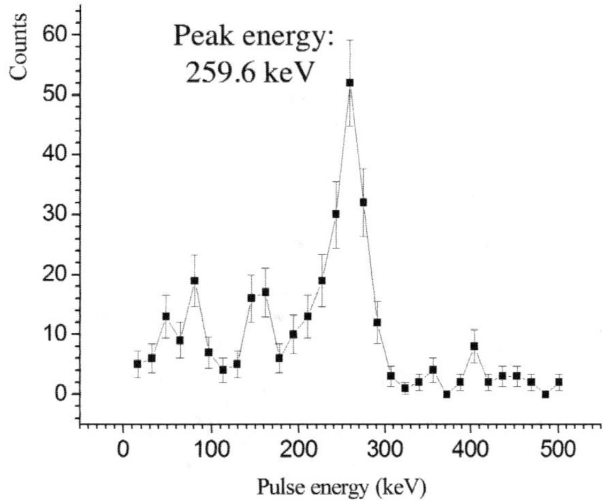

Figure 6.2. The gamma peak corresponding to a nuclear electron capture by a proton. The measurement methods are described in section 10.

details of this time correlation analysis are given in section 10. Therefore the gamma peak cannot originate from any radioactive isotope contamination. Such short duration of the gamma peak emission is characteristic of a free e_n^- particle's short half-life.

Protons are the main electron capture capable nuclei in our experiment, and therefore the capture of nuclear electrons may be written as $p^+ + e_n^- \rightarrow n$. Figure 6.2 shows the measured gamma peak, which implies E_b=259.6 keV binding energy between a proton and a nuclear electron. For ordinary electrons, the $p^+ + e^- \rightarrow n$ reaction is endothermic by E_{ec}=-782.4 keV. We again use equation 6.1 to calculate the nuclear electron mass, and obtain 1553 keV.

Third estimation of the $\mathbf{e_n^-}$ mass: Our precise estimate for the e_n^- mass is the average of the above two measurements, that is $m_{en}c^2$=1553.5 keV.

7. An estimation of the 1.5 MeV lepton half-life in free-particle state

At the JINR institute in Dubna, Vladimir A. Nikitin analyzed pair creation tracks in a bubble chamber, where particle-antiparticle pairs were generated from energetic photons. Upon the analysis of 7000 such tracks, he found nine anomalous lepton tracks. His analysis of these tracks revealed that they are produced by 9±2 MeV particles, that eventually decay into an electron or positron. The 9 MeV lepton is yet another electron-like particle that has been previously unknown. In two cases, the decay event was captured on the track photo. Figure 7.1 shows such a particle pair creation event and the subsequent decay of the negatively charged lepton particle. As can be seen in figure 7.1, the decay process is actually a two-step decay, and an electron is produced upon the second decay step. This electron's track is highlighted by the red circle. The bubble chamber is under 1.5 T magnetic field, and thus we can calculate the electron's momentum from its track radius: its track data yields 1.925 MeV/c momentum. The challenge is to identify the mass of the short-lived intermediate particle.

As can be seen in figure 7.1, the blue circle fitted track of the intermediate particle has three times higher radius than the electron track, and thus its momentum is p_i=5.78 MeV/c.

The track of the 9 MeV lepton is highlighted by the dashed ellipse on figure 7.1. Before its decay, its track is slightly elliptic, which indicates a Larmor precession; i.e. we are observing its

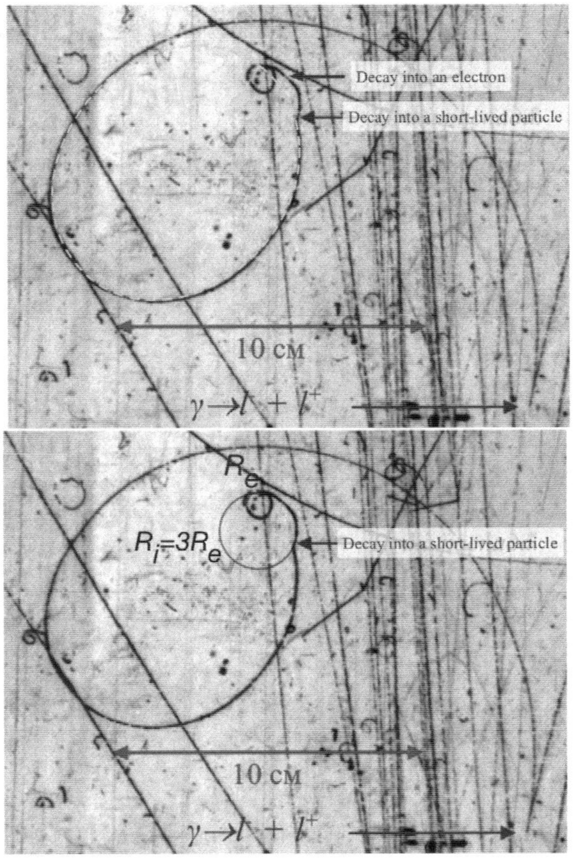

Figure 7.1. Heavy lepton pair creation in a bubble chamber. The 9 MeV lepton-antilepton pair is created at the bottom right corner. The other two purple arrows indicate the two decay events. The yellow dashed ellipse shows the elliptic track fitting of the 9 MeV lepton. The blue and red circles show the circular track fitting of the intermediate particle and final electron, respectively. Photograph provided by V. A. Nikitin.

circular track from a non-perpendicular angle. We use the large elliptic semi-axis to calculate the incoming lepton's momentum, and obtain p_l=21.13 MeV/c. Using the 9 MeV lepton mass value, we obtain a total energy of E_l=23 MeV. During the first decay event, the energy and momentum difference between the incoming heavy lepton and the outgoing intermediate particle are carried away by an anti-neutrino.

Fourth estimation of the $\mathbf{e_n^-}$ mass: In this first decay event, the angle between the incoming and outgoing particles is 39°. With respect to the incoming lepton momentum, the outgoing perpendicular momentum is $p_\perp = p_i \sin(39°)$=3.63 MeV/c. The momentum difference in the parallel direction is $p_\parallel = p_l - p_i \cos(39°)$=16.64 MeV/c. From the $p_\perp$ and $p_\parallel$ components we calculate the total neutrino momentum: p_ν=17 MeV/c. Since the neutrino mass is negligible, its energy is E_ν=17 MeV. We finally estimate the energy and mass of the intermediate particle:

$$E_i = E_l - E_\nu = 6\, MeV$$

$$m_i c^2 = \sqrt{E_i^2 - (p_i c)^2} = 1.6\, MeV$$

Journal of Physics: Conference Series **2482** (2023) 012019 doi:10.1088/1742-6596/2482/1/012019

This intermediate particle mass is very close to 1.55 MeV, and therefore we can identify the short lived intermediate particle with the e_n^- lepton. As far as we know, figure 7.1 contains the first photograph of an e_n^- track, although this interpretation must be confirmed by more observations of such events.

Considering the high kinetic energies of the above discussed particles, the intermediate e_n^- speed is close to the speed of light. We can thus make an order of magnitude estimation of its half-life. Considering its track length of about 1.5 cm, it decays after 5×10^{-11} seconds in the laboratory frame. Taking into account the Lorentz boost factor of 4, the e_n^- half-life is in the range of 10^{-11} seconds.

8. A summary of e_n^- properties

Summarizing our results so far, table 2 compares the electron against the newly identified e_n^- lepton.

	e^-	e_n^-
Charge	-1	-1
Mass	511 keV	1553.5 keV
Half-life	stable	$\sim 10^{-11}$ s

Table 2. A comparison between the e^- and e_n^- particles.

The exact ratio between the e_n^- and e^- particle masses is 3.04. In figure 1.1, we illustrated a similar mass tripling pattern among certain mesons. The specific ratio between the J/ψ and ϕ meson masses is 3.038, which is very close to 3.04. Similarly, the ratio between the Υ and J/ψ masses is also very close to this 3.04 value. Perhaps this ratio can be derived from some universal principle, which is valid for both leptons and mesons.

9. Towards a deeper understanding of the neutron

9.1. A single-particle neutron model

The above-discussed experimental data lead to the conclusion that the neutron comprises a positive and a negative elementary charge. How to understand the presence of a negative elementary charge within a proton-like particle? One may wonder whether the neutron is a meta-stable composite of e_n^- and p^+particles. In section 3 we showed that the deuteron's negative charge has 0.4 fm charge radius: this size is comparable to the overall neutron radius, which is around 1 fm. In our view, the neutron's positive and negative charges cannot be considered separate particles from the perspective of spin measurement or momentum uncertainty, because their internal fields mostly overlap. Only in response to very high frequency radiation, which cause Compton scattering, do these charges behave as distinct particles.

For the above reason, we consider the neutron to be a single particle. Within this single neutron particle, the electric field originates from the two well-defined spherical charge surfaces of its positive and negative charges. This electric field is non-zero inside the neutron, and zero beyond the fm-scale range. These intertwined charges generate a shared magnetic field, which is the essential feature of a single-particle neutron model. The neutron's meta-stability means that it eventually splits into bound e_n^- and p^+particles, as discussed in section 2, and then the e_n^- lepton decays into an electron 10^{-11} s later.

One arrives at the single-particle model also from spin entanglement considerations. As derived in the second appendix of [14], three particles with individually measurable spins cannot be in isotropic spin entanglement. In the neutron, the positive and negative charge spins have a fixed relative orientation, which generates the neutron's constant net magnetic moment. A proton's and a neutron's magnetic moments sum up to the deuteron's magnetic moment, which is also constant. It means that the deuteron's three charges all spin in a fixed relative orientation. Therefore, the deuteron's three elementary charges cannot be particles with individually measurable spins. This indicates that these charges are not distinct particles, i.e. the deuteron's negative charge is not a separate particle from the positive charges. Since the deuteron's negative charge originates from proton-neutron fusion, the neutron's negative charge is not a distinct particle either.

The neutron and deuteron do not have any known excited states. This absence of any excited states is yet another signature of a single particle, despite comprising 2 and 3 elementary charges, respectively.

9.2. The neutron's internal structure

Although we are still in the early stages of understanding the neutron, there is already sufficient experimental data to map out its internal constellation. On the basis of proton experimental data, reference [14] derives the proton's internal structure, which is illustrated on the left side of figure 9.2. Reference [14] derives the following proton parameter set: its major toroidal radius is 0.841 fm, its minor (poloidal) radius is 0.421 fm, and the positive charge's spherical radius is 0.0015 fm.

The neutron's negative charge is imposed over this proton structure. With the advancement of electron-proton scattering measurements, it has become possible to directly map out the neutron's radial charge distribution. Such radial charge distribution data is measured for example at JLAB [24], and is visualized in figure 9.1. The RMS (Root Mean Square) radius value of the positive charge is 0.8 fm, which differs by only 4% from the calculated proton toroidal radius of 0.841 fm. Figure 9.1 conveys the important information that the radial distribution of the neutron's positive and negative charges nearly overlaps; this means that the neutron's negative charge is also at approximately 0.841 fm radial distance from the neutron's center.

The more precise radial distance of the neutron's negative charge can be extracted from so-called "neutron charge radius measurements". Such experiments compare the mean radial distance of the positive and negative charges, and yield $r_n^2 = r_+^2 - r_-^2$=-0.11 fm^2 [31, 32]. Plugging in the above mentioned r_+=0.841 fm value, we obtain r_-=0.904 fm.

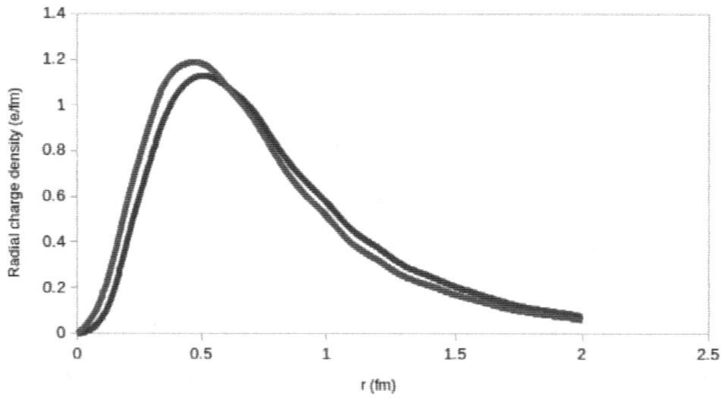

Figure 9.1. The radial density of the neutron's positive (red) and negative (blue) charges, from JLAB measurements [24].

The data of section 3 revealed that the deuteron's negative charge has about 0.4 fm charge radius, and we propose that the neutron's negative charge has the same spherical radius. This proposition is motivated by the observation that a proton's and a neutron's magnetic moments sum up to the deuteron's magnetic moment. This means that during $^1H + n \rightarrow {}^2H$ fusion the negative charge merely depolarizes between the two positive charges.

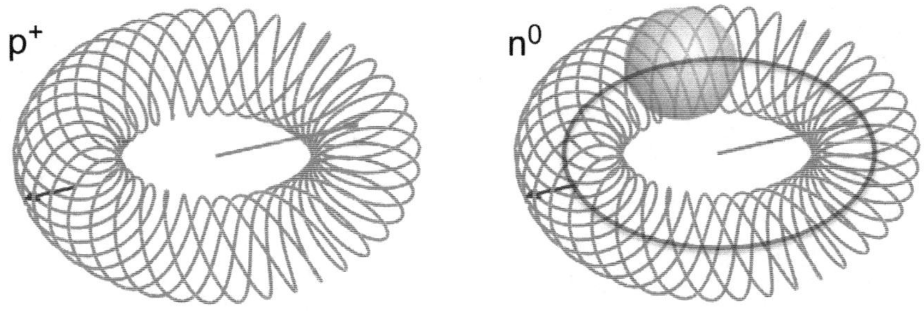

Figure 9.2. An illustration of the toroidal proton structure (left) and the neutron structure (right). The toroidal curve represents the Zitterbewegung path of the neutron's positive charge, and the blue ring represents the Zitterbewegung path of its negative charge. The blue sphere represents its negative charge. Purple arrow: 0.841 fm, blue arrow: 0.421 fm.

Summing up the above information, the neutron's negative charge has approximately 0.4 fm spherical radius, and its center is located at 0.904 fm radial distance from the neutron's center. Fitting the negative charge's spherical radius into the 0.421 fm poloidal radius of the proton structure, we arrive at the neutron geometry illustrated on the right side of figure 9.2[1]. We arrived at this proposed neutron structure by a direct interpretation of experimental data.

We demonstrate the predictive power of this neutron model by calculating the neutron's magnetic moment. The neutron's net magnetic moment comes from two contributions: the $\mu_+ = \mu_p = 2.793\mu_N$ magnetic moment of the toroidal proton structure, and the μ_- contribution of the negative charge. Here, μ_N is the nuclear magneton: its value corresponds to circular Zitterbewegung at $r_N = r_C \frac{m_e}{m_p}$=0.2103 fm radial distance, where r_C is the reduced Compton radius, m_e is the electron mass, and m_p is the proton mass. A point-like negative charge's Zitterbewegung circulation at r_-=0.904 fm radial distance implies a magnetic moment contribution of $\mu_- = -\mu_N \frac{r_-}{r_N} = -4.299\mu_N$. Since the negative charge is not point-like, we also need to take its anomalous magnetic moment factor into account. The derivation of the anomalous magnetic moment factor can be found in chapter 6 of [25], where the $g = \left(1 - \frac{r_{charge}}{2\pi r_{ZBW}}\right)^{-1}$ formula is derived from the basic principles of general relativity. Thus the more accurate μ_- result is:

$$\mu_- = -4.256\mu_N g_- = -4.299\mu_N \left(1 - \frac{0.4}{2\pi \cdot 0.904}\right)^{-1} = -4.625\mu_N$$

Finally, we estimate the neutron's magnetic moment as the sum of the positive and negative charges' contribution:

$$\mu_+ + \mu_- = -1.832\mu_N$$

This estimation matches rather well the neutron's experimentally measured $\mu_n = -1.913\mu_N$ magnetic moment. One may formulate the experimental μ_- value as:

$$\mu_{-,exp} = \mu_n - \mu_p = -4.706\mu_N$$

Our neutron model thus yields a μ_- value that 98% matches the measured one.

[1] As the negative charge's center moves on a ring-shaped Zitterbewegung path, its surface stays within the toroidal volume that is traced out by the positive charge.

9.3. Neutron stabilization by a high electric potential

We saw in section 1.2 that the electrostatic potential is a control parameter of nuclear beta decay. Here, we look into the role of electrostatic potential through the example of ^{17}O. The ^{17}O nucleus is an interesting case for the study of neutron dynamics: it contains a neutron-like sub-particle which is the most similar to a free neutron.

The ^{16}O nucleus comprises four alpha sub-particles. When a neutron is captured by ^{16}O, where will it reside? Considering that the charge radii of ^{16}O and ^{17}O are the exactly same, the captured neutron must be located in the nuclear center, surrounded by four alpha sub-particles. Since the magnetic moment of ^{16}O is zero, the magnetic moment of ^{17}O can be attributed to the captured particles. Interestingly, the magnetic moment of ^{17}O is nearly the same as the neutron's magnetic moment. Specifically, the ^{17}O magnetic moment is $-1.894\mu_N$, which 99% matches the neutron's $-1.913\mu_N$ magnetic moment. These data indicate that the ^{17}O nucleus contains a freely rotating neutron at its center, as schematically illustrated in figure 9.3[2].

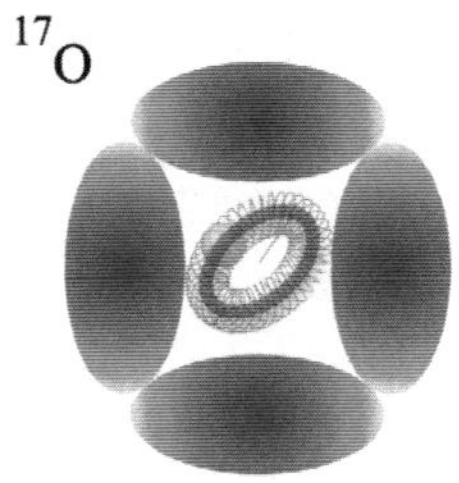

Figure 9.3. The ^{17}O nuclear structure. A neutron is in the center position, the red ellipses illustrate alpha sub-particles.

^{17}O is a stable particle. The neutron at its center cannot decay, because the electron capture energy of ^{17}F is 2.76 MeV; the positive sign of electron capture energy means that the energy needed to remove an electron from the nucleon's location exceeds the 0.87 MeV maximum electron kinetic energy shown in figure 2.1. In other words a neutron becomes stabilized by being at a sufficiently high electrostatic potential, which prevents the electron's departure.

9.4. Neutron destabilization by a strong magnetic field

We now revisit the discussion of ^{9}Be excited states. As noted in section 5, the 2429 keV excitation of ^{9}Be decays mainly by emitting a proton and an electron, in contrast to the neutron-emitting 1684 keV and 2780 keV excitations. Therefore, the break-up of the neutron at 2429 keV excitation is controlled by some other parameter than just the excitation energy.

The ^{9}Be ground state has $J = \frac{3}{2}-$ rotational quantum number designation. Here, the "-" sign denotes the currently popular conjecture of a "negative parity state", which means that the nuclear wavefunction supposedly changes its sign upon the mirror reflection of spatial coordinates. The above-mentioned 2429 keV excitated state has $J = \frac{5}{2}-$ rotational quantum number designation. Recognizing this pattern as a rotational series, the next rotational state is $J = \frac{7}{2}-$, and the corresponding excitation energy is 6380 keV. The rotational Hamiltonian's eigenvalues are:

$$E_J = J\left(J+1\right)\frac{\hbar^2}{2I} \tag{9.1}$$

[2] While the ^{4}He charge radius is 1.7 fm, the ^{17}O charge radius is less than twice this size; it is only 2.7 fm. Four alpha subparticles thus do not fit into ^{17}O without deformation, and therefore they are represented by ellipses in figure 9.3.

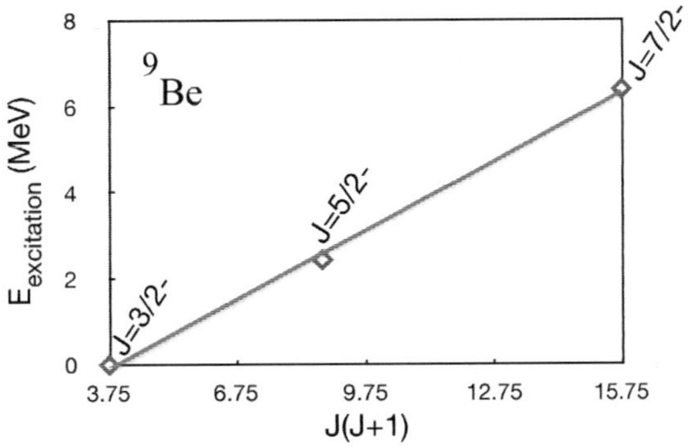

Figure 9.4. The rotational states of ^{9}Be

where E_J is the rotational energy eigenvalue at the rotational quantum number J, and I is the moment of inertia. Figure 9.4 shows the excitation energies as a function of $J(J+1)$. The three data points fit rather well onto a single line, which corresponds to the rotational excitations of a rigid structure. The data shown in figure 9.4 yields $\frac{\hbar^2}{2I}$=534 keV. Considering the 2.5 fm nuclear charge radius of ^{9}Be, this moment of inertia matches quite well the rotation of two alpha sub-particles around each other.

As the positively charged alpha sub-particles rotate around their center of mass, their rotation produces a strong magnetic field. The strength of this magnetic field is proportional to the rotational quantum number J because the moment of inertia remains constant. Since the neutron-like sub-particle breaks up in the $J = \frac{5}{2}-$ and $J = \frac{7}{2}-$ states, a sufficiently strong magnetic field appears to destabilize the neutron. Quantitatively, the magnetic moment values of the various ^{9}Be states are given in table 3. Its first two columns are literature data. Its last two columns are our predictions, calculated from the rotating alpha particles' J states. One may observe from table 3 values that, as a first approximation, the neutron remains stable if the magnetic moment of neighboring nucleons is $<\mu_N$, but becomes unstable if the magnetic moment of neighboring nucleons is $>\mu_N$.

	n	^{9}Be	^{9}Be* (2429 keV)	^{9}Be* (6380 keV)
μ	-1.913μ_N	-1.177μ_N	-0.686μ_N	-0.196μ_N
μ-μ_n	0	+0.736μ_N	+1.227μ_N	+1.717μ_N

Table 3. The magnetic moment of the neutron and various ^{9}Be states, in nuclear magneton units. The last row shows the magnetic moment values without the neutron's contribution.

9.5. The electric binding of neutron decay products

According to the virial theorem, the kinetic energy of a lepton that is electrically bound to a heavier proton is $E_{kinetic} = E_p\frac{\gamma_L}{\gamma_L+1}$, where E_p is the electric potential energy gain and γ_L is the Lorentz boost factor. Using this virial condition together with equation 2.2, we get the following set of equations for the emitted e_n^- lepton:

$$m_nc^2 - m_pc^2 = \gamma_L m_{en}c^2 - E_p$$

$$(\gamma_L - 1)\, m_{en}c^2 = E_p\frac{\gamma_L}{\gamma_L + 1}$$

After expressing E_p in terms of the other parameters, we obtain:

$$m_n c^2 - m_p c^2 = \frac{m_{en} c^2}{\gamma_L}$$

Using the measured $m_{en}c^2$=1553.5 keV value in the above equation, we obtain γ_L=1.2011. Finally, we can calculate the potential and binding energies:

$$E_p = \frac{\gamma_L^2 - 1}{\gamma_L} m_{en} c^2 = 572.5\, keV$$

$$E_{binding} = E_p - (\gamma_L - 1)\, m_{en} c^2 = 260.1\, keV$$

The obtained 99.8% precise match against the binding energy measurement shown in figure 6.2 demonstrates that the neutron decay products are indeed electrically bound during the e_n^- particle's short lifetime. The above calculated potential energy gain implies that the mean $p^+ - e_n^-$ distance is just 2.5 fm.

10. Experimental methods

Figure 10.1 illustrates the experimental setup. The electrode gap with a voltage of 20 kV is placed in a jet of finely dispersed drops of distilled water, supplied from the humidifier by an air stream. This jet of fine droplets is created by high-frequency mechanical vibrations. The pressure in the discharge zone is equal to atmospheric pressure. The discharge gap electrodes are located along the flow of a water-air jet of fine droplets. The flow moves from the cathode to the anode. After passing through the discharge zone, the water-air jet is directed to a ventilation tube, and is released into the atmosphere.

A gamma spectrometer is located above the discharge gap, and consists of a NaI crystal and a Photo-Electrom Multiplier (PhEM). The spectrometer was calibrated by two peaks of the ^{241}Am isotope, which emits gamma rays at 26.3 keV and 59.6 keV energies. These instruments are placed in a lead box with a volume of 15×15×15 cm^3, assembled from 5 cm thick lead blocks.

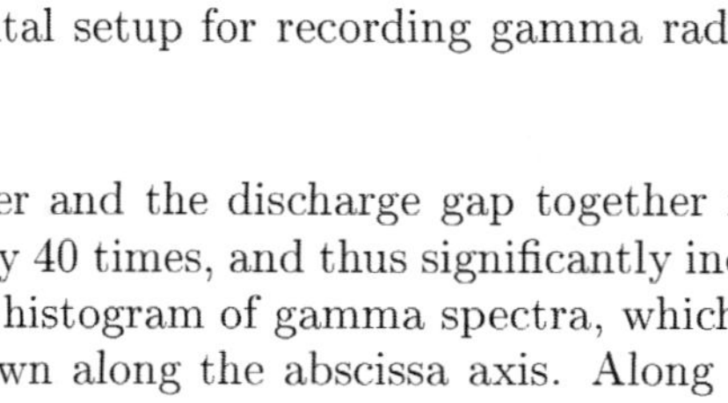

Figure 10.1. The experimental setup for recording gamma radiation during spark discharges in a water-air environment.

By placing the spectrometer and the discharge gap together into a lead box we reduce the gamma background signal in by 40 times, and thus significantly increase the signal-to-noise ratio. Figures 6.2 and 10.2 show the histogram of gamma spectra, which were recorded for 55 seconds. The gamma ray energy is shown along the abscissa axis. Along the ordinate axis, the number of pulses of electromagnetic radiation with a given energy are shown. A peak with an energy of 259.6 keV is clearly visible on figure 10.2. At that 259.6 keV energy, 53 pulses were registered in 55 seconds. When the same spark discharges were done in dry air atmosphere, i.e. without a jet of fine water droplets, no gamma signals were detected above the background.

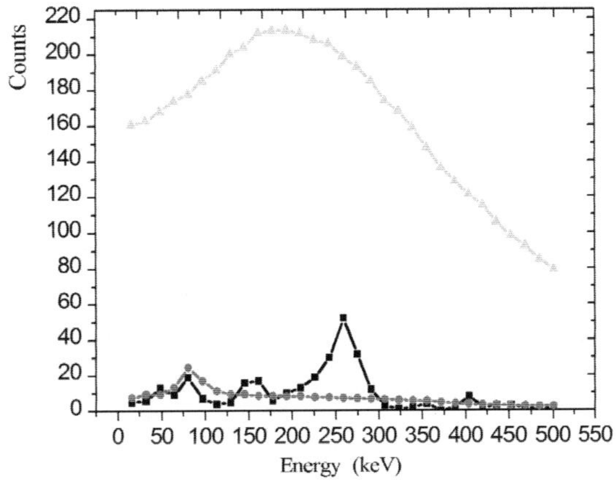

Figure 10.2. A comparison of gamma spectrograms. The black curve was recorded during a 20 kV spark discharge in water-air environment with the recorder located in a lead box, the red curve is the background signal on the recorder located in a lead box while the discharge is turned off, the green curve is the background signal on the recorder located outside the lead box while the discharge is turned off.

Figure 10.2 shows that the background level inside the lead chamber (red curve) is about 10 times lower than the registered peak signal. To the left of the 259 keV peak, another gamma peak with energy ranging from 125 to 175 keV is visible. This side peak provides additional confirmation that the 259 keV peak value is correct, since it is the Compton shoulder of an electron-scattered gamma ray quantum, with the incoming energy of 259 keV.

The green curve of figure 10.2 shows the gamma background in the laboratory area outside the lead chamber. The background outside the lead chamber is about 190 pulses per 55 seconds at 259 keV energy, which exceeds by 3.4 times the peak of the black curve. Therefore, the use of lead box enclosure is essential.

We now address the statistical analysis of the distribution shown in figure 6.2. A direct statistical analysis of figure 6.2 data is complicated by the fact that the NaI crystal has poor spectral resolution. To the left of the 259 keV peak is the Compton shoulder, which distorts the normal distribution of the main peak. Figure 10.3 shows the right branch of the 259 keV peak, which is located far from the Compton peak. This allows us to reasonably apply the normal distribution statistics to the ensemble of experimental points. The red curve shows a normal distribution fitted to the experimental points[3]. It can be seen that the experimental distribution is well approximated by a normal curve centered at 259.6 keV, with a mean squared deviation of 34.7 keV^2. We note that the standard deviation in our experiment is larger than the ±1 keV deviation shown in figure 6.1, which was recorded by a higher precision gamma spectrometer. Insofar as the gamma peaks of figures 6.1 and 6.2 are signatures of analogous e_n^- capture phenomenon, we can take the higher precision of ±1 keV as the error margin of table 2 data.

We also investigate the time correlation between the gamma ray signals and the electric discharge sparks. To do this, we use an oscilloscope to register gamma ray signals. In addition, a small radio antenna was also used, located at a distance of 0.5 m from the lead chamber. The oscillograms coming from the two channels of the oscilloscope are jointly analyzed: i.e. the signal

[3] this normal distribution was obtained using the Origin6 software

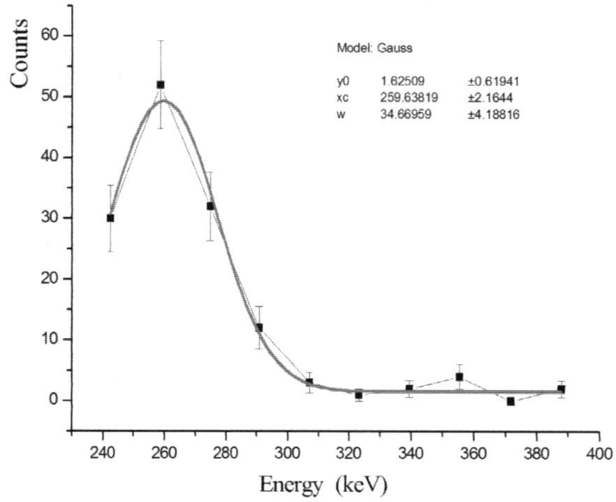

Figure 10.3. The normal distribution curve along the right branch of the 259.6 keV peak. $y0$ is the zero level of the Gaussian along the ordinate axis, xc is the location of the center point, and w is the mean square deviation.

from the PhEM and the signal from the radio antenna. The result of this experiment is a set of oscillograms obtained on the oscilloscope by repeatedly registering short pulses from the NaI detector and from the radio antenna (fast frame mode). The oscillogram recording was triggered when two conditions were met by the gamma signal: i) the duration of the signal is more than 1.5 microseconds, and ii) the signal amplitude is more than 50 mV[4]. Such double condition for starting the oscillogram recording confidently ensures that it starts when the gamma ray quantum passes through the NaI crystal.

Spark discharges generate a bunch of high-frequency oscillations, which were recorded by the radio antenna located 0.5 m away. Each oscillogram shows 25 microseconds before and after the trigger condition. Considering that the sparking frequency is about 10 discharges per second, the probability that a gamma ray quantum and an electromagnetic oscillation signal would both randomly show up in one oscillogram is estimated to be 5×10^{-4}. That is, if the discharge and the formation of a gamma ray quantum are not correlated, then it would take several thousand oscillograms in order to see both signals (i.e from the antenna and from the PhEM) registered in the same oscillogram.

An exemplary oscillogram record is shown in figure 10.4. The green curve shows the signal from the radio antenna; it shows the registration of electromagnetic field oscillations at -9.6 microseconds. It is at this moment that an electric discharge occurs between the two electrodes. These oscillations are almost completed after 5 microseconds, i.e. this is the duration of the spark discharge. The blue curve shows the signal of PhEM; it also reacts to the electromagnetic field like the radio antenna, and shows an oscillating signal at the same time. But 6 microseconds after the spark signal a unipolar negative signal appears on the blue curve, typical for a gamma ray quantum. As explained in the previous paragraph, the probability of a random coincidence between these two signal types is extremely small. We found similar correlated signals on 26% of the recorded oscillograms. Keeping in mind that the oscillogram recording is triggered by any $>$200 keV gamma ray, i.e. not only the peak signal, we obtain a causal relationship between the

[4] this roughly corresponds to a gamma ray quantum of 200 keV

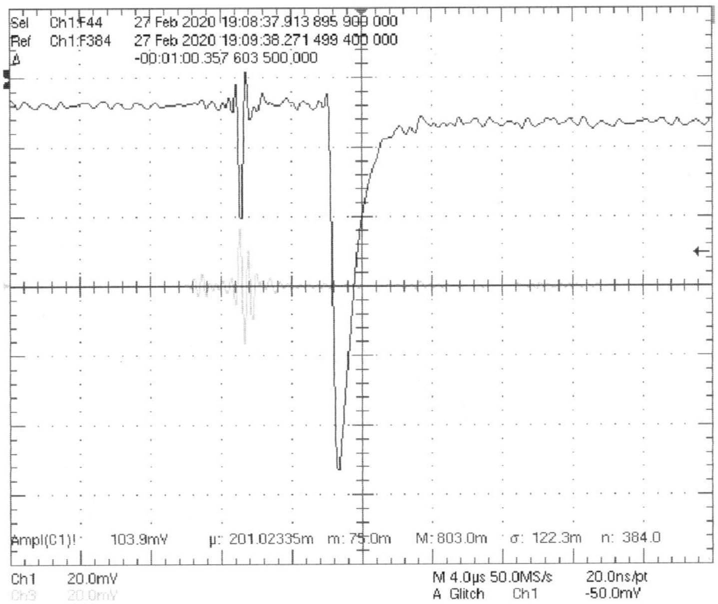

Figure 10.4. An exemplary oscillogram. The blue curve is the PhEM signal, and the green curve is the radio antenna signal. The large negative dip is the signature of a gamma ray quantum.

spark discharges and the 259 keV gamma ray signal.

In summary, the above described analysis proves that the 259 keV gamma peak is statistically significant and it is correlated with spark discharge events in the water-air medium.

We note that this experiment is difficult to replicate because it requires not only thick radiation shielding, but also a large gamma ray sensor and a very short distance between the sensor and the spark discharge location.

11. Conclusions

In conclusion, we presented numerous experimental evidences of a new lepton particle, which is a distinct particle from an ordinary electron. We measured its mass to be 1553.5 keV. It decays into an ordinary electron after just $\sim 10^{-11}$ seconds.

As this 1.5 MeV lepton appears upon neutron decay, it sheds new light on the neutron's internal structure. We proposed a neutron structure that is based on a direct interpretation of experimental data. We conclude that the neutron is a single particle, which is composed of a positive and a negative elementary charge.

Acknowledgements

The work of Andras Kovacs is financially supported by the European Union funded CleanHME project (grant agreement no. 951974). The authors thank Vladimir A. Nikitin for providing the bubble chamber track data.

References

[1] A. Pais "Inward Bound", Clarendon Press (1986), p. 401.

[2] F. Halzen et al "Quarks & Leptons: An introductory course in modern particle physics", John Wiley and Sons (1984)

[3] M. Gell-Mann "A schematic model of baryons and mesons", Physics Letters, Volume 8, 3 (1964)
[4] J. D. Bjorken et al "Inelastic Electron-Proton and γ-Proton Scattering and the Structure of the Nucleon", Physical Review, Volume 185, 5 (1969)
[5] J. Kuti et al "Inelastic Lepton-Nucleon Scattering and Lepton Pair Production in the Relativistic Quark-Parton Model", Physical Review D, Volume 4, 11 (1971)
[6] W. L. Stubbs "The Nucleus of Atoms: One Interpretation", CreateSpace Independent Publishing Platform (2018)
[7] R. L. Jaffe "Where Does the Proton Really Get Its Spin?", Physics Today (1995)
[8] C. van der Leun et al "The deuteron binding energy", Nuclear Physics A, Volume 380.2 (1982)
[9] G. Vassallo et al "The electron and Occam's razor", Journal of Condensed Matter Nuclear Science, Volume 25 (2017)
[10] B. M. Steinetz et al "Experimental Observations of Nuclear Activity in Deuterated Materials Subjected to a Low-Energy Photon Beam", arXiv:1704.00694 (2017)
[11] T. Faestermann et al "Indications for a bound tetraneutron", Physics Letters B, Volume 824 (2022)
[12] Nuclear data tables at www-nds.iaea.org
[13] M. H. MacGregor "The Enigmatic Electron", El Mac Books, 2nd edition (2013)
[14] G. Vassallo et al "The proton and Occam's razor", Proceedings of the 2022 International Association for Relativistic Dynamics (IARD) conference, Journal of Physics - Conference Series (2023)
[15] M. Jung et al "First Observation of Bound-State β^- Decay", Physical Review Letters, Volume 69.15 (1992)
[16] J. S. Nico "Neutron beta decay", Journal of Physics G: Nuclear and Particle Physics, Volume 36.10 (2009)
[17] M. Oettingen "The Application of Radiochemical Measurements of PWR Spent Fuel for the Validation of Burnup Codes", Energies, Volume 15 (2022)
[18] J. S. Kim et al "Analysis of high burnup pressurized water reactor fuel using uranium, plutonium, neodymium, and cesium isotope correlations with burnup", Nuclear Engineering and Technology, Volume 47 (2015)
[19] S. Focardi et al "Evidence of electromagnetic radiation from Ni-H Systems", proceedings of the ICCF-11 International Conference on Condensed Matter Nuclear Science, Marseille, France (2004)
[20] A. V. Gurevich et al "Strong Flux of Low-Energy Neutrons Produced by Thunderstorms", Physical Review Letters, Volume 108 (2012)
[21] L. P. Babich "Thunderstorm neutrons", Physics-Uspekhi, Volume 62.10 (2019)
[22] T. Enoto et al "Photonuclear reactions triggered by lightning discharge", Nature, Volume 551 (2017)
[23] H. Abele "The neutron. Its properties and basic interactions", Progress in Particle and Nuclear Physics, Volume 60 (2008), Pages 1–81
[24] The DOE/NSF Nuclear Science Advisory Committee "The Frontiers of Nuclear Science - A Long Range Plan" (2007)
[25] A. Kovacs et al "Unified Field Theory and Occam's Razor: Simple Solutions to Deep Questions", World Scientific (2022)

[26] J. Chadwick et al "The nuclear photoelectric effect", Proceedings of the Royal Society of London, Volume 151.873 (1935)

[27] E. Storm et al "Photon cross sections from 1 keV to 100 MeV for elements Z=1 to Z=100", Nuclear Data Tables, Volume A7 (1970)

[28] H. Arenhövel et al "Photodisintegration of the deuteron: a review of theory and experiment", Springer (2012)

[29] D. Babusci et al "Deuteron photo-disintegration with polarised photons in the energy range 30–50 MeV", Nuclear Physics A, Volume 633.4 (1998)

[30] Y. Torizuka et al "Entangled pairs of 2p atoms produced in photodissociation of H_2 and D_2", Physical Review A, Volume 99.6 (2019)

[31] H. Atac et al "Measurement of the neutron charge radius and the role of its constituents", Nature communications, Volume 12.1 (2021)

[32] B. Heacock et al "Pendellösung interferometry probes the neutron charge radius, lattice dynamics, and fifth forces", Science, Volume 373.6560 (2021)

The Proton and Occam's Razor

Giorgio Vassallo[1], **Andras Kovacs**[2]
[1]Università degli Studi di Palermo (Dipartimento di Ingegneria), Palermo, Italy
[2]BroadBit Energy Technologies, Espoo, Finland

E-mail: giorgio.vassallo@unipa.it

Abstract. Otto Stern's 1933 measurement of the unexpectedly large proton magnetic moment indicated to most physicists that the proton is not a point particle. At that time, many physicists modeled elementary particles as point particles, and therefore Stern's discovery initiated the speculation that the proton might be a composite particle. In this work, we show that despite being an elementary particle, the proton is an extended particle. Our work is motivated by the experimental data, which we review in section 1.

By applying Occam's Razor principle, we identify a simple proton structure that explains the origin of its principal parameters. Our model uses only relativistic and electromagnetic concepts, highlighting the primary role of the electromagnetic potentials and of the magnetic flux quantum $\Phi_M = {}^h\!/_e$. Unlike prior proton models, our methodology does not violate Maxwell's equation, Noether's theorem, or the Pauli exclusion principle.

Considering that the proton has an anapole (toroidal) magnetic moment, we propose that the proton is a spherical shaped charge that moves at the speed of light along a path that encloses a toroidal volume. A magnetic flux quantum $\Phi_M = {}^h\!/_e$ stabilizes the proton's charge trajectory. The two curvatures of the toroidal and poloidal current loops are determined by the magnetic forces associated with Φ_M. We compare our calculations against experimental data.

keywords
Aharonov-Bohm electrodynamics, magnetic flux quanta, nuclear anapole moment, Occam's razor, proton model, unified field theory, vector potential, Zitterbewegung.

1. Motivation

1.1. *A brief history of the proton model*

Before the 1970s, most scientists viewed the proton as an elementary particle. Starting from the 1970s, scientists working with high energy particle colliders proposed that protons and neutrons are not elementary particles, but comprise smaller sub-particles. According to their model, a proton and a neutron both comprise three quark sub-particles. The existence of quarks has been suggested initially in the 1960s, based on the theoretical efforts by Gell-Mann to model baryons and mesons [13], which were observed in a great variety during high energy nuclear experiments. The momentum distribution of particles emerging from a high-energy collision is characterized by the F_2 structure function[1]. Feynman's proposition was that the F_2 structure

[1] A detailed explanation of the $F_1(x)$ and $F_2(x)$ structure functions can be found for example in [24]. In these functions, the variable x measures the fraction of the nucleon's longitudinal momentum carried by the struck

Content from this work may be used under the terms of the Creative Commons Attribution 3.0 licence. Any further distribution of this work must maintain attribution to the author(s) and the title of the work, journal citation and DOI.
Published under licence by IOP Publishing Ltd

function probes the internal momentum distribution of sub-particles; for a particle comprising N sub-particles, its F_2 structure function must peak at $x = \frac{1}{N}$. Gell-Mann's original quark theory thus required the F_2 momentum distribution to peak at $x = \frac{1}{3}$. However, as will be shown in section 1.3, this is not the case because the experimentally observed F_2 data peaks at $x = \frac{1}{9}$. This deviation from the required peak at $x = \frac{1}{3}$ was explained away via the hypothesis that the three quarks originally thought to form the proton are the so-called "valence quarks", which are swimming in the background of "sea quarks" [6]. These so-called sea quarks are a collection of quark-antiquark pairs, radiated by the three valence quarks. However, the calculations of 1970s still showed that the valence quarks together with the sea quarks only accounted for 54% of the proton's momentum [16]. A further hypothesis was added to supplement the momentum shortfall of the quarks; chargeless particles called gluons were introduced into the proton model [25]. Since gluons have no electric charge, the thinking was that they are there, but the electrons probing the proton in deep inelastic scattering cannot see them. These hypothesized gluons were assigned the missing proton momentum, and the resulting proton model became the quark-gluon model that it is today. Despite the absence of any direct quark observation, the quark-gluon model gained popularity during the 1970s, and remained embraced by most theoretical physicists ever since.

According to the 1970s model of "valence quarks" swimming in the background of "sea quarks and gluons", there seemed to be an angular momentum deficit with respect to the measured angular momentum of the proton, and therefore the presence of "virtual strange quarks" was also postulated during the 1990s [20].

In the context of quark theory, the fulfillment of $F_2(x) = 2xF_1(x)$ relation around $x = \frac{1}{3}$ means that an individual quark's spin is detectable[2]. Since the proton's magnetic moment measurements yield a constant value of $\mu_p = 2.793\mu_N$, quark proponents postulated that the three valence quarks are always spin correlated and that the sea quarks' spin contribution always sums up to zero. The $\mu_p \approx 3\mu_N$ relation has been interpreted as the almost parallel orientation of the three valence quark spins. It was pointed out to pioneering quark proponents that their requirement is in contradiction with the Pauli exclusion principle. This issue lead Oscar Greenberg to postulate in 1964 that quarks also have "color charge"; the purpose of this color charge hypothesis was to remove the contradiction with respect to the Pauli exclusion principle.

1.2. Experimental counter-evidence to the quark model

Although the quark-based model was inspired by the great variety of mesons, the proposed quark masses do not add up the masses of observed mesons. According to quark proponents, this is explained by a negative binding energy between quarks: any particle's valence quarks masses are only a small percentage of the total particle mass, with the bulk of the particle mass coming from virtual particles which represent the binding force: i.e. virtual quarks and gluons. Moreover, the valence quark : virtual quark : gluon mass ratio is allowed to vary from particle to particle in order to match the observed masses. Now what is the physical meaning of negative binding energy? By definition, negative binding energy means a metastable bound state. This model implies that individual quarks should be easily observable upon the break-up of their metastable binding. However, quark proponents also postulated that these metastable bonds between quarks can never be dissociated. There is a fundamental contradiction between the hypothesis of metastable quark binding and the hypothesis of unbreakable quark bonds.

particle, evaluated in the Breit frame. The $F_1(x)$ function at a given x is interpreted as $\frac{1}{2}$ of the likelihood of scattering from a particle which, in the Breit frame, has longitudinal momentum fraction x of the proton. In case of scattering from spin-$\frac{1}{2}$ particles, $F_2(x) = 2xF_1(x)$.

[2] The $F_2(x) = 2xF_1(x)$ equation is referred to as the Callan-Gross relation. Scattering experiments have indeed observed this Callan-Gross relation, at least within the $x = [0.25, 0.75]$ range.

Proton-antiproton reactions provide rather direct counter-evidence. Figure 1.1 shows traces of a proton-antiproton reaction event, highlighting the produced pion tracks. According to the quark model, a proton-antiproton pair comprises six quarks. After a partial annihilation of quark-antiquark pairs, there can be up to four remaining quarks, which may be organized into two pions. However, figure 1.1 shows at least eight pions emerging from the annihilation event, which contradicts the quark model. A quark model proponent may try to explain this phenomenon by assuming that the kinetic energy of the incoming antiproton was converted into the production of numerous pion-antipion pairs just prior to its annihilation. However, such an explanation is refuted by reference [7], whose authors exposed a nuclear emulsion to antiprotons, and then analyzed the resulting tracks in the emulsion. Their discussion of figure 2 in reference [7] clearly states that the antiproton first came to a rest in the emulsion, and then produced at least five pions upon annihilation with a proton. Such large number of pions emerging from proton-antiproton reactions is impossible under the quark-antiquark annihilation model.

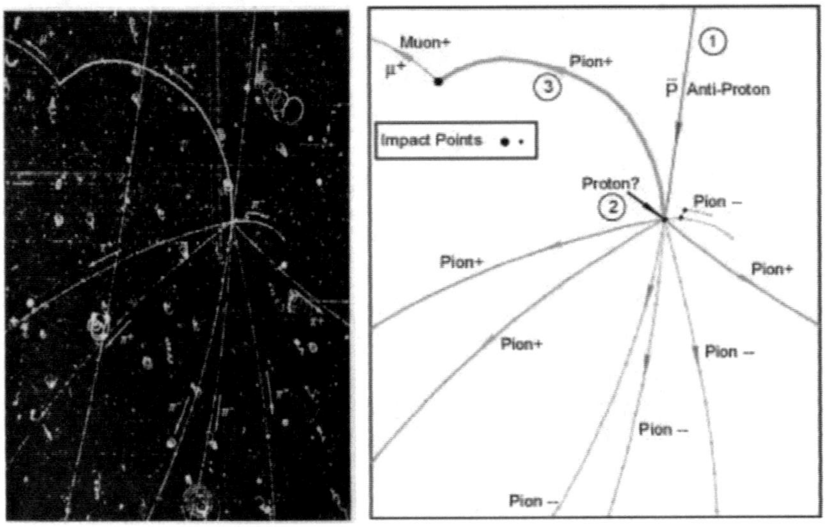

Figure 1.1. Proton-antiproton annihilation event. Left: bubble chamber photograph. Right: diagram of the photo, identifying the particles created by the annihilation event. Source: Lawrence Berkeley National Laboratory Science Photo Library - photograph K003/4377.

According to the quark model, the proton and neutron both comprise three quarks, only differing in one quark type. However, a recent work [23] establishes that the neutron comprises a positive and a negative elementary charge, which again invalidates the quark-gluon model.

Finally, we mention that the postulated spin correlation among the three valence quarks leads to a mathematical paradox, that will be discussed in section 6.

1.3. A re-interpretation of high-energy particle collision data

Considering the above outlined problems with the quark-based proton model, one may wonder about the origin of the F_2 momentum-distribution data recorded in high-energy collisions.

The production of particle-antiparticle pairs is a well established phenomenon of high-energy collisions. Therefore, an incoming energetic electron may produce muon-antimuon pairs upon scattering. Also, an incoming electron may be energized into a muon upon scattering. It is thus pertinent to consider a relationship between the F_2 data and the short-lived particles produced in preceding scattering events.

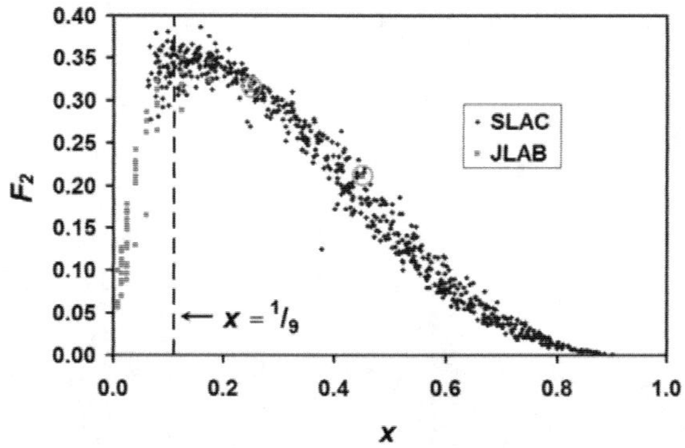

Figure 1.2. Combined SLAC and JLAB data of F_2 momentum distribution measurements from electron-proton scattering, reproduced from [37].

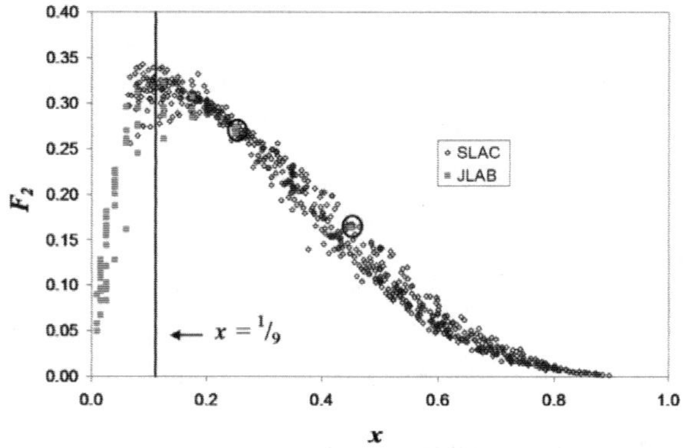

Figure 1.3. Combined SLAC and JLAB data of F_2 momentum distribution measurements from electron-deuteron scattering, reproduced from [37]. This scattering data shows the same F_2 momentum distribution as in the electron-proton case.

Reference [37] presents a thorough analysis of high energy scattering data from measurements performed at the Stanford Linear Accelerator Center (SLAC), Thomas Jefferson National Accelerator Facility (JLAB), and Hadron Electron Ring Accelerator (HERA). As shown in figures 1.2 and 1.3, the combined SLAC and JLAB data of F_2 momentum distribution measurements shows clearly, without any curve fittings, that their F_2 values peak in the vicinity of $x = \frac{1}{9}$. The JLAB F_2 values at $x = 0.45$ and $x = 0.25$ (circled) show that JLAB data integrates well with the original SLAC F_2 data. Is the use of single-variable $F_2(x)$ distribution justified, i.e. are the energy and momentum exchange sufficiently high for convergence? The use of $F_2(x)$ rather than $F_2(x, Q^2)$ is justified because the SLAC data was shown to satisfy Bjorken scaling, i.e. for $x{>}0.2$, the F_2 values are essentially the same for a given x regardless of the Q amount of energy transferred between the scattering particles. In this $x{>}0.2$ region, Q^2 values range from 0.6 to about 30 GeV^2. Regarding the $x{<}0.2$ region, the JLAB Q^2 values shown in Figure 1.4 are nearly the same as the SLAC Q^2 values listed in Appendix A.1 of [37], which makes the two results directly comparable.

One may wonder why this F_2 peak at $x = \frac{1}{9}$ doesn't show up in any other literature? By 1973, mainstream theorists have essentially embraced the quark-gluon model as adequately describing the structure of the proton. Most attempts to explain the SLAC scattering results any other

x	Q^2	F_2 - p	F_2 - d	x	Q^2	F_2 - p	F_2 - d
0.009	0.034	0.056	0.0492	0.04	0.287	0.2027	0.2002
0.009	0.051	0.0616	0.058	0.04	0.353	0.2244	0.2077
0.009	0.086	0.0997	0.0896	0.04	0.37	0.2288	0.2139
0.015	0.059	0.0696	0.0669	0.04	0.371	0.2186	0.2155
0.015	0.095	0.0842	0.0831	0.04	0.38	0.2102	0.2231
0.015	0.098	0.0961	0.0935	0.04	0.421	0.2416	0.2268
0.015	0.112	0.0876	0.0966	0.06	0.18	0.1641	0.1616
0.015	0.127	0.1058	0.1073	0.06	0.479	0.2622	0.2563
0.015	0.144	0.1114	0.1129	0.06	0.491	0.2617	0.2702
0.015	0.151	0.1216	0.1186	0.06	0.543	0.2751	0.2609
0.015	0.164	0.1253	0.1227	0.06	0.633	0.2863	0.2958
0.015	0.172	0.1118	0.1286	0.08	0.456	0.265	0.2451
0.025	0.067	0.0883	0.0834	0.08	0.617	0.2935	0.2752
0.025	0.092	0.096	0.0953	0.08	0.619	0.296	0.2767
0.025	0.104	0.104	0.0994	0.08	0.799	0.3128	0.295
0.025	0.113	0.1069	0.1024	0.08	0.818	0.3227	0.3122
0.025	0.14	0.1251	0.1208	0.125	0.588	0.2876	0.2609
0.025	0.186	0.1469	0.1441	0.125	0.797	0.3179	0.2873
0.025	0.195	0.1312	0.1439	0.125	1.032	0.3319	0.2952
0.025	0.212	0.1675	0.1545	0.125	1.056	0.3491	0.3228
0.025	0.222	0.1593	0.1568	0.175	1.029	0.3242	0.2846
0.025	0.24	0.178	0.1656	0.175	1.045	0.3235	0.2939
0.025	0.252	0.1696	0.1745	0.175	1.365	0.3447	0.3072
0.025	0.253	0.153	0.1601	0.25	1.332	0.3126	0.2673
0.025	0.287	0.1669	0.1814	0.25	1.761	0.3183	0.2744
0.04	0.133	0.1295	0.128	0.45	2.275	0.2104	0.1638
0.04	0.273	0.2038	0.1876				

Figure 1.4. F_2 structure function values for the proton (F_2 - p) and the deuteron (F_2 - d) as a function of x and Q^2 from the JLAB E99-118 deep inelastic scattering experiments [38].

way had ended, and the bulk of the theoretical effort focused on enhancing the quark model. Different versions of the quark-gluon plasma model predict either constant or rising F_2 values as $x \to 0$. In the 1990s, when HERA experiments began producing data in this $x<0.1$ range, it was assumed that HERA filled the low-x gap left by SLAC, even though its data was generated from scatterings with Q^2 values tens to hundreds of times higher than the SLAC data. Many theorists could not resist the temptation of mixing non-comparable data in this low-x region of the F_2 curve, and mistakenly proclaimed experimental support for their quark-gluon model. Reference [32] is a typical example of such erroneous data analysis. Around 2000, the JLAB experiment began producing scatterings with comparable Q^2 values to the SLAC experiment. By that time, the erroneous blending of high-Q^2 HERA data with low-Q^2 SLAC data was already a consensus procedure for obtaining the proton's F_2 curve, and mainstream theorists had no interest in pointing out their colleagues' mistakes or discussing the implications of the JLAB experiment.

Upon dividing the proton's mass by 9, we obtain approximately the muon mass. This match with the F_2 peak location at $x = \frac{1}{9}$ suggests that it may correspond to electron scattering from a muon or antimuon that was produced in some preceding scattering event. Our interpretation implies that one should also find a peak corresponding to electron-electron scattering in the very low-x region because an incoming electron may also collide with a previously scattered other electron. Upon the analysis of HERA experiments, reference [37] indeed identifies yet another F_2 momentum distribution peak near $x = \frac{1}{1836}$, which corresponds to the electron mass.

Since there is no F_2 peak at $x = \frac{1}{3}$, the fulfillment of Callan-Gross relation around $x = \frac{1}{3}$ simply means that the energetic electron is mainly scattering from the proton, which is a spin-$\frac{1}{2}$ particle.

In summary, the F_2 momentum distribution data shows signatures of electron-muon and electron-electron scattering. Consequently, the quark model is contradicted by all experimental data. The absence of a reasonable proton model motivates us to explore the proton's internal structure.

Antiprotons are generated in sufficiently energetic collisions between light and heavy nuclei. Would a violent collision create complex structures involving many sub-particles? That would be very unlikely, and it is moreover favored by Occam's razor principle to firstly explore simple

Journal of Physics: Conference Series **2482** (2023) 012020 doi:10.1088/1742-6596/2482/1/012020

proton structures. We thus investigate whether a relatively simple proton model exists, which would match its experimentally observed properties.

2. How large is the proton?

2.1. The proton's spherical charge radius

Any particle's Compton scattering cross-section is given by the Klein-Nishina formula, where one parameter is the spherical charge radius. Upon fitting the electron's experimental Compton scattering cross-section to the Klein-Nishina formula, in the 0.5 MeV photon energy range, one obtains 2.82 fm spherical charge radius. This 2.82 fm electron charge radius is referred to as the classical electron radius in the scientific literature.

Is the same method applicable for determining the proton's spherical charge radius? Figure 2.1 shows the proton's scattering cross-section in the 1 GeV photon energy range, which corresponds to the proton mass. There are numerous peaks in the scattering data of figure 2.1; these correspond to photo-production of new particles. Experimental measurements determined that the largest peak around 300 MeV corresponds to the photo-production of two neutral pions, while the peak around 700 MeV corresponds to the photo-production of a pion and an η meson. In contrast to the electron case, the scattering cross-section is now a sum of particle photo-production and Compton scattering processes. Nevertheless, we can make an estimation of the proton's spherical charge radius.

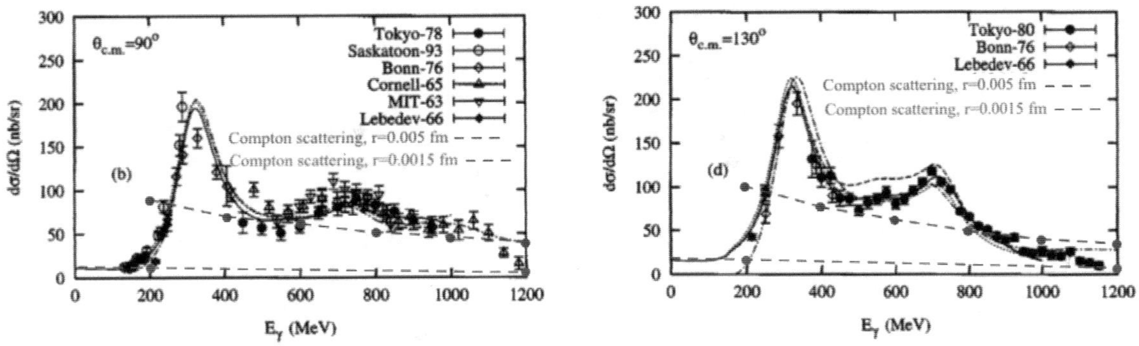

Figure 2.1. The proton's interaction cross-section with high frequency radiation, reproduced from [28]. The horizontal scale shows the incoming photon energy, the left and right panels show the cross-section at 90° and 130° scattering angles, respectively. The red and and blue dashed lines show the Compton scattering cross-section for the indicated spherical charge radius values.

The dashed lines on figure 2.1 show the Compton scattering cross-section at $5 \cdot 10^{-18}$ m and at $1.5 \cdot 10^{-18}$ m spherical charge radius values. With $5 \cdot 10^{-18}$ m radius, the Compton scattering cross-section becomes larger than the experimental values in the <200 MeV and >1000 MeV regions. Therefore, the true radius is smaller than $5 \cdot 10^{-18}$ m. In contrast, with $1.5 \cdot 10^{-18}$ m radius value the Compton scattering cross-section converges to the experimental values both in the <200 MeV and >1000 MeV regions. Therefore, light scattering measurements indicate that the proton's spherical charge radius is approximately $1.5 \cdot 10^{-18}$ m.

2.2. The proton's apparent Zitterbewegung radius

Numerous experiments aim to precisely measure the proton's so-called "charge radius", which is defined as the mean radius value of its charge distribution. High-energy electron-proton scattering experiments are one class of such measurements.

As shown in table 1, one of the earliest scattering analysis based proton charge radius extraction was published in 1963: it comprises a systematic review of scattering experiments performed up to that date, and its authors calculated a $0.805 \cdot 10^{-15}$ m charge radius value. By the early 2000s, the consensus mean proton radius value increased to $0.875 \cdot 10^{-15}$ m, but reference [27] re-analyzes the involved measurements and claims to have found a systemic error which caused over-estimations.

Recent measurements converge around the $0.84 \cdot 10^{-15}$ m mean radius value, and claim very small error margins of only $(5-8) \cdot 10^{-18}$ m.

This $0.84 \cdot 10^{-15}$ m mean radius value is several orders of magnitude larger than the above identified $r_{cp} < 5 \cdot 10^{-18}$ m parameter. To understand the physical meaning of the $0.84 \cdot 10^{-15}$ m radius value, we again use the analogy of electron scattering experiments. When an electron interacts with high frequency light, its scattering cross section is given by the Klein-Nishina formula, and such scattering data reveals the electron's 2.82 fm spherical charge radius. When an electron interacts with low frequency light, its scattering cross section is given by the Thomson scattering formula, and such scattering data reveals the electron's 386 fm Zitterbewegung radius. In the scientific literature this electron Zitterbewegung radius is also referred to as the electron's "reduced Compton radius". By analogy, we associate the proton's $0.84 \cdot 10^{-15}$ m radius value as an approximation of the major radius of the torus enclosed by the proton charge trajectory.

Publication year	Mean proton radius value	Reference
1963	0.805 ± 0.011 fm	[17]
2016	0.840 ± 0.016 fm	[14]
2020	0.831 ± 0.019 fm	[42]
2021	0.847 ± 0.008 fm	[10]
2022	0.840 ± 0.005 fm	[27]

Table 1. Electron-proton scattering analysis based mean proton radius measurements.

Besides the electron-proton scattering analysis, there are also spectroscopic methods for the proton's charge radius calculation [15]; all spectroscopic estimate the impact of non-zero proton radius on the electrostatic potential of the electron's wavefunction. Table 2 shows the results of recent proton radius measurements, based on spectroscopic methods.

Publication year	Involved particles	Charge radius	Reference
2017	e^-, p^+	0.8335 ± 0.0095 fm	[4]
2019	e^-, p^+	0.833 ± 0.01 fm	[5]
2020	e^-, p^+	0.8483 ± 0.0038 fm	[15]

Table 2. Spectroscopic analysis based mean proton radius measurements.

Tables 1 and 2 show remarkably similar values. Omitting the 1963 data, the remaining recent measurements average out to $r_{mean} = 0.839 \pm 0.007$ fm.

2.3. Charge density measurements

With the advancement of electron-proton scattering measurements, it has become possible to directly map out the proton's radial charge distribution. Such radial charge distribution data is measured for example at JLAB [8], and is visualized in figure 2.2. This distribution's average RMS (Root Mean Square) value is 0.8 fm, which implies only 4% deviation from the data of tables 1 and 2. Figure 2.2 conveys the important information that the proton charge is located mainly within 0.2-1 fm from its center. Any realistic proton model must yield a similar radial distribution range.

One can also calculate the radius within which 50% of the proton charge is contained. As can be seen in figure 2.2, the radius value at the 50% cumulative charge is $r_{phc} \simeq 0.627$ fm. Here, the "phc" index refers to the proton half-charge radius. An interpretation of this experimental value will be addressed in section 5.5.

The r_{mean} and r_{phc} values are two distinct measures of the proton size. The relevant measure to use depends on the context.

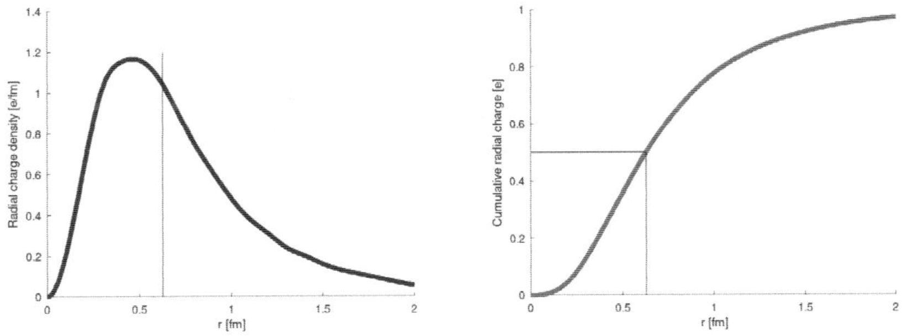

Figure 2.2. The proton's radial charge distribution, according to JLAB measurements [8]. Left: the proton's radial charge density. Right: the cumulative proton charge contained within a given radius. In both charts, the black vertical line indicates the radius $r_{phc} \simeq 0.627$ fm which contains 50% of the total charge.

2.4. The proton's electric polarizability radius
Electric polarizability measurements represent yet another proton size measurement method. Recent measurements by this method are reported in references [26, 12]: their authors obtain 1.2−1.3 fm proton size. This size is significantly larger than the above-mentioned Zitterbewegung radius measurements, and the origin of such discrepancy has not been understood in preceding works.

3. Methodology
In this work, we explore an electromagnetic proton structure which is in accordance with Maxwell's equation. Our methodology is based on the recently published electron model [22], which explains what an electron is made of, why it has a spin, and what the origin of the quantum mechanical wavefunction is. A main conclusion of [22] is that the electron mass comprises electromagnetic field energy. Given that a high-frequency electromagnetic wave can produce an electron-positron pair, while traveling through a sufficiently strong electric field, the ideas of [22] are quite natural. More specifically, by calculating the electric field energy around the electron's 2.82 fm spherical charge radius, one obtains 255.5 keV, which is exactly half of the electron mass. As explained in [22], the other half of the electron mass is magnetic field energy. These two electromagnetic energy types continuously induce each other. Such dynamics is completely analogous to the perpetual induction within an electromagnetic wave, which may give birth to the electron-positron pair.

The 255.5 keV magnetic energy of an electron corresponds to the circular Zitterbewegung of its spherical charge. Such circular Zitterbewegung generates the electron spin. The constant value of the electron spin follows from the constant speed of its circular Zitterbewegung; as discussed in [22], it directly follows from gaugeless Maxwell's equation that the Zitterbewegung speed is the speed of light.

As outlined in table 3, our methodology is in accordance with all fundamental physical laws. In comparison, the quark-based methodology has multiple drawbacks: i) the quark model violates foundational laws, such as Maxwell's equation or Noether's theorem, ii) as explained in section 1, the quark model lacks any experimental evidence, and iii) the quark model is contradicted by the commercial Nuclear Magnetic Resonance (NMR) technology. The implications of proton NMR data will be discussed in section 6.

	Quark-based proton model	**This work**
Violates Maxwell's equation?	Yes (point-like charges)	No
Neglects infinite quantities?	Yes (renormalization)	No
Violates Noether's theorem?	Yes (virtual particles)[3]	No
Relation to NMR measurements?	Contradicts p^+ NMR data	Explains p^+ NMR
Radius calculations can be verified?	No (too lengthy/complex)	Yes

Table 3. A comparison between the quark-based proton model and our present work.

At first it seems natural to use the exact same model for describing both the electron and proton, scaling the given particle's dimensions by the appropriate particle mass. The advantages of a simple ring-shaped proton model were indeed pointed out by David L. Bergman in his paper "The Real Proton" [2]. This approach, while works well for muons, introduces unacceptably large errors if naively used for proton modeling. Firstly, the magnetic moment of such a simple model is equal to the nuclear magneton μ_N, while the experimental proton magnetic moment value is approximately 2.79 times larger. Secondly, as discussed in section 2.2, the proton's experimental Zitterbewegung radius value is 0.839 ± 0.007 fm, while the "scaled positron" model yields a 0.2103 fm Zitterbewegung radius from the $e^+ : p^+$ mass ratio. The following sections present a simple proton model that overcomes these large discrepancies while fully maintaining the conceptual framework introduced in this section.

4. Gaugeless electrodynamics

It's important to note that, at the Compton scale, certain quantized physical values appear dimensionless in natural units. The elementary charge value $e = \pm\sqrt{\alpha}$, its magnetic flux $\Phi_M = {}^{2\pi}/_{e}$, Zitterbewegung speed $c = 1$, and angular momentum $\hbar = 1$ *cannot be separated but are different characteristics of the same physical entity.* As already pointed out in [22, 11], a non-linear dynamic equation can be derived when the Maxwell's equation and the Proca equation are considered to apply simultaneously. This equation essentially describes, using the language of spacetime Clifford Algebra $Cl_{3,1}(\mathbb{R})$, the behavior of an elementary charge that always moves at the speed of light and is subjected to a magnetic centripetal force that is responsible for the curvatures of its Zitterbewegung trajectory. Therefore, the electromagnetic four-potential can be seen as the field, a "*Materia Prima*", from which the physical entities that we call "particles" are generated. It is therefore reasonable to universally apply this approach to all charged elementary particles.

Leaving behind the experimentally paradoxical hypothesis of electromagnetic gauges [36, 9, 35], we do not assume the presence of any electromagnetic gauge, and arrive at the simplest form of Maxwell's equation [22, 11, 3]: $\boldsymbol{\partial}^2 \boldsymbol{A}_\square = 0$. The $\boldsymbol{A}_\square$ notation refers to the electromagnetic four-potential $\boldsymbol{A}_\square = \boldsymbol{A} + \gamma_t V$. The electric charges and currents then correspond to a scalar field on a spherical surface. As required by Maxwell's equation, this charged surface

[3] The quark-based model assigns over 98% of the proton mass to virtual particles.

is moving at light speed, and is characterized by a vector potential $\boldsymbol{A}$, an electric potential V, a current $I = \alpha A/2\pi$, a mechanical momentum $m\boldsymbol{c} = e\boldsymbol{A}$, and an angular speed $\omega = eA$. Respecting both magnetic and electric Aharonov-Bohm relations, the charge's electromagnetic four potential $\boldsymbol{A}_\square = \boldsymbol{A} + \gamma_t V$ is a nilpotent vector $\left(\boldsymbol{A}_\square^2 = 0\right)$. These laws may be considered as a powerful tool for modeling the structure and properties of elementary charged particles. Prior to [22, 11], gaugeless electrodynamics has been already introduced and explored by other authors [1, 18, 19, 21, 29, 30, 31, 33, 34, 39, 41]. Most of these preceding works introduce the electromagnetic scalar field as an additional entity besides charges and currents, rather than the entity that actually produces the apparent charges and currents. A notable exception is the work of Giuliano Bettini [3] that recognizes the electromagnetic sources as the partial derivatives of the scalar field.

The elementary charge is characterized by a simple Lagrangian $\mathscr{L}$ that defines the action $\mathscr{S}$:

$$\mathscr{L} = e\boldsymbol{A} \cdot \boldsymbol{c} - eV \tag{4.1}$$

$$\mathscr{S} = \int \mathscr{L} dt$$

The stationary action condition $\delta\mathscr{S} = 0$ is a consequence of the Aharonov-Bohm relations

$$\mathscr{S} = \int \left(e\boldsymbol{A} \cdot \boldsymbol{c} - eV\right) dt = e \int \boldsymbol{A} \cdot \boldsymbol{dl} - e \int V dt = 0$$

$$\delta\mathscr{S} = 0$$

As $\boldsymbol{A}$ and $\boldsymbol{c}$ are parallel vectors for a freely moving charge, it's possible to substitute the dot product with the product of their modulus:

$$\mathscr{L} = eAc - eV = eA\frac{dl}{dt} - eV$$

If the radius of the charge's Zitterbewegung trajectory is r, the differential of the displacement dl can be substituted by the product $rd\varphi$:

$$dl = rd\varphi$$

$$\mathscr{L} = eAr\frac{d\varphi}{dt} - eV$$

Consequently, the following simple conditions guarantee that the action $\mathscr{S}$ is always zero:

$$eAr = \hbar = 1$$

$$\frac{d\varphi}{dt} = eV = eA = \frac{1}{r}$$

$$r^{-1} = \frac{d\varphi}{dt} = \omega = m$$

In natural units, the elementary particle's mass-energy is equal to its Zitterbewegung angular speed, to the inverse of its Zitterbewegung radius, and to the value its Zitterbewegung momentum eA.

5. Proton model

5.1. Proton geometric structure

We develop our proton model in agreement with the above considerations. While the natural choice for a proton model is a simple "scaled positron" model, it leads to some unacceptable discrepancies with the experimental data, that we have already pointed out.

Measurements of the proton's anapole magnetic moment have been claimed since 1997 [40]. In such experiments, electron-proton coupling interactions are used for mapping out the proton's various magnetic modes. Since the anapole magnetic moment is generated by a toroidal charge current, these experiments suggest that the proton's charge moves on a toroidal surface.

We therefore consider a model where the stationary proton charge follows a toroidal Zitterbewegung trajectory, similar to a toroidal coil winding. The toroidal volume enclosed by the proton charge trajectory has a minor (poloidal) proton radius r_{pp}, which remains to be determined.

At the Compton radius scale, the magnetic flux quantum h/e induces a centripetal magnetic force that constrains an elementary particle to follow either a circular Zitterbewegung path (positron case) or helicoidal Zitterbewegung path (proton case).

In this article, we describe two ways of applying our methodology to a toroidal proton model. These two approaches agree on the spherical charge radius value, but yield slightly different toroidal charge radius values. One approach, which represents the perspective of one author, is described in sections 5.2-5.5 of this article[4]. This first approach yields a toroidal charge radius that coincides with the measured r_{50c}=0.624 fm value. The other approach, which represents the perspective of the other author, is described in the first appendix. This second approach yields a toroidal charge radius that coincides with the measured $r_{mean} = 0.839$ fm value. We invite readers to analyze these two approaches, and debate the pros and cons for each one.

5.2. The proton's electromagnetic energy and charge radius

The proton charge is assumed to follows a closed helicoidal trajectory similar to a toroidal coil winding. The main geometric parameters of this model, such as its spherical charge radius or its toroidal and poloidal radii, can be found by starting from the proton mass-energy value and imposing the quantization of the angular momentum and of the experimental magnetic moment value.

The electric energy W_{pE} of the proton is calculated by integrating the energy density of its electric field down to its spherical charge radius r_{cp}:

$$W_{pE} = \frac{e^2}{8\pi} \int_{r_{cp}}^{\infty} \frac{1}{r^4} \cdot 4\pi r^2 dr = \frac{e^2}{2} \int_{r_{cp}}^{\infty} \frac{1}{r^2} dr = -\frac{e^2}{2} \left. \frac{1}{r} \right|_{r_{cp}}^{\infty} = \frac{e^2}{2r_{cp}} = \frac{\alpha}{2r_{cp}}$$

We assume that the electric energy is equal to one half of the proton mass, as required by Maxwell's equation for any electromagnetic wave. Introducing r_{pp} as the reduced Compton wavelength of the proton,

$$W_{pE} = \frac{m_p}{2} = \frac{1}{2r_{pp}}$$

and we calculate $r_{pp} = m_p^{-1} = 1.06578893 \cdot 10^{-9}\, eV^{-1} \left(0.2103089103 \cdot 10^{-15} m\right)$, as well as the proton charge radius:

$$\frac{\alpha}{2r_{cp}} = \frac{1}{2r_{pp}}$$

[4] This approach is the perspective of Giorgio Vassallo

$$r_{cp} = \alpha r_{pp}$$

$$r_{cp} = \frac{\alpha}{m_p} \simeq 7.777437549 \cdot 10^{-12}\, eV^{-1} \left(1.534698267 \cdot 10^{-18} m\right)$$

The other half of the proton mass comprises its magnetic energy W_{pM}:

$$W_{pM} = \frac{1}{2}\Phi_M I_p = \frac{1}{2} \cdot \frac{2\pi}{e} \cdot \frac{\alpha}{2\pi} A_p = \frac{1}{2} e A_p = \frac{m_p}{2}$$

The charge radius r_{cp} implies a potential V_p at the surface of the charge. Consequently it is in agreement with the electric Aharonov-Bohm 5.1equation if we assume that the charge spans an angle $d\varphi$ in a time dt moving at light speed $c = 1$:

$$d\varphi = \frac{cdt}{r_{pp}} = \frac{dt}{r_{pp}}$$

$$V_p = \frac{e}{r_{cp}}$$

$$d\varphi = eV_p dt \tag{5.1}$$

$$\frac{d\varphi}{dt} = eV_p = \frac{e^2}{r_{cp}} = \frac{\alpha}{r_{cp}} = \frac{1}{r_{pp}} = m_p$$

5.3. Proton torus aspect ratio

Assuming that the torus volume enclosed by the proton charge trajectory has a minor radius r_{pp} equal to the proton reduced Compton wavelength $\lambda_p/2\pi$ and that, using natural units, is equal to the inverse of the proton mass [2], it's possible to find the major radius imposing the quantization of the proton angular momentum L_p and the experimental value of the magnetic moment μ_p

$$L_p = m_p v_{pt} r_{pt} = \hbar = 1$$

$$v_{pt} = \frac{r_{pp}}{r_{pt}}$$

Multiplying the toroidal component I_{pt} of the proton current by the enclosed area πr_{pt}^2 we get the proton magnetic moment:

$$I_{pt} = I_p v_{pt} = \frac{\alpha}{2\pi} A_p v_{pt} = \frac{\alpha}{2\pi} A_p \frac{r_{pp}}{r_{pt}}$$

$$I_{pt} \pi r_{pt}^2 = \mu_p$$

$$\frac{\alpha}{2\pi} A_p \frac{r_{pp}}{r_{pt}} \pi r_{pt}^2 = \mu_p$$

remembering that $\alpha = e^2$ and that $eA_p = m_p = r_{pp}^{-1}$, we can write

$$\frac{e}{2} r_{pt} = \mu_p$$

$$r_{pt} = \frac{2\mu_p}{e} = 2.976586476 \cdot 10^{-9}\, eV^{-1} \left(0.5873608214 \cdot 10^{-15} m\right)$$

$$\mu_N = \frac{e}{2m_p}$$

$$\frac{r_{pt}}{r_{pp}} = v_{pt}^{-1} = \frac{2\mu_p}{e} m_p = \frac{\mu_p}{\mu_N} = \sqrt{\frac{39}{5}}$$

This means that the torus aspect ratio $\eta = \sqrt{39/5}$ is a value that is equal to the ratio of proton's magnetic moment and the nuclear magneton μ_N.

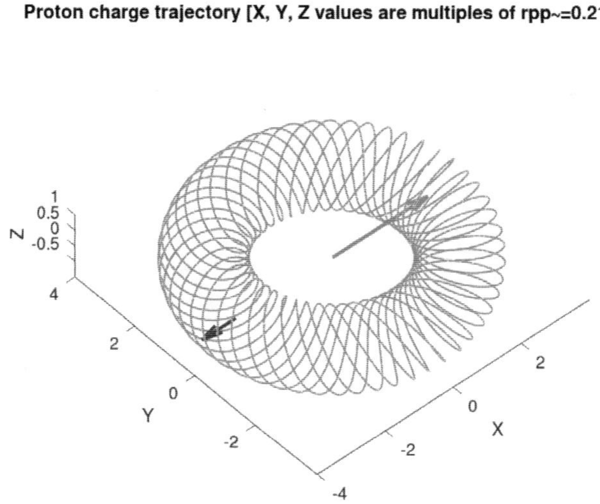

Figure 5.1. An illustration of the toroidal proton geometry. The brown curve is the Zitterbewegung trajectory, the blue arrow represents the poloidal proton radius (r_{pp}), and the purple arrow represents the toroidal proton radius (r_{pt}). The X, Y, Z values are multiples of $r_{pp} \simeq 0.21$ fm.

The toroidal component of a charge displacement equal to one Compton wavelength λ_p is equal to $\lambda_p v_{pt}$. The aspect ratio η implies that the proton charge travels along a path length of 39 λ_p after 5 turns around the torus center:

$$39\lambda_p v_{pt} = 5 \cdot 2\pi r_{pt}$$

$$39 \cdot 2\pi r_{pp} v_{pt} = 5 \cdot 2\pi r_{pt}$$

$$\frac{r_{pp}}{r_{pt}} v_{pt} = \frac{5}{39}$$

$$v_{pt} = \frac{r_{pp}}{r_{pt}}$$

$$v_{pt}^2 = \frac{5}{39}$$

$$v_{pt} = \eta^{-1} = \sqrt{\frac{5}{39}}$$

5.4. Proton charge and Lorentz force

The proton's charge has a mechanical momentum $m_p\boldsymbol{c}$ equal to the product of the elementary charge and its vector potential:

$$e\boldsymbol{A}_p = m_p\boldsymbol{c}$$

The proton charge is subjected to the magnetic Lorentz force:

$$\boldsymbol{F}_p = e\boldsymbol{c} \times \boldsymbol{B}_p = e\frac{d\boldsymbol{A}_p}{dt} = m_p\frac{d\boldsymbol{c}}{dt}$$

$$F_p = m_p\frac{c^2}{r_{pp}} = \frac{m_p}{r_{pp}} = r_{pp}^{-2}$$

The force vector $\boldsymbol{F}_p$ has a component F_{pt} that is always directed towards the torus' center

$$\boldsymbol{F}_{pt} = e\frac{d\boldsymbol{A}_{pt}}{dt} = eA_p\frac{d\boldsymbol{v}_{pt}}{dt} = m_p\frac{d\boldsymbol{v}_{pt}}{dt}$$

$$F_{pt} = m_p\frac{v_{pt}^2}{r_{pt}} = \frac{v_{pt}^2}{r_{pp}r_{pt}} = \frac{r_{pp}^2}{r_{pt}^3 r_{pp}} = \frac{r_{pp}}{r_{pt}^3}$$

The magnetic flux density $\boldsymbol{B}_p$ seen by the proton charge is one half the averaged valued of the magnetic flux density obtained dividing the magnetic flux $\Phi_M = {}^{2\pi}/_{e}$ by an area equivalent to the toroid cross section πr_{pp}^2

$$B_p = \frac{1}{2}\frac{\Phi_M}{\pi r_{pp}^2} = \frac{2\pi}{e\pi r_{pp}^2} = \frac{1}{er_{pp}^2}$$

5.5. The "proton charge radius" interpretation

Observing the proton on time scales much larger than the time required for a complete Zitterbewegung turn around the torus center, which is $T_{toroidal} = {}^{2\pi r_{pt}}/_{v_{pt}}\ (\simeq 3.44 \cdot 10^{-23}s)$, the charge appears confined inside a radius R_p equal to the sum of the two torus radii, r_{pt} and r_{pp}, and the proton's spherical charge radius $r_{cp} = \alpha r_{pp}$:

$$r_{pmax} = r_{pt} + r_{pp} + r_{cp} \simeq 0.7992 \cdot 10^{-15}m$$

.

Recent measurements of the "proton radius", which are listed in tables 1 - 2, average out to the following value: $r_{mean} \simeq (0.839 \pm 0.007) \cdot 10^{-15}m$. This value is 30% larger than the r_{pt} value calculated in sections 5.2-5.5.

Remembering however that r_{mean} is defined as the mean radius value of the charge distribution, this value, for the toroidal model, should be computed by the Root Mean Square distance r_{prms} of the charge from the toroid center in a time $T \gg \frac{2\pi}{\omega_{pt}}$ and a path length $L_p = cT$

$$r_{prms} = \sqrt{\frac{1}{L_p}\int_0^{L_p} r^2 dl} \simeq 0.6239 \cdot 10^{-15}m$$

This value is about 26% smaller than the r_{mean} experimental value.

An other possibility is to compare the r_{prms} toroidal radius with the $r_{phc} \simeq 0.627 \cdot 10^{-15}m$ radius, within which half of the proton charge is contained, and that is shown in figure 2.2. This value, for the proton model, coincides with its r_{prms} radius:

$$r_{phc} = r_{prms}$$

In this case, the above proton radius calculation deviates less than 0.5% from the experimental one 2.2.

6. Proton spin and gyromagnetic factor

6.1. The proposed electromagnetic model based interpretation of proton spin

Analogously to the electron case, the absolute value L_p of the proton's toroidal angular momentum $\boldsymbol{L}_p$ is equal to the reduced Planck constant[5]:

$$L_p = \hbar$$

In the presence of an external magnetic field $\boldsymbol{B}_E$, we can write the vector L_p as the sum of two vectors. One vector is parallel to B_E, and the second one is orthogonal to it:

$$\boldsymbol{L}_p = \boldsymbol{L}_{p||} + \boldsymbol{L}_{p\perp}$$

$$L_{p||} = L_p \cos(\theta)$$

$$L_{p\perp} = L_p \sin(\theta)$$

where θ is the angle between the vectors $\boldsymbol{\mu}_p$ and $\boldsymbol{B}_E$. The proton is therefore subjected to a torque $\boldsymbol{\tau}$:

$$\tau = \left|\boldsymbol{\mu}_p \times \boldsymbol{B}_E\right| = \mu_p B_E \sin(\theta)$$

Consequently, the proton's toroidal structure will be in a Larmor precession, with angular frequency ω_{pP}:

$$\tau = \left|\frac{d\boldsymbol{L}_p}{dt}\right| = L_p \sin(\theta)\frac{d\phi}{dt} = L_p \sin(\theta)\,\omega_{pp}. \tag{6.1}$$

$$\mu_p B_E = \hbar\omega_{pp}$$

What we call "proton spin" s_p is the measured component of its angular momentum vector L_p along the external magnetic field B_E :

$$s_p = \hbar \cos(\theta)$$

Figure 6.1 illustrates the precessing proton structure under an external magnetic field $\boldsymbol{B}_E$.

The measurable angular momentum transitions are also universally quantized to $\hbar$ value. This implies the following for the quantization of the angle θ :

$$\triangle L_{p||} = \pm\hbar \Longrightarrow \theta \in \left\{\frac{\pi}{3}, \frac{2\pi}{3}\right\},\ \cos(\theta) = \pm\frac{1}{2}$$

$$s_p = \pm\frac{\hbar}{2}$$

[5] Under the approach of sections 5.2-5.5, the toroidal angular momentum is $\hbar$ according to the $\boldsymbol{L}_p = \boldsymbol{r}_{pt} \times m_p\boldsymbol{v}_{pt}$ formula, where m_p is the full proton mass. Under the approach of appendix 1, the toroidal angular momentum is $\hbar$ according to the $\boldsymbol{L}_p = \boldsymbol{r}_{pt} \times m_{pt}\boldsymbol{v}_{pt}$ formula, where m_{pt} is the mass component in the toroidal direction that corresponds to the toroidal current loop. See the explanation of appendix 1 about splitting the proton mass into the toroidal and poloidal current loop components.

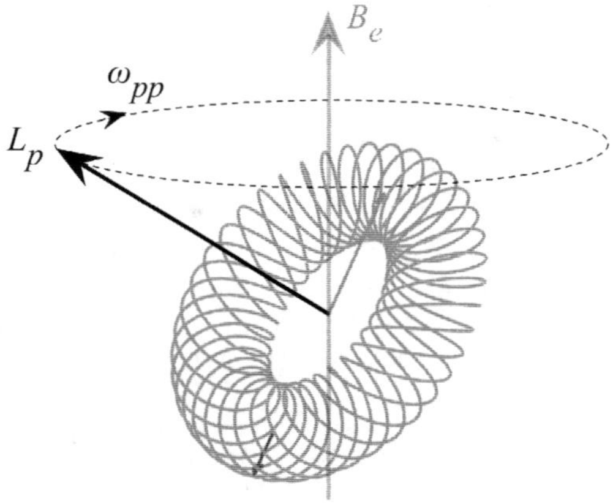

Figure 6.1. The Larmor precessing proton in external magnetic field $\boldsymbol{B}_E$. The proton's toroidal angular momentum vector $\boldsymbol{L}_p$ precesses with angular frequency ω_{pp}.

The two spin values are characterized by two different energy levels, E_L and E_H;

$$\theta = \frac{\pi}{3} \Longrightarrow E_L = -\hbar\omega_{pp}$$

$$\theta = \frac{2\pi}{3} \Longrightarrow E_H = \hbar\omega_{pp}$$

$$\triangle E = E_H - E_L = 2\hbar\omega_{pp}$$

This energy gap $\triangle E$ is equal to $\hbar\omega_{NMRp}$, where ω_{NMRp} is the proton's NMR angular frequency. Therefore:

$$\omega_{NMRp} = 2\omega_{pp} = 2\mu_p B_E = \frac{e}{m_p}\frac{\mu_p}{\mu_N} B_E \tag{6.2}$$

This linear relationship between the applied B_E magnetic field and the resulting $\triangle E$ energy gap is the basis of NMR technology.

The value of ω_{NMRp} can be written as a function of the gyromagnetic factor g_p:

$$\omega_{NMRp} = \frac{e}{2m_p} g_p B_E \tag{6.3}$$

The proton's gyromagnetic-factor g_p is therefore

$$g_p = 2\frac{\mu_p}{\mu_N} = 5.585696018$$

Our calculation precisely matches the CODATA value of the proton's experimentally measured gyromagnetic factor, which is 5.5856946893.

We note that all of the above applies completely analogously to the electron, whose angular momentum value is also $\hbar$. The exact same Larmor precession arises when the electron is placed under an external magnetic field $\boldsymbol{B}_E$, and thus the measured value of its angular momentum becomes $s_e = \pm\frac{\hbar}{2}$ [22]. This phenomenon is the basis of Electron Spin Resonance (ESR) technology: the measured energy gap $\triangle E$ is then equal to $\hbar\omega_{ESR}$, where $\omega_{ESR} = 2\mu_B B_E$ and μ_B is the Bohr magneton.

6.2. The quark model based interpretation of proton spin
When Otto Stern measured the proton's $\mu_p = 2.793\mu_N$ magnetic moment in 1933, most physicists assumed that this measured value is the absolute value of the proton's internal magnetic moment vector. The quark model based magnetic moment calculations were developed under this assumption. With the recent advent of NMR technology, the operators of NMR equipment have recognized that under applied magnetic field the proton is subjected to Larmor precession. However, quark proponents never revised their calculations, which do not consider quarks being in Larmor precession. Any Larmor precession implies that the absolute value of individual quarks' angular momentum vector must be larger than the $\frac{\hbar}{2}$ value assumed in those calculations. Therefore, the quark model based spin interpretation is fundamentally contradicted by the NMR technology.

The Callan-Gross relation, which was mentioned in section 1, implies that a quark's spin is individually measurable, in principle. Yet proton spin measurements always yield $\mu_p = 2.793\mu_N$. One may wonder why the hypothetical quarks' magnetic moments always add up to the same value of $2.793\mu_N$. To explain this constant value, the quark model based proton spin interpretation also requires that the three valence quarks remain in isotropic spin entanglement, which means that their individual spin measurements are always correlated. Specifically, n particles are said to be isotropically spin-correlated, if a measurement made in an *arbitrary* direction θ on *one* of the particles allows us to predict with certainty the spin value of each of the other $n-1$ particles for the same direction θ. Such spin-correlation is required to maintain a constant value of the measured proton magnetic moment. As mentioned in section 1, the color charge hypothesis was introduced to remove the contradiction between the 3-quark spin correlation and the Pauli exclusion principle. However, Paul O'Hara recently proved that Greenberg's postulate does not remove the contradiction with the Pauli exclusion principle: the isotropic spin entanglement of three particles is a mathematical impossibility if their spins are individually measurable. This mathematical contradiction holds regardless of the presence or absence of color charges. Paul O'Hara's proof can be found in the second appendix.

In summary, the quark model based proton spin interpretation involves two fundamental contradictions. Each of these contradictions invalidates the quark model.

7. Conclusions
We have presented a proton model that describes the physical origin of numerous proton parameters, such as its mass, its spin, charge radius, magnetic moment. Despite our model's simplicity, our calculations are in a relatively fair agreement with experimental values. The proton's spherical electric charge generates a Zitterbewegung current over a toroidal surface.

A consistent application of Maxwell's equation thus lead to the discovery of the proposed proton model. The strong similarities with the electron model suggest a universal applicability of fundamental physical laws. Both the electron and the proton comprise an electromagnetic wave, whose formulation can be derived by solving Maxwell's equation. These solutions must not neglect the effects of general relativity, as demonstrated in reference [22] and in our present work. Our proton mass calculation demonstrates that Maxwell's equation remains valid at least down to 10^{-18} m, which is the length scale of the proton's spherical charge radius.

Based on our results, the proton may regain its elementary particle status. The main difference between an electron and a proton is the topology of their Zitterbewegung: a circular Zitterbewegung current in the electron case and a toroidal Zitterbewegung current in the proton case. It remains to be understood why only these two topologies lead to a stable particle.

Acknowledgements
The authors thank Paul O'Hara for insightful discussions of the Pauli exclusion principle, William Stubbs for insightful discussions of high-energy electron-proton scattering data and Giuliano

Bettini for some essential suggestions.

Appendix 1: An alternative approach to the proton geometry calculation[6]

Proton geometry

We calculate the proton radii by applying the toroidal Zitterbewegung model, which was introduced in section 5.1 and is illustrated in figure 5.1.

The total electric energy of the proton is calculated by integrating the energy density of its electric field down to its spherical charge radius:

$$W_e = \frac{e^2}{32\pi^2\epsilon_0}\int_{r_{cp}}^{\infty}\frac{1}{r^4}\cdot 4\pi r^2 dr = \frac{e^2}{8\pi\epsilon_0}\int_{r_{cp}}^{\infty}\frac{1}{r^2}dr = -\frac{e^2}{8\pi\epsilon_0}\left.\frac{1}{r}\right|_{r_{cp}}^{\infty} = \frac{e^2}{8\pi\epsilon_0 r_{cp}}$$

In accordance with Maxwell's equation, the electric an magnetic fields induce each other within the proton. Therefore its electric and magnetic energies must be equal: $W_e = W_m$. From the 938.272 MeV proton mass value we get $W_e = W_m$ =469.136 MeV. We can now calculate the proton's spherical charge radius r_{cp}:

$$r_{cp} = \frac{e^2}{8\pi\epsilon_0 W_e} = 1.5347 \cdot 10^{-18}\ \text{m}$$

This calculated r_{cp} value is remarkably similar to the experimental value discussed in section 2.1.

To transform the circular Zitterbewegung model of the positron [22] into a toroidal geometry, the naive approach is to view the positron from rotating reference frame. Such reference frame transformation must take into account the relativistic Thomas precession effect which arises in a rotating reference frame. This effect reduces the apparent lab-frame speed of a circularly orbiting object in proportion to its Lorentz boost factor:

$$\beta_{lab} = \frac{\beta'}{\gamma_{lab}}$$

where $c\beta'$ is the true rotation speed, $c\beta_{lab}$ is the apparent rotation speed in the lab frame, and $\gamma_{lab} = \sqrt{1-\beta_{lab}^2}^{\,-1}$. When $\beta_{lab} = \frac{1}{\sqrt{2}}$, we get $\beta' = 1$: this limiting value corresponds to the true rotation speed being the speed of light. This result means that in the $\beta_{lab} < \frac{1}{\sqrt{2}}$ regime the toroidal charge distribution is in fact a rotating scaled positron because we can make a rotational change of reference frame which transforms the charge current back to the positron's ring shaped Zitterbewegung. The rotating scaled positron will eventually loose its rotational energy by interacting with other particles, and will thus transform back into an ordinary scaled positron. Therefore, the $\beta_{lab} < \frac{1}{\sqrt{2}}$ regime does not correspond to a stable proton particle. On the other hand, the limiting $\beta_{lab} = \frac{1}{\sqrt{2}}$ value stays invariant under any rotational reference frame transformation, and therefore it corresponds to a truly toroidal charge current, which retains the same geometry in any reference frame. Since the proton retains its basic properties in all reference frames, this $\beta_{lab} = \frac{1}{\sqrt{2}}$ value corresponds to its toroidal Zitterbewegung speed: $v_t = \frac{c}{\sqrt{2}}$.

It follows from Maxwell's equation that electromagnetic waves propagate at the speed of light, which means that the spherical charge moves at a Zitterbewegung speed which is always the speed of light [22]. In the toroidal geometry, the Zitterbewegung speed vector comprises toroidal and poloidal components, which are perpendicular to each other:

$$v_t^2 + v_p^2 = c^2$$

6 This approach is the perspective of Andras Kovacs

Since we already know the toroidal Zitterbewegung speed, we can calculate the poloidal one as well from he above relationship: $v_p = \frac{c}{\sqrt{2}}$.

The proton's toroidal and poloidal radii
The electron's and positron's circular Zitterbewegung structure is discussed in reference [22]. In order to determine the proton's toroidal and poloidal radii, we must briefly review the positron's magnetic energy and magnetic flux calculation. The positron's canonical momentum, generated by the vector potential A at its spherical charge surface, is $p = eA$. The corresponding positron angular momentum is $\Omega = eAr_{ZBW}$, where r_{ZBW} is the circular Zitterbewegung radius. The electron's and positron's Zitterbewegung radius is experimentally determined by Thomson scattering: $r_{ZBW} = 0.3861592676 \cdot 10^{-12}\ m$. This r_{ZBW} value is referred to in the scientific literature as the reduced Compton radius.

By setting $\Omega = \hbar$, we obtain the norm of the vector potential at the positron's spherical charge surface:

$$A = \frac{\hbar}{er_{ZBW}}.$$

Once we know the vector potential, it is possible to determine the magnetic flux produced by the rotating elementary charge by applying the circulation of the vector potential A:

$$\phi = \oint_\lambda A d\lambda = \int_0^{2\pi} \frac{\hbar}{er_{ZBW}} r_{ZBW} d\vartheta = 2\pi\frac{\hbar}{e} = \frac{h}{e} \approx 4.135\,667 \cdot 10^{-15}\ \mathrm{V \cdot s}$$

i.e. the magnetic flux crossing the Zitterbewegung loop is quantized. Now it is possible to calculate the magnetic energy stored in the positron current loop:

$$W_m = \frac{1}{2}\phi I_{positron} = \frac{1}{2} \cdot 2\pi\frac{\hbar}{e} \cdot \frac{ec}{2\pi r_{ZBW}} = \frac{\hbar c}{2r_{ZBW}} \approx 255.5\ \mathrm{keV}$$

which is equal to half the positron rest energy, thereby satisfying the $W_e = W_m$ requirement of an electromagnetic wave. This result demonstrates the correctness of setting the intrinsic angular momentum value to $\hbar$. We note that although the above equation refers to a static current loop, the result stays the same in the case of a circulating elementary charge. To see this, we evaluate the current interaction part of the electromagnetic Lagrangian density:

$$\mathscr{L}_{int} = \boldsymbol{J}_\triangle \cdot \boldsymbol{A}_\triangle = JA = \frac{I_{positron}}{\pi r^2_{charge}} \cdot \frac{\hbar}{er_{ZBW}} \approx 1.352\,604 \cdot 10^{27}\ \mathrm{J \cdot m^{-3}}$$

By integration over the volume described by the spherical charge trajectory, it is possible to recompute the positron's energy:

$$W_{positron} = \iiint_V JA\, dV = \frac{I_{positron}}{\pi r^2_{charge}} \cdot \frac{\hbar}{er_{ZBW}} \cdot 2\pi^2 r_{ZBW} r^2_{charge} = \phi I_{positron} \approx 511\ \mathrm{keV}$$

Considering the above expression, we can take the toroidal volume, and divide it into two halves. The spherical positron charge is in one of those halves, and thus the integration volume becomes half of the toroid volume, while the effective current between the integration endpoints is now twice as large. The integration result for $W_{positron}$ remains invariant. By repeating this halving of the toroidal volume segments, we see that the total magnetic energy remains invariant as we approach the circulating spherical charge scenario.

The recognition that the magnetic flux of a Zitterbewegung loop is quantized to $\frac{h}{e}$ is a central result of [22]. We also show that this magnetic flux quantization is equivalent to the electric charge quantization. Since the proton's charge is the elementary charge e, the $\frac{h}{e}$ magnetic flux quantization must hold true for the proton.

How is the proton's W_m =469.136 MeV magnetic field energy divided between its toroidal and poloidal current loops? The proton's magnetic moment measurement is in fact its toroidal magnetic moment measurement. For an elementary particle, its measured magnetic moment is given by the $\mu = \frac{e\hbar}{2m}$ formula, where the only non-constant factor is the particle mass. Since the proton mass is derived from electromagnetic induction, its excess toroidal magnetic moment is inversely proportional to its toroidal magnetic mass. Therefore the proton's toroidal magnetic energy is:

$$W_{mt} = W_m / \left(\frac{\mu_p}{\mu_N}\right) = \frac{469.136}{2.79285}\,\text{MeV} = 167.978\,\text{MeV}$$

The remaining poloidal magnetic field energy is:

$$W_{mp} = W_m - W_{mt} = 301.158\,\text{MeV}$$

Under the toroidal proton geometry there are two Zitterbewegung loops: a toroidal loop and a poloidal loop. Since both current loops are generated by the elementary charge e, the $\frac{h}{e}$ magnetic flux quantization holds for each current loop. Thus we can calculate the magnetic energy values by applying the $\frac{h}{e}$ magnetic flux quantization condition:

$$W_{mt} = \frac{1}{2}\phi I_{toroidal} = \frac{1}{2}\cdot 2\pi\frac{\hbar}{e}\cdot\frac{ev_t}{2\pi r_{pt}} = \frac{\hbar v_t}{2r_{pt}}$$

$$W_{mp} = \frac{1}{2}\phi I_{poloidal} = \frac{1}{2}\cdot 2\pi\frac{\hbar}{e}\cdot\frac{ev_p}{2\pi r_{pp}} = \frac{\hbar v_p}{2r_{pp}}$$

We thus evaluate the toroidal and poloidal Zitterbewegung radii from the above equations:

$$r_{pt} = \frac{\hbar v_t}{2W_{mt}} = 0.831\,\text{fm}$$

$$r_{pp} = \frac{\hbar v_p}{2W_{mt}} = 0.463\,\text{fm}$$

The obtained 0.831 fm toroidal radius value has a 99% match with the experimentally measured $r_{mean} = 0.839 \pm 0.007$ fm proton charge radius value.

To validate the consistency of our model, we check that the $\mu_p = I_{toroidal}\pi r_{pt}^2$ magnetic moment formula is fulfilled for the toroidal current loop. The above-discussed scaled positron model corresponds to the nuclear magneton:

$$\mu_N = I\pi r_{ZBW}^2 = \frac{ec}{2\pi r_{ZBW}}\pi r_{ZBW}^2 = \frac{ec}{2}r_{ZBW}$$

where $r_{ZBW} = r_{e+}\frac{m_{e+}}{m_p}$=0.2103 fm is the scaled positron's Zitterbewegung radius, while the positron's Zitterbewegung radius and mass values r_{e+}=386.16 fm and m_{e+}=511 keV. We now evaluate the proton's magnetic moment according to the current loop formula:

$$\mu_p = I_{toroidal}\pi r_{pt}^2 = \frac{e\left(c/\sqrt{2}\right)}{2\pi r_{pt}}\pi r_{pt}^2 = \frac{ec}{2\sqrt{2}}r_{pt}$$

$$\mu_p/\mu_N = \left(cr_{pt}/\sqrt{2}\right)/\left(cr_{ZBW}\right) = \frac{1}{\sqrt{2}}\frac{0.831}{0.2103} = 2.794$$

which precisely matches with the experimental μ_p/μ_N ratio. Comparing the above magnetic moment equation with the magnetic energy based r_{pt} calculation, one arrives at the same $\mu_p/\mu_N = r_{pt}/\left(\sqrt{2}r_{ZBW}\right)$ formulation in either case: i.e. these are two ways of expressing the same physics.

The toroidal proton geometry implies that the circulating proton charge is radially distributed between $r_{pt} - r_{pp} - r_{cp}$ and $r_{pt} + r_{pp} + r_{cp}$ distance from its center, i.e. its charge reaches up to 1.296 fm radial distance. This 1.296 fm radius precisely matches the electric polarizability measurements based radius value, that was introduced in section 2.4. Indeed, it is logical to interpret the electric polarizability based radius as the proton charge's furthest distance from its center because one can polarize a charge distribution only within that range where it is physically present.

In summary, we calculated the proton's r_{cp}, r_{pt}, and r_{pp} radii without any parameter fittings, and found that each of them matches well with experimental data.

Appendix 2: Isotropic spin entanglement[7]

By definition, n particles are said to be isotropically spin-correlated (ISC), if a measurement made in an *arbitrary* direction on *one* of the particles allows us to predict with certainty the spin value of each of the other $n-1$ particles for the same direction.

Essentially, to show that ISC states exist only for $n = 2$, it is sufficient to prove that it is impossible to have three such particles. The impossibility of three ISC particles also excludes the possibility of $n \geq 3$ ISC particles.

Suppose that an ISC state exists for $n = 3$. We demonstrate in the following paragraphs that this assumption leads to a mathematical contradiction.

In the interest of clarity, assume without loss of generality that the three ISC particles are such that they are detected to be in $(+,+,+)$ correlation for an arbitrary measurement direction. Later we will generalize the proof to any other correlation type. Define the x axis along this arbitrary direction, and define the z axis in any orthogonal direction to x. We will perform further spin measurements in the $x-z$ plane. Spin measurements in orthogonal directions are statistically independent. Although we know a given particle spin to be $|+\rangle$ along the x axis, a subsequent spin measurement along the z axis of the apparatus gives $\frac{1}{2}$ probability of measuring $|-\rangle$ state. In general, a spin state in direction 2θ with respect to the x axis, given that it is in the state $|+\rangle$ with respect to the x axis, can be constructed from the rotation R and is given by $R|+\rangle = \cos\theta\,|+\rangle - \sin\theta\,|-\rangle$. Therefore, in direction 2θ the probability of measuring $|+\rangle$ state is $\cos^2\theta$ and of measuring $|-\rangle$ is $\sin^2\theta$. Taking the $(x, 2\theta)$ direction with respect to two spin correlated particles, the joint probabilities are $P(+,+) = \frac{1}{2}\cos^2\theta$ and $P(+,-) = \frac{1}{2}\sin^2\theta$. Similarly, for the ket $|-\rangle$, $R|-\rangle = \sin\theta\,|+\rangle + \cos\theta\,|-\rangle$ and the joint probabilities are $P(-,-) = \frac{1}{2}\cos^2\theta$ and $P(-,+) = \frac{1}{2}\sin^2\theta$. In principle, if three ISC particles exist, a sequence of spin correlated measurements in the directions $2\theta_1$, $2\theta_2$, $2\theta_3$ can be performed on the three entangled particles. Let $(s_1(\theta_1), s_2(\theta_2), s_3(\theta_3))$ represent each particle's observed spin values in the three different directions. Recall that the above stated spin correlation implies that if any particle is measured to be in the $s_i(\theta_i) = |+\rangle$ spin state, the probability of measuring an other particle in the $s_j(\theta_j) = |-\rangle$ spin state becomes $\frac{1}{2}\sin^2(\theta_j - \theta_i)$.

Given that $s_n(\theta_n) = |\pm\rangle$ for each n, there exists only two possible values for each measurement, which we associate with "spin-up" and "spin-down" respectively. Hence, for three measurements there are a total of 8 possibilities. In particular,

$$\{(+,+,-),(+,-,-)\} \subset \{(+,+,-),(+,-,-),(-,+,-),(+,-,+)\}$$

implies the following probability relationship:

[7] This appendix is authored by Paul O'Hara

$$P\{(+,+,-),(+,-,-)\} \leq P\{(+,+,-),(+,-,-),(-,+,-),(+,-,+)\}$$

Consider the meaning of various subsets in the above inequality:

- The $\{(+,+,-),(+,-,-)\}$ subset is interpreted as follows: we measured the spin of particle 1 to be in $|+\rangle$ state and particle 3 to be in $|-\rangle$ state. The corresponding probability is $\frac{1}{2}\sin^2(\theta_3-\theta_1)$.
- The $\{(+,+,-),(-,+,-)\}$ subset is interpreted as follows: we measured the spin of particle 2 to be in $|+\rangle$ state and particle 3 to be in $|-\rangle$ state. The corresponding probability is $\frac{1}{2}\sin^2(\theta_3-\theta_2)$.
- The $\{(+,-,-),(+,-,+)\}$ subset is interpreted as follows: we measured the spin of particle 1 to be in $|+\rangle$ state and particle 2 to be in $|-\rangle$ state. The corresponding probability is $\frac{1}{2}\sin^2(\theta_2-\theta_1)$.

Substituting the above terms into the above inequality, we arrive at

$$\frac{1}{2}\sin^2(\theta_3-\theta_1) \leq \frac{1}{2}\sin^2(\theta_3-\theta_2) + \frac{1}{2}\sin^2(\theta_2-\theta_1)$$

which is Eugene Wigner's interpretation of Bell's inequality. Taking $\theta_3-\theta_2=\theta_2-\theta_1=\frac{\pi}{6}$ and $\theta_3-\theta_1=\frac{\pi}{3}$ gives $\frac{1}{2}\geq\frac{3}{4}$, which is a contradiction. Therefore, three particles cannot all be in the same spin state with probability 1.

Remark: The proof of the above theorem was worked out for $(+,+,+)$ or $(-,-,-)$ type spin correlation. To generalize the proof, suppose that the ISC particles are measured to be $(+,-,+)$ along an arbitrary measurement direction. Then the spin outcomes in the three different directions θ_1, θ_2, θ_3 can be written as:

$$\{(+,-,-),(+,+,-)\} \subset \{(+,-,-),(+,+,-),(-,-,-),(+,+,+)\}$$

Essentially, this means that we flipped the $|+\rangle$ to $|-\rangle$ to represent the state of particle 2. Applying the same probability argument as before, but noting that $P\{(+,-,-),(-,-,-)\} = \frac{1}{2}\cos^2(\theta_3-\theta_2)$, the inequality becomes

$$\frac{1}{2}\sin^2(\theta_3-\theta_1) \leq \frac{1}{2}\cos^2(\theta_3-\theta_2) + \frac{1}{2}\cos^2(\theta_2-\theta_1)$$

Then upon taking $\theta_3-\theta_2=\theta_2-\theta_1=\frac{\pi}{2}-\frac{\pi}{6}$ and $\theta_3-\theta_1=\pi-\frac{\pi}{3}$ gives as before $\frac{1}{2}\geq\frac{3}{4}$, which is a contradiction.

Appendix 3: Physical constants in Natural Units

Conversion constants for natural units:

$1.9732898\cdot10^{-7}\ m \simeq 1eV^{-1}$ length
$6.5821220\cdot10^{-16}\ s \simeq 1eV^{-1}$ time
$2.99792458\cdot10^{8}\ ms^{-1} = 1$ speed
$1.5192669\cdot10^{15}\ Hz \simeq 1eV$ frequency
$8.1193997\cdot10^{-13}\ N \simeq 1eV^{2}$ force
$1.8755460\cdot10^{-18}\ C = 1$ charge

Relevant physical constants in natural units:

$h = 2\pi$ Planck's constant $(6.62607015 \cdot 10^{-34} JHz^{-1})$
$\hbar = h/2\pi = 1$ reduced Planck's constant
$\varepsilon_0 = \frac{1}{4\pi}$ vacuum permittivity
$\mu_0 = 4\pi$ vacuum magnetic permeability
$c = 1$ light speed in vacuum $(2.99792458 \cdot 10^8 ms^{-1})$
$\alpha^{-1} \simeq 137.036$ inverse of the fine structure constant
$e = \pm\sqrt{\alpha} \simeq 0.085424546$ $(1.602176634 \cdot 10^{-19} C)$ elementary charge
$\mu_N \simeq 4.552225759 \cdot 10^{-11}\, eV^{-1}$ $(5.0507837461 \cdot 10^{-27} JT^{-1})$ nuclear magneton
$\mu_{pC} \simeq 1.271367397 \cdot 10^{-10}\, eV^{-1}$ $(1.41060679736 \cdot 10^{-26} JT^{-1})$ proton magnetic moment
$\frac{\mu_p}{\mu_N} \simeq 2.79284734463$ CODATA proton magnetic moment to nuclear magneton ratio
$m_p \simeq 0.93827208816 \cdot 10^9\, eV$ proton mass
$\lambda_p \simeq 6.696549362 \cdot 10^{-9}\, eV^{-1}$ $(1.32140985539 \cdot 10^{-15} m)$ proton Compton wavelength
$A_e \simeq 5.981875085 \cdot 10^6 eV$ norm of the vector potential of the electron charge
$V_e = A_e$ electric potential at surface of the electron's charge
$\Phi_M = \frac{h}{e} = 2\pi\alpha^{-1/2} \simeq 73.55246020$ elementary charge's magnetic flux
$m_e = \omega_e \simeq 0.51099895 \cdot 10^6 eV$ electron rest mass
$\omega_e = m_e$ electron's charge angular speed
$T_e = \frac{2\pi}{\omega_e}$ electron Zitterbewegung period
$r_e = \omega_e^{-1} \simeq 1.956951198 \cdot 10^{-6} eV^{-1}$ $(0.3861592676 \cdot 10^{-12} m)$ electron radius
$r_{ce} = \alpha r_e$ electron charge radius
$R_{p,exp} \simeq 4.264 \cdot 10^{-9}\, eV^{-1}$ $(0.8414 \cdot 10^{-15} m)$ "proton charge radius" CODATA value

Appendix 4: Proton model parameters

parameter set 1 (section 5):

$r_{pp} = m_p^{-1} = \frac{\lambda_p}{2\pi} \simeq 1.06578893 \cdot 10^{-9}\, eV^{-1}$ $(0.2103089103 \cdot 10^{-15} m)$ proton torus minor radius
$r_{pt} = \eta r_{pp} \simeq 2.976586476 \cdot 10^{-9}\, eV^{-1}$ $(0.5873608214 \cdot 10^{-15} m)$ proton torus major (toroidal) radius
$r_{cp} = \alpha r_{pp} \simeq 7.777437549 \cdot 10^{-12}\, eV^{-1}$ $(1.534698267 \cdot 10^{-18} m)$ proton spherical charge radius
$r_{prms} \simeq 3.16 \cdot 10^{-9}\, eV^{-1}$ $(0.624 \cdot 10^{-15} m)$ proton charge Root Mean Square radius
$r_{phc} = r_{prms}$ radius that contains half of the proton charge
$\eta = \frac{r_{pt}}{r_{pp}} = \sqrt{\frac{39}{5}} \simeq 2.792848009$ proton torus aspect ratio
$v_{pt} = \eta^{-1} c$ toroidal component of the charge speed c
$A_p = \frac{m_p}{e} \simeq 1.098363601 \cdot 10^{10}\, eV$ the absolute value of the vector potential at the proton charge
A_{pt} toroidal component of the vector potential A_p
$L_p = eA_{pt} r_{pt} = \hbar = 1$ the proton's toroidal angular momentum
$eA_{pt} r_{pt} \cos(\vartheta) = \pm 1/2\hbar$ $(\vartheta \in \{\pi/3, 2\pi/3\})$ measured proton spin
$\mu_p = \frac{r_{pt}}{2} e \simeq 2.792848009\, \mu_N$ proton model magnetic moment
$\Phi_p = \frac{h}{e} \simeq 73.55246020$ proton model magnetic flux quantum

parameter set 2 (appendix 1):

$r_{pp} \simeq 2.3465 \cdot 10^{-9}\, eV^{-1}$ $(0.463 \cdot 10^{-15} m)$ proton torus minor (poloidal) radius
$r_{pt} \simeq 4.2113 \cdot 10^{-9}\, eV^{-1}$ $(0.831 \cdot 10^{-15} m)$ proton torus major (toroidal) radius
$r_{cp} \simeq 7.777437549 \cdot 10^{-12}\, eV^{-1}$ $(1.534698267 \cdot 10^{-18} m)$ proton spherical charge radius
$v_{pt} = \frac{c}{\sqrt{2}}$ toroidal component of the charge speed c
$L_p = \hbar = 1$ the proton's toroidal angular momentum
$\hbar \cos(\vartheta) = \pm 1/2\hbar$ $(\vartheta \in \{\pi/3, 2\pi/3\})$ measured proton spin
$\mu_p \simeq 2.792848009\, \mu_N$ proton model magnetic moment
$\Phi_p = \frac{h}{e} \simeq 73.55246020$ proton model magnetic flux quantum

References

[1] P. Anastasovski, T. Bearden, C. Ciubotariu, W. Coffey, L. Crowell, G. Evans, M. Evans, R. Flower, S. Jeffers, A. Labounsky, B. Lehnert, M. Mészáros, P. Molnar, and S. Roy, *Aharonov-bohm effect as the basis of electromagnetic energy inherent in the vacuum*, Foundations of Physics Letters, 15 (2002), pp. 561–568.
[2] D. Bergman, *The real proton*, Foundations of Science, 3 (2000).
[3] G. Bettini, *Clifford Algebra, 3 and 4-Dimensional Analytic Functions with Applications. Manuscripts of the Last Century.*, viXra.org, Quantum Physics (2011), pp. 1–63. http://vixra.org/abs/1107.0060.
[4] A. Beyer, *The rydberg constant and proton size from atomic hydrogen*, Science, 358 (2017).
[5] N. Bezginov, *A measurement of the atomic hydrogen lamb shift and the proton charge radius*, Science, 365 (2019).
[6] J. D. Bjorken, *Inelastic electron-proton and gamma-proton scattering and the structure of the nucleon*, Physical Review, 185 (1969).
[7] O. Chamberlain, *Example of an antiproton-nucleon annihilation*, Physical Review, 102 (1956).
[8] T. D. N. S. A. Committee, *The frontiers of nuclear science - a long range plan*, (2007).
[9] P. Cote and M. Johnson, *Groupthink and the blunder of the gauges*, arXiv, (2011).
[10] Z. F. Cui, *Fresh extraction of the proton charge radius from electron scattering*, Physical Review Letters, 127 (2021).
[11] A. Di Tommaso and G. Vassallo, *Electron structure, Ultra-Dense Hydrogen and Low Energy Nuclear Reactions*, Journal of Condensed Matter Nuclear Science, 29 (2019), pp. 525–547.
[12] H. Fonvieille, *Virtual compton scattering and nucleon generalized polarizabilities*, Progress in Particle and Nuclear Physics, 113 (2020).
[13] M. Gell-Mann, *A schematic model of baryons and mesons*, Physics Letters, 8 (1964).
[14] K. Griffioen, C. Carlson, and S. Maddox, *Consistency of electron scattering data with a small proton radius*, Physical Review C, 93 (2016).
[15] A. Grinin, *Two-photon frequency comb spectroscopy of atomic hydrogen*, Science, 370 (2020).
[16] F. Halzen, *Quarks and leptons: An introductory course in modern particle physics*, John Wiley and Sons, (1984).
[17] L. N. Hand, *Electric and magnetic form factors of the nucleons*, Reviews of Modern Physics, 35 (1963).
[18] L. Hively and G. Giakos, *Toward a more complete electrodynamic theory*, International Journal of Signal and Imaging Systems Engineering, 5 (2012), pp. 3–10.
[19] L. Hively and A. Loebl, *Classical and extended electrodynamics*, Physics Essays, 32 (2019), pp. 112–126.
[20] R. L. Jaffe, *Where does the proton really get its spin?*, Physics Today, (1995).
[21] O. Keller and L. Hively, *Ohmura's extended electrodynamics: longitudinal aspects in general relativity*, Journal of Physics Communications, 3 (2019), p. 115002.
[22] A. Kovacs, G. Vassallo, P. O'Hara, F. Celani, and A. Di Tommaso, *Unified Field Theory and Occam's Razor*, World Scientific, 04 2022.
[23] A. Kovacs, V. Zatelepin, and D. Baranov, *Signatures of 1.5 MeV leptons in nuclear reactions*, Journal of Physics - Conference Series, (2023).
[24] S. E. Kuhn, *Nucleon structure functions: Experiments and models*, Proceedings of the 12th Annual HUGS at CEBAF, 231 (1998).
[25] J. Kuti, *Inelastic lepton-nucleon scattering and lepton pair production in the relativistic quark-parton model*, Physical Review D, 4 (1971).
[26] R. Li, *Measured proton electromagnetic structure deviates from theoretical predictions*, Nature, 611 (2022).
[27] Y. H. Lin, *New insights into the nucleon's electromagnetic structure*, Physical Review Letters, 128 (2022).
[28] A. I. L'vov and V. A. Petrun'kin, *Dispersion theory of proton compton scattering in the first and second resonance regions*, Physical Review C, 55 (1997).
[29] R. L. Mills, J. J. Farrell, and W. R. Good, *Unification of Spacetime, the Forces, Matter, and Energy*, Science Press, Ephrata, PA 17522, 1992.
[30] G. Modanese, *Generalized maxwell equations and charge conservation censorship*, Modern Physics Letters B, 31 (2016).
[31] ——, *Unified Field Mechanics II: Formulations and Empirical Tests - Covariant Formulation of Aharonov-Bohm Electrodynamics and Its Application to Coherent Tunneling*, 2018.
[32] E. Perez, *The quark and gluon structure of the proton*, Reports on Progress in Physics, 76 (2013).
[33] D. Reed, *Unravelling the potentials puzzle and corresponding case for the scalar longitudinal electrodynamic wave*, Journal of Physics: Conference Series, 1251 (2019).
[34] D. Reed and L. Hively, *Implications of gauge-free extended electrodynamics*, Symmetry, 12 (2020).
[35] H. Reiss, *Fundamental formulation of light-matter interactions revisited*, Physical Review A, 100 (2019).
[36] G. Rousseaux, *The gauge non-invariance of classical electromagnetism*, Annales de la Fondation Louis de

Broglie, 30 (2005).
[37] W. L. Stubbs, *The nucleus of atoms: One interpretation*, CreateSpace Independent Publishing Platform, (2018).
[38] V. Tvaskis, *Proton and deuteron f2 structure function at low q2*, Physical Review C, 81 (2010).
[39] K. J. van Vlaenderen, *A generalisation of classical electrodynamics for the prediction of scalar field effects*, arXiv, (2003).
[40] C. S. Wood, *Measurement of parity nonconservation and an anapole moment in cesium*, Science, 275 (1997).
[41] D. A. Woodside, *Three-vector and scalar field identities and uniqueness theorems in euclidean and minkowski spaces*, American Journal of Physics, 77 (2009), pp. 438–446.
[42] W. Xiong, *A high precision measurement of the proton charge radius at jlab*, PhD dissertation, Duke University, (2020).

Towards testing a dark matter candidate that emerges from the scalar ether theory

Mayeul Arminjon

Laboratory "Soils, Solids, Structures, Risks" (CNRS, Université Grenoble-Alpes, Grenoble-INP), Grenoble, France

E-mail: `Mayeul.Arminjon@3sr-grenoble.fr`

Abstract. According to a scalar theory of gravity with a preferred frame, electromagnetism in the presence of a gravitational field implies that there is an additional energy tensor, which might contribute to dark matter. The expression of this tensor is determined by a mere scalar p, that depends on the EM field and (for a weak field) on the Newtonian gravitational field. We briefly recall why this tensor arises and how the EM field in a galaxy can be calculated. The data fields that enter the PDE for the scalar field p oscillate very quickly in space and time, as does the EM field. This prevents integration of that PDE at the relevant galactic scale. Therefore, a homogenization of that PDE has to be operated. We discuss in some detail three possible ways of applying the homogenization theory to that PDE: time, space, or spacetime homogenization. The second and third ways may lead to feasible, albeit heavy calculations.

1. Introduction

Although an unseen, exotic form of matter seems to make up the dominant part in the mass of galaxies and clusters of galaxies, the problem of identifying what this "dark matter" could be made of has been and remains one of the big enigmas in contemporary physics. The reason for the present work is that it turns out [1] that, according to an alternative, relativistic theory of gravity with a preferred frame: In the presence of both a gravitational field and an electromagnetic (EM) field, there must indeed appear some exotic "interaction energy" E_{inter}, which should be distributed in space, and be gravitationally active. This has been found while searching [2, 3, 1] for a consistent formulation, in that theory of gravity, of the Maxwell equations in the presence of gravitation — the aim was thus not to invent a new candidate for dark matter.

That energy E_{inter} could nevertheless contribute to the dark matter. It depends on the EM field $(\mathbf{E}, \mathbf{B})$ and the gravity field. The theory of gravity that we refer to, based on a scalar field only, is called "scalar ether theory" or **SET**; see Ref. [4]. To check if E_{inter} might indeed build a "dark halo" around a galaxy, we should be able to compute the Interstellar Radiation Field (hereafter **ISRF**) as a solution of the Maxwell equations — because the equations that govern E_{inter} [1] contain the Maxwell field $(\mathbf{E}, \mathbf{B})$ and its first-order derivatives. However, the existing models for the ISRF focus on radiation transfer (e.g. [5, 6, 7, 8, 9, 10]) and do not produce an EM field $(\mathbf{E}, \mathbf{B})$, even less one that would be an exact solution of the Maxwell equations. Hence, we built from scratch a model that does produce an exact solution of the Maxwell equations [12]. That model is based on axial symmetry (of the galaxy and the ISRF) as a relevant

Content from this work may be used under the terms of the Creative Commons Attribution 3.0 licence. Any further distribution of this work must maintain attribution to the author(s) and the title of the work, journal citation and DOI.
Published under licence by IOP Publishing Ltd

approximation. It turns out that there exists an explicit representation for all axisymmetric solutions of the source-free Maxwell equations [11]. The model makes predictions for the Spectral Energy Density (SED) of the ISRF, that are close to those of the existing models [13]. Except for one thing: the new model predicts extremely high values of the SED on the galaxy's axis [14].

Our current work, of which this paper is a progress report, is as follows: Using the axisymmetric model of the ISRF in a galaxy as an exact Maxwell EM field, just mentioned, we seek to compute the interaction energy density field E_{inter} in the weak gravitational field of a galaxy. Beyond the EM field, E_{inter} depends also on the Newtonian potential U, more precisely on $\partial_t(\nabla U)$. The time derivative ∂_t is taken in the preferred frame ("ether") $\mathcal{E}$. Hence the field E_{inter} will depend on the velocity $\mathbf{V}$ of the center of the galaxy w.r.t. $\mathcal{E}$.

Before exposing our present work, in the next two sections we shall briefly summarize the foregoing steps [1, 11, 12].

2. Interaction tensor in SET

The equations of electrodynamics on general relativity (GR) simply rewrite those of special relativity (SR) by using the "comma goes to semicolon" rule: $_{,\nu} \rightarrow {}_{;\nu}$, i.e.: partial derivatives are replaced by covariant derivatives. That is not possible in SET, for the Dynamical Equation for the energy(-momentum-stress) tensor $\boldsymbol{T}$ is not generally $T^{\lambda\nu}_{\ \ ;\nu} = 0$ (which rewrites $T^{\lambda\nu}_{\ \ ,\nu} = 0$ valid in SR). In SET, the first group of the Maxwell equations is unchanged. The second group <u>was</u> obtained by applying the Dynamical Equation of SET to a charged medium in the presence of Lorentz force, assuming that (as is the case in GR):

$$\text{Total energy tensor } \boldsymbol{T} = \boldsymbol{T}_{\text{charged medium}} + \boldsymbol{T}_{\text{field}}. \tag{1}$$

The additivity (1) leads to a form of Maxwell's second group in SET [2, 15]. But that form of Maxwell's second group in SET predicts charge production/destruction at untenable rates, and hence has to be discarded [3]. The additivity assumption (1) is contingent and may be abandoned. It means introducing an "interaction" energy tensor $\boldsymbol{T}_{\text{inter}}$ such that

$$\boldsymbol{T}_{(\text{total})} = \boldsymbol{T}_{\text{charged medium}} + \boldsymbol{T}_{\text{field}} \ \underline{+\boldsymbol{T}_{\text{inter}}}\,. \tag{2}$$

In the work [1], it has been found that the assumption that $\boldsymbol{T}_{\text{inter}}$ should be Lorentz-invariant in the situation of special relativity (SR), i.e. when the "physical" spacetime metric $\boldsymbol{\gamma}$ is Minkowski's metric $\boldsymbol{\gamma}^0$ ($\gamma^0_{\mu\nu} = \eta_{\mu\nu}$ in Cartesian coordinates), leads unambiguously to the following definition in the general case:

$$T^{\mu}_{\text{inter}\ \nu} := p\,\delta^{\mu}_{\nu}, \qquad \text{or} \quad (T_{\text{inter}})_{\mu\nu} := p\,\gamma_{\mu\nu}. \tag{3}$$

(Note that this is generally-covariant.) It has also been shown that the additional equation made necessary by the additional unknown: the scalar field p, can then be consistently imposed to be the equation for *charge conservation.* In that way, the electrodynamics of SET is a closed system of PDE's that satisfies charge conservation. The "interaction energy"

$$E_{\text{inter}} := T^{00}_{\text{inter}} = p\gamma^{00} \tag{4}$$

may be regarded as (macroscopic) "dark matter", because (i) it is not localized inside usual matter: we have $p \neq 0$ at a generic point; (ii) it is gravitationally active: T^{00}_{inter} contributes to T^{00} which is the source of the gravitational field in SET; (iii) it is not usual matter (e.g. no velocity can be defined for the medium having $T^{\mu}_{\text{inter}\ \nu}$ as energy tensor, since, when considered in coordinates that are Cartesian at the event considered, the tensor (3) is invariant under local Lorentz transformations).

Moreover, the scalar PDE that determines the first approximation of the scalar field p in a given general EM field and in a given weak gravitational field with Newtonian potential U has been obtained. Its initial form is (Eq. (69) in Ref. [1]): [1]

$$(-G^{\mu\nu}\, p_{,\nu})_{,\mu} = -f, \tag{5}$$

where $G^{\mu\nu}$ (noted ${G_1}^{\mu\nu}$ in Ref. [1]) are the components of an antisymmetric spacetime tensor $\boldsymbol{G}$: the inverse tensor of the EM field tensor made with the components of $(\mathbf{E},\mathbf{B})$ — the EM field of the first approximation, that obeys the flat-spacetime Maxwell equations. In addition, in Eq. (5), we have

$$f := c^{-3}\left(e^i \partial_T U\right)_{,i}\left(1 + O\left(c^{-2}\right)\right), \tag{6}$$

in which e^i $(i = 1, 2, 3)$ are the components of a spatial vector $\mathbf{e}$ also made with the components of $(\mathbf{E},\mathbf{B})$. Using the antisymmetry of $G^{\mu\nu}$, Eq. (5) can be rewritten [1] as:

$$\left(\frac{\mathrm{d}p}{\mathrm{d}t}\right)_{\mathbf{u}} := \frac{\partial p}{\partial t} + \mathbf{u}.\nabla p = S, \tag{7}$$

in which the source field S and the spatial vector field $\mathbf{u}$ are given, both depending on $(\mathbf{E},\mathbf{B})$, and with the source S depending also on $\nabla \partial_T U$. Thus, in principle, p obtains by integrating S along the curves $\frac{\mathrm{d}\mathbf{x}}{\mathrm{d}t} = \mathbf{u}(t,\mathbf{x})$ [1].

3. Maxwell model of the ISRF

To check if E_{inter} might build a "dark halo", we must have the Interstellar Radiation Field in a galaxy (ISRF) as a Maxwell field. In Ref. [12], a model has been proposed to calculate the ISRF as an exact solution of the Maxwell equations. Note that this is of astrophysical interest, independently of the problem of checking what has just been mentioned. See e.g. Ref. [14], Sect. 2, for a short presentation of the model. Here we shall provide only the main elements of that model, as follows:

(i) We assume *axial symmetry.* This is a relevant approximation for many galaxies. The z axis is taken as the symmetry axis.

(ii) We consider the *source-free* Maxwell equations. Indeed, we want to describe the ISRF at a galactic scale, not in the stars (which are the primary source of the ISRF) or in their neighborhood.

(iii) We consider a finite set of frequencies (ω_j) $(j = 1, ..., N_\omega)$, thus a finite number of time-harmonic source-free Maxwell fields. The number of frequencies of a simulation, N_ω, determines how accurately the continuous spectrum is represented for that simulation.

(iv) We use a theorem [11] that expresses any time-harmonic axisymmetric source-free Maxwell field as the sum of two such fields, obtained from two scalar potentials A_z and A'_z.

[1] Equation (5) is valid in coordinates x^μ that are Cartesian for the flat spacetime metric γ^0 of SET and adapted to the preferred frame. Thus the time coordinate is $x^0 = cT$ with T the preferred time of the theory, and the spatial coordinates are Cartesian for the Euclidean spatial metric $\boldsymbol{g}^0$ of SET [1]. The l.h.s. of (5) is invariant under a Lorentz transformation of γ^0, but the r.h.s. is so only if that transformation is internal to the preferred frame.

Moreover, in the relevant "totally-propagating" case, each of these potentials depends only on a "spectrum" function of one variable, $S = S(k)$ with k the axial wave number. In addition, for a galaxy, the logic of the model (points (v) and (vi) below) leads one to assume that $A_z = A'_z$ [14], thus one spectrum function S_j for each frequency ω_j $(j = 1, ..., N_\omega)$.

(v) An axisymmetric galaxy is schematized as a finite set $\{\mathbf{x}_i\,;\ i = 1, ...i_{\max}\}$ of point-like "stars", given by their cylindrical coordinates ρ, ϕ, z, which are drawn with specific probability laws ensuring axisymmetry and representativity.

(vi) The model is fixed by the spectrum functions S_j $(j = 1, ..., N_\omega)$, approximated by their discretized values S_{nj} $(n = 1, ..., N)$. For each frequency ω_j, to determine the values S_{nj} $(n = 1, ..., N)$, we do a least-squares fitting of a sum of spherical potentials:

$$\Sigma_j := \sum_{i=1}^{i_{\max}} \varphi_{\mathbf{x}_i\,\omega_j}. \tag{8}$$

The potential $\varphi_{\mathbf{x}_i\,\omega_j}$ is the outgoing spherical solution of the wave equation with frequency ω_j, emanating from the star at point $\mathbf{x}_i$.

4. Calculation of u and S

In order to integrate Eq. (7), we have to be able to compute the values of the vector field $\mathbf{u}$ and the source S at any given event $(t, \mathbf{x})$. The explicit expressions of $\mathbf{u}$ and S were given in Ref. [1]. Setting $\Pi := \mathbf{E}.\mathbf{B}$, we get immediately from Eqs. (73)-(77) there, that [2]

$$\mathbf{u} = \frac{\mathbf{E} \wedge \nabla\Pi - (\partial_t\Pi)\mathbf{B}}{\mathbf{B}.\nabla\Pi}, \tag{10}$$

$$S = \frac{\mathrm{div}\,(\mathbf{e}\,\partial_t U)}{c^3\mathbf{B}.\nabla\frac{1}{\Pi}}, \tag{11}$$

where the spatial (3-)vector $\mathbf{e}$ (Eq. (68) in Ref. [1]) is easily shown to be

$$\mathbf{e} = \frac{1}{c\mu_0}\left(\mathbf{E} + \frac{c^2\mathbf{B}^2 - \mathbf{E}^2}{2\Pi}\mathbf{B}\right) \tag{12}$$

(here μ_0 is the vacuum permeability). Since $A'_{jz} = A_{jz}$ for the model EM field, it follows that here $c^2\mathbf{B}^2 - \mathbf{E}^2 = 0$, see Eq. (19) in Ref. [14]. Thus $\mathbf{e} = \mathbf{E}/(c\mu_0)$. With the free Maxwell equations, we have $\mathrm{div}\,\mathbf{E} = 0$, so the numerator of (11) is $\mathbf{E}.\nabla(\partial_t U)/(c\mu_0)$.

The calculation of $\mathbf{u}$ was numerically implemented, and tried with the model of the Milky Way, already investigated [12, 13, 14]. As could be expected in view of Eq. (10), the time variation of $\mathbf{u}$ is on the scale of the quasi-period of the EM field, and its space variation is on the scale of its wavelengths, thus both are extremely small as compared with the relevant scales for a galaxy. Therefore, it is numerically unfeasible to integrate on a galactic scale the PDE (7) for the scalar field p.

[2] We must assume that $\mathbf{E}.\mathbf{B} \neq 0$ [2]. It is not valid for the simple and most known solutions of the Maxwell equations: purely electric or purely magnetic fields, usual plane waves, dipole field [3]. However, it is generally true, because a real EM field is a combination of simple solutions and, if one adds two solutions $(\mathbf{E}_1, \mathbf{B}_1)$, $(\mathbf{E}_2, \mathbf{B}_2)$ of the standard Maxwell equations such that $\mathbf{E}_1.\mathbf{B}_1 = 0$ and $\mathbf{E}_2.\mathbf{B}_2 = 0$, then $(\mathbf{E}_1 + \mathbf{E}_2, \mathbf{B}_1 + \mathbf{B}_2)$ is also a solution, but

$$(\mathbf{E}_1 + \mathbf{E}_2).(\mathbf{B}_1 + \mathbf{B}_2) = \mathbf{E}_1.\mathbf{B}_2 + \mathbf{E}_2.\mathbf{B}_1, \tag{9}$$

which generally is not zero. In short, $\mathbf{E}.\mathbf{B}$ depends non-linearly on the field $(\mathbf{E}, \mathbf{B})$.

5. Homogenization of the PDE for the scalar field p

We want to obtain information on the field of interaction energy, thus on the scalar field p, at the macroscopic scale of a galaxy. But to do this, what is at our disposal is the PDE (5) or the PDE (7) — whereas either equation involves fields that vary on a very microscopic scale, e.g. Eq. (7) involves the fields $\mathbf{u}$ and S. This situation is typical of the homogenization theory, which has been developed mostly for periodic media — see e.g. Refs. [16, 17, 18]. The aim of this kind of theory is to get "homogenized" PDE's allowing one to describe the medium at the macroscopic scale. Note that for Eq. (7) the "medium" is, essentially, characterized by the pair of given heterogeneous fields $(\mathbf{u}, S)$. For Eq. (5), the "medium" is characterized by the pair of given heterogeneous fields $(\boldsymbol{G}, f)$. Actually, for a real galaxy, that "medium" is quasi-periodic rather than periodic, but the structure of the homogenized equations is the same for a quasi-periodic medium as for a truly periodic medium [18, 19]. The most important difference is that, in the first case, the elementary cell and/or the data fields within the cell, vary with the macroscopic position. (See in Subsects. 5.1–5.3 below for the relevant definitions of the elementary cell.) We note also that the homogenization theory has often been applied to the EM field, e.g. [17, 20], thus obtaining homogenized Maxwell equations, that apply to some specific situation — in the case of Refs. [17, 20], an anisotropic linear medium with permittivity and permeability tensors that vary with the position at a very small scale (and in a periodic way). However, these works do not seem to be relevant to the problem of homogenizing the PDE's (5) or (7). Two obvious points are: (i) a galaxy (thus a set of stars in the first place) cannot be regarded as a medium of the kind just mentioned. (ii) What has to be homogenized here, is the PDE for the scalar field p, not the Maxwell equations for the ISRF. Anyway, the ISRF in a galaxy does oscillate at the extremely small time and space scales mentioned above.

The homogenization technique considers two variables related by a small parameter ϵ: a "slow" variable, say s, which is the one that browses the medium at the macroscopic scale as is sought for, and the "quick" variable, s/ϵ, of which an $O(1)$ variation will browse the period of the medium, which is very small as evaluated in terms of the slow variable. Several different frames can thus be envisaged for the homogenization of the PDE's (5) or (7), depending on which set of variables is considered for the homogenization: the time variable, the space variables, or the spacetime variables.

5.1. *Time homogenization of Eq. (7)*

We will first illustrate the technique in the case of the time variable, thus we will outline the application to Eq. (7) of the *time homogenization* technique as proposed (for the mechanics of heterogeneous materials) by Guennouni [21]. The idea of the method is to consider two "separated" time scales: a "quick" time τ and a "slow" time t, with $\tau = t/T$, where $T \ll 1$ is the period of the quick variation (here that of the EM radiation field), when expressed in terms of the "slow" time variable t. Thus $T \ll t$ — here the galactic time scale: $t \simeq r/c$ with r a galactic distance, hence the ratio $t/T \simeq r/\lambda$ has a huge value $\simeq 10^{25}$: we have here a very good separation of scales. The elementary cell is one interval of periodicity, thus $[0, T]$ in terms of the slow variable t, and $[0, 1]$ in terms of the quick variable $\tau = t/T$. Formally, one assumes that the given fields $\mathbf{u}$ and S, as well as the boundary values for p (which shall not be precised), have the form

$$\mathbf{u} = \mathbf{u}(\mathbf{x}, t, \tau) = \lambda(t, \tau)\mathbf{u}^*(\mathbf{x}), \quad S = S(\mathbf{x}, t, \tau) = \mu(t, \tau)S^*(\mathbf{x}), ... \tag{13}$$

where λ and μ are τ-periodic of period 1.

Setting

$$\mathbf{u}^T(\mathbf{x}, t) := \mathbf{u}\left(\mathbf{x}, t, \frac{t}{T}\right), ..., \tag{14}$$

one has a boundary value problem π^T for p, depending smoothly on T. Its solution field $p^T(\mathbf{x},t)$ defines a field depending on two time variables:

$$p(\mathbf{x},\, t,\, \tau) := p^T(\mathbf{x},t) \quad \text{with } T = \frac{t}{\tau}. \tag{15}$$

The total time derivative is

$$\frac{\mathrm{d}p^T}{\mathrm{d}t} = \frac{\partial}{\partial t}\left(p\left(\mathbf{x}, t, \frac{t}{T}\right)\right) = \frac{\partial p}{\partial t} + \frac{1}{T}\frac{\partial p}{\partial \tau}. \tag{16}$$

One states an asymptotic expansion as $T \to 0$ for the unknown field:

$$p^T(\mathbf{x},t) = p_0(\mathbf{x},t,\tau)T^0 + p_1(\mathbf{x},t,\tau)T + O(T^2) \quad \left(\tau = \frac{t}{T}\right), \tag{17}$$

where p_0 and p_1 are τ-periodic of period 1. Note that, since λ and μ in Eq. (13) are assumed τ-periodic, the given fields $\mathbf{u}$ and S are "already developed": Eq. (13) provides their expansions, those expansions having only the T^0 term. Using (16) and (17) in (7) and identifying powers, we get

$$\text{order } T^{-1}: \frac{\partial p_0}{\partial \tau} = 0 \; i.e. \; p_0 = p_0(\mathbf{x},t), \quad \text{order } T^0: \frac{\partial p_0}{\partial t} + \frac{\partial p_1}{\partial \tau} + \mathbf{u}.\nabla p_0 = S. \tag{18}$$

Averaging the last equation over the period T, according to

$$\bar{f}(\mathbf{x},t) := \int_0^1 f(\mathbf{x},t,\tau)\mathrm{d}\tau, \tag{19}$$

yields

$$\frac{\partial p_0}{\partial t} + \bar{\mathbf{u}}.\nabla p_0 = \bar{S}, \tag{20}$$

the sought-for time-averaged equation. Note that, from $(18)_1$, it follows that $p_0 = \bar{p}$. Thus, the time-homogenized equation (20) differs from the "microscopic" starting equation (7) merely by the fact that time-averaged fields are substituted for the fields entering (7), both for the given fields $\mathbf{u}$ and S and for the unknown field p. This was not *a priori* obvious: a naive averaging of Eq. (7) would leave us with $\overline{\mathbf{u}.\nabla p}$ in the place of $\bar{\mathbf{u}}.\nabla p_0 \equiv \bar{\mathbf{u}}.\nabla \bar{p}$.

Thus, the application of the time homogenization technique to Eq. (7) is quite straightforward, at least formally. (We did not investigate the convergence as $T \to 0$.) However, the data fields that are obtained after the time-averaging, $\bar{\mathbf{u}}$ and $\bar{S}$, have still the very rapid *spatial* variation, at the scale of the typical wavelength. As a result: as it is the case for the starting equation (7), again it is numerically unfeasible to integrate Eq. (20) on a galactic scale. This has been checked numerically, after having implemented the calculations of the time averages of $\mathbf{u}$ and S. We give some indications about the latter calculations below.

5.1.1. Time-averaging $\mathbf{u}$ *and* S The Maxwell model of the ISRF provides each of the six components of $\mathbf{E}$ and $\mathbf{B}$ in the form

$$F^{(q)}(t,\mathbf{x}) = \mathcal{R}e\left(\sum_{j=1}^{N_\omega} C_j^{(q)}(\mathbf{x})e^{-\mathrm{i}\omega_j t}\right) \qquad (q = 1, ..., 6). \tag{21}$$

Then, $\Pi := \mathbf{E}.\mathbf{B}$ expands on the $e^{-\mathrm{i}(\omega_j+\omega_k)t}$'s and $e^{-\mathrm{i}(\omega_j-\omega_k)t}$'s. Hence, $\mathbf{E}\wedge\nabla\Pi$, $(\partial_t\Pi)\mathbf{B}$, and $\mathbf{B}.\nabla\Pi$, which enter (10) and (11), expand on the $e^{-\mathrm{i}\Omega_\kappa t}$'s and $e^{-\mathrm{i}\Psi_\kappa t}$'s, with $\kappa = (j,k,m)$, $\Omega_\kappa = \omega_j+\omega_k+\omega_m$, $\Psi_\kappa = \omega_j+\omega_k-\omega_m$. Those three fields time-average to zero, because $\omega_j+\omega_k-\omega_m$ does not vanish, not any more than (of course) $\omega_j+\omega_k+\omega_m$. But equations (10) and (11) for $\mathbf{u}$ and S involve *ratios* of these fields. Thus we have to compute the time average of a ratio of two trigonometric polynomials: [3]

$$Q(t) = \frac{\mathcal{R}e\left(\sum_j C_j e^{-\mathrm{i}\omega_j t}\right)}{\mathcal{R}e\left(\sum_k D_k e^{-\mathrm{i}\omega_k t}\right)} = \frac{\sum_j C_j e^{-\mathrm{i}\omega_j t} + C_j^\star e^{\mathrm{i}\omega_j t}}{\sum_k D_k e^{-\mathrm{i}\omega_k t} + D_k^\star e^{\mathrm{i}\omega_k t}}$$

$$= \sum_j \frac{1}{\sum_k \frac{D_k}{C_j} e^{-\mathrm{i}(\omega_k-\omega_j)t} + \frac{D_k^\star}{C_j} e^{\mathrm{i}(\omega_k+\omega_j)t}} + \frac{1}{\sum_k \frac{D_k}{C_j^\star} e^{-\mathrm{i}(\omega_k+\omega_j)t} + \frac{D_k^\star}{C_j^\star} e^{\mathrm{i}(\omega_k-\omega_j)t}}. \tag{22}$$

We compute that, to an often quite good approximation,

$$\left\langle \frac{1}{a+\sum_\kappa b_\kappa e^{-\mathrm{i}\Omega_\kappa t}} \right\rangle \simeq \frac{1}{a}. \tag{23}$$

With this approximation, we get

$$\bar{Q} \simeq \sum_j \frac{1}{\frac{D_j}{C_j}} + \frac{1}{\frac{D_j^\star}{C_j^\star}} = 2\sum_j \mathcal{R}e\left(\frac{C_j}{D_j}\right). \tag{24}$$

5.2. Space homogenization of Eq. (7)

The (space) homogenization of a PDE governing a medium with a fine periodic microstructure has been studied notably in Refs. [16, 17], followed by very numerous works. Here we will outline the space homogenization of Eq. (7) by adapting the work of Caillerie [18]. We assume that the given fields $\mathbf{u}$ and S are defined in the spatial elementary cell Y, a rectangular parallelepiped with sides parallel to the Cartesian axes, as functions of the "quick" spatial variable $\mathbf{y}$, and can be extended to Y-periodic functions of $\mathbf{y}$ defined in the whole space. This means essentially that each of them, along with its derivatives, takes the same value at corresponding points on opposite faces of the parallelepiped Y. As in the time homogenization described above, one considers a boundary value problem for (7) depending on a small parameter ϵ (in the place of T) — ϵ can be considered as the size of the cell (its dimension in the direction of the first coordinate axis, say), that size being expressed in terms of the slow space variable $\mathbf{x}$ such that $\mathbf{y} = \mathbf{x}/\epsilon$. To do so one defines [4]

$$\mathbf{u}^\epsilon(\mathbf{x},t) = \mathbf{u}\left(\frac{\mathbf{x}}{\epsilon},t\right), \tag{25}$$

and the like for S. As in Eq. (17), one states an asymptotic expansion as $\epsilon \to 0$ for the solution field:

$$p^\epsilon(\mathbf{x},t) = p_0(\mathbf{x},\mathbf{y},t)\,\epsilon^0 + p_1(\mathbf{x},\mathbf{y},t)\,\epsilon + O(\epsilon^2), \qquad \mathbf{y} = \frac{\mathbf{x}}{\epsilon}, \tag{26}$$

where p_0 and p_1 are Y-periodic in the quick space variable $\mathbf{y}$. Similarly with Eq. (16), the total spatial derivatives are given by

$$\frac{\mathrm{d}p^\epsilon}{\mathrm{d}x^i} = \frac{\partial}{\partial x^i}\left(p\left(\mathbf{x},\frac{\mathbf{x}}{\epsilon},t\right)\right) = \frac{\partial p}{\partial x^i} + \frac{1}{\epsilon}\frac{\partial p}{\partial y^i}. \tag{27}$$

[3] The indices in Eq. (22) browse in general a different set as compared with the set $\{1,...,N_\omega\}$ of the frequencies of the EM field.

[4] This is somewhat different from Eq. (13) that was adapted to Eq. (7) from the presentation of the time-homgenization technique in Ref. [21]. However, it is equivalent in practice, because actually we have indeed just one field $\mathbf{u}$, not a family of fields. Thus Eq. (25) defines such a family $(\mathbf{u}^\epsilon)$ from the data of the unique field $\mathbf{u}$.

Then, inserting (26) in (7) using (27), and identifying powers, we get

$$\text{order } \epsilon^{-1} : \frac{\partial p_0}{\partial y^j} u^j = 0, \quad \text{order } \epsilon^0 : \frac{\partial p_0}{\partial t} + \left(\frac{\partial p_0}{\partial x^j} + \frac{\partial p_1}{\partial y^j} \right) u^j = S. \tag{28}$$

We note that, in contrast with the time homogenization case (see Eq. $(18)_1$), inserting the expansions (26) into the PDE (7) does not necessarily imply that p_0 is independent of the quick variable $\mathbf{y}$. [5] However, it is consistent to *assume* that p_0 is independent of $\mathbf{y}$: clearly, Eq. $(28)_1$ is then automatically satisfied. Another difficulty as compared with the time homogenization case, as well as with the works [18, 21], is the following one. The volume integration, over the elementary cell Y, of the main (zero-order) equation, Eq. $(28)_2$, does not in the most general case leave us with a PDE for the averaged fields. Assuming that p_0 is independent of $\mathbf{y}$ implies obviously that

$$\overline{p_0}(\mathbf{x}, t) := \frac{1}{|Y|} \int_Y p_0(\mathbf{x}, \mathbf{y}, t) \, \mathrm{d}\mathbf{y} = p_0(\mathbf{x}, t) \tag{29}$$

(where $|Y|$ is the volume of Y), so that $P := p_0 = \overline{p_0}$ is indeed an averaged field. With this, the integration of Eq. $(28)_2$ over Y gives us

$$|Y| \left(\frac{\partial P}{\partial t} + \overline{u^j} \frac{\partial P}{\partial x^j} \right) + \int_Y \mathbf{u}.\nabla_{\mathbf{y}} p_1(\mathbf{x}, \mathbf{y}, t) \, \mathrm{d}\mathbf{y} = |Y| \, \overline{S}(\mathbf{x}, t). \tag{30}$$

The integral on the l.h.s. may be rewritten as

$$\int_Y \mathbf{u}.\nabla_{\mathbf{y}} p_1(\mathbf{x}, \mathbf{y}, t) \, \mathrm{d}\mathbf{y} = \int_{\partial Y} \mathbf{u}.\mathbf{n} \, p_1 \mathrm{d}\Sigma - \int_Y p_1 \mathrm{div}_{\mathbf{y}} \mathbf{u} \, \mathrm{d}\mathbf{y} = - \int_Y p_1 \, \mathrm{div}_{\mathbf{y}} \mathbf{u} \, \mathrm{d}\mathbf{y}, \tag{31}$$

the last equality resulting from the fact that p_1 and $\mathbf{u}$ are Y-periodic in $\mathbf{y}$, so that $\mathbf{u}.\mathbf{n} \, p_1$ takes opposite values at corresponding points on opposite faces of Y. However, the integral on the rightmost side of (31) has no apparent reason to vanish in general. If it can be neglected, we can rewrite Eq. (30) as

$$\left(\frac{\mathrm{d}P}{\mathrm{d}t} \right)_{\overline{\mathbf{u}}} := \frac{\partial P}{\partial t} + \overline{u^j} \frac{\partial P}{\partial x^j} = \overline{S}(\mathbf{x}, t), \tag{32}$$

an equation for the averaged fields. Assuming that the same equation is valid for the relevant case of quasi-periodic fields $\mathbf{u}$ and S, it would then remain to implement the calculation of the spatially-averaged fields $\overline{\mathbf{u}}$ and $\overline{S}$ and the integration of Eq. (32).

5.3. Spacetime homogenization of Eq. (5)

Equation (5) for p has exactly the same form as the stationary heat conduction equation for the temperature θ (see Eq. (18) in Ref. [18]), except for the fact that here we have a spacetime equation instead of a spatial equation. That is, the tensor $\boldsymbol{G}$ is a spacetime second-order tensor, that plays the role played in Ref. [18] by the spatial second-order tensor of thermal conductivity, $\boldsymbol{K}$. Therefore, we can easily adapt the work [18]. Again, one considers a boundary value problem for (5) depending on a small parameter ϵ: ϵ can be considered as the size of the

[5] In the case of the stationary heat conduction equation for the temperature θ, studied in Ref. [18], the fact that θ_0 is independent of the quick variable is not obtained directly (in contrast with the equation studied in Ref. [21], as well as for Eq. (7) with the time homogenization). Nevertheless, this fact does follow from the expanded equation of the lowest order, by using the weak form of this equation together with the periodic boundary conditions and the fact that the quadratic form associated with $\boldsymbol{K}$ is coercive (D. Caillerie, private communication). Here the argument valid for the heat equation does not apply and, clearly, Eq. $(28)_1$ does not necessarily imply that $\nabla_{\mathbf{y}} p_0 = 0$.

spacetime elementary cell Υ, that size being expressed in terms of the slow spacetime variable $\mathbf{X} = (t, \mathbf{x})$. To do so one defines, similarly to Eq. (21) of Ref. [18]:

$$\boldsymbol{G}^{\epsilon}(\mathbf{X}) = \boldsymbol{G}\left(\frac{\mathbf{X}}{\epsilon}\right), \tag{33}$$

and the like for f. (The latter is a minor difference with Ref. [18], in which f is considered independent of ϵ.) As in Eqs. (17) and (26), one states an asymptotic expansion as $\epsilon \to 0$ for the solution field (cf. Eq. (22) of Ref. [18]):

$$p^{\epsilon}(\mathbf{X}) = p_0\,(\mathbf{X}, \mathbf{Y})\,\epsilon^0 + p_1\,(\mathbf{X}, \mathbf{Y})\,\epsilon + O(\epsilon^2), \qquad \mathbf{Y} = \frac{\mathbf{X}}{\epsilon}, \tag{34}$$

where p_0 and p_1 are Υ-periodic in the quick spacetime variable $\mathbf{Y} = (\tau, \mathbf{y})$. A difference with the heat conduction case is that the inverse EM field tensor $\boldsymbol{G}$ is antisymmetric, instead of being symmetric as is the conductivity tensor $\boldsymbol{K}$. Therefore the quadratic form associated with $\boldsymbol{G}$ is identically zero. This prevents us from using the argument used in the former case (see Footnote 5) to show that the first term in the expansion (here p_0) is independent of the quick variable, here $\mathbf{Y}$. As in Subsect. 5.2, we nevertheless still may assume that $p_0 = p_0(\mathbf{X})$. Then, just the same calculations of Ref. [18] do apply, substituting $\boldsymbol{G}$ for $\boldsymbol{K}$, p for θ, and defining, as in Eq. (20) of Ref. [18]:

$$\mathbf{q} := -\boldsymbol{G}.\nabla_{\mathbf{X}} p. \tag{35}$$

In particular, the following expanded equation is derived (Eq. (25b) in Ref. [18]):

$$\mathrm{div}_{\mathbf{X}} \mathbf{q}_0 + \mathrm{div}_{\mathbf{Y}} \mathbf{q}_1 + f = 0, \tag{36}$$

with (Eqs. (26a) and (26b) in Ref. [18]):

$$\mathbf{q}_0 = -\boldsymbol{G}.\left(\nabla_{\mathbf{X}} p_0 + \nabla_{\mathbf{Y}} p_1\right), \tag{37}$$

$$\mathbf{q}_1 = -\boldsymbol{G}.\nabla_{\mathbf{X}} p_1. \tag{38}$$

(The $\nabla_{\mathbf{Y}} p_2$ term does not appear, for we stop the expansion (34) at p_1.) By integrating (36) on the spacetime cell Υ, the following homogenized equation is derived (Eq. (27) in Ref. [18]):

$$\mathrm{div}_{\mathbf{X}} \langle \mathbf{q}_0 \rangle + \langle f \rangle = 0, \tag{39}$$

with the angular brackets denoting the average over the spacetime cell:

$$\langle A \rangle (\mathbf{X}) := \int_{\Upsilon} A(\mathbf{X}, \mathbf{Y})\mathrm{d}\mathbf{Y} \Big/ \int_{\Upsilon} \mathrm{d}\mathbf{Y}. \tag{40}$$

Moreover, as with the heat conduction case, the average field $\langle \mathbf{q}_0 \rangle$ is calculated by using a homogenized tensor (Eq. (45) in Ref. [18]):

$$\langle \mathbf{q}_0 \rangle = -\boldsymbol{G}^{\mathrm{H}}.\nabla_{\mathbf{X}} p_0, \tag{41}$$

the latter having the form (Eq. (46) in Ref. [18]):

$$\boldsymbol{G}^{\mathrm{H}} = \langle \boldsymbol{G}.(1 + \nabla_{\mathbf{Y}} \chi) \rangle. \tag{42}$$

Note that, with Eq. (41), the homogenized PDE (39) has just the same form as the starting "microscopic" equation (5). However, the homogenized tensor $\boldsymbol{G}^{\mathrm{H}}$ is not in general the average of the microscopic tensor $\boldsymbol{G}$: in Eq. (42), the spacetime vector χ is the solution of a boundary value problem on the spacetime cell Υ (transposed from Eq. (31) in Ref. [18]). Due to this fact, the application of this method to the actual numerical solution of Eq. (5) looks heavy at first sight: in the relevant quasi-periodic case, we would have to solve that boundary value problem on a representative cell neighbouring an integration point in the macroscopic spacetime domain considered, and this, *a priori*, for any integration point — though possibly less often, depending on the variability that would be found for $\boldsymbol{G}^{\mathrm{H}}$.

6. Conclusion

In the alternative gravity theory "SET", electromagnetism in the presence of gravitation demands to introduce an additional energy tensor $\boldsymbol{T}_{\mathrm{inter}}$, depending on a scalar field p. This exotic energy tensor might contribute to dark matter. To check this, we developed a model of the interstellar radiation field that provides it as an exact Maxwell field. Then we may calculate the fields $\mathbf{u}$ and S that determine p and $\boldsymbol{T}_{\mathrm{inter}}$ through the PDE (7). But the very quick variation of $\mathbf{u}$ and S, both in time and in space, makes it unfeasible to integrate (7) on a galactic scale.

Therefore, we discussed the application of the homogenization theory. Together with the choice of the precise form for the PDE: either (5) or (7), each choice of a set of variables for the homogenization: time variable, space variables, or spacetime variables, determines a specific homogenization process.

- For Eq. (7) with the time variable alone ("time homogenization"), the PDE stays unchanged but with time-averaged fields: Eq. (20). However, the data fields that are obtained after the time-averaging, $\bar{\mathbf{u}}$ and $\bar{S}$, have still the very rapid *spatial* variation, at the scale of the typical wavelength. As a result, it is numerically unfeasible to integrate Eq. (20) on a galactic scale.
- For Eq. (7) with the space variables alone ("spatial homogenization"), an equation for the spatially-averaged unknown field $P = \overline{p_0}$ is obtained if the integral (31) can be neglected, and that equation again is the same as the starting equation, though with spatially-averaged data fields $\overline{\mathbf{u}}$ and $\overline{S}$, Eq. (32). This equation seems to be a tractable one for an integration at the galactic scale — provided the spatial averaging at a given time also sufficiently smooths out the time variation.
- Equation (5) for p has precisely the same form as the stationary heat conduction equation for the temperature studied in Ref. [18], except for the fact that here it is a spacetime equation instead of a spatial equation. Hence, many results of Ref. [18] translate immediately into results for the "spacetime homogenization" of Eq. (5). The homogenized equation, Eq. (39) with (41), has the same form as the starting equation (5), but here the homogenized tensor $\boldsymbol{G}^{\mathrm{H}}$ is not the average of the microscopic tensor $\boldsymbol{G}$. The calculation of $\boldsymbol{G}^{\mathrm{H}}$ involves the solution of a boundary value problem on the spacetime cell. In the relevant quasi-periodic case, we would have to solve that boundary value problem at many different points in the macroscopic spacetime domain considered.

Acknowledgments

The author is grateful to Denis Caillerie for a helpful exchange.

References

[1] Arminjon M 2018 On the equations of electrodynamics in a flat or curved spacetime and a possible interaction energy *Open Phys.* **16** 488-498

[2] Arminjon M 2016 Continuum dynamics and the electromagnetic field in the scalar ether theory of gravitation *Open Phys.* **14** 395-409

[3] Arminjon M 2017 Charge conservation in a gravitational field in the scalar ether theory *Open Phys.* **15** 877-890

[4] Arminjon M 2006 Space isotropy and weak equivalence principle in a scalar theory of gravity *Braz. J. Phys.* **36** 177-189

[5] Draine B T 1978 Photoelectric heating of interstellar gas *Astrophys. J. Suppl. Ser.* **36** 595-619

[6] Mathis J S, Mezger P G and Panagia N 1983 Interstellar radiation field and dust temperatures in the diffuse interstellar matter and in giant molecular clouds *Astron. Astrophys.* **128** 212-229

[7] Chi X and Wolfendale A W 1991 The interstellar radiation field: a datum for cosmic ray physics *J. Phys. C: Nucl. Part. Phys.* **17** 987-998

[8] Gordon K D, Misselt K A, Witt A N and Clayton G C 2001 The DIRTY model. I. Monte Carlo radiative transfer through dust *Astrophys. J.* **551** 269-276

[9] Robitaille T P 2011 HYPERION: an open-source parallelized three-dimensional dust continuum radiative transfer code *Astron. Astrophys.* **536** A79, 17 pages

[10] Popescu C C, Yang R, Tuffs R J, Natale G, Rushton M and Aharonian F 2017 A radiation transfer model for the Milky Way: I. Radiation fields and application to High Energy Astrophysics *Mon. Not. Roy. Astr. Soc.* **470** 2539–2558

[11] Arminjon M 2020 An explicit representation for the axisymmetric solutions of the free Maxwell equations *Open Phys.* **18** 255-263

[12] Arminjon M 2021 An analytical model for the Maxwell radiation field in an axially symmetric galaxy *Open Phys.* **19** 77-90

[13] Arminjon M 2021 Spectral energy density in an axisymmetric galaxy as predicted by an analytical model for the Maxwell field *Adv. Astron.* **2021**, 5524600, 13 pages

[14] Arminjon M 2021 Interstellar radiation as a Maxwell field: improved numerical scheme and application to the spectral energy density *Preprint* HAL-03341905

[15] Arminjon M 2017 On continuum dynamics and the electromagnetic field in the scalar ether theory of gravitation *J. Phys.: Conf. Series* **845** 012014, 9 pages

[16] Bensoussan A, Lions J-L and Papanicolaou P 1978 *Asymptotic Analysis for Periodic Structures* (Amsterdam: North Holland)

[17] Sanchez-Palencia E 1980 *Non-Homogeneous Media and Vibration Theory* (Lectures Notes in Physics No. 127, Berlin: Springer)

[18] Caillerie D 2012 Homogénéisation des matériaux à structure périodique, Ecole d'été *Méthodes Asymptotiques en Mécanique*, Quiberon, September 2012. Text available at http://mam.ida.upmc.fr/Telechargements/cours-Caillerie.pdf

[19] Auriault J-L 2011 Heterogeneous periodic and random media. Are the equivalent macroscopic descriptions similar? *Int. J. Eng. Sci.* **49** 806-808

[20] Barbatis G and Stratis I G 2003 Homogenization of Maxwell's equations in dissipative bianisotropic media *Math. Meth. Appl. Sci.* **26** 1241-1253

[21] Guennouni T 1988 Sur une méthode de calcul de structures soumises á des chargements cycliques : l'homogénéisation en temps *Math. Model. Num. Anal.* **22** 417-455

IARD 2022 IOP Publishing
Journal of Physics: Conference Series **2482** (2023) 012022 doi:10.1088/1742-6596/2482/1/012022

Weyl and Majorana for Neutral Particles

Valeriy V. Dvoeglazov
UAF, Universidad Autónoma de Zacatecas, México
E-mail: valeri@fisica.uaz.edu.mx

Abstract. We compare various formalisms for neutral particles. It is found that they contain unexplained contradictions. Next, we investigate the spin-1/2 and spin-1 cases in different bases. Next, we look for relations with the Majorana-like field operator. We show explicitly incompatibility of the Majorana anzatzen with the Dirac-like field operators in both the original Majorana theory and its generalizations. Several explicit examples are presented for higher spins too. It seems that the calculations in the helicity basis only give mathematically and physically reasonable results.

1. Weyl Formalism.

The Weyl formalism is just a massless limit of the Dirac equation:

$$[i\gamma^\mu \partial_\mu]\Psi(x) = 0\,. \tag{1}$$

Of course, it can be re-written in the 2-component forms:

$$[p_0 + \sigma \cdot \mathbf{p}]\chi(x) = 0\,, \quad [p_0 - \sigma \cdot \mathbf{p}]\phi(x) = 0\,. \tag{2}$$

However, if we apply the Noether theorem to the Lagrangian

$$\mathcal{L} = \frac{i}{2}[\bar{\Psi}\gamma_\mu \partial^\mu \Psi - \partial^\mu \bar{\Psi}\gamma_\mu \Psi] \tag{3}$$

we obtain the current operator

$$J_\mu = \bar{\Psi}\gamma_\mu \Psi\,, \tag{4}$$

as in the massive case. So, it is doubtful that we can use the massless limit of the Dirac equation for neutral particles.

In Refs. [2]-[6] we considered the procedure of construction of the field operators *ab initio* (including for neutral particles). The Bogoliubov-Shirkov method has been used.

In the present article we investigate the spin-1/2 and spin-1 cases in different bases. The Majorana theory of the neutral particles is well known [7]. We look for relations of the Dirac-like field operator to the Majorana-like field operator. It seems that the calculations in the helicity basis give mathematically and physically reasonable results.

Content from this work may be used under the terms of the Creative Commons Attribution 3.0 licence. Any further distribution of this work must maintain attribution to the author(s) and the title of the work, journal citation and DOI.
Published under licence by IOP Publishing Ltd

2. The Spin-1/2.

Usually, everybody uses the following definition of the field operator [8] in the pseudo-Euclidean metrics:

$$\Psi(x) = \frac{1}{(2\pi)^3}\sum_h \int \frac{d^3\mathbf{p}}{2E_p}[u_h(\mathbf{p})a_h(\mathbf{p})e^{-ip_\mu\cdot x^\mu} + v_h(\mathbf{p})b_h^\dagger(\mathbf{p})e^{+ip_\mu\cdot x^\mu}]\,, \tag{5}$$

as given *ab initio*. The momentum-space 4-spinors ($u-$ and $v-$) satisfy the equations: $(\hat{p}-m)u_h(p) = 0$ and $(\hat{p}+m)v_h(p) = 0$, respectively; the h is the polarization index. It is easy to prove from the characteristic equations $Det(\hat{p}\mp m) = (p_0^2 - \mathbf{p}^2 - m^2)^2 = 0$ that the solutions should satisfy the energy-momentum relations $p_0 = \pm E_p = \pm\sqrt{\mathbf{p}^2+m^2}$ for both $u-$ and $v-$ solutions.

The general scheme of construction of the field operator has been given in [9]. In the case of the $(1/2,0)\oplus(0,1/2)$ representation we have:

$$\Psi(x) = \frac{1}{(2\pi)^3}\int dp\, e^{ip\cdot x}\tilde{\Psi}(p)\,. \tag{6}$$

We know the condition of the mass shell: $(p^2-m^2)\tilde{\Psi}(p) = 0$. Thus, $\tilde{\Psi}(p) = \delta(p^2-m^2)\Psi(p)$. After simple transformations we obtain

$$\begin{aligned}\Psi(x) &= \frac{1}{(2\pi)^3}\sum_h\int\frac{d^3\mathbf{p}}{2E_p}\theta(p_0)\Big[u_h(p)a_h(p)|_{p_0=E_p}e^{-i(E_pt-\mathbf{p}\cdot\mathbf{x})}+\\ &+ u_h(-p)a_h(-p)|_{p_0=E_p}e^{+i(E_pt-\mathbf{p}\cdot\mathbf{x})}\Big]\end{aligned}$$

During the calculations we had to represent $1 = \theta(p_0)+\theta(-p_0)$ above in order to get positive- and negative-frequency parts. We did not yet assumed, which equation does this field operator (namely, the $u-$ spinor) satisfy, with negative- or positive- mass and/or $p^0 = \pm E_p$. We should transform $u_h(-p)$ to the $v_h(p)$ 4-spinor. The procedure is the following one [1, 2].

In the Dirac case we should assume the following relation in the field operator:

$$\sum_{h=\pm 1/2} v_h(p)b_h^\dagger(p) = \sum_{h=\pm 1/2} u_h(-p)a_h(-p)\,, \tag{7}$$

which is compatible with the "hole" theory and the Feynman-Stueckelberg interpretation. We need $\Lambda_{\mu\lambda}(p) = \bar{v}_\mu(p)u_\lambda(-p)$. By direct calculations, we find

$$-mb_\mu^\dagger(p) = \sum_\lambda \Lambda_{\mu\lambda}(p)a_\lambda(-p)\,. \tag{8}$$

Hence, $\Lambda_{\mu\lambda} = -im(\boldsymbol{\sigma}\cdot\mathbf{n})_{\mu\lambda}$, $\mathbf{n} = \mathbf{p}/|\mathbf{p}|$, and

$$b_\mu^\dagger(p) = +i\sum_\lambda(\boldsymbol{\sigma}\cdot\mathbf{n})_{\mu\lambda}a_\lambda(-p)\,. \tag{9}$$

Multiplying (7) by $\bar{u}_\mu(-p)$ we obtain

$$a_\mu(-p) = -i\sum_\lambda(\boldsymbol{\sigma}\cdot\mathbf{n})_{\mu\lambda}b_\lambda^\dagger(p)\,. \tag{10}$$

The equations are self-consistent.

The details of the helicity basis are given in Refs. [10, 11]. However, in this helicity case we have:

$$\Lambda_{hh'}(p) = \bar{v}_h(p)u_{h'}(-p) = i\sigma^y_{hh'}\,. \tag{11}$$

So, someone may argue that we should introduce the creation operators by hand in every basis.

It is well known that "*particle=antiparticle*" in the Majorana theory [7]. So, in the language of the quantum field theory we should have:

$$b_\mu(E_p, \mathbf{p}) = e^{i\varphi} a_\mu(E_p, \mathbf{p})\,. \tag{12}$$

Usually, different authors use $\varphi = 0, \pm\pi/2$ depending on the metrics and on the forms of the 4-spinors and commutation relations. So, on using (9) and the above-mentioned postulate we come to:

$$a^\dagger_\mu(p) = +ie^{i\varphi}(\boldsymbol{\sigma}\cdot\mathbf{n})_{\mu\lambda}a_\lambda(-p)\,. \tag{13}$$

On the other hand, on using (10) we make the substitutions $E_p \to -E_p$, $\mathbf{p} \to -\mathbf{p}$ to obtain

$$a_\mu(p) = +i(\boldsymbol{\sigma}\cdot\mathbf{n})_{\mu\lambda}b^\dagger_\lambda(-p)\,. \tag{14}$$

The totally reflected (12) is $b_\mu(-E_p, -\mathbf{p}) = e^{i\varphi}a_\mu(-E_p, -\mathbf{p})$. Thus,

$$b^\dagger_\mu(-p) = e^{-i\varphi}a^\dagger_\mu(-p)\,. \tag{15}$$

Combining with (14), we come to

$$a_\mu(p) = +ie^{-i\varphi}(\boldsymbol{\sigma}\cdot\mathbf{n})_{\mu\lambda}a^\dagger_\lambda(-p)\,, \tag{16}$$

and

$$a^\dagger_\mu(p) = -ie^{i\varphi}(\boldsymbol{\sigma}^*\cdot\mathbf{n})_{\mu\lambda}a_\lambda(-p)\,. \tag{17}$$

This contradicts with the above equation unless we have the preferred axis in every inertial system.

Next, we can use another Majorana anzatz $\Psi = \pm e^{i\alpha}\Psi^c$ with usual definitions

$$\mathcal{C} = e^{i\vartheta_c}\begin{pmatrix} 0 & i\Theta \\ -i\Theta & 0 \end{pmatrix}\mathcal{K}\,, \quad \Theta = \begin{pmatrix} 0 & -1 \\ 1 & 0 \end{pmatrix}\,. \tag{18}$$

Thus, on using $\mathcal{C}u^*_\uparrow(\mathbf{p}) = iv_\downarrow(\mathbf{p})$, $\mathcal{C}u^*_\downarrow(\mathbf{p}) = -iv_\uparrow(\mathbf{p})$ we come to other relations between creation/annihilation operators

$$a^\dagger_\uparrow(\mathbf{p}) = \mp ie^{-i\alpha}b^\dagger_\downarrow(\mathbf{p})\,, \tag{19}$$

$$a^\dagger_\downarrow(\mathbf{p}) = \pm ie^{-i\alpha}b^\dagger_\uparrow(\mathbf{p})\,, \tag{20}$$

which may be used instead of (12). Due to the possible signs $\pm$ the number of the corresponding states is the same as in the Dirac case that permits us to have the complete system of the Fock states over the $(1/2, 0) \oplus (0, 1/2)$ representation space in the mathematical sense.[1] However, in this case we deal with the self/anti-self charge conjugate quantum field operator instead of the self/anti-self charge conjugate quantum states. Please remember that it is the latter that answers for the neutral particles.

[1] Please note that the phase factors may have physical significance in quantum field theories as opposed to the textbook nonrelativistic quantum mechanics, as was discussed recently by several authors.

We conclude that something is missed in the foundations of both the original Majorana theory and its generalizations in the $(1/2,0)\oplus(0,1/2)$ representation.

We define the *self/anti-self* charge-conjugate 4-spinors in the momentum space [12]:

$$\mathcal{C}\lambda^{S,A}(\mathbf{p}) = \pm\lambda^{S,A}(\mathbf{p})\,, \qquad (21)$$
$$\mathcal{C}\rho^{S,A}(\mathbf{p}) = \pm\rho^{S,A}(\mathbf{p})\,. \qquad (22)$$

Such definitions of 4-spinors differ, of course, from the original Majorana definition in x-representation:

$$\nu(x) = \frac{1}{\sqrt{2}}(\Psi_D(x) + \Psi^c_D(x))\,, \qquad (23)$$

$C\nu(x) = \nu(x)$ that represents the positive real $C-$ parity field operator. However, the momentum-space Majorana-like spinors open various possibilities for description of neutral particles (with experimental consequences, see [13]). For instance, "for imaginary C parities, the neutrino mass can drop out from the single β decay trace and reappear in $0\nu\beta\beta$, a curious and in principle experimentally testable signature for a non-trivial impact of Majorana framework in experiments with polarized sources."

Thus, in the accustomed basis the explicit forms of the 4-spinors of the second kind $\lambda^{S,A}_{\uparrow\downarrow}(\mathbf{p})$ and $\rho^{S,A}_{\uparrow\downarrow}(\mathbf{p})$ are:

$$\lambda^S_\uparrow(\mathbf{p}) = \frac{1}{2\sqrt{E_p+m}}\begin{pmatrix} ip_l \\ i(p^-+m) \\ p^-+m \\ -p_r \end{pmatrix}, \lambda^S_\downarrow(\mathbf{p}) = \frac{1}{2\sqrt{E_p+m}}\begin{pmatrix} -i(p^++m) \\ -ip_r \\ -p_l \\ (p^++m) \end{pmatrix}$$
$$\lambda^A_\uparrow(\mathbf{p}) = \frac{1}{2\sqrt{E_p+m}}\begin{pmatrix} -ip_l \\ -i(p^-+m) \\ (p^-+m) \\ -p_r \end{pmatrix}, \lambda^A_\downarrow(\mathbf{p}) = \frac{1}{2\sqrt{E_p+m}}\begin{pmatrix} i(p^++m) \\ ip_r \\ -p_l \\ (p^++m) \end{pmatrix} \qquad (24)$$

In this basis one has

$$\rho^S_\uparrow(\mathbf{p}) = -i\lambda^A_\downarrow(\mathbf{p})\,,\ \rho^S_\downarrow(\mathbf{p}) = +i\lambda^A_\uparrow(\mathbf{p})\,, \qquad (25)$$
$$\rho^A_\uparrow(\mathbf{p}) = +i\lambda^S_\downarrow(\mathbf{p})\,,\ \rho^A_\downarrow(\mathbf{p}) = -i\lambda^S_\uparrow(\mathbf{p})\,. \qquad (26)$$

The $\lambda-$ and $\rho-$ spinors are connected with the $u-$ and $v-$ spinors by the following formula:

$$\begin{pmatrix} \lambda^S_\uparrow(\mathbf{p}) \\ \lambda^S_\downarrow(\mathbf{p}) \\ \lambda^A_\uparrow(\mathbf{p}) \\ \lambda^A_\downarrow(\mathbf{p}) \end{pmatrix} = \frac{1}{2}\begin{pmatrix} 1 & i & -1 & i \\ -i & 1 & -i & -1 \\ 1 & -i & -1 & -i \\ i & 1 & i & -1 \end{pmatrix}\begin{pmatrix} u_{+1/2}(\mathbf{p}) \\ u_{-1/2}(\mathbf{p}) \\ v_{+1/2}(\mathbf{p}) \\ v_{-1/2}(\mathbf{p}) \end{pmatrix}, \qquad (27)$$

provided that the 4-spinors have the same physical dimension.

We construct the field operators on using the procedure above with $\lambda^S_\eta(p)$. Thus, the difference is that 1) instead of $u_h(\pm p)$ we have $\lambda^S_\eta(\pm p)$; 2) possible change of the annihilation operators, $a_h \to c_\eta$. Apart, one can make corresponding changes due to normalization factors. Thus, we should have

$$\sum_{\eta=\pm 1/2} \lambda^A_\eta(p) d^\dagger_\eta(p) = \sum_{\eta=\pm 1/2} \lambda^S_\eta(-p) c_\eta(-p)\,. \qquad (28)$$

We find surprisingly:

$$d^\dagger_\eta(p) = -\frac{ip_y}{p}\sigma^y_{\eta\tau} c_\tau(-p)\,, \quad c_\eta(-p) = -\frac{ip_y}{p}\sigma^y_{\eta\tau} d^\dagger_\tau(p)\,. \tag{29}$$

The bi-orthogonal anticommutation relations are given in Ref. [12]. See other details in Ref. [14, 15]. Concerning with the P,C and T properties of the corresponding states see Ref. [15] in this model.

The above-mentioned contradictions may be related to the possibility of the conjugation which is different from that of Dirac. Both in the Dirac-like case and the Majorana-like case ($c_\eta(p) = e^{-i\varphi} d_\eta(p)$) we have difficulties in the construction of field operators.

3. The Spin-1.

We use the results of Refs. [17, 16, 18] in this Section. The polarization vectors of the standard basis are defined [19]:

$$\epsilon^\mu(\mathbf{0},+1) = -\frac{1}{\sqrt{2}}\begin{pmatrix}0\\1\\i\\0\end{pmatrix}, \quad \epsilon^\mu(\mathbf{0},-1) = +\frac{1}{\sqrt{2}}\begin{pmatrix}0\\1\\-i\\0\end{pmatrix}, \tag{30}$$

$$\epsilon^\mu(\mathbf{0},0) = \begin{pmatrix}0\\0\\0\\1\end{pmatrix}, \quad \epsilon^\mu(\mathbf{0},0_t) = \begin{pmatrix}1\\0\\0\\0\end{pmatrix}. \tag{31}$$

The Lorentz transformations are ($\widehat{p}_i = p_i/|\mathbf{p}|$):

$$\epsilon^\mu(\mathbf{p},\sigma) = L^\mu{}_\nu(\mathbf{p})\epsilon^\nu(\mathbf{0},\sigma)\,, \tag{32}$$

$$L^0{}_0(\mathbf{p}) = \gamma\,,\; L^i{}_0(\mathbf{p}) = L^0{}_i(\mathbf{p}) = \widehat{p}_i\sqrt{\gamma^2-1}\,,\; L^i{}_k(\mathbf{p}) = \delta_{ik} + (\gamma-1)\widehat{p}_i\widehat{p}_k\,. \tag{33}$$

Hence, for the particles of the mass m we have:

$$u^\mu(\mathbf{p},+1) = -\frac{N}{\sqrt{2}m}\begin{pmatrix}-p^r\\ m+\frac{p^1p^r}{E_p+m}\\ im+\frac{p^2p^r}{E_p+m}\\ \frac{p^3p^r}{E_p+m}\end{pmatrix}, \quad u^\mu(\mathbf{p},-1) = \frac{N}{\sqrt{2}m}\begin{pmatrix}-p^l\\ m+\frac{p^1p^l}{E_p+m}\\ -im+\frac{p^2p^l}{E_p+m}\\ \frac{p^3p^l}{E_p+m}\end{pmatrix}, \tag{34}$$

$$u^\mu(\mathbf{p},0) = \frac{N}{m}\begin{pmatrix}-p^3\\ \frac{p^1p^3}{E_p+m}\\ \frac{p^2p^3}{E_p+m}\\ m+\frac{(p^3)^2}{E_p+m}\end{pmatrix}, \quad u^\mu(\mathbf{p},0_t) = \frac{N}{m}\begin{pmatrix}E_p\\-p^1\\-p^2\\-p^3\end{pmatrix}. \tag{35}$$

N is the normalization constant for $u^\mu(\mathbf{p},\sigma)$. They are the eigenvectors of the parity operator ($\gamma_{00} = \mathrm{diag}(1\;-1\;-1\;-1)$):

$$\hat{P}u_\mu(-\mathbf{p},\sigma) = -u_\mu(\mathbf{p},\sigma)\,, \quad \hat{P}u_\mu(-\mathbf{p},0_t) = +u_\mu(\mathbf{p},0_t)\,. \tag{36}$$

It is assumed that they form the complete orthonormalized system of the $(1/2, 1/2)$ represntation, $\epsilon^*_\mu(\mathbf{p}, 0_t)\epsilon^\mu(\mathbf{p}, 0_t) = 1$, $\epsilon^*_\mu(\mathbf{p}, \sigma')\epsilon^\mu(\mathbf{p}, \sigma) = -\delta_{\sigma'\sigma}$.

The helicity operator should act as:

$$\frac{(\mathbf{S}\cdot\mathbf{p})}{p}\epsilon^\mu_{\pm1} = \pm\epsilon^\mu_{\pm1}\,, \quad \frac{(\mathbf{S}\cdot\mathbf{p})}{p}\epsilon^\mu_{0,0_t} = 0\,. \tag{37}$$

The eigenvectors are in the helicity basis:

$$\epsilon^\mu_{+1} = \frac{1}{\sqrt{2}}\frac{e^{i\alpha}}{p}\begin{pmatrix} 0 \\ \frac{-p^1p^3+ip^2p}{\sqrt{(p^1)^2+(p^2)^2}} \\ \frac{-p^2p^3-ip^1p}{\sqrt{(p^1)^2+(p^2)^2}} \\ \sqrt{(p^1)^2+(p^2)^2}\end{pmatrix}, \ \epsilon^\mu_{-1} = \frac{1}{\sqrt{2}}\frac{e^{i\beta}}{p}\begin{pmatrix} 0 \\ \frac{p^1p^3+ip^2p}{\sqrt{(p^1)^2+(p^2)^2}} \\ \frac{p^2p^3-ip^1p}{\sqrt{(p^1)^2+(p^2)^2}} \\ -\sqrt{(p^1)^2+(p^2)^2}\end{pmatrix} \tag{38}$$

$$\epsilon^\mu_0 = \frac{1}{m}\begin{pmatrix} p \\ \frac{E}{p}p^1 \\ \frac{E}{p}p^2 \\ \frac{E}{p}p^3\end{pmatrix}, \quad \epsilon^\mu_{0_t} = \frac{1}{m}\begin{pmatrix} E_p \\ p^1 \\ p^2 \\ p^3\end{pmatrix}. \tag{39}$$

The normalization is the same as in the standard basis. The eigenvectors $\epsilon^\mu_{\pm1}$ are not the eigenvectors of the parity operator $(\gamma_{00}R)$ of this representation. However, the $\epsilon^\mu_{1,0}$, $\epsilon^\mu_{0,0_t}$ are. Various-type field operators are possible in this representation. Let us remind the procedure to get them.

$$\begin{aligned} A_\mu(x) &= \frac{1}{(2\pi)^3}\int d^4p\,\delta(p^2-m^2)e^{-ip\cdot x}A_\mu(p) = \\ &= \frac{1}{(2\pi)^3}\sum_\lambda\int\frac{d^3\mathbf{p}}{2E_p}[\epsilon_\mu(p,\lambda)a_\lambda(p)e^{-ip\cdot x} + \epsilon_\mu(-p,\lambda)a_\lambda(-p)e^{+ip\cdot x}]\,. \end{aligned} \tag{40}$$

We should transform the second part to $\epsilon^*_\mu(p,\lambda)b^\dagger_\lambda(p)$ as usual. In such a way we obtain the states which are considered to be the charge-conjugate states. In this Lorentz group representation the charge conjugation operator is just the complex conjugation operator for 4-vectors. We postulate

$$\sum_\lambda \epsilon_\mu(-p,\lambda)a_\lambda(-p) = \sum_\lambda \epsilon^*_\mu(p,\lambda)b^\dagger_\lambda(p)\,. \tag{41}$$

Then we multiply both parts by $\epsilon^\mu(p,\sigma)$, and use the normalization conditions for polarization vectors.

In the $(\frac{1}{2},\frac{1}{2})$ representation we can also expand (apart of the equation (41)) in a different way. For example,

$$\sum_\lambda \epsilon_\mu(-p,\lambda)c_\lambda(-p) = \sum_\lambda \epsilon_\mu(p,\lambda)d^\dagger_\lambda(p)\,. \tag{42}$$

From the first definition we obtain:

$$\begin{pmatrix} b^\dagger_{0_t}(p) \\ -b^\dagger_{+1}(p) \\ -b^\dagger_0(p) \\ -b^\dagger_{-1}(p)\end{pmatrix} = \sum_{\mu\lambda}\epsilon^\mu(p,\sigma)\epsilon_\mu(-p,\lambda)a_\lambda(-p) = \sum_\lambda \Lambda^{(1a)}_{\sigma\lambda}a_\lambda(-p) =$$

$$= \begin{pmatrix} -1 & 0 & 0 & 0 \\ 0 & \frac{p_r^2}{p^2} & -\frac{\sqrt{2}p_z p_r}{p^2} & \frac{p_z^2}{p^2} \\ 0 & -\frac{\sqrt{2}p_z p_r}{p^2} & -1+\frac{2p_z^2}{p^2} & +\frac{\sqrt{2}p_z p_l}{p^2} \\ 0 & \frac{p_z^2}{p^2} & +\frac{\sqrt{2}p_z p_l}{p^2} & \frac{p_l^2}{p^2} \end{pmatrix} \begin{pmatrix} a_{00}(-p) \\ a_{11}(-p) \\ a_{10}(-p) \\ a_{1-1}(-p) \end{pmatrix} . \quad (43)$$

Possibly, we should think about modifications of the Fock space in this case. Alternatively, one can think to introduce several field operators for the $(\frac{1}{2}, \frac{1}{2})$ representation. The Majorana-like anzatz is compatible for the 0_t time-like polarization state only in this basis of this representation. However, the corresponding matrices Λ^2 in the helicity basis are different. Here they are:

$$\begin{pmatrix} b^\dagger_{0_t}(p) \\ -b^\dagger_{+1}(p) \\ -b^\dagger_0(p) \\ -b^\dagger_{-1}(p) \end{pmatrix} = \sum_{\mu\lambda} \epsilon^\mu(p,\sigma)\epsilon_\mu(-p,\lambda)a_\lambda(-p) = \sum_\lambda \Lambda^{(2a)}_{\sigma\lambda} a_\lambda(-p) =$$

$$= -\begin{pmatrix} 1 & 0 & 0 & 0 \\ 0 & e^{2i\alpha} & 0 & 0 \\ 0 & 0 & 1 & 0 \\ 0 & 0 & 0 & e^{2i\beta} \end{pmatrix} \begin{pmatrix} a_{00}(-p) \\ a_{11}(-p) \\ a_{10}(-p) \\ a_{1-1}(-p) \end{pmatrix} , \quad (44)$$

and

$$\begin{pmatrix} d^\dagger_{0_t}(p) \\ -d^\dagger_{+1}(p) \\ -d^\dagger_0(p) \\ -d^\dagger_{-1}(p) \end{pmatrix} = \sum_{\mu\lambda} \epsilon^{\mu^*}(p,\sigma)\epsilon_\mu(-p,\lambda)c_\lambda(-p) = \sum_\lambda \Lambda^{(2b)}_{\sigma\lambda} c_\lambda(-p) =$$

$$= -\begin{pmatrix} 1 & 0 & 0 & 0 \\ 0 & 0 & 0 & -e^{-i(\alpha-\beta)} \\ 0 & 0 & 1 & 0 \\ 0 & -e^{+i(\alpha-\beta)} & 0 & 0 \end{pmatrix} \begin{pmatrix} c_{00}(-p) \\ c_{11}(-p) \\ c_{10}(-p) \\ c_{1-1}(-p) \end{pmatrix} . \quad (45)$$

This is compatible with the Majorana-like anzatzen. Of course, the same procedure can be applied in the construction of the quantum field operator for $F_{\mu\nu}$.

The solutions of the Weinberg-like equation

$$[\gamma^{\mu\nu}\partial_\mu\partial_\nu - \frac{(i\partial/\partial t)}{E} m^2]\Psi(x) = 0 . \quad (46)$$

are found in Refs. [17, 20, 21, 22]. Here they are:

$$u_\sigma(\mathbf{p}) = \begin{pmatrix} D^S(\Lambda_R)\xi_\sigma(\mathbf{0}) \\ D^S(\Lambda_L)\xi_\sigma(\mathbf{0}) \end{pmatrix} , \; v_\sigma(\mathbf{p}) = \begin{pmatrix} D^S(\Lambda_R\Theta_{[1/2]})\xi^*_\sigma(\mathbf{0}) \\ -D^S(\Lambda_L\Theta_{[1/2]})\xi^*_\sigma(\mathbf{0}) \end{pmatrix} = \Gamma^5 u_\sigma(\mathbf{p}), \quad (47)$$

$$\Gamma^5 = \begin{pmatrix} 1_{3\times3} & 0_{3\times3} \\ 0_{3\times3} & -1_{3\times3} \end{pmatrix} , \quad (48)$$

where D^S is the matrix of the $(S, 0)$ representation of the spinor group $SL(2, c)$. In the $(1, 0) \oplus (0, 1)$ representation the procedure of derivation of the creation operators leads to somewhat different situation:

$$\sum_{\sigma=0,\pm1} v_\sigma(p)b^\dagger_\sigma(p) = \sum_{\sigma=0,\pm1} u_\sigma(-p)a_\sigma(-p) , \quad \text{hence } b^\dagger_\sigma(p) = 0 . \quad (49)$$

However, if we return to the original Weinberg equations $[\gamma^{\mu\nu}\partial_\mu\partial_\nu \pm m^2]\Psi_{1,2}(x) = 0$ with the field operators:

$$\Psi_1(x) = \frac{1}{(2\pi)^3}\sum_\mu \int \frac{d^3\mathbf{p}}{2E_p}[u_\mu(\mathbf{p})a_\mu(\mathbf{p})e^{-ip_\mu\cdot x^\mu} + u_\mu(\mathbf{p})b^\dagger_\mu(\mathbf{p})e^{+ip_\mu\cdot x^\mu}], \tag{50}$$

$$\Psi_2(x) = \frac{1}{(2\pi)^3}\sum_\mu \int \frac{d^3\mathbf{p}}{2E_p}[v_\mu(\mathbf{p})c_\mu(\mathbf{p})e^{-ip_\mu\cdot x^\mu} + v_\mu(\mathbf{p})d^\dagger_\mu(\mathbf{p})e^{+ip_\mu\cdot x^\mu}], \tag{51}$$

we obtain

$$b^\dagger_\mu(p) = [1-2(\mathbf{S}\cdot\mathbf{n})^2]_{\mu\lambda}a_\lambda(-p)\,, \tag{52}$$

$$d^\dagger_\mu(p) = [1-2(\mathbf{S}\cdot\mathbf{n})^2]_{\mu\lambda}c_\lambda(-p)\,. \tag{53}$$

The applications of $\overline{u}_\mu(-p)u_\lambda(-p) = \delta_{\mu\lambda}$ and $\overline{u}_\mu(-p)u_\lambda(p) = [1-2(\mathbf{S}\cdot\mathbf{n})^2]_{\mu\lambda}$ prove that the equations are self-consistent. This situation signifies that in order to construct the Sankaranarayanan-Good field operator (which was used by Ahluwalia, Johnson and Goldman [21]) we need additional postulates. One can try to construct the left- and the right-hand side of the field operator separately each other. In this case the commutation relations may also be more complicated.

Repeating the above procedures, on using (52) and the Majorana postulate, we come to:

$$a^\dagger_\mu(p) = +e^{+i\varphi}[1-2(\mathbf{S}\cdot\mathbf{n})^2]_{\mu\lambda}a_\lambda(-p)\,. \tag{54}$$

On the other hand, on using the inverse relation, namely, that for $a_\mu(-p)$, we make the substitutions $E_p \to -E_p$, $\mathbf{p} \to -\mathbf{p}$ to obtain

$$a_\mu(p) = +[1-2(\mathbf{S}\cdot\mathbf{n})^2]_{\mu\lambda}b^\dagger_\lambda(-p)\,. \tag{55}$$

The totally reflected Majorana anzatz is $b_\mu(-E_p,-\mathbf{p}) = e^{i\varphi}a_\mu(-E_p,-\mathbf{p})$. Thus,

$$b^\dagger_\mu(-p) = e^{-i\varphi}a^\dagger_\mu(-p)\,. \tag{56}$$

Combining with (55), we come to

$$a_\mu(p) = +e^{-i\varphi}[1-2(\mathbf{S}\cdot\mathbf{n})^2]_{\mu\lambda}a^\dagger_\lambda(-p)\,, \tag{57}$$

and

$$a^\dagger_\mu(p) = +e^{+i\varphi}[1-2(\mathbf{S}^*\cdot\mathbf{n})^2]_{\mu\lambda}a_\lambda(-p)\,. \tag{58}$$

In the basis where S_z is diagonal the matrix S_y is imaginary [19]. So, $(\mathbf{S}^*\cdot\mathbf{n}) = S_x n_x - S_y n_y + S_z n_z$, and $(\mathbf{S}^*\cdot\mathbf{n})^2 \neq (\mathbf{S}\cdot\mathbf{n})^2$ in the case of $S=1$. So, we conclude that there is the same problem in this point, in the aplication of the Majorana-like anzatz, as in the case of spin-1/2. Similarly, one can proceed with (53).

Meanwhile, the attempts of constructing the self/anti-self charge conjugate states failed in Ref. [12]. Instead, the $\Gamma^5 S^c_{[1]}-$ self/anti-self conjugate states have been constructed therein.

4. Conclusions.

Thus, some equations are incompatible each other after the application of the Majorana anzats. Moreover, the Majorana-like formalisms may indicate on the existence of preferred axis. We conclude that something is missed in the foundations of both the Weyl theory, the original Majorana theory and its generalizations. Similar problems exist in the theories of higher spins.

Acknowledgments
I acknowledge discussions with colleagues at recent conferences.

References

[1] V. V. Dvoeglazov, Hadronic J. Suppl. **18**, 239 (2003).
[2] V. V. Dvoeglazov, Int. J. Mod. Phys. B**20**, 1317 (2006).
[3] V. V. Dvoeglazov, in *Einstein and Others: Unification.* Ed. V. V. Dvoeglazov. (Nova Sci. Pubs., Hauppauge, NY, USA, 2015), p. 211.
[4] V. V. Dvoeglazov, Z. Naturforsch. A **71**, 345 (2016).
[5] V. V. Dvoeglazov, SFIN **22**-A1, 157 (2009); J. Phys. Conf. Ser. **284**, 012024 (2011); Bled Workshops **10**-2, 52 (2009); ibid. **11**-2, 9 (2010); in *Beyond the Standard Model. Proceedings of the Vigier Symposium.* (Noetic Press, Orinda, USA, 2005), p. 388.
[6] V. V. Dvoeglazov, Bled Workshops **14**-2, 199 (2013).
[7] E. Majorana, Nuovo Cim. **14**, 171 (1937).
[8] C. Itzykson and J.-B. Zuber, *Quantum Field Theory* (McGraw-Hill Book Co., 1980).
[9] N. N. Bogoliubov and D. V. Shirkov, *Introduction to the Theory of Quantized Fields.* 2nd Edition. (Nauka, Moscow, 1973).
[10] H. M. Rück and W. Greiner, J. Phys. G: Nucl. Phys. **3**, 657 (1977).
[11] V. V. Dvoeglazov, Int. J. Theor. Phys. **43**, 1287 (2004).
[12] D. V. Ahluwalia, Int. J. Mod. Phys. A**11**, 1855 (1996).
[13] M. Kirchbach, C. Compean and L. Noriega, Eur. Phys. J. A**22**, 149 (2004).
[14] V. V. Dvoeglazov, Nuovo Cim. A**108**, 1467 (1995).
[15] V. V. Dvoeglazov, Mod. Phys. Lett. A**12**, 2741 (1997); J. Phys. Conf. Ser. **1010**, 012011 (2018).
[16] S. Weinberg, *The Quantum Theory of Fields. Vol. I. Foundations.* (Cambridge University Press, Cambridge, 1995).
[17] Yu. V. Novozhilov, *Introduction to Elementary Particle Physics* (Pergamon Press, 1975).
[18] V. V. Dvoeglazov, in *Photon: Old Problems in Light of New Ideas.* Ed. V. V. Dvoeglazov (Nova Sci. Pubs., Huntington, NY, USA, 2000).
[19] D. A. Varshalovich, A. N. Moskalev and V. K. Khersonskiĭ, *Quantum Theory of Angular Momentum* (World Scientific, Singapore, 1988), §6.2.5.
[20] A. Sankaranarayanan and R. H. Good, Jr., Nuovo Cim. **36**, 1303 (1965).
[21] D. V. Ahluwalia, M. B. Johnson and T. Goldman, Phys. Lett. B**316**, 102 (1993).
[22] V. V. Dvoeglazov, Int. J. Theor. Phys. **37**, 1915 (1998).

Time dispersion in quantum electrodynamics

John Ashmead
Visiting Scholar, University of Pennsylvania, USA
E-mail: jashmead@seas.upenn.edu

Abstract. If we use the path integral approach, we can write quantum electrodynamics (QED) in a way that is manifestly relativistic. However the path integrals are confined to paths that are on mass-shell. What happens if we extend QED by computing the path integrals over all paths in energy momentum space, not only those on mass-shell? We use the requirement of covariance to do this in an unambiguous way. This gives a QED where the time/energy components appear in a way that is manifestly parallel to the space/momentum components: we have dispersion in time, entanglement in time, full equivalence of the Heisenberg uncertainty principle (HUP) in time to the HUP in space, and so on. Entanglement in time has the welcome side effect of eliminating the ultraviolet divergences. We recover standard QED in the long time limit. We predict effects at scales of attoseconds. With recent developments in attosecond physics and in quantum computing, these effects should be detectable. Since the predictions are unambiguous and testable the approach is falsifiable. Falsification would sharpen our understanding of the role of time in QED. Confirmation would have significant implications for attosecond physics, quantum computing and communications, and quantum gravity.

1. Introduction

"Look, I don't care what your theory of time is. Just give me something I can prove wrong." – Nathan Gisin at the 2009 Feynman Festival in Olomouc

Is quantum electrodynamics fully relativistic? Quantum electrodynamics (QED) can be developed in a large number of ways. Perhaps the most common is the Hamiltonian/canonical momentum approach. Time plays a special role in this formalism (in defining the canonical momenta) so it is not clear that this approach is completely relativistic. However the canonical momentum formalism is equivalent to the Feynman path integral formulation. And in the Feynman path integral formulation, QED is developed in a relativistic way.

However perhaps even the Feynman path integral formulation is not as fully relativistic as it might be. In it the paths are limited to on mass-shell paths. Consider the simplest possible propagator, the propagator for a massive spinless field:

$$\Delta\left(p_\mu\right) = \frac{\imath}{p^\mu p_\mu - m^2 + \imath\varepsilon} \tag{1}$$

Consider the infinitesimal $\imath\epsilon$. We construct the Feynman diagrams by doing integrals in four momentum $\int d^4p$ over the propagators. The $\imath\epsilon$ identifies one of these four integrals not as a normal but as a contour integral. Say we make the contour integral the one over energy. When we look in detail at this – done in the text – we see that for any fixed value of the three

Content from this work may be used under the terms of the Creative Commons Attribution 3.0 licence. Any further distribution of this work must maintain attribution to the author(s) and the title of the work, journal citation and DOI.
Published under licence by IOP Publishing Ltd

momentum $\vec{p}$ the value of the fourth component E is fixed by the value of the residues at the poles, typically $E \to E_{\vec{p}} \equiv \sqrt{m^2 + \vec{p}^2}$. The effect is to fix the paths to only the on-shell paths.

We can generalize the paths to include off-shell paths as well. By replacing the contour integral with a normal integral we can include paths that vary in four dimensions. For instance we can write E as $E_{\vec{p}} + \delta E$ and include paths which vary over all values of δE. Letting the paths vary in all four dimensions simultaneously is arguably more in keeping with the "spirit" of relativity, more fully relativistic.

But of course the question is not whether this is more fully in keeping with the spirit of relativity but does including off-shell paths in the path integrals give a more accurate description of nature?

Our goal here is to put this question in a way that is falsifiable with current technology.

If dispersion in time/energy is real, why has it not already been seen? QED has been confirmed to extraordinary precision in a wide variety of experiments, to the point where there is a wikipedia page on "Precision Tests of QED". If such dispersion in energy (and therefore time) is present, wouldn't we have already seen indications of this?

The most obvious estimate of the scale at which such effects should be seen is the Bohr radius a_0 divided by c; the time it would take a photon to cross an atom. This is of order attoseconds: $\frac{a_0}{c} \approx .177as$. This is at the edge of current experimental technology so technically within reach. But it is small enough that associated effects are unlikely to be seen if not specifically looked for.

Factors that make it less likely that dispersion in time would be seen by accident include:

(i) Calculations in QED are normally done by taking the limit as time goes to $\pm\infty$. This will naturally tend to obscure effects at attosecond scale.

(ii) Averaging over many interactions – i.e. shining beams against targets – will tend to average out effects in time.

(iii) It is not something which is expected, so therefore less likely to be seen. The effects of dispersion in time might be hiding within the error bars in some existing data sets.

Objective What we are going to do here is to treat include off-shell and on-shell paths on the same basis when computing the Feynman diagrams and see what breaks. Do we encounter an unavoidable contradiction on the one hand? or can we formulate experimental tests of this idea on the other?

Our objective is to force the question; to extend the paths in QED off-shell in a way that is:

(i) Manifestly covariant,

(ii) Consistent with observation and experiment,

(iii) Self-consistent,

(iv) Has no free parameters,

(v) Falsifiable with current technology.

Literature The work here has its starting point in the path integral approach as originated by Stueckelberg and Feynman [1, 2, 3, 4, 5, 6] and as further developed in [7, 8, 9, 10, 11, 12, 13, 14, 15]. This work is specifically part of the Relativistic Dynamics approach as developed by Horwitz, Fanchi, Piron, Land, Collins, and others [16, 17, 18, 19, 20, 21, 22, 23].

We are also much indebted to general reviews of the role of time in quantum mechanics: [24, 25, 26, 27, 28, 29, 30].

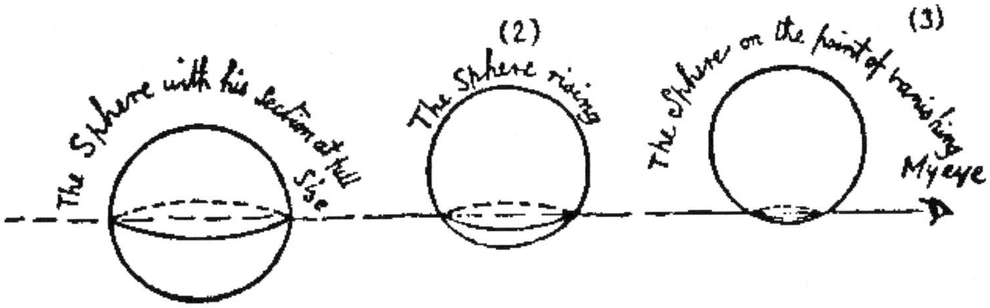

Figure 1. Flatland: A Romance of Many Dimensions – Edwin A. Abbott [52]

And we have taken considerable advantage of the extraordinary literature for QED. References particularly helpful here include [31, 32, 33, 34, 35, 36, 37, 38, 39, 40, 12, 41, 42, 43, 44, 45, 46, 47, 48].

We use the path integral formalism here. Texts on QED typically include a chapter on path integrals. There is also a considerable literature on them in their own right, as [8, 9, 49, 10, 11, 13, 14, 15].

In previous work we have looked at time dispersion in the single particle case [50] (paper A) and at the specific problems created in doing time-of-arrival measurements [51] (paper B). The investigation here extends this work to QED. The extension to QED is necessary to extend the results to high energies/short times, critical for falsification.

Overview In Edwin Abbott's charming 1884 romance Flatland, "A Sphere" – a visitor to Flatland – explains how three dimensions work to "A Square", an inhabitant of Flatland. A Sphere uses rotational symmetry to help take A Square from the idea of a circle or a square to the idea of a sphere or a cube.

We take a similar approach here. We will start with the established rules in standard quantum mechanics (SQM): quantum mechanics applied along the three space dimensions, but time treated classically. We use covariance to extend the rules of SQM to include time as an observable (TQM).

In the early days of quantum mechanics, classical techniques were extended to apply to quantum problems. For instance Feynman used the classical Lagrangian to develop his sum over paths in space. We are extending his approach to include paths in time as well.

Essentially we are completing the square between special relativity and quantum mechanics, adding quantum effects to time on the one side, adding time aspects to quantum mechanics on the other.

Strategy We use path integrals. These are simple, they require only a few basic ingredients: paths, a Lagrangian, a procedure for summing over the paths weighted by the Lagrangian.

In the single particle case, we promote the paths from three space dimensions (3D) to time plus the three space dimensions (4D), while leaving the Lagrangian unchanged.

In QED we again take the paths – now seen as successive values of a field – and promote them from being fields in three space to being fields in four. We are again able to keep the Lagrangian and the rest of the machinery of QED unchanged. In particular, we are able to develop the Feynman rules in TQM in a way that is clearly parallel to the rules in SQM. We get manifest covariance by construction.

As a result, conformity to existing results for SQM can be verified in a straight-forward way. And we can easily pick out experimental tests to look for the differences.

The major sections are:

(i) *Time dispersion in QED.* We use the requirement of covariance to extend the rules of quantum mechanics to include time on the same basis as space.

(ii) *Applications.* We apply these rules to the free case, to the simplest possible scattering case, to a simple mass correction loop, and then to the simplest tree diagrams (Møller, Bhabha, and Compton scattering).

(iii) *Experimental tests.* We propose a specific experiment to test the Heisenberg uncertainty principle in time/energy. The experiment is non-trivial, but appears within reach of current technology. It is, further, only one of many possible tests.

One surprising result is that the simple mass correction loop is convergent without regularization. We might expect that with an additional dimension to integrate over, the usual loop diagrams would become still more divergent, perhaps even unrenormalizable. But instead the combination of dispersion in time and entanglement in time keeps the loop diagrams finite.

In general, any time dependent system monitored by time sensitive detectors should show small but definite effects of dispersion in time. In addition to the Heisenberg uncertainty principle (HUP) in time/energy we can look at forces of "anticipation and regret", diffraction in time, entanglement in time, corrections to existing loop and bound state predictions at short times, and so on.

Summary In general, we expect to see the effects of dispersion in time at scales of attoseconds and less. With recent developments in attosecond physics and in quantum computing, these effects should now be visible. The most dramatic are those involving the HUP in time. The hypothesis is therefore falsifiable in practice.

Since the promotion of time to an operator is done by a straightforward application of agreed and tested principles of quantum mechanics and relativity, falsification will have implications for our understanding of those principles.

Confirmation will have implications for attosecond physics, quantum computing and communications, and quantum gravity.

2. Time dispersion in Quantum Electrodynamics

"The rules of quantum mechanics and special relativity are so strict and powerful that it's very hard to build theories that obey both." – Frank Wilczek [53]

In this section we work out the rules for extending quantum electrodynamics to include time as an operator. We work up the ladder of complexity till we have all the necessary pieces in place.

(i) Time dispersion and the single particle. We work out the Schrödinger equation for a single particle.

(ii) Spin zero propagator. We work out the Feynman propagator for a spin zero particle with mass greater than zero.

(iii) Photon propagator. We work out the Feynman propagator for photons.

(iv) Dirac propagator. We work out the Feynman propagator for fermions.

(v) Interactions. We note that the usual vertex terms are already TQM-ready.

Because of the need to carefully distinguish between SQM and TQM versions of otherwise familiar objects, some care is required in the notation. We describe our choices in Appendix A.

2.1. Time dispersion and the single particle

We start with the single particle case. The treatment here is largely based on paper A, but reworked to prepare for QED.

We develop the path integral approach for the single particle: we use the standard path integral approach, but with the usual paths generalized from three to four dimensions.

2.1.1. Clock time We start with clock time, defined operationally as what clocks measure: Alice with a stop watch or perhaps the laboratory clock on the wall or a carefully tended Cesium clock. This is also referred to as laboratory time, as in Busch [54] and others. We reserve the letter τ for this.

Note this is *not* the proper time. For a single particle, the use of proper time for the particle would give similar results to those here. However there is no way to extend the proper time approach to QED: the proper time for each massive particle will be in general different, while for photons it will be identically zero.

We give an alternate definition of τ towards the end of this subsection.

2.1.2. Paths Normally in single particle path integrals the paths vary in space but not in time. At each clock tick, the path will be assigned a specific triad of space coordinates. To get the amplitude to go from a starting point A to an endpoint B we will consider the set of all paths from A to B, weighting each by the action. We usually do the sum by breaking up the clock time from A to B into N time steps with each tick of size:

$$\varepsilon \equiv \frac{T}{N} \tag{2}$$

At the end we will take the limit $N \to \infty$. The path is defined by its space coordinates at each clock tick. To sum over the paths, we sum over the associated measure:

$$\mathcal{D}\vec{x} \equiv \prod_{n=0}^{N} d\vec{x}_n \tag{3}$$

Now we extend the paths to include time:

$$\pi\left(\vec{x}_\tau\right) \to \pi\left(t_\tau, \vec{x}_\tau\right) \tag{4}$$

We refer to the time dimension used in this way as coordinate time t, with its properties defined with respect to space by covariance.

The resulting paths are in four dimensions. They curve around in time, so can dart into the future or the past. To be sure, the sum over the paths is in general dominated by the classical paths, whose behavior is more sedate.

We extend the measure to include the sum over 4D paths:

$$\mathcal{D}x \equiv \prod_{n=0}^{N} dt_n d\vec{x}_n \tag{5}$$

2.1.3. Kernel Our primary object is to compute the kernel to go from A to B. This is given by the sum over all paths, weighted by the action, defined as the integral of the Lagrangian along each path:

$$K_\tau\left(x''; x'\right) = \int \mathcal{D}x_\tau \exp\left(\imath \int_0^\tau d\tau' \mathcal{L}\left[x_\tau, \dot{x}_\tau\right]\right) \tag{6}$$

Choice of Lagrangian We need a Lagrangian which is manifestly covariant, which correctly models the behavior of a particle in an electro-magnetic field, and which works equally well in 3D and 4D. We will use the following Lagrangian, which we have from Goldstein [55] and also from Feynman [31]:

$$\mathcal{L}\left[x_\tau, \dot{x}_\tau\right] = -\frac{1}{2}m\dot{x}^\mu \dot{x}_\mu - q\dot{x}^\mu A_\mu\left(x\right) - \frac{m}{2} \tag{7}$$

Convergence of the calculation Path integrals are normally computed by starting at a specific time, then integrating slice-by-slice. For this to make sense, the individual integrals have to converge.

Convergence is normally forced by adding small convergence factors, i.e. rewriting the mass as $m \to m + \imath\epsilon$. Such tricks are not usable here as they generally break covariance. However if we are integrating against a Gaussian test function (GTF), these tricks are not needed in the first place. The GTF itself will keep each step convergent. GTFs are completely general: by using Morlet wavelet analysis we can decompose any normalizable wavelet into sums over GTFs (further discussed in Appendix B).

This give us convergence.

Result In paper A [50] we derive the explicit form of the free kernel:

$$K_\tau\left(x;x'\right) = -\imath\frac{m^2}{4\pi^2\tau^2}e^{-\frac{\imath m}{2\tau}\left(x-x'\right)^2 - \imath\frac{m}{2}\tau} \tag{8}$$

or in momentum space:

$$K_\tau\left(p;p'\right) = \exp\left(\imath\frac{E^2 - \vec{p}^2 - m^2}{2m}\tau\right)\delta^4\left(p - p'\right) \tag{9}$$

This matches the non-relativistic kernel found in introductory quantum mechanics textbooks except that it now includes paths in time. From the momentum form, we can see that deviations off-shell will be punished.

2.1.4. Schrödinger equation One normally derives the path integral formula from the Schrödinger equation, see for instance Schulman or Kleinert [8, 14]. However we can also start with the path integral formula and get the Schrödinger equation by taking the short time limit of the path integral expression and running their derivations "in reverse". We get:

$$\imath\frac{\partial\psi_\tau}{\partial\tau}\left(t,\vec{x}\right) = -\frac{1}{2m}\left(\left(\imath\partial_\mu - qA_\mu\left(t,\vec{x}\right)\right)\left(\imath\partial^\mu - qA^\mu\left(t,\vec{x}\right)\right) - m^2\right)\psi_\tau\left(t,\vec{x}\right) \tag{10}$$

or in momentum space:

$$-2m\imath\frac{\partial\psi_\tau}{\partial\tau} = \left(\left(p_\mu - qA_\mu\right)\left(p^\mu - qA^\mu\right) - m^2\right)\psi_\tau \tag{11}$$

This is formally identical to the Feynman-Stueckelberg equation in the Relativistic Dynamics literature [3, 18, 19, 20, 23].

While the equation 10 is formally the same as the Feynman-Stueckelberg equation, the interpretation and use of the equation here is distinct. We will therefore refer to this as the FS/T: the Feynman-Stueckelberg equation in the TQM context.

Note also the resemblance to the non-relativistic Schrödinger equation:

$$\imath\frac{\partial\psi}{\partial\tau} = \frac{1}{2m}\left(\vec{p} - q\vec{A}\right)^2\psi \tag{12}$$

The only difference is that we have added a term that represents dispersion in time:

$$-\frac{1}{2m}\left(\left(\imath\partial_t - q\Phi\left(t,\vec{x}\right)\right)\left(\imath\partial^t - q\Phi\left(t,\vec{x}\right)\right)\right)\psi_\tau\left(t,\vec{x}\right) \tag{13}$$

We will modify the FS/T slightly as part of the extension to QED below.

2.1.5. Long, slow approximation If the dependence on clock time is weak, we get the familiar Klein-Gordon equation with minimal substitution:

$$\left(\imath\partial_\mu - qA_\mu\right)\left(\imath\partial^\mu - qA^\mu\right)\psi - m^2\psi = 0 \tag{14}$$

Is it reasonable to assume that the dependence on clock time is weak? That is that:

$$\imath\frac{\partial\psi}{\partial\tau} \approx 0 \tag{15}$$

or more specifically that the expectation value of dependence on clock time is small:

$$\langle\psi|\,\imath\frac{\partial}{\partial\tau}\,|\psi\rangle \approx 0 \tag{16}$$

Effects of dependence on clock time of order picoseconds To see the relevant scale, we estimate the clock frequency ϖ_p:

$$\varpi_p \sim -\frac{E^2 - \vec{p}^2 - m^2}{2m} \tag{17}$$

We are using ϖ rather than ω for the clock frequency to distinguish it clearly from the usual frequency ω. We will modify the definition of ϖ slightly below, again as part of the extension to QED (equation 53).

In the non-relativistic case E is of order mass plus kinetic energy:

$$E \sim m + \frac{\vec{p}^2}{2m} \tag{18}$$

so we have:

$$E^2 - \vec{p}^2 - m^2 \sim \left(m + \frac{\vec{p}^2}{2m}\right)^2 - \vec{p}^2 - m^2 = \left(\frac{\vec{p}^2}{2m}\right)^2 \tag{19}$$

This is just the kinetic energy, squared. In an atom the kinetic energy is of order the binding energy:

$$\frac{\vec{p}^2}{2m} \sim eV \tag{20}$$

So the numerator is of order eV squared. But the denominator is of order MeV. Therefore we can estimate the clock frequency ϖ_p as:

$$\varpi_p \sim \frac{(eV)^2}{MeV} \sim 10^{-6}eV \tag{21}$$

Energies of millionths of an electron volt $10^{-6}eV$ correspond to times of order millions of attoseconds 10^6as or picoseconds, a million times longer than the natural time scale of the effects we are looking at. Therefore the long, slow approximation (LSA) is reasonable.

Over long times, the clock frequency term will tend to reinforce on-shell components of the wave function with respect to the off-shell components. It is not so much that the off-shell components vanish, it is that averaged over nanoseconds, as by a slow detector, the off-shell components will average out to approximately zero. We then get what may look like a long, slow collapse of the wave function.

Over short times, we will treat the effects of the dependence on clock time as relatively less significant.

2.1.6. Meaning of laboratory time We now have two different kinds of time in play: coordinate time and clock time. We can reduce the ontological overhead of TQM by combining them. To do this, we take the clock time as the average over coordinate time over the rest of the universe $\mathcal{U}$:

$$\tau \equiv \langle \mathcal{U} | t \, | \mathcal{U} \rangle \tag{22}$$

We are effectively dividing the wave function of the universe into two parts, the small part we are focused on and the large part which is us, the laboratory, and the rest of the universe. We now re-define the clock time as the expectation of the coordinate time of the large part.

Therefore the properties of the clock time are those associated with an expectation value over an Avogadro's number of particles. In particular, it does not go backwards, as such fluctuations are wildly unlikely for the usual statistical dynamics reasons. As with a crowd, composed of individuals, but with the dynamics of the crowd very different from the dynamics of the individual. So we have:

(i) Defined clock time in terms of laboratory clocks.

(ii) Defined the extension of paths to coordinate time using clock time and covariance.

(iii) Worked out rules for quantum mechanics with coordinate time. Coordinate time is now an operator in the same way as the three space dimensions are operators.

(iv) Then turned around and defined clock time as the coordinate time operator applied to the laboratory – and the rest of universe, if it comes to that.

With this, laboratory time is not only not an operator, it is not even a parameter, it is merely an expectation value of the fundamental operator t. Therefore Pauli's theorem [56, 25] does not apply to it. We will continue to use clock time as a short hand for equation 22.

This is a significant variation of the work here from the literature in the Relativistic Dynamics program. In that, the parameter we have been calling clock time is an additional parameter which is introduced because various other parts of the problem then become more tractable. Here it is fixed: defined operationally by clocks and defined theoretically as the average over the coordinate time. This eliminates τ as a degree of freedom. The fewer the degrees of freedom the more falsifiable.

In TQM we have only one time, the coordinate time. The clock time is derivative, useful as scaffolding to get the analysis started, but dispensable once the analysis is in place. The clock time applies in full force only to macroscopic ensembles. The coordinate time represents the underlying reality.

2.1.7. Choice of laboratory frame So we understand what is meant by laboratory time in Alice's lab. But what if Bob is working in his laboratory moving at relativistic speeds relative to Alice's? Whose time should we use?

If the speeds are not too great, we can argue the effects will be of second order so may be neglected on a first attack.

However in the interests of achieving a clean and complete treatment we note we can define an invariant reference frame, to whose judgments both Alice and Bob must defer. (This is analogous to the way we can work in the center-of-mass frame, take advantage of the resulting simplicity, and then transform back to a specific laboratory frame at the end.)

In [57] Weinberg shows we can use Einstein's equations of general relativity to define an appropriate energy-momentum tensor of local spacetime. See Appendix D for specifics.

Since this is an energy-momentum tensor, we can use it to define a "local rest frame of spacetime" or $\mathcal{V}$ (for vacuum) frame. We take $\mathcal{V}$ as the required common frame. The defining laboratory time is therefore the clock time in this frame. Alice and Bob can agree on this, then perform the necessary Lorentz transforms from and to their respective frames confident they will make the same physical predictions.

2.2. Spin zero propagator

Having established a foundation in the single particle case, we extend TQM to the case of a massive spin zero particle. This is the core case for managing the transition from single particle quantum mechanics to QED. The photon and fermion cases will turn out to be relatively straightforward extensions of this.

We start with the SQM form. We use as a starting point the careful and detailed treatment in Klauber's text [45], but adapt his notation and techniques to the requirements of TQM. We give only the key "twists and turns".

For SQM and then for TQM, we look at:

(i) the free solutions and their associated Fock space,

(ii) the field operators constructed as sums over the free solutions,

(iii) the propagator constructed as a sum over the field operators.

2.2.1. Spin zero propagator in SQM We start with the Lagrangian for SQM:

$$\mathcal{L}^S \equiv \partial_\tau \phi \partial^\tau \phi - \nabla\phi\nabla\phi - m^2\phi^2 \tag{23}$$

The corresponding Euler-Lagrange equation is the Klein-Gordon equation:

$$\left(\partial_\tau \partial^\tau - \nabla^2 + m^2\right)\phi_\tau(x) = 0 \tag{24}$$

The free solutions of this are:

$$\phi_\tau^{(\vec{k})}(\vec{x}) \sim \exp\left(-\imath\omega_{\vec{k}}\tau + \imath\vec{k}\cdot\vec{x}\right), \omega_{\vec{k}} \equiv \sqrt{m^2 + \vec{k}^2} \tag{25}$$

Fock space The corresponding spin zero Fock space is built up in the usual way as appropriately symmetrized combinations of the free single particle solutions:

$$\phi_{\vec{k}}(\vec{x}) = \frac{1}{\sqrt{V}}\exp\left(\imath\vec{k}\cdot\vec{x}\right) \tag{26}$$

We are using box normalization to a volume V here. This is useful for dimension checking. We will shift back and forth freely between box and continuous normalization.

We use the occupation number representation for Fock space:

$$\left|\{n_{\vec{k}}\}\right\rangle \tag{27}$$

where n is an integer from zero to infinity and the wave functions are fully symmetric. The creation and annihilation operators are defined by their effects on Fock space:

$$a_{\vec{k}}\left|n_{\vec{k}}\right\rangle = \sqrt{n_{\vec{k}}}\left|n_{\vec{k}} - 1\right\rangle, a_{\vec{k}}^\dagger\left|n_{\vec{k}}\right\rangle = \sqrt{n_{\vec{k}} + 1}\left|n_{\vec{k}} + 1\right\rangle \tag{28}$$

with the 3D commutators being:

$$\left[a_{\vec{k}}, a^{\dagger}_{\vec{k}'}\right] = \delta^3\left(\vec{k} - \vec{k}'\right) \tag{29}$$

All other commutators are zero. We make *no* use of the usual interpretation of Fock space in terms of harmonic oscillators. The creation and annihilation operators are defined entirely by their effects on Fock space.

Field Operators Now we build up the spin zero field operators as sums over the free single particle solutions in the interaction picture. We have the sums over the positive frequency components on the left and negative frequency components on the right:

$$\phi^S_\tau\left(\vec{x}\right) = \sum_{\vec{k}} \frac{1}{\sqrt{2V\omega_{\vec{k}}}}\left(a_{\vec{k}} \mathrm{e}^{-\imath\omega_{\vec{k}}\tau + \imath\vec{k}\cdot\vec{x}} + a^{\dagger}_{\vec{k}} \mathrm{e}^{\imath\omega_{\vec{k}}\tau - \imath\vec{k}\cdot\vec{x}}\right) \tag{30}$$

We mark SQM parts with a superscript S. The normalization factor $\frac{1}{\sqrt{2\omega_{\vec{k}}}}$ corresponds to the convention of normalizing beams to energy/volume (see for instance Feynman [31]).

Feynman propagator The SQM Feynman propagator is defined as the time-ordered vacuum expectation value of two of these field operators. This definition is key to evaluating the S matrix as a sum over Feynman diagrams:

$$\imath\Delta^S_{\tau_x\tau_y}\left(\vec{x} - \vec{y}\right) \equiv \left\langle 0 \left| T\left\{\phi^S_{\tau_x}\left(\vec{x}\right), \phi^S_{\tau_y}\left(\vec{y}\right)\right\}\right| 0\right\rangle \tag{31}$$

T is the time-ordering operator; if $\tau_y < \tau_x$, then the y operator is on the right and vice versa. We break ϕ up into its positive and negative frequency parts:

$$\phi^S_\tau\left(\vec{x}\right) = \phi^{S+}_\tau\left(\vec{x}\right) + \phi^{S-}_\tau\left(\vec{x}\right) \tag{32}$$

In a vacuum expectation value, the only non-zero terms are those with an annihilation operator a on the left and a creation operator $a^\dagger$ on the right. As a result most of the terms vanish.

When $\tau_x > \tau_y$ the only non-zero term is:

$$\langle 0 | \phi^{S+}_x\left(\vec{x}\right) \phi^{S-}_y\left(\vec{y}\right) | 0 \rangle \tag{33}$$

By taking advantage of:

$$0 = -\langle 0 | \phi^{S-}_y\left(\vec{y}\right) \phi^{S+}_x\left(\vec{x}\right) | 0 \rangle \tag{34}$$

we can rewrite this in terms of the commutator:

$$\imath\Delta^{S+}_{xy}\left(\vec{x} - \vec{y}\right) = \langle 0 | \left[\phi^{S+}_x\left(\vec{x}\right), \phi^{S-}_y\left(\vec{y}\right)\right] | 0 \rangle \tag{35}$$

We expand the operators and use the commutators. To simplify the calculations, we shift from discrete sums to continuous integrals by replacing $\Sigma \to \int, V \to (2\pi)^3$. We use the vacuum product $\langle 0 \mid 0 \rangle = 1$. We are left with a pure number:

$$\imath\Delta^{S+}_{xy}\left(\vec{x} - \vec{y}\right) = \frac{1}{(2\pi)^3} \int d\vec{k} \frac{e^{-\imath\omega_{\vec{k}}\tau_{xy} + \imath\vec{k}\cdot(\vec{x}-\vec{y})}}{2\omega_{\vec{k}}} \tag{36}$$

The same development on the negative frequency side gives:

$$\imath\Delta^{S-}_{xy}\left(\vec{x} - \vec{y}\right) = \frac{1}{(2\pi)^3} \int d\vec{k} \frac{e^{\imath\omega_{\vec{k}}\tau_{xy} - \imath\vec{k}\cdot(\vec{x}-\vec{y})}}{2\omega_{\vec{k}}} \tag{37}$$

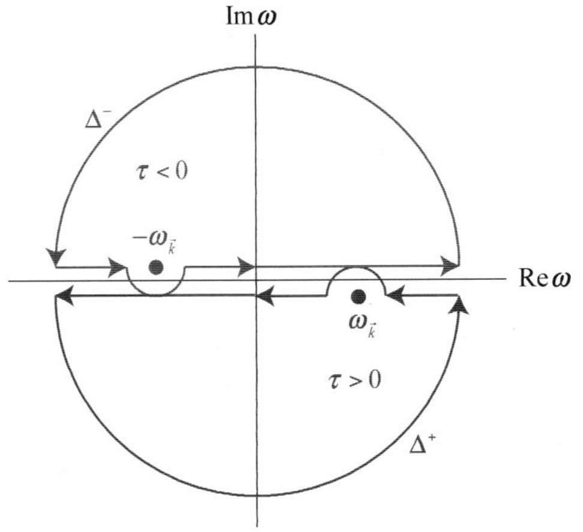

Figure 2. Contour integral for Feynman propagators – after Klauber [45]

We could also show this by interchanging $x \leftrightarrow y$.

We see clearly here that time and space are being treated differently: there is no integral over the energy coordinate ω; the value of $\omega_{\vec{k}}$ is fixed by $\vec{k}$ rather than being allowed to roam. This implies no dispersion in ω – and therefore none in time.

Combining the positive and negative parts we get the full spin zero propagator in SQM:

$$\imath\Delta^{S}_{xy}\left(\vec{x}-\vec{y}\right)=\frac{1}{\left(2\pi\right)^{3}}\int d^{3}\vec{k}\frac{\mathrm{e}^{-\imath\omega_{\vec{k}}\tau_{xy}-\imath\vec{k}\cdot\left(\vec{x}-\vec{y}\right)}}{2\omega_{\vec{k}}}\theta\left(\tau_{xy}\right)+\frac{\mathrm{e}^{\imath\omega_{\vec{k}}\tau_{xy}+\imath\vec{k}\cdot\left(\vec{x}-\vec{y}\right)}}{2\omega_{\vec{k}}}\theta\left(-\tau_{xy}\right)\tag{38}$$

We refer to this as the "unpacked form". Effectively it carries positive frequency components into the future; negative into the past.

Now we turn this into something that is more covariant in appearance. We start with the positive frequency side. We replace the integrand with a contour integral:

$$\frac{e^{-\imath\omega_{\vec{k}}\tau}}{2\omega_{\vec{k}}}\theta\left(\tau\right)=\frac{\imath}{2\pi}\int_{-\infty}^{\infty}d\omega\frac{e^{-\imath\omega\tau}}{\left(\omega-\omega_{\vec{k}}+\imath\varepsilon\right)\left(\omega+\omega_{\vec{k}}\right)}\tag{39}$$

with a pole at $\omega=\omega_{\vec{k}}-\imath\epsilon$. This gives the positive frequency side as:

$$\imath\Delta^{S+}_{xy}\left(\vec{x}-\vec{y}\right)=\frac{\imath}{\left(2\pi\right)^{4}}\int_{-\infty}^{\infty}d\omega\int d\vec{k}\frac{e^{-\imath\omega\tau_{xy}+\imath\vec{k}\cdot\left(\vec{x}-\vec{y}\right)}}{\left(\omega-\omega_{\vec{k}}+\imath\varepsilon\right)\left(\omega+\omega_{\vec{k}}\right)}\tag{40}$$

The negative frequency part can be replaced by a contour integral in the same way. We take the pole at $\omega=-\omega_{\vec{k}}+\imath\varepsilon$:

$$\frac{e^{\imath\omega_{\vec{k}}\tau}}{2\omega_{\vec{k}}}\theta\left(-\tau\right)=\frac{\imath}{2\pi}\int_{-\infty}^{\infty}d\omega\frac{e^{-\imath\omega\tau}}{\left(\omega-\omega_{\vec{k}}\right)\left(\omega+\omega_{\vec{k}}-\imath\varepsilon\right)}\tag{41}$$

Inserting this back in the previous expression we get:

$$\imath\Delta^{S-}_{xy}\left(\vec{x}-\vec{y}\right)=\frac{\imath}{\left(2\pi\right)^{4}}\int_{-\infty}^{\infty}d\omega\int d\vec{k}\frac{e^{-\imath\omega\tau_{xy}+\imath\vec{k}\cdot\left(\vec{x}-\vec{y}\right)}}{\left(\omega-\omega_{\vec{k}}\right)\left(\omega+\omega_{\vec{k}}-\imath\varepsilon\right)}\tag{42}$$

We could also just flip $x \leftrightarrow y$ again. Combined form spelled out:

$$\imath\Delta_{xy}^{S}(\vec{x}-\vec{y}) = \frac{\imath}{(2\pi)^4}\int d\omega d\vec{k}\frac{e^{-\imath\omega\tau_{xy}+\imath\vec{k}\cdot(\vec{x}-\vec{y})}}{\omega^2-\vec{k}^2-m^2+\imath\varepsilon} \tag{43}$$

The momentum space part is:

$$\imath\Delta_{\omega}^{S}\left(\vec{k}\right) = \frac{\imath}{\omega^2-\vec{k}^2-m^2+\imath\varepsilon} \tag{44}$$

This is the propagator for the original Klein-Gordon equation. If we apply the Klein-Gordon equation (24) in momentum space we have:

$$\left(\omega^2-\vec{k}^2-m^2\right)\Delta_{\omega}^{S}\left(\vec{k}\right) = 1 \tag{45}$$

The $+\imath\epsilon$ gives the propagator Feynman boundary conditions.

Energy is only a virtual 4th dimension So we can see that the propagator includes only on-shell paths. For instance, fixing the momentum $\vec{k}$ at any point fixes the energy ω. The integral over ω sees only the poles in the contour integral; it does not see the whole of ω space. Given this, the usual practice of referring to the particles associated with these propagators as virtual is correct. Our goal in building TQM is, in a certain sense, to make them real.

2.2.2. Spin zero propagator in TQM We do the same thing for TQM, replacing all 3D functions with 4D functions. The dependence on clock time and the normalization will require a bit of thought. We start with the TQM plane waves. They have four coordinates rather than three:

$$\phi_{w,\vec{k}}(t,\vec{x}) = \frac{1}{\sqrt{TV}}\exp\left(-\imath wt+\imath\vec{k}\cdot\vec{x}\right) \tag{46}$$

The T is the length of a box in time, starting well before anything interesting happens, and finishing well after everything interesting is done. It represents box normalization in time. The Fock space is built up of appropriately symmetrized products of these. It is given in the occupation representation by:

$$|\{n_k\}\rangle \tag{47}$$

where n is an integer from zero to infinity and the wave functions are fully symmetric.

The creation and annihilation operators are defined by their effects in Fock space:

$$a_k|n_k\rangle = \sqrt{n_k}|n_k-1\rangle, a_k^{\dagger}|n_{\vec{k}}\rangle = \sqrt{n_k+1}|n_k+1\rangle \tag{48}$$

with 4D commutators:

$$\left[a_k, a_{k'}^{\dagger}\right] = \delta^4(k-k') \tag{49}$$

All other commutators are zero. Again, per discussion above, we make *no* use of the usual interpretation in terms of harmonic oscillators.

Field Operators Now we extend the SQM field operator to TQM. For a first cut we take this as:

$$\phi_\tau(t,\vec{x}) = \sum_{w,\vec{k}} \frac{1}{\sqrt{TV}} \frac{1}{\sqrt{2\omega_{\vec{k}}}} \left(a_{w,\vec{k}} \mathrm{e}^{-\imath\varpi_k\tau - \imath wt + \imath\vec{k}\cdot\vec{x}} + a^\dagger_{w,\vec{k}} \mathrm{e}^{\imath\varpi_k\tau + \imath wt - \imath\vec{k}\cdot\vec{x}} \right) \tag{50}$$

The two questions here are:

(i) What should we use for ϖ_k?

(ii) And what should we use for $\frac{1}{\sqrt{2\omega_{\vec{k}}}}$? This depends to a considerable extent on the answer to the first question, so we will tackle that first.

Dependence on clock time We need a way to define the dependence on clock time – the clock energy – in a way that works equally well for massive spin zero particles, fermions, and photons. This is a non-trivial problem. For instance, the previous formula for the single particle ϖ_p has a $1/m$ factor which makes it unsuitable for use with photons. It is also unclear how best to extend the single particle approach to the Dirac equation: for instance, should antiparticles use the same sign for ϖ_p as particles do? We require an approach which lets us treat all kinds of particles uniformly.

To do this, recall we are using the local rest frame of spacetime as our reference point. What if we argue that the momentum of a particle should be understood not as an absolute but as relative to the average four momentum of the vacuum $\mathcal{P}$?

We do this by replacing the particle's four momentum k with its four momentum relative to $\mathcal{P}$; $k \to k - \mathcal{P}$. With this ansatz we rewrite the Klein-Gordon equation as:

$$\left((k-\mathcal{P})^2 - m^2 \right) \psi = 0 \tag{51}$$

By working in the rest frame of local spacetime $\mathcal{V}$ (as above) we reduce spacetime's four momentum to just its energy $\mathcal{P} \to \left(\mathcal{E}, \vec{0}\right)$. We replace its energy $\mathcal{E}$ with its complementary time operator $\mathcal{E} \to \imath\frac{\partial}{\partial\tau_V}$. The laboratory time is defined as $\tau \equiv \langle t \rangle$ so is *also* going opposite to the time of the vacuum so $\imath\frac{\partial}{\partial\tau_V} \to -\imath\frac{\partial}{\partial\tau}$:

$$\left(k^2 - m^2\right)\psi = -2w\imath\frac{\partial}{\partial\tau}\psi \tag{52}$$

We refer to this approach as the Machian hypothesis. We give a more detailed treatment in Appendix D.2. This is the FS/T for the single particle case (equation 11) with the substitution $m \to w$ or in coordinate space $m \to \imath\frac{\partial}{\partial t}$. Looking forwards, we will see this gives reasonable results for photons and fermions as well. In energy momentum space we can write the clock time dependence of the wave function as $\psi \sim \exp\left(-\imath\varpi_k\tau\right)$ to get:

$$\imath\frac{\partial}{\partial\tau}\psi = \varpi_k\psi, \varpi_k \equiv -\frac{w^2 - \vec{k}^2 - m^2}{2w} \tag{53}$$

In general we expect that off-shell components will tend to average out: $\langle w \rangle \approx \omega_{\vec{k}}$. Therefore we can expand the clock frequency in terms of $\delta w \equiv w - \omega_{\vec{k}}$:

$$\varpi_k \approx -\delta w + \frac{(\delta w)^2}{2\omega_{\vec{k}}} \tag{54}$$

Using ω as the complementary variable to τ, we write in this equation in terms of clock frequency ω and coordinate energy w:

$$\left(w^2 - \vec{k}^2 - m^2 + 2\omega w\right)\psi = 0 \tag{55}$$

From the analysis of the LSA above, we expect that the $2\omega w$ term will have little effect at short times, only coming into its own at longer times, i.e. on the "legs" of the Feynman diagrams.

For the rest of the text, we will treat the Machian hypothesis as a formal hypothesis useful for achieving a consistent treatment. In Appendix D.3 we take a quick look at some of the implications of treating this hypothesis as real.

Normalization There are two requirements for the normalization. The first is that the resulting propagator should be a propagator for (55), that in momentum space it should look like:

$$\imath\Delta_\omega(k) \sim \frac{\imath}{w^2 - \vec{k}^2 - m^2 + 2\omega w} \tag{56}$$

The second is that it should obey Feynman boundary conditions, specifically its dependence on clock time should go as:

$$\exp(-\imath\varpi_k\tau)\,\theta(\tau) + \exp(\imath\varpi_k\tau)\,\theta(-\tau) \tag{57}$$

This is required so that the construction of the S matrix in TQM will go in parallel to the construction in SQM. These two requirements strongly constrain the normalization. We can meet them if we take:

$$\frac{1}{\sqrt{2\omega_{\vec{k}}}} \to \frac{1}{\sqrt{2w}} \tag{58}$$

We therefore take as the TQM operator:

$$\phi_\tau(t,\vec{x}) = \sum_{w,\vec{k}} \frac{1}{\sqrt{TV}}\frac{1}{\sqrt{2w}}\left(a_{w,\vec{k}}\mathrm{e}^{-\imath\varpi_k\tau - \imath wt + \imath\vec{k}\cdot\vec{x}} + a^\dagger_{w,\vec{k}}\mathrm{e}^{\imath\varpi_k\tau + \imath wt - \imath\vec{k}\cdot\vec{x}}\right) \tag{59}$$

The differences from the original guess are the normalization $\frac{1}{\sqrt{2w}}$ and the precise definition of ϖ.

Propagator We can now derive the unpacked form of the propagator in close parallel to the derivation for SQM. The propagator is defined by:

$$\imath\Delta_{xy}(x-y) = \langle 0\left|T\left\{\phi_x(x), \phi_y(y)\right\}\right|0\rangle \tag{60}$$

We are using the same conventions and approach as for SQM, but generalizing $\vec{x} \to x, \vec{y} \to y$.

We again break the wave function into its positive and negative frequency parts:

$$\phi_x(x) = \phi_x^+(x) + \phi_x^-(x) \tag{61}$$

As with SQM, for $\tau_x > \tau_y$ most terms are zero. We are left with only:

$$\langle 0\left|\phi_x^+(x)\,\phi_y^-(y)\right|0\rangle \tag{62}$$

We rewrite this in terms of the commutator:

$$\imath\Delta_{xy}^+(x-y) = \langle 0\left|\left[\phi_x^+(x), \phi_y^-(y)\right]\right|0\rangle \tag{63}$$

which we write in turn as the integral:

$$\imath\Delta_{xy}^{+}\left(x-y\right)=\frac{1}{\left(2\pi\right)^{4}}\int d^{4}k\frac{e^{-\imath\varpi_{k}\tau_{xy}-\imath k(x-y)}}{2w} \tag{64}$$

As with SQM, for $\tau_x < \tau_y$ we can get the results for the propagator by interchanging $x \leftrightarrow y$:

$$\imath\Delta_{xy}^{-}\left(x-y\right)=\frac{1}{\left(2\pi\right)^{4}}\int d^{4}k\frac{e^{\imath\varpi_{k}\tau_{xy}+\imath k(x-y)}}{2w} \tag{65}$$

We flip the sign of k to line this up with the positive frequency side. Since ϖ_k is odd in w, it flips sign as well:

$$\imath\Delta_{xy}^{-}\left(x-y\right)=-\frac{1}{\left(2\pi\right)^{4}}\int d^{4}k\frac{e^{-\imath\varpi_{k}\tau_{xy}-\imath k(x-y)}}{2w} \tag{66}$$

The result is the full propagator:

$$\imath\Delta_{xy}\left(x-y\right)=\frac{1}{\left(2\pi\right)^{4}}\int d^{4}k\frac{\mathrm{e}^{-\imath\varpi_{k}\tau_{xy}-\imath k(x-y)}}{2w}\theta\left(\tau\right)-\frac{\mathrm{e}^{-\imath\varpi_{k}\tau_{xy}-\imath k(x-y)}}{2w}\theta\left(-\tau\right) \tag{67}$$

The second term differs in overall sign and in the sign of the clock frequency from the second term for SQM.

We now rewrite the integrand in terms of a contour integral over ω. We do this first to match as closely as possible the development in SQM and secondly to let us write the S matrix expansion in a way that gives us conservation not only of coordinate energy but also of clock energy. We use the representations of the Heaviside unit step function:

$$\begin{aligned}\theta\left(\tau\right)&=\lim_{\varepsilon\to0^{+}}-\frac{1}{2\pi\imath}\int_{-\infty}^{\infty}\frac{1}{\omega+\imath\varepsilon}e^{-\imath\omega\tau}d\omega\\ \theta\left(-\tau\right)&=\lim_{\varepsilon\to0^{+}}\frac{1}{2\pi\imath}\int_{-\infty}^{\infty}\frac{1}{\omega-\imath\varepsilon}e^{-\imath\omega\tau}d\omega\end{aligned} \tag{68}$$

We write the propagator in terms of ω:

$$\exp\left(-\imath\varpi_{k}\tau\right)\theta\left(\tau\right)-\exp\left(-\imath\varpi_{k}\tau\right)\theta\left(-\tau\right)=\frac{\imath}{2\pi}\int_{-\infty}^{\infty}d\omega\exp\left(-\imath\omega\tau\right)\left(\frac{1}{\omega-\varpi_{k}+\imath\varepsilon}+\frac{1}{\omega-\varpi_{k}-\imath\varepsilon}\right) \tag{69}$$

which implies:

$$\imath\Delta_{\omega}\left(k\right)=\frac{1}{2w}\left(\frac{\imath}{\omega-\varpi_{k}+\imath\varepsilon}+\frac{\imath}{\omega-\varpi_{k}-\imath\varepsilon}\right) \tag{70}$$

We multiply out the $2w$ in the denominator to get:

$$\imath\Delta_{\omega}\left(k\right)=\frac{\imath}{w^{2}-\vec{k}^{2}-m^{2}+2w\omega+2w\imath\varepsilon}+\frac{\imath}{w^{2}-\vec{k}^{2}-m^{2}+2w\omega-2w\imath\varepsilon} \tag{71}$$

where both sides are inverses of equation 55, with Feynman boundary conditions.

Conservation of clock energy in TQM The unpacked form makes the physical meaning more transparent: the normalization is obvious, the direction in time is obvious, the fact that the expression is a relativistic invariant is obvious.

But the packed form does show the propagator as a function of ω, the Fourier transform of the clock time. The packed form always travels with an implicit:

$$\exp\left(-\imath\omega\tau\right) \tag{72}$$

which will be used in the inverse Fourier transform back to clock space.

This has considerable practical advantages. At a typical vertex, if we have for example an incoming external line, an outgoing line, and a exchanged photon the associated packed propagators will give factors of:

$$\exp\left(\imath\varpi_{out}\tau\right)\exp\left(\mp\imath\omega_{\gamma}\tau\right)\exp\left(-\imath\varpi_{in}\tau\right) \tag{73}$$

where the factors of $\imath\epsilon$ make sure the sign of the photon part is correct. If we have an overall integral over clock time to $\mp\infty$, then we get integrals of the form:

$$\int\limits_{-\infty}^{\infty} d\tau\exp\left(\imath\varpi_{out}\tau\right)\exp\left(\mp\imath\omega_{\gamma}\tau\right)\exp\left(-\imath\varpi_{in}\tau\right) \tag{74}$$

associated with each vertex. And after we have done all these integrals we have conservation of clock energy at each vertex plus an overall conservation of clock energy for the diagram. In SQM:

$$\delta\left(\sum\Omega_{out}-\sum\Omega_{in}\right) \tag{75}$$

in TQM:

$$\delta\left(\sum\varpi_{out}-\sum\varpi_{in}\right) \tag{76}$$

This works for both SQM and TQM; the math is the same. This is helpful in practical calculations.

But it is also a crutch: it depends in a critical way on being able to take the limits of the integral over clock time to infinity, and therefore limits the applicability of the S matrix in SQM to long times. If the limits are for short times, then the conservation of clock energy will be at best approximate. This is troubling in SQM.

But it is not a problem in TQM: in TQM only the *coordinate energy* is real, the clock energy – like its companion clock time – is ultimately a statistical variable. If it fluctuates a bit here and there, well that is to be expected when you are dealing with statistical variables.

We finesse this problem here by looking only at diagrams where the limits of the clock time integrals may be taken to $\pm\infty$ but which still allow a direct comparison of TQM to SQM.

2.2.3. Long and short time scales The TQM propagator is significantly more complex than the SQM equivalent. The problem is that we are treating time at two different levels: the low level quantum realm where coordinate time is fully symmetric with space and the higher level macroscopic realm of clock time, laboratories, and observers. The propagator, like its complementary equation, is a bridge between two disparate realms of analysis. In practice this can be difficult to work with.

From the LSA the clock term:

(i) is small: $|\varpi_k| << \omega_{\vec{k}}$,

(ii) averages to zero: $\langle \varpi_k \rangle \approx 0$, and

(iii) only takes effect over longer terms i.e. picoseconds.

At short times we expect it will not play much of a role, Therefore it is convenient to split the analysis into long (picoseconds) and short (attosecond) times (with femtosecond times left for negotiation).

Long time scales It will normally take quite a few picoseconds for a wave packet to get from the interaction zone to the detector, arguably enough time for the clock frequency to play a significant role in shaping the wave packet. For instance, if there is decoherence en route, the on-shell terms will be preferentially favored over the off-shell. And if the detector itself is not sensitive to sub-picosecond changes, the detector will be unlikely to see off-shell components of the wave function.

We have therefore a natural way to understand how the wave packet evolves from what is initially a fully four dimensional wave packet (as it leaves the interaction zone) to what appears to be on-shell at it registers at a detector.

Short time scales In SQM the combination of clock frequency and clock time give a clear direction in time: $\omega > 0 \Leftrightarrow \theta(\tau), \omega < 0 \Leftrightarrow \theta(-\tau)$. But in TQM, at sub-picosecond times, we have $\exp(-\imath\varpi_k\tau) \approx 1$ and the clock frequency approximately zero, as likely to be negative as positive. The clock time/clock frequency pair no longer provides reliable directionality in time.

Nor should it.

The clock time is defined as the expectation over the coordinate time, only valid at longer times and for statistical assemblies. Backported to extremely short times and small numbers of individual particles, the use of clock time is suspect. Just as it is improper to infer from the macroscopic behavior of a gas the details of the motion of a specific molecule within it.

Recall our fundamental hypothesis, that coordinate time is to be completely defined by covariance and the rules for the three space dimensions. In SQM the expectation of the three momentum gives the direction in space: $\langle p_x \rangle > 0$ implies we are going in the positive x direction, $\langle p_x \rangle < 0$ that we are going in the negative x direction, and so on. Therefore if we have $\langle w \rangle > 0$ we should be going forwards in time; if $\langle w \rangle < 0$ backwards in time.

(Admittedly this latter case is perhaps less often seen in the laboratory. For discussions of what this might look like in practice see Schulman [58] and also Greenberger and Svozil [59]).

So at short times, the clock time/clock frequency should have little or nothing to do with the direction in time. That should be defined by the wave function itself.

Therefore at short times, we approximate the exponential of the clock time as one. Since the left side is forwards in clock time and the right side backwards, we add the two to get the short-time propagator. (We take $\theta(0) \equiv 1/2$ to have exactly $\theta(\tau) + \theta(-\tau) = 1$).

We therefore take as our propagator for attosecond times:

$$\imath\Delta_\omega^{\mathrm{A}}(k) \approx \frac{\imath}{w^2 - \vec{k}^2 - m^2} \tag{77}$$

With space time form:

$$\imath\Delta_\tau^{\mathrm{A}}(x) = \frac{\imath}{(2\pi)^5}\int d^4k d\omega \frac{\mathrm{e}^{-\imath\omega\tau - \imath kx}}{w^2 - \vec{k}^2 - m^2} \tag{78}$$

We will refer to this as the attosecond propagator, tagging it by a superscript A to make this clear. We expect it will start to fail at picosecond and greater times.

So at attosecond times, we have *no* imposed direction in time. We have no dependence in the propagator on clock time, not even via an $\imath\epsilon$. And we have a natural map to the SQM propagator: $w \to \omega$.

The attosecond propagator directly addresses the question posed in the introduction: what do we get if we apply the replacement $\omega_{\vec{k}} \to w$ to the Feynman propagators? And replace the contour integrals with real ones (by dropping the $\imath\epsilon$'s)?

Quantum energy and quantum time We define the "quantum energy" as the difference between the coordinate energy and the energy value (the classical energy) associated with the parallel SQM calculation. For free particles this is $\delta w \equiv w - \omega_{\vec{k}}$. More generally we can write the quantum energy as the coordinate energy less the value expected from SQM: $\delta E \equiv E - E^S$. This latter definition works within Feynman diagrams as well, where E^S is the "virtual energy" or the energy associated with a virtual particle.

We define in parallel the "quantum time" as the difference between the coordinate time and the clock time: $\delta t \equiv t - \tau$.

We are primarily focused on the quantum energy here, but the quantum time has its uses as well. Both serve as measures of the difference between TQM and SQM. And in that sense summarize the effect we are looking for.

Summary At sub-picosecond times we can use the attosecond time propagator. On the legs, at longer times, we will use the FS/T equation (52) and the associated single particle solutions.

This division makes sense when we are looking at high speed scattering experiments, where the interaction zone is at attosecond scale, but then the products of the interactions take journeys that can be nanoseconds or longer.

In more complex cases we may need to fall back on the full propagator.

Next we develop TQM versions of photon and fermion propagators. The polarization and spin parts will turn out to be relatively minor complications from a TQM point-of-view; we have just navigated the trickier parts of the analysis. Once these are ready, we will turn to applications.

2.3. Photon propagator

We use the same approach here as for spin zero case. The addition of polarization turns out to be an inessential complication from the point of view of TQM.

2.3.1. Photons in SQM

Fock space We have as the basis functions:

$$\frac{1}{\sqrt{2V\omega_{\vec{k}}}}\varepsilon_r^\mu\left(\vec{k}\right)\exp\left(-\imath\omega_{\vec{k}}\tau + \imath\vec{k}\cdot\vec{x}\right) \tag{79}$$

$$\omega_{\vec{k}} \equiv \left|\vec{k}\right| \tag{80}$$

with the polarization vectors:

$$\begin{aligned} \varepsilon_1 &= \begin{pmatrix} 1 & 0 & 0 & 0 \end{pmatrix}, \varepsilon_2 = \begin{pmatrix} 0 & 1 & 0 & 0 \end{pmatrix}, \\ \varepsilon_3 &= \begin{pmatrix} 0 & 0 & 1 & 0 \end{pmatrix}, \varepsilon_4 = \begin{pmatrix} 0 & 0 & 0 & 1 \end{pmatrix} \end{aligned} \tag{81}$$

We build up the associated Fock space from these as above; the only difference is that the Fock space labels have a polarization index as well:

$$\left|\{m_{r\vec{k}}\}\right\rangle \tag{82}$$

We have the usual creation and annihilation operators, indexed by polarization as well as the three space momenta: $a_{r\vec{k}}, a^{\dagger}_{r'\vec{k}'}$. Commutators:

$$\left[a_{r\vec{k}}, a^{\dagger}_{r'\vec{k}'}\right] = \delta_{rr'}\delta^3\left(\vec{k}-\vec{k}'\right) \tag{83}$$

All other commutators are zero.

Field Operators The field operators include a sum over the polarization vectors as well as over the three space momenta:

$$A_{\tau}^{(S)\mu}(\vec{x}) \equiv \sum_{r,\vec{k}} \frac{1}{\sqrt{2V\omega_{\vec{k}}}}\varepsilon_r^{\mu}\left(\vec{k}\right)\begin{pmatrix} a_r\left(\vec{k}\right)\exp\left(-\imath\omega_{\vec{k}}\tau + \imath\vec{k}\cdot\vec{x}\right) \\ +a_r^{\dagger}\left(\vec{k}\right)\exp\left(\imath\omega_{\vec{k}}\tau - \imath\vec{k}\cdot\vec{x}\right)\end{pmatrix} \tag{84}$$

Propagator The propagator is again defined as the vacuum expectation value of the time ordered product of two field operators:

$$\imath D_{21}^{(S)\mu\nu}(\vec{x},\vec{y}) \equiv \langle 0|\, T\left\{A_x^{\mu}(\vec{x}), A_y^{\nu}(\vec{y})\right\}|0\rangle \tag{85}$$

And using the same methods as earlier (see also the more detailed treatment in [45]) we get the unpacked SQM photon propagator:

$$\imath D_{\tau}^{(S)\mu\nu}(\vec{x}) = -\imath g^{\mu\nu}\int \frac{d\vec{k}}{2\omega_{\vec{k}}}\begin{pmatrix}\exp\left(-\imath\omega_{\vec{k}}\tau + \imath\vec{k}\cdot\vec{x}\right)\theta(\tau) \\ +\exp\left(\imath\omega_{\vec{k}}\tau - \imath\vec{k}\cdot\vec{x}\right)\theta(-\tau)\end{pmatrix} \tag{86}$$

We use the same approach as with the SQM spin zero case to rewrite the propagator in terms of an integral over clock frequency:

$$\imath D_{\tau}^{(S)\mu\nu}(\vec{x}) = \frac{-\imath g^{\mu\nu}}{(2\pi)^4}\int d\omega d\vec{k}\frac{\exp\left(-\imath\omega\tau + \imath\vec{k}\cdot\vec{x}\right)}{\omega^2 - \vec{k}^2 + \imath\varepsilon} \tag{87}$$

$$\imath D_{\omega}^{(S)\mu\nu}\left(\vec{k}\right) = \frac{-\imath g^{\mu\nu}}{\omega^2 - \vec{k}^2 + \imath\varepsilon} \tag{88}$$

The same comments about the virtual character of the SQM particles earlier apply here as well.

2.3.2. Photons in TQM From the perspective of TQM, the SQM photon is a hybrid of 3D and 4D approaches: the time coordinate is clock time, but the vector field is basically a four dimensional object, requiring no adjustment to promote it to TQM.

In Lorenz gauge the individual components of the vector potential obey the Klein-Gordon equation. Therefore in TQM the application of the Machian hypothesis gives:

$$-2w\imath\frac{\partial}{\partial\tau}A^{\nu} = \left(w^2 - \vec{k}^2\right)A^{\nu} \tag{89}$$

with 4D solutions:

$$A_{\tau}^{(k)\mu}(x) = \frac{\varepsilon_r^{\mu}(k)\exp\left(-\imath\varpi_k\tau - \imath kx\right)}{\sqrt{2TVw}} \tag{90}$$

and clock frequency as defined above (with the exception of no mass term):

$$\varpi_k \equiv -\frac{w^2 - \vec{k}^2}{2w} \tag{91}$$

Note that the polarization part is the same in the SQM and TQM free wave functions.

Fock space In TQM the Fock space is built up from the 4D solutions, appropriately symmetrized:

$$|\{m_{rk}\}\rangle \tag{92}$$

The index r to the polarization vectors is unchanged, but the three vector $\vec{k}$ is promoted to a four vector $w, \vec{k}$. We have the 4D commutation relations in complete parallel:

$$\left[a_{rk}, a^{\dagger}_{r'k'}\right] = \delta_{rr'}\delta^4\left(k - k'\right) \tag{93}$$

All other commutators are zero.

Field Operators The field operator is similarly uncomplicated:

$$A^{\mu}_{\tau}\left(t, \vec{x}\right) \equiv \sum_{r,w,\vec{k}} \frac{1}{\sqrt{2TVw}} \varepsilon^{\mu}_{r}\left(w, \vec{k}\right) \begin{pmatrix} a_{rk} \exp\left(-\imath\varpi_k\tau - \imath wt + \imath\vec{k}\cdot\vec{x}\right) \\ +a^{\dagger}_{rk} \exp\left(\imath\varpi_k\tau + \imath wt - \imath\vec{k}\cdot\vec{x}\right) \end{pmatrix} \tag{94}$$

Feynman propagator The Feynman propagator is the vacuum expectation value of the time ordered product of the TQM vector potentials:

$$\imath D^{\mu\nu}_{\tau} \equiv \langle 0 \left| T\left\{A^{\mu}_{\tau}\left(x\right), A^{\nu}_{\tau}\left(y\right)\right\}\right| 0\rangle \tag{95}$$

By the same procedure as for spin zero (equation 67):

$$\imath D^{\mu\nu}_{\tau}\left(x\right) = -\frac{\imath g^{\mu\nu}}{\left(2\pi\right)^4} \int \frac{dwd\vec{k}}{2w} \left(\exp\left(-\imath\varpi_k\tau - \imath kx\right)\theta\left(\tau\right) - \exp\left(-\imath\varpi_k\tau + \imath kx\right)\theta\left(-\tau\right)\right) \tag{96}$$

in momentum space:

$$\imath D^{\mu\nu}_{\tau}\left(k\right) \equiv -\imath\frac{g^{\mu\nu}}{\left(2\pi\right)^4}\left(\frac{\exp\left(-\imath\varpi_k\tau\right)\theta\left(\tau\right)}{2w} - \frac{\exp\left(-\imath\varpi_k\tau\right)\theta\left(-\tau\right)}{2w}\right) \tag{97}$$

We use the same approach as with the TQM spin zero case to rewrite the propagator in terms of an integral over clock frequency:

$$\imath D^{\mu\nu}_{\tau}\left(k\right) \equiv \frac{-\imath g^{\mu\nu}}{\left(2\pi\right)^5} \int_{-\infty}^{\infty} d\omega \frac{\exp\left(-\imath\omega\tau\right)}{2w} \begin{pmatrix} \frac{\imath}{w^2 - \vec{k}^2 + 2w\omega + 2w\imath\varepsilon} \\ + \frac{\imath}{w^2 - \vec{k}^2 + 2w\omega - 2w\imath\varepsilon} \end{pmatrix} \tag{98}$$

In momentum space:

$$\imath D^{\mu\nu}_{\omega}\left(k\right) = \frac{-\imath g^{\mu\nu}}{2w}\left(\frac{\imath}{w^2 - \vec{k}^2 + 2w\omega + 2w\imath\varepsilon} + \frac{\imath}{w^2 - \vec{k}^2 + 2w\omega - 2w\imath\varepsilon}\right) \tag{99}$$

or:

$$\imath D_{\omega}^{\mu\nu}\left(k\right) = -\imath g^{\mu\nu}\Delta_{\omega}\left(k\right) \tag{100}$$

So the photon propagator is a polarization wrapper for the spin zero propagator. This point considerably simplifies the subsequent analysis. From the arguments above, we have also the attosecond form:

$$\imath D_{\omega}^{(A)\mu\nu}\left(k\right) = -\imath g^{\mu\nu}\frac{\imath}{w^2 - \vec{k}^2} \tag{101}$$

2.4. Dirac propagator

We now extend TQM to include the Dirac propagator (Dirac [60]).

2.4.1. Dirac propagator in SQM

Dirac equation We start by reviewing the Dirac equation in a way that will prepare for the extension to TQM. Dirac equation in SQM:

$$\left(\imath\gamma_o\partial_\tau + \imath\vec{\gamma}\cdot\nabla - m\right)\psi = 0 \tag{102}$$

We use standard choices for γ^μ:

$$\gamma^0 \equiv \begin{pmatrix} I & 0 \\ 0 & -I \end{pmatrix}, \gamma^{1,2,3} \equiv \begin{pmatrix} 0 & \sigma^{1,2,3} \\ -\sigma^{1,2,3} & 0 \end{pmatrix} \tag{103}$$

This has positive solutions (particles):

$$\psi_{1,2}^{(S)\vec{p}}\left(\tau,\vec{x}\right) = u_{1,2}^S\left(\vec{p}\right)\exp\left(-\imath E_{\vec{p}}\tau + \imath\vec{p}\cdot\vec{x}\right) \tag{104}$$

$$u_1^S = \sqrt{\frac{E_{\vec{p}}+m}{2m}}\begin{pmatrix} 1 \\ 0 \\ \frac{p^3}{E_{\vec{p}}+m} \\ \frac{p^1+\imath p^2}{E_{\vec{p}}+m} \end{pmatrix}, u_2^S = \sqrt{\frac{E_{\vec{p}}+m}{2m}}\begin{pmatrix} 0 \\ 1 \\ \frac{p^1-\imath p^2}{E_{\vec{p}}+m} \\ -\frac{p^3}{E_{\vec{p}}+m} \end{pmatrix} \tag{105}$$

and negative solutions (anti-particles):

$$\psi_{3,4}^{(S)\vec{p}}\left(\tau,\vec{x}\right) = v_{2,1}^S\exp\left(\imath E_{\vec{p}}\tau - \imath\vec{p}\cdot\vec{x}\right) \tag{106}$$

$$v_2^S = \sqrt{\frac{E_{\vec{p}}+m}{2m}}\begin{pmatrix} \frac{p^3}{E_{\vec{p}}+m} \\ \frac{p^1+\imath p^2}{E_{\vec{p}}+m} \\ 1 \\ 0 \end{pmatrix}, v_1^S = \sqrt{\frac{E_{\vec{p}}+m}{2m}}\begin{pmatrix} \frac{p^1-\imath p^2}{E_{\vec{p}}+m} \\ \frac{-p^3}{E_{\vec{p}}+m} \\ 0 \\ 1 \end{pmatrix} \tag{107}$$

These solutions define a Fock space:

$$\left|\left\{n_{s\vec{p}}\right\}\right\rangle \tag{108}$$

where $n_{r\vec{p}}$ the occupation number is 0 or 1; s indexes u_1, u_2, v_2, v_1, and the product wave functions are anti-symmetric under interchange. As with spin zero, we define creation and annihilation operators to navigate Fock space. The $c^\dagger, c$'s create and annihilate the particles; the $d^\dagger, d$'s the anti-particles. Their anti-commutation relations are:

$$\left[c_{s\vec{p}}, c^{\dagger}_{s'\vec{p}'}\right]_{+} = \left[d_{s\vec{p}}, d^{\dagger}_{s'\vec{p}'}\right]_{+} = \delta_{ss'}\delta^{3}\left(\vec{p} - \vec{p}'\right) \tag{109}$$

with all other anti-commutators zero.

Field Operators We define the field operators in parallel with those in the spin zero case, but with anti-commutators for commutators. The normalization is given by the $\sqrt{\frac{m}{VE_{\vec{p}}}}$. $c, c^{\dagger}$ are the annihilation and creation operators for $u^{S}_{1,2}$; $d, d^{\dagger}$ the annihilation and creation operators for $v^{S}_{1,2}$. The resulting field operator is:

$$\psi^{S}_{\tau}\left(\vec{x}\right) = \sum_{s=1}^{2}\sum_{\vec{p}}\sqrt{\frac{m}{VE_{\vec{p}}}}\begin{pmatrix} c_{s\vec{p}}u^{S}_{s}\left(\vec{p}\right)\exp\left(-\imath E_{\vec{p}}\tau + \imath\vec{p}\cdot\vec{x}\right) \\ +d^{\dagger}_{s\vec{p}}v^{S}_{s}\left(\vec{p}\right)\exp\left(\imath E_{\vec{p}}\tau - \imath\vec{p}\cdot\vec{x}\right)\end{pmatrix} \tag{110}$$

We define adjoints $\bar{u} \equiv u^{\dagger}\gamma_{0}, \bar{v} \equiv v^{\dagger}\gamma_{0}, \bar{\psi} \equiv \psi^{\dagger}\gamma_{0}$:

$$\bar{\psi}^{S}_{\tau}\left(\vec{x}\right) = \sum_{s=1}^{2}\sum_{\vec{p}}\sqrt{\frac{m}{VE_{\vec{p}}}}\begin{pmatrix} d_{s\vec{p}}\bar{v}^{S}_{s}\left(\vec{p}\right)\exp\left(-\imath E_{\vec{p}}\tau + \imath\vec{p}\cdot\vec{x}\right) \\ +c^{\dagger}_{s\vec{p}}\bar{u}^{S}_{s}\left(\vec{p}\right)\exp\left(\imath E_{\vec{p}}\tau - \imath\vec{p}\cdot\vec{x}\right)\end{pmatrix} \tag{111}$$

Dirac propagator The derivation of the Feynman propagator runs in parallel to the derivation for spin zero, except for replacing the simple wave functions with spinors and the commutators with anti-commutators. We define the Feynman propagator as time-ordered vacuum expectation value:

$$\imath S^{S}_{xy}\left(\vec{x} - \vec{y}\right) \equiv \left\langle 0\left|T\left\{\psi^{S}_{x}\left(\vec{x}\right), \bar{\psi}^{S}_{y}\left(\vec{y}\right)\right\}\right|0\right\rangle \tag{112}$$

Inside a vacuum sandwich, only the $cc^{\dagger}, dd^{\dagger}$ terms survive. This lets us rewrite the propagator in terms of the anti-commutators using the same procedure as for the spin zero case. We have the particle case when $\tau_{x} > \tau_{y}$:

$$\imath S^{S+}_{xy}\left(\vec{x} - \vec{y}\right) = \left\langle 0\left|\left[\psi^{S+}_{x}\left(\vec{x}\right), \bar{\psi}^{S-}_{y}\left(\vec{y}\right)\right]_{+}\right|0\right\rangle \tag{113}$$

and the anti-particle case when $\tau_{y} > \tau_{x}$:

$$\imath S^{S-}_{xy}\left(\vec{x} - \vec{y}\right) = -\left\langle 0\left|\left[\bar{\psi}^{S+}_{y}\left(\vec{y}\right), \psi^{S-}_{x}\left(\vec{x}\right)\right]_{+}\right|0\right\rangle \tag{114}$$

Then for the first term we have:

$$\imath S^{S+}_{xy}\left(\vec{x} - \vec{y}\right) = \frac{m}{\left(2\pi\right)^{3}}\int d\vec{p}\sum_{s=1,2}u^{S}_{s}\left(\vec{p}\right)\bar{u^{S}}_{s}\left(\vec{p}\right)\frac{\exp\left(-\imath E_{\vec{p}}\tau_{xy} + \imath\vec{p}\cdot\left(\vec{x} - \vec{y}\right)\right)}{E_{\vec{p}}} \tag{115}$$

We use:

$$\frac{\not{p}^{S} + m}{2m} = \sum_{s=1,2}u^{S}_{s}\left(\vec{p}\right)\bar{u}^{S}_{s}\left(\vec{p}\right) \tag{116}$$

to get the positive frequency part of the propagator as:

$$\imath S^{S+}_{xy}\left(\vec{x} - \vec{y}\right) = \frac{1}{\left(2\pi\right)^{3}}\int d\vec{p}\left(\not{p}^{S} + m\right)\frac{\exp\left(-\imath E_{\vec{p}}\tau_{xy} + \imath\vec{p}\cdot\left(\vec{x} - \vec{y}\right)\right)}{2E_{\vec{p}}} \tag{117}$$

We define $\not{p}^S \equiv \gamma_0 E_{\vec{p}} - \vec{\gamma} \cdot \vec{p}$ to distinguish this construction from $\not{p} \equiv \gamma_0 E - \vec{\gamma} \cdot \vec{p}$, needed for TQM.

For the negative frequency part we use:

$$\frac{\not{p}^S - m}{2m} = \sum_{s=1,2} v_s^S(\vec{p})\, \bar{v}_s^S(\vec{p}) \tag{118}$$

to get the negative frequency part of the propagator as:

$$\imath S_{xy}^{S-}(\vec{x} - \vec{y}) = -\frac{1}{(2\pi)^3} \int d\vec{p}\, (\not{p}^S - m) \frac{\exp\left(\imath E_{\vec{p}} \tau_{xy} - \imath \vec{p} \cdot (\vec{x} - \vec{y})\right)}{2E_{\vec{p}}} \tag{119}$$

Putting the two pieces back together:

$$\imath S_{xy}^{S}(\vec{x} - \vec{y}) = \frac{1}{(2\pi)^3} \int d\vec{p} \begin{pmatrix} (\not{p}^S + m) \frac{\exp\left(-\imath E_{\vec{p}} \tau_{xy} + \imath \vec{p} \cdot (\vec{x} - \vec{y})\right)}{2E_{\vec{p}}} \theta(\tau_{xy}) \\ -(\not{p}^S - m) \frac{\exp\left(\imath E_{\vec{p}} \tau_{xy} - \imath \vec{p} \cdot (\vec{x} - \vec{y})\right)}{2E_{\vec{p}}} \theta(-\tau_{xy}) \end{pmatrix} \tag{120}$$

This is again the "unpacked form":

$$\imath S_\tau^S(\vec{x}) = \frac{\imath}{(2\pi)^4} \int d\omega d\vec{p} \frac{\not{p}^S + m}{\omega^2 - \vec{p}^2 - m^2 + \imath\varepsilon} \exp(-\imath\omega\tau + \imath \vec{p} \cdot \vec{x}) \tag{121}$$

As with the spin zero case, once the three space momenta have been picked, the value of the energy component is forced: there is no dispersion in energy.

The momentum space form is:

$$\imath S_\tau^S(\vec{p}) = \imath \frac{\not{p}^S + m}{\omega^2 - \vec{p}^2 - m^2 + \imath\varepsilon} \tag{122}$$

or:

$$\imath S_\omega^S(\vec{p}) = \imath \left(\not{p}^S + m\right) \Delta_\omega^S(\vec{p}) \tag{123}$$

2.4.2. Dirac propagator in TQM

Dirac equation We now add in the time coordinate. We again apply the Machian hypothesis, this time to get the TQM form of the Dirac equation:

$$\left(\not{p} - \not{\mathcal{P}} - m\right) \psi_\tau = 0 \tag{124}$$

As in the spin zero case:

$$(\not{p} - m)\, \psi_\tau = \mathcal{P}\psi_\tau, \mathcal{P}\psi_\tau \to \mathcal{M}\psi_\tau \to \imath\gamma_0 \frac{\partial}{\partial \tau_V} \psi_\tau \to -\imath\gamma_0 \frac{\partial}{\partial \tau} \psi_\tau \tag{125}$$

The result is the TQM form of the Dirac equation:

$$(\not{p} - m)\, \psi_\tau = -\imath\gamma_0 \frac{\partial}{\partial \tau} \psi_\tau \tag{126}$$

Dirac propagator To get the corresponding propagator we use the Machian hypothesis to form an ansatz for the TQM propagator then plug this ansatz in the equation to verify it is correct.

First we observe that Dirac's approach had its roots in the Klein-Gordon equation: it was an attempt to get a version of the Klein-Gordon equation which did not have negative energy solutions. We can see this directly from the Dirac equation. Apply $\not p + m$ to it using $\not p \not p = p^2$:

$$(\not p + m)(\not p - m)\psi = \left(p^2 - m^2\right)\psi = 0 \tag{127}$$

This is still true in TQM:

$$\left((\not p - \not{\mathcal{P}}) + m\right)\left((\not p - \not{\mathcal{P}}) - m\right)\psi = \left((p - \mathcal{P})^2 - m^2\right)\psi = 0 \tag{128}$$

We extend (122) in the obvious way: replacing 3D with 4D objects and replacing p with $p - \mathcal{P}$:

$$\imath S_\tau(p) = \int d\omega \frac{(\not p - \not{\mathcal{P}}) + m}{\left((p - \mathcal{P})^2 - m^2\right) + \imath\varepsilon} \exp(-\imath\omega\tau) \tag{129}$$

We can see that this propagator is the inverse of the TQM Dirac equation 128.

To see more clearly what this means we again apply the Machian hypothesis, this time to get ω:

$$\mathcal{P} \to \left(\mathcal{M}, \vec{0}\right) \to \left(\imath\frac{\partial}{\partial\tau_V}, \vec{0}\right) \to \left(-\imath\frac{\partial}{\partial\tau}, \vec{0}\right) \to \left(-\omega, \vec{0}\right) \tag{130}$$

We get:

$$\imath S_\tau(p) = \int d\omega \frac{(\not p + \gamma_0\omega) + m}{E^2 - \vec{p}^2 - m^2 + 2E\omega + \omega^2 + \imath\varepsilon} \exp(-\imath\omega\tau) \tag{131}$$

We get a term proportional to ω in the numerator and terms proportional to ω, ω^2 in the denominator. We drop the ω^2 term as on the grounds that if the $E\omega$ term is small, the ω^2 term is going to be exceptionally small. The term linear in ω in the numerator will be replaced by ϖ when we do the contour integral. This in turn will average to zero to lowest order. We will drop this term for now as well.

Therefore we take as the lowest order TQM propagator for fermions:

$$\imath S_\omega(p) = \imath \left(\begin{array}{c} \dfrac{(\not p + m)}{E^2 - \vec{p}^2 - m^2 + 2\omega E + 2E\imath\varepsilon} \\ + \dfrac{(\not p + m)}{E^2 - \vec{p}^2 - m^2 + 2\omega E - 2E\imath\varepsilon} \end{array} \right) \tag{132}$$

or in "unpacked" form:

$$\imath S_\tau(p) = \frac{\imath}{(2\pi)^4} \left(\frac{e^{-\imath\varpi_p\tau}}{2E}\theta(\tau) - \frac{e^{-\imath\varpi_p\tau}}{2E}\theta(-\tau) \right) \tag{133}$$

From the arguments above, we have the attosecond form:

$$\imath S_\omega^{A}(p) = \imath \frac{(\not p + m)}{E^2 - \vec{p}^2 - m^2} \tag{134}$$

We have implicitly promoted $E_{\vec{p}} \to E$ in the various elements of the spinor formalism:

$$u^S(\vec{p}) \to u(p), v^S(\vec{p}) \to v(p), \sqrt{\frac{m}{E_{\vec{p}}}} \to \sqrt{\frac{m}{E}} \tag{135}$$

2.5. Vertexes

2.5.1. QED interaction in SQM For interactions, we take the Fock space as the product of the photon and fermion Fock spaces. Here it is enough to consider one species of fermion, say electrons. For photon and electrons we have:

$$\left|\left\{n_{s\vec{p}}^{e}\right\}\right\rangle\left|\left\{m_{r\vec{k}}^{\gamma}\right\}\right\rangle \tag{136}$$

The interaction term is:

$$-e\bar{\psi}_{\tau}^{S}\left(\vec{x}\right)A_{\tau}^{(S)\nu}\left(\vec{x}\right)\gamma_{\nu}\psi_{\tau}^{S}\left(\vec{x}\right) \tag{137}$$

The dependence on the coordinates is already absorbed into the definition of the Feynman diagram, so this is reduced to sums over terms of the form:

$$-e\bar{\psi}^{S}\left(\vec{p}'\right)A_{\tau}^{(S)\nu}\left(\vec{k}\right)\gamma_{\nu}\psi^{S}\left(\vec{p}\right) \tag{138}$$

The $\psi^{S}\left(\vec{p}\right)$ are built up over sums of the $u_{s}^{S}\left(\vec{p}\right), v_{s}^{S}\left(\vec{p}\right)$. For an electron $e = -\left|e\right|$.

2.5.2. QED interaction in TQM The full Fock space in TQM is also a product of the fermion and photon Fock spaces:

$$\left|\left\{n_{sp}^{e}\right\}\right\rangle\left|\left\{m_{sk}^{\gamma}\right\}\right\rangle \tag{139}$$

The vertex term is:

$$-e\bar{\psi}_{\tau}\left(t,\vec{x}\right)A_{\tau}^{\nu}\left(t,\vec{x}\right)\gamma_{\nu}\psi_{\tau}\left(t,\vec{x}\right) \tag{140}$$

Again, the dependence on the coordinates is already absorbed into the definition of the Feynman diagram, so this is reduced to sums over terms of the form:

$$-e\bar{\psi}\left(p'\right)A_{\tau}^{\nu}\left(k\right)\gamma_{\nu}\psi\left(p\right) \tag{141}$$

The $\psi\left(p\right)$ are built up over sums of the $u_{s}\left(p\right), v_{s}\left(p\right)$.

2.5.3. Conservation of momentum at a vertex In SQM the integrals over the 3D plane waves give conservation of three momentum at each vertex:

$$\delta^{3}\left(\vec{p}_{out}-\vec{p}_{in}\right) \tag{142}$$

As noted earlier, we get conservation of clock energy at a vertex from the integrals over clock time. For this to work we need to take the limit as $\tau \to \pm\infty$. If we are looking at short time S matrix elements – legitimate if infrequent – we will not get exact conservation of clock energy at each vertex, but only an approximation thereof, corresponding to *not* taking the limit in:

$$\lim_{T\to\infty}\int_{-T}^{T}\frac{d\tau}{2\pi}\exp\left(-\imath\left(\omega_{out}-\omega_{in}\right)\tau\right)=\lim_{T\to\infty}\frac{\sin\left(\left(\omega_{out}-\omega_{in}\right)T\right)}{\pi\left(\omega_{out}-\omega_{in}\right)}=\delta\left(\omega_{out}-\omega_{in}\right) \tag{143}$$

If the time range T is sufficiently small the difference between the *sin* and the δ function could become noticeable. So we see that in SQM the clock energy is not present on exactly the same basis as the three space momenta.

To a reasonable first approximation, TQM may be thought of as SQM with a fourth space dimension. So we get conservation of 4D momentum at each vertex:

$$\delta^4 \left(p_{out} - p_{in}\right) \tag{144}$$

If we able to take the limit as clock time goes to infinity, we get a fifth conservation condition at each vertex and over the entire Feynman diagram:

$$\delta \left(\sum \varpi_{out} - \sum \varpi_{in}\right) \tag{145}$$

but we can easily do without this at shorter times.

To keep the comparisons between SQM and TQM as straightforward as possible, we will finesse this problem by looking primarily at the scattering of GTFs which only intersect for very short times. We can therefore extend the integrals to infinity at will since, except over the short interaction zone, the integrand is zero.

2.5.4. Normalization factors To go from SQM to TQM we changed the normalization on the field operators from $\frac{1}{\sqrt{2\omega_{\vec{k}}}}$ to $\frac{1}{\sqrt{2w}}$ and from $\sqrt{\frac{m}{E_{\vec{p}}}}$ to $\sqrt{\frac{m}{E}}$.

At any vertex, the field operators either connect to a propagator or to an outside line. If they connect to a propagator their normalization factor joins up with the normalization from the other side to give normalizations which are accounted for in the calculation of the propagators:

$$\frac{1}{\sqrt{2\omega_{\vec{k}}}}\frac{1}{\sqrt{2\omega_{\vec{k}}}} \to \frac{1}{2\omega_{\vec{k}}}, \frac{1}{\sqrt{2w}}\frac{1}{\sqrt{2w}} \to \frac{1}{2w}, \ldots \tag{146}$$

If on the other hand they are connected to an external line the normalization is passed to the overall definition of the S; we get the familiar exterior factors, as:

$$\prod^{\substack{exterior\\boson}} \sqrt{\frac{1}{2\omega_{\vec{k}}}}, \prod^{\substack{exterior\\boson}} \sqrt{\frac{1}{2w}}, \ldots \tag{147}$$

In SQM there is a conventional understanding that on the legs the virtual particles will drop off and we will be left with the true 3D on-shell wave function. But in TQM there is no such transition: the wave packets on the legs are fully four dimensional in character.

So ultimately, in TQM if we consider the "Feynman diagram of the universe" we can divide it up into S matrices as we please. Where the legs of two S matrices meet, the two associated exterior factors will join to help form the propagator for the leg. There is no physical transition from the system *under* examination to the system *doing* the examination; it is a matter of convention and names where and how we choose to make the divisions.

3. Applications

> "In other words, we are trying to prove ourselves wrong as quickly as possible, because only in that way can we find progress." – Richard P. Feynman [61]

With the basic tools built, we apply them to a starter set of cases:

(i) Free particles. The creation, propagation, and detection of wave functions which are dispersed in time presents specific questions.

(ii) Spin-zero particle exchange. This serves as the first real test of how TQM plays out in practice.

(iii) Simple mass correction loop. We show that renormalization is not needed in TQM; the combination of dispersion in time and entanglement in time keep the loop integrals finite.

(iv) The three basic second order scattering diagrams in QED:

(a) Møller scattering. Electron-electron scattering by photon exchange.
(b) Bhabha scattering. Electron-positron scattering by photon exchange and by photon creation/annihilation.
(c) Compton scattering. Electron-photon scattering.

This starter set is sufficient to expose a fair number of experimental possibilities; we work out the general Feynman rules in Appendix C.

3.1. Free particles

"nothing can be created out of nothing" – Lucretius [62]

To see the effects associated with dispersion in time we need to look at wave functions which have finite dispersion in time/energy. Plane waves and δ functions are equally unsuitable. Gaussian test functions (Appendix B) are well-suited as models for this: they are easy to work with and they are completely general – any wave function may be represented as a sum over GTFs using Morlet wavelet analysis.

In SQM the use of GTFs is a convenience, but in TQM the use of GTFs is mandatory: the convergence of the path integrals in the single particle case and in loop diagrams depends on the use of GTFs.

To lay a correct foundation for the examination of scattering problems in TQM we will need to look at the initial values, the propagation, and the detection – the birth, life and death – of GTFs.

3.1.1. Initial wave function The first problem is how do we create a GTF of known dispersion in time/energy? This problem is solved in paper A. There we showed that if we are given the norm, average momentum, and average dispersion in momentum we can get a maximum entropy estimate of the corresponding GTF in energy using the method of Lagrange multipliers.

As maximum entropy estimates tend to be robust, we have what we need for falsifiability.

We summarize for use here the results of paper A. We assume we have a description of the wave function in momentum space with the expectation, average relativistic mass, and dispersion in relativistic mass defined:

$$\begin{aligned} &\langle 1\rangle = 1\\ &\bar{E} \equiv \langle E\rangle = \sqrt{m^2 + \langle\vec{p}\rangle^2}\\ &\left\langle E^2\right\rangle = \left\langle m^2 + \vec{p}^2\right\rangle = m^2 + \left\langle\vec{p}^2\right\rangle \end{aligned} \tag{148}$$

$\bar{E}$ is the average relativistic mass.

By the method of Lagrange multipliers the corresponding GTF in energy is:

$$\varphi_0(E) = \sqrt[4]{\frac{1}{\pi\sigma_E^2}} e^{\imath(E-\bar{E})\tau_0 - \frac{(E-\bar{E})^2}{2\sigma_E^2}} \tag{149}$$

$$\sigma_E^2 = 2(\Delta E)^2 = 2\left(\left\langle E^2\right\rangle - \bar{E}^2\right) \tag{150}$$

With this estimate of the GTF in energy we can get the corresponding GTF in time by taking the Fourier transform. In this case we have:

$$\varphi_0(t) \equiv \sqrt[4]{\frac{1}{\pi\sigma_t^2}} e^{-\imath\bar{E}t - \frac{(t-\tau_0)^2}{2\sigma_t^2}}, \sigma_t = \frac{1}{\sigma_E} \tag{151}$$

The dispersions in energy are likely to be of order eV, if we are starting with atomic wave functions. Assuming we are starting with minimum uncertainty packets (expected) the uncertainty in time will be the inverse of this:

$$\Delta t = \frac{1}{\Delta E} \tag{152}$$

and therefore of order attoseconds.

We will refer to this as the "entropic estimate". It will not pick up any complex structure: to the entropic estimate, everything looks like an s state. But it should get the order-of-magnitude of the time part of the wave function right, which is the critical marker for falsifiability. If no more specific estimate is available, it should do.

3.1.2. Propagation of free wave functions in clock time In general, the three or four dimensional GTFs in momentum space are solutions of the corresponding free equation with appropriate choices of the clock frequency. If we have the wave function as time zero as a function of momentum then we have it at time τ later. In SQM:

$$\varphi_\tau^S(\vec{p}) = \varphi_0^S(\vec{p}) \exp\left(-\imath E_{\vec{p}}\tau\right) \tag{153}$$

In TQM:

$$\varphi_\tau(E,\vec{p}) = \varphi_0(E,\vec{p}) \exp\left(-\imath\varpi_p\tau\right) \tag{154}$$

These two results are exact.

The square root in the definition of the relativistic mass for SQM and the $1/2w$ factor in the definition of the clock frequency for TQM can make these two expressions difficult to work with analytically. A quadratic approximation offers useful insight. We will use this approach when analyzing the loop corrections, for instance. Note that $\varpi_p = 0$ when we are exactly on-shell, increasing quadratically as we get far enough off-shell:

$$\varpi_p = -\frac{\left(\bar{E}+\delta E\right)^2 - \bar{E}^2}{2\left(\bar{E}+\delta E\right)} \approx -\delta E + \frac{(\delta E)^2}{2\bar{E}} \tag{155}$$

We expect therefore that if TQM wave packets start on-shell (per the entropic estimate) they will tend to stay approximately on-shell subsequently. This means we can take the initial wave packets in a scattering experiment as on-shell. We will take advantage of this in the next subsection.

3.1.3. Detection of the wave function The issues associated with the detection of a wave function in time were discussed at length in paper B [51]. To keep this work self-contained we give a summary.

Effect of convoluted paths The paths in TQM are significantly more complex than those in SQM. In particular while paths in SQM can go left and right, up and down, forwards and back, paths in TQM can also go into the future and into the past. We expect these excursions will be of order attoseconds and be centered around the classical or SQM values. (This will keep us from attracting the unwelcome attention of the time police.)

When we calculate the paths in TQM we typically calculate the wave function at the detector as a function of clock time:

$$\varphi_\tau(t,x) \tag{156}$$

This gives the sum not of all paths to t, x but only of those paths with length τ. To get the full amplitude at t, x we will need to sum over paths of all lengths:

$$\varphi(t, x) = \int d\tau \varphi_\tau(t, x) \tag{157}$$

We are, as required, taking coordinate time as fundamental with clock time playing a secondary role, here letting us categorize the paths by length.

The problem of measurement

> "To re-emphasize, from a broader perspective, the main argument that is being articulated in this section is that quantum measurements, in the interferometric and polarization domains, can be described without resorting to the concept of the collapse of the wave function or the collapse of the probability amplitude." – p185 Duarte and Taylor [63]

The fundamental problem here is that ultimately we have to treat the detectors (and even the observers) as also quantum mechanical systems. They are, after all, made of atoms, and atoms are unavoidably quantum systems. There is no such thing as a classical atom – in classical mechanics electrons spiral into the nucleus in a minute fraction of a second due to loss of energy from Larmor radiation [64].

There is of course an extraordinary literature on the "problem of measurement". This problem is not a central focus of this work. In paper B, where this problem was a central focus, we took as a starting point an analysis by Marchewka and Schuss [65, 66, 67, 68] who made a strong argument (from probability conservation) that the probability current correctly gives the detection rate.

In a certain sense this merely postpones the problem: to compute the probability current you must first compute the wave function, which means that you have to first solve the problem of the interaction of the particle's wave function with that of the detector – including absorption, reflection, loss, emission, emission followed by re-absorption, and so on. Rather a lot to consider.

However if you are prepared to posit an ideal detector, with no loss or lag, you can use the probability current at the detector as giving a reasonable estimate of the detection rate as a function of time. The beauty of this approach, in the current context, is that it:

(i) is easy to use,
(ii) should give a reasonable first estimate in general,
(iii) is typical of what workers often do, so clearly not that bad in practice,
(iv) and most importantly here, works the same way for SQM and TQM, so creates no bias.

In SQM We take a detector in the y-z plane placed at $x = L$. For a simple GTF headed, say, left to right with momentum p_0, the detection rate is:

$$D_\tau^S = \frac{p_0}{m} \rho_\tau(x)\big|_{x=L} \tag{158}$$

$\rho_\tau(x)$ has a dispersion in space given by σ_x. But it is the resulting dispersion of the detection rate D_τ^S in clock time that we are interested in here. This is:

$$\sigma_\tau^S = \frac{1}{m v_0 \sigma_x} \bar{\tau} \tag{159}$$

This is proportional to $\bar{\tau}$ the average time of flight: the greater the average time of flight the greater the uncertainty in clock time at the detector. It is inversely proportional to σ_x because of diffraction (see Appendix B.3.2). And it is inversely proportional to the speed of the packet: the slower the packet the more spread out it will be by the time it reaches the detector. The packet functions like a train; the slower the train the longer the time from when the engine arrives at the station till the time when the caboose does.

The corresponding uncertainty in clock time, the number we are most interested in, is:

$$\Delta_{\tau}^{S} = \frac{1}{\sqrt{2}}\sigma_{\tau}^{S} \tag{160}$$

We are seeing uncertainty in clock time, but it is *entirely* the product of dispersion in space. As Busch and other workers have pointed out, in SQM there is a HUP for time/energy, but it is not on the same basis at all as the HUP for momentum/space.

In TQM In TQM we will have dispersion in time in addition to the dispersion in space. Suppose the initial wave function is a product of time and space parts:

$$\varphi_{\tau}(t,x) = \varphi_{\tau}^{T}(t)\,\varphi_{\tau}^{S}(x) \tag{161}$$

We may be getting the wave function in time from the entropic estimate. And we are ignoring the more general possibility that time and space are entangled. Using again the probability current, now in TQM, we showed that the detection rate is now the product of the time and space parts:

$$D_{\tau}(t) = D_{\tau}^{S}\rho_{\tau}^{T}(t) \tag{162}$$

where the probability distribution in coordinate time as a function of clock time is:

$$\rho_{\tau}^{T}(t) \approx \sqrt{\frac{1}{\pi\sigma_{\tau}^{(T)2}}}e^{-\frac{1}{\sigma_{\tau}^{(T)2}}(t-\tau)^{2}}, \sigma_{\tau}^{T} \equiv \frac{\bar{\tau}}{m\sigma_{t}} \tag{163}$$

σ_{τ}^{T} is proportional to the average time to the detector. It is inversely proportional to the initial dispersion in coordinate time σ_{t}. To get the full detection rate at coordinate time t we have to sum over the clock time:

$$D(t) = \int d\tau D_{\tau}^{S}\rho_{\tau}^{T}(t) \tag{164}$$

This is a convolution of the S and the T distributions. The total dispersion squared is the sum of the squares of the dispersions:

$$\sigma_{t}^{2} = \sigma_{\tau}^{(S)2} + \sigma_{\tau}^{(T)2} \tag{165}$$

This is a "nice" result in that if we compute the dispersion in clock time in SQM and the dispersion from the time part of TQM we need only to add them to get the total dispersion. If we monitor the dispersion of arrival times long enough, we should see – to whatever level of statistical certainty is desired – whether the dispersion is better described by SQM or by SQM + TQM.

Slow train problem The main problem here is the $1/v_{0}$ factor in the dispersion from space. From the entropic estimate we expect that the dispersions in space and time will in general be of the same order, $\sigma_{x} \sim \sigma_{t}$. But if the wave packet is going at non-relativistic speeds, the effects of the space part will be far greater than those from time, making it hard to pick out any effects from TQM.

There are (at least) four possible solutions to the slow train problem:

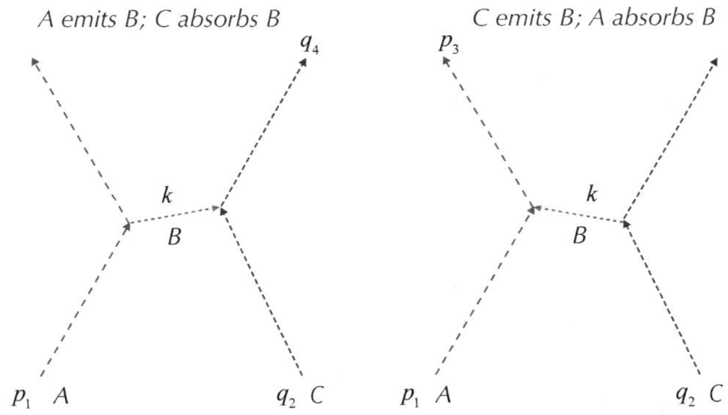

Figure 3. An A and a C particle exchange a B particle

(i) Use a faster train. Get v_0 closer to one. This in fact is part of the motivation for this paper, as we need QED to work with high speeds.
(ii) Wait longer for the train. Collect enough data points that even a small difference becomes statically significant to whatever level is desired.
(iii) Break the train up into its individual cars. If we do this, we can see the dispersion in time on a per car basis, which should make it stand out more. Or in less figurative language, run the wave packet through a strong magnetic field. Discussed further below in subsubsection 3.4.1.
(iv) Diffract the train. (This is a quantum train.) Use the Heisenberg uncertainty principle in time. We use a second wave packet, chosen to be narrow in time, as effectively a single slit in time. The single slit in time acts as a single slit in space would, diffracting the particle with correspondingly increased uncertainty in time-of-arrival.

3.1.4. Summary At this point we have achieved a measure of falsifiability.
(i) We write the TQM prediction of the uncertainty in time at the detector as $(\Delta t)^2_D$,
(ii) We write the SQM prediction of the uncertainty in (clock time) at the detector as $(\Delta\tau)^{(S)2}_D$.
(iii) Our signal is the difference, meaning that part of the uncertainty in time at the detector which is not accounted for by SQM:

$$(\Delta t)^{(T)2}_D \equiv (\Delta t)^2_D - (\Delta\tau)^{(S)2}_D \tag{166}$$

For this to be well-defined, we need a good estimate of the initial wave function in coordinate time. This is why the entropic estimate is critical: it gives a simple but robust estimate of the initial uncertainty in coordinate time.

However this does not yet take us to feasibility. We see that for non-relativistic wave packets, the SQM prediction for uncertainty in time at the detector may be much larger than the additional contribution from TQM; therefore the space contribution may drown out the time contribution.

To get from falsifiability in principle to falsifiability in practice we need a way to increase the relative size of the time contribution.

To do this we look next at the simplest possible scattering problem. This continues our development of TQM and it defines the basic effects we can use to see its effects.

3.2. *Spin zero scattering*

3.2.1. Objective We now look at the simplest possible scattering problem. We work with a toy model consisting of three spin zero particles: A, B and C. A and C both have mass m. They do not interact directly, but only by exchanging B's. B's have mass μ. A's and C's are stand-ins for fermions, B is a stand-in for photons. We will refer to this as the ABC model.

The very considerable practical advantage of looking at spinless particles is that most of the effects that distinguish TQM from SQM have nothing to do with spin or polarization.

The field operators for A's are for *SQM:*

$$A_\tau^S(\vec{x}) \equiv \sum_{\vec{p}} \frac{1}{\sqrt{2VE_{\vec{p}}}} \begin{pmatrix} a_{\vec{p}} \exp\left(-\imath E_{\vec{p}}\tau + \imath\vec{p}\cdot\vec{x}\right) \\ +a_{\vec{p}}^\dagger \exp\left(\imath E_{\vec{p}}\tau - \imath\vec{p}\cdot\vec{x}\right) \end{pmatrix} \tag{167}$$

and for *TQM:*

$$A_\tau(t,\vec{x}) \equiv \sum_{p} \frac{1}{\sqrt{2TVE}} \begin{pmatrix} a_p \exp\left(-\imath\varpi_p\tau - \imath px\right) \\ +a_p^\dagger \exp\left(+\imath\varpi_p\tau + \imath px\right) \end{pmatrix} \tag{168}$$

B and C's are the same.

The interaction Lagrangian has the form:

$$V = \lambda\frac{A^2B}{2} + \lambda\frac{C^2B}{2} \tag{169}$$

with specific terms of the form:

$$\frac{\lambda\left(a+a^\dagger\right)\left(b+b^\dagger\right)\left(a+a^\dagger\right)}{2} + \frac{\lambda\left(c+c^\dagger\right)\left(b+b^\dagger\right)\left(c+c^\dagger\right)}{2} \tag{170}$$

We will assume λ is small, so that perturbation theory makes sense. We will look at the case where an A and C exchange a B.

To help make the diagrams more readable, we identify the A's by p's, B's by k's, and C's by q's. We use $\bar{E}_{1,2,3,4}, \omega$ for the clock energies for SQM. We use $E_{1,2,3,4}, w$ for the coordinate energies and $\varpi_{1,2,3,4}, \varpi_k$ for the clock energies in TQM. A will start with momentum p_1, finish with p_3; C will start with q_2, finished with q_4; and the exchanged B will have momentum k.

We will work in the center-of-mass frame. For definiteness we will assume the particles are approaching each other along the x-axis and are scattered along the y-axis. Results for the z-axis follow from symmetry. The coordinate system is given by $(x,y,z) = r(\sin\theta\cos\varsigma, \sin\theta\sin\varsigma, \cos\theta)$. In the case of GTFs, we will work in the center-of-time frame as well: arrange for $\tau = 0$ to correspond to the time of closest approach of the opposing GTFs.

We assume we have detectors well off to the left and right, placed at $x = \pm L$, running in the y,z plane, and time-sensitive. And further we will assume that the detectors are far enough away that the position of the detector divided by time-of-flight from the interaction zone is a good proxy for the velocity and therefore for the momentum. With this arrangement it is easy to map detector time and position to momentum. To first order:

$$p_x = \gamma m v_x, p_y = \gamma m v_y, E = \gamma m \tag{171}$$

For SQM $v_x = \frac{L}{\tau}, v_y = \frac{y}{\tau}$; for TQM $v_x = \frac{L}{t}, v_y = \frac{y}{t}$. We will compute the energy component indirectly, by using the three velocity to compute $\gamma = \frac{1}{\sqrt{1-\vec{v}^2}}$. We will look first at the case where the initial wave functions are defined as plane waves; then as GTFs.

We have conservation of clock energy as well, for both SQM and TQM. For plane waves it is customary to take the limit as the integral over clock time goes to $\pm\infty$, which in turn gives conservation of clock energy. As noted earlier, for GTFs we argue that if the interaction time is small an integration over the interaction will be zero outside of that time, so that the $\pm$ limits may be extended to $\pm\infty$ without affecting the value of the integral. So we can again take the limit as clock time goes to $\pm\infty$ and therefore have conservation of clock energy.

3.2.2. Plane waves We can see the plane wave case as Møller scattering without the spin or polarization. In terms of the Mandelstam variables [69] we are working in t-channel, with the energy-momentum of the exchanged particle being given by $p_3 - p_1$ or $q_4 - q_2$:

$$t = (p_1 - p_3)^2 = (q_2 - q_4)^2 \tag{172}$$

The S matrix is given in each case as a product of the appropriate δ functions, normalization factors, and the matrix element. We start with the matrix element as the interesting bit. In this simple case, this is λ^2 times the propagator for the exchanged particle.

SQM For SQM the propagator is:

$$\Delta^S_\omega\left(\vec{k}\right) = \frac{1}{\omega^2 - \vec{k}^2 - \mu^2 + \imath\varepsilon} \tag{173}$$

We can compute this from the values of the external momenta and the δ functions at each vertex. If the A particle emits the B, then we have:

$$\left(\omega, \vec{k}\right) = \left(\bar{E}_3 - \bar{E}_1, \vec{p}_3 - \vec{p}_1\right), \left(\bar{E}_4, \vec{q}_4\right) = \left(\bar{E}_2 + \omega, \vec{q}_2 + \vec{k}\right) \tag{174}$$

while if the C particle emits the B we have:

$$\left(\omega, \vec{k}\right) = \left(\bar{E}_4 - \bar{E}_2, \vec{q}_4 - \vec{q}_2\right), \left(\bar{E}_3, \vec{p}_3\right) = \left(\bar{E}_1 + \omega, \vec{p}_1 + \vec{k}\right) \tag{175}$$

In both cases we have overall conservation of momentum:

$$\left(\bar{E}_3 + \bar{E}_4, \vec{p}_3 + \vec{q}_4\right) = \left(\bar{E}_1 + \bar{E}_2, \vec{p}_1 + \vec{q}_2\right) \tag{176}$$

This forces the magnitude of the three vectors for both A and C to be the same before and after the interaction. From conservation of clock energy:

$$2m^2 + \vec{p}_1^{\,2} + \vec{q}_2^{\,2} = 2m^2 + \vec{p}_3^{\,2} + \vec{q}_4^{\,2} \tag{177}$$

In the center-of-mass frame:

$$\vec{p}_1^{\,2} = \vec{q}_2^{\,2}, \vec{p}_3^{\,2} = \vec{q}_4^{\,2} \tag{178}$$

Combining these two results we get:

$$2m^2 + 2\vec{p}_1^{\,2} = 2m^2 + 2\vec{p}_3^{\,2} \Rightarrow \vec{p}_1^{\,2} = \vec{p}_3^{\,2} \tag{179}$$

and from this we see there is no clock energy available for the unfortunate B. As the propagator is not being integrated over, we may drop the $\imath\epsilon$. We have:

$$\Delta^S_\omega\left(\vec{k}\right) = -\frac{1}{\vec{k}^2 + \mu^2} \tag{180}$$

We set $\vec{p} \equiv \vec{p}_1 = -\vec{q}_2$ and write:

$$\vec{k} = |\vec{p}|\left(\cos\left(\theta\right), \sin\left(\theta\right), 0\right) \tag{181}$$

so the propagator is:

$$\Delta_{\omega}^{S}\left(\vec{k}\right) = -\frac{1}{2\vec{p}^2\left(1-\cos\left(\theta\right)\right)+\mu^2} \tag{182}$$

and the matrix element $(-\imath\lambda)^2$ times this:

$$\mathcal{M}^{S} = \frac{\lambda^2}{2\vec{p}^2\left(1-\cos\left(\theta\right)\right)+\mu^2} \tag{183}$$

The external factors are:

$$N^{S} = \sqrt{\frac{1}{2\bar{E}_1}}\sqrt{\frac{1}{2\bar{E}_2}}\sqrt{\frac{1}{2\bar{E}_3}}\sqrt{\frac{1}{2\bar{E}_4}} \tag{184}$$

In the center-of-mass frame they are all equal, so there is just one $\bar{E}$. The δ functions are:

$$D^{S} = (2\pi)^4\delta\left(\left(\bar{E}_3+\bar{E}_4\right)-\left(\bar{E}_1+\bar{E}_2\right)\right)\delta^3\left(\left(\vec{p}_3+\vec{q}_4\right)-\left(\vec{p}_1+\vec{q}_2\right)\right) \tag{185}$$

The full S matrix is:

$$S^{S} = D^{S}N^{S}\mathcal{M}^{S} \tag{186}$$

The probability is given by the square of this. The four δ functions make clear that this is a formal object; it will not acquire meaning until we integrate over the GTFs.

TQM The calculation for the plane wave for TQM is the same. We will take the SQM values as the starting point. Recall the quantum energy δw is defined as the difference between the coordinate energy and the SQM energy. Since the SQM energy is zero, the quantum energy is just the coordinate energy w itself.

We have from the entropic estimate that the average energy in the initial value of a TQM wave function will match that in SQM, so we have:

$$E_1 = \bar{E}_1, E_2 = \bar{E}_2 \Rightarrow \delta E_1 = 0, \delta E_2 = 0 \tag{187}$$

From this we also get $\varpi_1 = \varpi_2 = 0$. Or to put it another way, wave functions in TQM start on-shell. We have four δ functions in momentum plus one in clock energy at each vertex, forcing the four momenta for the exchanged B in the same way as with SQM.

If the A particle emits the B, then we have:

$$E_3 = \bar{E} - w, \vec{p}_3 = \vec{p}_1 - \vec{k}; E_4 = \bar{E} + w, \vec{q}_4 = -\vec{p}_1 + \vec{k} \tag{188}$$

while if the C particle emits the B we have:

$$E_3 = \bar{E} + w, \vec{p}_3 = \vec{p}_1 + \vec{k}; E_4 = \bar{E} - w, \vec{q}_4 = -\vec{p}_1 - \vec{k} \tag{189}$$

In both cases we have overall conservation of momentum:

$$E_3 + E_4 = E_1 + E_2, \vec{p}_3 = -\vec{q}_4 \tag{190}$$

The only remaining ambiguity is whether we have $\varpi_3 = \varpi_4 = 0$ or not, whether the two outgoing particles are forced to be on-shell or not. To check this we expand in powers of w.

We have for the two external clock energies if A emits the B:

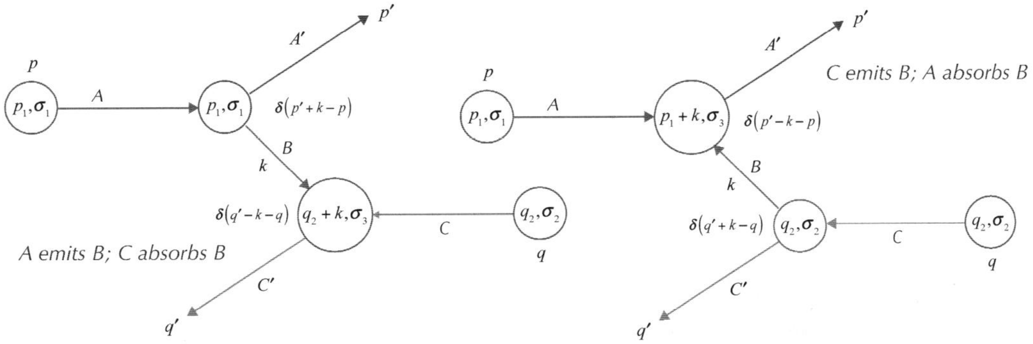

Figure 4. Calculation of dispersion in exchange of B particle

$$\varpi_3 = -\frac{\left(\bar{E} - w\right)^2 - \left(\vec{p}_1 - \vec{k}\right)^2 - \mu^2}{\bar{E} - w}, \varpi_4 = -\frac{\left(\bar{E} + w\right)^2 - \left(\vec{p}_1 - \vec{k}\right)^2 - \mu^2}{\bar{E} + w} \quad (191)$$

and the same thing with $w \to -w$ if C emits the B. Therefore the sum of the two, in either case, is:

$$-\frac{\left(\bar{E} - w\right)^2 - \left(\vec{p}_1 - \vec{k}\right)^2 - \mu^2}{2\left(\bar{E} - w\right)} - \frac{\left(\bar{E} + w\right)^2 - \left(\vec{p}_1 - \vec{k}\right)^2 - \mu^2}{2\left(\bar{E} + w\right)} = 0 \quad (192)$$

We already know, since the incoming lines are on-shell, that:

$$\left(\bar{E}\right)^2 - \left(\vec{p}_1\right)^2 - \mu^2 = 0 \quad (193)$$

Using this and clearing the fractions we get:

$$2\bar{E}w^2 = 0 \Rightarrow w = 0 \quad (194)$$

So here the quantum energy w of the exchanged particle is zero just as the classical energy is. And therefore the clock energies ϖ_3, ϖ_4 are also zero. And therefore both outgoing lines are on-shell. To put it another way, an analysis using on-shell plane waves shows no effects of TQM. We will need to work with GTFs to see the effects of dispersion in time/energy.

3.2.3. Gaussian test functions

Calculation of the propagation of the GTFs through the diagram Our initial wave functions will be direct products of GTFs in x, y, z and (in TQM) t. The value of the t wave function will be inferred from the other three parts using the entropic estimate.

For GTFs the calculation is essentially the same for all three or four components of the momentum, which we will represent as p, q, p', q'.

We will start by assuming we are dealing with fixed input GTFs, functions of p, q centered on p_1, q_2 with dispersions σ_1, σ_2. We wish to compute the amplitude to measure the outgoing momenta with values p', q'.

We can compute this by integrating over the S matrix from the plane wave case. We express this in a generic notation:

$$\psi\left(p',q'\right)=\int dpdqdkS\left(p',q';p,q\right)\varphi_1\left(p\right)\varphi_2\left(q\right) \tag{195}$$

First we assume the incoming GTFs are narrow in energy/momentum, that the dispersions is a small fraction of the corresponding momenta:

$$\frac{\sigma_{|\vec{p}|}}{\vec{p}}\ll 1 \tag{196}$$

We refer to this as the "narrow beam" approximation. Our GTFs are more rounded than simple plane waves but each is still relatively focused on a specific value of the momentum. They inhabit a kind of halfway house between plane wave and GTF.

This lets us make some useful approximations for the normalizations and for the matrix elements. In particular we can approximate each external factor by its average. For SQM we can ignore the variation caused by the GTF:

$$\sqrt{\frac{m}{E_{\vec{p}_{1,2}}}}\approx\sqrt{\frac{m}{\bar{E}_{1,2}}} \tag{197}$$

for TQM we can ignore both this and variation caused by the quantum energy ($\delta E\equiv E-E_{\vec{p}}$):

$$\sqrt{\frac{m}{E_{1,2}}}\approx\sqrt{\frac{m}{E_{\vec{p}_{1,2}}}}\approx\sqrt{\frac{m}{\bar{E}_{1,2}}} \tag{198}$$

By the same argument, for SQM we replace the technically correct $k_i+\delta k_i$ in the denominator of the matrix element by k_i:

$$\frac{\imath}{\left(\vec{k}+\delta\vec{k}\right)^2+\mu^2}\to\frac{\imath}{\left\langle\vec{k}^2\right\rangle+\mu^2} \tag{199}$$

For TQM we do the same:

$$\frac{\imath}{\left(\delta w\right)^2-\left(\vec{k}+\delta\vec{k}\right)^2+\mu^2}\to\frac{\imath}{\left\langle\vec{k}^2\right\rangle+\mu^2} \tag{200}$$

This means we can pull out from inside the integral a common factor (the same for SQM and TQM) of:

$$\bar{S}\equiv\prod_{i=1}^{4}\sqrt{\frac{m}{\bar{E}_i}}\frac{\imath}{\vec{k}^2+\mu^2} \tag{201}$$

leaving the integral as purely an integral over the dispersions in the GTFs and the δ functions at the vertexes.

There are three of these integrals for SQM plus one more – over energy – for TQM:

$$I\left(p'q'\right)=\int dpdqdk\begin{pmatrix}\delta\left(p'+k-p\right)\delta\left(q'-k-q\right)\theta\left(\tau\right)\\+\delta\left(p'-k-p\right)\delta\left(q'+k-q\right)\theta\left(-\tau\right)\end{pmatrix}\varphi\left(p\right)\varphi\left(q\right) \tag{202}$$

where the sign of τ depends on whether we are looking at the case where A emits B and C absorbs it or where C emits a B and A absorbs it.

What we have here is a "stick-and-cloud" approximation. The plane wave analysis gives the average result – the stick – then the integrals over the dispersions pull in the quantum effects we are interested in here, the clouds. With the δ functions we have only one real integral to do.

For the first pair, we do the integral over p first. This turns $\varphi_1(p) \to \varphi_1(p'+k)$. Next we do the integral over k and the second δ function. This leaves:

$$\psi^{(1)}(p', q') = \int dq \varphi_1(p'+k)\varphi_2(q'-k) \tag{203}$$

This is a convolution of two Gaussians. This results in a single Gaussian of the form:

$$\exp\left(-\frac{((p'+q') - (p_1+q_2))^2}{2\sigma_3^2}\right) \tag{204}$$

with dispersion:

$$\sigma_3^2 = \sigma_1^2 + \sigma_2^3 \tag{205}$$

What this is saying is that the resulting amplitude is a GTF centered on the average value of the total initial momentum, with a total dispersion which is a Pythagorean sum over the two initial dispersions. The second pair of δ functions goes exactly the same way, but in reverse.

At this point we need to check the counting. Each vertex has a factor of $1/2$, for a total of $1/4$. We can have the B emit from the first vertex or the second and from A or from C. This is a counter-balancing factor of four, for a total value of one. This is as expected. Therefore the final distribution is still given by equation 204.

Compare to conservation of energy with plane waves We define:

$$\delta p \equiv p' - \bar{p}, \delta q \equiv q' - \bar{p} \tag{206}$$

giving:

$$\delta p + \delta q = p' + q' - (p+q) \tag{207}$$

and:

$$\psi(\delta p, \delta q) = \exp\left(-\frac{(\delta p + \delta q)^2}{2\sigma_3^2}\right) \tag{208}$$

So the incoming total value of the momentum in question has an average of $\bar{p}$, but is dispersed around that average. The outgoing value is also dispersed around the average, with the dispersion given by the Pythagorean sum.

The full outgoing wave function is the product of three of these functions for SQM; the product of four of these functions – the three from SQM and a fourth on the energy axis – for TQM. Defining:

$$\varphi_{3,4}^{\mu}(\delta p_\mu, \delta q_\mu) \equiv \exp\left(-\frac{(\delta p_\mu + \delta q_\mu)^2}{2\left(\sigma_\mu^{(1)2} + \sigma_\mu^{(2)2}\right)}\right) \tag{209}$$

we get for SQM:

$$\varphi_{3,4}^{S}(\vec{p}', \vec{q}') = \bar{S}(\vec{p}_3, \vec{q}_4; \vec{p}_1, \vec{q}_2) \prod_{i=1}^{3} \varphi_{3,4}^{i}(\delta p_i, \delta q_i) \tag{210}$$

and for TQM we have the product of the energy part with this:

$$\varphi_{3,4}(p', q') = \varphi_{3,4}^{\mu=0}(\delta p_0, \delta q_0)\varphi_{3,4}^{S}(\vec{p}', \vec{q}') \tag{211}$$

For our purposes the SQM part is the carrier; the energy GTF is the signal.

Distribution in three space for both SQM and TQM The distribution in the three space dimensions is the same for both SQM and TQM. We pick a specific scattering angle θ.

Suppose we have in the p_θ direction the final dispersion $\hat{\sigma}_\theta$:

$$\hat{\sigma}_\theta^2 = \hat{\sigma}_\theta^{(1)2} + \hat{\sigma}_\theta^{(2)2} \tag{212}$$

then we have σ_θ, the space dispersion in the θ direction, as:

$$\sigma_\theta^2 = \frac{1}{\hat{\sigma}_\theta^2} \tag{213}$$

Then from the previous section the effective dispersion in clock time is:

$$\sigma_\tau^{(S)2} = \frac{1}{\bar{p}^2 \sigma_\theta^2} \bar{\tau}^2 = \frac{\hat{\sigma}_\theta^{(1)2} + \hat{\sigma}_\theta^{(2)2}}{\bar{p}^2} \bar{\tau}^2 \tag{214}$$

Distribution in coordinate time/energy for TQM TQM we have a distribution in energy as well:

$$\rho(E) \sim \exp\left(-\frac{(\delta E_3 + \delta E_4)^2}{\left(\sigma_E^{(1)2} + \sigma_E^{(2)2}\right)}\right) \tag{215}$$

This is centered on the average in energy. Therefore if δE for one of the outgoing legs is positive, the mostly likely value of δE for the other is its negative. But what does δE mean in operational terms? The most direct way to determine this is to run the outgoing particles through a detector that responds to energy. But to be consistent with the treatment so far, we will convert this into time-of-arrival measurements.

If $\delta E > 0$ then the particle is traveling "hot" and will arrive a bit early at the detector. For a fixed angle θ, if we have the expected time-of-arrival as $\bar{\tau}$, then we will measure a δt which is "early", so negative. The other particle is most likely to be traveling exactly that amount "cold" and therefore to arrive (to first order) exactly $-\delta t$ "late", so positive.

We don't have an absolute distribution for either particle, what we have is a correlated distribution: if, say, the C particle exactly on time, then the A particle will have a distribution in time centered on $\bar{\tau}$, with dispersion:

$$\rho(t) = \exp\left(-\frac{(\delta t)^2}{\sigma_t^2}\right), \sigma_t^2 = \frac{1}{\sigma_E^2}, \sigma_E^2 = \sigma_E^{(1)2} + \sigma_E^{(2)2} \tag{216}$$

If the C particle is a bit off-center, the A particle's own distribution will shift in the opposite direction. So we have the distribution in coordinate time as we leave the interaction zone.

Diffraction with respect to coordinate time in TQM There is one more piece to the puzzle: we have the dispersion in coordinate time as we leave the interaction zone. But this is not the dispersion in coordinate time at the detector, some temporal distance in the particle's future. Referring to the appendix on GTFs , we have the probability density at the detector as:

$$\rho_\tau^T(t) = \sqrt{\frac{1}{\pi\sigma_t^2\left(1 + \frac{\bar{\tau}^2}{E^2\sigma_t^4}\right)}} \exp\left(-\frac{(t-\bar{\tau})^2}{\sigma_t^2\left(1 + \frac{\bar{\tau}^2}{E^2\sigma_t^4}\right)}\right) \tag{217}$$

with uncertainty in time as:

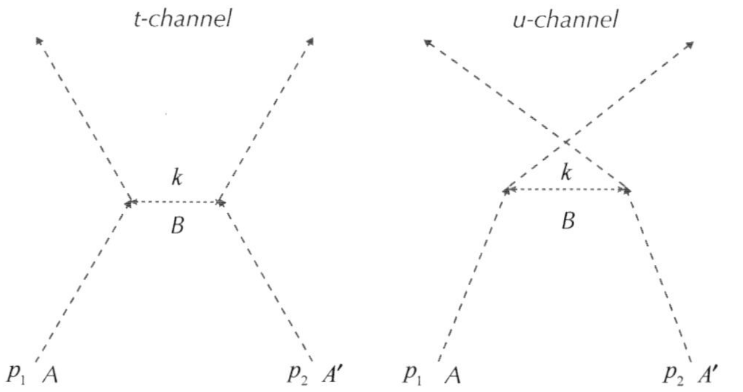

Figure 5. Two A particles exchange a B particle

$$(\Delta t)^2 \equiv \langle t^2 \rangle - \bar{\tau}^2 = \frac{\sigma_t^2}{2}\left|1 + \frac{\bar{\tau}^2}{\bar{E}^2\sigma_t^4}\right| \to \frac{\bar{\tau}^2}{2\bar{E}^2\sigma_t^2} \tag{218}$$

so both the TQM part and the SQM part scale as $\bar{\tau}$, as one would expect intuitively. But the TQM contribution to the uncertainty in time at the detector is inversely proportional to the initial dispersion in coordinate time. The smaller we can make the post-interaction σ_t, the greater the dispersion in time at the detector. How to make σ_t small?

Single slit in time Suppose we think of the particle C as a gate in time, a single slit in time, with A being the particle passing through the slit. We need C narrow in time, so we make its dispersion in energy large, enough to dominate the sum in (216). We can do this by setting its original dispersion in three momentum large and relying on the entropic estimate to set the dispersion in energy large as well. Now the dispersion in energy in A (post-interaction) is large and its dispersion in coordinate time is therefore small: A will be strongly localized in time. We will have the desired small σ_t.

By judicious manipulation of the uncertainty from SQM – reducing it by increasing the velocity in space and by reducing the dispersion in momentum – while at the same time using the shortest possible wave function for C, we should be able to use C as a time gate, and get a workable ratio of:

$$\frac{\Delta_t^{(T)2}}{\Delta_t^{(T)2} + \Delta_\tau^{(S)2}} \tag{219}$$

enough to make the effects of TQM visible if it is correct. Or *prove* it is false, if it is not correct.

3.2.4. Scattering of indistinguishable particles If we replace the C particle with a second A particle we have two indistinguishable particles to deal with. We therefore cannot tell whether the particle we detect at location 3 is the one that started at location 1 or the one that started at location 2. The diagram on the left is again the t-channel as above. The crossed diagram on the right is associated with the u-channel:

$$u = (p_1 - q_4)^2 = (p_2 - q_3)^2 \tag{220}$$

The sum of the t and u channel diagrams gives the total amplitude. The core matrix element will go as:

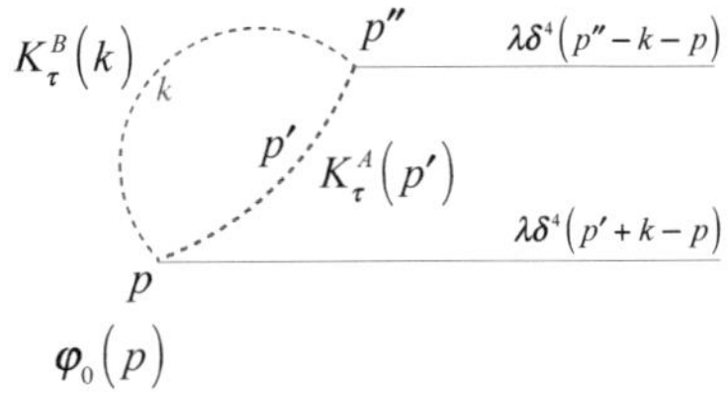

Figure 6. Simple mass correction loop

$$\lambda^2\left(\frac{1}{(\vec{p}_3-\vec{p}_1)^2+\mu^2+\imath\varepsilon}+\frac{1}{(\vec{p}_3-\vec{p}_2)^2+\mu^2+\imath\varepsilon}\right) \tag{221}$$

Since we are dealing with bosons, the relative sign is positive and the two amplitudes add. We see we have symmetry under the interchange of $1 \leftrightarrow 2$ as expected.

By our fundamental hypothesis this rule must apply in TQM for time and the three space coordinates as a whole. To take advantage of this, consider a starting wave function which is symmetric under the interchange of both time and space, but anti-symmetric under each separately. This is not that hard to arrange: we showed in paper A that if the left side has, say, large dispersions in time and space, the right side small – like one of those comedy duos with a tall wide partner and a short thin one – then the total wave function will be the sum of a part which is symmetric in both time and space and a part which is anti-symmetric in time and space separately but symmetric in the combination.

The anti-symmetric part will then give rise to a probability distribution which has a part symmetric in space and time separately and a part anti-symmetric in space and in time, but still symmetric in the combination. For instance it might be symmetric going from $+y, +t \rightarrow -y, -t$, but anti-symmetric going $+y \rightarrow -y$. Identical yet opposite considerations apply to fermions.

The result is we will see small apparent violations of space (anti-)symmetry in TQM where it is mandated in SQM. As we suspect that this effect is of the second or third order of practicality, we do not pursue it further here. But it is suggestive of the large number of experimental possibilities that TQM generates.

3.2.5. Primary effects Judging by this simple example, the first order of corrections from TQM comes less from the propagators and normalization factors; more from the simple presence of time dispersion in the initial wave functions. There are three kinds of effects we might look for:

Increased dispersion in time/energy These are expected typically small, of scale attoseconds. However they are expected present in all cases.

Diffraction in time In a scattering experiment, we can use one particle, narrowly focused in time, to act as a kind of "single slit in time" experiment with respect to another. This provides a direct test of the HUP in time/energy, and appears to be the most promising of the candidate experiments.

Entanglement in time/energy Anti-symmetry in time is an example of this. These in some ways the most interesting, but may be a bit subtle for use in a first attack on the problem.

3.3. Simple loop diagram

3.3.1. Ultraviolet divergences In the founding days of QED the problem of the loop divergences was one of the most troubling. If you look at essentially any loop diagram in QED, the associated integral diverges. For instance, the loop diagram for an electron to emit and then absorb a photon is divergent.

We will work, as above, with our simple ABC model which has the same problem and for the same reason. Consider the amplitude for an A particle to emit a B particle and then absorb it. We assume a starting momentum for the A particle of p and integrate over all possible values of the B particle's four momentum k:

$$L = \int d\omega d\vec{k} \frac{\imath}{\omega^2 - \vec{k}^2 - \mu^2 + \imath\varepsilon} \frac{\imath}{(E-\omega)^2 - \left(\vec{p} - \vec{k}\right)^2 - m^2 + \imath\varepsilon} \tag{222}$$

This integral does not converge. If we focus on the high momentum part of the integral, the integrand goes as $1/k$. If we impose a high momentum cutoff Λ, the integral diverges logarithmically:

$$\int^{\Lambda} \frac{dk}{k} \sim \ln(\Lambda) \tag{223}$$

This was very troubling to the founders of QED. This particular form of the loop shows up as simple corrections to the mass of a particle, so would be expected small. And these infinities show up in all the basic loop diagrams.

Renormalization Now suppose we were to compare the value of this loop diagram taken at a specific value of p to the value of the same loop diagram taken at a slightly different starting value $p' = p + \delta p$:

$$L' = \int d\omega d\vec{k} \frac{\imath}{k^2 - \mu^2 + \imath\varepsilon} \frac{\imath}{(p + \delta p - k)^2 - m^2 + \imath\varepsilon} \tag{224}$$

The difference between the two loop diagrams goes as:

$$L' - L = \frac{\partial L}{\partial p}\delta p \tag{225}$$

The derivative picks up an extra factor of the A propagator in the denominator:

$$\frac{\partial L}{\partial p} \to -\frac{2\imath(p-k)}{\left((p-k)^2 - m^2 + \imath\varepsilon\right)^2} \tag{226}$$

causing the integrand to acquire an extra factor of $1/k$:

$$\frac{1}{k^4} \to \frac{1}{k^5} \tag{227}$$

and thereby making the integral converge. So that while the original loop is infinite, the differences between nearby loops are finite. Since we do not actually measure any mass directly, but only by comparison – every scale needs a weight on the left side and on the right – we are not strictly required to compute the absolute value of any loop but only the difference between it and another. The calculations are tricky (see for example [33, 38, 45, 47]) but give not only finite results but extraordinarily good agreement with experiment.

The comparison program – referred to as renormalization – has been worked out to a high degree of sophistication. It can be done to all orders, for not only QED but for the entire Standard Model, and in a way that is covariant. So we have a procedure that makes no sense (why are these basic loop diagrams infinite?) yet agrees perfectly and to many places with experiment.

This problem would appear to be even more severe in the case of TQM. TQM has one more dimension to integrate over, so we would expect its divergences to be linear, perhaps even beyond the reach of a renormalization program. This would not keep us from using TQM as a way to develop a program to develop interesting experiments involving time and quantum mechanics, but it would make clear that TQM could not be an entirely satisfactory theory in its own right.

3.3.2. Simple loop calculation We are not going to show that TQM is fully renormalizable. But we will show that simple loop diagrams in TQM – we will focus on the A, B mass correction diagram – are in fact finite and do not require regularization in the first place, provided they are done in a way that is consistent with the spirit of TQM. It is still true that we never measure a mass or any other quantity in isolation but only in comparison with other measurements – but the associated integrals are not only finite but small.

The key is to take advantage of two points:

(i) We have to start with a physically realistic wave function, one which is normalizable itself. Our GTFs will work well for this.

(ii) In computing each step in a loop diagram we have to take advantage of the entanglement in time of that step with the previous. This lets each step take advantage of the finite character of the previous steps to make sure that it stays finite as well. This entanglement in time is not available in SQM, since it is a quantum feature not available to any theory that treats time classically.

We will do the calculation in two stages:

(i) Calculate the value of the loop for a fixed clock time.

(ii) Take the Fourier transform of the fixed clock time result, to get the actual mass correction.

The attosecond form of the propagator is not suitable for this. This is one of the places where that approximation breaks down. But the full, "unpacked" propagator works. We compute the loop from past to future, so need only the positive time branches of the propagators for the A and B particles. Since we are only interested in the questions "is the loop integral finite?" and "does the loop integral make sense?" we ignore the squared coupling constant and an overall factor of $\imath$. We are after the loop, the whole loop, and nothing but the loop.

We start as usual with a GTF. We again use the narrow beam approximation, assuming that for all four momentum components $\frac{\sigma}{|p|} \ll 1$.The propagator for the A particle is:

$$K_\tau^A(p) = \frac{1}{2E}\exp(-\imath\varpi_p\tau), \varpi_p = -\frac{E^2 - \vec{p}^2 - m^2}{2E} \tag{228}$$

The propagator for the B particle is:

$$K_\tau^B(k) = \frac{1}{2w}\exp(-\imath\omega_k\tau), \varpi_k = -\frac{w^2 - \vec{k}^2 - \mu^2}{2w} \tag{229}$$

For fixed clock time the loop integral is:

$$L_\tau(p) = \int d^4k\hat{K}_\tau^A(p-k)\hat{K}_\tau^B(k)\hat{\varphi}_0(p-p_0) \tag{230}$$

We are using the sample GTF (B.11) from the GTF appendix.

In the narrow beam approximation we may estimate the normalization factor using:

$$E \to E_0 \equiv \sqrt{m^2 + \vec{p}_0^2}, w \to \mu \tag{231}$$

Since these factors are in the denominator, this approximation will make the integral more divergent rather than less. Since they are now constant, we can pull them out of the integral. For the rest of the analysis we will work with a re-scaled loop integral $L \to 4E_0\mu L$.

We will take the same step with respect to the factors of clock time in the exponentials:

$$\varpi_k^B \approx -\frac{w^2 - \vec{k}^2 - \mu^2}{2\mu}, \varpi_p^A \approx -\frac{E^2 - \vec{p}^2 - m^2}{2E_0}, \tag{232}$$

This will tend to make the arguments of the exponentials oscillate less, again making the integrals more divergent rather than less.

The loop integral is a pair of convolutions in momentum space. If we shift to coordinate space, these convolutions will turn into the coordinate space equivalents. The GTF is:

$$\varphi_0(x) = \sqrt[4]{\frac{1}{\pi^4 \det(\Sigma)}} \exp\left(-\imath p_0 x - \frac{(x - x_0)^2}{2\Sigma}\right), \Sigma = \hat{\Sigma}^{-1} \tag{233}$$

The coordinate space forms of the kernels are (compare to Appendix B.3.2):

$$\begin{aligned} K_\tau^A(x_1; x) &= -\imath \tfrac{E_0^2}{4\pi^2\tau^2} \exp\left(-\tfrac{\imath E_0}{2\tau}(x_1 - x)^2 + \imath \tfrac{m^2}{2E_0}\tau\right) \\ K_\tau^B(x_1; x) &= -\imath \tfrac{\mu^2}{4\pi^2\tau^2} \exp\left(-\tfrac{\imath \mu}{2\tau}(x_1 - x)^2 + \imath \tfrac{\mu}{2}\tau\right) \end{aligned} \tag{234}$$

The product equals a single coordinate space kernel:

$$K_\tau^{(M)}(x_1; x) = -\imath \frac{M^2}{4\pi^2\tau^2} \exp\left(-\imath \frac{M}{2\tau}(x_1 - x)^2 + \imath \frac{M}{2}\tau\right) \tag{235}$$

with a modified mass $M \equiv E_0 + \mu$, a prefactor $-\imath \frac{1}{4\pi^2} \frac{E_0^2\mu^2}{\tau^2 M^2}$, and a post-factor $\exp\left(-\imath \frac{M}{2}\tau + \imath \frac{m^2}{2E_0}\tau + \imath \frac{\mu}{2}\tau\right)$. In the limit as $\mu \to 0$ we have $M \to E_0$.

So the loop term in coordinate space is (without prefactor and post-factor):

$$L'_\tau(x_1) = \int d^4x K_\tau^{(M)}(x_1; x) \varphi_0(x) \tag{236}$$

This is just the integral to advance a wave function of mass M a distance τ in time, so we have:

$$L'_\tau(x_1) = \varphi_\tau^{(M)}(x_1) \tag{237}$$

We have a correction that shows a spread in time, but at the slightly slower rate associated with the slightly larger mass M. The full correction is greater at shorter clock times, as one would expect. Now we transform back to momentum space:

$$\begin{aligned} L'_\tau(p_1) &= \int d^4p \hat{K}_\tau^{(M)}(p_1; p) \hat{\varphi}_0(p) \\ \hat{K}_\tau^{(M)}(p_1; p) &= \exp\left(\imath \tfrac{p^2 - M^2}{2M}\tau\right) \delta^4(p_1 - p) \end{aligned} \tag{238}$$

So the loop correction for fixed clock time (folding back in the prefactor and post-factor):

$$L_\tau(p_1) = -\imath \frac{1}{4\pi^2} \frac{m^2\mu^2}{M^2\tau^2} \exp\left(\imath \frac{p_1^2}{2M}\tau + \imath \frac{m^2}{2E_0}\tau + \imath \frac{\mu}{2}\tau\right) \hat{\varphi}_0(p_1) \tag{239}$$

At this point the value of the loop correction at a particular value of p is independent of the specific shape of the incoming wave function. We are therefore free to drop the initial wave function from the analysis. We rewrite the loop without the initial wave function, but fold back in the trailing factor:

$$L_{\tau}(p) = -\imath \frac{1}{4\pi^2} \frac{m^2\mu^2}{M^2\tau^2} \exp\left(-\imath \varpi_p^M \tau\right), \varpi_p^M \equiv -\frac{p^2}{2(E_0+\mu)} - \frac{m^2}{2E_0} - \frac{\mu}{2} \tag{240}$$

The $\varpi_p^M \to \varpi_p$ as $\mu \to 0$. The initial GTF acted in a way parallel to the regularization factors often employed in SQM, but it comes organically out of the calculation, it is not supplied by hand.

3.3.3. Fourier transform The actual mass correction is given by the Fourier transform of the loop integral. Now that we have the loop integral for a specific value of the clock time we can take the Fourier transform with respect to τ. We have the value of the Fourier transform of the core element:

$$\mathcal{FT}\left[\frac{\exp\left(-\imath \varpi_p^M \tau\right)}{\tau^2}\right] = -\sqrt{\frac{\pi}{2}}\left|\omega - \varpi_p^M\right| \tag{241}$$

and therefore of the loop:

$$\hat{L}_{\omega}(p) = \imath \frac{1}{4\pi^2} \frac{m^2\mu^2}{M^2} \sqrt{\frac{\pi}{2}}\left|\omega - \varpi_p^M\right| \tag{242}$$

The loop correction is zero if $\omega = \varpi$. However in TQM this is a set of measure zero: the 4D wave functions will form a cloud around the average, so the correction will be proportional to the uncertainty in the coordinate energy:

$$\left\langle\left|\omega - \varpi_p^M\right|\right\rangle \sim \left\langle \frac{(\delta E)^2}{2(E_0+\mu)} \right\rangle \approx \frac{\sigma_E^2}{4(E_0+\mu)} \tag{243}$$

So although the initial wave function dropped out of the loop calculation, its influence, like the smile of the Cheshire cat, is still felt.

3.3.4. Implications This is a simple result for a toy model. But it was achieved without artifice. We did not have to add regularizing or convergence factors by hand to the integrals; the initial wave function acted as the convergence factor. And the results make sense: the correction is small, and still smaller when the mass of the B is smaller.

This is perhaps not that surprising. In classical electrodynamics comparable calculations, as of the self-energy of the electron, suffered from linear divergences. The transition from classical electrodynamics to QED softened these divergences from linear to logarithmic.

TQM represents a further step from the classical to the quantum mechanical approaches. We are applying quantization procedures used in space and applying them in time as well. We see a further "softening" of the integrals, to the point where they are finite.

Therefore we have addressed our motivating question: is TQM ruled out as a theory in its own right by not being renormalizable?

And we have provided additional evidence to suggest that TQM is worth exploring experimentally.

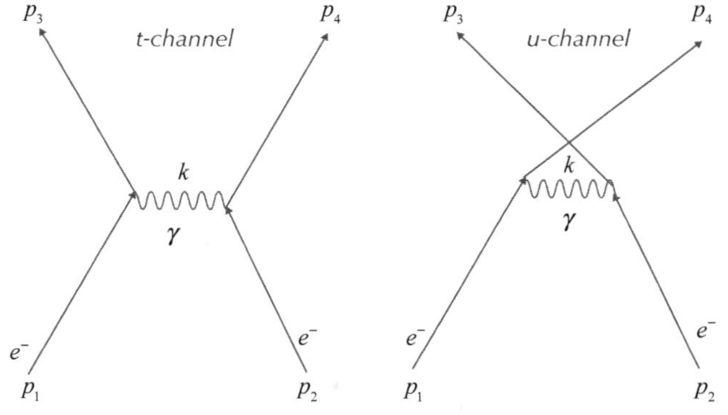

Figure 7. Møller scattering

3.4. Second order-of-magnitude tree diagrams

We discuss here the second order tree diagrams. There are three:

(i) Møller scattering: two electrons exchange a photon.
(ii) Bhabha scattering: an electron and a positron either exchange a photon or else annihilate and then recreate themselves.
(iii) Compton scattering: a photon and an electron scatter.

We see photons and fermions as composed of a spin zero component times a polarization vector or times a spinor. To compute the associated S matrices, we have to look at the input wave functions, the external normalization factors, the polarization vectors, and the spinors. We are primarily looking for additional dispersion in time-of-flight measurements.

If we use the narrow beam approximation, we will be able to ignore the effects of the changes to the external normalization factors, spinors, and polarization vector to lowest order. The SQM part of the diagram will function, as above, largely as a carrier. The principle effects of TQM will be:

(i) The general presence of dispersion in time, leading to additional uncertainty at a detector.
(ii) Diffraction in time, as in the HUP in time/energy.
(iii) Entanglement in time, as in the effects of anti-symmetry in the time part of the wave function.

3.4.1. Møller scattering Møller scattering is the term for electron-electron scattering via the exchange of an electron. From our point of view Møller scattering is a relatively minor generalization of the spin zero scattering problem in subsection 3.2.4. The differences between the SQM and TQM versions of this are in the external factors, the photon propagator, and the spinors at the vertex.

We use Mandelstam variables t, u to characterize the interaction:

$$\begin{aligned} t &= (p_1 - p_3)^2 = (p_2 - p_4)^2 \\ u &= (p_1 - p_4)^2 = (p_2 - p_3)^2 \end{aligned} \tag{244}$$

The heart of the S matrix is the photon propagator. Here the mapping from SQM to SQM is simple, if we use the attosecond propagator:

$$\imath D^{(S)\mu\nu}_{(\cdot)}\left(\vec{k}\right) = \frac{-\imath g^{\mu\nu}}{\omega^2 - \vec{k}^2 + \imath\varepsilon} \rightarrow \imath D^{(A)\mu\nu}_{(\cdot)}\left(k\right) = \frac{-\imath g^{\mu\nu}}{w^2 - \vec{k}^2} \tag{245}$$

We are mapping $\omega \to w$ (coordinate frequency to clock frequency). In SQM, $\omega \to 0$, so the propagator is just:

$$\imath D_{\omega}^{(S)\mu\nu}\left(\vec{k}\right) = \frac{\imath g^{\mu\nu}}{\vec{k}^2} \tag{246}$$

The same analysis as in the spin zero case gives same result for the TQM propagator, so it too is:

$$\imath D_{\omega}^{(A)\mu\nu}\left(k\right) = \frac{\imath g^{\mu\nu}}{\vec{k}^2} \tag{247}$$

And therefore the values of the t and u variables are the same for both SQM and TQM:

$$\frac{1}{t} \approx \frac{1}{t^S}, \frac{1}{u} \approx \frac{1}{u^S} \tag{248}$$

where the SQM context merely means we replace coordinate energy w by its average ω, which is zero.

In the external factors we have to map $E_{\vec{p}} \to E$:

$$\sqrt{\frac{m}{E_{\vec{p}}}} \to \sqrt{\frac{m}{E}} \tag{249}$$

This is where the narrow beam approximation is useful. We are assuming that the variation in energy and the other components of the four momentum are no more than, say, 1% of the average for each component. Therefore if we are looking for order-of-magnitude changes only, we can use the SQM external factor for TQM.

The vertex contribution is more complex. To go from the ABC model to QED in SQM we have:

$$\imath\lambda ABA, \imath\lambda CBC \to -\imath e\bar{\psi}A^{\nu}\gamma_{\nu}\psi \tag{250}$$

To then go from SQM to TQM we replace $E_{\vec{p}} \to E$ in the spinors:

$$-\imath e\bar{\psi}^S\left(\vec{p}'\right)A^{\nu}\left(\vec{k}\right)\gamma_{\nu}\psi^S\left(\vec{p}\right) \to -\imath e\bar{\psi}\left(p'\right)A^{\nu}\left(k\right)\gamma_{\nu}\psi\left(p\right) \tag{251}$$

The ψ's are sums over the u, v's. The u, v's depend on E via factors of $\sqrt{\frac{E+m}{2m}}, \frac{1}{E+m}$. We can expand these in power series of δE:

$$\begin{aligned} \sqrt{\frac{E+m}{2m}} &= \sqrt{\frac{E_{\vec{p}}+m}{2m}} + \frac{\delta E}{2m\sqrt{E_{\vec{p}}+m}} - \frac{(\delta E)^2}{16m\sqrt{E_{\vec{p}}+m}^3} + \ldots \\ \frac{1}{E+m} &\approx \frac{1}{E_{\vec{p}}+m} - \delta E\frac{1}{\left(E_{\vec{p}}+m\right)^2} + (\delta E)^2\frac{1}{\left(E_{\vec{p}}+m\right)^3} - \ldots \end{aligned} \tag{252}$$

In the narrow beam approximation we drop the first order and higher terms in δE so that $u, v \approx u^S, v^S$. The implication is that the matrix elements in TQM are the same as in SQM:

$$S \approx S^S \tag{253}$$

All the usual trace-tricks will work the same way: use the SQM procedures replacing $E_{\vec{p}} \to E$ throughout, then drop the quantum energy component $\delta E \equiv E - E_{\vec{p}}$ leaving us right back where we started with the SQM case.

This implies that the first order effects are, as with the spin zero case, a function of the GTFs, rather than the propagator, spinors, or the normalizations.

If we may write the incoming wave functions as the direct product of a spinor part and a GTF, then we have for any SQM calculation the same formula for the TQM signal as with the spin zero case. We have to lowest order reduced the Møller calculation to the ABC one.

Samurai versus pirate Now that we are dealing with charged particles we can employ some additional techniques to help prove TQM effects do not exist. Consider a charged particle going through a magnetic field. It will have a radius of curvature given by:

$$r = \frac{mv}{qB} = \frac{p_x}{eB} \tag{254}$$

so the faster it is going, the greater the radius of curvature. So the larger p_x, the less the path of the particle is bent. This provides us a way to address the "slow train" problem. We can start with the FS/T equation for a single particle, using the minimal substitution to include the magnetic and electric fields. We ignore spin:

$$\imath E \frac{\partial \psi_\tau}{\partial \tau} = -\frac{\left(\left(p_\mu - q A_\mu \right) \left(p^\mu - q A^\mu \right) - m^2 \right)}{2} \psi_\tau \tag{255}$$

We can use this equation to calculate the motion of a TQM wave packet as function of x, y, t. We can use the Klein-Gordon equation with the minimal substitution to calculate the motion of the particle in the SQM case.

Suppose the particle is going in the x direction and the magnetic field is in the z direction. The force from the magnetic field will be in the y direction. The slower particles will have a smaller radius of curvature, so will be pushed further in the y direction. They will also arrive later, so if the particles are being bent to the right and time is being tracked going up, the main trace will go to the right and up. It will look a bit like a sword trace.

Since the magnetic field has no effect on the time part, the dependence on p_x and on E are separate. If SQM is true, then the dispersion in time at each y position will be small. If TQM is true, then the dispersion in time at each y position will be greater, depending on the specifics significantly greater. For SQM the sword trace will be narrow, looking like the thin precise scar left by a skilled samurai's katana. But if TQM is true, then the sword trace will be broader, looking more like the undisciplined scar left by a pirate's cutlass.

3.4.2. Bhahba scattering In Bhabha scattering we look at an electron scattering from a positron. They can either do this by exchanging a photon, with results very like those for the Møller case, or they can interact by annihilating with emission of a photon, which then decays into an electron-positron pair.

There are two slight differences from the Møller case for our purposes. The exchanges are described by the s and t channels:

$$\begin{aligned} s &= (p_1 + p_2)^2 = (p_3 + p_4)^2 \\ t &= (p_1 - p_3)^2 = (p_2 - p_4)^2 \end{aligned} \tag{256}$$

This is the first point at which the Machian hypothesis has come up as a possible issue. Recall this gave us a well-defined value for the magnitude and sign of the clock frequency. The symmetry between particle and anti-particle means that the magnitude of the clock frequency can't plausibly change. The only reasonable possibility is that the sign changes for anti-matter. However for a normal wave packet the clock frequency will be averaged over values of the coordinate energy centered on the on-shell case. Therefore the sign of the clock frequency will be averaged away.

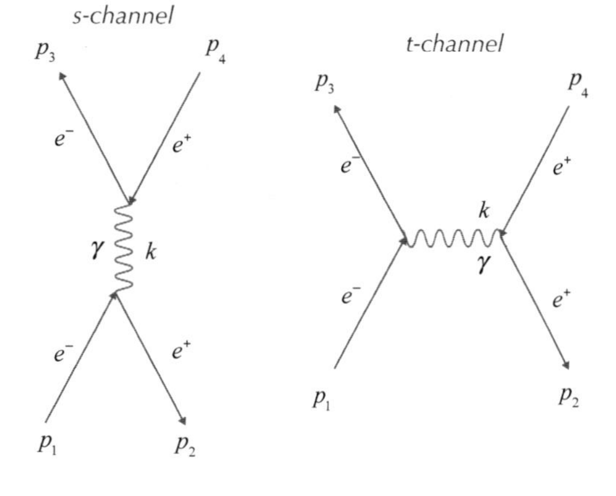

Figure 8. Bhabha scattering

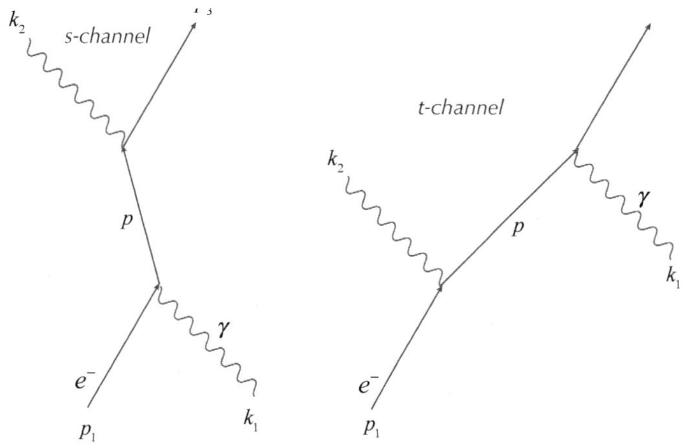

Figure 9. Compton scattering

The other point is that for the first time, the photon propagator will be carrying a non-zero energy component, in the pair annihilation / pair creation branch (s-channel):

$$\imath D_{\omega}^{(S)\mu\nu}\left(\vec{k}\right) = \frac{-\imath g^{\mu\nu}}{\omega^2 - \vec{k}^2 + \imath\varepsilon} \to \imath D_{\omega}^{(A)\mu\nu}(k) = \frac{-\imath g^{\mu\nu}}{(\omega + \delta w)^2 - \vec{k}^2} \approx \frac{-\imath g^{\mu\nu}}{\omega^2 - \vec{k}^2} \tag{257}$$

Here again we invoke the narrow beam approximation to argue that the dependence on δw in the denominator of the propagator will be averaged out in first order and small in second.

And we are left with again the position that all significant dependence on TQM will be in the GTF part of the wave function, there will be no first order dependence on the spinor parts.

The results for the GTFs are essentially the same as with the ABC case: the two final dispersions in energy will be the averages of the initial dispersions in energy. If one of the two incoming particles has a much greater dispersion in energy, it will dominate the result. This particle will then act as a narrow gate in time with respect to the other, with results already discussed.

Journal of Physics: Conference Series **2482** (2023) 012023 doi:10.1088/1742-6596/2482/1/012023

3.4.3. Compton scattering The exchanges are described by the s and t channels:

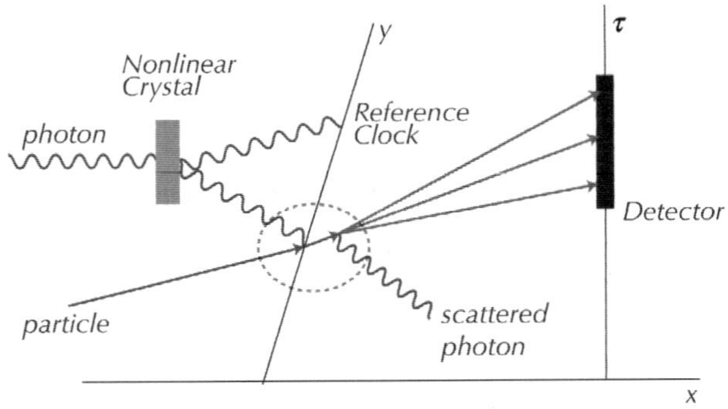

Figure 10. Test of Heisenberg uncertainty principle in time

$$\begin{aligned} s &= (p_1 + k_1)^2 = (p_3 + k_2)^2 \\ t &= (p_1 - k_2)^2 = (k_1 - p_3)^2 \end{aligned} \tag{258}$$

This is our first chance to see the fermion propagator in action:

$$\imath S_\omega^S(\vec{p}) = \imath \frac{\left(\not{p}^S + m\right)}{E^{(S)2} - \vec{p}^2 - m^2} \to \imath S_\omega^{\mathrm{A}}(p) = \imath \frac{(\not{p} + m)}{E^2 - \vec{p}^2 - m^2} \tag{259}$$

We see that the situation is not much different than with the two previous cases. The difference between $\not{p}^S$ and $\not{p}$ drops out in the narrow beam approximation, as does the difference between E^S and E in the denominator. We are left with SQM as carrier and the TQM GTF in time/energy the signal.

The main point of interest here is that as our technology for creating short photon pulses is now extremely sophisticated [70, 71], the changes of using a short pulse of light as a "narrow gate in time" should be good. We discuss this next.

4. Experimental Tests

"In so far as a scientific statement speaks about reality, it must be falsifiable: and in so far as it is not falsifiable, it does not speak about reality." – Karl Popper [72]

Our goal in this investigation has not been to argue that TQM is a correct extension of QED but rather that it is falsifiable, to give Gisin and his peers "something they can prove wrong".

There is no question that due to the small size of expected effects the associated experiments will be difficult. But at the same time there is a compensating variety of experiments: any time dependent system monitored by time sensitive detectors should show small but definite effects of dispersion in time.

Here we look at one possible experiment, define a figure of merit for such experiments, and then review experimental possibilities in general.

4.1. Heisenberg uncertainty principle in time/energy

"Ah, but a man's reach should exceed his grasp, Or what's a heaven for?" – Robert Browning

We sketch out a simple candidate solution for using the HUP in time/energy to falsify TQM.

(i) Suppose we start with a photon with a narrow width in time (as in Lindner's classic experiment [73]).
(ii) We send this through a non-linear crystal to split the photon into two equal-but-opposite photons, as in tests of Bell's theorem.
(iii) We send one of these to a reference clock to provide a start point for a time-of-flight measurement.
(iv) We use the other as a narrow gate in time, per above.
(v) We arrange for this photon to scatter an electron via Coulomb scattering.
(vi) We measure the electron's time-of-arrival at a detector.
(vii) We compute the time-of-flight as the difference between the time-of-arrival and the reference time.
(viii) We do this enough times to build up a time-of-flight distribution which will either conform to the predictions of TQM or falsify those.

Dr. Klag was kind enough to point out that this will not, in fact, work. The photons that are split by non-linear crystals in Bell's theorem tests are much longer in time than those Lindner used. As we probably need ultra-short photons, this is likely to be a problem.

Still this does show the essential elements of a realistic experiment:

(i) We need a time-of-flight, so need in general both a start time and a time of arrival.
(ii) We need something that can act as a narrow gate in time.
(iii) We are likely to need many many data points.

We next propose a "figure of merit" to pick among candidate solutions.

4.2. Figure of merit

"by recording single electron detection events diffracting through a double-slit, a diffraction pattern was built up from individual events." – Bach et al [74]

The primary effect of dispersion in time will be to increase the uncertainty in time from what would otherwise be predicted by standard QED. The fundamental metric is equation 166, repeated here for convenience:

$$(\Delta t)_D^{(T)2} \equiv (\Delta t)_D^2 - (\Delta t)_D^{(S)2} \tag{260}$$

The term on the left is the signal $(\Delta t)_D^{(T)2}$: the uncertainty squared predicted by TQM $(\Delta t)_D^2$ minus the uncertainty squared predicted by SQM $(\Delta t)_D^{(S)2}$.

If we are looking at time-of-flight as the prediction, then the distribution in time-of-flight will build up one event at a time. If we assume, say, Gaussian predictions for both TQM and SQM we will expect in general a wider, flatter Gaussian for TQM.

Assume we pick a degree of confidence, say the traditional five sigmas (giving a one in 3.5 million chance that the distribution was assigned incorrectly). There are many different statistical tests for making this sort of discrimination between two Gaussian distributions; we will assume one appropriate to the specific situation has been chosen.

The chosen degree of confidence will in turn imply a minimal sample size N to achieve it. Now suppose our apparatus can run T tests per second. The number of seconds to achieve the targeted level of confidence is then:

$$S = \frac{N}{T} \tag{261}$$

Our proposed figure of merit is the log to the base ten of the number of seconds required to achieve a five sigma level of confidence that TQM is falsified:

$$M \equiv \log_{10}(S) \tag{262}$$

If we need 100,000 tests and can run one test per second, then our figure of merit is:

$$\log_{10}(100000) = 5 \tag{263}$$

We choose a log scale because we expect there will be considerable variation in the efficiency of various experimental arrangements. In this way, arrangements that generate a stronger signal can be fairly compared to those that generate more tests per second and so on. The smaller S is the better of course. If S is longer than the mean-time-between-failure (MTBF) of the apparatus, then the specific experiment is not practical. And of course if it is longer than the duration of the associated grant, that too will be a problem.

One advantage of having a reasonably well-defined figure of merit is that this makes it easier for an AI system e.g. [75] to compare experimental possibilities. AI systems have the specific advantage in this case that they know nothing about time and are therefore less likely to be distracted by preconceptions.

4.3. TQM as an experiment factory

> "Henceforth space by itself, and time by itself, are doomed to fade away into mere shadows, and only a kind of union of the two will preserve an independent reality" – Hermann Minkowski [76]

With respect to the falsifiability of TQM, the small size of the basic effect may be compensated for by the large number of experimental possibilities. If quantum mechanics should in fact be extended in the time direction then essentially any time dependent apparatus with time dependent detectors may provide a possible line of attack. By hypothesis, *all* quantum mechanical phenomena seen in space – interference, diffraction, uncertainty, entanglement, tunneling, ... – apply in time as they do in space.

Effects of TQM on the legs We have focused here on the applications within QED. This is needed to treat interactions correctly. However there are many interesting effects at the single particle level.

These may described using the FS/T equation for a spinless particle (52):

$$\left((E - q\Phi)^2 - \left(\vec{p} - q\vec{A}\right)^2 - m^2\right)\psi_\tau = -2E\imath\frac{\partial\psi_\tau}{\partial\tau} \tag{264}$$

or the TQM Dirac equation (126):

$$\left(\not{p} - q\not{A} - m\right)\psi_\tau = -\imath\gamma_0\frac{\partial}{\partial\tau}\psi_\tau \tag{265}$$

We include the vector potential via the minimal substitution. We can also make use of the TQM equation for the free vector potential (89):

$$\left(w^2 - \vec{k}^2\right)A^\nu = -2w\imath\frac{\partial}{\partial\tau}A^\nu \tag{266}$$

All three equations are defined with reference to the rest frame of the vacuum $\mathcal{V}$, so are invariant. As noted, these can often be usefully simplified by rewriting the energy in terms of the quantum energy: $E = E^S + \delta E$ where E^S is the value of the energy in SQM, and then assuming that δE is small (as in the "narrow beam" approximation).

Reuse of existing SQM results We will often start with an existing solution in SQM, perhaps a hard-earned part analytic/part numeric approximation specific to a complex mashup of lab-built and off-the-shelf tech. In many cases we will be able to reuse these pre-existing solutions.

Assume we can use such a pre-existing solution to compute a time-of-arrival distribution. The uncertainty in time-of-arrival is defined as:

$$\Delta_{\tau}^{(S)2} \equiv \int_{-\infty}^{\infty} d\tau \left(\tau - \langle\tau\rangle^{S}\right)^{2} \rho_{\tau}^{S}, \langle\tau\rangle^{S} \equiv \int_{-\infty}^{\infty} d\tau \tau \rho_{\tau}^{S} \tag{267}$$

We can also use the existing SQM solution to compute a first order estimate of the corresponding GTF in coordinate time/energy by using the entropic estimate. This will usually be much less sophisticated than our existing SQM solution. But for falsification we are interested primarily in order-of-magnitude numbers. We therefore use the estimated GTF in time to infer the corresponding probability distribution in time $\rho_{\tau}(t)$.

The total uncertainty in coordinate time will be given by:

$$\Delta_t^2 \equiv \int_{-\infty}^{\infty} dt \rho(t) (t - \langle t\rangle)^2, \langle t\rangle \equiv \int_{-\infty}^{\infty} dt \rho(t) t, \rho(t) \equiv \int_{-\infty}^{\infty} d\tau \rho_{\tau}(t) \tag{268}$$

The difference between these two metrics is our signal. By this method existing SQM results for time-of-flight and the like may be converted into tests of TQM.

Use of combinations of approaches Further as suggested in the Samurai and Pirate experiment 3.4.1, a combination of approaches may be useful. Post-filtering of the results of an interaction to make the effects of dispersion in time more obvious; pre-filtering to get a cleaner result are possible approaches.

4.3.1. Three classes of effects The primary effects of TQM may be categorized as the common, the dramatic, and the subtle.

Common By the fundamental hypothesis the effects of dispersion in time are omnipresent. In a detector they will show as additional uncertainty in time-of-flight measurements. Within a quantum system, they will act as forces of "anticipation and regret": causing interactions to start sooner and last longer than otherwise would be the case.

But by the initial estimate they are expected small, of order attoseconds or less. There are several ways to address this problem:

(i) Statistical approaches are implicit in the figure of merit.
(ii) Scattering through a crystal is another approach; all of the atoms of the crystal can act in combination to achieve an effect. Therefore we can try sending a beam through a time crystal [77, 78, 79, 80, 81, 82, 83, 84, 85]; look for diffraction effects specifically from TQM.

Dramatic Diffraction effects, especially those associated with the Heisenberg uncertainty principle in time/energy, would appear to present the most dramatic possibilities. In SQM a single slit in time *clips* the wave function in time: the narrower the gate, the less the dispersion in time-of-flight. But in TQM a single slit in time *diffracts* the wave function in time: the narrower the gate, the greater the dispersion in time-of-flight. The effects go in opposite directions, so the contrast may, in principle, be set *arbitrarily* great.

Journal of Physics: Conference Series **2482** (2023) 012023 doi:10.1088/1742-6596/2482/1/012023

Subtle While it is natural to propose tests using SQM wave functions as carrier, the TQM part as signal; by the fundamental hypothesis a wave function is always to be understood as fully entangled in time and space. Several lines of attack arise out of this including:

(i) Effects of duality:
  (a) The single particle Lagrangian we started with (equation 7) is symmetric under the interchange of $t, \Phi \leftrightarrow x, A_x$. So the effects of sending a charged particle though a time-varying electric field are dual to sending a charged particle through a space-varying magnetic field. For instance we can start with the Aharonov-Bohm effect with respect to magnetic fields [86] and ask if the Aharonov-Bohm effect with respect to the electric field [87] has possibilities.
  (b) In general, Maxwell's equations are symmetric under an interchange of electric and magnetic fields, of the time-space and space-space components of the Maxwell stress tensor $F_{\mu\nu}$. We can look for effects associated with this.
  (c) And of course we can look at any experiment in space and ask if we interchange $t \leftrightarrow x$, do any interesting possibilities present themselves?

(ii) Effects of anti-symmetry in time:
  (a) For bosons: create a wave function which is anti-symmetric in both time and space parts, then look for anti-symmetry under reflection solely in the space part.
  (b) For fermions: create a wave function which is anti-symmetric in the time part but symmetric in the space part, then look for symmetry under reflection solely in the space part.
  (c) And in both cases we can look directly for anti-symmetry in time.

(iii) We can look at EPR effects in time, Bell's theorem tests in time, Greenberger–Horne–Zeilinger experiments, and so on.

(iv) We can look for the effects of tunneling in time, which may have interesting practical applications in steganography (the art of concealing the true message within an apparently innocent one).

4.3.2. Variations on existing fundamental tests of quantum mechanics Another, bottom up, line of attack is to take existing lists of foundational tests of quantum mechanics, to see if there is an "in time" variation for specific tests. Possible starting points are Lamoreaux [88], Ghose [89], and the three hundred or so experiments detailed in Auletta [90]. Examples are the "single slit in time" (as in the text), the "double slit in time" (as Lindner op cit), and so on.

And of course any experiment that explicitly mentions time, i.e. the Delayed Choice Quantum Eraser [91, 92] is a possible starting point.

4.3.3. What if TQM is confirmed but with qualifications? This in our own opinion is the most likely result. We are pushing classical mechanics in a quantum direction, historically a profitable line of attack. And we are making aggressive use of established principles, also often effective.

But we are extrapolating from a 3D theory to a 4D one, from shadow to substance. Any such extrapolation has ambiguities, whether done by a CAT scan or a philosopher.

The figure of merit provides a convenient way to categorize deviations from the simple extrapolation proposed by TQM. The points of maximal deviation provide a natural guide for followup experiments.

4.3.4. What if TQM is falsified? On the other hand, if TQM is falsified, then the falsification itself should suggest further experimental possibilities. The most obvious falsification would be that we do not in fact see the dispersion in time/energy that is the central prediction. There are two main possibilities here:

(i) There is a frame in which the violation is maximal. This would be a preferred frame, anathema to relativity, and of great interest to the contrarian experimenter.

(ii) The absence is uniform across frames. The restriction of the paths to on-shell paths is confirmed. This would perhaps not be entirely in the spirit of relativity. But it would at a minimum help clarify the relationship between the HUP in energy/time and the HUP in space/momentum. For instance it would show how to transform between these two in one frame and in another. As the precise relationship between these two has been the subject of a considerable literature (e.g.[93, 94, 95, 96, 54, 97]) this would be interesting as well.

4.3.5. There are no null experimental results Therefore there are no null experiments. Either we will have a variety of novel phenomena to explore or our understanding of the role of time/energy in QED will be deepened.

5. Discussion

"Anything that is not compulsory is forbidden." – Murray Gell-Mann [98]

The main problem here has been to extend QED to include time while keeping it consistent with all that has gone before. The approach has been to use the path integral formulation but keep everything but the paths themselves the same. We then extended the paths in a way that is manifestly covariant.

While there may be alternative ways to the same end, the requirement of manifest covariance should cause the results to be consistent to the first order of magnitude. Since the relevant scales, of attoseconds, are now accessible by experiment the hypothesis that the wave function should be extended in time is therefore falsifiable.

Evidence for confirmation Obviously no one writes a paper as long as this to define a hypothesis that has no chance of being confirmed. We argue that the odds of it being confirmed are perhaps even better than that given various advantages:

(i) The treatment of the time/energy and space/momentum coordinates is manifestly symmetric. Simply being able to do this is interesting.

(ii) We have a clear explanation of why time normally appears asymmetric at the level of the observer (due to statistical effects at the scale of Avogadro's number) while still at the particle level being completely symmetric.

(iii) We have a treatment which goes smoothly from the single to the multiple particle cases.

(iv) We do not see the ultra-violet divergences. We still have to normalize the loops, but we no longer have to regularize them: that drops out of the formalism.

(v) And we have Gell-Mann's principle: what is not forbidden is compulsory. If there is not a conservation principle or symmetry rule forbidding dispersion in time, it would be surprising *not* to see dispersion in time.

Implications of falsification As noted, if we falsify TQM we would at a minimum get a clearer understanding of the relationship between the HUP in energy/time and the HUP in space/momentum, especially with regard to the way they transform from one frame to another.

Implications of confirmation High speed chemical and biological interactions, i.e. attosecond scale, should show effects of time dispersion. For instance, if molecules can sense into the future, it may affect their ability to find optimal configurations.

There are potential applications for quantum communication and quantum computing. With TQM we have an additional channel to use for calculation/communication but also an additional channel to act as a source of decoherence.

The implications for quantum gravity are particularly interesting: with manifest covariance, elimination of the ultra-violet divergences, some recent work by Horwitz, and earlier work by Verlinde, we appear to have all the pieces needed to construct a complete, covariant, and convergent theory of quantum gravity. Leaving the question of the odds of this being correct to one side, we note that recent advances in technique mean such a theory has a reasonable chance of being falsifiable as well. We explore this in slightly more detail in the Appendix D.3.

Conclusion Any time dependent quantum phenomena viewed at a sufficiently short time scale (attoseconds or less) and with sufficiently time sensitive detectors should either display novel phenomena along the time/energy axis or at a minimum deepen and make more precise our understanding of the role of time in quantum mechanics.

Acknowledgments
I thank my long time friend Jonathan Smith for invaluable encouragement, guidance, and practical assistance.

I thank Ferne Cohen Welch for extraordinary moral and practical support.

I thank Martin Land, L. P. Horwitz, James O'Brien, Tepper Gill, Petr Jizba, Matthew Trump, and the other organizers of the International Association for Relativistic Dynamics (IARD) 2018, 2020, and 2022 Conferences for encouragement, useful discussions, and hosting talks on the papers in this series in the IARD conference series. I also thank my fellow participants in the 2022 conference – especially Pascal Klag, Bruce Mainland, Luca Smaldone, Howard Perko, Philip Mannheim, Alexey Kryukov, Ariel Edery, and others – for many excellent questions and discussions.

I thank Steven Libby for several useful conversations and in particular for insisting on the extension of the original ideas to the high energy limit and therefore to QED.

I thank Larry Sorensen for many helpful references. I thank Ashley Fidler for helpful references to the attosecond physics literature.

I thank Avi Marchewka for an interesting and instructive conversation about various approaches to time-of-arrival measurements.

I thank Hou Yau for an interesting discussion of variations on these themes and for directing my attention to his paper[99].

I thank Asher Yahalom for insisting that a better explanation of the clock time than simply what clocks measure was required

I thank Thomas Cember for useful clarifications of several points in the argument.

I thank the reviewer who drew my attention to Horwitz's work on gravity [100].

I think Danko Georgiev of the journal Quanta for very practical suggestions and advice.

I thank Y. S. Kim for organizing the invaluable Feynman Festivals, for several conversations, and for general encouragement and good advice.

I thank Catherine Asaro, Julian Barbour, Gary Bowson, Howard Brandt, Daniel Brown, Ron Bushyager, John G. Cramer, J. Ferret, Robert Forward, Fred Herz, J. Peřina, Linda Kalb, A. Khrennikov, David Kratz, Andy Love, Walt Mankowski, O. Maroney, John Myers, Paul Nahin, Marilyn Noz, R. Penrose, Stewart Personick, V. Petkov, H. Price, Matt Riesen, Terry Roberts, J. H. Samson, Lee Smolin, L. Skála, Arthur Tansky, R. Tumulka, Joan Vaccaro, L. Vaidman, A. Vourdas, H. Yadsan-Appleby, S. Weinstein, and Anton Zeilinger for helpful conversations over the years.

I thank the organizers of several QUIST, DARPA, Perimeter Institute conferences I've attended and the very much on topic conferences *Quantum Time* in Pittsburgh in 2014 and *Time and Quantum Gravity* in San Diego in 2015.

And none of the above are in any way responsible for any errors of commission or omission in this work.

Appendix A. Notation

We are using natural units throughout.

Since the text usually alternates between SQM and TQM sections, the meaning of an object should often be clear from context. In general SQM objects will have three vectors (i.e. $\vec{p}$) as arguments and TQM objects will have four vectors (i.e. p) as arguments. Where necessary we use a superscript S to mark an SQM object, a superscript T to mark a purely time object, and absence of an explicit mark to indicate a fully relativistic (i.e. TQM object). For example:

$$\begin{aligned} \psi_\tau (t, \vec{x}) &= \psi_\tau^T (t)\, \psi_\tau^S (\vec{x}) \\ \psi_\tau (E, \vec{p}) &= \psi_\tau^T (E)\, \psi_\tau^S (\vec{p}) \end{aligned} \tag{A.1}$$

In paper A we used an over-bar and over-tilde for the same markings, but we had to abandon that usage because it conflicted with the use of an over-bar to mark adjoint spinors in the Dirac equation. We still use the occasional over-bar to indicate average as $\bar{E} \equiv \left\langle \sqrt{m^2 + \vec{p}^2} \right\rangle$. We are putting the clock time at the bottom right in the position of a traditional index, in honor of its frequent use as an index in the time-slicing used to compute path integrals. To further streamline the frequent references to clock time we replace τ_1 with just the 1. And if we are looking at differences between two times we just put the two indexes in:

$$K_{\tau_1} \to K_1; K_{\tau_2;\tau_1} \to K_{2;1} \to K_{21} \tag{A.2}$$

The complementary variable for clock time is always ω; the complementary variable for coordinate time may be E, w, k_0, p_0 depending on context. It is usually obvious when a particular object refers to momentum or to coordinate space. When it might not be obvious, we use an over-hat to mark the momentum space form, i.e. $\hat{\phi}(p)$ for a four dimensional plane wave or $\hat{\sigma}_x \equiv \sigma_{p_x}$ for the dispersion in p_x. We use the Greek letter ϖ for the clock energy/clock frequency (energy and frequency are the same in natural units of course):

$$\varpi_k \equiv -\frac{w^2 - \vec{k}^2 - \mu^2}{2w} \tag{A.3}$$

We define $E_{\vec{p}}$ as the relativistic mass $E_{\vec{p}} \equiv \sqrt{m^2 + \vec{p}^2}$.

We use a superscript A to tag specific propagators as Attosecond, primarily meant for use at attosecond scales (77). We use ϕ for plane waves, φ for Gaussians, and ψ for general wave functions.

Appendix B. Gaussian Test Functions

> "That's a great deal to make one word mean," Alice said in a thoughtful tone.
> "When I make a word do a lot of work like that," said Humpty Dumpty, "I always pay it extra."
> – Lewis Carroll *Through the Looking Glass* 1871

Appendix B.1. Uses of GTFs
By Gaussian test functions (GTFs) we mean functions of the general form:

$$\varphi_0(x) = \sqrt[4]{\frac{1}{\pi\sigma^2}} e^{\imath p_0 x - \frac{(x-x_0)^2}{2\sigma^2}} \tag{B.1}$$

We generally take them as normalized to one. We refer to the σ as the dispersion. The uncertainty in the associated dimension is given by the dispersion divided by $\sqrt{2}$:

$$\Delta x \equiv \sqrt{\left\langle (x - \langle x \rangle)^2 \right\rangle} = \frac{\sigma}{\sqrt{2}} \tag{B.2}$$

We can get a rough approximation of any normalizable wave function by using the GTF with the same uncertainty:

$$\varphi_0^{(\text{TYPICAL})}(x) \equiv \sqrt[4]{\frac{1}{2\pi(\Delta x)^2}} e^{\imath \langle p_0 \rangle x - \frac{(x - \langle x_0 \rangle)^2}{4(\Delta x)^2}} \tag{B.3}$$

(i) We are making this use of GTFs when we estimate the initial wave functions using the entropic estimate of the uncertainty in subsection 3.1.1. This corresponds loosely to a statistician's use of the mode to summarize a population.
(ii) By using Morlet wavelet analysis, we can represent any normalizable wave function as a sum over Gaussians (see [101]).
(iii) Since the GTFs in momentum space are exact solutions of the various free equations in TQM (and in SQM for that matter) we can use them as starting functions in perturbation expansions.
(iv) A final and perhaps surprising use here is that the use of normalizable functions – whether typical GTFs or sums over GTFs – is critical for ensuring convergence of path integrals in general and loop diagrams in particular.

Appendix B.2. Starting GTFs
Appendix B.2.1. In momentum and space In p_x:

$$\varphi_0(p_x) = \sqrt[4]{\frac{1}{\pi\sigma_{p_x}^2}} e^{-\imath\left(p_x - p_x^0\right)x_0 - \frac{\left(p_x - p_x^0\right)^2}{2\sigma_{p_x}^2}} \tag{B.4}$$

In x:

$$\varphi_0(x) = \sqrt[4]{\frac{1}{\pi\sigma_x^2}} e^{\imath p_0 x - \frac{(x-x_0)^2}{2\sigma_x^2}} \tag{B.5}$$

where $\sigma_{p_x} = \frac{1}{\sigma_x}$. We can simplify slightly by taking $\hat{\sigma}_x \equiv \sigma_{p_x}$. The y, z GTFs are the same, replacing $x \to y, z$ and $p_x \to p_y, p_z$. In momentum space:

$$\varphi_0^S(\vec{p}) = \sqrt[4]{\frac{1}{\pi^3 \det\left(\hat{\Sigma}^S\right)}} e^{-\imath(\vec{p}-\vec{p}_0)\cdot\vec{x}_0 - (\vec{p}-\vec{p}_0)\cdot\frac{1}{2\hat{\Sigma}^S}\cdot(\vec{p}-\vec{p}_0)}, \hat{\Sigma}^S \equiv \begin{pmatrix} \hat{\sigma}_x^2 & 0 & 0 \\ 0 & \hat{\sigma}_y^2 & 0 \\ 0 & 0 & \hat{\sigma}_z^2 \end{pmatrix} \tag{B.6}$$

and in coordinate space:

$$\varphi_0^S(\vec{x}) = \sqrt[4]{\frac{1}{\pi^3 \det(\Sigma^S)}} e^{\imath \vec{p}_0 \cdot \vec{x} - (\vec{x}-\vec{x}_0) \cdot \frac{1}{2\Sigma^S} \cdot (\vec{x}-\vec{x}_0)}, \Sigma^S = \begin{pmatrix} \sigma_x^2 & 0 & 0 \\ 0 & \sigma_y^2 & 0 \\ 0 & 0 & \sigma_z^2 \end{pmatrix} = \frac{1}{\hat{\Sigma}^S} \tag{B.7}$$

Appendix B.2.2. In energy and time We get the time and energy forms taking $p_x \to E, x \to t$ and complex conjugating. If we take the wave function in energy as:

$$\varphi_0(E) \equiv \sqrt[4]{\frac{1}{\pi \sigma_E^2}} e^{\imath (E-E_0) t_0 - \frac{(E-E_0)^2}{2\sigma_E^2}} \tag{B.8}$$

we have the wave function in time as:

$$\varphi_0(t) \equiv \sqrt[4]{\frac{1}{\pi \sigma_t^2}} e^{-\imath E_0 t - \frac{(t-t_0)^2}{2\sigma_t^2}} \tag{B.9}$$

where $\sigma_E = \frac{1}{\sigma_t}$.

Appendix B.2.3. In time/energy and space/momentum We can get four dimensional wave functions by taking the direct product of the wave functions in time/energy and space/momentum. In four momentum space:

$$\varphi_0(E, \vec{p}) = \varphi_0^T(E) \varphi_0^S(\vec{p}) \tag{B.10}$$

spelled out:

$$\varphi_0(p) = \sqrt[4]{\frac{1}{\pi^4 \det(\hat{\Sigma})}} e^{\imath (p-p_0)_\mu x_0^\mu - (p-p_0)^\mu \frac{1}{2\hat{\Sigma}^{\mu\nu}} (p-p_0)^\nu}, \hat{\Sigma}^S \equiv \begin{pmatrix} \hat{\sigma}_t & 0 & 0 & 0 \\ 0 & \hat{\sigma}_x^2 & 0 & 0 \\ 0 & 0 & \hat{\sigma}_x^2 & 0 \\ 0 & 0 & 0 & \hat{\sigma}_z^2 \end{pmatrix} \tag{B.11}$$

In coordinate space:

$$\varphi_0(t, \vec{x}) = \varphi_0^T(t) \varphi_0^S(\vec{x}) \tag{B.12}$$

spelled out:

$$\varphi_0(x) = \sqrt[4]{\frac{1}{\pi^4 \det(\Sigma)}} e^{-\imath p_0^\mu x_\mu - (x-x_0)^\mu \frac{1}{2\Sigma^{\mu\nu}} (x-x_0)^\nu}, \Sigma = \begin{pmatrix} \sigma_t & 0 & 0 & 0 \\ 0 & \sigma_x^2 & 0 & 0 \\ 0 & 0 & \sigma_x^2 & 0 \\ 0 & 0 & 0 & \sigma_z^2 \end{pmatrix} = \frac{1}{\hat{\Sigma}} \tag{B.13}$$

We are treating time and space as disentangled. We can entangle by generalizing the dispersion matrix Σ to be an arbitrary positive definition matrix.

Appendix B.3. GTFs as a function of clock time
We look at the evolution of the GTFs as a function of clock time.

Appendix B.3.1. Momentum space

Non-relativistic and SQM GTFs The behavior of the momentum space GTFs is simple. In the non-relativistic case:

$$\varphi_\tau^{NR}(\vec{p}) = \exp\left(-\imath\frac{\vec{p}^2}{2m}\tau\right)\varphi_0^S(\vec{p}) \tag{B.14}$$

The GTFs in SQM are similar:

$$\varphi_\tau^S(\vec{p}) = e^{-\imath\sqrt{m^2+\vec{p}^2}\tau}\varphi_0^S(\vec{p}) \tag{B.15}$$

If the dispersion in three space momenta is not too great we may write:

$$\sqrt{m^2+\vec{p}^2} \approx \bar{E}_0 + \frac{(\vec{p}-\vec{p}_0)^2}{2E_0}, \bar{E}_0 \equiv \sqrt{m^2+\vec{p}_0^2}, \vec{p}_0 \equiv \langle\vec{p}\rangle \tag{B.16}$$

Note we are taking the average relativistic mass $\bar{E}_0$ as the reference point, not the bare mass m. This means that the utility of the approximation can survive to much greater velocities; all that is required is that the dispersion of the momentum be small relative to the average momentum. As $\bar{E}_0 \to m$ we get the non-relativistic form, modulo a constant and therefore uninteresting overall factor of $\exp\left(-\imath\bar{E}_0\tau\right)$. Note also that the diagonal form of the clock time dependence is a result of using direct product GTFs in the three space dimensions.

For TQM the behavior is, of course, a bit more complex. We have:

$$\varphi_\tau(p) = \exp\left(\imath\frac{E^2-\vec{p}^2-m^2}{2E}\tau\right)\varphi_0(p) = \exp\left(\imath\frac{E^2-\vec{p}^2-m^2}{2E}\tau\right)\varphi_0^T(E)\varphi_0^S(\vec{p}) \tag{B.17}$$

So even though the energy and the three momentum parts start disentangled, they become entangled as a result of the $\frac{\vec{p}^2}{2E}$ term. For now, we will deal with this by again assuming that the dispersions in energy/momentum are not that great, so that it makes sense to write $\delta E = E - \bar{E}_0$ and therefore:

$$-\frac{E^2-\vec{p}^2-m^2}{2E} = \varpi_p = -\frac{\left(\bar{E}_0+\delta E\right)^2 - (\vec{p}_0+\delta\vec{p})^2 - m^2}{2\left(\bar{E}_0+\delta E\right)} \approx -\delta E + \frac{(\delta E)^2 - 2\vec{p}\cdot\delta\vec{p} - (\delta\vec{p})^2}{2\bar{E}_0} \tag{B.18}$$

We can then divide the TQM GTF into the energy part:

$$\varphi_\tau(E) = \sqrt[4]{\frac{1}{\pi\sigma_E^2}}\exp\left(\imath\delta E(\tau+t_0) - \imath\frac{(\delta E)^2}{2\bar{E}_0}\tau - \frac{(\delta E)^2}{2\sigma_E^2}\right) \tag{B.19}$$

and momentum space part:

$$\varphi_\tau^{TQM}(\vec{p}) = \exp\left(\imath\frac{2\vec{p}\cdot\delta\vec{p}+(\delta\vec{p})^2}{2\bar{E}_0}\tau\right)\varphi_0^S(\vec{p}) \tag{B.20}$$

The space part in TQM does not evolve in quite the same way as the space part in SQM; it seems better in practice to compare SQM as a whole with TQM as a whole.

Appendix B.3.2. Coordinate space forms

Non-relativistic GTFs We start with the non-relativistic form:

$$\varphi_\tau^{NR}(\vec{x}) = \varphi_\tau^{NR}(x)\,\varphi_\tau^{NR}(y)\,\varphi_\tau^{NR}(z) \tag{B.21}$$

We look at the x direction [102, 8]:

$$\varphi_\tau^{NR}(x) = \sqrt[4]{\frac{1}{\pi\sigma_x^2}}\sqrt{\frac{1}{f_\tau^{(x)}}}e^{\imath p_x^{(0)}x - \frac{1}{2\sigma_x^2 f_\tau^{(x)}}(x-x_\tau)^2 - \imath\frac{p_x^{(0)2}}{2m}\tau}, f_\tau^{(x)} = 1 - \imath\frac{\tau}{m\sigma_x^2} \tag{B.22}$$

with average position in x:

$$x_\tau = x_0 + v_x^{(0)}\tau, v_x^{(0)} = \frac{p_x^{(0)}}{m} \tag{B.23}$$

We have y, x the same. The corresponding probability density is:

$$\rho_\tau^{NR}(\vec{x}) = \rho_\tau^{NR}(x)\,\rho_\tau^{NR}(y)\,\rho_\tau^{NR}(z) \tag{B.24}$$

And again focusing on the x direction:

$$\rho_\tau^{NR}(x) = \sqrt{\frac{1}{\pi\sigma_x^2\left(1+\frac{\tau^2}{m^2\sigma_x^4}\right)}}\exp\left(-\frac{(x-x_\tau)^2}{\sigma_x^2\left(1+\frac{\tau^2}{m^2\sigma_x^4}\right)}\right) \tag{B.25}$$

Notice the kink in behavior at the point where $\frac{\tau}{m\sigma_x^2} \approx 1$. At this turning point, the uncertainty goes from being proportional to σ to being proportional to τ/σ.

$$(\Delta x)^2 \equiv \left\langle x_\tau^2 \right\rangle - \left\langle x_\tau \right\rangle^2 = \frac{\sigma_x^2}{2}\left|1 + \frac{\tau^2}{m^2\sigma_x^4}\right| \tag{B.26}$$

So if the dispersion starts small it will end large. This may be understood in terms of the HUP in space/momentum. If the initial uncertainty in space is small, the uncertainty in momentum goes as $1/\sigma_x$ so is correspondingly large. Given a bit of time (clock time here) the large dispersion in momentum causes the wave function to spread out in space, creating correspondingly large dispersion in space. We may think of this as diffraction at work. The behavior for the SQM GTFs is essentially the same.

TQM GTFs The time part is in close parallel to the x part:

$$\varphi_\tau^T(t) = \sqrt[4]{\frac{1}{\pi\sigma_t^2 f_\tau^t}}e^{\imath E_0 t_0 - \frac{(t-t_\tau)^2}{2\sigma_t^2 f_\tau^t}}, f_\tau^t \equiv 1 + \frac{\imath\tau}{E_0\sigma_t^2} \tag{B.27}$$

with:

$$t_\tau = t_0 + v_t\tau, v_t = \gamma = \frac{E_0}{m} \tag{B.28}$$

Probability density:

$$\rho_\tau^T(t) = \sqrt{\frac{1}{\pi\sigma_t^2\left(1+\frac{\tau^2}{E_0^2\sigma_t^4}\right)}}\exp\left(-\frac{(t-t_\tau)^2}{\sigma_t^2\left(1+\frac{\tau^2}{E_0^2\sigma_t^4}\right)}\right) \tag{B.29}$$

Again note the shift in uncertainty in time at the point where $\frac{\tau}{E_0\sigma_t^2} \approx 1$.

$$(\Delta t)^2 \equiv \langle t^2 \rangle - t_\tau^2 = \frac{\sigma_t^2}{2} \left| 1 + \frac{\tau^2}{E_0^2 \sigma_t^4} \right| \tag{B.30}$$

For TQM, the HUP in time/energy is fully equivalent to the HUP for space/momentum.

TQM Kernels In the analysis of the mass loop correction we will work directly with the kernels in the narrow beam approximation. We ignore the normalization factor of $\frac{1}{2\bar{E}_0}$ here. We developed this in some detail in paper A. In momentum space:

$$\hat{K}_\tau (p; p') = \delta^{(4)} (p - p') \exp \left(-\imath \frac{E^2 - \vec{p}^2 - m^2}{2\bar{E}_0} \tau \right) \tag{B.31}$$

and in coordinate space:

$$K_\tau (x; x') = -\imath \frac{\bar{E}_0^2}{4\pi^2 \tau^2} e^{-\imath \bar{E}_0 \frac{(t-t')^2}{2\tau} + \imath \bar{E}_0 \frac{(\vec{x}-\vec{x}')^2}{2\tau} - \imath \frac{m}{2} \tau} \tag{B.32}$$

Appendix C. Feynman rules

In the text we develop the Feynman rules for a few simple cases; here we show we can extend the Feynman rules to all orders. This is not a given. The problem is that in most treatments of the path integral, the Hamiltonian part acts like a kind of locomotive pulling the sum over paths forward. With TQM there is no natural equivalent to the Hamiltonian, since there is no dependence of the Lagrangian on clock time and therefore no canonical momenta and therefore no non-trivial Hamiltonian.

To see this in the simplest case, consider the Lagrangian for a free spin zero massive particle above (equation 23). The obvious choice for the corresponding TQM Lagrangian is:

$$\mathcal{L} [\phi] = \frac{1}{2} \partial_t \phi \partial_t \phi - \frac{1}{2} \nabla \phi \nabla \phi - \frac{m^2}{2} \phi^2 \tag{C.1}$$

In fact our requirement of complete covariance does not give us any real alternatives. The derivatives with respect to time must be with respect to coordinate time if they are to form part of a four vector with the ∇ operator. We can write the action as either the integral of d^4x or the integral of $d\tau d^4x$ over this. But we still have no dependence on clock time in the Lagrangian. Therefore this Lagrangian goes not give us an equation of motion. It has no kinetic energy term; it is all potential. Therefore while we can usually develop path integrals with the Hamiltonian or the Lagrangian approaches, with TQM we *must* use the Lagrangian formulation.

Other major differences are that:

(i) Most 3D objects are promoted to 4D.

(ii) The treatment of clock time has a different character; it still orders the Dyson series and the like, but most of the dependence on time is via the dependence on coordinate time.

(iii) We are primarily interested in interactions over short times: the effects we are interested in are of order attoseconds. Over picoseconds – or still more glacial intervals – the effects of time dispersion are likely to be averaged out.

Appendix C.1. Dyson series

Appendix C.1.1. Dyson series in SQM We start with the familiar case of the Dyson expansion for the S matrix in SQM, in the interaction picture:

$$S_{fi} = \langle f| \exp\left(-\imath \int\limits_{\tau_i}^{\tau_f} d\tau \int d\vec{x}\mathcal{H}_I\right)|i\rangle \tag{C.2}$$

with interaction Hamiltonian:

$$H_I \equiv \int d\vec{x}\mathcal{H}_I \tag{C.3}$$

We expand the exponential in a power series:

$$S_{fi} = \sum_{n=0}^{\infty} \frac{(-\imath)^n}{n!} \int\limits_{\tau_i}^{\tau_f} d\tau_1 \int\limits_{\tau_i}^{\tau_f} d\tau_2 \ldots \int\limits_{\tau_i}^{\tau_f} d\tau_n H_I(\tau_1) H_I(\tau_2) \ldots H_I(\tau_n) \tag{C.4}$$

and then time order the individual terms, which disposes of the $1/n!$:

$$S_{fi} = \sum_{n=0}^{\infty} (-\imath)^n \int\limits_{\tau_i}^{\tau_f} d\tau_1 \int\limits_{\tau_i}^{\tau_f} d\tau_2 \ldots \int\limits_{\tau_i}^{\tau_f} d\tau_n T\left\{H_I(\tau_1) H_I(\tau_2) \ldots H_I(\tau_n)\right\} \tag{C.5}$$

We then use Wick's theorem to replace the difficult-to-work-with time ordered terms with normal ordered terms plus "contractions" – the Feynman propagators. The sum over all topologically distinct ways of doing this is the sum over all Feynman diagrams.

The handling of clock time requires particular attention in this context. For instance at each vertex, the integrals over three space induce in a natural way a δ function in three momentum. The usual δ function over clock energy is produced by the taking the S matrix to run from $\tau = -\infty \to +\infty$. This guarantees conservation of clock energy, treated as the fourth component of a four vector, at each vertex and also for the S matrix overall.

But this conservation law is purchased at the expense of limiting the domain of applicability in a serious way. If we want to apply the theory to short times (as in this investigation) we have to accept that conservation of laboratory energy may be approximate. So we have an unnatural limitation of the domain of applicability of the theory or else have to take our chances with conservation of laboratory energy. In the text we are able to finesse the problem by focusing on the interaction of individual wave packets. They are set to interact for only short periods of time, so we can let the limits of clock time integrals go to $\pm\infty$ without changing the numeric value of the integral. But the problem remains in the general case.

Appendix C.1.2. Dyson series in TQM Wick's theorem is general, applying equally to SQM and TQM. It can be used in TQM in the same way as in SQM, letting the time ordered elements in the Dyson series be expressed as sums over products of normal ordered operators and of propagators. The topology and symmetry of the two series are the same.

However the individual objects go in general from 3D to 4D. The initial and final wave functions go from 3D plane waves to 4D and the propagators change as described in the text. At each vertex we go from an integral over clock time and three space dimensions to an integral over clock time, coordinate time, and the three space dimensions. The coordinate time and the three space dimensions are part of a four vector, the clock time is treated as a separate object (but really the time coordinate of the rest frame of the vacuum, per discussion).

For the most part the effect is as if we had simply added a 4th space dimension to each object in the series, one that comes in with opposite sign with respect to the three space dimensions. With this it is possible to write out each term in the Dyson series and solve for any specified problem to any required order, with one further change. We will assume, as with the SQM case, that we are working in the interaction picture. Therefore we need only consider the interaction potential; the rest of the dependence on clock time is contained in the basis states.

The TQM Dyson series for the S matrix is then:

$$S_{fi} = \langle f| \exp\left(\imath \int_{\tau_i}^{\tau_f} d\tau \int dt d\vec{x} \mathcal{L}_I \right) |i\rangle \tag{C.6}$$

The interaction Lagrangian is:

$$\mathcal{L}_I = -V_I \tag{C.7}$$

so:

$$\mathcal{H} = -\mathcal{L} = V_I = \mathcal{H}_I \tag{C.8}$$

The two sign flips of the potential cancel out and we are left with the same arguments for the exponential, albeit with four dimensions to integrate over rather than three:

$$\int d\vec{x} \mathcal{H}_I \to \int dt d\vec{x} \mathcal{H}_I \tag{C.9}$$

Appendix C.2. Feynman rules in TQM

Therefore we can write the Feynman rules for TQM down by inspection, with the substitutions noted. The basic topology is unchanged, the symmetry factors are unchanged. The external factors are adjusted per discussion; the external δ functions go from three in three momentum plus clock energy to four in four momentum plus clock energy.

Appendix C.2.1. S matrix The S matrix in SQM is:

$$S_{fi} = \delta_{fi} + (2\pi)^4 \delta\left(\Omega_f - \Omega_i\right) \delta^3\left(\vec{P}_f - \vec{P}_i\right) \left(\prod^{\text{all external bosons}} \sqrt{\frac{1}{2V\omega_{\vec{k}}}} \right) \left(\prod^{\text{all external fermions}} \sqrt{\frac{m}{VE_{\vec{p}}}} \right) \mathcal{M}^S \tag{C.10}$$

while the S matrix in TQM is the formally similar:

$$S_{fi} = \delta_{fi} + (2\pi)^5 \delta\left(\varpi_f - \varpi_i\right) \delta^4\left(P_f - P_i\right) \left(\prod^{\text{all external bosons}} \sqrt{\frac{1}{2TVw}} \right) \left(\prod^{\text{all external fermions}} \sqrt{\frac{m}{TVE}} \right) \mathcal{M} \tag{C.11}$$

In both cases the amplitude $\mathcal{M}$ is a sum over all topologically distinct terms:

$$\mathcal{M} = \sum_{n=0}^{\infty} \mathcal{M}^{(n)} \tag{C.12}$$

In TQM the raw interaction vertex is unchanged:

$$-e\gamma^\mu \tag{C.13}$$

The sign rules are the same as in SQM. For our purposes, writing out the usual Feynman diagrams while adding an additional -1 for each flip of identical fermions suffices.

Appendix C.2.2. Incoming and outgoing wave functions The handling of the incoming and outgoing wave functions is different. The incoming wave functions in SQM are generally taken as plane waves. In TQM we have to use GTFs. In SQM both incoming and outgoing legs are taken as on-shell. In TQM the 4D wave functions oscillate around the on-shell values, now taken as averages rather than absolutes, more guidelines than rules.

To go from SQM to TQM we replace the relativistic mass $E_{\vec{p}}$ in each spinor with the coordinate energy E, but leave them otherwise unchanged. So each spinor becomes a function of all four components of the momentum:

(i) Initial electron: $u_s^S(\vec{p}) \rightarrow u_s(p)$
(ii) Each final electron: $\bar{u}_s^S(\vec{p}) \rightarrow \bar{u}_s(p)$
(iii) Each initial positron: $\bar{v}_s^S(\vec{p}) \rightarrow \bar{v}_s(p)$
(iv) Each final positron: $v_s^S(\vec{p}) \rightarrow v_s(p)$

The polarization vectors for the photons are unchanged. We change the argument from a three vector to a four vector to indicate we are changing the context.

(i) Each initial photon: $\varepsilon_{r,\mu}\left(\vec{k}\right) \rightarrow \varepsilon_{r,\mu}(k)$
(ii) Each final photon: $\varepsilon_{r,\mu}\left(\vec{k}\right) \rightarrow \varepsilon_{r,\mu}(k)$

Appendix C.2.3. Propagators The propagators are different, as noted in some detail in the text. The apparently covariant character of the SQM propagators falls apart under close examination: there is dispersion in the x, y, z directions but not in t. As a result in SQM the intermediate particles are virtual; in TQM they are real.

Photon propagator in SQM (88):

$$\imath D_\omega^{(S)\mu\nu}\left(\vec{k}\right) = \frac{-\imath g^{\mu\nu}}{\omega^2 - \vec{k}^2 + \imath\varepsilon} \tag{C.14}$$

Fermion propagator in SQM (122):

$$\imath S_\omega^S(\vec{x}) = \imath\frac{\omega\gamma_0 - \vec{p}\cdot\vec{\gamma} + m}{\omega^2 - \vec{p}^2 - m^2 + \imath\varepsilon} \tag{C.15}$$

The $\imath\epsilon$ is used to pick out the contours so that positive frequencies are associated with forwards in time; negative with backwards in time. In TQM, at least in this initial analysis, we have found it helpful to break out the forward and backward components and deal with them separately.

Photon propagator in TQM (99):

$$\imath D_\omega^{\mu\nu}(k) = \frac{-\imath g^{\mu\nu}}{2w}\left(\frac{\imath}{w^2 - \vec{k}^2 + 2w\omega + 2w\imath\varepsilon} + \frac{\imath}{w^2 - \vec{k}^2 + 2w\omega - 2w\imath\varepsilon}\right) \tag{C.16}$$

with:

$$\varpi_k \equiv -\frac{w^2 - \vec{k}^2}{2w} \tag{C.17}$$

and short time limit:

$$\imath D_\omega^{(A)\mu\nu}(k) = -\imath g^{\mu\nu}\frac{\imath}{w^2 - \vec{k}^2} \tag{C.18}$$

Feynman propagator for fermions (132):

$$\imath S_\omega\left(p\right)=\imath\frac{\left(\not p+m\right)}{E^2-\vec{p}^2-m^2+2\omega E+2E\imath\varepsilon}+\imath\frac{\left(\not p+m\right)}{E^2-\vec{p}^2-m^2+2\omega E-2E\imath\varepsilon} \tag{C.19}$$

with short time limit:

$$\imath S_\omega^{(A)}\left(p\right)=\imath\frac{\not p+m}{E^2-\vec{p}^2-m^2} \tag{C.20}$$

Appendix C.2.4. Vertexes

SQM SQM vertex (138):

$$-e\bar{\psi}^S\left(\vec{p}'\right)A^{(S)\nu}\left(\vec{k}\right)\gamma_\nu\psi^S\left(\vec{p}\right) \tag{C.21}$$

The vertex is accompanied by δ functions in the three space momenta:

$$\delta^3\left(\vec{p}_{out}-\vec{p}_{in}\right) \tag{C.22}$$

and in clock energy if we are taking the limit as $\tau\pm\infty$:

$$\delta\left(\sum\Omega_{out}-\sum\Omega_{in}\right) \tag{C.23}$$

TQM TQM uses the same vertex except for the obvious replacement of three vector by four vector (141):

$$-e\bar{\psi}\left(p'\right)A^\nu\left(k\right)\gamma_\nu\psi\left(p\right) \tag{C.24}$$

The vertex is accompanied by δ functions in the four momenta:

$$\delta^4\left(p_{out}-p_{in}\right) \tag{C.25}$$

and in clock energy if we are taking the limit as $\tau\pm\infty$:

$$\delta\left(\varpi_{out}-\varpi_{in}\right) \tag{C.26}$$

Appendix D. The rest frame of the vacuum

"If you look long enough into the void, the void begins to look back through you."
– Frederick Nietzsche

Appendix D.1. Energy-momentum of spacetime

In the development in the text we have to pick a specific laboratory frame to define the clock time τ; we have therefore dependence on that choice. For laboratories going at non-relativistic velocities, the corrections to clock time will result in a correction to a correction, therefore not of first order, therefor not essential to falsifiability.

Still we would like to define TQM in a completely frame independent way. We noted in paper A that we can make an invariant choice of frame by taking advantage of an observation from Weinberg [57]. Consider the Einstein field equations:

$$G_{\mu\nu}\equiv R_{\mu\nu}-\frac{1}{2}g_{\mu\nu}R=-8\pi GT_{\mu\nu} \tag{D.1}$$

Rewrite as:

$$\left(G^{\mu\nu} + 8\pi G T^{\mu\nu}\right)_{;\nu} = 0 \tag{D.2}$$

We may use this to associate an energy momentum tensor with local spacetime. Define:

$$g_{\mu\nu} = \eta_{\mu\nu} + h_{\mu\nu} \tag{D.3}$$

where $h_{\mu\nu}$ vanishes at infinity but is not assumed small. The part of the Ricci tensor linear in h is:

$$R^{(1)}_{\mu\nu} \equiv \frac{1}{2}\left(\frac{\partial^2 h^\lambda_\lambda}{\partial x^\mu \partial x^\nu} - \frac{\partial^2 h^\lambda_\mu}{\partial x^\lambda \partial x^\nu} - \frac{\partial^2 h^\lambda_\nu}{\partial x^\lambda \partial x^\mu} + \frac{\partial^2 h_{\mu\nu}}{\partial x^\lambda \partial x_\lambda}\right) \tag{D.4}$$

The exact Einstein equations may be written as:

$$R^{(1)}_{\mu\nu} - \frac{1}{2}\eta_{\mu\nu} R^{(1)\lambda}_{\lambda} = -8\pi G\left(T_{\mu\nu} + t_{\mu\nu}\right) \tag{D.5}$$

where $t_{\mu\nu}$ is defined as quadratic in h and higher:

$$t_{\mu\nu} \equiv \frac{1}{8\pi G}\left(R_{\mu\nu} - \frac{1}{2} g_{\mu\nu} R^\lambda_\lambda - R^{(1)}_{\mu\nu} + \frac{1}{2}\eta_{\mu\nu} R^{(1)\lambda}_{\lambda}\right) \tag{D.6}$$

Weinberg then argues we may interpret $t_{\mu\nu}$ as the energy-momentum of the gravitational field itself.

If we can associate an energy momentum with spacetime, we may define a local rest frame with respect to that energy momentum tensor. We will refer to this as $\mathcal{V}$, the rest frame of the vacuum. If we define clock time with respect to this frame, we have an invariant definition of the clock time. Presumably this invariant frame is in free fall. A laboratory in Near Earth Orbit would do. More practically, we can make the associated calculations to correct for our earth bound laboratory being in an accelerated frame.

Appendix D.2. The four dimensional Schrödinger equation in the rest frame of the vacuum
Weinberg was working in terms of spacetime, an essentially classical concept. Now let us replace spacetime with a quantum vacuum. We will assume that it is full of real (rather than virtual) particles in a statistical ensemble.

With this context we return to the 4D Schrödinger equation:

$$\imath\frac{\partial}{\partial \tau}\psi = -\frac{p^2 - m^2}{2m}\psi \tag{D.7}$$

We define the energy momentum operators in the rest frame of the vacuum as:

$$\mathcal{E} \equiv \imath\frac{\partial}{\partial \tau_{\mathcal{V}}}, \overrightarrow{\mathcal{P}} \equiv -\imath\vec{\nabla}_{\mathcal{V}} \tag{D.8}$$

with four momentum:

$$\mathcal{P} = \left(\mathcal{E}, \vec{\mathcal{P}}\right) \tag{D.9}$$

We are using capital script letters for values and operators associated with the vacuum. We form the invariant $\mathcal{Q}^2$ for the invariant difference between the foreground and the vacuum momentum:

$$\mathcal{Q}^2 \equiv (\mathcal{P} - p)^2 \tag{D.10}$$

Suppose that the Klein-Gordon equation should really not be written as an absolute but as relative to the local spacetime. This is clearly reasonable. Then the Klein-Gordon equation becomes, in an invariant form:

$$\mathcal{Q}^2\psi = m^2\psi \tag{D.11}$$

Any realistic quantum vacuum will be some sort of quantum soup of photons and other bosons, electrons and other fermions. But at core the free equations for all of these are the Klein-Gordon equation with a light flavoring of spin/polarization added. We therefore take the vacuum as obeying an averaged Klein-Gordon equation of its own:

$$\left(\mathcal{P}^2 - \mathcal{M}^2\right)|\mathcal{V}\rangle = 0 \tag{D.12}$$

using $\mathcal{P}, \mathcal{M}$ as conveniently vague local averages over the mass-energy of the vacuum. Write the system of vacuum plus particle as a direct product:

$$|\psi\rangle\,|\mathcal{V}\rangle \tag{D.13}$$

We expand the Klein-Gordon equations:

$$\left(\mathcal{P}^2 - 2p\mathcal{P} + p^2 - m^2 - \mathcal{M}^2\right)|\psi\rangle\,|\mathcal{V}\rangle = 0 \tag{D.14}$$

The purely vacuum part cancels by assumption $\left\langle \mathcal{P}^2 - \mathcal{M}^2 \right\rangle \approx 0$. Further we choose to work in the rest frame of the vacuum so that $\mathcal{P} \to \mathcal{E}, \vec{0}$:

$$\langle 2E\mathcal{E}\rangle = \left\langle p^2 - m^2 \right\rangle \tag{D.15}$$

The energy of the vacuum $\mathcal{E}$ is, in coordinate space, given by the time operator of the vacuum $\mathcal{E} \equiv \imath\frac{\partial}{\partial\tau_{\mathcal{V}}}$:

$$\imath\frac{\partial}{\partial\tau_{\mathcal{V}}} = \frac{p^2 - m^2}{2E} \tag{D.16}$$

Since we are looking at the difference between the energy operator of the particle and the vacuum we therefore also need to look at the difference between the energy operator of the laboratory and the vacuum (the laboratory itself is after all nothing but particles). Therefore the clock time or laboratory time is to be understood as the *negative* of the time operator of the vacuum:

$$\frac{\partial}{\partial\tau_{\mathcal{V}}} = -\frac{\partial}{\partial\tau} \tag{D.17}$$

Therefore the correct 4D Schrödinger equation is (if we are in the $\mathcal{V}$ frame):

$$p^2 - m^2 = -2E\imath\frac{\partial}{\partial\tau} \tag{D.18}$$

So the slow drift of the observed system's wave function with respect to the observer is tracked by the cross-term of observed and observer wave functions. The 4D Schrödinger equation reduces to the non-relativistic form (equation 10) using $E \to m$.

Note use of the Machian hypothesis has recovered the FS/T, which we built starting from the single particle path integral approach. It is striking that these two distinct approaches agree.

Appendix D.3. Implications for quantum gravity

In the text proper we treat the Machian hypothesis as a formal hypothesis, useful for extending QED in time in a self-consistent way. However if we are prepared to accept this hypothesis as physical, at least for purpose of argument, there are some interesting implications for quantum gravity.

In the text proper we show in TQM we have:

(i) a fully covariant treatment of time in QED,

(ii) and the elimination of the ultra-violet divergences.

These are two of the principal barriers in the way of getting to quantum gravity.

Further, as noted in the introduction, TQM is a part of the Relativistic Dynamics program so TQM can draw on the extensive Relativistic Dynamics literature. In particular we can take advantage of Horwitz's extension of Relativistic Dynamics to General Relativity [100, 103, 104].

The Horwitz approach does not itself supply a mechanism. However, consider the conventional practice of dropping disconnected diagrams in Feynman diagrams. What if these terms should be seen not as disconnected but as connected to the vacuum?

Consider the mass terms in particular. They typically have a form like:

$$\frac{m^2}{2}\left(a_k a_{-k} + a_k a_k^\dagger + a_k^\dagger a_k + a_k^\dagger a_{-k}^\dagger\right) \tag{D.19}$$

All four of the terms conserve momentum. The leftmost describes pair annihilation; the rightmost pair creation. These are typically thrown away as part of the process of throwing out "disconnected diagrams". What unconscionable waste!

Perhaps a pair annihilation term is really describing two particles descending to the vacuum, to flit round there for a short time, then return as particles spontaneously appearing from the vacuum via the pair creation term. The discarded terms might represent a kind of quantum friction with the vacuum.

This is consistent with Weinberg's approach, providing a mechanism for the exchange of energy/momentum between foreground and vacuum. And consistent with Verlinde's entropic gravity approach [105, 106], which encourages us to treat spacetime as itself a statistical system.

Of course, these provide only "elements" of a theory of quantum gravity. There may be zero, one, or multiple acceptable ways of combining these elements.

To motivate such an effort consider this:

(i) explains the hierarchy problem: "quantum friction" is a qualitatively different mechanism; it is not surprising it would be much weaker than the electromagnetic, weak, or strong interactions.

(ii) treats the mass in foreground and background the same way: the pair creation term in the foreground is also a pair annihilation term in the vacuum, and vice-versa. This suggests an interesting perspective on the equivalence principle.

(iii) resolves the information paradox [107]: as spacetime is "nothing but" the quantum vacuum, information can transfer to it, hide out for a bit, then escape as during the process of black hole evaporation or the like. Total information in foreground plus background is still expected constant, as per the various no-cloning, no-deleting theorems.

(iv) provides a mechanism by which gravity could act as a source of decoherence, as in the Penrose Interpretation [108, 109, 110].

But the primary advantage that such a theory naturally couples interactions at the quantum scale with effects at the scale of the universe. This opens up interesting experimental and observational possibilities.

For instance the foreground (i.e. the particles we observe) should be at a higher energy and less disordered than the vacuum. So we expect a continuous transfer of energy and information from foreground to vacuum over time. We expect this transfer would be a monotonically increasing function of the energy of the foreground, more rapid within a supernova than in the gaps between the stars.

If we take the zero-zero component of the mass-energy tensor of the vacuum as providing a time scale, this transfer could look like an expansion of the vacuum (as it acquires a greater proportion of the total energy). To lowest order this might therefore look like a general expansion of spacetime (continuously increasing dark energy).

Further if the increased energy in the vacuum is identified as dark matter we can predict that the amount of dark matter in a galaxy will be proportional to the time the galaxy has existed (i.e. small for new borns) and to the mass of the galaxy.

Therefore, for essentially no work, we have two qualitative predictions for the evolution of the universe. And we have an interesting line of attack on the problem of quantum gravity. The associated technical problems are obviously non-trivial (e.g. what does the total Lagrangian look like?), but any well-defined and testable hypothesis has considerable value. At a minimum, this should suggest interesting experiments, especially now that we can see quantum effects of gravity (as in Bothwell et al [111]). For candidate quantum gravity experiments see for instance [112, 113, 114, 115, 116].

References

[1] Stueckelberg E C G 1941 *Helv. Phys. Acta.* **14** 51
[2] Stueckelberg E C G 1941 *Helv. Phys. Acta.* **14** 322–323
[3] Feynman R P 1948 *Rev. Mod. Phys.* **20**(2) 367–387
[4] Feynman R P 1949 *Phys Rev* **76** 749–759
[5] Feynman R P 1949 *Phys Rev* **76** 769–789
[6] Feynman R P 1950 *Physical Review* **80** 440–457
[7] Feynman R P 1961 *Quantum Electrodynamics* (New York: W. A. Benjamin, Inc)
[8] Schulman L S 1981 *Techniques and Applications of Path Integrals* (New York: John Wiley and Sons, Inc.) ISBN 0 471 16610 3
[9] Rivers R J 1987 *Path Integral Methods in Quantum Field Theory* (Cambridge [Cambridgeshire] ; New York: Cambridge University Press) ISBN 0521259797 URL `http://www.loc.gov/catdir/description/cam032/86026861.html` `http://www.loc.gov/catdir/toc/cam032/86026861.html`
[10] Swanson M 1992 *Path Integrals and Quantum Processes* (Academic Press, Inc.) ISBN 0 12 678945 2
[11] Kashiwa T, Ohnuki Y and Suzuki M 1997 *Path Integral Methods* (Clarendon Press; Oxford University Press) ISBN 0198517718
[12] Huang K 1998 *Quantum Field Theory: from Operators to Path Integrals* (New York: Wiley) ISBN 0471141208 (cloth alk. paper) URL `http://www.loc.gov/catdir/bios/wiley047/98013437.html`
[13] Zinn-Justin J 2005 *Path Integrals in Quantum Mechanics* (Oxford: Oxford University Press)
[14] Kleinert H 2009 *Path Integrals in Quantum Mechanics, Statistics, Polymer Physics, and Financial Markets* (New Jersey: World Scientific) ISBN 9789814273558 9814273554
[15] Feynman R P, Hibbs A R and Styer D F 2010 *Quantum Mechanics and Path Integrals* (Mineola, N.Y.: Dover Publications) ISBN 9780486477220 0486477223 URL `http://www.loc.gov/catdir/enhancements/fy1006/2010004550-d.html`
[16] Horwitz L P and Piron C 1973 *Helvetica Physica Acta* **46**
[17] Fanchi J R and Collins R E 1978 *Found Phys* **8** 851–877
[18] Fanchi J R 1993 *Found. Phys.* **23**
[19] Fanchi J R 1993 *Parameterized Relativistic Quantum Theory* (*Fundamental Theories of Physics* vol 56) (Kluwer Academic Publishers)
[20] Land M C and Horwitz L P 1996 *ArXiv e-prints* (*Preprint* `hep-th/9601021v1`) URL `http://lanl.arxiv.org/abs/hep-th/9601021v1`
[21] Horwitz L P 2006 *Physics Letters A* **355** 1–6 (*Preprint* `quant-ph/0507044`) URL `http://arxiv.org/abs/quant-ph/0507044`
[22] Fanchi J R 2011 *Found Phys* **41** 4–32

[23] Horwitz L P 2015 *Relativistic Quantum Mechanics* Fundamental Theories of Physics (Springer Dordrecht Heidelberg New York London)
[24] Dirac P A M 1958 *General Principles of Quantum Mechanics* 4th ed International series of monographs on physics (Oxford, Clarendon Press)
[25] Pauli W 1980 *General Principles of Quantum Mechanics* (Springer-Verlag) URL `http://www.springer.com/us/book/9783540098423`
[26] Schulman L S 1997 *Time's Arrows and Quantum Measurement* (New York: Cambridge University Press) ISBN 0 521 56775 0
[27] Zeh H D 2001 *The Physical Basis of the Direction of Time* (Berlin: Springer-Verlag)
[28] Muga J G, Sala Mayato R and Egusquiza I L 2002 *Time in Quantum Mechanics* (Berlin; New York: Springer) ISBN 3540432949
[29] Muga J G, Sala Mayato R and Egusquiza I L 2008 *Time in Quantum Mechanics - Vol 2* (Berlin; New York: Springer-Verlag) ISBN 3540734724 9783540734727 0075-8450 ;
[30] Callender C 2017 *What Makes Time Special?* (Oxford University Press)
[31] Feynman R P 1961, 1998 *Quantum Electrodynamics* Advanced Book Classics (New York: Addison-Wesley) ISBN 0-201-36075-6
[32] Bjorken J D and Drell S D 1965 *Relativistic Quantum Mechanics* (New York: McGraw Hill)
[33] Bjorken J D and Drell S D 1965 *Relativistic Quantum Fields* (New York: McGraw Hill)
[34] Sakurai J J 1967 *Advanced Quantum Mechanics* (Addison-Wesley) ISBN 0 201 06710 2
[35] Ramond P 1990 *Field theory* (Addison Wesley) ISBN 0 201 54611 6
[36] Kaku M 1993 *Quantum Field Theory: a Modern Introduction* (New York: Oxford University Press) ISBN 0195076524 (acid-free paper) URL `http://www.loc.gov/catdir/enhancements/fy0604/92027704-d.html` `http://www.loc.gov/catdir/enhancements/fy0604/92027704-t.html`
[37] Greiner W and Reinhardt J 1994 *Quantum Electrodynamics* 2nd ed (Berlin: Springer-Verlag) ISBN 0387580921 (U.S.) URL `http://www.loc.gov/catdir/enhancements/fy0902/95119144-d.html`
[38] Peskin M E and Schroeder D V 1995 *An Introduction to Quantum Field Theory* (Reading, Mass.: Addison-Wesley Pub. Co.) ISBN 0201503972 URL `http://www.loc.gov/catdir/enhancements/fy0831/89027170-b.html` `http://www.loc.gov/catdir/enhancements/fy0831/89027170-d.html`
[39] Weinberg S 1995 *Quantum Theory of Fields: Foundations* vol I (Cambridge University Press) ISBN 0 521 55001 7
[40] Weinberg S 1995 *Quantum Theory of Fields: Modern Applications* vol II (Cambridge University Press) ISBN 0 521 55002 5
[41] Greiner W 2000 *Relativistic Quantum Mechanics: Wave Equations* 3rd ed (Berlin: Springer) ISBN 3540674578 (softcover) URL `http://www.loc.gov/catdir/enhancements/fy0815/00041960-d.html`
[42] Itzykson C and Zuber J B 2005 *Quantum Field Theory* (Mineola, N.Y.: Dover Publications) ISBN 0486445682 (pbk.) URL `http://www.loc.gov/catdir/enhancements/fy0625/2005053026-d.html`
[43] McMahon D 2008 *Quantum Field Theory Demystified* (New York, N.Y.: McGraw-Hill) ISBN 9780071543828 (pbk.) 0071543821 URL `http://www.loc.gov/catdir/toc/fy0804/2008299146.html`
[44] Zee A 2010 *Quantum Field Theory in a Nutshell* (Princeton, N.J.: Princeton University Press) ISBN 9780691140346 (hardcover alk. paper)
[45] Klauber R D 2013 *Student Friendly Quantum Field Theory* 2nd ed (Sandtrove Press)
[46] Lancaster T and Blundell S 2014 *Quantum Field Theory for the Gifted Amateur* 1st ed (Oxford University Press) ISBN 9780199699339 (pbk.)
[47] Schwartz M D 2014 *Quantum Field Theory and the Standard Model* (Cambridge University Press)
[48] Schwichtenberg J 2020 *No-Nonsense Quantum Field Theory* (No-Nonsense Books)
[49] Khandekar D C, Lawande S V and Bhagwat K V 1993 *Path-integral Methods and their Applications* (Singapore: World Scientific Publishing Co. Pte. Ltd.)
[50] Ashmead J 2019 *Journal of Physics: Conference Series* **1239** 012015 URL `https://iopscience.iop.org/article/10.1088/1742-6596/1239/1/012015`
[51] Ashmead J 2021 *Journal of Physics: Conference Series* **1956** 012014 (*Preprint* `2101.10512`) URL `https://doi.org/10.1088/1742-6596/1956/1/012014`
[52] Abbott E A 1884 *Flatland: A Romance of Many Dimensions* (Seeley and Co. of London)
[53] Wilczek F 2008 *The Lightness of Being: Mass, Ether, and the Unification of Forces* (New York, NY: Basic Books) ISBN 9780465003211 (alk. paper)
[54] Busch P 2001 *The Time Energy Uncertainty Relation* (Berlin: Springer-Verlag) pp 69–98 Lecture Notes in Physics
[55] Goldstein H 1950 *Classical Mechanics* (Reading, MA: Addison-Wesley)

[56] Pashby T 2014 *Time and the Foundations of Quantum Mechanics* Ph.D. thesis Dietrich School of Arts and Sciences, University of Pittsburgh URL http://d-scholarship.pitt.edu/21183/1/Pashby2014TimeFoundationsQM.pdf
[57] Weinberg S 1972 *Gravitation and Cosmology: Principles and Applications of the General Theory of Relativity* (New York: John Wiley and Sons, Inc.) ISBN 0 471 92567 5
[58] Schulman L S 1999 *Phys.Rev.Lett.* **83** 5419–5422 (*Preprint* cond-mat/9911101) URL https://arxiv.org/pdf/cond-mat/9911101.pdf
[59] Greenberger D M and Svozil K 2005 *Quo Vadis Quantum Mechanics?* ed Elitzur A, Dolev S and Kolenda N (Berlin: Springer Verlag) (*Preprint* https://arxiv.org/abs/quant-ph/0506027)
[60] Dirac P A M 1928 *Proceedings of the Royal Society of London. Series A: Containing Papers of a Mathematical and Physical Character* **117** 610–624 URL http://www.jstor.org/stable/94981
[61] Feynman R P 1965 *The Character of Physical Law* (Modern Library) ISBN 0 679 60127 9
[62] Lucretius 55 BCE *The Nature of the Universe* (Penguin Books)
[63] Duarte F J and Taylor T S 2021 *Quantum Entanglement Engineering and Applications* (IOP Publishing) URL file:///.file/id=6571367.169809146
[64] Olsen J D and McDonald K T 2017 *URL reference* (*Preprint* https://www.physics.princeton.edu/ mcdonald/examples/orbitdecay.pdf) URL https://www.physics.princeton.edu/ mcdonald/examples/orbitdecay.pdf
[65] Marchewka A and Schuss Z 1998 *Phys.Lett.* **A240** 177–184 (*Preprint* quant-ph/9708034) URL https://arxiv.org/abs/quant-ph/9708034
[66] Marchewka A and Schuss Z 1999 *Phys. Lett. A* **65** 042112 (*Preprint* quant-ph/9906078) URL https://arxiv.org/abs/quant-ph/9906078
[67] Marchewka A and Schuss Z 1999 *ArXiv e-prints* (*Preprint* quant-ph/9906003) URL https://arxiv.org/abs/quant-ph/9906003
[68] Marchewka A and Schuss Z 2000 *Phys. Rev. A* **61** 052107 (*Preprint* quant-ph/9903076) URL https://arxiv.org/abs/quant-ph/9903076
[69] Mandelstam S 1958 *Phys. Rev.* **112**(4) 1344–1360 URL https://link.aps.org/doi/10.1103/PhysRev.112.1344
[70] Luis Plaja Ricardo Torres A Z (ed) 2013 *Attosecond Physics: Attosecond Measurements and Control of Physical Systems* 1st ed (*Springer Series in Optical Sciences* vol 177) (Springer-Verlag Berlin Heidelberg) ISBN 9783642376221 (hard cover : alk. paper)
[71] Lin C D, Le A T, Jin C and Wei H 2018 *Attosecond and Strong-Field Physics* (Cambridge University Press)
[72] Popper K R 1968 *The Logic of Scientific Discovery* (New York: Harper & Row)
[73] Lindner F, Schätzel M G, Walther H, Baltuška A, Goulielmakis E, Krausz F, Milošević D B, Bauer D, Becker W and Paulus G G 2005 *Physical Review Letters* **95** 040401 URL http://link.aps.org/doi/10.1103/PhysRevLett.95.040401 http://lanl.arxiv.org/abs/quant-ph/0503165
[74] Bach R, Pope D, Liou S H and Batelaan H 2013 *New Journal of Physics* **15** 033018 URL http://stacks.iop.org/1367-2630/15/i=3/a=033018
[75] Nichols R, Mineh L, Rubio J, Matthews J C F and Knott P A 2018 *ArXiv e-prints* (*Preprint* 1812.01032) URL https://arxiv.org/pdf/1812.01032.pdf
[76] Minkowski H 1908 *Space and Time* (Dover)
[77] Shapere A and Wilczek F 2012 *Phys. Rev. Lett.* **109**(16) 160402 URL https://link.aps.org/doi/10.1103/PhysRevLett.109.160402
[78] Wilczek F 2012 *Phys. Rev. Lett* **109** 160401 ISSN 0031-9007
[79] Else D V, Bauer B and Nayak C 2016 *Phys. Rev. Lett.* **117** 090402 (*Preprint* 1603.08001) URL https://arxiv.org/pdf/1603.08001.pdf
[80] Giergiel K, Kosior A, Hannaford P and Sacha K 2018 *ArXiv e-prints* (*Preprint* 1805.05718) URL https://arxiv.org/pdf/1805.05718
[81] Sacha K and Zakrzewski J 2018 *Rep. Prog. Phys.* **81** 016401 (*Preprint* 1704.03735) URL https://arxiv.org/pdf/1704.03735.pdf
[82] Sacha K 2018 *Time Crystals* (Springer)
[83] Khemani V, Moessner R and Sondhi S L 2019 *ArXiv e-prints* (*Preprint* 1910.10745) URL https://arxiv.org/pdf/1910.10745.pdf
[84] Nakatsugawa K, Fujii T, Saxena A and Tanda S 2017 *ArXiv e-prints* (*Preprint* 1711.10179) URL https://arxiv.org/pdf/1711.10179.pdf
[85] Niemi A J 2021 *ArXiv e-prints* (*Preprint* 2109.06091) URL https://arxiv.org/pdf/2109.06091.pdf
[86] Aharonov Y and Bohm D 1959 *Phys. Rev.* **115** 485–491
[87] Weder R 2011 *Journal of Mathematical Physics* **52** 052109 (*Preprint* 1006.1385v2) URL http://scitation.aip.org/content/aip/journal/jmp/52/5/10.1063/1.3592150

[88] Lamoreaux S K 1992 *Intl. J. of Mod. Physics* **7** 6691–6762
[89] Ghose P 1999 *Testing Quantum Mechanics on New Ground* (Cambridge: Cambridge University Press)
[90] Auletta G 2000 *Foundations and Interpretation of Quantum Mechanics: In the Light of a Critical-Historical Analysis of the Problems and of a Synthesis of the Results* (Singapore: World Scientific Publishing Co. Pte. Ltd.)
[91] Song Ma X, Kofler J and Zeilinger A 2014 *ArXiv e-prints* (*Preprint* 1407.2930v1)
[92] Marlow A and Wheeler J A 1978 *The 'Past' and the 'Delayed Choice' Double-Slit Experiment* (New York: Academic Press) pp 9–48
[93] Heisenberg W 1930 *The Physical Principles of the Quantum Theory* (Chicago: University of Chicago Press)
[94] Aharonov Y and Bohm D 1961 *Phys. Rev.* **122** 1649–1658 URL https://link.aps.org/doi/10.1103/PhysRev.122.1649
[95] Hilgevoord J 1996 *American Journal of Physics* **64** 1451–1456
[96] Hilgevoord J 1998 *American Journal of Physics* **66** 396–402
[97] Aharonov Y and Rohrlich D 2005 *Quantum Paradoxes: Quantum Theory for the Perplexed* (Wiley-VCH)
[98] Gell-Mann M 1956 *Il Nuovo Cimento* **4**
[99] Yau H Y 2020 *Int. J. Quant. Info.* **18** 1941016 URL https://arxiv.org/pdf/2111.01534.pdf
[100] Horwitz L P 2019 *The European Physical Journal Plus* **134** 313 (*Preprint* 1810.09248) URL https://arxiv.org/abs/1810.09248
[101] Ashmead J 2012 *Quanta* **1** ISSN 1314-7374 URL http://quanta.ws/ojs/index.php/quanta/article/view/5
[102] Merzbacher E 1998 *Quantum Mechanics* (New York: John Wiley and Sons, Inc.)
[103] Horwitz L P 2020 *The European Physical Journal Plus* **135** 479 URL https://doi.org/10.1140/epjp/s13360-020-00446-0
[104] Horwitz L P 2021 *The European Physical Journal Plus* **136** 32 URL https://doi.org/10.1140/epjp/s13360-020-00967-8
[105] Verlinde E P 2011 *JHEP* **1104** 029 (*Preprint* 1001.0785) URL https://arxiv.org/pdf/1001.0785.pdf
[106] Verlinde E P 2017 *SciPost Phys.* **2** 016 (*Preprint* 1611.02269) URL https://arxiv.org/pdf/1611.02269.pdf
[107] Susskind L 2008 *The Black Hole War: My Battle with Stephen Hawking to Make the World Safe for Quantum Mechanics* (New York: Little, Brown) ISBN 9780316016407 0316016403 URL http://www.loc.gov/catdir/enhancements/fy0804/2007048355-b.html http://www.loc.gov/catdir/enhancements/fy0804/2007048355-d.html
[108] Ekert A, Jozsa R, Penrose R and Penrose R 1998 *Philosophical Transactions of the Royal Society of London. Series A: Mathematical, Physical and Engineering Sciences* **356** 1927–1939 (*Preprint* https://royalsocietypublishing.org/doi/pdf/10.1098/rsta.1998.0256) URL https://royalsocietypublishing.org/doi/abs/10.1098/rsta.1998.0256
[109] Penrose R 1996 *General Relativity and Gravitation* **28** 581–600
[110] Penrose R 2014 *Foundations of Physics* **44** 557–575 URL https://doi.org/10.1007/s10701-013-9770-0
[111] Bothwell T, Kennedy C J, Aeppli A, Kedar D, Robinson J M, Oelker E, Staron A and Ye J 2022 *Nature* **602** 420–424 URL https://doi.org/10.1038/s41586-021-04349-7
[112] Christodoulou M, Biagio A D and Martin-Dussaud P 2020 *ArXiv e-prints* (*Preprint* 2007.08431) URL https://arxiv.org/pdf/2007.08431.pdf
[113] Huggett N, Linnemann N and Schneider M 2022 *ArXiv e-prints* (*Preprint* 2205.09013) URL https://arxiv.org/abs/2205.09013
[114] Polino E, Polacchi B, Poderini D, Agresti I, Carvacho G, Sciarrino F, Biagio A D, Rovelli C and Christodoulou M 2022 *ArXiv e-prints* (*Preprint* 2207.01680) URL https://arxiv.org/pdf/2207.01680.pdf
[115] Christodoulou M, Biagio A D, Howl R and Rovelli C 2022 *ArXiv e-prints* (*Preprint* 2207.03138) URL https://arxiv.org/pdf/2207.03138.pdf
[116] Marshall W, Simon C, Penrose R and Bouwmeester D 2003 *Physical Review Letters* **91** 130401 URL http://lanl.arxiv.org/abs/quant-ph/0210001

Journal of Physics: Conference Series **2482** (2023) 012024 doi:10.1088/1742-6596/2482/1/012024

Group contraction in physics: History and potential

Jaroslav Knap
Department of Physics
Faculty of Nuclear Sciences and Physical Engineering, CTU, Prague, Czech Republic
E-mail: knapjaro@fjfi.cvut.cz

Abstract. This article serves as a review of the mathematical tool of group contraction with an emphasis on physical applications. It was implicitly understood that some theories contain others as a special case (e.g., the relation between special relativity and Newtonian mechanics), but until the 1950s there was no firm mathematical grounding for such relations. We discuss the historical development of these concepts, their application to symmetry groups of space and time, and their relevance to the phenomenon of spontaneous symmetry breaking. Finally, we close by proposing a speculative chronology of space-time symmetries in the early universe.

1. Introduction

Physical theories are, in essence, an attempt to explain observed physical phenomena by deducing them as a consequence of some smaller set of fundamental principles. Typically, they are framed as mathematical theories, as mathematics provides us with a good language for describing the relations of relevant entities.

A change in theory may be required as observational power increases (e.g., with the discovery of telescopes, microscopes, X-rays, etc.) or new situations are encountered. Sometimes the needed change is minor, such as the addition of a new element to the periodic table; other times, such as in the case of special relativity and Newtonian mechanics, the discovery necessitates a complete replacement of a theory.

In the second case, we are left with an "obsolete" theory; however, this theory is still useful, as after all, it was used to explain some phenomena. It is intuitively understood that the older theory is in some sense a special case of the newer theory. The question then arises: is it possible to describe the relationship between the two theories via some mathematical mechanism?

It turns out that the answer is "yes" in at least some cases. The case we are specifically interested in is that of SR and Newtonian mechanics, where the relation between the theories is described via group contraction of the underlying symmetry groups. If we have theories that are defined via their symmetry groups, we can relate them using this method, provided the groups involved satisfy certain relations.

2. History

Before delving into the mathematics involved and its applications, it is useful to review the history of the usage of groups in physics, specifically in the discussion of space-time symmetries.

The first implicit usage of symmetry groups when describing physics was by G. Galilei in 1632, when in his "*Discourses and Mathematical Demonstrations Relating to Two New Sciences*" [1] he stated his principle of relativity. This concept of relativity of motion arose from the observation

Content from this work may be used under the terms of the Creative Commons Attribution 3.0 licence. Any further distribution of this work must maintain attribution to the author(s) and the title of the work, journal citation and DOI.
Published under licence by IOP Publishing Ltd

that motions can be superimposed on one another. Effectively what he was proposing was invariance of physical laws under Galilean boosts, i.e. equivalency of inertial frames of reference, today this would be the group $\mathbb{R}^3 \rtimes SO(3)$. What was lacking in his description was any reference to or codification of absolute space and time.

That was done later by I. Newton in 1687 in his seminal work "*Philosophiæ Naturalis Principia Mathematica*" [2] where he elaborated on the ideas introduced by Galilei, introducing absolute space and time, and developed Newtonian mechanics, in a sense the first physical theory in "modern" understanding. The enlarged symmetry group of this theory was then $\mathbb{R}^4 \rtimes \left(\mathbb{R}^3 \rtimes SO(3)\right)$, today known as Galilean group.

This was the prevailing paradigm until 1905, when the tensions between then new theory of electromagnetism and classical mechanics were resolved by A. Einstein in his article "*On the Electrodynamics of Moving Bodies*" [3], which overturned and superseded Newtonian mechanics and replaced the Galilean boosts with relativistic ones, effectively changing the symmetry group from Galilean to Poincaré.

Very soon after publishing of the article it was intuitively understood that Newtonian mechanics represents low velocity limit of the new theory (or infinite speed of light limit, as the relevant quantity is $\frac{v}{c}$), however despite this understanding there was no proper mathematical method to relate these two theories. On the level of individual formulas, it was always possible to expand into a power series and then keep only the lowest relevant order terms; however, this was mathematically unsatisfying.

This changed in 1951 when I. Segal published an article "*A class of operator algebras which are determined by groups*" [4] where he first proposed the idea of a limiting procedure for groups and its application to symmetry groups of physical theories. These ideas were developed further by E. Inönü and E. P. Wigner who two years later, in 1953, published the now classical article "*On the contraction of groups and their representations*" [5], where they described a specific mathematical method for contracting one group into another. It was immediately clear that this mechanism can be seen as a way to link different theories and to describe theory change.

Further developments of this concept were made, e.g. by E. J. Saletan [6], however we will concentrate on the Inönü-Wigner mechanism as it is sufficient for our purposes.

3. Inönü-Wigner group contraction

In this section, we go over the necessary steps to describe and illustrate the group contraction procedure, we follow similar explanations in [7], [8] and [9].

The setting in which Inönü-Wigner contraction operates are continuous symmetries described by Lie groups G_i, with the easiest way to approach it being via studying transformation of the associated Lie algebras $\mathfrak{g}_i$.

Suppose then we have a Lie algebra $\mathfrak{g}$ associated with group G, with J_i as the chosen basis of the vector space of algebra. Commutation relations can be written in the given basis as

$$[J_i, J_k] = \sum_{k=1}^{n} f_{ijk} J_k \quad i, j = 1, \cdots, n \tag{1}$$

with f_{ijk} being the structure constants of the group. Additionally the basis must satisfy *Jacobi identity*, which puts constraint on the structure constants

$$\sum_{c=k}^{n} f_{jlk} f_{ikm} + f_{lik} f_{jkm} + f_{ijk} f_{lkm} = 0. \tag{2}$$

Suppose now we introduce a *contraction parameter* ϵ and redefine the basis elements $J_i \to J_i^{(\epsilon)}$ such that the following is satisfied:

- the infinite sequence $[J_i]^\epsilon$ and its corresponding structure constants $[f_{ijk}]^{(\epsilon)}$ are known
- the limit $\lim_{\epsilon\to 0}[f_{ijk}]^\epsilon = [f_{ijk}]^0$ exists for all i, j, k and is consistent under under Jacobi identity

If the above conditions are satisfied, then the structure constants $[f_{ijk}]^0$ generate a new Lie algebra $\mathfrak{g}'$ (and so also Lie group G') which is called the *contraction* of $\mathfrak{g}$.

It turns out that there are conditions which describe when/how can the contraction be performed for given Lie algebra. Specifically, if $\mathfrak{g}$ has a subalgebra $\mathfrak{h}$ such that

$$\mathfrak{g} = \mathfrak{h} + \mathfrak{p} \tag{3}$$

$$[\mathfrak{h},\mathfrak{h}] \subset \mathfrak{h}, \quad [\mathfrak{h},\mathfrak{p}] \subset \mathfrak{p}, \quad [\mathfrak{p},\mathfrak{p}] \subset \mathfrak{h} + \mathfrak{p} \tag{4}$$

We can then explicitly re-parametrize generators from the subspace $\mathfrak{p}$ as $\mathfrak{p}' = \epsilon\mathfrak{p}$. This does not fundamentally change the algebra, as $\mathfrak{g}_\epsilon$ is isomorphic to $\mathfrak{g}$, since the commutators are

$$[\mathfrak{h},\mathfrak{h}] \subset \mathfrak{h}, \quad [\mathfrak{h},\mathfrak{p}'] \subset \mathfrak{p}', \quad [\mathfrak{p}',\mathfrak{p}'] \subset \epsilon^2(\mathfrak{h} + \mathfrak{p}) \tag{5}$$

However, if we now perform the limiting procedure $\epsilon \to 0$, the commutators in the singular limit become

$$[\mathfrak{h},\mathfrak{h}] \subset \mathfrak{h}, \quad [\mathfrak{h},\mathfrak{p}'] \subset \mathfrak{p}', \quad [\mathfrak{p}',\mathfrak{p}'] = 0. \tag{6}$$

From this, it is clear that $\mathfrak{g}_0$ is no longer isomorphic to $\mathfrak{g}$, as the subspace $\mathfrak{p}'$ has become an abelian subalgebra after the limit. The algebra $\mathfrak{g}_0$ is clearly the algebra of the contracted group.The structure of the resulting group is that of semi-direct product, i.e. $G' = G_1 \rtimes G_2$ for some groups G_1, G_2.

It is good to notice two things about the procedure: one, the dimension of the symmetry group is preserved under the contraction. This means that we cannot, in this way, relate theories that have different numbers of symmetry generators. Two, the algebra of the contracted group contains an abelian subalgebra, and hence the presence of a subspace of commuting symmetry generators can be taken as a sign of possible group contraction. These observations will be important in the later discussion of the symmetries of space-time.

3.1. *Example: SO(3) → ISO(2)*

To finish this section, allow a demonstration of the procedure on a simple case. The most common example is the contraction of 3D rotation group $SO(3)$ to a group of isometries of a plane $ISO(2)$.

As the generators for the $SO(3)$ we will take the usual selection J_i

$$[J_1, J_2] = J_3, \quad [J_2, J_3] = J_1, \quad [J_3, J_1] = J_2. \tag{7}$$

Using the method described previously, we select for the subalgebra $\mathfrak{h}$ a one dimensional space generated by J_3, the complementary space $\mathfrak{p}$ is then generated by the remaining J_1, J_2. We now rescale $\mathfrak{p}$ via the contraction parameter ϵ

$$j_1 = \epsilon J_1, \quad j_2 = \epsilon J_2, \quad j_3 = J_3. \tag{8}$$

The commutators now change as

$$[j_1, j_2] = \epsilon^2 j_3, \quad [j_2, j_3] = j_1, \quad [j_3, j_1] = j_2, \tag{9}$$

which after the limit $\epsilon \to 0$ leads to a new Lie algebra

$$[j_1, j_2] = 0, \quad [j_2, j_3] = j_1, \quad [j_3, j_1] = j_2. \tag{10}$$

This Lie algebra characterizes the group $ISO\,(2) = SO\,(2) \rtimes \mathbb{R}^2$, with the generators j_1, j_2 being the translations and j_3 the rotation around the axis perpendicular to the plane.

We note that we can assign a geometric interpretation to the contraction parameter, and these can be interpreted physically as well. In this particularly simple case we can describe ϵ as the inverse radius of a sphere $1/R$. As the radius of this imagined sphere grows, the neighbourhood of any point starts to increasingly resemble a plane, and at the limit of an infinite radius, it becomes (geometrically) indistinguishable.

4. Symmetries of space-time

After exploring the basics of the historical context and the mathematical underpinnings of the group contraction, it is now time to apply the method to some physical theories.

Specifically, we will trace the evolution of the assumed symmetry group of space and time, through the lens of group contraction. From looking at the historical examples, we will also try to see if we can deduce something about potential future developments in this area.

As mentioned previously, the first proposal of a symmetry of space and time (even if not using these exact terms) came from Galilei in 1638 [1]. This was invariance under Galilean boosts and spatial rotations, $\mathbb{R}^3 \rtimes SO(3)$. To complement our understanding, we extend this by addition of space and time translations [2], to obtain full *Galilean group* $\mathbb{R}^4 \rtimes \left(\mathbb{R}^3 \rtimes SO(3)\right)$. The non-zero commutation relations are as follows:

$$\frac{1}{\mathrm{i}}\,[C_i, H] = P_i \tag{11}$$

$$\frac{1}{\mathrm{i}}\,[L_{ij}, C_k] = \delta_{ik} C_j - \delta_{jk} C_i, \quad \frac{1}{\mathrm{i}}\,[L_{ij}, P_k] = \delta_{ik} P_j - \delta_{jk} P_i \tag{12}$$

$$\frac{1}{\mathrm{i}}\,[L_{ij}, L_{jk}] = \delta_{ik} L_{jl} - \delta_{il} L_{jk} - \delta_{jk} L_{il} + \delta_{jl} L_{ik} \tag{13}$$

where the generators are

- H - time translation
- P_i - space translation
- $L_i j$ - spatial rotations
- C_i - Galilean boost

We can immediately notice that the Galilean boosts form an abelian subalgebra, hinting that perhaps we can obtain this group from another via the contraction procedure. Indeed, if we take Poincaré group $\mathbb{R}^{3,1} \rtimes SO\,(3,1)$ and take the slowness parameter $1/c$ to zero, we do get (inhomogeneous) Galilean group.

However, we do need to point out that such an approach gets appropriate results only in the classical setting, once we move over to the quantum setting, there appear effects that leave imprints in the non-relativistic limit that are not described by the Galilean group [10]. It is then necessary to include additional generator M, and to take care in the choice of representation. For our discussion and arguments, however, it is sufficient to consider the Galilean group.

Explicitly, we can consider generators of the Lorentz group $SO(3,1)$ J_i, B_i that obey commutation relations

$$\frac{1}{\mathrm{i}}\,[J_i, J_j] = \epsilon_{ijk} J_k, \quad \frac{1}{\mathrm{i}}\,[J_i, B_j] = \epsilon_{ijk} B_k, \quad \frac{1}{\mathrm{i}}\,[B_i, B_j] = -\epsilon_{ijk} B_k \tag{14}$$

with J_i begin the spatial rotations, and B_i the Lorentz boosts. If we now introduce the inverse speed of light as the limit parameter (the "slowness"), we define

$$J_i^{(c)} = J_i, \quad B_i^{(c)} = \frac{B_i}{\mathrm{i}c}. \tag{15}$$

Taking now the limit $c \to \infty$ (or equivalently $1/c \to 0$) we obtain commutation relations

$$\frac{1}{\mathrm{i}}\left[J_i, J_j\right] = \epsilon_{ijk} J_k, \quad \frac{1}{\mathrm{i}}\left[J_i, B_j^{(0)}\right] = \epsilon_{ijk} B_k^{(0)}, \quad \frac{1}{\mathrm{i}}\left[B_i^{(0)}, B_j^{(0)}\right] = 0, \tag{16}$$

which are those of a homogeneous Galilean group.

We could have also speculated immediately from the form of the group that it can be obtained via contraction, as it has a semi-direct product structure, which does appear in groups resulting from contraction procedures.

Historically speaking, this relations of these groups, and hence also of Newtonian mechanics and special relativity was only discovered after the theory itself was. However, we can also look at it from another angle: if the mechanism was known prior to the formulation of special relativity, we could have speculated whether there is some physical parameter that is sufficiently large (or small), so that the effective symmetry group of low velocity mechanics is Galilean and the full symmetry group is different. Experimentally, of course, until there was observational evidence for the finite speed of light, there was no reason to consider such a parameter to play a role in the kinematical group of mechanics.

Taking this second viewpoint, we can ask if perhaps there are also other finite parameters that are currently outside of our observational bounds that we are not considering (e.g., a finite invariant length scale).

With this in mind, let us look at the Poincaré group in more detail. It is again a 10 dimensional group, which does have a semi-direct product structure, so it could be a contraction of another group. Its commutation relations are

$$\frac{1}{\mathrm{i}}\left[M_{\mu\nu}, P_\rho\right] = \eta_{\mu\rho} P_\nu - \eta_{\nu\rho} P_\mu \tag{17}$$

$$\frac{1}{\mathrm{i}}\left[M_{\mu\nu}, M_{\rho\sigma}\right] = \eta_{\mu\rho} M_{\nu\sigma} - \eta_{\mu\sigma} M_{\nu\rho} - \eta_{\nu\rho} M_{\mu\sigma} + \eta_{\nu\sigma} M_{\mu\rho} \tag{18}$$

with

- P_μ - space-time translations
- $M_{\mu\nu}$ - spatial rotations and boosts (spatio-temporal rotations)

Note that we can relate $M_{\mu\nu}$ to explicit generators of rotations and boosts as $J_i = \frac{1}{2}\epsilon_{ijk} M_{jk}$, $B_i = M_{0i}$. From this, we also see that spatial and temporal translations form an abelian subgroup, another hint that the group is a potential result of a contraction procedure. Specifically, Poincaré group can be obtained as a contraction from either the de Sitter group $SO(4,1)$ or the anti-de Sitter group $SO(3,2)$, via sending their scalar curvature Λ to zero.

Which of these two groups is more worthy of our attention? From the perspective of the author, it makes more sense to pick the de Sitter group and focus on it for the following reasons:

- positive cosmological constant (i.e. "dark energy")
- inflationary space-times are approximately de Sitter during inflation [11]
- de Sitter relativity is a mathematically appropriate version of doubly special relativity (in the sense that there is no explicit Lorentz symmetry breaking) [12]

The De Sitter group $SO(4,1)$ is again a 10-dimensional group, with commutation relations simplified to

$$\frac{1}{\mathrm{i}}\left[M_{AB}, M_{CD}\right] = \eta_{BC}M_{AD} - \eta_{BD}M_{AC} - \eta_{AC}M_{BD} + \eta_{AD}M_{BC}. \tag{19}$$

The contraction to the Poincaré group can be constructed as follows: let us define a new basis

$$\Pi_\mu = \frac{1}{l}M_{5\mu} \tag{20}$$

with the rest staying the same. Commutation relations can then be rewritten as

$$\frac{1}{\mathrm{i}}\left[\Pi_\mu, \Pi_\nu\right] = \frac{1}{l^2}M_{\mu\nu} \tag{21}$$

$$\frac{1}{\mathrm{i}}\left[M_{\mu\nu}, \Pi_\rho\right] = \eta_{\mu\rho}\Pi_\nu - \eta_{\nu\rho}\Pi_\mu \tag{22}$$

$$\frac{1}{\mathrm{i}}\left[M_{\mu\nu}, M_{\rho\sigma}\right] = \eta_{\mu\rho}M_{\nu\sigma} - \eta_{\mu\sigma}M_{\nu\rho} - \eta_{\nu\rho}M_{\mu\sigma} + \eta_{\nu\sigma}M_{\mu\rho}. \tag{23}$$

Sending now the pseudo-radius l to infinity, we obtain exactly the commutation relations of the Poincaré group, with $\Pi_\mu \to P_\mu$ becoming the translation generators.

From the commutation relations above, we can see that the de Sitter group has neither a semi-direct product structure nor a non-trivial abelian subgroup, so we could in principle end our discussion here.

However, we would like to take a step further by investigating transformations that can naturally extend the Sitter group, namely operations that act transitively on the Sitter space [13].

These de Sitter "translations" can be written as a combination of translations and special conformal transformations, i.e.

$$\partial_\mu - \frac{1}{4l^2}\left(2\eta_{\mu\tau}x^\tau x^\sigma - x^2\delta^\sigma_\mu\right)\partial_\sigma, \tag{24}$$

where $l2$ denotes the de Sitter pseudo-radius and $x_m u$ some specific coordinates. This object on its own is not a de Sitter generator, as its action on space-time leads to a conformal rescaling of the metric, transformation which is not from the de Sitter group. If we demand that the symmetry group of space-time include transformations under which it is transitive, we must include both the usual translations and the special conformal transformations.

This inclusion forces us to also include the dilation generator in order to close the commutation relations. This extensions then moves us to a group $ISO(4,1)$ a 15-dimensional group of isometries of $SO(4,1)$. This group has again the semi-direct product structure, and can be obtained as a contraction of the conformal group $SO(4,2)$, in a similar fashion to the preceding contractions.

Why should we be interested in the conformal group, if we are discussing symmetries of space-time? We see several reasons for this

- conformal group $SO(4,2)$ is the largest group of symmetries preserving causal ordering in 4D space-time
- other physical theories have natural scale symmetry (gauge theories, massless theories), and this can be promoted to full conformal symmetry under a broad set of conditions (in $d=2$ there is *Zamolodchikov-Polchinski theorem* [14])
- CMB is nearly scale-invariant, suggesting scale invariant (or conformal invariant) theories are suitable for description of very early universe [11]

Considering these arguments persuasive enough to explore this direction, let us take a closer look at conformal group. Its commutation relations can be compactly written as

$$\frac{1}{\mathrm{i}}\left[M_{AB}, M_{CD}\right] = \eta_{BC} M_{AD} - \eta_{BD} M_{AC} - \eta_{AC} M_{BD} + \eta_{AD} M_{BC} \tag{25}$$

in the exact same form as the de Sitter ones (which should not be surprising, as both are pseudo-rotation groups). We can also consider an alternative basis that is related to the usual generators as,

$$L_{\alpha\beta} = M_{\alpha\beta}, \quad D = M_{56} \tag{26}$$
$$P_\alpha = M_{\alpha 5} + M_{\alpha 6}, \quad K_\alpha = M_{\alpha 5} - M_{\alpha 6} \tag{27}$$
$$\alpha, \beta = 0, 1, 2, 3, \quad \alpha = 0 \equiv A = 4. \tag{28}$$

Again, we can see from the structure of the group and from the commutation relations that there is no obvious need for further extension. We could again consider transformations acting on the space as in the de Sitter case, but from a physical perspective, we are interested in 3+1 dimensional spacetimes, and any further extension would take us away from that (as $SO(4,2)$ is the conformal group of 3+1 dimensional space-time).

Additionally, now we are left with a problem. The present-day universe is well described by considering only 10 space-time symmetries (local), yet the conformal group has 15. How can we reduce this number? One answer is spontaneous symmetry breaking, and it turns out that group contraction has a relationship to it.

5. Group contraction and symmetry breaking

In this section we recall relation of group contraction to symmetry breaking patters, as developed in [15] and [16].

One of the important realizations of 20th century physics was that symmetry in a system need not be realized linearly, i.e., in the Wigner-Weyl realization, but instead can be realized non-linearly in what is known as the Nambu-Goldstone realization. This is also colloquially known as spontaneous symmetry breaking (SSB) [17], as a particular vacuum state is invariant only under a certain subgroup H of the symmetry group G, i.e., it breaks the symmetry, with the rest of the former symmetries now transforming it into a different vacuum state.

It is also a well known fact that spontaneous symmetry breaking leads to the appearance of massless Goldstone modes (which can be composite fields in the case of dynamical symmetry breaking). These fields are either scalar, spinorial, or vector, depending on whether the broken symmetry is internal, super, or space-time. The number of these Goldstone modes corresponds to the number of broken generators, i.e. to the difference between the dimensions of the symmetry group G and the symmetry group of the vacuum state H,

$$\#\, Goldstone\, modes = \dim G - \dim H \tag{29}$$

It turns out that the "remnant" symmetry group H combined with the abelian Goldstone modes (which generate field translations) must be a contraction of the original symmetry group G [15]. This then provides us with a mechanism to calculate the massless Goldstone modes. Also (and more importantly for us) it provides us with a mechanism to change the dimension of the contracted group, provided we ignore the "abelianized" field translations.

This relationship between symmetry breaking and group contraction provides us with a tool for describing the relationship of space-time symmetry groups in a unified manner. We can assume that the conformal group is spontaneously broken (e.g., by the introduction of scale, breaking the dilatation generator), and the left-over non-trivial part is the de Sitter group, with the remaining 5 generators being abelianized and representing field translations

Journal of Physics: Conference Series **2482** (2023) 012024 doi:10.1088/1742-6596/2482/1/012024

6. Hierarchy of space-time symmetry groups

Now that we have discussed the relation between symmetry breaking and group contraction, we have all the required ingredients ready. In section 4 we have progressed from the least to the most general symmetry group. However, in universe it would appear the other way around.

We would propose the following timeline of space-time symmetry groups:

$$\text{Conformal} \rightarrow \text{de Sitter} \rightarrow \text{Poincare} \rightarrow \text{Galilei}$$

In the pre-inflationary and inflationary universe we propose that the symmetry group of space-time was the conformal group $SO(4,2)$. Due to the fact that the present day universe is neither conformal nor scale invariant, this part of the symmetry was then spontaneously broken down to the de Sitter group $SO(4,1)$, with the symmetry breaking pattern determined by the contraction, as described in the previous sections.

We further suggest that the true "local" symmetry of space-time is the de Sitter, not the Poincaré. The transition from de Sitter to Poincaré (and from Poincaré to Galilei) are then not true transitions, but merely approximation artifacts due to limited observational capabilities.

As the shift to Poincaré happened when we had other theoretical signs and further experimental evidence of the finite speed of light, similarly, we propose that there is an invariant length scale that is currently out of observational bounds. If this invariant scale exists, the symmetry group of space-time would be naturally de Sitter, not Poincaré.

Another kind of theoretical argument in support of an invariant length (or energy) scale would be that the scale on which quantum gravity becomes relevant should play a fundamental role. An example could be that a photon with a wavelength of Planck length should collapse into a black hole (based on classical understanding), however in different reference frames its wavelength could be longer due to red-shift. As a result, the presence of a black hole would then be observer-dependent, leading to contradictions within the theory. Eventual quantum theory of gravity should be able to resolve this issue, and treating some length or energy scale as invariant would lead to resolution.

Additionally, we could argue that the structure of the Poincaré group combined with the knowledge and history of group contraction also points in the direction of further refinement being necessary.

7. Summary and conclusion

In this note, we have reviewed the historical context and mathematical footing of group contraction, with an emphasis on the physical applications and interpretation of the procedure.

We show that group contraction can be used to provide a natural way to link different theories, provided a link exists between their symmetry groups. Additionally, we show how this procedure is related to symmetry breaking pattern.

From a physical standpoint, we used this to investigate the relationship of historical space-time symmetry groups (Poincaré and Galilei) and then extended that to propose the de Sitter group for consideration. Additionally, we further extended this to the conformal group, for which discussion of the link between group contraction and symmetry breaking was necessary.

We used this to propose that the true symmetry group of space-time in the early universe was the conformal group, which was spontaneously broken down to the de Sitter group. The current understanding of the Poincaré group as the symmetry group is thus an observational artifact resulting from limited experimental and observational capabilities, similar to how the Galilean group was considered a symmetry of space-time until it was superseded by the Poincaré group due to advances in mathematical tools and other theories.

Acknowledgments
Author would like to thank his thesis advisor Petr Jizba for consultations that helped shape the the work, and the organizing committee of the IARD conference.
This work has been supported by the grant SGS22/178/OHK4/3T/14.

References
[1] Galilei G 1638 *Discourses and Mathematical Demonstrations Relating to Two New Sciences*
[2] Newton I 1687 *Philosophiæ Naturalis Principia Mathematica*
[3] Einstein A 1905 On the Electrodynamics of Moving Bodies *Annalen der Physik* **17** 891-921
[4] Segal I E 1951 A class of operator algebras which are determined by groups *Duke Math. J.* **18** 221
[5] Inönü E, Wigner E P 1953 On the Contraction of Groups and Their Representations *Proc. Natl. Acad. Sci.* **39** (6) 510-24
[6] Saletan E J 1961 Contraction of Lie Groups *J. Math. Phys* **2** (1) 1-21
[7] Inönü E 1964 *Contractions of Lie Groups and their representations* Group theoretical concepts in elementary particle physics (Gordan and Breach) 391-402
[8] Gilmore R 2008 *Lie Groups, Physics, and Geometry* (Cambridge University Press, London)
[9] Hermann R 1966 *Lie Groups for Physicists* (W. A. Benjamin, Amsterdam)
[10] Greenberger D M 2001 *Inadequacy of the Usual Galilean Transformation in Quantum Mechanics*
[11] Planck Collaboration 2020 Planck 2018 results. I. Overview and the cosmological legacy of Planck *A&A* **641** A1
[12] Aldrovandi R, Beltran Almeida J P and Pereira J G 2007 de Sitter special relativity *Class. Quant. Grav.* **24** 1385-1404
[13] Pereira J G, Sampson A and Savi L L 2013 de Sitter transitivity, conformal transformations and conservation laws *Int. J. Mod. Phys.* D **23** (4)
[14] Polchinski J 1988 Scale and conformal invariance in quantum field theory *Nucl. Phys.* B **303** 226-36
[15] De Concini C, Vitiello G 1976 Spontaneous breakdown of symmetry and group contractions *Nucl. Phys.* B **116** 141-56
[16] Blasone M, Jizba P and Vitiello G 2010 *Quantum Field Theory and Its Macroscopic Manifestations* (World Scientific & ICP, London)
[17] Beekman A J, Rademaker L and van Wezel J 2019 An introduction to spontaneous symmetry breaking *SciPost Phys. Lect. Notes* **11**

The measurement in classical and quantum theory

Alexey A. Kryukov

Department of Mathematics & Natural Sciences, University of Wisconsin-Milwaukee, USA

E-mail: kryukov@uwm.edu

Abstract. The Bohigas-Giannoni-Schmit (BGS) conjecture states that the Hamiltonian of a microscopic analogue of a classical chaotic system can be modeled by a random matrix from a Gaussian ensemble. Here, this conjecture is considered in the context of a recently discovered geometric relationship between classical and quantum mechanics. Motivated by BGS, we conjecture that the Hamiltonian of a system whose classical counterpart performs a random walk can be modeled by a family of independent random matrices from the Gaussian unitary ensemble. By accepting this conjecture, we find a relationship between the process of observation in classical and quantum physics, derive irreversibility of observation and describe the boundary between the micro and macro worlds.

1. Introduction

Given the fundamental character of the topics discussed in the paper, it is necessary to place the paper in the context of existing research and to explain prerequisites for the obtained results. First and foremost, standard quantum mechanics is assumed throughout the paper. The equation of motion is the usual Schrödinger equation, with no extra terms. In proving the initial results related to the connection of Schrödinger and Newtonian dynamics, the state of the system that represents a macroscopic particle is assumed to be constrained to a certain submanifold in the space of states. This may seem to contradict the Schrödinger dynamics of the system. However, this step is only used to understand the geometry of the constraint and is later replaced by a condition on the Hamiltonian of the system. Second, the presented results are mathematical theorems proved in the paper and therefore, given the assumptions in the theorems, the results follow rigorously and need not be questioned, unless the validity of the proof itself is in question. What is open to discussion is the nature of the assumptions and the physical interpretation of the derived results.

Now, the only assumption of essence used in proving the results is related to a specific form of the Hamiltonian of a macroscopic particle in natural surroundings and a microscopic particle under a measurement. This assumption is stated in the conjecture (**RM**). It claims that the Hamiltonian of a microscopic particle whose position is measured, as well as a macroscopic particle in the natural surroundings can be represented by a random matrix from a Gaussian unitary ensemble and that random matrices considered at different moments of time are independent. Of course, (**RM**) is not BGS. In particular, BGS is about stationary states and the distribution of eigenvalues of the Hamiltonian. On the contrary, (**RM**) is about a non-stationary stochastic process on the space of states. At the same time, both conjectures claim that under certain physical conditions on the system the Hamiltonian of the system can be represented by a random matrix. This is the point of connection of the paper with the BGS conjecture. It is

Content from this work may be used under the terms of the Creative Commons Attribution 3.0 licence. Any further distribution of this work must maintain attribution to the author(s) and the title of the work, journal citation and DOI.
Published under licence by IOP Publishing Ltd

important to point out that there is no attempt to prove or disprove the BGS conjecture, or to justify it in any way in the paper. Rather, the validity of the BGS conjecture confirmed in numerous observations is used to argue in favor of (**RM**). One may disagree with the argument, but given that (**RM**) is the only assumption of essence used in proving the theorems in the paper, and the statements of theorems have far reaching consequences to physics, the approach deserves our full attention.

Finally, what are the results obtained in the paper? How are they related to previous research? What is the main merit of the paper? The coherent states used in identifying the classical phase space manifold in the paper are well known and widely used in quantum optics and quantum physics in general. The fact that when the state is coherent, the motion of the system is closest to the classical motion and, in the limit, may become identical to it is not surprising. The literature related to the subject ranges from textbook results in quantum optics, e.g., [1] to research in quantum gravity (see [2] for a historic context, main ideas and some references). A semiclassical approximation obtained by constraining the state to remain Gaussian during the evolution [3] is known to be useful in modeling a range of quantum systems over a time interval.

However, the current paper is not concerned with a semiclassical approximation. It deals with the full-fledged Schrödinger evolution with the Hamiltonian that satisfies (**RM**). The state of the system under such evolution does not stay coherent, but performs a random walk in the space of states. This walk is used in the paper to find a new link between measurement in classical and quantum mechanics. Namely, the normal distribution typical for classical observations and the Born rule valid in the micro-world turn out to follow from the Schrödinger equation with the Hamiltonian satisfying (**RM**). It becomes possible to specify on this basis the boundary between the micro and macro worlds. It becomes also possible to explain irreversibility of measurement. Other essential results are the derivation of the classical behavior of macroscopic systems and an explanation of why the cat states of macroscopic systems are not observed in nature. This in turn provides an explanation of the role played by measuring devices in quantum theory and how the quantum theory, while remaining linear, may avoid using superpositions of states of macroscopic bodies.

As stated before, the derived results are mathematical consequences of the assumption (**RM**). They are new and are not directly based on previous research. The measurement problem and physics of measuring devices is of course a topic of significant research of its own. However, as far as the author is aware, the existing body of literature related to quantum measurement was never concerned with (**RM**). The theory of decoherence [4] probably came closest and is possibly most relevant to the topics considered in the paper. However, in classification of the decoherence theory, we are dealing here with a "fake" decoherence, as the evolution remains unitary. Furthermore, the paper deals with evolution of the state and not the density matrix of a system. The irreversibility of a measurement is the result of the Hamiltonian lacking time-reversal symmetry rather than a "leakage" of information into the environment. Similarly, the dynamical collapse theories surveyed in [5, 6] use non-linear modifications of the Schrödinger equation, while only linear evolution is considered in the paper. Accordingly, we only refer to general surveys in both of these fields of research with the goal of providing a context for the derived results. Let us mention the Ornstein-Uhlenbeck process of stochastic relaxation of Hermitian matrices first considered by Dyson [7, 8]. Although not directly related to the stochastic process considered here, it provides an interesting example of a process that leads to the Gaussian ensemble used in the paper.

To make the presentation self-contained, simple proofs of the previously published theorems (**A**) and (**B**), and propositions (**P2**) and (**P3**) are provided. The reader is referred to the cited publications for details.

Journal of Physics: Conference Series **2482** (2023) 012025 doi:10.1088/1742-6596/2482/1/012025

2. The measurement in classical and quantum theory

The linear nature of quantum mechanics poses a persistent problem for reconciliation of the classical and quantum mechanics. The superposition principle is foreign to classical physics. We don't see macroscopic objects in two different places or the cat being alive and dead at the same time. However, such states are commonplace in the microworld. The notable attempts to resolve the situation include non-linear modifications of the Schrödinger equation to account for the transition to states observed in a measurement, the De Broglie-Bohm theory of classical particles lead by a pilot-wave, an appeal to many coexisting worlds representing the components of a superposition and the decoherence program aiming to explain how superpositions decohere to probabilistic mixtures. None of these attempts is generally accepted as successful in resolving the problem. There is no experimental evidence for needing to modify the Schrödinger equation. The simultaneous presence of quantum and classical trajectories in the pilot-wave theory seems redundant. The many-world approach fails to explain the world that is real and unique to us. The decoherence program derives the laws of probability valid for macroscopic bodies but fails to account for a specific outcome of a measurement.

On the other hand, the Newtonian and Schrödinger dynamics have a simple relationship that does not seem to be known nor used in the above-mentioned attempts. Namely, the Newtonian dynamics can be identified with a constrained Schrödinger dynamics. The latter is similar to the Newtonian dynamics of a constrained system, e.g., a bead on a wire. However, since the Schrödinger equation describes the dynamics in a Hilbert space of states, the constraint is imposed on the state of the system. Consider the subset $M^{\sigma}_{3,3}$ of the Hilbert space $L_2(\mathbb{R}^3)$, formed by the states

$$\varphi(\mathbf{x}) = g_{\mathbf{a},\sigma}(\mathbf{x})e^{i\mathbf{px}/\hbar}, \tag{1}$$

where

$$g_{\mathbf{a},\sigma} = \left(\frac{1}{2\pi\sigma^2}\right)^{3/4} e^{-\frac{(\mathbf{x}-\mathbf{a})^2}{4\sigma^2}} \tag{2}$$

is the Gaussian function of a sufficiently small variance $2\sigma^2$ centered at a point $\mathbf{a}$ in the Euclidean space $\mathbb{R}^3$ and $\mathbf{p}$ is a fixed vector in $\mathbb{R}^3$. The state $\varphi(\mathbf{x})$ represents a narrow wave packet with group velocity $\mathbf{p}/m$, where m is the mass of the particle. Let's identify the set of all pairs $(\mathbf{a}, \mathbf{p})$ with the classical phase space $\mathbb{R}^3 \times \mathbb{R}^3$ of possible positions $\mathbf{a}$ and momenta $\mathbf{p}$ of a particle. The map $\Omega : (\mathbf{a}, \mathbf{p}) \longrightarrow g_{\mathbf{a},\sigma}e^{i\mathbf{px}/\hbar}$ identifies then the classical phase space with the submanifold $M^{\sigma}_{3,3}$ of $L_2(\mathbb{R}^3)$. The equivalence classes of states in $L_2(\mathbb{R}^3)$ differing only by a constant phase factor $e^{i\alpha}$ form the projective space CP^{L_2}. Under the equivalence relation, the embedded manifold $M^{\sigma}_{3,3}$ becomes a submanifold of CP^{L_2}, denoted here by the same symbol.

The six-dimensional manifold $M^{\sigma}_{3,3}$ is embedded into the space of states in a very special way. There are no vectors in the Hilbert space $L_2(\mathbb{R}^3)$ orthogonal to all of $M^{\sigma}_{3,3}$. Instead, the points of $M^{\sigma}_{3,3}$ represent an overcomplete basis in the Hilbert space [12]. Furthermore, the projective space CP^{L_2} possesses the Fubini-Study metric, induced by the embedding of CP^{L_2} into the sphere S^{L_2} of unit-normalized states in $L_2(\mathbb{R}^3)$ furnished itself with the induced round Riemannian metric. For any two vectors ξ, η in $L_2(\mathbb{R}^3)$ tangent to the sphere S^{L_2} and the corresponding vectors $X = (\mathrm{Re}\xi, \mathrm{Im}\xi)$ and $Y = (\mathrm{Re}\eta, \mathrm{Im}\eta)$ in the realization $L_{2R}(\mathbb{R}^3)$ of $L_2(\mathbb{R}^3)$, the Riemannian metric G_{φ} on the sphere is defined by

$$G_{\varphi}(X, Y) = \mathrm{Re}(\xi, \eta). \tag{3}$$

Perhaps surprisingly, the metric induced on the submanifold $M^{\sigma}_{3,3}$ of CP^{L_2} with 2σ as a unit of length turns out to be the ordinary Euclidean metric. In other words, the map $\Omega : \mathbb{R}^3 \times \mathbb{R}^3 \longrightarrow CP^{L_2}$ is an isometric embedding of the Euclidean space into the space of states. The manifold $M^{\sigma}_{3,3}$ can be also furnished with a compatible linear structure, making it isomorphic to the Euclidean space $\mathbb{R}^3 \times \mathbb{R}^3$ [9–13].

We can now state the first theorem. Here and later, it is assumed that the value of the parameter σ is sufficiently small for the linear approximation of the potential $V(\mathbf{x},t)$ to be valid within intervals of length σ.

> **(A)** *The Newtonian dynamics of an arbitrary mechanical system is the Schrödinger dynamics of that system with the state constrained to the classical phase space submanifold of the space of states of the system. Furthermore, the Schrödinger dynamics is the only unitary evolution that reduces under the constraint to the Newtonian one.*

Note that for the purpose of this paper we only need the proof of the first part of the theorem and only in the case of a single particle. A complete prove can be found in [11].

Proof. A simple way to prove the theorem is by using the variational principle. The variation of the functional

$$S[\varphi]=\int\overline{\varphi}(\mathbf{x},t)\left[i\hbar\frac{\partial}{\partial t}-\widehat{h}\right]\varphi(\mathbf{x},t)d^3\mathbf{x}dt \tag{4}$$

with the Hamiltonian $\widehat{h}=-\frac{\hbar^2}{2m}\Delta+V(\mathbf{x},t)$ yields the Schrödinger equation for φ. For the states φ constrained to the manifold $M^{\sigma}_{3,3}$, this functional reduces to the classical action

$$S=\int\left[\mathbf{p}\frac{d\mathbf{a}}{dt}-h(\mathbf{p},\mathbf{a},t)\right]dt, \tag{5}$$

where $h(\mathbf{p},\mathbf{a},t)=\frac{\mathbf{p}^2}{2m}+V(\mathbf{a},t)+const$ is the Hamiltonian function for the particle. We used here the fact that for a sufficiently small σ, the terms $V(\mathbf{a},t)$ and $\int g^2_{\mathbf{a},\sigma}(\mathbf{x})V(\mathbf{x},t)d^3\mathbf{x}$ are arbitrarily close to each other. In fact, as σ approaches 0, the terms $g^2_{\mathbf{a},\sigma}$ form a delta sequence. It follows that the variation of the functional (4) subject to the constraint that φ belongs to $M^{\sigma}_{3,3}$ yields Newtonian equations of motion. □

The theorem provides us with a geometric condition for the transition from Schrödinger to Newtonian dynamics. Note that we are not trying to change the Schrödinger equation by imposing the constraint. Rather, the upcoming theorems will demonstrate that the constraint itself can be traced back to the Schrödinger dynamics with a proper Hamiltonian. Theorem **(A)** will play an important role in this derivation. To proceed, let us first work on a deeper understanding of geometry and physics behind the theorem. The reader who is more interested in the applications of the theorem, may skip propositions **(1)** and **(2)** below.

(1) The transition from Schrödinger to Newtonian dynamics can be expressed in terms of transition from quantum commutators to Poisson brackets. A simple calculation with the above $h(\mathbf{p},\mathbf{a},t)$ and $\widehat{h}$, and φ in $M^{\sigma}_{3,3}$ yields the following:

$$(\varphi,\tfrac{1}{i\hbar}[\widehat{\mathbf{x}},\widehat{h}]\varphi)=\{\mathbf{a},h\}, \tag{6}$$

and

$$(\varphi,\tfrac{1}{i\hbar}[\widehat{\mathbf{p}},\widehat{h}]\varphi)=\{\mathbf{p},h\}. \tag{7}$$

The brackets on the right side of (6) and (7) are the usual Poisson brackets. Consider the linear vector fields $x_\varphi=\widehat{\mathbf{x}}\varphi$ and $p_\varphi=\widehat{\mathbf{p}}\varphi$ on $L_2(\mathbb{R}^3)$ associated with the operators of position and momentum. For a given state φ, let $\overline{\mathbf{x}}$ and $\overline{\mathbf{p}}$ be the expected values of the operators on this state. The components $x_{\varphi\perp}=(\widehat{\mathbf{x}}-\overline{\mathbf{x}}I)\varphi$ and $p_{\varphi\perp}=(\widehat{\mathbf{p}}-\overline{\mathbf{p}}I)\varphi$ are tangent to the sphere S^{L_2} at φ and orthogonal to the fibres $\{\varphi\}$ of the fibre bundle $CP^{L_2}=S^{L_2}/S^1$. Therefore, they are tangent to the projective space CP^{L_2} itself. Moreover, the vector fields $x_{\varphi\perp}$ and $p_{\varphi\perp}$ constrained to the submanifold $M^{\sigma}_{3,3}$ of CP^{L_2} are tangent to it. The integral curves of these fields are lines of constant position and momentum that provide the coordinate grid in the classical phase space $\mathbb{R}^3 \times \mathbb{R}^3$. This can be seen directly by comparing the right hand sides of the equations

$$(\widehat{\mathbf{x}} - \overline{\mathbf{x}}I)\varphi = (\mathbf{x} - \mathbf{a})\varphi \tag{8}$$

and

$$(\widehat{\mathbf{p}} - \overline{\mathbf{p}}I)\varphi = -\frac{i\hbar}{2\sigma^2}(\mathbf{x} - \mathbf{a})\varphi, \tag{9}$$

valid for φ in $M^{\sigma}_{3,3}$, with the gradients $\nabla_{\mathbf{a}}\varphi$ and $\nabla_{\mathbf{p}}\varphi$. The latter gradients represent vectors tangent to the lines of constant values of $\mathbf{p}$ and $\mathbf{a}$ through a point of the projective manifold $M^{\sigma}_{3,3}$.

We have the following proposition.

(P1) *At any point* $(\mathbf{a}, \mathbf{p})$ *of the phase space* $\mathbb{R}^3 \times \mathbb{R}^3$*, the derivatives* $\frac{d\mathbf{a}}{dt}$ *and* $\frac{d\mathbf{p}}{dt}$ *of a Newtonian motion are projections at the point* $\Omega(\mathbf{a}, \mathbf{p})$ *of the velocity* $\frac{d\varphi}{dt}$ *of the Schrödinger evolution onto the coordinate lines of the phase space submanifold* $M^{\sigma}_{3,3}$.

Proof. Using the Ehrenfest theorem written in terms of the vector fields $x_{\varphi\perp}$ and $p_{\varphi\perp}$ and (6) and (7), we have for the initial state φ in $M^{\sigma}_{3,3}$ at $t = 0$:

$$2\mathrm{Re}\left(\frac{d\varphi}{dt}, x_{\varphi\perp}\right) = \{\mathbf{a}, h\} = \frac{d\mathbf{a}}{dt} \tag{10}$$

and

$$2\mathrm{Re}\left(\frac{d\varphi}{dt}, p_{\varphi\perp}\right) = \{\mathbf{p}, h\} = \frac{d\mathbf{p}}{dt}. \tag{11}$$

□

(2) Furthermore, it is possible to decompose the initial velocity of state of a particle in a potential V with the initial state in $M^{\sigma}_{3,3}$ into four physically meaningful orthogonal components. The two components tangent to the phase space submanifold $M^{\sigma}_{3,3}$ were already identified with the usual classical velocity $\mathbf{v} = d\mathbf{a}/dt$ and acceleration $\mathbf{w} = d\mathbf{v}/dt = -\nabla V/m$ of the particle. The other two components orthogonal to $M^{\sigma}_{3,3}$ are the *phase velocity* (projection of $d\varphi/dt$ onto the unit vector $-i\varphi$) and the *velocity of spreading* of the wave-packet (projection of $d\varphi/dt$ onto the unit vector in the direction $id\varphi/d\sigma$).

(P2) *The norm of the total velocity* $d\varphi/dt$ *at* $t = 0$*, i.e., the speed of motion of the initial state* φ*, is given by the formula*

$$\left\|\frac{d\varphi}{dt}\right\|^2 = \frac{\overline{E}^2}{\hbar^2} + \frac{\mathbf{v}^2}{4\sigma^2} + \frac{m^2\mathbf{w}^2\sigma^2}{\hbar^2} + \frac{\hbar^2}{32\sigma^4 m^2}. \tag{12}$$

When the state is constrained to the manifold $M^{\sigma}_{3,3}$*, the first term (the phase velocity squared) and the last term (the velocity of spreading squared) disappear. The only surviving terms are associated with the classical velocity and acceleration and the motion follows the classical Newtonian dynamics in the phase space.*

Proof. The decomposition (12) is obtained by projecting the velocity $\frac{d\varphi}{dt}$ of state under the Schrödinger evolution with an arbitrary Hamiltonian $\widehat{h} = -\frac{\hbar^2}{2m}\Delta + V(\mathbf{x}, t)$ at a point $\varphi = g_{\mathbf{a},\sigma}e^{i\mathbf{p}\mathbf{x}/\hbar}$ in the classical phase space $M^{\sigma}_{3,3}$ onto an orthonormal set of vectors specified by changing the parameters, $\mathbf{a}, \mathbf{p}, \sigma$ that define $\{\varphi\}$ and the phase parameter θ of a possible constant phase factor $e^{-i\theta}$ of φ. Calculation of the classical space components of $\frac{d\varphi}{dt}$ at an arbitrary point φ in $M^{\sigma}_{3,3}$ yields

$$\mathrm{Re}\left(\frac{d\varphi}{dt}, -\widehat{\frac{\partial\varphi}{\partial a^{\alpha}}}\right)\Bigg|_{t=0} = \frac{p^{\alpha}}{2m\sigma}, \tag{13}$$

where the hat here and in other calculations denotes normalization. For the momentum space components of $\frac{d\varphi}{dt}$ at φ we similarly obtain, assuming that σ is small enough to make the linear approximation of $V(\mathbf{x})$ valid:

$$\mathrm{Re}\left(\frac{d\varphi}{dt}, \widehat{\frac{\partial\varphi}{\partial p^{\alpha}}}\right)\Bigg|_{t=0} = \frac{mw^{\alpha}\sigma}{\hbar}, \quad where \quad mw^{\alpha} = -\left.\frac{\partial V(\mathbf{x})}{\partial x^{\alpha}}\right|_{\mathbf{x}=\mathbf{a}}. \tag{14}$$

The components (13) and (14) are tangent to $M^{\sigma}_{3,3}$ and orthogonal to the fibre $\{\varphi\}$. The component of the velocity $\frac{d\varphi}{dt}$ due to change in σ (spreading) is orthogonal to the phase space $M^{\sigma}_{3,3}$ and the fibre $\{\varphi\}$, and is equal to

$$\mathrm{Re}\left(\frac{d\varphi}{dt}, i\widehat{\frac{d\varphi}{d\sigma}}\right) = \frac{\sqrt{2}\hbar}{8\sigma^2 m}. \tag{15}$$

The component of the velocity parallel to the fibre $\{\varphi\}$ is the expected value of energy divided by $\hbar$:

$$\mathrm{Re}\left(\frac{d\varphi}{dt}, -\widehat{\frac{d\varphi}{d\theta}}\right) = \frac{1}{\hbar}(i\widehat{h}\varphi, i\varphi) = \frac{\overline{E}}{\hbar}. \tag{16}$$

Calculation of the norm of $\frac{d\varphi}{dt} = \frac{i}{\hbar}\widehat{h}\varphi$ at $t = 0$ gives

$$\left\|\frac{d\varphi}{dt}\right\|^2 = \frac{\overline{E}^2}{\hbar^2} + \frac{\mathbf{p}^2}{4m^2\sigma^2} + \frac{m^2\mathbf{w}^2\sigma^2}{\hbar^2} + \frac{\hbar^2}{32\sigma^4 m^2}, \tag{17}$$

which is the sum of squares of the found components. This completes a decomposition of the velocity of state at any point φ in $M^{\sigma}_{3,3}$. □

Similar results are true for systems of n particles [11]. In this case, the transition to classical Newtonian dynamics happens when the state of the system is constrained to the phase space submanifold $M^{\sigma}_{3n,3n} = M^{\sigma}_{3,3} \otimes ... \otimes M^{\sigma}_{3,3}$ in the space of states of the system. That is, the constrained state is the product of states of the n particles in the system with the state of each particle constrained to a copy of $M^{\sigma}_{3,3}$. Although not directly used in the paper, the results **(1)** and **(2)** provide an insight into the theorem (**A**) and help a deeper understanding of the upcoming theorems.

(3) The identification of the classical phase space of a particle with the manifold $M^{\sigma}_{3,3}$ yields a useful metric relationship. Let M^{σ}_{3} denote the submanifold of $M^{\sigma}_{3,3}$ of the Gaussian states $g_{\mathbf{a},\sigma} = \left(\frac{1}{2\pi\sigma^2}\right)^{3/4} e^{-\frac{(\mathbf{x}-\mathbf{a})^2}{4\sigma^2}}$. The map $\omega : \mathbf{a} \longrightarrow g_{\mathbf{a},\sigma}$, $\omega(\mathbb{R}^3) = M^{\sigma}_{3}$ identifies the submanifold M^{σ}_{3} of CP^{L_2} with the classical space $\mathbb{R}^3$. We then have the following proposition.

(P3) *Let* $\theta(g_{\mathbf{a},\sigma}, g_{\mathbf{b},\sigma})$ *be the distance from* $g_{\mathbf{a},\sigma}$ *to* $g_{\mathbf{b},\sigma}$ *in* M^{σ}_{3} *in the Fubini-Study metric on the space of states* CP^{L_2} *and let* $(\mathbf{a}-\mathbf{b})^2$ *be the square of the Euclidean distance between the corresponding points* $\mathbf{a}$ *and* $\mathbf{b}$ *in* $\mathbb{R}^3$*. Then*

$$e^{-\frac{(\mathbf{a}-\mathbf{b})^2}{4\sigma^2}} = \cos^2\theta(g_{\mathbf{a},\sigma}, g_{\mathbf{b},\sigma}) \tag{18}$$

Likewise, for arbitrary states $\varphi = g_{\mathbf{a},\sigma} e^{i\mathbf{px}/\hbar}$ *and* $\psi = g_{\mathbf{b},\sigma} e^{i\mathbf{qx}/\hbar}$ *in the classical phase space* $M^{\sigma}_{3,3}$, *the distances in the Euclidean phase space and the space of states are related by*

$$e^{-\frac{(\mathbf{a}-\mathbf{b})^2}{4\sigma^2} - \frac{(\mathbf{p}-\mathbf{q})^2}{\hbar^2/\sigma^2}} = \cos^2\theta(\varphi, \psi). \tag{19}$$

Proof. The metric relationship (18) follows from the inner product of two states in M^{σ}_3:

$$(g_{\mathbf{a},\sigma}, g_{\mathbf{b},\sigma}) = e^{-\frac{(\mathbf{a}-\mathbf{b})^2}{8\sigma^2}}. \tag{20}$$

This expression squared is equal to the right hand side of (18) by the definition of the Fubini-Study distance between states in CP^{L_2}. The result (19) is obtained in a similar way by evaluating the Fourier transform of a Gaussian function along the way. □

The relation (18) implies an important connection between probability distributions of multivariate random variables valued in the classical space and the space of states. Namely, consider a random variable φ with values in the space of states CP^{L_2}. Because the classical space M^{σ}_3 is a submanifold of CP^{L_2}, we can restrict φ to take values in M^{σ}_3. We then have the following theorem.

(B) *Suppose the conditional probability of* φ *given that* φ *is in the classical space is described by the normal distribution. Suppose the probability* P *of transition between any two states depends only on the distance between the states. Then* P *is given by the Born rule, i.e.,* $P = |(\varphi, \varphi_0)|^2$, *where* φ_0 *is the initial and* φ *is the observed state. The opposite is also true: For states in the classical space the Born rule yields the normal distribution of the observed position.*

So, the normal probability distribution is the Born rule in disguise!

Proof. In fact, (18) establishes the needed connection between the normal distribution and the Born rule for the states in M^{σ}_3. By assumption, the probability of transition between states depends only on the Fubini-Study distance between them. Since the Fubini-Study distance between states in M^{σ}_3 takes on all possible values, the validity of the Born rule for the states in M^{σ}_3 signifies its validity for arbitrary states. See [10] for details. □

These results establish a deep connection between Newtonian and Schrödinger dynamics and between the classical space and classical phase space, and the submanifolds M^{σ}_3 and $M^{\sigma}_{3,3}$ of the space of states. Based on these results we put forward the following embedding hypothesis:

(EH) *The constructed mathematical embedding of the classical space and classical phase space into the space of states and the resulting identification of Newtonian dynamics with the constrained Schrödinger dynamics are physical. That is, the classical space, phase space and the Newtonian dynamics of a system are not only fully derivable from, but also physically originate in the Schrödinger dynamics in the Hilbert space of states of the system.*

The hypothesis asserts that the provided mathematical derivation of the classical Newtonian dynamics from the Schrödinger dynamics of the system is the true physical correspondence between classical and quantum systems.

Let us point out that this embedding hypothesis is not required for validity of the theorems in the paper. For this, the mathematical isometric embedding of $M^{\sigma}_{3,3}$ into CP^{L_2} is sufficient. Viewed this way, the results that will be derived can be considered curious mathematical consequences of the already verified isometry of the embedding and a single forthcoming conjecture (**RM**). The next few paragraphs preceding (**RM**) can be then looked at as a physical motivation of the conjecture. Whatever point of view is preferred by the reader, the results will provide an advance in our understanding of the transition from quantum to classical theory and of the process of measurement.

To validate the hypothesis, or to motivate the forthcoming conjecture, let us show that the hypothesis is consistent with all observed classical and quantum phenomena. This involves showing that all observable classical phenomena can be derived from the corresponding quantum phenomena by constraining the state to the classical phase space submanifold in the space of states. This also involves physically explaining the constraint itself. Now, an arbitrary deterministic classical motion is described by the Newtonian equations of motion. The theorem (**A**) demonstrates that these equations follow from the Schrödinger equation with the constraint. Therefore, given the constraint, the hypothesis is consistent with an arbitrary deterministic classical motion. To validate the hypothesis, it remains then to: (1) verify its consistency for the motion of particles described in statistical mechanics, specifically, for the Brownian motion, and (2) explain the origin of the constraint from the Schrödinger dynamics itself. The rest of the paper will be dedicated to these two tasks.

Instead of diving into the subject of a quantum version of the Brownian motion, note that a macroscopic Brownian particle in a medium can be considered a classical chaotic system. For instance, the motion of the particle through an appropriate lattice of round obstacles provides a classical chaotic realization of the Brownian motion [14]. A lively discussion of the stochastic, deterministic chaotic, and regular characterizations of the Brownian motion can be found in [17–19]. With the view that the Brownian particle in a medium is a chaotic system comes the applicability of the BGS-conjecture to the system [16]. It asserts that the Hamiltonian of the corresponding quantum system can be represented by a random matrix.

Random matrices were originally introduced into quantum mechanics by Wigner [15] in a study of excitation spectra of heavy nuclei. Wigner reasoned that the motion of nucleons within the nucleus is so complex that the Hamiltonian of the system can be modeled by a random matrix from an ensemble that respects symmetries of the system but otherwise contains no additional information. It was later discovered that correlations in the spectrum of random matrices possess a remarkable universality in being applicable to a large number of quantum systems. That includes nuclear, atomic, molecular and scattering systems, chaotic systems with few degrees of freedom as well as complex systems, such as solid state systems and lattice systems in the field theory. This wealth of experimental evidence suggests that all quantum systems whose classical counterpart is chaotic exhibit random matrix statistics. This is the essence of the BGS-conjecture.

As mentioned, from the possibility to attribute chaotic character to Brownian motion and the BGS conjecture it follows that the Hamiltonian of the quantum analogue of the Brownian motion at any time is given by a random matrix. On the physical grounds, we can also claim that the random matrices that represent the Hamiltonian at two different moments of time must be independent. This leads us to the following conjecture:

> **(RM)** *The quantum-mechanical analogue of the Brownian motion can be modeled by a random walk of state in the space of states of the system. The steps of the random walk without drift satisfy the Schrödinger equation with the Hamiltonian represented by a random matrix from the Gaussian unitary ensemble (GUE). The matrices representing the Hamiltonian at two different times are independent and belong to the same ensemble.*

Note that the Schrödinger equation with the Hamiltonian represented by a random matrix describes evolution of the state and not of the density matrix for the system. In that sense, it is analogous to the Langevin equation for the position of a Brownian particle rather than the diffusion equation for the probability density function. Note also, that even though both, (**RM**) and BGS conjectures deal with random matrices, they should not be confused with each other. In particular, while BGS deals with spectra of Hamiltonians, (**RM**) defines a non-stationary stochastic process. A mathematically inclined reader may skip the motivation for (**RM**) and consider it an assumption used in the forthcoming theorems. In fact, this is the only important assumption needed for validity of the theorems in the paper.

First of all, let us prove that the random walk in (**RM**) yields the Brownian motion in $\mathbb{R}^3$. More precisely, we have the following theorem.

> (**C**) *The random walk described in* (**RM**) *but conditioned to stay on the submanifold* M_3^g *in the space of states yields a random walk on* $\mathbb{R}^3$ *that approximates the Brownian motion of a particle in a medium.*

Proof. In fact, a general Schrödinger evolution with Hamiltonian $\widehat{h}$ can be thought of as a sequence of steps connecting the points $\varphi_{t_0}, \varphi_{t_1}, ...$ in the space of states. For small time intervals $\Delta t = t_k - t_{k-1}$, the state φ_{t_N} at time t_N is given by the time ordered product

$$\varphi_{t_N} = e^{-\frac{i}{\hbar}\widehat{h}(t_N)\Delta t} e^{-\frac{i}{\hbar}\widehat{h}(t_{N-1})\Delta t} ... e^{-\frac{i}{\hbar}\widehat{h}(t_1)\Delta t} \varphi_{t_0}. \tag{21}$$

Suppose the evolution of the state φ of a particle is constrained to the classical space submanifold M_3^g. The points $\varphi_{t_0}, \varphi_{t_1}, ...$ belong then to the submanifold M_3^g and the steps can be identified with translations in the classical space. This is to say that for each k, the operator $\widehat{h}(t_k)$ acts as the generator of translation by a vector ξ_k in $\mathbb{R}^3$, so that $\widehat{h}(t_k) = \xi_k \widehat{\mathbf{p}}$, where $\widehat{\mathbf{p}}$ is the momentum operator. Because all operators of translation commute with each other, the equation (21) yields the following expression:

$$\varphi_{t_N}(\mathbf{x}) = \varphi_{t_0}(x - \xi_1 \Delta t - \xi_2 \Delta t - ... - \xi_N \Delta t). \tag{22}$$

That is, the initial state is simply translated by the vector

$$\mathbf{d} = \sum_{k=1}^{N} \xi_k \Delta t \tag{23}$$

in $\mathbb{R}^3$. Now, the probability distribution of steps $-\frac{i}{\hbar}\widehat{h}(t_{k+1})\varphi_{t_k}$ in the tangent space $T_{\varphi_k} M_3^g$ must be the conditional probability distribution of steps for the Hamiltonian satisfying (**RM**) under the condition that the steps take place in $T_{\varphi_k} M_3^g$. From the properties of the random matrix, it follows that ξ_k are independent and identically normally distributed random vectors, so that the equation (23) defines a random walk with Gaussian steps in $\mathbb{R}^3$. This is known to approximate the Brownian motion in $\mathbb{R}^3$ and yield the normal distribution of the position vector $\mathbf{d}$ at any time t, proving the claim. □

Note that the theorem would not be true if the Hamiltonian were a random matrix in the Gaussian orthogonal ensemble in place of the unitary ensemble. In fact, the momentum operator in the proof is Hermitian but not orthogonal. Now, the Gaussian unitary ensemble corresponds to systems that are not invariant under time reversal [20,21]. Therefore, the fact that we obtained in (**C**) a diffusion process on the submanifold M_3^g is tied to the fact that the Hamiltonian is not invariant under time reversal.

By the extension of a random walk on $\mathbb{R}^3$ to the space of states CP^{L_2}, we understand a walk in CP^{L_2} that satisfies (**RM**) and that reduces to the original random walk on $\mathbb{R}^3$ when conditioned to stay on M_3^g. From (**C**), we know that an extension of the walk with Gaussian steps on $\mathbb{R}^3$ exists. We claim that such an extension is *unique*. In fact, because the random walk conditioned to stay on M_3^g must be Gaussian, the entries of the matrix of the Hamiltonian associated with the directions tangent to M_3^g at a point are distributed normally. But the probability distribution of a single entry of the matrix defines the corresponding Gaussian unitary ensemble. Because of that, the random walk with Gaussian steps in $\mathbb{R}^3$ defines a unique random walk satisfying (**RM**). In what follows, the random walk of the state of a particle in (**RM**) will always be assumed to be this unique extension of the random walk with Gaussian steps that approximate the Brownian motion of the particle. We are free to make this choice as long as the resulting extension satisfies our needs. At the same time, this choice makes sense physically, because a displacement of the particle in $\mathbb{R}^3$ results in the like displacement of its state in M_3^{σ}, tying the two motions together.

The theorems (**A**) and (**C**) prove the consistency of the hypothesis (**EH**) for the deterministic Newtonian dynamics and the Brownian motion. Our goal now is to explain the origin of the constraint to the classical space and phase space submanifolds in the Hilbert space of states. For this, we need to establish properties of the random walk on the space of states in the conjecture (**RM**) and establish its applicability to the process of measurement.

(**D**) *The probability distribution of steps of the random walk specified in* (**RM**) *is isotropic and homogeneous. That is, for all initial states* $\{\varphi\}$ *in the space of states* CP^H*, the vector* $d\varphi = -\frac{i}{\hbar}\widehat{h}\varphi dt$ *is a normal random vector in the tangent space* $T_{\{\varphi\}}CP^H$*, with spherical distribution. The probability of reaching a certain state during the walk depends only on the distance between the initial and the final state.*

Proof. To prove the theorem, note that because for any t the matrix of $\widehat{h}$ is in GUE, the probability density function $P(\widehat{h})$ of $\widehat{h}$ is invariant with respect to conjugations by unitary transformations. That is, $P(U^{-1}\widehat{h}U) = P(\widehat{h})$ for all unitary transformations U acting in the Hilbert space of states. Also, for all unitary transformations U that leave $\{\varphi\}$ unchanged and therefore all U that act in the tangent space $T_{\{\varphi\}}CP^H$, we have

$$(U^{-1}\widehat{h}U\varphi, v) = (\widehat{h}U\varphi, Uv) = (\widehat{h}\varphi, Uv), \tag{24}$$

where v is a unit vector in $T_{\{\varphi\}}CP^H$. It follows that

$$P(\widehat{h}\varphi, v) = P(\widehat{h}\varphi, Uv), \tag{25}$$

where P is the probability density of the components of $\widehat{h}\varphi$ in the given directions. By a proper choice of U, we can make Uv to be an arbitrary unit vector in $T_{\{\varphi\}}CP^H$, proving the isotropy of the distribution. On the other hand, for all unitary operators V in H and a unit vector v in $T_{\{\varphi\}}CP^H$, we have

$$P(\widehat{h}\varphi, v) = P(V^{-1}\widehat{h}V\varphi, v) = P(\widehat{h}V\varphi, Vv). \tag{26}$$

Because $V\varphi$ is an arbitrary state and Vv is in the tangent space $T_{\{V\varphi\}}CP^H$, we conclude with the help of (24) that the probability density function is independent of the initial state of the system, proving the homogeneity of the distribution. The components of the vector $\widehat{h}\varphi dt$ are independent, because the entries of the matrix of $\widehat{h}$ are independent. It follows that $-\frac{i}{\hbar}\widehat{h}\varphi dt$ is a normal random vector with spherical distribution. Finally, because different steps of the walk are independent, the probability of reaching a certain state during the walk may depend only on the distance between the initial and the final state. □

The fact that the components of the vector $\widehat{h}\varphi dt$ are independent signifies that the probability distribution of steps conditioned to take place in M_3^{σ} is the same as the probability distribution of these steps without the condition. In particular, the random walk obtained in (**C**) is the usual random walk on $\mathbb{R}^3$, independent of the embedding of $\mathbb{R}^3$ into CP^{L_2}.

Let us show now that conjecture (**RM**) provides a consistent approach to the process of observation of the position of a single particle in the classical and quantum mechanics alike.

(**E**) *The Schrödinger evolution with the Hamiltonian that satisfies* (**RM**) *models the measurement of the position of a particle in classical and quantum physics. Namely, under the constraint that the random walk that approximates the evolution of state* *takes place on the submanifold M_3^g, the probability distribution of the position random vector is normal. Without the constraint, the probability to find the initial state φ_0 at a point φ is given by the Born rule. The transition from the initial to final state is time-irreversible.*

Proof. In fact, according to (**C**), the random walk that satisfies (**RM**) but is conditioned to stay on M_3^g is the random walk with Gaussian steps on $\mathbb{R}^3$. The latter random walk considered over the time interval of observation yields the normal distribution of the position random vector. On the other hand, according to (**D**), the probability P of reaching a state φ from the initial state φ_0 by means of a unconstrained random walk that satisfies (**RM**) depends only on the distance between the states. From (**B**), it then follows that P is given by the Born rule: $P = |(\varphi, \varphi_0)|^2$. Also, the choice of the Gaussian unitary ensemble corresponds to time-irreversible systems [21]. □

Note that the normal distribution of the position agrees with observations in the macro world. It is also consistent with the central limit theorem applied to describe the cumulative effect of uncontrollable fluctuations from the mean in a series of measurement outcomes. It follows that the model satisfies the basic properties of observation of the position in classical and quantum mechanics.

Quantum mechanics does not explain how and when the deterministic Schrödinger evolution is replaced by probabilistic evolution, whose outcomes obey the Born rule. It does not explain why macroscopic bodies are never observed in superpositions of position eigenstates. The wave-function collapse models [5, 6] aim to answer these questions by introducing a non-linear stochastic modification of the Schrödinger equation. It is assumed that the modified equation must be non-linear to be able to suppress superpositions of states. It is also assumed that the modified equation must vary with a change in the observed quantity. However, here the Born rule was derived from the linear Schrödinger equation with the Hamiltonian satisfying (**RM**). Moreover, it was derived for all observed quantities at once, without needed to change the equation of motion. Is there a contradiction?

The modified Schrödinger equations in the collapse models make the state of the system converge to an eigenstate of the measured quantity, usually the position of a particle. An equation like that must break an arbitrary initial superposition of states, so it must be non-linear. Under the Schrödinger evolution with the Hamiltonian satisfying the conjecture (**RM**), the state of the system does not converge to a position eigenstate. The equation does not suppress superpositions. It makes the state wander around the space of states and ensures that the probability of reaching a particular neighborhood in the space is given by the Born rule. That explains why the two approaches do not contradict each other.

We are now ready to investigate why the state of a macroscopic body is confined to the classical space submanifold of the space of states. The key to this is that the Brownian motion and the motion of state under a measurement are now derived from the same dynamics. We know that large particles do not experience a Brownian motion. That is because the total force acting on any such particle from the particles of the medium is nearly zero. As a result, in the absence of an external potential, the particle remains at rest in the medium. A similar mechanism explains why the state of a particle may be confined to the submanifold M_3^g:

(**F**) *Suppose the Hamiltonian of a particle in the natural surroundings satisfies the conjecture* (**RM**). *Then, the state of the particle is constrained to the submanifold M_3^g precisely when the induced Brownian motion of the particle described in* (**C**) *vanishes. More precisely, under* (**RM**), *the boundary between the quantum and the classical occurs for particles that satisfy the following two conditions. The particles must be sufficiently large in size so that their Brownian motion in the natural surroundings is observable. At the same time, the particles must not be too large, when the Brownian motion* *trivializes, the state initially in M_3^g becomes constrained to M_3^g and the Schrödinger evolution becomes Newtonian motion.*

Proof. Recall that the spaces M_3^g and $\mathbb{R}^3$ are metrically identical. Recall also that the random walk in (**RM**) is the unique extension of the random walk in $\mathbb{R}^3$ that approximates the Brownian motion of the particle in the surroundings. Consider a particle whose initial state is in M_3^g. Suppose the particle is sufficiently large in size so that the Brownian motion of the particle in the natural surroundings is negligible. Therefore, the motion of state in the directions tangent to M_3^g is negligible. From (**D**), we know that for the Hamiltonian satisfying (**RM**), the probability distribution of steps of the random walk in the space of states of the particle is isotropic. Because the probability of steps in the directions tangent to M_3^g vanishes, the same holds true for any other direction tangent to the space of states at the same point. Therefore, in the absence of an external potential, the motion of state of the particle in the space of states is trivial. When an external potential is applied to the particle, the two middle terms in the decomposition (12) may appear. However, these terms can only contribute to the motion of state in the direction tangent to the classical phase space submanifold $M_{3,3}^g$. Therefore, the state will remain constrained to the submanifold. The theorem (**A**) asserts then that the particle in such a state will move in accord with the Newtonian dynamics.

On the contrary, suppose the particle is sufficiently small, but not too small, so that the interaction between the particle and the surroundings cannot be ignored and results in a noticeable Brownian motion of the particle. By the isotropy of the probability distribution of steps of the random walk of state, a displacement of the particle away from M_3^g is then equally likely. Such a displacement would mean that the particle is now in a superposition of states of different positions. If $\mathbf{a}$ is the initial position and $\mathbf{l}$ is the observed displacement of the particle in M_3^g during the measurement, then the states $g_{\mathbf{a},\sigma}$ and $g_{\mathbf{a+l},\sigma}$ are distinguishable in the experiment. It means that the superpositions of these states can be observed as well, indicating that we are dealing with a quantum system. □

We see how the properties of the Hamiltonian in the conjecture (**RM**) are responsible for the fact that the state of a macroscopic particle driven by the Schrödinger dynamics with this Hamiltonian is constrained to the classical phase space manifold. One might be tempted to say that the resulting "freezing" of the state is in agreement with the quantum Zeno effect for the particle whose position is continuously measured. However, the essential difference is that the result is derived here from the unitary evolution generated by a random Hamiltonian without ever needing to involve the projection operators.

As an example, let us find the displacement of a particle of radius $1mm$ in still air in normal conditions. The estimate of the diffusion coefficient of a macroscopic spherical particle of radius $1mm$ in still air in normal conditions is based on the Stokes-Einstein relationship

$$D = \frac{k_B T}{6\pi\eta r}, \tag{27}$$

where D is the diffusion coefficient, r is the radius of the particle, η is the dynamic viscosity, T is temperature of the medium and k_B is the Boltzmann constant. Using the room temperature and the known value of dynamic viscosity $\eta \sim 10^{-5} N \cdot s/m^2$, we get $D \sim 10^{-12} m^2/s$. The variance of the x-coordinate of position of the particle is given by $\overline{x^2} = 2Dt$. If we scatter visible light off the particle to determine its position, the time interval of observation can be estimated to be as short as $10^{-13}s$. This amounts to the displacement of the order of $10^{-12}m$. The accuracy of measurement is limited by the wavelength $\lambda \sim 10^{-5}m$. The Fubini-Study distance between Gaussian states that are $10^{-12}m$ apart in M_3^g with $\sigma \sim 10^{-5}m$ is calculated via (18) and is about $10^{-7}rad$. These displacements are too small to be observed in the described experiment. The state of the particle remains "frozen". The particle is constrained to M_3^g and no superpositions of states of a different position of the particle can be observed.

Journal of Physics: Conference Series **2482** (2023) 012025 doi:10.1088/1742-6596/2482/1/012025

Consider now a system consisting of a microscopic particle P and a macroscopic device D capable of measuring the position of P. The particle and the device form a two-particle system whose state belongs to the product Hilbert space $H = H_P \otimes H_D$. We have the following result.

(G) *Suppose the initial state of the system consisting of a microscopic particle and a macroscopic device in the natural surroundings is a product state with the state of the device in the manifold M_3^g. Suppose also that the conjecture* (**RM**) *holds true. Then, during the evolution, the state of the system remains in a product form with the state of macroscopic device confined to M_3^g and evolving classically.*

Proof. Consider the motion of the corresponding classical system. Let say, D is a cloud chamber, small enough to be considered a material point, when treated classically. The macroscopic device interacts with the surroundings. It also interacts with the particle. However, in classical physics, the effect of the particle on the device can be neglected. It follows that we can apply (**RM**) to the device itself. The interaction of the device with the surroundings results in a Brownian motion of the device. When the device is sufficiently large, its Brownian motion is trivial and the device is at rest in the lab system. By applying (**RM**), we conclude that the state of the device positioned initially in M_3^g can be treated independently of the state of the particle and is at rest in the space of state CP^{H_D}. On the other hand, a small macroscopic particle placed in the medium of the chamber would undergo a Brownian motion. By applying (**RM**) to this case, we conclude that the state of the corresponding quantum system will perform a random walk in the space CP^{H_P}. According to (**E**), the probability to find the particle at a particular point of $\mathbb{R}^3$ during the walk is given by the Born rule.

The conjecture (**RM**) is valid by assumption. Using the conjecture, we managed to predict the evolution of states of the device and the measured particle. This can only be the case when the state of the particle-device system remains separable throughout the evolution. Furthermore, according to (**F**), the state of the device will be confined to the submanifold M_3^g and in the presence of an external potential will be evolving in accord with the Newtonian dynamics. Under these conditions, the system remains in a product state with both factors being able to change. In particular, the position of the particle can be mechanically recorded with no entanglement between the particle and the device ever appearing or needing to appear in the process. □

3. Implications for the notion of objective reality

Theorem (**G**) provides an explanation of the issue of an apparent inconsistency of the description of quantum mechanics by macroscopic observers. The issue was pointed out first by Wigner [22] and has picked up a significant interest in recent times. Wigner has demonstrated that the rules of quantum mechanics applied universally may lead to contradictory accounts of observations performed by two observers. In his thought experiment [22], an observer (Wigner's friend) makes a measurement on a system and another observer (Wigner himself) observes the laboratory, where his friend performs the measurement. The contradiction is seen in the observers getting two different states after the friend has performed the measurement, but before the result is communicated to Wigner. This, as well as more recent thought experiments based on a similar scenario [23], raise the question of when the collapse of the state actually occur, what does the consciousness of the observers have to do with the collapse, and, ultimately, is there such a thing as an objective reality that all observers can agree on.

A possible reason for the contradiction in the accounts of Wigner and the friends in the Wigner's friend-type thought experiments is the assumption that a macroscopic observer or the whole lab may exist in a superposition of classical states. Without this assumption the no-go result in [23] doesn't hold and the original Wigner's friend paradox disappears. Note that the experiment reported in [24], in which the photons played the role of an observer is consistent with this conclusion. The reality as "described" by microscopic particles is different than the one the macroscopic observers are aware of. Consequently, many researchers, including the authors of [23] think that quantum mechanics is not directly applicable to macroscopic bodies. This is also the thinking behind the modifications of the Schrödinger equation with the goal of suppressing the unobserved superpositions, and accounting for the transition to classicality and the measurement results [5, 6].

The results proved in this paper offer an alternative solution to the problem. The Schrödinger evolution may still be valid for macroscopic bodies, but we need to be careful in identifying the Hamiltonian of the corresponding system. Because of the unavoidable interaction of macroscopic bodies with the surroundings, the Hamiltonian cannot have the same form as before. In some ways, the macroscopic bodies are constantly "measured" by the surroundings. At the same time, measuring a microscopic particle requires proper conditions. For instance, the needed conditions may be provided by a photographic plate, a cloud chamber, or a high-intensity radiation. It was assumed here that the Hamiltonian of a macroscopic particle in the natural surroundings and of a microscopic particle whose position is measured is represented by a random matrix that satisfies (**RM**). This assumption turned out to be sufficient to explain why no superpositions of classical states of macroscopic bodies are observed, to obtain irreversibility of a measurement and to derive the Born rule. Without superpositions of macroscopic observers and the lab, the observations by Wigner and the friends become consistent and objective reality is restored.

References

[1] Klauder, J., & Sudarshan, E., *Fundamentals of quantum optics* (W.A. Benjamin, New York, 1968).
[2] Martin-Dussaud, P., Searching for Coherent States: From Origins to Quantum Gravity. *Quantum* **5** 390 (2021).
[3] Heller, E., Frozen Gaussians: A very simple semiclassical approximation. *J. Chem. Phys.* **75** 2923 (1981).
[4] Joos, E., Giulini, D., Kiefer, C., Kupsch, J., and Stamatescu, I., *Decoherence and the Appearance of a Classical World in Quantum Theory* (Springer-Verlag, Berlin Heidelberg, 2003).
[5] Bassi, A. and Ghirardi, G. Dynamical Reduction Models. *Physics Reports* **379** 257 (2003).
[6] Bassi, A., Lochan, K., Satin, S., Singh, T. and Ulbricht, H. Models of wave-function collapse, underlying theories, and experimental tests. *Rev. Mod. Phys.* **85** 471 (2013).
[7] Dyson, F., A Brownian-motion model for the eigenvalues of a random matrix. *J. Math. Phys.* **3** 1191 (1962).
[8] Fyodorov, I., Introduction to the Random Matrix Theory: Gaussian Unitary Ensemble and Beyond. *Lond. Math. Soc. Lect. Notes* **Ser.322** 31 (2005).
[9] Kryukov, A. On the motion of macroscopic bodies in quantum theory. *J. Math. Phys.* **58**, 082103 (2017).
[10] Kryukov, A. On observation of position in quantum theory. *J. Math. Phys.* **59** 052103 (2018).
[11] Kryukov, A. Mathematics of the classical and the quantum. *J. Math. Phys.* **61** 082101 (2020).
[12] Kryukov, A. Linear algebra and differential geometry on abstract Hilbert space. *Int. J. Math. Math. Sci.* **2005** 2241 (2005).
[13] Kryukov, A. Geometric derivation of quantum uncertainty. *Phys. Lett.* A **370** 419-422 (2007).
[14] Szasz, D., editor. *Hard Ball Systems and the Lorentz Gas.* (Springer-Verlag, Berlin, 2000).
[15] Wigner, E. P. On the statistical distribution of the widths and spacings of nuclear resonance levels. *Proc. Cambridge Philos. Soc.* **47** 790 (1951).
[16] Bohigas, O., Giannoni, M. J. and Schmit, C. Characterization of Chaotic Quantum Spectra and Universality of Level Fluctuation Laws. *Physical Review Letters* **52** 1 (1984).
[17] Gaspard, P., Briggs, M. E., Francis, M. K., Sengers, J. V., Gammon, R. W., Dorfman, J. R. and Calabrese, R. V. Experimental evidence for microscopic chaos. *Nature* **394** 865-868 (1998).
[18] Grassberger, P. and Schreiber, T. Microscopic chaos from brownian motion? *Nature* **401** 875-876 (1999).
[19] Cecconi, F., Cencini, M., Falcioni, M. and Vulpiani, A. Brownian motion and diffusion: from stochastic processes to chaos and beyond. *Chaos* **15**(2) 26102 (2005).
[20] Dyson, F., The threefold way. Algebraic structure of symmetry groups and ensembles in quantum mechanics. *J. Math. Phys.* **3** 1199 (1962).
[21] Wigner, E.P. *Group theory and its application to quantum mechanics of atomic spectra.* (Academic Press, New York, 1959).
[22] Wigner, E.P. Remarks on the mind-body question. In: Good,I.J., editor. *The Scientist Speculates*, (Heinemann, London, 1961)
[23] Frauchiger, D. and Renner, R. Quantum theory cannot consistently describe the use of itself. *Nature Communications* **9** 3711 (2018).

[24] Proietti, M., Pickston, A., Graffitti, F., Barrow, P., Kundys, D., Branciard, C., Ringbauer, M. and Fedrizzi, A. Experimental test of local observer independence. *Science Advances* **5** 9832 (2019).

Torsion Geometry 5-Fold Symmetry, Anholonomic Phases, Klein Bottle Logophysics, Chaos, Resonance: Applications Towards a Novel Paradigm for the Neurosciences and Consciousness

Diego Lucio Rapoport
Department of Science and Technology
Universidad Nacional de Quilmes (ret.) Buenos Aires, Argentina.
E-mail: diego.rapoport@gmail.com

Abstract. We discuss the torsion geometries as the universal dynamical setting for the five-fold symmetry and its relation to nonorientable surfaces of selfreference embodying a supradual logophysics, rooted in the Möbius strip and Klein Bottle. We frame the discussion in terms of image-schemas in cognitive semantics and their disruption stemming from supraduality and nonorientability. We present the relation with anholonomic phases, chaos and the brain-mind as an integrated dynamical system. We discuss nonorientability as the characteristic transcendental metapattern of resonant connection, pattern formation and recognition. We present the torsion geometry and nonorientability in psychophysics and the neurosciences. We discuss the homology of the torsion geometry of physical space or spacetime with that of the unconscious modelization by the brain-mind of the kinematics of objects in physical space and in the perception of music, elaborating on the Principle of Complementarity of cognitive psychology due to Shepard. We elaborate on the phenomenological construal of experience, the world as experience. We discuss the supraduality of the liminal states of consciousness and the basis of awareness in memory. We introduce the cognitive psychology foundations of memory on nonorientability and its relations to the short-memory space and its quantumlike nature, and the hyperbolicity of the psychophysics of vision. We discuss the chaotic behaviour of dynamical systems as a manifestation of supraduality as their nonorientability. We propose a basis for the source of consciousness, the Hard Problem, given by the principles of selfreference and hetero-reference which generate the Klein Bottle supradual logophysics. We present the supradual logophysics of neuron cytoskeletal structures, its relation to torsion, resonances, topological and geometrical phases and the microtubule dynamics in terms of nonlinear buckling patterns and nonorientability, and still the torsion geometry of the irreversible thermodynamical processes supporting interactions-at-a-distance. We discuss the primal relation of torsion, nonorientability and memory, particularly arising in the buckling of microtubules. We discuss the primal morphogenesis of the cell as a tensegrity structure, torsion and the indiscernibility of elastodynamics, electromagnetic and gravitational wavefronts as morphogenetic fields. We discuss nonorientability as the metaform pattern of connection and resonance, particularly of interaction-at-a-distance. We apply it to a topological allosteric effect mediated by electromagnetic fields. We discuss the topological chemistry paradigm, particularly of organic chemistry where conformation superposition is crucial, its relation to anaesthetics and its application to the ORC OR theory of Penrose and Hameroff for consciousness as arising from the collapse of the wave function which we relate to the multiconformation in the topological chemistry paradigm.We propose a model of continuous signal processing in digital terms which allows an optimal reconstruction -from digital to continuous- in terms of the Nyquist-Shannon theorem whose constraints on frequencies and bandwidths is naturally interpreted in the terms of the 2:1 harmonics of nonorientability in the Möbius strip or Klein Bottle. We discuss the non-dual logic of the tubulin code and the resonator nature of the neuronal cytoskeleton. We propose that the topological phases appearing in the cytoskeleton may correspond to the topological anholonomity, namely, nonorientability, as was elicited empirically by the Bandyopadhyay group, the existence of anholonomity corresponding to the 360° rotation characteristic of the topological anholonomity, Möbius strip. We propose the Matrix Logic representation of the Klein Bottle logic as the basis for microtubule computations, and discuss the relation to quantum computation, topological entanglement and the topological coherence/decoherence Klein Bottle cycle. We also relate it to holography, the brain's integration and Velmans' Reflexive Monism extended by Rapoport. We discuss the multiconformation orientable and nonorientable resonators electronic devices and antennas as classical-quantum realisations of microtubule coherent behaviour. We discuss the relation of life and consciousness and the topological nonorientable embodiment of memory. We introduce the dynamical reduction program for the collapse of the quantum state in terms of the torsion stochastic differential geometry of the quantum geometry of Quantum Mechanics as projective space, and particularly the stochastic extension of the Schroedinger equation to account for the coupling of quantum system and environment, say a measurement apparatus.

Content from this work may be used under the terms of the Creative Commons Attribution 3.0 licence. Any further distribution of this work must maintain attribution to the author(s) and the title of the work, journal citation and DOI.
Published under licence by IOP Publishing Ltd

Thus we identify a realisation of the dynamical reduction program which is based on the torsion geometry, yet supports an agent-free collapse of the quantum state, as an objective albeit random process due to quantum fluctuations. We shall discuss the present supradual logophysics in several aspects such as 1) microtubule structure and dynamics, 2) the orientable and nonorientable conformations of organic molecules, 3) the relation with anaesthetics and the altered states of consciousness and 4) confront them with the tenets of Penrose & Hameroff ORC OR theory for the origin of consciousness, not only the topological chemistry but their choice of a superposition of null torsion as in General Relativity, which renders a trivial selfreferentiality, with the topological chemistry multiconformations which requires non-null torsion. The nontrivial selfreferentiality is both proper to the mind and the torsion geometry, be that on the physical, chemical, perceptual or cognitive domains. All in all, we propose that consciousness is neither based on panpsychism nor the collapse of quantum states, nor exclusively on the control of superposition of conformations of organic chemistry, but rather on the torsion-nonorientable geometry-topology based on the principles of selfreference and hetero-reference, operating in all domains: that of meaning,the mental-anatomical-physiological domain, structural processes of matter energy and in-formation, be that physical, chemical, biological, cognitive and perception. We present a remarkable connection between them. Our approach rather than multidisciplinary will be transdisciplinary, a possibility supported on supraduality - transcending the dual-logic based logophysics- from which the theory will follow in a rather smooth way.

1. Introduction

Given the universality of the Golden Mean it is quite remarkable that its association with the universal dynamical geometries of nonlinear vortical morphology, namely the torsion geometries, is unacknowledged. This is still quite remarkable given that one of the most basic example of torsion geometries is none other than the Lie groups of continuous symmetries which sustains much of theoretical physics, originally introduced by Sophus Lie to give a structural support for perception (Pribram [196]).These geometries are basic to physics, cosmology, biology, chemistry, perception, cognition, music, anatomy-physiology, phenomenology as the world as created/experienced – rather than conceived as given data. These geometries introduce morphogenesis as well as pattern formation and pattern recognition through their vortical nonlinearity. We remark, pattern formation operates on Nature at large, particularly the brain-mind, while pattern recognition is proper to the latter. The Klein Bottle being the metaform for both (Rapoport,[228,231-234,236], Carlsson [31,32]), it is natural to conceive vortical motions as the dynamical metaform of Nature associated to the Golden Mean and the Fibonacci series, as shown by John Bell Pettigrew in *Design in Nature,* published in 1907 [18]. Actually, D'Arcy Thompson Wentworth, considered the father of mathematical biology was mentored by Bell Pettigrew. Although he curated his collection and succeeded him at St Andrews, he failed to acknowledge the primality of vortices but partially and almost entirely he did so with Bell Pettigrew's contribution to his own work [13]. The fact that torsion geometry is tied to the pentagon allows to elicit another remarkable extensions of the 5-fold symmetry. Indeed, already in the Principle of Complementarity in cognitive psychology (Shepard [250-253]), torsion supports the association of cognition, torsion and nonorientable topology, as crucial to the ongoing mental formation of models of the kinematics of extended objects (Rapoport [236]). Nonorientability as in the surfaces of selfreference, Möbius strip, Klein Bottles, not only support this modeling, but are also basic to embody memorization, mental and physiological modeling of visual and somatosensory experience (Rapoport [222,226,228,229,231-236]). The defining characteristic one-sidedness of nonorientable surfaces supports the integration of Outside and Inside. Thus they transcend the duality inherent to Aristotelian-Boolean logic [230-236]. Another surface of selfreference, yet orientable, the 2d-torus, provides a banal selfreference and a topology for dualism (Rapoport [235]). Thus these nonorientable surfaces embody the disposition for cognitive capabilities by providing the metaforms –mother-form or form of forms- of a supradual logophysics which further supports a transdisciplinary unification of the material world of matter and energy, information, the psyche, as well as the imaginal domain and that of sign processes (Rapoport [236]) associated to cognition as meaning. Returning to torsion as the topological shape instrumental for embodying memorization, the Golden Mean appears coding brain harmonics as the quantum of thought associated with short-memory span (see Weiss &Weiss [280]) related to the quantization of subjective visual space, to be discussed below.

The first purpose of this article is to briefly present the pentagonal torsion geometries and its association to nonorientability and the anholonomic geometro-topological phases pervasive to physics, to set a more comprehensive bearing for the 5-fold symmetry. Remarkably, this realises Felix Klein's Erlangen program (see Sharpe & Chern [111]) and unseparably the gestaltic modeling construal of the world as experience, to which we shall also refer in this setting. Furthermore, it is a geometry of chance prophesied by Pascal, which considers Brownian motion as a geometry with continuous nondifferentiable paths with a stochastic character, Stochastic Differential Geometry (Elworthy [47], Ikeda & Watanabe, [89], Rapoport [201,206,209]), developed in theoretical physics (Rapoport [200-210, 216-221,227]) and fluid-dynamics (Rapoport [201-217]). In particular this was elaborated by Rapoport for the Schroedinger equation on space or spacetime as torsion geometries associated to Brownian motion, or the Schroedinger equation for open -i.e.non-isolated- systems, say subjected to measurements, evolving in the so-called quantum geometry of complex projective space [218,219]. Both the linear and nonlinear Schroedinger equations reveal the underlying torsion geometry and its relations to Brownian motion. Moreover, this is revealed as well in the linear and nonlinear Dirac-Hestenes for spinor-operator fields of Relativistic Quantum Mechanics, and its equivalence to the Maxwell equation, in the setting of Clifford bundles (Rapoport [207,208,217]). In Stochastic Differential Geometry the torsion field describes the average velocity while the Brownian fluctuations describing the fluctuations are related to a diffusion matrix related to a square-root form of a metric which can be positive definite [201,203,206]or non-positive definite as in Einstein's Relativity theory, or Minkowski space (Oron & Horwitz [180]). Here the Fokker-Planck operator of the diffusion process is the Laplacian operator of a Riemann-Cartan-Weyl covariant derivative, with trace-torsion (Rapoport [201]). The notion of a non-differentiable structure of spacetime at Planck scale as the geometry of quantum fluctuations and its relevance to the measurement problem of Quantum Mechanics has become the physical basis for developing schemes for the so called collapse of the wavefunction and its role as a form of "quantum gravity" related to the alleged emergence of consciousness as in (Penrose & Hameroff [74,75,192]). Remarkably enough, rather than torsion what is considered is the derivative curvature field, which at Planck scale would grow rather than a flat structure being the case (Percival [188]). However, independently of these considerations of a Stochastic Differential Geometry as the geometry of fluctuations, 't Hooft has shown [87] that the singularities of a "quantum foam" have for event horizon a Möbius strip arising from the antipodal identification of projective space, as described by Rapoport, which is the case of visual perception and the arousal of the sense of selfhood, selfconsciousness (Rapoport [236]).

With respect to philosophy in relation to which science has mostly developed as unrelateable forms of cognition and of probing the world and phenomena at large [230] the supradual Klein bottle logophysics and their interweaving by 'gluing' several Klein Bottles, namely HyperKlein Bottles as forms of multi-distinctions proper to complexity ([230,236]), realise Heidegger's claim (see Malpas [155]) that phenomenology is topologically grounded (see Rapoport[230,236]; Rosen [241]). It provides for a transdisciplinarian unification of science [223], knowledge systems at large, music (Mazzola [159], Tymozcko [274], Payeron [185]) and its cognition (Merrick,[320]).

As second -yet the major- topic of this article, we shall make the case of the transdisciplinarian unification of science in terms of this supradual logophysics, extending and applying the previous discussion to develop a novel paradigm for the neurosciences and consciousness, based on the principles of selfreference and hetero-reference proper to the foundations of torsion geometry and nonorientability, and to cognition.

2. Nonorientability and torsion geometries: the 5-fold symmetry

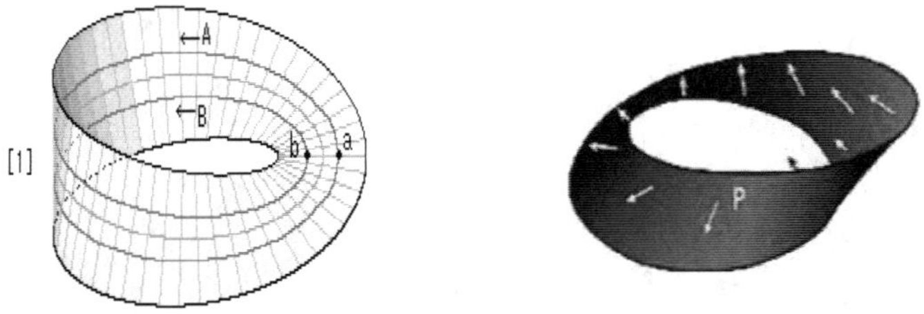

Figure 1 (Möbius strip; left figure from Rapoport, [228] CC):The green line is a centre line. Suppose the red line extends counterclockwise from point to turn back and arrives at point b which is the opposite side of point, so to return to b again another complete 360° turn is required. The same is the case of the red line which extends one more round, thus gets back to the starting point a. The red line never went across the centre line. Nevertheless the red line runs both sides of the centre line. CC

Another way of expressing this is what is *above* the centre line turns to become *below*, and the converse as well: Top and Bottom are continuously intertransformable. Top and Bottom are the elements of one of the most basic universal image-schemas -as conceived in cognitive linguistics after Johnson and Lakoff [96,139]. While dualism in an orientable surface assigns to this image-schema a fixed dual categorization, the nonorientabiity of the surface belies their dual character. Indeed, the red line looks like parallel lines when we see them partially. But it is not true. Route A and route B are actually a single route. This belies the top/down duality. Maurits Cornelis Escher represented this supraduality with his famous *Ascending and Descending* litograph. The green line looks a median or a central reservation on a road, a safe place from collision with whatever transits outside of it. However, it does not divide Möbius strip into two: thus the supraduality just expressed; or non-duality, would we stick to the habit of dualism as primal. Together with this, what we may describe as whatever is placed inside turns to be placed outside, and the converse. So another universal image-schema which is construed as a dual categorical divide is also belied. This is another manifestation of the supradual nature of Möbius strip. Philosophers Wittgenstein and Bachelard contested the dual Inside/Outside categorization., usually phrased as the CONTAIN image-schema (see Wittgenstein, IV 35, 4023 [282], Bachelard [9]). The notion developed in cognitive linguistics is that our thoughts, our conceptualization and theoretizations at large, are anchored in figures of speech denoting a spatial disposition, and these image-schemas -as they are called- are cross-culture universals. CONTAIN -and dualism, at large- is paradoxically both eminent and transparent in our construal of the representation of the world as already manifest in language, actually hegemonic based on the dual onto-epistemology, after Aristotelian-Boolean logic. In system theory, CONTAIN introduces a divide of form and content. In mathematics set theory this image-schema established its formalization which were further applied to attempt the axiomatical formalization of mathematics as a whole, the program faltered, proved unrealizable. It is problematic indeed, to consider any language as if exterior to its context -understood literally, a complex of physical, biological, cognitive, psychological, social, or economic conditions-, producing at times unresolvable paradoxes as Russell's invocation of an infinite set that contains itself. CONTAIN is indeed a very important image-schema, transcendental in establishing the dual onto-epistemology, i.e. based on Aristotelian logic. Its fundamental laws or principles are the principle of identity, the principle of non-contradiction and the principle of no third value, tertium non datur -but true and false, for its two possible values. We shall this discuss further with respect to Quantum Cognition, the biophysics of action potentials and to chaotic systems.

CONTAIN pertains either to the real space where we suffer on kicking a stone, the abstract space of the mathematician and physicist, and the imaginal space of a prophet, poet, shaman or dreamer. Whereas the latter two may experience as if real that Inside and Outside are connected, this is unlikely to the previous formalizing subjects. This is a privilege which on delivery to this world from the maternal womb every newborn experiences, with no due need of immediate reflection about the transformation. However, the fecunded ovum starting from a cleavage on its surface, turns outside-in, the blastopore invagination as a first step of development towards the maturation of a viviparous organism. As a stage of this inward flexing and folding development of a morphomechanics of tissue surfaces, the exterior surface of the ovum turns outside-in migrating to constitute the nervous system which the later outgoing brain finalises turns outwards to become the eyes, the progeny of a turning Outside-Inside-Outside development. Invertebrate *Hydra* makes of its -in principle – eternal being this perpetual transformation. Most of the matter in the Universe is produced by supernovae remnants of collapsing stars. Cassiopeia A for one such star turned Outside-Inside ([232]). CONTAIN image-schema and the cosmologies constructed on its terms appear banal in terms of such transformations. An ideally elastic 2-sphere might be turned inside-out with no discontinuities nor creases to produce another 2-sphere with the previous Inside and Outside interchanged, through mediating nonorientable surfaces such as the Klein Bottle or Morin's surface (Smale's Paradox [4,158,230]). In distinction with the ideally elastic 2sphere, the ovum requires a discontinuity [230,231]. We are to more surprises:

If we cut the Möbius strip along the green line, its length becomes doubled. It plays a crucial role in the process of duplication and transcription of DNA (Rapoport [233]).There is no opposite lane on Möbius strip. This is the topological origin of the 2:1 harmonic. It is also the origin of a protoform of Newton's Third Law, albeit it does not require, in distinction with Newton's formulation, a dualistic assumption, nor the instantaneous causality that this law implicitly assumes; causality, if any, is embodied by the 2:1 resonance produced by the 180º twist. Indeed, consider a normal (i.e. perpendicular) vector to Möbius strip; if we move it along any curve in a 360° turn as before, we would find it pointing in the opposite turn will return the vector to the samepoint further coinciding with its original configuration. Rather than having an action and a reaction as distinct, in the Möbius strip and Klein Bottle, the opposite and equal modulus of normal vectors is a resultant of the 2:1 harmonic produced by the nonorientable torsion, not an hypothesis for the foundations of physics at large. Thus it appears that this harmonic is more fundamental to physics than Newton's Third Law. Also, this is the case of the origin of the 4π symmetry of relativistic quantum mechanics (Rapoport [228]). The Möbius strip is a topological (Berry)-phase (an)holonomy. We shall return to this below.Would we cut and unfold the Möbius strip while keeping the length of the edge with notches, it turns into a trapezoid and the length and slope of the notches lines are distorted; see figure 4. This distortion is of great physical import. Indeed, while the fundamental rule of the sum of vectors, say forces, on Euclidean space is valid in the case of the parallelogram, in the case of Möbius strip unfolded i.e. the trapezoid, this is no longer the case. In fact, this property is shared by spaces with torsion, in which parallelograms don't close so that the sum of two vectors does not yield the diagonal which is ill-defined. Thus, instead in the non-closure of the parallelogram formed by the vectors u_P and v_P at P and their parallel transport, $u_R^{\parallel}$ at R and $v_Q^{\parallel}$ at Q, a fifth side T(u,v) closes the infinitesimal vectors. Prolonging T(u,v) to the line joining P and Q, we have the trapezoid whose completion is depicted in red, the inferior triangle at the right. Would we start with u_P and v_P of equal length then we can form a regular pentagon by scaling T(u,v) by a factor of the lengths of u_P (or v_P), $|v_P|$, and as $|v_P|T(u,v)/|T(u,v)|$ whose length coincides with theirs. (For the latter normalization a metric compatible with the parallel transport is required (Rapoport [201]).Thus the regular pentagon, i.e. all sides of equal length- is formed by considering torsion and defining parallel translation in terms of it, and further a metric as described. An example of this was introduced by the geometrization of dislocations of crystals by Kondo at the 1940s. In dislocation theory the torsion is called the Burgers vector, and as it will turn out, is a geometrical anholonomy.

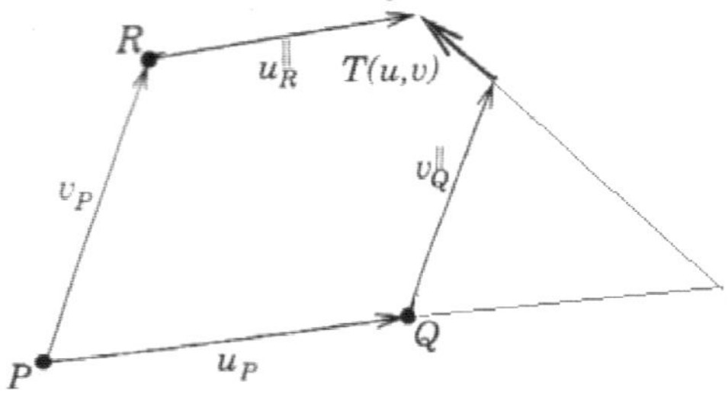

The infinitesimal parallelogram of vectors drawn in black does not close, the torsion being the fifth side closing the otherwise open parallelogram. We prolong it to draw the inferior rhs triangle in red and thus the trapezoid cnaturally introduced by torsion.

Figure 2 (From Rapoport [228] CC):The five-fold symmetry of torsion geometry. Notice that the torsion can be introduced with either chirality, clockwise or anticlockwise¸ which introduces a minus sign to distinguish them. As a torsion tensor or differential 2-form this makes of them noncommutative, specifically anticommutative: on transposing the covariant indices of the torsion tensor or as vector-valued differential form of degree 2, a minus sign is introduced . Noncommutation, say as in the product of matrices, is the crucial property which supports Quantum Mechanics, and is key to Quantum Cognition, the application of the mathematical formalism to model cognitive processes such as information decision process, concepts and conceptual reasoning, human judgment, and perception, cognition in general [2,29,37,38,100,101]. This arises from the contextuality of Quantum Mechanics as a theory of physics and of cognitive processes. The Klein Bottle is a body of contextuality, to be introduced below.

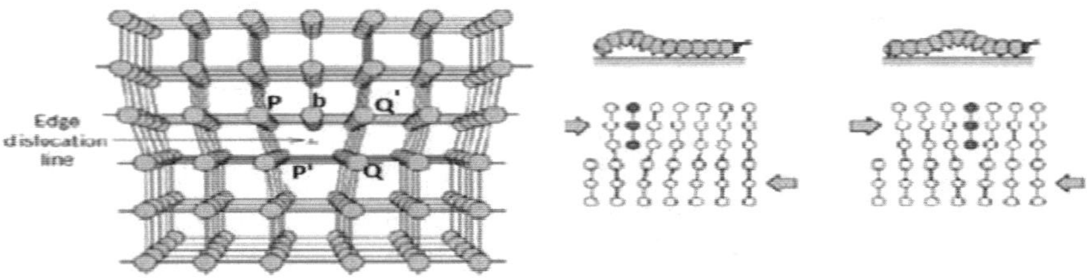

Figure 3 (From [228] CC): We introduce torsion by shear in a lattice; in the rhs by a caterpillar moving in the surface of a perfect crystal. In the lhs we have a perfect lattice, as is the case of a discrete space-time rendering of the homogeneous spacetime of General Relativity, but in the central area in which a dislocation is the case. We see then the transition from parallelograms that close (null torsion as in General Relativity), to non-closing and the formation of a fifth side, the torsion field. The figure above shows the meaning of torsion. In an otherwise perfect crystal (whose vertices are described in red), i.e. free of inhomogeneities an edge dislocation is produced. This is done either by removal (i.e., introduction of singularities) of atoms of the crystal, as the figure shows, or by introducing extra material; both are inhomogeneities. Torsion can be introduced by shear (i.e., the relative motion of two planes) as figure 3 shows, on the plane itself. (The other class of dislocations are the screw dislocations involving a shear on the direction transversal to the plane, which acting together with a shear dislocation we get the figure of DNA). Think of a caterpillar which moves a lattice a step at a time, and the shear produces the torsion of the crystal; this shear produces a vortical motion on the vertical plane to the shearing plane under the mixing of layers for small that it can be. Another analogy is that of a rug, which moving in the perfect background of the homogeneous crystal; local changes affect the whole structure; the analogy strikingly applies to the crease (the folded rug) formation in the gastropore invagination in Embryology (Rapoport, 2011c, 2014,2016a,b). It can also be produced by a hole in the surface, producing an embryological expansion wave. Thus, it is an action-dependent participative geometry introduced in terms of inhomogeneities by the subject, say the caterpillar, or more basically the photon.

This stands in stark contrast with the *homogeneous* situation of a Cartesian ideal geometry exterior to the subject, which corresponds in the continuum limit in which the atoms of the crystal approach indefinitively, i.e. the continuum hypothesis of Einstein, which due to the lack of a singularity, it corresponds to the zero torsion metric-based geometry of General Relativity (Cartan [33],Rapoport & Sternberg [200,204];Hehl et.al. [81-83], Vargas [276]). In short, to have loci, a geometrical reference at all and in the first place – quite literally! -self-referentially dislocations are needed: These are the inhomogeneities that put up a geometrical locus. The most basic dislocation is produced by a photon, a quantum particle, a singularity of an electromagnetic field, or still, a low dimensional submanifold of the wavefront eikonal propagation of light rays (Rapoport [220,221]); *no duality of wave and particle*. They play an important role not only as topological anholonomities to be discussed below but as a kind of square root of the null state (the "mind Apeiron", or vacuum state of the mind) in Matrix Logic as Penrose twistors [220,221,223] The parallelograms where inhomogeneities are present do not close,while in the perfect crystal they do close indicating a self-referential trivial loop: by default is indeed a loop, i.e. closed without mediations. This is also the case of the metric-derived spacetime geometry of General Relativity. Instead, in the former case a pentagon is produced. The fifth newborn side at the upper right side of the centre, joining Q' with b, is the torsion –the self-referential mediator; it is necessary and sufficient to the effect of completing the self-referential closure of the otherwise closed parallelogram; we shall explain this further below. As the figure 4 shows in the centre, the torsion appears producing a trapezoid with vertices P, Q, P' and Q', since the upper side is now longer than the lower one. Yet trapezoids are themselves Möbius strips cut transversally, say along the line on the Möbius strip below joining P (P') and Q (Q') below, and laid open on a plane, which by identifying the opposite lateral sides PP', QQ', on a previously 180° twist of the trapezoid, re-establishes thus the Möbius strip, as shown below:

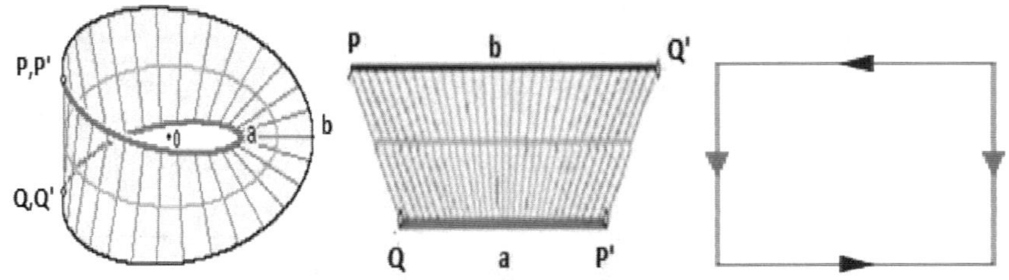

Figure 4 (From [228] CC): The Möbius strip, its unfoldment (as the completion of the dislocated parallelogram in fig. 2 above and the identifications producing the Klein Bottle, respectively.

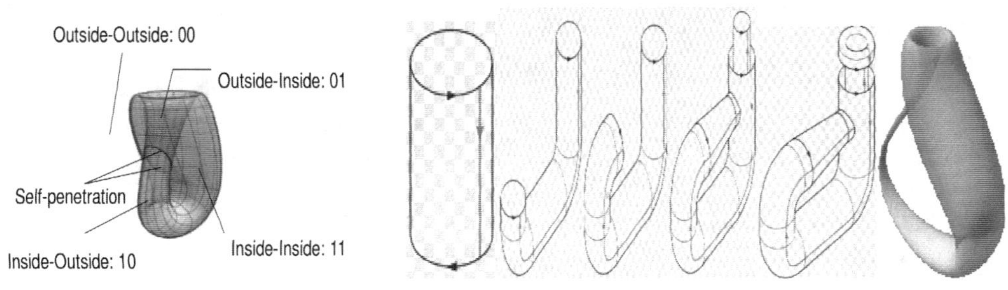

Figure 5. *Left* (From [224] CC) Natural logic of 4 states (the Klein Bottle Logic, KBL) which has two digits representation: Inside-Inside (represented as 11), Inside-Outside (represented by 10), Outside-Inside (represented by 01) and Outside-Outside (represented by 00). The mediation states arise from self-penetration absent in the mechanical-dual-membrane of biological cells and in the Cartesian conception of object-in-space-before-subject. They are paradoxical states associated to time waves

(Rapoport [224,234]), as introduced in terms of non-dual de Morgan logic by Kauffman ([112]). They express the imaginary logic values proposed by Spencer-Brown in his calculus of indications [303] as reentrant states of the primal distinction: torsion. These time waves are crucial to Cantor's diagonal construction for his argument of a continuum, which due to his unacknowledged choice of Boolean logic he was unaware of.They underly the paradoxical unending tail of the dyadic expansion of most real numbers (Hellerstein, [304], [226,231]); for an update on these issues see (Kauffman et.al. [131])

From the KBL appears the genetic code with its 64 elementary codons, and furthermore the genome and its association to the Dirac algebra of Quantum Mechanics. The latter is also based on the KBL reinterpretation of the recursive Universal Rewrite System (NURS) of Rowlands [243] . Since NURS topologically is based on the Klein Bottle sustaining a coupling of a fermion singularity with the rest of the universe as its dual element, this coupling evokes a boson interaction of the fermion with the universe for its dual. Indeed, in NURS bosons of spin 0 and spin 1 can be produced by fermion-antifermion combinations while boson condensates can be formed in terms of fermion-fermion combinations. In other words, the Klein Bottle with its 2:1 harmonics supports fermions and bosons. We shall return to this upon discussing the cytoskeletal neuron architecture and action potentials.

Right: (sequential construction of the Klein Bottle; rendered with Mathematica 8 using the parameterization provided by Robert Israel; uploaded by Wridgers. Creative Commons SA BY 3.0 File: Klein bottle translucent.png, Wikipedia)

Möbius strip although it surmounts the dualism of CONTAIN it is still embedded in Euclidean 3d space, as contained in it. The Klein Bottle is entirely different: rather than contained in what seems ambient space, it is selfcontained through the selfpenetration. Thus we see that dualism as supported by CONTAIN is inappropriate to theoretization vis-à-vis this selfpenetrating surface. However, in practical terms the Klein Bottle can be construed in ambient 3d space by using liquid crystals further solidified as exposed at the London Museum of Science, in the crystal figures by artist-extraordinaire Alan Bennett, HyperKlein Bottles.The latter introduce image-schemas which transcends hierarchy through heterarchies, functional organisations which disrupt the iterated nesting of CONTAIN as in the Babushkas Russian dolls [230,236].

Klein Bottle can be represented in 4d or 5d removing its selfpenetration, thus serving as the metaform for both pattern recognition and pattern formation (Carlsson), which is basic to the patterns of the brain-mind formed by eigenstates harmonics of the brain connectome (Atasoy et.al. [8], Rapoport & Perez [234], Rapoport [236])

Notably, the most basic example of a manifold which bears a torsion geometry is a Lie group of continuous symmetries! The so-called structural coefficients which describe the operations of infinitesimal symmetries, the Lie Algebra, stand for the components of the torsion tensor ([232]). It seems that conceptual astraying from the geometry of Lie symmetries which lay at the very basis of theoretical physics, in particular since the inception of gauge theories, has produced a cognitive dissonance that made of metrics, i.e., distances, as the basic structure of the geometrization of physics after Einstein, rather than symmetries. A Poincaré group gauge theory of gravitation with torsion and spin was somewhat exceptional (Goenner [67], Hehl et.al. [81-83]). So, given that symmetries provide the basic geometry of physics, as well as chemistry, which also is the case of biology, in addition of vortices as the dynamical form of torsion, we are led to examine what are the inherent symmetries of the Möbius strip and the Klein Bottle. While elementary dual image-schemas CONTAIN and ON - the latter as in the expression "on the table" pointing to above/below categorization- are surmounted by the Möbius strip and Klein Bottle, the left/right categorization in terms of specular symmetry is also compromised by considering chirality as in enantiomeric pairs. Handedness is no longer global. This is of crucial importance to chemistry and biology, as discovered by Pasteur, which we shall refer to (Rapoport [228]) for further graphical information.

3. The induction of selforganisation and nonorientability

The topological entanglement supported by nonorientabiity is crucial to the overall coherence of genomes, in which separate genomic configurations are interlinked in time and in space, establishes a coherence in time and in space as a form of non-Darwinian evolution linked to palindromic sections. This operates through resonant harmonics given by the proportions of codons to mirror-codons, which is almost perfectly equal to one, and to the proportion of the 32 most frequent codon-mirror codons pairs to the 32 less frequent pairs -discovered by Perez [113], which is 2 to 1; both being the signature of non-orientability (Rapoport [224,233]). The Golden Mean plays a crucial role in genomes and in harmonic principles at large, for the material world, cognition and its experience, the phenomenology of music [320]. Genomes are liquid crystals, omnipresent in biological systems, in which nonorientability is crucial to their coherence (Bouligand [24,25]).

The remarkable fact is that the topology of liquid crystals is amenable to control, and to the induction of topological changes. These can be produced by the expedience of introducing into them colloidal particles; thus topological changes arise due to the incompatibility of constraints, namely the anchoring conditions on the particle surfaces with the alignment imposed by the cell boundaries, or at *large distances* (Machon and Alexander [153,154]; Melle [164]; Musevic [171]). So, forms which are pervasive to organic chemistry such as knotted defects, Möbius strips and molecular motors, can be induced. This amounts to a general principle of nature for the topological generation of *complexity* as intricate structures in which nonorientable surfaces act as a kind of "glue", simply by inducing them through emplacing colloidal particles on cholesteric or in nematic liquid crystals, such as DNA (Chow [36]). The sole factor appears to be the nonorientable topology of the director field of the crystal, which underlies the formation of these intricate structures in terms of topological dislocations (generically, torsion geometries). It operates through the elastical adaptation of the director field to the extraneous colloidal particles, as in the previous references. But rather than the non-orientable topology being erased, more intricate robust structures appear to be the case, still carrying as their progeny, the nonorientable structure which make them possible. Thus, complexity –as intricate structures *progenies* of foldedness, though no genomes are involved- appears to be related to non-orientability and its elastic deformation to ensure its preservation under disturbances!

The addition of particles as inductors of orders can be interpreted in very general terms as the introduction to asymmetry to a system which otherwise already has an ordered structure producing constraints to those which may already be acting. The existence and relevance of constraints in any form, which usually appear as boundary conditions, are crucial to selforganisation and life [231,232].

4. Berry topological phases, (an)holonomities, torsion and nonorientability

The paradigm of theoretical physics as spacetime or state-space geometries (say, the projective Hilbert space of quantum mechanics), which in General Relativity was predicated in terms of homogeneity and geodesic motion, the latter as an extension of linear free fall motion, is clearly invalid vis-à-vis the omnipresence of vortical nonlinear motion in Nature. First of all, vortices are indissociable from singularities which stand as organising centres. Furthermore, the rotational motion establishes cyclicity which is selfreferential and yet hetero-referential since the cyclical motion also refers to the centre. So we have two principles which support vortical inhomogeneities: Namely selfreference and hetero-reference, to be discussed below. They are basic to cognition and awareness, and to perception in the first place as the pinwheel hypercolumn structure of the primary visual cortex. Thus, in the primal stage of processing the visual system has for functional topology the Klein Bottle (Swindale [266] Tanaka [268, 269]). Furthermore the primary visual cortex arises from selforganisation as a π-periodic structure (Miller [128]) which we related to the above hypercolumnar Klein Bottle functional architecture ([234,236]) and was independently interpreted as produced by a Berry quantum phase (Marcer & Rowlands [156]): The Klein Bottle as the topological anholonomity given by gluing two opposite chirality Möbius strips already mentioned [228]. Secondly, vortices cannot be identified as objects occupying space or space-time, say as in classical physics or General Relativity; vortices are geometries, dynamic as such. Furthermore they are eminently relational and processual, and they do not produce a space or spacetime as a container with absolute "Inside" and "Outside " categories. Instead these are connected dynamically, at times the differences, if any, are blurred or undefinable, uncertain: generically, vortices mix Insides and Outsides. Yet the geometry of inhomogeneity which embodies vortices is that of torsion, and ultimately at the Planck scale level, associated to quantum spin (Ross [244]), with spin-density for the macroscopic source of torsion in the extended theory of gravitation (Hehl et.al. [81-83]). Torsion stands as the fundamental geometry of physics, for which cyclical processes are the most fundamental (Dolce [45]). Particularly torsion geometry is the case of i) quantum physics (Rapoport, as already indicated), ii) the fundamental interactions as spacetime symmetries rather than internal symmetries (Rapoport &Tilli [202]), iii) fluid and magneto-hydrodynamics -the most pervasive form of matter in the cosmos, iv) biology (Rapoport, [224,225,229,230, 232-236,238]), v) the topological chemistry paradigm, vi) in the Klein Bottle form of the Mendeleev Periodic Table of Elements which is the unfoldment of the Golden spiral as a standing wave (Boeyens & Lavendis [22]), vii) perception and cognition [213,220,223], and viii) crucial to music and its experience [159,251,274]. Torsion geometry is the case of fluids as the paramount matter organisations of the cosmos at all scales, particularly liquid crystals. The latter have nonorientable topologies pervasive to biology [225, 231,232], genomes (Rapoport, [233]), cholesterics and nematics (Machon [153,154)], and coherent water domain crucial to cell metabolism (Ling [135], Rapoport [225]). They are also crucial to the topological chemistry paradigm (Mezey [118], Flapan [51], Bonchev 136, Sokolov [258]) , and to the topological entanglement of molecules having multiple conformations (Putz & Ori [197])

We have already introduced geometrical torsion as the natural geometry of nonorientability, yet we wish to examine this in a different guise: As a primal form of "panpsychism", more properly of pansemiosis -the world constituted by signs processes eliciting meaning-, which is proper to the supradual logophysics through the action of referral to. However, where the psyche, if any, though unnecessary in general, is related to selfreference and hetero-reference, as already discussed. Indeed, it appears as the completion of a space or spacetime dislocation which closes a fractured cycle taken around a singularity, thus blending selfreference (to produce the closure of the path around a singularity) and hetero-reference (as the alter of the singularity). Selfreference and hetero-reference are the two fundamental cognitive principles, however present as a vortical process of geometrical comparison which Nature itself produces. Alternatively it appears from cutting open non-transversally the Möbius strip to render either the trapezoid as the crystal relaxed description of the homogenized dislocated parallelogram or the pentagon, described in the figures 2 and 3 above (see Batterman [16], Figs. 3 & 4 where it is identified as an anholonomy). Thus we can identify torsion as the primal geometrical anholonomy to be discussed below.

The name anholonomy is a major source of confusion since some authors call it as holonomy. In theories of gravitation torsion is related to the so called anholonomic nonintegrable systems of reference, and coherently with this in the realm of theoretical physics at large,it has been called anholonomy. The term has been derived from classical mechanics, The terms ''holonomy'' and ''anholonomy'' derive from the classical mechanics of systems evolving under certain constraints. If the constraint is integrable, i.e. the evolution of the system does not depend on the particular path, and leads to a reduction in the number of degrees of freedom, it is called ''holonomic.'' Nonintegrable constraints, i.e. path dependent motions, are called''anholonomic'' or ''nonholonomic" an in gravitation or quantum physics. Geometers apparently do not respect this distinction calling anholonomies ''holonomies.'' Berry took this reversal of usage to be ''a barbarism" [288,289]. The name anholonomy as called by the physicists which have mostly remained. They are also called alternatively quantum phases, geometric phases or Berry phases. He discovered that in the cyclic trajectories in the quantum state-geometry given by projective Hilbert space adding to it a circular U(1)-phase symmetry , antipodal quantum states carry a phase. This phase is elicited upon completion of the cylic path in which an initial state is transported by a rule of parallel transport, the torsion appearing as the end result. Due to its its rendering of an understanding of seemingly disparate phenomenon in terms of a unifying concept, the discovery of Berry's phase is considered fundamental to physics (Lyre [152]). More of this below.

A cursory definition of anholonomy is as follows. We transport a system on a selfreferential– i.e., closed –path in the abstract space of states of a system to further discover the fact that the state variables return to their initial values ,and that there is no local rate of change for quantity S. However, it appears there is a global change in S's value at the end of the closed path in the complete system of states. Thus, in manifest physical appearance there is a difference with the initial state on returning to it through a closed loop in state-space. This final state is called the anholonomity. Two observations are in order. Firstly is that the notion of transport is more precisely of parallel transport acting on what is being transported, say a state of a system or a vector in spacetime say a crystal. It is a relational construct.The path upon which the parallel transport can be differentiable or continuous non-differentiable. The latter is the case of Stochastic Differential Geometry, the torsion geometry of Brownian motion already introduced, elaborating on Edward Nelson's conception of the Schroedinger equation arising from random fluctuations of the vacuum and Newton's second law with acceleration derived from the *mean* velocity of the random motions [175], the torsion drift vectorfield [201,206,208-211]. The notion of parallel transport in the setting of torsion geometries was introduced by Elie Cartan. It is called a Cartan or affine connection [33]. Its centrality is apparent in being the core instrument in theoretical physics in the modern extension of Klein's Erlangen Program (Sharpe & Chern [111]). It further extends to the psyche through the Principle of Complementarity of Shepard ([250-253]): Namely the unconscious modeling of the mind by introjection of the physical world in terms of representations which are further conditioned by embodied constraints [236]. Secondly a cyclic character about which the transport is carried out, as in the case of torsion and nonorientability. Thirdly, the ultimate cognitive operation is of establishing a difference and furthermore, second-order differences arising from the primal one (Johansen [294]). In other words, the primal perception *and* principle for creation, in the first place, is the perception of a distinction, a difference, its metaform the Klein Bottle Second-order differences allow for the eliciting of meaning and understanding, as proposed by cybernetist Gregory Bateson.

Let us identify primal forms of geometric-topological anholonomities in the very structure of the supradual Klein Bottle logophysics. Indeed, the primal topological anholonomy is embodied as the 2:1 harmonics of the Möbius strip and Klein Bottle decurring from their nonorientability. We recall Fig.1: When we take a vector normal to the surface at some point of a Möbius strip points *upwards* (action) and we further transport longitudinally this vector to the same point placed on the "other" side after a 360° percourse completing thus half a closed path, the normal vector now points *downwards* (reaction). Another way of understanding the anholonomity of the Möbius strip is as follows. Given a point on the vortex as centre of coordinates would we consider the radius vector from the centre to the point on the surface, then after a complete 360° rotation the radius returns to the same point but the normal vector does not, thus the anholonomity of 360° degrees of the Möbius strip. Another complete turn of 360° will complete the path to close it and yield the same upper-pointing vector. Möbius strip is the case of a topological anholonomity, and is founded on non-abstract-physical space terms, not a state-space, which nevertheless is the case. Actually, this state-space is given by Möbius strip and the normal vectors to it, and the parameter space is subsumed by two values, +1 and -1. They describe the upper-pointing and downward-pointing direction of normal vectors. Thus the nonorientability allows for a characterization either as a "real"-space or state-space anholonomity, somewhat of a Platonian archetype which very much manifests as real. The latter issue of the abstract or real character of anholonomities is a subject of discussion (Lyre [152]; Bliokh et.al. [21]).

As for the space or spacetime anholonomity we have already identified the torsion, say of a dislocated crystal, a viscous fluid or a quantum system, the latter being the velocity field and the logarithmic differential of the wave function, respectively. It provides for the selfreferential closure of the path around a singularity, which when mapped into a relaxed configuration elicits the torsion (Burgers vector) as the anholonomity (Batterman [16]).

In the quantum geometry of state-spaces, projective Hilbert space, the path transports a wave function which acquires a phase upon completion by closure. Yet this does not depend on the closed path itself in distinction of the configuration "real"-space. However, the quantum geometry does have a singularity which is the excluded null state. Furthermore, antipodal identification of quantum states produces nonorientability (Bliokh et.al. [21], Rapoport [228]). Both singularities are usually unacknowledged, whereas nonorientability has lately been found omnipresent. More of this below.

It has been lately acknowledged -and presently taken a widespread momentum- the relevance of the quantum anholonomy and its omnipresence in quantum physics, electromagnetism, acoustics, pressure waves, electronic devices, sound, gravity waves, fluids, crystals, etc. However, this has proceeded as the unacknowledgement of the geometrical anholonomity of ("real") space or spacetime torsion, rather than state-space, say in dislocated crystals as elementary examples of torsion geometry, or quantum fields associated to linear and non-linear Schroedinger and Dirac equations [207,208,217-219]. They reveal the unity of continuity and singularities as discontinuities, much sought for in the search of a theory of quantum gravity. (Yet seemingly impossible to achieve in terms of the metric geometries of General Relativity which assume a continuous spacetime and null torsion; these assumptions imply no selfreference nor primal singularities, by default, though solutions to the field equations show otherwise). Furthermore they have brought to the fore the fundamental relevance of "non-conventional" (viz. nonorientable) topologies. On the other hand the torsion geometries of fluid and magneto-hydrodynamics, quantum physics and thermodynamics which was revealed in the works by Rapoport have not been acknowledged. Furthermore, torsion has appeared in the qualitative studies of nonlinear systems, particularly in blow-ups, chaos and bifurcations (Yong Wu and Yi Lin [286];Rapoport [228,232,233]). For example, electromagnetic fields with Möbius strip polarizations were first discovered in optics (Freund [54], Bauer et.al. [17], Andrews [7], Bliokh et.al. 21), to later extend this discovery to tidal ocean waves, sound, fluids and particularly turbulence, gases, water-surfaces, gravitational waves, which "show that polarization singularities and Möbius strips are also ubiquitous for them. In contrast to well-studied electromagnetic polarizations associated with the motion of abstract field vectors, acoustic wave polarizations correspond to real-space trajectories of the medium particles" (Bliokh et.al. [21]). The robust quality of these topological singularities is relevant.Thus the abstract state-spaces of quantum mechanics in terms of which quantum anholonomies were first discovered, were anticipated by spacetime ("real-space") effects such as the Aharonov-Bohm effect associated to the primality of electromagnetic potentials, actually torsion potentials as appearing in the trace-torsion differential 1-form of Cartan-Weyl type (Rapoport [202,203-210,212,217-220]) and in the Schroedinger equation for open systems upon measurement of quantum geometry state-space (Rapoport [218]). They were complemented by a plethora of space or spacetime quantum anholonomies associated to such elementary phenomena such as geometrical-optics eikonal propagation of light, sound waves (Deymier & Runge [43]*)*, fluid and pressure waves, phonon vibrations on lattices, metamaterials, etc. The quantum anholonomy has shown its omnipresence (Bliokh et.al. [2]1) to which we can add the universal phenomena of resonance of harmonic systems which has nonorientabilty and quantum and classical anholonomies for their very basis (Dembowski et.al. [122], Deymier & Runge [43], Xu et.al. [287]). It has been empirically verified that resonance arises from tuning harmonic systems through the mediation of nonorientable states (Shelton [249], Xu et.al. [287]).

In *Steps to an ecology of mind,* Bateson's posthumous work on cybernetics, general systems theory and the foundations of biosemiotics, he envisaged the existence of a universal transcendental metaform, a "pattern which connects" (Brier [27]). We identified nonorientability as the characteristic property of the connection pattern (Rapoport,[236]). As already mentioned, the Klein Bottle is the metaform of both pattern formation and pattern recognition in Nature, and in the brain-mind connectome as an harmonic system (Atasoy et.al. [8], Rapoport & Perez [234])

5. Anholonomies, nonorientability and chaos in the brain-mind dynamics

Nonorientability plays a crucial role in the theory of dynamical systems; we already mentioned the blow-ups of nonlinear systems which lead to chaotic attractors. Due to the complexity of nonlinearity which manifests in the asymptotic morphologies given by their attractors, particularly those that arise as periodic orbits that makeup the unstable manifolds, this is perforce a qualitative theory. Topological invariants are basic to characterize the complexity of the invariant manifolds of these attractors. The reason is simple: stretching, folding, twisting, tearing, squeezing are the topological transformations which nonlinear processes undergo in their phase spaces. However, the characterization of the topological properties are much involved. One such topological invariant is an anholonomy, the so called local torsion (Gilmore & Lefranc [61]). Indeed, consider a normal vector to a periodic orbit of the solution flow of a dynamical system. As the flow evolves the base of the vector moves along the orbit, the tip rotates along the direction of the flow. We are already familiar with this, the Möbius strip. However, in this dynamical system flow, when the initial state returns to its starting point, the vector rotates through an angle, Θ,precisely Θ/Π, the local torsion around the periodic orbit. The value of the local torsion is generally not an integer. The latter are relevant to topology, nonorientability and complexity. Say, if the local torsion around a closed period-n orbit is an even integer, then it is possible for two orbits of period n to entwine the original period-n orbit. Otherwise, if the local torsion is an odd integer, the original period-n orbit can be entwined by an orbit that closes after two cycles, i.e. a period-2n orbit. This notion of anholonomity can be extended to nonlinear dissipative systems undergoing Hopf bifurcations with a continuous symmetry (Ning & Haken [176]), which characteristically produce nonorientable unstable periodic attractors.

Would we consider dynamical systems described by a vectorfield on 3d-Euclidean space, i.e. a field of vectors in 3d space which evolve in time as a parameter, namely describable by a system of ordinary differential equations. Since detailed description is not the issue but rather qualitative models which are useful to capture the essential dynamics in the simplest low-dimensional model. Classical examples are the Van der Pol equation, which models a simple electronic circuit exhibiting oscillations, and the Lorenz system capturing the essence of complexity in weather models. Very often, as is also the case with these examples, the resulting model is three-dimensional due to 3d-vectorfields being the lowest-dimensional continuous time models that can exhibit extremely complicated behaviour, including chaos. To understand the dynamics of such a vectorfield, we search for attractors, their basins of attraction and in particular, the boundaries of these attracting domains. Two-dimensional invariant manifolds often act as basin boundaries, and it is important to know for understanding the qualitative aspects which are the very core of the theory of dynamical systems where they are located and their shapes. They are called separating manifolds, as they separate the state-space into two invariant regions. Indeed, trajectories of the solution flow of the dynamical system falling on one side of the manifold are trapped into to this region since they cannot move through an invariant manifold. However, not every 2d-invariant manifold is a separating manifold. Indeed, a separating manifold must have two sides, namely an inside and an outside and it is the case that the nonorientability of Möbius strip and Klein Bottle unite them.Thus invariant manifolds may be nonorientable and thus nonseparating, and this bears the signature of a supradual logophysics. The particular feature of Euclidean 3d-space is that any nonorientable manifold embedded in it is topologically equivalent to a Möbius strip. Already in 4d this is no longer the case. In (Mindlin &Solari [166]) 2d surfaces contained in the 3d invariant manifolds of periodic orbits were found, clear signatures of both orientable and nonorientable surfaces, invariant tori and Klein bottles in a 4d flow, respectively. Thus the torus and the Möbius strip appear conspicuously while for 4d nonlinear dynamical systems is the twisted torus and the Klein Bottle (Charó et.al. [34,35]). The complex topology of attractors was considered initially as "fragile anomalies", but this was meant in relation to Cantor dust structures, rather than nonorientability, which was much disregarded.

This is of importance to the brain-mind would we consider it as a metastable chaotic system, having electroencephalograms as the wave patterns for the stable functioning, which are organized as a series of 2:1 Möbius strip harmonic series (Rapoport [236)]. The chaotic attractors which represent the 2:1 arms of the Hopf bifurcation say of the logistic map, elicit Möbius strips and Klein Bottles which are pervasive to chaotic attractors, the unstable periodic orbits appearing in the models by Lorenz, Rössler and others (Crawford & Omohundro [40], Meiske and Schneider [163]). This transition between chaotic attractors of the nonlinear dynamics of neural circuits are conceived as the functional support for the multistable switching between different perceptions or behaviours, say as in the Necker cube or Gestalt figures, supporting the transition of the brain dynamics between different oscillatory states. This can be conceived as the core of the dynamical systems approach to the neurosciences. Thus the nonlinear metastable dynamics of the brain-mind produced by the cross-influence between the electroencephalogram's octave rhythmic progression, manifests as a sequential transition between non-orientable attractors, providing the basis for the self-organized control of a "communication-through-coherence" (Battaglia et.al. [15]), resonance. Indeed, these different attractors would correspond to different dynamical states of a fixed structural network. However, the metastable behaviour from which the sequential chaotic attractors appear is conceived as a single 'complex' attractor, where the complexity refers to the shape of the manifold drawn by the trajectory, which can be characterized as a set of connected sub-manifolds, each capable of sequestering the trajectory for a limited period of time (Rapoport [236]). They are connected to, or embedded in a larger surface in a way such that the trajectory will find this connection and a new transient will emerge as the trajectory moves off to another submanifold. The previous considerations can be extended with René Thom's Catastrophe theory. Following the Aristotelian revindication of morphology this theory claims the universality of some forms, to account for the sudden abrupt transitions of nonlinear systems between regimes as universal, following small changes in additional parameters. However, examined in a larger parameter space, Catastrophe Theory reveals that such bifurcation points tend to occur as part of well defined qualitative geometrical-topological structures. Despite the number of discontinuities in Nature appear infinite in variety, Thom showed that the graphs of these processes could be categorized into a few basic shapes. The jump phenomena occur as the state of the system approaches local phase-space manifolds (attractors) with different (non)orientability properties. In these cases the so-called invariant manifolds describe the asymptotic- in- time geometry of the periodic oscillations, known as Instantaneous Centre Manifolds. The instantaneity is related to the discontinuity in sudden regime change emerging as the change of topology, in terms of orientability to nonorientability and the converse. They show the archetypical form of infinity or 8, the lemniscal projection of the Klein Bottle, of the unstable periodic orbits of chaotic attractors, say of Rössler and Lorenz. Current discoveries indicate the relevance of knots (Kaufmann [126]) to the topologies of chaotic attractors (Charó et.al. [34,35]) suggested in (Gilmore & Letelier [61]), following a theorem by Birman and Williams.

6. On the logophysical topological nature of the brain-mind: emotions, Markov blankets and free energy principle

Nonorientability does not only appear in the chaotic attractors of the dynamical theory mathematical treatment of neurons. Rapoport identified it in the vision of a subject which operates through projectivity, as initiated by the Renaissance Masters which is crucial to the constitution of selfhood: the *topology of consciousness* (Rapoport [236]). The young Sigmund Freud was a neurophysiologist in the making, whose reflections on the psyche have become integral to the current developments of Biological Psychology (Panksepp [182] ,Solms 259-262), and the neurosciences at large. In his *Project for a Scientific Psychology,* Freud associated the brain's resting state as constituting a spatiotemporal structure supported by the brain's neuronal structure for what he conceived as a "psychological structure" of the ego for which he suggested a topological consideration (Northoff [177]). Freud referred to this spatiotemporal anchoring of the psyche as "It must lie on the borderline *between* [our emphasis] inside and outside; it must be turn outwards toward the external world and must envelop the other psychical systems" (quoted by Solms and Panksepp, [126]). It is as if Freud unbeknownst on the Klein Bottle had indicated its role as the integrated agency, psyche. Later on Jacques Lacan temptatively pursued this further, alternatively choosing the 2-torus, which supports a banal form of selfreference [239], and the Klein Bottle, yet with deep intuition indicating the importance of topological knots to the psyche (Miller [165], Hewitson [85], Ragland [199]). This anticipated crucial notions of the psyche's operations which appear to indicate the Klein Bottle as the metaform for the spatialization of the psyche elaborated by Rapoport. Phenomenological theoretizations calling for anchoring phenomenology on topology, such as the works by Heidegger further elaborated by Malpas, with Stern, Rosen and Rapoport's identification of nonorientability for grounding phenomenology, for the latter not restricted to physical space. Striving to find a quantum theory for the brain-mind, particularly in terms of quantum field theory of open systems (Vitiello [278]), Globus referred to observational practices which appear to indicate the spatiotemporal anchoring of the body-mind already displayed by the novelist Marcel Proust. Globus' proposal for the scientific study of consciousness [66] was to extend the third-person account of the subject-observer by the praxis of first-order account. The latter was also proposed in the Radical Neurophenomenology and biogenetic structuralism initiated by Laughlin et.al. [141]. It was later pursued by Varela [276], Petitmengin [189]. Globus [66] pointed out to the ultimate spatial-temporal referral of consciousness somewhat treading on Heidegger's conception for anchoring his phenomenology (Malpas [155]) . The third-person account is the signature of objectivity practiced by science curtailed by the lack of a supradual principle, requiring nevertheless to be completed by a first-order account proper of the selfreference of consciousness. The notion pledged by Globus is the "in-betweeness" of phenomenology -just like Freud-, the world as experience, which he stops short of linking with the Klein Bottle (personal communication to the author) yet linked to the "tilde" and "non-tilde" states of Vitiello [278], Jibu and Yasue [123]; see also [120]. An important comment: as a field anthropologist Laughlin remarked the intervention of rhythmic and resonance elements inducing altered states of consciousness arising in rituals [142], and stressed their crucial character to consciousness rather singularly. Nowadays has resurfaced in the neurosciences the appropriateness of claiming a musical nature to the living brain (Bandyopadhyay [14]), which we suggested to be based on the Klein

Bottle metaform of pattern formation and pattern recognition at the brain connectome in terms of harmonics [234] which may appear as cymatic patterns [8].

The experiences associated with the cognition of music have a signifying richness and ample spectra perhaps unique and nonreproducible by any other sensorial stimulus but sound, besides its abstractness perhaps comparable to the seemingly most abstract mathematics, say topoi and category theory (Mazzola [159]). Returning to the anthropologist Laughlin's interest for music and its embodiment induced by brain entrainment through repetition and mimicry of animal motion through ritualisytic performance, later validated by mirror neuron research, it has been ventured that it is music which presents those consciousness states: They are brought to the mind by music and this is so in all states of consciousness, mostly in states of unawareness (Addis [1]). In other words, Addis claims that music is the rather unique agency which fully activates through the body-brain the spectra of consciousness states and particularly emotions, which are conceived both in the neurosciences and Biological Psychology after Panksepp [182], as the most elementary qualia. The latter conclusion was reached on the studies of the anatomical and physiological evidence that led to conclude that the primary role of consciousness is to scan the proper body states rather than the states of the seemingly independent Exterior world (Solms & Friston [308]). In other words, the primary role of consciousness is interoceptive, directed to the mind's embodiment. The most elementary qualia require no cognitive nor perceptual representations, they are the so-called called affections: hunger, lust, surprise or curiosity, particularly aroused by sound. This is a radical departure of consciousness as primed on abstractions.
The experience of music is both a first-person and third-person issue, in the latter case providing for a universal form of intersubjectivity, yet one which is anchored subjectively, in the sense of interoceptiveness.

The first-person account is very much anchored through emotions, to which we shall return briefly. Remarkably this notion of the embodied mind integrating interoprioception with exteroception is crucial to the body-mind homeostasis-directed cybernetics. Freud's conceptualization of the psyche has influenced the cognitive neuroscience, for which the primal form of consciousness is grounded on emotics operating through the limbic system (Damasio [41,42], Solms [261,262]). More recently Craig has identified the middle insula as crucial to homeostatic sentience, and interoception and exteroception integration, leaning on the motor aspect -rather than feelings- in the motor-perception cycle which Turvey identified as a Möbius strip [273] and whose somatotopographic representation is supported by the Klein Bottle [119,281]. See Craig [316.]

The spatialization structure in the neurosciences are referred to the mainly unconscious modeling activity of the psyche in terms of a Markov blanket, a 4-state logophysics proposed by Friston [55] which was identified as the Klein Bottle by Rapoport ([234,236]), further identifying the somatosensory and visual systems Klein Bottle topographic representations of the body periphery through touch and vision, respectively, as key Markov blankets integrations.

These considerations have arised through the search for the minimal conditions to ascribe feelings to an entity, to be alive. A fundamental property of biological self-organising systems is their tendency to resist the second law of thermodynamics. Friston characterized this functional property as emerging in the conditions of ergodic random dynamical system that naturally arises within any such system that possesses a Markov blanket. They are characterized by systems consisting in two sets: ('sensory' and 'active' states. They further influence each other in a selfreferential cycle fashion: [active] external states [Outside–Outside; Klein Bottle-wise, as in the other states of the blanket] cause sensory states which influence—but are not influenced by—internal states [Outside-Inside], while internal states [Inside–Inside] cause active states which influence—but are not influenced by—external [Inside–Outside] states" (Rapoport [236])

Together with a free energy principle the Markov blankets are the core of the developments in neurosciences conceiving the brain-mind as operating through a Bayesian inferential process Actually Solms identified any living biological system enclosed by a boundary as a Markov blanket ([262]; see also [317]). Such is the case of the living cell organised with its membrane which is not amenable to identification as a dual-logic gate (Rapoport [225], Ling [135]).The cell's cytoplasm appear to operate through the "like likes like" principle, introduced by Richard Feynman, which is crucial to the cell's metabolism (Pollack [193])], Ling [135]). and to the formation of coherent water domains. Of course,

Feynman's principle expresses non-duality. This substantiates Spencer-_Brown's conceptualization of worlds arising from a reentering system defined by a primal distinction/boundary [303], a reentering which cannot be supported by a 2d surface other than the selfpenetrating Klein Bottle (Rapoport[231,232]). Boundaries rather than operating in a dualistic setting as merely defining the system by the CONTAIN image-schema, as that what remains Inside -and thus a somewhat passive object- and what is left Outside, are constitutive to the systems bounded by them rather than mere tokens of separation (Celliers [311]).

The nonorientability decurring from antipodal identification of projectivity in the case of the visual field, is crucial to the topology of consciousness, as we already referred to. In Algebraic Topology the notion of blowup is introduced as a geometric transformation which replaces a subspace of a given space with all the directions pointing out of that subspace. For instance, the replacement of a point by the projectivized tangent space at that point. Remarkably we can generate the Möbius strip as the blow-up of the origin of R^2 replacing the plane with the projectivized tangent space. This is related to the pinhole singularities of the visual cortex (Petitot [190] and its periodic architecture with its Klein Bottle topology, and to the purported origin of the Universe through a Big Bang [236]. However, the underlying conception proper to this is the so called resolution of singularities in Algebraic Topology, which conceives it as zooming on a point, rather than an explosion. There is nothing that is that has not already been, as another form, perhaps. No ex-nihilo. As already observed the event-horizon of the Planck-scale singularities of the vacuum elicit non-orientability as the topology of the infinitesimal boundary of the experientable. In short, the speculation of a subquantum void elicits a frontier to cognition, which has the signature of nonorientability.

Whatever, this reflects the notion of spacetime as a product of the mind, conceived as a "geometric agent", which Piaget identified at the early stage of the child's cognitive development. Kant: "Space is a necessary a priori intuition, that lies at the root of every awareness. One cannot imagine its negation, although one can imagine it to be empty. It is the necessary condition for the possibility of perception, and not a consequence of perception, it is an intuition a priori, the necessary basis for external perceptions". (p. 38) (quoted from Northoff [177]) While space may be an existential precondition to perception, the denial of the space-construal nature by the embodied mind has been disproved.

7. The Principle of Complementarity in cognitive psychology, torsion and nonorientability in the homology of Inside and Outside representations

Shepard's Principle of Complementarity conceives visual experience as arising from a priori embodied structures, constraints operating on the construal of experience. Constraints arise from perception as framed by the visual system, an ecological psychology. The crucial constraints arise from the perceived locally Euclidean 3-dimensionality of space and of the objects placed as if contained in it. This appears in principle as the usual conception of Cartesian space as a container, yet framed by perception. However, objects appear through two-dimensional surfaces of their boundary, both of them as characteristics of the perceived environment and our actions on them. So rather than staying on the 3-dimensionality of objects, perception arise from their boundaries, and these are 2-dimensional. We can conceive of this in terms of the complementarity of form, related to symmetry, with function, related to asymmetry, first suggested by Pierre Curie. Biologist cum crystallologist Lima de Faria upholded this principle for the very foundation for biology as selforganisation, and he argued that it may carry from the level of elementary particles to higher forms [148].

This complementarity is already a Klein Bottle embodiment, as the right hand side rectangular figure (see fig.3) shows its formation: two equally directed sides -the symmetry aspect- produce the cylindrical shape, and the identification of the other directed sides -the asymmetrical aspect- produces in turn the torsion towards the selfpenetration. Steven Rosen has called this a synsymmetry (Rosen [241]). Without the asymmetrical aspect of synsymmetry Nature would be timeless, inert, functionless, and without the symmetrical aspect the morphological regularity would not stand to shape it, as if inert, yet in pregnant openness to change.

According to this, phenomenological space is 2-dimensional rather than 3-dimensional, alike in the holography principle of physics [313], but embodied , for fundamental difference. Remarkably, the visual receptive fields of the cortex are mathematically modeled as Gabor functions, of holography theory. The built-in symmetry of these functions, as appear in their phases, is the Klein Bottle, 2-dimensional. In other words: Phenomenology is related to 2-dimensional construal of experience, in which holography is sustained by Gabor functions of receptive fields, yet to be able to perform pattern recognition the selfpenetration of the Klein Bottle needs to be removed, and this is carried out in 4d or 5d (Rapoport [228]; Rapoport & Perez 2018 [234]).

At the basis of this mental construal which is further related to the physiologics of visual perception, we retrieve torsion and nonorientability as the geometry-topology of the unconscious ongoing modelization of the living brain of the kinematics of objects (Shepard [250,252,253], Carlton and Shepard [49]). Actually, Shepard departed from the experience of music: the Tritone Paradox motion inside an octave which moves as in Escher's *Ascending and Descending* pivoting on the tritone halfway [251], a Möbius strip [228])). The former was further pursued by Risset in the paradox of rhythm perception (Risset [237],Vernooij et.al. [277])). Shepard further discovered that this geometro-topological principle extended to the unconscious modelization by the brain-mind of the kinematics of objects offered to the subject.

Remarkably, there is an homology of the geometry of dynamics of space or spacetime and the geometry of phenomenology, the world as experienced, particularly in the mathematical modelization of the living brain-mind. Phenomenology points out to symmetries, nonorientability through the Cartan torsion geometry that underlies the Erlangen Program as of XXth century (Sharpe and Chern [111]), and the very organisation of the visual cortex with its π-periodic hypercolumnar structure underlying the Klein Bottle and operating as the visual receptive fields [234]). This is further supported by the human heart which rather than a pump it operates as a Möbius strip torsional motion, as elicited by Francisco Torrent Guasp; see Kocika et.al. [104]). Or still further, the body architecture of fascia -the connective tissue system [129] made of liquid crystals which extends from the cellular cytoskeleton to connect the whole body and covers all tissues, muscle and bones, as lemniscal one-dimensional projections of either the Klein Bottle or Möbius strip. We already mentioned the somatosensory topographic map of touch with its Klein Bottle topology, which models on the primary somatosensory cortex the skin periphery supporting thus the unity of action and perception, an homology which comprises the lemniscal trace of the gait of mammals, flying and liquid inhabiting species. The homology extends to the distortions of the perception of space and time, through subjective Lorentz transformations, which show that "the subjective psychophysical laws of motion perception are relative to the observer's reference frame and that the Lorentz transformations govern the relationships between time, distance, and velocities of perceived events" (Caelli et.al. [114]). The visual system psychophysics does not comply with the Galilean transformations but appears to repose on hyperbolicity. Remarkably, there is a connection between hyperbolicity and nonorientability as shown in (Rapoport [228])).

Furthermore, there is a physiological connection between the subjective space-time and quantum physics structures through what is conceived as the visual component of short-time memory, VSTM. The capacity of VSTM is set both by information load and by number of objects: A racing driver or a fighter pilot may reach a limiting velocity of processing of information in which there is a perceived time dilation and space contraction which upon the subject's attention fixation on the phenomenon it may turn to be the final instant before crushing into an obstacle.

The point is that short-time memory is quantized as discovered by Juan Pascual-Leone [184], a former disciple of Jean Piaget. He pioneered cognitive psychology, and in studying child development claimed that topology is a primer on framing the cognition of children. According to Piaget, at the early states of cognitive developments the child puts up elementary spatial representations which *"are none other than those relations which the geometricians tell us are of a primitive character, forming that part of geometry called Topology, foreign to notions of rigid shapes, distances, and angles, or to mensuration and projective relations"* (see Piaget & Inhelder [319]). This does not disappear with the child's development. Indeed, spatial representations that form the elements of perception are not "based on a fixed homogeneous space in the sense of geometry these representations have a dynamic situation-dependent nature involving topological, projective and metrical aspects involving a complex structure of parameters and interactions with other primitives" (Mausfeld, p.86 [115]) which participate with is conceived as an innate system of organisation which support/produce appearances. This system is attributed to the brain as a "geometrical engine" (Koenderik [105]). The fact that they are mostly transparent to the final outcome operates a dimensional reduction, a process of integration which rather than working through discrimination reduces the information to a small number of parameters referred to as dimensions (Pang et.al. [183]). They early appeared in the pioneering visual research of Hubel and Wiesel who primed object orientation as experienced in the primal visual cortex where visual information is represented: topologically, a Klein Bottle. In the dynamical systems approach to the neurosciences the dimensional reduction is referred to the low dimensional attractors, which we discussed in a previous section. This also suggests that together with the spikes, neuron action potentials attribution of a dual logic, there is a coordination which in that paradigm is related to mitigation of the discrete character attributed to them and the necessity of reducing noise. We shall return to this below.

However, psychologists have discovered that attention has " *a quantitative dimension in terms of the maximum number of items to which a person can attend at one time"* (Weiss & Weiss [280]). The limits were indeed the case, yet rather than fixed and uncontextual,they have a plastic character subjectable to learning and other factors. In distinction with symmetries as the logophysical constraints in the unconscious modeling of object kinematics, in the case of attention it is the range of the number of objects or mental states which an observer can attend to, which also led to the so-called magic rule of 7 plus minus 2. Pascual Leone discovered that this space of objects subjected to attention appears to follow up an Einstein-Bose statistics, it is a bosonic-like space of objects vis-à-vis attention [280]. From this followed the proposal by Weiss & Weiss "... *that memory span has to be understood as the quantum of action of thought. In fact, this quanta of action represent macroscopic ordered states in the sense of quantum mechanics*". VSTM representations are stored by means of sustained firing of action potentials, which corresponds to a supradual logophysics, to be discussed below. So the visual space on which subjective relativity operates through VSTM is quantized: Unlike the theory of spacetime of objects as contained in physical "real" space, where in principle the homogeneity and continuity is posited, the visual space of experience as entertained by attention, does not need to search for a physical quantum basis: The singularity is the viewing subject, very much alike in the very definition of torsion when we supersede the primality of physical space as if exterior to our nature. *The subject as the primal distinction,* as offered by Spencer-Brown [303], and identified as a torsion geometry by Rapoport [231].

8. The hard problem of consciousness and the torsion of cognition

We have presented the relation of the five-fold structure to dynamical torsion geometries and nonorientable topologies pervasive to Nature. Rather than a purely algorithmic static structure it turned out to be related to a vortical universal dynamics which comprises both the material and cognitive domains. Particularly, so called *chaotic systems appear to indicate a supradual logophysics*, an issue which has been disregarded. These topologies embody harmonics and thus transpired that Nature is fundamentally related to harmonic principles, as already arises in genomes, in the brain connectome, and in the Principle of Complementarity of cognitive psychology, further substantiated in (Rapoport [236]).The mysterious nature of resonance has turned out to be supported on nonorientability,which is also evident in the dynamical system models of neurons, the polyrhythmicity of motion as reflected by the brain-mind, and the topographic maps of the visual and somatosensory modes. Actually the latter somatosensory topography which is crucial to the constitution of the 'body image' does so by further integration of the projection zones of the other sensory modalities, themselves consisting of topological maps of different sensory organs (Solms and Panksepp, [126]; Craig [316]). Thus, rather than a Klein Bottle topology for the somatosensory modality, this further integration indicates a HyperKlein Bottle. The supradual logophysics is keenly related to selfconsciousness, on the one hand by the 'body image' of selfhood, as experienced by the subject contemplating a distant point at the horizon [236]). While the so called Hard Problem of consciousness as the quest to identify the 'source of consciousness' points to emotions, as sustained by Biological Psychology after Panksepp and upholded in the neurosciences (Solms & Turnbull [259]), there is much empirical evidence that resonance is the 'gluing' phenomenon producing the synchronicity of the brain-mind as a whole, and of many-minds (Hunt & Schooler [88]). This biological base for emotions somehow contrasts with the prevalent notion that consciousness stems from complex computation among a network of simple neurons which each receives and integrates synaptic inputs to a threshold which performs a kind of implicit gate to a dual-logic firing/non-firing logic, which as we shall discuss below, is currently disproved. The other approach to understand consciousness queries on the origin of cognition as how it originates from cognitive processes accompanied or driven by phenomenal conscious experience and subjective feelings, referred to by philosophers as 'qualia'. This is the so qualified "Hard Problem" of consciousness. Whatever the difficulty is, we face the problem of understanding the nature of distinctions between awareness and the major form of consciousness. After all, "qualia" do not disappear in consciousness at all, whatever the form or degree. We see the redness of an apple while awaken or dreaming, not to mention the intensity of qualia in other altered states of consciousness. Much of what is unavailable to awareness and performs in seemingly autonomic form which never is the case as the so-called Autonomous Nervous Systems – a component of the peripheral nervous system. It regulates involuntary physiological processes , say heart rate, blood pressure, etc.; it is not isolated from the environment through its connection to the Central Nervous System. Its coupling to it appears to be embodied as a Klein Bottle topology as the somatosensory and visual systems, cogent to the unity of action and perception (Turvey [273]). The same problem arises in attempting to define Life, from the perspective of dualism [230], since it already denies a continuity from the physical-chemical level to the biological one, as viruses show to be mediating states which remarkable have a remarkable geometric complexity. Distinctions between conscious and unconscious are not amenable to be conceived in dual terms, as we shall meet upon discussing anaesthetics as well as other states of altered states of consciousness, those of enhancement of its functionality. To the query posed by the Hard Problem, Biological Psychology has pointed out to emotions as having the inherent quality from which experience can be anchored to. Solms: "An explanation of experience will never be found in the function of vision—or memory, for that matter—or in any function that is not inherently experiential. The function of experience cannot be inferred from perception and memory, but it *can* be inferred from feeling… affection" (Solms, [261,262]). As for the function of affects " feeling enables complex organisms to register—and thereby to regulate and prioritize through thinking and voluntary action—deviations from homeostatic settling points *in unpredicted contexts"*. So they cannot be conceived as performing in terms of dual logic, rather analogics appears to be the case. As uniquely elaborated by Antonio Damasio, from his clinical work with brain-body damages or amputations, emotions also provide for the basis of the most complex forms of consciousness, as presented in his extraordinary *The Strange Order of Things* ([41,42]). Consciousness is primally interoceptive, it registers the state of the subject, not of the object world. "The sentient subject is first and foremost an affective subject. Only then can it experience perceptual and cognitive representations... The subject of consciousness is primary. The secondary (perceptual and cognitive) form of consciousness is achieved only when the subject of consciousness feels its way into its perceptions and cognitions, which are unconscious in themselves" (Solms , ibid). This selfreferential nature of elementary consciousness does not breed in isolation but rather in relation to the environment, in mammals in the maternal uterus and interacting with it through the umbilical chord nurturing matter and through sound, as the most fundamental form of sensing in direction to alterity. The complex mammal-to-be-born with relation to the maternal being is a HyperKlein Bottle. Resonance as the most fundamental primarily exteroceptive form of communication of the feeling of what happens as referred to the Outside Other.

Emotions are supported by selfreference and hetero-reference, self and alter-directed and/or experienced, supradually. Empathy in its wide spectra appears to be related to resonance, as the experience of music shows to be the case. While a form of panpsychism which evades identification might be attributed to Nature (Skrbina [257],) the supradual logophysics of the Klein Bottle is based on the principles of selfreference and hetero-reference, both torsion and nonorientability, *as if independently of the existence of a subjective agency*. It is known in neurophysiology that selfreference is fundamental to the brain-mind, and as the mirror-neuron phenomenology shows it is undissociable from hetero-reference: There is no Self without Other, as the so called mirror neurons phenomenology has exposed. Interoception is unseparable from our perspectival alter-related mutual construal be that of reality, the imaginal or ideational domains, mimetics, having resonance and empathy for its vehicles (Baiano et.al. [10], Rizzollati et.al. [238], Ferrari et.al. [50]): Its chameleonic nature a disproof of the hegemonic attribution of the primality of the dual logophysics.

As for the origin of emotions has been proposed that the interoception scanning primary intention of consciousness establish a sense of value related to the preservation of existence. Spinoza gave it a name, conatus [308]. It is a selfreferential process which elicits intentionality, selfreferentially: preservation of being. It is being sensed as the different states of homeostasic regulation. The dynamics of this process is embodied as a Markov blanket [55,307], which we already identified as a Klein Bottle architecture.

9. The supradual phenomenology of neuronal action potentials and topological phases

An important role of the receptive field -whose Gabor function representation first introduced in holography has a phase with a builtin Klein Bottle symmetry (Tanaka [268,269])- is the enabling of cell-to-cell communication through the action potentials. All of information of the outside world arrives through patterns of action potentials created at one of five sensory organs. Action potentials occurs when the membrane potential of a specific cell location rapidly rises and falls; the current paradigm attribute this to ion current transport. This depolarization in turn causes adjacent locations to similarly depolarize. Action potentials occur in several types of animal cells, called excitable cells, which include neurons, muscle cells, endocrine cells and in some plant cells. Action potentials support the propagation of signals along the neuron's axon toward synaptic boutons, the axon terminals, small swellings situated at the ends of an axon. Synaptic boutons are typically the sites where synapses with other neurons are found, and neurotransmitters are stored here to communicate with other neurons via these synapses. These signals can then connect with other neurons at synapses, or to motor cells or glands.

Let us start by noting that the neural network operation is not a matter of separate elements. This already dispels the notion that a dual gate phenomenology could support this connectivity, it rather elicits a contextuality, which already the Klein Bottle embodies. Indeed, a typical neuron receives inputs from thousands other neurons through the contacts on its dendritic tree called synapses. The inputs produce electrical transmembrane currents that change the membrane potential of the neuron. Synaptic currents produce changes, the so called postsynaptic potentials (PSPs). Small currents produce small PSPs; larger currents produce significant PSPs that can be amplified by the voltage-sensitive channels embedded in the neuronal membrane and lead to the generation of an action potential or spike – an abrupt and transient change of membrane voltage that propagates to other neurons via a long protrusion called an axon. Such spikes are the main means of communication between neurons. Spikes tune the time for cognition and manage the times of firing, so they perform as time-operators.

We continue by dispelling the notion that complex behaviour may be necessarily related to a complex network of neurons, a connectome dynamics (Atasoy et.al. [8]), but rather is based from an ecological integration of the living organism with its environment. Such an integration is the Klein Bottle embodiment. A paramount example is *Paramecium,* a unicellular organism, which swims in its habitat. *Paramecium* is a single spiking "neuron"; there is no connectome to this organism, and yet it can accomplish a variety of ecologically relevant tasks. It performs and controls complex navigation in crowded multisensory environments, practices accordingly social behaviour, sexual mating, and spatial memory might also be the case. Its motions with multiple cilia -mainly constituted by microtubules- is vortical. However, this does not arise from the complexity of the cell but "rather from the interaction between this spiking cell and the environment, together with the exploratory properties conferred by the pseudo-random nature of the effect of a spike [referring to the trial-and-error pragmatics of *Paramecium* in search for nurture]. This highlights the importance of embodiment and coupling with the environment, which are increasingly appreciated in cognitive science and philosophy of mind" (Brette [26])

In general neurons do not fire on their own; they fire as a result of incoming spikes from other neurons. If wished their identity as action is not entirely proper but a resultant of the endogenous *and* exogenous influence. As already said, there is no Self without Alter. In other words the CONTAIN image-schema appears to be inapplicable to the individual neuron in a collective of neurons . Hence, neurons are said to fire and yet they can be characterized *as being fired.* This, we claim, is the signature of the Klein Bottle supradual logophysics. Thus, there is an inherent "circularity" to the behaviour of neurons, which being associated to a joint selfreference and hetero-reference, it is associated to a

HyperKlein Bottle cybernetics, rather than a dual logic associated to the membrane and its electrochemical potential. Yet, not only a collective of neurons are involved in the spike formation and propagation, neurotransmitters are required for this propagation. The usual paradigm claims that action potentials are caused by ion currents which diffuse inside the neuron and when reaching and surpassing a threshold the spike is produced.

The mathematical model for action potentials as arising from ionic currents is the Hogdkin-Huxley model (HHM), which describes neuronal dynamics in terms of activation and inhibition of voltage-gated conductances. An important hypothesis of the HHM is that neurons are dynamical systems, and thus described by a system of ordinary nonlinear differential equations. HHM is used together with the claim that the continuous action potentials can be reduced to sudden appearance of peaks called "spikes" to posit the dual logic; this is done in terms of the individual neuron presumably upon surmounting a threshold voltage which thus fires producing the spike or otherwise stays quiescent not producing it at all. In HHM biological neurons are dual-logic gates working through an "integrate-and-fire" scheme based on comparison, a difference-producing-differences process, after Bateson: Firstly, |a cell body (soma) and several branched dendrites receive and integrate synaptic inputs as action potentials. Then, a comparison between the action potential and the integrated threshold potential at the axons initiation segments takes place, when the latter reaches the former value then a dual all-or-nothing "spike", or action potential as output and transmitted across the axon. The issue left unresolved by this dual logophysics is what makes the comparison, in the first place. In closer examination it further produces an infinite regress proper to its first-order cybernetics based on dual logic onto-epistemology [234,236].

However, further examination of this model does not support its in dual logic preconception, which is rather hegemonic in the sciences (Rapoport [230]). Izhikevich: "The irony is that the Hodgkin-Huxley model does not have a well-defined threshold; it does not fire all-or-none spikes; and it is not an integrator, but a resonator, it prefers inputs having certain frequencies that resonate with the frequency of subthreshold oscillations of the neuron" (p.2 [91]). As for establishing an adequate methodology, Fitzhugh concluded his mathematical modeling of threshold phenomena by saying that "...many neuronal properties are invariant under continuous, one-to-one transformations of the coordinates of phase space and fall within the domain of topology...the most logical procedure in the description of a complex biological might be to characterize the topology of its phase space, then to establish a set of physically identifiable coordinates in the space, and finally to fit differential equations to the trajectories" [315]. Upon examining the HHM attractors, nonorientable twists were discovered in the phase space chaotic attractors (Guckenheimer & Labouriau [70]). Rather than subthreshold, a conceptual hindrance of the assumption of dual logophysics, the more appropriate term would be subcellular or cytoskeletal.

However, it has been empirically verified that neurons operate as resonators (see Izhikevich [91]); hence, dual logic is not the case for the basis of neuronal phenomenology. Yet this resonator character turned out to be the core of what is conceived as a novel paradigm for the neurosciences elaborated by Anirban Bandyopadhyay and collaborators, the Hindu group [14,64,246,254-256]. Already in 2014 Hameroff and Penrose [75] appealed to the empirical findings of this group as support for their ORC OR theory, and very recently further promoted them for a novel paradigm for the neurosciences. Their elaboration is quite distinct with the present Klein Bottle logophysics, disregarding the torsion geometry and topological issues, not mentioning the topological phases in the cytoskeleton and the shape of action potentials, nor the deep connections to cognition and perception, topological chemistry and nonorientability of liquid crystals, which are crucial to the supradual Klein Bottle logophysics.

Already visual receptive fields do not operate in terms of dual logic: the consideration of context such as central locus and periphery elicits a non-dual logic by which neurons are in a state of superposition of firing and non-firing, inhabitation and activation superposed, yielding a 4-state Klein Bottle logophysics (Rapoport [228]).

Redirecting our attention to the so called cytoskeletal structures and processes we first remark their association to the filamentary structures of the neuron. Focusing the attention on the spikes in disregard of the phenomena being produced on the subthreshold and before actual spikes firing is very much in ignorance of the complexity and information associated to it. Indeed, the current spike dual-logic based paradigm ignores all components beneath the neuron membrane - fundamentally the cytoskeleton, tagging them as inconsequential: the prejudice of the neural membrane as a dual gate is hegemonic. We turn to this now.

The cytoskeleton is basically made of proteins, the most remarkable proteins are the molecular motors, the so called motor molecules which play a crucial role in the cytoskeletal dynamics. Indeed, motor proteins operate transporting membrane-enclosed organelles or changing the shape of structures built from cytoskeletal filaments (Alberts et.al. [5]). In other words, in the latter case motor proteins operate as topological agencies. They bind to a polarized cytoskeletal filament and use the energy derived from repeated cycles of ATP hydrolysis to move steadily along it. Dozens of different motor proteins coexist in every eukaryotic cell. They differ in the type of filament they bind to (either microtubules or actin), the direction in which they move along the filament, and what they carry.

The cytoskeleton is a protein lattice which has three kinds of filaments, microtubule made of tubulin, neurofilaments, and microfilaments (actin). Microtubules are hollow cylindrical tubes with outer diameter of 25 nm and inner diameter of 14 nm with thirteen linear tubulin chains ('protofilaments') aligning side-to-side to form the hollow microtubule Given the large surface area inside and outside the tube, ordered water domains are considered to be bound to microtubules. These ordered water domains are crucial to metabolism and to the overall coherence of organisms [79,80,135,168-170,193]. Due to their uniaxial symmetry rather than having the topology of the abstract projective space (Machon and Alexander [153,154]), given that organisms are bounded rather than extending to infinity, they are characterized by Möbius strips due to the antipodal identifications of points at the boundaries supporting nonorientability (Rapoport [228]). Remarkably, in the case of vision this supports the establishment of a sense of selfhood ([236]). In other words, already coherent water domains have the selfreferential nonorientable shaping as a form of protoconsciousness.

Using the quantum field theory for condensed matter systems it has been argued that microtubule cavities with water in their vicinity could support superradiance, i.e. collective infrared photon emission by water molecules known as superradiance. Pollack showed that coherent water domains are produced by electromagnetic absorption by regular water, and upon electromagnetic emission a cycle of coherence/decoherence of water is produced. Also a coupling between the water molecules inside the cavity of the microtubule and the tubulins that build up the microtubule walls may produce superradiance (Pollack [193]). In the case of coherent water domain, the supradual like likes like principle operates towards their formation, hence we may venture whether this also applies to the allosterics of water in the coupling of tubulin aggregating into microtubules, as a shaping typical of aromatic molecules behaviour.

It has been argued that the ordered water domains environment of microtubules could produce a dissipationless transfer of energy on neuronal microtubules due to the collective coherent behaviour. This transport would be supported by electromagnetic sine-Gordon solitons (Georgiev [59]). They are solutions to a non-linear Schroedinger equation, and as such associated to a torsion geometry of vacuum Brownian fluctuations (Rapoport [219]). Remarkably sine-Gordon solitons support Möbius strips as the edge of knots in the 2-torus phase-space of coupled oscillators [116]. They appear as the dynamical system attractors of the "mental" rotations discovered in the cognition of music and in the nonlinear dynamical systems of coupled oscillators say of two tonically spiked neurons , whose behaviours is represented by the strange attractors of their dynamics (Izhikevich, [91]). We shall return to this below.

Since superradiance is not exclusive to microtubule interaction with the coherent water domains- indeed it is the case of genomes as real-space nonorientable configurations [233]- Georgiev argues that it further requires to examine the interaction of microtubules terminals with MAPs (microtubule applied proteins) or motor proteins. We shall see later that this produces nonlinear elasticity buckling patterns as a solitary wave on a Möbius strip surface.

Cytoskeletons conform a tensegrity structure, a scaffold-like structure made of proteins intermittently folding and unfolding, assembling and disassembling (Ingber [117]). In neurons microtubules prevail and are uniquely stable; their collapse is related to the evanescence of consciousness, say under anaesthesia. In the general case microtubules are highly dynamic as they alternate in a cycle of alternative depolymerization and polymerization, known as dynamic instability. In this sense they resemble the superradiance cycle of formation and disgregation of coherent water domains by alternative absorption and emission of electromagnetic radiation already introduced. Notably Voeikov proposed this cycle as the very signature of the origin of life [279]. Some specific MAPs bind to the terminal sector of microtubules thus controlling their dynamics and attachments to other cellular structures. Other specific MAPs attach to microtubules along their entire length such as α-tubulin and β-tubulin thus modulating microtubule functionality. The molecular patterns generated by them conform the so called tubulin code which control microtubule behaviour. Remarkably, the tubulin code does *not* operate as a dual switch, but rather analogically as a fine regulator (Janke & Magiera [93]). In other words, the dual logophysics of microtubule functionality *is not* the case.

The notion of the cell as a tensegrity architecture provides a model to understand the processes subjacent to cell shape, movement and cytoskeletal mechanics, cells sensing and response to mechanical forces as well as cell differentiation and tissue development. Thus they perform as a cognitive-like structure which has adaptative behaviour. This is particularly notable in the behaviour of ciliate colonies -cilia are microtubules- for which the vortical metaform of motion as a shape of in-formation acting as the logophysical agency at some level being reproduced at another level of organisation: They manifest a natural intelligence as a mimetic reproduction of vortices (Ben Jacob [19]).This isomorphism surges from the common nonorientability of genome sections [233], the lemniscal 1d-projections of Möbius strip inscribed as the gait patterns in different media -swimming in the former case [18], nonorientability of sound and electromagnetic waves [7 ,43], and still in the topological multiconformation of organic molecules, to be discussed below. The latter is basic to ORC OR theory in the superposition of protein conformations.

Filaments look alike nano-wires which are densely packed in neurons, and as a tensegrity structure they are somewhat reductively considered as providing a tensile resisting skeleton. Thus, they resemble electronic devices such as resonators, which upon assembling on Möbius strip shape, the signals transmitted by them are somewhat liminal to the classical-quantum usual divide, to be later discussed. Tensegrities model the mechanical behaviour of the cell as it emerges from physical interactions among the different molecular filament systems that form the cytoskeleton and its role orienting much of the cell's metabolic and signal transduction machinery. Furthermore it elicits that a single physical control parameter, cell shape distortion through gradual variations can switch cells between distinct gene programs (e.g. growth, differentiation and apoptosis). This process can be viewed as a biological phase transition, in which mechanical distortion of cells and of their cytoskeleton through cell surface integrin receptors profoundly affect cell behaviour.

That the cytoplasm is such a tensegrity structure having tensile and compressive elements is well established and particularly microtubules are under compression due to tension generated by actin filaments and the so called cytomusculature (Hameroff, [73]). Microtubules form a scaffoldlike selfsupportive structure without contact among them. Furthermore, there is a complex torsion architecture organized in terms of microtubules with contractile actin filaments spirally winding around them, actually with the two opposite directions alike a spinning top which may cause rotational oscillations. The figure that appears is of a cytoskeletal tensegrity dynamical structure "twisting back and forth, even "rockin' and rollin'!" (Hameroff, p.170 [73]). Currently, this complex structure associated to resonances has been empirically probed by Bandyopadhyay and collaborators, using novel scanning techniques.

10. Introducing the torsion dynamics of microtubules

Astrophysicist N.A.Kozyrev following his discovery of the existence of a nonlocal field of stars along their past, present and future percourse in time as an holographic-like anticipative memory (Kozyrev, [108-110]; Lavrentiev [143-145], Eganova [46], Korotaev [106]) which he attributed to a spinning torsion field, he studied the behaviour of spinning tops. Seemingly, Kozyrev was not aware of the joint work of Einstein and Cartan introducing torsion to spacetime geometry (Debever [298]) which failed to identify the significance of torsion as a gravitational field element due to the omission of macroscopic spin (Goenner [67]). Kozyrev's findings of anomalous behaviour of weight variations of spinning tops were reproduced by Hayasaka and Takeuchi [78]. Remarkably, Kozyrev's initial hypothesis was that the existence of an irreversible thermodynamical process as the local manifestation of the action-at-a-distance. His conception of time was radically different to the usual time-parameter descriptor of change. According to Kozyrev in these irreversible processes what manifests is that time has a stufflike character: On the one hand being heterogeneous its manifestation depends on what the system it acts upon is like; secondly it establishes a primality of cause-effect, so time is more about *timing* (i.e. a time-operator) and its form is vortical, and thus the effect of this timing is the functionality and form of the system itself. Alike a vortex it selforganizes viz-a-viz the environment and the morphing effect is the system itself. A similar conception in relation to torsion and blow-ups of nonlinear systems was developed by Soucheng OuYang and Jeffrey Yi Lin Forest (see [286]).

For a proof of Kozyrev's findings of a torsion field which operates as an anticipative holographic memory it only takes to look at the centre of our galaxy, the Milky Way, to find a megaparsec size Möbius strip of light (Molinari et.al. [291]). In Bohm's conception of the quantum world as structured in terms of an "explicit" and "implicit" orders in terms of the so-called quantum potential, which is related to the electromagnetic torsion potential under the eikonal light-ray constraint and also arising in the Schroedinger equation as a torsion field, Bohm and Hiley- invoked a "prespace" [292,294]. This prespace is a byproduct of the nonlinear interference wave patterns from which in-formation surges by beaming ligthwaves upon them. They described this prespace as topological, and algebraically by Clifford algebras [28].The shapes of the latter whenever their phases possess an inhomogeneous distribution- which is as well the case of holograms, are non-orientable Möbius strips (Freund [54], Bauer et.al. [17], Bliokh et.al. [21]).

Would we consider the living cell, say a neuron, its cytoskeleton, water and interactions with vacuum fluctuations and still producing and absorbing electromagnetic waves and vibrations, we find a very complex fractal tensegrity architecture in which different periodicities operate in a ample range of frequencies and resonances elicit the overall coherence. Actually the investigation of this architecture has led to conceive the cytoskeleton as a time-crystal by which the cytoskeleton is under perpetual change in a pattern which is periodically repeated [14]. In any given cycle the system may be in a phase in a non-equilibrium state, in which the system appears to be in a thermodynamical irreversible state.

So returning to Kozyrev's conception of time as an operator causing an irreversible process as it morphs in different vortical structures such as proteins, tubulin, microtubules, we can further examine the torsion geometry of this.

In the study of the torsion geometry of thermodynamical irreversible processes and of Brownian motions (Rapoport [210,227]), or still in the equivalence of the Maxwell equation of electromagnetism and the Dirac-Hestenes equations of relativistic quantum mechanics upon examining the Helmholtz decomposition (extended to differential forms on manifolds, the deRham-Kodaira-Hodge theorem) of the trace-torsion differential one-form (or its metric- conjugate vector field) we identified the terms related to irreversibility. It is the non-exact gradient trace-torsion terms: a divergenceless co-closed term and an harmonic term. The harmonic term is both divergenceless and closed, its differential is null. These can be also thought as electromagnetic one-forms, the harmonic term has a zero-field, alike an Aharonov-Bohm potential term [206,208,217]. These terms may support topological anholonomities and are further related to the phenomenon of resonance as interaction-at-a-distance. Furthermore, what this elicits is a dynamical torsion given by differential one-forms, like the electromagnetic potentials which in fact is a trace-torsion Cartan-Weyl 1-form, say Q, yet whose electromagnetic field dQ -d being the differential operator- is a differential 2-form, say F as in electromagnetism, which is null [84]: In other words a dynamical torsion producing null "curvature". This is in stark contrast with null torsion metric geometry with non vanishing curvature which is postulated by Penrose and Hameroff in their ORC OR theory. We claim that this choice of geometry -as in General Relativity- defined by metrics, to be entirely unrelated to what Hameroff described as the rock-and-roll microtubule having tubulin helices morphing inhomogeneously as an invertible gyroscope either supported on any of its extremes, producing an inhomogeneous vortex at the same time.

It is this behaviour which we suggest to be the underlying process for the collapse of microtubules subjected to tubulin loads, while in the present setting the coherence is no longer an action-at-a-distance; more of this below. In other words, the collapse has to do with a thermodynamical irreversible process in which rotations indeed intervene as Hameroff early observed to be the case of microtubules dynamics. Alternatively can be conceived as the vorticity turbulence producing term of the velocity of a viscous fluid satisfying the Navier-Stokes equations of fluid-dynamics, or the equations of magneto-fluid-dynamics [211-217]. This is in stark contrast with null torsion metric geometry with non vanishing curvature which is postulated by Penrose and Hameroff in their ORC OR theory, to be discussed below. We early suggested that the continuity of the cytoskeleton with the extracellular matrix together with the confinement of the nonorientable coherent water domains basic to cellular metabolism, rather than making of a cell's membrane a semipermeable dual gate it performs an integration supported by local nonorientable membrane topology. According to Ingber, integrin receptors, critical mechanosensitive molecules and cellular components integrate the cell's Inside and Outside. So we can identify the *nonorientability* of Möbius strip elasticity as a supradual protoform of agency, a topological "protoconsciousness", though not necessarily ascribed to a living agent, but a function arising from the interaction of the microtubule with the environment, particularly by additional load producing a change of shape and order. This is still more patent upon considering that crystalline ribbons made of a compound of niobium and selenium, NbSe3 - non ascribable to the constituents of a living structure- in view of its inherent rigidity, which would be expected to prevent it from either bending or twisting, may conform as Möbius strips (Tanda et.al. [270]).

For a proof of concept, we note that selfassembled viral membranes were shown to have scallop-rodlike units selforganising into Möbius conformations at the membrane edge (see Gibaud et.al. [60]). We must recall that viruses are crystal structures *intermediary* to geometrical lifeless structure and living organisms. Indeed they carry genetic material, reproduce and evolve ascribed to natural selection. Alternatively, in the self-organising theory, free of Darwinian natural selection attribution, which is conspicuous to crystal structures, the natural selection process can be left aside, as argued by Lima de Faria. However viruses lack the key characteristics, such as cell structure, that are generally considered necessary criteria for defining life.

Given the importance of proteins to the cytoskeleton, we proceed to the shapes and regularities which are elicited by them. Palindromic sequences are abundant in proteins. Palindromes, say "neuquen", "ana", "somanydynamos", are topologically supported by collapsed 2-tori and Möbius strips, are crucial to genomes and their evolution (Rapoport [233,234]). Studies have shown that reversing the sequence leads to the production of the same 3d protein fold or still its enantiomer (Sheari et.al. [117]). This is most remarkable since it indicates that the 3d structure of proteins may be related to a lower dimensional -actually a 2d Möbius strip- metaconfiguration, the untwisted orientable case corresponding to the symmetric former case. Thus, the principle of "multiple identity" or, if wished "split personality" appears also to be the case of proteins, contradicting thus the principle of identity basic to dual logic.

This is still more notable following the topological studies of protein domains in terms of relational representation of protein domains using the locations of the backbone atoms and the hydrogen bonds, which allows a construal of a combinatorial object – a so-called *fatgraph.* Despite that the space of possible protein structures appears vast and continuous, and the complexity of the relationship between primary, secondary and tertiary structure levels, fatgraphs allow for a *discrete* representation from which a 2d-*surface* appears as the ultimate representation. This surface can be topologically represented as gluing 2-tori (viz bagel-shaped) and Klein Bottles. The former represents a banal selfreferential figure where CONTAIN is structured surrounding an excluded central singularity in a rigid relation -a hole which is part of the whole yet as if excluded (Rapoport [235]). The latter embodies the dynamical contextualization of protein domains vis-à-vis the environment relatable to the instability of protein conformations. In other terms, a juxtaposition of duality and supraduality, of permanence and mutability. The gluing of orientable 2-torus and Klein Bottles that lie as the topological structure of proteins elicits the multiconformation of proteins as possible modifications between the two forms of organisation vis-à-vis the environment: the closed organisation given by the tori, the open-closed given by the Klein Bottle elements, and the possibility intertransformations due to electron delocalization. Given that the Klein Bottle through its Hadamard matrix representation is the single operator which allows to transcend classical computations it is these elements which support this transcendence as quantum computations (Aharonov [3]). Furthermore, since the Klein Bottle operates on its own superposition states to produce Boolean states as a topological decoherence, as much as reiterating its action now on the Boolean states recovers its own coherence as topological superposition, then these nonorientable elements of protein domains control not only the quantum computations but also the transformations to dual logic elements and futher the reconstitution of quantum computation in the second phase of the cycle. Actually, as observed in (Hameroff et.al., [74]) "protein conformational dynamics suggest that protein conformational states—fundamental information units ('bits') in biological systems—are governed by quantum events, and are thus perhaps akin to quantum bits ('qubits') as utilized in quantum computation … signaling, communication and conductivity in MTs, and theoretical models have predicted both classical and quantum information processing in MTs". As related to their relevance "Quantum processes such as electron tunneling, delocalization, and superposition could, therefore, couple to nuclear locations and influence conformational changes, particularly if the quantum processes are collective and self-organized. Proteins may be designed to amplify these quantum processes. Within proteins ideal sites for electron delocalization are hydrophobic pockets (sites of anaesthetic effect), in particular those containing aromatic rings like that of tryptophan" (ibid.). We shall return to this upon discussing ORC OR theory.

Tubulin dimers are arranged not only in vortical shaped protofilament chains forming thus the microtubules. They further form a particular skewed hexagonal lattice whose crystal structure gives rise to helical winding patterns with regular repeat intervals. Increasingly steep winding patterns complete one cycle around the cylinder at a certain number of tubulins above where the cycle started. These numbers, 3, 5, 8, 13, 21 etc. follow the well known Fibonacci series and whose sequential ratios F_N / F_{N+1} tend to approximate the Golden Mean, reproducing thus the five-fold symmetry which we identified already as a torsion geometry anholonomy. From the association between torsion and nonorientability presented in §2, it is *coherent to claim that topological anholonomities, i.e. nonorientability would appear in microtubular conformations*. We shall discuss below a Möbius strip buckling pattern as a solitary wave arising from the nonlinear elasticity response of (nonorientable) microtubules which would ensue from the coupling of tubulin to the anaesthetic factor.

The association of microtubules with a torsion geometry and nonorientability points to the supradual Klein Bottle logic, in contrast to Hameroff and Watt' s take that microtubules operate as switching Boolean matrices processing binary bits states as they move through microtubules and proteins associated to them [72]. Rather than Boolean matrices and further recalling the analogical nature of the tubulin code, this points out to Matrix Logic, which has fuzzy logic, Boolean logic and quantum logic for subcases. Matrix Logic is the matrix form of the Klein Bottle logic introduced by August Stern, it is homologous to Quantum Mechanics for 2-state systems [265]. Rather than Matrix Logic being circumscribed to logical operators these are jointly logical-quantum operators. Quantum statements are interchangeably expressed as logical connections. It is noteworthy to recall that Orlov showed that Quantum Mechanics can be derived from Boolean logic expressed as matrices ([179]) . In Matrix Logic invertible operators are also considered; their eigenvalues are not restricted to the real numbers proper to measurements processes of quantum mechanical systems but may be complex numbers. Indeed, these logical-quantum operators are not necessarily self-adjoint, the latter restriction as in Quantum Mechanics for closed systems which is the case of Orlov's rendering of Quantum Mechanics from Boolean logic in matrix form. Non selfadjoint operators appear in the case of quantum open system, say the cytoskeleton. First we discuss some crucial relations of this to Quantum Information Theory.

11. The fundamental role of the Klein Bottle in Quantum Information theory and a theory of consciousness

The Klein Bottle plays a core role in Quantum Information theory central to quantum computation, through its 2x2 matrix representation (up to a scaling factor): the Hadamard matrix. This theory is constructed from elementary quantum building blocks, namely qubits, rather than bits as the unit of classical information, and logical gates. In quantum computation the gates are drawn from a universal set of gates, namely, a set which can be used to perform general quantum computations (Nielsen & Chang [174]). It is the Hadamard gate which supports transcending Boolean logic. In fact, in quantum computation it is the single logical gate that transcends the computation of all Boolean functions through the Toffoli reversible gate,which together with the Hadamard gate constitute a universal set of logical gates (Aharonov [3]).Thus, they constitute the logical gates for which any possible operation on a quantum computer can be expressed as a finite sequence of them. The notion that neuron microtubules would support quantum computations has been raised by Hameroff and Penrose, yet they were mainly concerned with microtubules as a Boolean gate architecture and the Hadamard gate was not considered in their models [72-76,192]. This is quite remarkable since it is the Hadamard gate which may bring a single qubit into a superposition state, the latter being the qubits which their theory of consciousness deals with. Furthermore as already noted, the tubulin code is not dual logic.

12. Topological folding, nonlinear elasticity in the buckling of microtubules, memory and the origin of life

Remarkably there is a close relation between the Möbius strip as a physical space topology and nonlinear elasticity, which has been unnoticed by Penrose & Hameroff, which we turn to discuss. The phenomenon of buckling of structures, say tubulin or microtubules, as in the Euler instability, does not correspond to linear load induction but to a sudden instability due to a load of tubulin that may lead to the collapse of the structure (de Pablo et.al. [297]). The response of an elastic Möbius strip to shear stresses is intrinsically nonreciprocal and nonlinear, although the local stress-strain relation is linear. A direct consequence of non-orientability is that no stress distribution can yield homogeneous shear deformations over a Möbius strip. The topological anholonomity constrains the ribbon to remain undeformed at one point whatever the magnitude of the applied stress. The associated buckling patterns were found to propagate as solitary waves on a Möbius strip; this is a zero-energy cost process. Furthermore the Möbius strip's elasticity may support several conformations and posses a form of mechanical memory as the static deformations encode their stress history. According to Bartolo & Carpentier ([12]): " Nonorientable surfaces offer a paradigmatic example of static mechanical memory. Information is coded and stored by the temporal variations of the stress. Information is read measuring the shear angle and deleted releasing the applied stresses". Thus, for the elastic Möbius strip, the coding of *memory as stuff-stress* reduces to a *single* variable, the shear angle: As in Occam's Razor, the principle of parsimony is the case. At this point we characterize that while memory might be stufflike as it is embodied as nonorientable shape, most basically it is a function, very much alike consciousness is (Solms [260]). Microtubules have a notorious involvement in learning and memory (Woolf [285]). In this multistability of buckling patterns of elastic Möbius strips there is a common behaviour with regards to the hybrid organic aromatic molecules which are present in the neuron, which may also assume different shapes according to π-electron delocalization, leading to both orientable and nonorientable conformations, to be discussed below.

In the mathematical formalization of cognition in the wake of Shepard's Principle of Complementarity, Leighton noted that memory cannot be formalized only in terms of symmetries and the Erlangen Program. Indeed, in the investigation of the psychological relationship between shape and time, Leyton argued compellingly that shape is used by the mind to recover the past and as such it forms a basis for memory, for which Rapoport argued that torsion is the basis for any memorization (Leyton [147]). Leyton identified a twist on a plane as the primal shape embodiment of memory. However, what was left unmentioned by Leighton is that this primeval coding can further produce nonorientability, establishing thus a permanence of the memory produced by the twisted folding of a plane, as noticed in (Rapoport, [236]). In other words, the function of memory in producing shape is introduced by nonorientability, embodied as shape, though it might not be notorious, just like consciousness is not necessarily so. As for the role of memory in biological evolution and development, any development is governed by the tension between the persistence of the past, and an anticipation of the future as proposed in the Klein Bottle logophysics [225,229]. Similarly Stuart Kauffman proposed that in biology evolution in the present is somewhat conditioned by the future future as possibility [295]. So any development is governed by the tension between the persistence of the past, and an anticipation of the future. This points out to Heidegger's notion of time-space, time as a nonlinear shape operating as mediating past and future through the present [298]. Where this shape of time-space is most notorious is in the experience of music yet it also transpires in literature, the so called chronotopes of Bakhtin. One such chronotope is Dante's description of the trip to Hell in search of Beatrice, a Möbius strip journey (Bethea [314].)

To make further the case of topological folding and biology, they are the basis for the *symbolic meaning* establishment of living organisms. On the one hand they are crucial to the morphomechanics of the establishment of their bodyplans during development [232,233]. Furthermore they are crucial to the symbol-matter complementarity of biological systems [257,312]. Pattee identified the relevance of folding to the timing process crucial to metabolism [290]. Indeed, he described folding as operating to transform the passive symbolic gene expressions into the dynamic rate control of enzymes. Pattee: " Folding transforms what are essentially rate-independent syntactically coded sequences into rate-dependent functional controls. Protein folding is a highly parallel process with so many degrees of freedom that is difficult to model even on supercomputers". Consequently, Rapoport followed by identifying topological folding as the operation that turned the material world to the symbolic world of organisms, the very origin of life [235]. *As in the origin of life, the basis of consciousness is, in essence, what we claim to be the case*. Meaning would then have an origin, in the interplay of matter and symbols, both for life in the first place and for consciousness, later for evolution, conceived in terms of folding and palindromic sectors of genomes [233].

That this may apply to the elasticity of microtubules, actually to the transition of the awaken state to the (partially) unconscious state as under anaesthetics, would be another proof of the homology of the mathematical geometric-topological formalization of the physical world and that of the unconscious modeling by the mind be that of the kinematic of objects or the "motions" arising in music perception (Rapoport, [236]). This memory supports the mental modelization of the kinematics of regular and irregular objects, by which an homology of the modelization of dynamics of point systems on spacetime are interiorized as the mental modelization both in terms of torsion geometry. Already Shepard not only discovered that the unconscious kinematical modelization required torsion geometries but also that nonorientability came with it by terminating the twist with a topological identification, a memory stabilization as already discussed. Furthermore Shepard discovered that "mental" rotations were inherent to the cognition of music, as in the Tritone Paradox [251]. However he suggested that "mental" rotations had no real substrate in the brain, say the neuronal dynamics. However, the nonlinear dynamical systems modeling of motor activity, say of coupled oscillators, where behaviour is represented as the strange attractors of this dynamics, disproved Shepard's claim. Indeed, in the case of polyrhythmic motor performance (Jagacinski et.al. [92]; Turvey [273]) and the rotational dynamics of population of motor cells first discovered by Churchland (Lebedev et.al. [146]) their attractors are nonorientable. Topological phases, retrieved.

13. Anaesthetics, the persistence of consciousness, the analogics of the tubulin code and nonorientability in the topological chemistry paradigm and the multiconformation of aromatic molecules

Hameroff came to collaborate with Penrose from his work in anesthesiology. He based his theoretization on the evidence that upon administering anaesthetics to patients undergoing surgery, the evanescence of the consciousness present at the awaken state holds as anaesthesia blocks consciousness and memory while sparing non-conscious brain activities. Certainly most of the brain activity is beyond awareness (Solms [259-262], Panksepp [182], Solms and Panksepp [126]). However, the previous description of the anaesthetic state as a duality of awareness/unconscious-state is not the case. Particularly emotions are not entirely suppressed in the anaesthetic state, the most common being anxiety, panic, helplessness, etc (Gidron et.al. [65]). Dreams do not subside neither, being much alike dreams while sleeping, and their experience appears to be unrelated to the depth of the anaesthetic state (Scheinen et.al. [247]). Neither memory is inactive, uttered words and conversations can be registered and expressed after recovery (Moerman et.al. [167]). The feeling of what happens may also be the case, in particular in near-death-experiences (Lopez et.al. [150]). Since emotions are considered the primal form if not the very source for consciousness as selfawareness further associated to hetero-reference -as the mimicry elicited by the mirror neuron phenomenology (Rizzolatti et.al.[238]), the attribution of consciousness as emerging from the physical process of the reduction of a quantum state under experience, say a measurement, is left wanting. Indeed, it is hard to conceive this physicalist reductionism: It attributes to this reduction to a single eigenstate the extraordinary physical source for consciousness of an agent-independent measurement what in fact is the emergence of an ampler selfawareness which already emotions elicit, fundamentally as interoceptive, *rather than exteroceptive as in the measurement process* discussed in physics. (Interoceptive homeostasic scanning regulation may involve measurements operating in a Klein Bottle cybernetics [234]). Furthermore, under certain conditions this interoceptive attention can be literally turned inside-òut, becoming a real world if not hyper-real, the focusing and control of attention playing a crucial role (Laughlin et.al. [140-142]). Hameroff acknowledges this problem and offers for possible explanation: "... abrupt transition from unconscious, or subconscious possibilities to consciousness, and causal selection of particular actions and perceptions, may be seen as "quantum state reduction," or "collapse of the wavefunction" (Hameroff & Penrose [76]). In other words, anaesthesia which elicits the "dark side of consciousness" -as Hameroff calls it- actually comes with the need of showing that its "darkness" is not to be confused with "blackout", at all. This is treating by "normalizing" the situation: both the dark side of consciousness and its non complete evanescence are all related to collapse of the wave function. We shall return to this below.

Remarkably all living molecules are susceptible to anaesthesia and plants are known to counter stress by producing anaesthetics. Actually plants are capable of memory and display different forms of intelligence (Trevawas [271]).This may be prompted by the alterations introduced by anaesthetics to the existing oscillatory modes in proteins which may tend to alter microtubule polymerization and function. The organising behaviour of macromolecular entities, such as tubulin, indicates the existence of a complex and very fast information processing activity. High -frequency neural oscillations have been associated with conscious states, while low-frequency activity has been associated with unconscious states (Cariani [30]). Coherence theories of anaesthesia suggest that general anaesthesia act by disrupting coherent neuronal activity in critical brain structures. They do by preventing the protein conformational changes which participate in the normal functioning of neurons [75,76] . In other words, anaesthesia blocks the multiconformational states of molecules, only the one in action is preserved. Thus it supresses at the level of the awaken state of awareness the supraduality operating these changes.

However, anaesthetics binding to tubulin in neurons disrupt the microtubules at their extreme or sidewise, which otherwise they are stable for a lifetime, while for the other protein structures they are susceptible to their intermittent assembling and disassembling (Karp [97]). The latter is naturally conceived as the breakdown of microtubules by buckling, a core instability to nonlinear elasticity theory.

Given the importance of organic chemistry to the cytoskeletal dynamics which is upholded as basic to the novel paradigm for the neurosciences proposed by Hameroff [76] in terms of aromatic arrangements crucial to anaesthetics, it is important to observe that not only the usual Hückel 2-torus-like aromatics can be the case but rather there is an alternative Möbius strip-like aromatics, also called Hückel- Möbius aromatics. In terms of molecular orbital theory these compounds have in common a monocyclic array of molecular orbitals in which there is an odd number of out-of-phase overlaps, the opposite pattern compared to the aromatic character of Hückel systems. The nodal plane of the orbitals, viewed as a ribbon, is a Möbius strip, rather than a 2-torus, hence the name (Rainer [198)]. While originally it was considered that Hückel- Möbius aromatics was unknown, in the last forty years starting with the synthesis of Möbius strip shaped molecules, a host of such molecules have appeared and furthermore the existence of multistable conformations transiting from nonorientability to orientability and the converse. The nonorientability provides extra stabilization and with this molecules are less reactive than those without this shape (see Janowicz [94]). The existence of multiple conformations as stereoisomers lays at the very basis of Topological Chemistry (Mezey [118]), and the synthesis of catenanes, rotaxanes, knotted molecules such as the Borromean ring, etc., of molecular motors in short, for which Möbius strips may be used to that effect (Breault, [305]). This paradigm led to the bestowal of the 2016 Nobel Prize in Chemistry to J.P. Sauvage, J. Fraser Stoddart and B. Feringa.This coincided with the 2016 bestowal of the Nobel Prize in Physics for the development of topological approach to condensed matter physics and exotic matter to D. Thouless, F. Duncan, M. Haldane and J. Michael Kosterlitz ([107])

We recall that hybridization is the concept used for explaining molecular shapes in the usual approach, of mixing atomic orbitals into new *hybrid orbitals* (with different energies, shapes, etc. than the component atomic orbitals) suitable for the pairing of electrons to form chemical bonds in valence bond theory. As already mentioned, molecules may appear as having multiple conformations, or a "split personality" in a colloquial dualistic description: on the one hand orientable conformations may turn to be non-orientable, and the converse can also be the case (Rzepa [245]; Sokolov [258]; Stapien et.al. [263]). It is the hybrid orbitals which are considered to support the "split personality" whereby, say graphene, coexists as in the normal *and* the Möbius strip configurations . Graphene is considered to be a peculiar form of aromatic molecule (Popov et.al. [130]).This double configuration is more generally encountered in aromatics which started with Kekule's model of benzene, which is a case of Hückel-Möbius aromatics as well as Hückel aromatics, i.e. the cyclic ring 2-torus shaped planar molecules.

14. The topological chemistry paradigm and superposition in ORC OR theory

With respect to ORC OR the most basic example of multiple conformations and of the switching between them, the "split personality" is that of tubulin itself: it leads to the quantum superposition from which this theory departs from. Hameroff: "... since London forces are quantum mechanical, the electron pair may occupy both states, and the protein exist in a quantum superposition of both open and closed conformations. By going from bottom (quantum superposition) to top (one particular conformation) the protein functions as a qubit" [76].

So, tubulins may operate as two different conformations upon coupling with a pair of quantum coupled electrons in an internal hydrophobic pocket. Each tubulin may occupy two classical conformations (a "black"-open state and a "white"-closed state"), two classical bit states) or exist in quantum superposition of both conformational states (a qubit). The former two states correspond to Boolean states, the latter to a quantum superposition, typical of the Klein Bottle superposition as nonorientable topology, lies at the very basis of ORC OR. We shall discuss this below.

Early in 1973 Sokolov reviewing the nascent paradigm of topological chemistry showed that a modified form of the allosteric effect could be realised in molecules by an electron delocalization mediated by Van der Waals-London forces which could be rendered as a change of electron density shape with a migration of an electron as if a Klein Bottle outside/inside or inside/outside motion. Sokolov proposed that the allosteric coupling of molecules rather than a lock-and-key coupling it operates in terms of topological changes as discussed above. It is to be superseded by a dynamical shape accommodation (Tripathi and Bankaitis [272]. Cosic and Cosic further identified resonant electromagnetic energy transfer at specific frequency for the specific activation and interaction for macromolecules such as proteins, DNA and RNA as the crucial driving force for macromolecules (protein, DNA and RNA) activation and interaction (Cosic & Cosic [39]). We already discussed nonorientability supporting resonance, Hameroff recognized the importance of resonance as the main interaction with the zero-point fluctuations (Penrose & Hameroff [75,76,192]) and electron delocalization in the anaesthetic state and the phenomenon of consciousness (Hameroff et.al. [74]): He observed that anaesthetic gas molecules act in hydrophobic (lipid-like, water-excluding) regions within critical target proteins...by weak van der Waals London forces between the anaesthetic and nonpolar amino acid groups (the same type of endogenous interactions occurring between non-polar amino acid groups in the absence of anaesthetics)". Furthermore "anaesthetics retard electron mobility—the movement of free electrons in a corona... [B]y forming their own London force attractions in hydrophobic pockets, anaesthetics may inhibit electron mobility required for protein dynamics, quantum superposition and consciousness". In other words, anaesthetics control by retarding or impeding the electron delocalization that operate on aromatic molecules quintaessential to topological shaping and the possible orientable to nonorientable reconformations. Nonanaesthetics may be understood as occupying hydrophobic pockets without altering electron mobility, and convulsants as forming cooperative van der Waals interactions, which promote excessive electron mobility and protein dynamics in excitatory proteins.

We observed that biophotons emission could be produced due to change of conformation by re-organisation of chemical bonding, which is known in biology as chemiluminescence [225,229]. Furthermore, chemiluminescence is recognized to be the basic process for syntropic – rather than entropic-organisation of organisms using stored energy of chemical machines, and that is the resonant energy communication of quantum entanglement in organisms (Ho [79,80]). McClare argued that this can produce using a chemical reaction a form for producing useful work. Namely a "molecular energy" mechanism, which depends upon the rapidity of resonant energy-transfer (Mc Clare [160])

In ORC OR it is this quantum superposition which while in superposition, tubulins communicate/ compute with entangled tubulins in the same microtubule, and in other MTs in the same neuron. Quantum states of tubulin/MTs are isolated/ protected from environmental decoherence by biological mechanisms which are related to liquid crystals which may have nonorientable configurations, such as phases of actin gelation, coherent water domains where the polar ordering produce a Möbius strip, and alike tubulin they assemble and disassemble by absorption and emission of light, respectively, coherent pumping and topological quantum error correction.

15. Consciousness states, warps, multiconformation molecules and anaesthetics, Klein Bottle supraduality and Eastern Philosophy

The lingering question is: if after all, whatever the kind of states of consciousness are, under deep anaesthesia or awaken, if in all cases the reduction to a single state is ascribed to the "source of consciousness" what is the source for the differences between them? Would we consider anaesthesia as a special case of induction of a blockage of the possibility of recognition of sensorial experience as if arising from a given exterior world, together with the blockage of transforming this into action due to the induced immovilization?

We are taken to contrast anaesthesia with the experience of sensory deprivation. Say, by immersion into a lightless soundless ambience (incubation, as were known the long periods of sojourns in dark caves (Kingsley [102]) where motion was allowed, the ingestion of hallucinogens, or still the submersion in a water tank with no further restrictions. What is reported in these situations is that rather consciousness receding, fading or becoming impoverished, exactly the opposite is eventuated. Their importance to the transformations of civilizations and human consciousness has been highlighted by several scholars [102,103].

These transitional situations are called warps, and what warps elicit is a supradual logophysics. For instance the hypnagogic warp from the waking phase to the dream phase of consciousness and the inverse *hypnopompic* warp from the dream phase to the waking phase. Paraphrasing Benny Shanon, a cognition psychologist who travelled across South America to experience entheogens for understanding the psyche, what is elicited in these experiences is the "antipodes of the mind" being unified [248]. They involve the autonomic nervous system (ANS) as it channels energy to the sensorium, and does so through two systems considered complimentary: the excitation system and the relaxation system (Gellhorn [57] Gellhorn and Kiely [58]). It is the combination of these systems and the evocation of symbols what elicits the complexity of awareness as a cycle of meaning. This latter cycle also operates in supradual terms (Rapoport [236]). With respect to anaesthetics (consciousness reduced) and hallucinogens (consciousness enhanced), they act in opposite ways (Hameroff et.al., [74]), which we shall discuss below.

Warps are conceived as "discrete phases, a phase cognized strip of unfolding experience dominating for some time the sensorium, has a causal influence on the organisation of the phase succeeding it, and a causal influence on the end of the phase preceding it" (Laughlin et.al. [142]) . Thus, they are keenly related to memory, and they have an anticipative character, to the effect that perception has an apperceptive nature: Prediction is involved in the selfreferential interoceptive homeostasic regulation and related to a Bayesian inference process and the Markov blankets. We already discussed their Klein Bottle topology.

Laughlin et.al.: "The *awareness* of experience involves cognition about self and world. The root meaning of "awareness" is the same as "wary" and connotes careful attention to the world and detection of danger. Because the definitive characteristic of awareness is recollection, remembering, or recognition of patterns in experience, awareness obviously refers in part to a role played by knowledge in experience. Further, since the *recursive* [my emphasis] quality of segments of experience may be cognized as recurrent, awareness itself is organized into different "states" of consciousness" (p.109; ibid.). So awareness is related to memory, and the selfreferential re-enactment of its "contents". Thus if nonorientability would arise associated to the neuron cytoskeleton, then memory as an inherent behaviour of these material structures would be embodied as nonorientability of the cytoskeleton dynamical tensegrity structure. Upon collapse of such microtubules the Möbius strip shaped soliton arising as the pattern produced by a nonorientable microtubule buckling would be the memory in a zero-energy process [12]. Thus, through another warp it may in turn be evoked and in this conception "a minimal reentrainment across warps is all that is required for integration of phases into some semblance of a cognizable continuity" (Laughlin et.al., p.121 [142]). The experiences reified by warps do not "lie" in another branch of the universe as Penrose & Hameroff put it [75,76,192]. McGilchrist qualifies the latter invocation as the dominance of the either/or dual logic action proper to the analytical function of the brain left hemisphere over the intuitive direct cognition holistic integration by the right hemisphere [312].

From the topological chemistry paradigm. taking in account the already discussed Van der Waals-London forces operating the multistable topological shaping the research on both the reduced-consciousness anaesthetic states and the enhanced-consciousness-psychedelic states, say Ayahuasca, LSD, DMT (Dimethyltryptamine), we can phrase it like Hameroff et.al. which spell it out in terms of dependences rather than associations [74]. The different states of consciousness as elicited both by the anaesthetic and psychedelic states appear to be related "to quantum processes in hydrophobic pockets. Furthermore these quantum processes are inhibited by anaesthetics which impair electron mobility in van der Waals London forces", i.e. related to the metamorphical condition of aromatic multiconformations. Further: "The same processes are enhanced (hallucinations, but also enlightenment) in the presence of psychedelic drugs, those with indole rings donating electron resonance energy to indole rings in tryptophan within the hydrophobic pocket, forming a collective quantum state. Consciousness depends on quantum states of electrons within hydrophobic pockets in a class of brain proteins". In other words, as already stated anaesthetics block the quantum processes and computations -which ultimately depend on the Klein Bottle-Hadamard operator- by constraining the electron delocalization and thus multiconformation metamorphical topological transformation. In distinction with the latter, the altered-enhanced-consciousness states induced by psychedelic conformations or meditative techniques appear to enhance electron delocalization towards the formation of coherent collective states. Thus they induce a metamorphical supradual logophysics of multiconformation topological transformations proper to the two versions of aromatic molecules. One of these enhanced yet seemingly rare state is called in the Eastern traditions by different names: Nirvana, Satori, Turiya, perhaps others we are not informed of.

Notably, the shamanistic and Shinto roots of Japanese culture, and the later inception of Zen Buddhism have known and continue to practice meditational states. They allow for the first-person account of consciousness which is mostly decried by science as falling into the figments of our imagination, a practice which anyway has been and continues to be the hallmark of the curiosity and enthusiasm of the human condition. They procure to achieve the state of "supreme unification", which has the supradual logophysics at its core. Hagiwara: "the work of the Japanese Zen master Hakuin [Hakuin Ekaku (白隠 慧鶴 1685-1768)] utilized the idea of the Möbius strip (and the similarly paradoxical Klein bottle) in his expression of satori (nirvana). …a crucial visual expression of both Hakuin's satori and the Buddhist Dharma (Law)" (Hagiwara [71]). In the Indian *Advaita Vedånta* philosophy there is a fourth state of consciousness other than the three states of awakeness, dream-filled sleep and deep sleep: beyond deep sleep, Turiya. It is discussed in Verse 7 of the Mandukya Upanishad [178]:

"Not inwardly cognitive, nor outwardly cognitive, not both-wise cognitive, not a cognition-mass, not cognitive, not non-cognitive, unseen, with which there can be no dealing, ungraspable, having no distinctive mark, non-thinkable, that cannot be designated, the essence of assurance, of which is the state of being one with the Self the cessation of development, tranquil, benign, without a second, such they think is the fourth. He is the Self (Atman). He should be discerned" (Olivelle).

Hence, Turiya is transcendent and immanent, the metaform for cognition and perception, the metaform for both pattern formation and recognition, yet neither Outside-Outside nor Inside-Inside, neither Inside-Outside and Outside-Inside ("nor even both together"); beyond perception, invisible, ungraspable and without distinctions, and still supports all the other states of consciousness, the Klein Bottle.

To resume this section: prompting consciousness as related to microtubule dynamics primarily elicited through anaesthetics has brought us to examine states of consciousness other than the awaken state. They appear to indicate a supradual logophysics where memory is highlighted as a primary function. This will be discussed in terms of microtubule dynamics and buckling patterns below.

17. The topological chemistry paradigm and the origin of life and meaning.
Returning to the supraduality that the multiple "split personality"conformations embody, this is relevant in regards to biosemiotics and the characterization of life as related to artifact-making, for which molecular motors -particularly the cytoskeletal protein motors- and DNA may play the role of paradigmatic examples. According to biosemiotician Barbieri this is setup in terms of the Inside/Outside categorical divide and still in the alledgedly linear character of the geometry of the biological digital codes, and its digital character. Barbieri: "…molecular biology has discovered that the production of genes and proteins requires not only catalysts but also templates. The catalysts join the subunits together by chemical bonds, and the templates provide the order in which the subunits are assembled. It is precisely that order that determines biological specificity, the most important characteristic of life, and that order comes from a molecule that is outside the assembled molecule. This is precisely the characteristic that divides spontaneous objects from artifacts. In spontaneous and in catalyzed processes, the order of the components comes from within the molecules, i.e., is determined by internal factors, whereas in genes and proteins it comes from without, from an external template" ([318], p.11) *:*

Actually it was biosemiotician Hoffmeyer who first suggested that life appeared as a lipid membrane surface produced an Inside and Outside in an asymmetric relation vis-à-vis an autocatalytic system of compounds, namely the elements of the cytoskeleton and water (see Rapoport [232]). Its protoconsciouness described as "the never-ending *interest* of the insides into their outsides or, in other words, cellular *aboutness* [the term introduced by philosopher Brentano, considered by phenomenological philosophy as a protoform of consciousness]...this "interest" should be understood as a property that ultimately was derived from the primordial membrane itself".

However, as we have just seen and it turned out to be the case of genomes, rather than linearity we have that Nature introduces memory through the torsion encoded as nonorientability and develops its genomic codification in palindromic sequences related to nonlinear non-orientable surfaces, and the digital character of the codifications is derived from the Klein Bottle logic [234]; we shall later discuss the possibility of reconstructing the continuous signal by adequate data sampling of a discrete series. Furthermore, what chemical topology has shown is that these molecular motors can be synthetized from the application of Möbius-like twists that *anticipate* life, so that Barbieri's account appears to be amenable to be understood in terms of the non-orientable surfaces of self-reference, rather than requiring an elaborate distinction between life and non-life as discussed in (Rapoport, [230]): The dualistic ontology cannot actually integrate what it fractured, ab initio. We have already discussed the semiotic sign-processes origin of life and memory, as well as the interpretation of codes in terms of topological nonorientable folding. If life is identified as artifact-making, clearly chemical topology shows that this is unprecise –or still, ontologically mistaken- with regards to the actual non-dual ontology and logophysics that operates as an *ontopoiesis*, *the supradual logos as the source of creativity (*Rapoport, [230]). The semiotic agency that develops as complex structures required for life exist in the chemical domain in terms of *non-orientability* while in biology the multisystem integration of an organism requires an overall control which penetrates in the particular subsystems and selforganises Klein Bottle wise (Neumann [173] ; Rapoport [234]). Chemical topology dispels the claim of a divide between the "analogue world of chemistry and the digital world of life." Still, instead of envisaging the Interior/Exterior duality, the interplay of templates, genes, proteins and catalysts is to be understood in terms of a HyperKlein Bottle cybernetics [230 -234,236].

Examining the development of structure and tissue differentiation in cells, it was found to be mediated by an active cytoskeleton propagating alike light ways in terms of the Fermat Principle (Gordon [13]). We proposed that the cytoskeleton as a tensegrity has a physical support on optical vortical electromagnetic or sound waves network operating as emmiting and absorbing antennas, with nonorientable polarizations interacting with the coherent liquid crystal nonorientable topology of the intracellular coherent water domains (Rapoport [225,229]). Remarkably, Hameroff has proposed that acupuncture practiced in Traditional Chinese Medicine may be related to holography having microtubules operating as tuned resonators carrying electromagnetic pulses which by either stimulating or relaxation with a needle applied to the upper epidermis induce membrane ion currents to the effect of reaching a state of homeostasis . (He further suggested that Chi, the Chinese energy field would be coherent photic energy from the Sun and stars which is refracted by the upper epidermis (stratum corneum) further resonating with the microtubules network). We shall further discuss the notion of microtubules as electronic resonator devices and antennas.

18. Cytoskeletal dynamics, morpho-logic, topological phases and nonorientabiity

A coherent network is not exclusive to the brain-mind, genomes with their dynamical nonorientable dynamics resonating with electromagnetic and sound vortices with nonorientable phases or amplitudes, respectively, as if a coordinated orchestra of resonant topological phases. This tensegrity is the basis for a supradual logophysics rather than a dual-logophysics (Rapoport [225,229]). In cellular cytoplasm tensegrity structures are basically produced in terms of geodesic paths which continuously absorb-produce tension, so they are considered in terms of elastodynamics (Gordon [132]), though electromagnetic fields with eikonal light rays propagation can also be a natural interpretation; they develop topological anholonomities (Bliokh et.al. [21]). These differentiation waves support tissue differentiation and meiosis [225,229]. Since the wavefront propagation of gravitational and electromagnetic waves (light rays under the eikonal condition) in matter coincide as proved by Fock in his theory of gravitation [53] , and the latter coincide with the wavefront sets of elastodynamics (Bóna and Slawinski [23]), as wavefronts define the main morphological features, the actual physical content is, in general principles, secondary. *In principle, either the gravitational, electromagnetic and elastodynamical interpretations are undiscernible morphological-wise.* Actually, whatever the interpretation of the physical nature of the waves might be, as long as is consistent with the stuff of the object , environmental interrelations and the morphologies, they are methodologically valid. Certainly, it is hard to conceive the empirical possibility of discerning which is the case.

However, recent experimental settings have allowed to show the filaments deep inside the neural branches check the potential difference across the junctions by sending and receiving electromagnetic signals (Ghosh et.al. [64]; Agrawal et.al. [133], Singh et.al. [254-256]). Filaments and four ordered structures inside a neuron fire a thousand times faster than the ionic current. Filamentary communication is less known to the ionic signal transmission whereas the two are intimately linked through time domains. Novel scanning methods were introduced by the Hindu group, dielectric resonance microscopy (DRM), to operate in two-time domains, milliseconds and microseconds simultaneously, for the first time. It was found that four ordered structures in the cytoskeletal filaments exchange energy ~250 μs before a neuron fires, editing spike-time-gap—key to the brain's cognition. Also the control of the neuron process was achieved, either by deterring from firing above a threshold or initiating a fire by wirelessly pumping electromagnetic signals. The experimental discoveries "support a new dielectric *resonator* model of neuron functioning in various time domains, thus suggesting the dynamic anatomy of electrical activity as information-rich!". Furthermore, a fundamental aspect that has been elicited by the resonator frequencies is that the cytoskeleton operates as a fractal architecture of triplets conformed by triplets supporting quantum topological phases. While tubulin and microtubules are effective insulators with low conductance upon applying to them an alternating electric current at specific "resonant" frequencies, their conductance become extremely high and microtubules excellent conductors. Excitation at these resonant frequencies causes microtubules to assemble extremely rapidly; Fröhlich-type condensation is invoked to describe them [56,151]. Remarkably the same superconductive-like behaviour is the case of Möbius strip electronic resonator devices and antennas to be discussed below.These resonant frequencies show selfsimilar conductance patterns repeating every 3 or so orders of magnitude, over 15 orders of magnitude in the brain, from the quantum world to the electro-encephalogram ("EEG"). From terahertz through gigahertz, megahertz, kilohertz and hertz, conductance at each of these frequencies showed self-similar "triplet-of-triplets" resonance patterns

To resume, there is more to the neurons that ion transport: Indeed, that there is a subcellular cytoskeletal phenomenology to neurons which is subjacent to the process of polarization of neurons and their ensuing depolarization. This cytoskeletal behaviour is continuous, wavelike and thus neurons are conceived as continuous dynamical systems, described by nonlinear differential equations, for which their stability and the asymptotic state known as attractors are crucial. However, rather than perception being continuous or discrete, exclusive or which is the case of dual Boolean logic, the Klein Bottle logophysics integrates both. We have discussed this already in terms of the warps transitions of awareness states. Notably in morphological terms the abrupt nondifferentiable form of a spike as a wave is a point singularity, i.e. formless, which according to the current paradigm further propagates on the neuron membrane. In other words, since this signal is formless, it cannot generate in-formation as morphogenesis. It stands for the "it out of bit" digital dual-logophysics proposed by Wheeler (see Rapoport [234,236], Floridi [52]). In the terms of semiotics for which signs are architectonical organisation of packets of energy (Taborsky [267]) the singular dotlike packet of the spike as a morphless architecture cannot elicit any meaning at all, it bears no in-formation, no induction of morphology! Actually the identification of topological folding together with a membrane as precursor of life was in terms of folding being a semiotical meaning-conveyor sign as expounded by Hoffmeyer. As the neuronal action potentials are the primal signs of the brain-mind, be that in its preponderantly unconscious operation and its selfawareness mode, the process of meaning assignment cannot be generated nor sustained by a pointlike formless signal. The attribution of a non-differentiable Dirac-delta like ion-based spike has been brought into empirical examination, departing from the notion that there is much more to the neuronal behaviour the ion currents which is based in the continuous so-called subthreshold phenomenology.

19. Nonorientability and the discrete optimal representation of continuous signals: The topographic maps of the sensorium as a discrete sampling reconstruction of a continuous signal

The dual-logic spike paradigm conceives the number of spikes as the discrete variable for a "neural coding" such that objective properties of the world are communicated to the brain in the form of spikes. But the pointlike nature of spikes in this paradigm cannot support any meaning at all, as already discussed. Moreover, the details of the cytoskeletal phenomenology have been obliterated although they support the behaviour of the neuron. In contradistinction perceptual systems construe models introjecting from relations between sensory signals and actions, which are subjected to constraints which determine the outcome of the underlying physical and chemical laws, which have a nonorientable topology for signature. In the visual and somatosensory systems their topographic representation of periphery-to-cortex-to-periphery relations conform a Klein Bottle logophysics as elicited by Werner &Whitsel for the somatosensory system [119,281]. Neurons in the hippocampus represent the rat's position mapping (Mac Naughton et.al. [161,162]) producing a Möbius strip (Rapoport, [236]).

Brette: "Neural codes are inadequate for this purpose because they are unstructured and therefore unable to represent relations…The coding metaphor [following Shannon's statistical theory of communication wrongly deemed information-wise] tries to fit the dynamic, circular [selfreferential Klein Bottle-wise], and distributed causal structure of the brain into a linear chain of transformations between observables, but the two causal structures are incongruent" [26].

Indeed, nonlinearity is incongruent with discreteness, unless we represent the former as a discrete representation of a continuous signal whose maximal frequency, f_{max} is incorporated into a sample rate that permits a discrete sequence of *samples* to capture all the information from a continuous-time signal of finite bandwidth. This is guaranteed by the Nyquist–Shannon sampling theorem in the field of signal processing [157]. It serves as a fundamental bridge between continuous time signals and discrete-time signal processing. Indeed, if the continuous signal has a maximal frequency, say f_{max} a sufficient sample-rate processing is therefore anything larger than $2f_{max}$. Equivalently, for a given sample rate f_s perfect reconstruction is guaranteed possible for a bandlimit less than half the sample rate.

We can consider this duplication, the 2:1 harmonics, as a representation of a continuous signal performed on a non-orientable Möbius strip or a Klein Bottle with length $2f_{max}$, on which we take samples distributed on it that exceed its length. They need not be distributed homogeneously.

Whether the nature of perception is discrete or continuous is unsettled (van Rullen & Koch [137]). Brain function involves oscillations at various frequencies. This could imply that perception and cognition operate periodically, as a succession of cycles mirroring the underlying oscillations. In the case of EEGs we already saw that they operate as a system of 2.1 harmonics conceivable as overlayed

Möbius strips with cross-interaction. Also, the cytoskeleton appears to be organized as a fractal organisation with specific frequencies which appear to organize as chronomes, time structures (Rapoport [226]).

Contrary to earlier views of discrete perception as a unitary sampling rhythm, contemporary evidence points not to one but several rhythms of perception. Would we consider the discrete case, then it is in principle possible to construe a representation of it as if a continuous signal, optimally, by the previous prescription and the metaform of such construal is again, a Möbius surface, as if the continuous model would be the original signal.

This reconstitution by considering the sampled discrete signal as if disposed on a Möbius strip or a Klein Bottle -the 2:1 harmonic being also the case- leads us to consider the Klein Bottle topologies of the visual and somatosensory topographies as such general reconstruction process. In this process the Fourier transform plays the crucial role (see Pribram [125,196]; de Valois [306]) that already the Klein Bottle invariance of the phase of the cortical receptive fields shows that the Klein Bottle as metaform of pattern formation and pattern recognition would be preserved along the reconstruction process. In the contrary perception and cognition would be impossible.

20. Morphology of spikes as revealed by resonance and topological anholonomities

Returning to morphology as the natural analogical codification as in-formation, shape supporting meaning, spikes could not be pointlike. Such a singular signal to be transmitted along an axon, would not survive, it would dissipate instantaneously (Ghosh et.al. [64]). An actual form is required to this effect, "the spike should have the appearance of a ring, encompassing the diameter of a cylindrical axon or dendron…" (Ghosh et.al.., [64])). In other words, the form of the sign should somewhat be accommodated by the architecture which supports the sign as its transmission, in this case a cylinder, the form of the axon or dendrite, as an electrical circuit the connecting "wires" of the neuronal architecture. The suggested form actually observed is a ring, moving pairwise: two confronting rings; see fig.1 in [64]. Furthermore, they are associated to a 3D electric field, which in homogeneous media appear as Möbius strip polarizations transversal to the propagation of the field.

Bandyopadhyay and collaborators used nanotechnology at ambient temperatures to measure quantum resonance conduction in microtubules as a function of applied alternating current frequency. They found self-similar conductance patterns at specific frequencies every few orders of magnitude, i.e., in terahertz, gigahertz, megahertz, and kilohertz, equivalent to the 10^{-4} seconds. These results also were claimed by Hameroff & Penrose to confirm the ORC OR theory.

Action potentials and the cytoskeleton, protein and microtubule morphogenesis of the living brain appear to selforganise as topologies associated to Berry phases related to resonances of vortical dynamics (Bandyopadhyay [14]). They pertain to an intricate cellular tensegrity architecture which responds to alternate currents resonating in ranges which reach terahertz values. This confirmed the hypothesis that neurons are resonators in which the complex architecture of the subcellular level resonate to a now identified spectra of frequencies. Each resonant frequency produces a phase shift: "while passing through a microtubule nanowire, the ac signal shifts its phase by $n\pi$=4, (n = 0,1,2 . . .). The sum of phase shifts for a set of resonance frequencies forming a group is 360°, so completing a phase cycle is not incidental" (Ghosh et.al., [64]). Indeed, 360° is the topological (an)holonomy of the Möbius strip, as already shown. Rather than action potentials being morphless, points, the Hindu group was able to identify them as pairs of luminous annular shapes, two electromagnetic optical vortices sitting above the terminating dendrites. The nonorientability suggested by these empirical findings of luminous structures point to characterize each of them as a "solid" Möbius strip, the pinched-torus Klein Bottle. Remarkably it appears in the modelization of visualization by the moving subject while watching a distant moving point, the so called parallax vision, to be presented elsewhere by the present author (see **https://commons.wikimedia.org/wiki/File:Pinched_Torus_Klein_bottle.jpg**).

21. Classical-quantum resonator devices as metaphors of the neuron cytoskeletal architecture

We return to the discussion of nonorientability supporting boson-fermion systems. This is generic as can be seen in 1d and 2d electronic and microwave devices, a special class of transmission lines of finite extent with well defined boundary conditions such as coupled resonators systems embedded on a Möbius strip. As a proof of concept, on producing such a nonorientable embedding of a linear system turns to behave like an infinite yet compact transmission line with remarkable superconductive-like behaviour. These systems are low-pass filters with remarkable superconductive properties. They have been used to design antennas, magnetic resonance and nanodevices, since they allow for efficient compact packing producing an infinite transmission line with half the resonance frequency of the original untorsioned line (Pond [194,195], Rodhe et.al. [239]). They enable, in principle, a Tesla Coil for global wireless transmission of electricity and high-temperature superconductors.They give rise to boson-fermion symmetries, where the non-orientable 4π symmetry are formal patterns of annular resonators with the 2π symmetries (Ballon [11], Hamilton et.al. [77]): Spinor fields arising with 2d devices.

Together with the findings of the Hindu group, this suggests that the cytoskeletal neuron architecture, conceptually at least, can be conceived as a recursive architecture of resonators which may functionally operate alternating in the nonorientable and annular ring forms: This is an alternative to Hameroff & Penrose ORC OR theory to be discussed below. A resonator fractal architecture turned indeed to be the case though nonorientability was not mentioned at all [14], nor by them.

This differs with the quantum reductionism of Hameroff & Penrose who try to anchor consciousness in a rather remarkable superposition, on the one hand the microtubule structures of the cytoskeletal neuronal structures -which we extensively showed to be anchored in a supradual logophysics- and on the other an unknown "quantum gravitation", which these authors invoke as a metaphor for a model which we describe below. They further invoke the neuron dual fire/non-fire paradigm while they focus their attention on the subthreshold cytoskeleton continuous phenomenology of integrate-and-fire neurons where the resonator character of neurons is central and related to a supradual principle. They invoke the findings by the Hindu group on the multifrequency resonant behaviour of the elements of the neuron cytoskeleton as supporting evidence of their ORC OR theory. Remarkably they refrain from mentioning the vortical structures and the topological phases arising in physics and revealed by this group in their investigations of the action potentials arising from the subthreshold cytoskeleton phenomenology. They chose to posit the appearance of consciousness as arising from the decoherence of the purported quantum mechanical phenomenology which they refer to massive particles which somehow alter the coherence of the quantum fluctuations to produce the actual world of experience, one of the many branched world of possible states, consciousness the emergent agency. For them the collapse of the wave function is an agency-independent physical phenomenon, thus the "objective reduction" (OR) qualification, which further "provides the bridge between quantum and classical physics". They depart from extending the Schroedinger equation modified by random terms associated to quantum fluctuations, which somehow are resolved into decoherence by the microtubules.

That quantum physics is indeed the case is evidently associated to the electromagnetic character of the action potentials. We recall that Dirac's equation of relativistic quantum mechanics, both the linear and nonlinear forms, for a Dirac-Hestenes spinor field, and Maxwell equation -as a single equation- for the electromagnetic field, as written in a Clifford bundle, are equivalent, under the proviso of a rotation on the 2d spin-plane [207,208,217]. The latter is not necessarily a 2d-plane, but in principle can be a nonorientable Möbius strip as arises in optical vortices with nonorientable polarizations (Bauer et.al., [17], Bliokh et.al. [21], Freund [54]).

22. Supradual Klein Bottle logophysics, quantum cognition and Matrix Logic

We already discussed the nondual analogic of the tubulin code.The topological phases emerging from the resonant subthreshold phenomenology appears as a supradual logophysics related to the elementary brain phenomenon, which also is the case of cognition and the unconscious modeling by the brain-mind of object-kinematics and the perception of music. Action potentials rather than morphless nondifferentiable spikes are pairs of electromagnetic annular-like optical vortices related to topological phases.

We propose otherwise, incorporating the supradual logophysics of the subthreshold phenomenology, which is supported by topological phases *and* the quantum-mechanical-likeness of the mind, in the sense considered by Quantum Cognition, to be discussed below.

Neuropsychologist cum anthropologist Jacobo Grinberg-Zylberbaum proved that macroscopic entanglement of minds is the case particularly when strong emotions are involved, as elicited in pair of lovers undergoing meditation [68,69]. So the quantum-likeness of the brain is not to be looked upon as exclusively brainwise.

This leads us to topological entanglement, the Klein Bottle. The non-Boolean logic of the mind, operating inferentially, deductively and memorywise -the inversion of operators for backtracking cognitive state- is indeed noncommutative: the Matrix Logic based on the Klein Bottle, after August Stern [265]. In the former, the mind operators as 2x2 matrices operating on cognitive bra-ket states as vectors on a plane. These operators may be invertible and in distinction with the standard physical quantum operators they may be non-hermitean, so real eigenvalues proper of measurement of physical processes are superseded by complex values. They can be written as quantum field creation and annihilation operators, and some are nilpotent, non trivial square roots of the null operator. Particularly they arise as quaternionic light rays giving rise to torsion fields with 3d polarizations represented as Penrose's twistors (Rapoport, [223,220,226]).

The most fundamental noncommutativity in this logic is that TRUE and FALSE operators are noncommutative. This introduces a torsion given by the commutator, which we shall denote as M,

$$[\text{TRUE},\text{FALSE}] = \text{TRUE} \,.\, \text{FALSE} - \text{FALSE} \,.\, \text{TRUE} = \text{TRUE} - \text{FALSE} = \text{M}.$$

The dots in the first identity stand for matrix multiplication. M is the fundamental operator of Matrix Logic. Nonlinear multiplication is turned to additive structure in the third expression. M is nilpotent, $\text{M}^2 = 0$. This commutator, i.e. M = TRUE – FALSE, is the fifth side of the pentagon given by TRUE and FALSE for the other sides of a dislocated parallelogram, alike u and v in Fig.2. Here TRUE = $\begin{pmatrix} 0 & 0 \\ 1 & 1 \end{pmatrix}$, FALSE = $\begin{pmatrix} 1 & 1 \\ 0 & 0 \end{pmatrix}$. Both TRUE and FALSE are idempotent, i.e. $\text{TRUE}^2 = \text{TRUE}$ and $\text{FALSE}^2 =$ FALSE, and so they stand for projection operators.

In addition to the previous antisymmetric form, we have a symmetric form, which together with M becom the two fundamental cognitive operators of Matrix Logic. It is

$$\{\text{TRUE}, \text{FALSE}\} = \text{TRUE} \,.\, \text{FALSE} + \text{FALSE}.\, \text{TRUE} = \text{FALSE} + \text{TRUE}.$$

In M we find the two possible orientations, plus 1 and minus 1, *superposed* as either the Möbius strip or Klein Bottle. Matrix Logic is isomorphic to 2-state Quantum Mechanics. Alike Lie groups as torsion manifolds where the commutator of the elements of the Lie algebra yields the structure coefficients, the torsion tensor, the torsion of cognitive space appears as a non-null commutator. The five-fold symmetry of cognitive space. The importance of this lack of commutativity cannot be overstressed: it introduces the notion of a cognitive operator associated to this topological superposition -vis. nonorientability, a logical derivative and topological charge associated to cognitive loops. This is the domain of the mind, for which knots and their tying and untying are fundamental. Rather than this being separated from the geometry of the mind, which is torsion, they are united. We recall that the brain is qualified as a geometric "engine", referring to the visual system and its ecological integration. However as visual cognition working through projectivity, as introduced by the Renaissance Masters, is integrated to the mind by antipodal *topological* identification which produces a Möbius strip, from which selfhood arises unconsciously [236]. This is also the case of the hippocampal spatial mapping of the environment by rats, as elicited by Moser & Moser with the Norwegian University of Science and Technology neuroscience group (Mc Naughton et.al. [161,162]). It also is the case of the unconscious modeling of the kinematics of objects, or still the cognition of music, as elicited by the Principle of Complementarity of cognitive psychology. Remarkably, in the quantum reductionism that aims to anchor consciousness as emerging from Quantum Mechanics, the topological nature of the mind's operations and its relation to torsion are unconsidered altogether. Yet as introduced in this article, it is in the nonorientable and the torsion primal anholonomities where this logophysical fusion of the mind and the physical -in the ample sense- domains operate in the construal of experience. This further extends to the seemingly abstract topological phases, which are no less "real" than the case of physical space or spacetime, say the Aharonov-Bohm phenomenon, which actually is a trace-torsion differential 1-form. Topological charges is the case of the hypercolumnar architecture of the primal visual cortex and its Klein Bottle topology. It is also the case of condensed matter physics and particularly liquid crystals, the matter of which the brain is made of, and genomes too where nonorientability is eminent in its palindromic sectors (Rapoport [233]).

23. The unbroken information phase as nonorientable topology, Matrix Logic and Quantum Mechanics, the nonorientable topological form of coherence/decoherence cycle and holography

In Matrix Logic information is considered as something physically real, no different that space and time, with thinking as its manifestation [265,220,230-236]. In classical physics we resort to classical probability and real numbers. In quantum physics complex numbers are required to describe probability amplitudes and phases, real and imaginary components respectively. But the mind-brain is neither quantum nor classical, but a distinct phase, its "matter" is information, as form-inducing (which we called in-formation) and as data which anyway have a form [31,32,231-234]: *the information phase*. This requires for its description the so called denktors, thinking states, which use both complex probabilities and amplitudes. However, as is the case of the topological paradigm of physics as appears in condensed and soft matter physics, Matrix Logic deals with topological charges, as already the pinwheel architecture of the primal visual system shows to be the case, i.e. the very periodic architecture of the visual cortex, whose topology is the Klein Bottle. The information phase appears as a transition from orientability to nonorientability, as already the unconscious modelling of the kinematics of objects or music cognition requires: a torsion on a plane further glued as the very basis for the appearance of memory as discussed before relative to the Complementary Principle of cognitive psychology. Or still, the appearance of memory of buckling patterns of nonlinear elasticity in microtubules. Upon performing the twist and further stabilizing by the topological identification of the sides of a priorly orientable surface the information phase is constituted. It is free of constraints as imposed by relativistic consideration. All points or events in the priorly bilateral surface become united independently of distance be that of space or time: a topological form of holism and holography as well. We already described though non-formally the relation between the Klein Bottle, i.e, the information phase on the one hand and classical logic as arises in classical physics, say with the validity of Newton's Third Law. We also indicated the primal role of the in-formation phase-Klein Bottle- in producing the classical dual logophysics, i.e. the phase of classical physics in terms of the principles of dual logic, and the further reconstitution of the former from the latter in a selfreferential cycle. This begs for the reformulation of the discussions on the classical world as the topological decoherence of the information phase.

In Matrix Logic logophysical operators, L, act on bra states $<q| = (\bar{q} \quad q)$ and kets $|q> = (\bar{q}\ q)^{tr}$, L|q> with $\bar{q}$ + q= 1. Thus, TRUE| q> = |1>, FALSE |q> = |0>, with |0> = $(1\ 0)^{tr}$ and |1>= $(0\ 1)^{tr}$, the true and false states.The scalar true and false values of dual Boolean logic correspond to the eigenvalues of these operators. They are orthogonal. Further consider the orthogonal quantum basis: $|S_+> = (1\ 1)^{tr}$ and $|S_->= (1-1)^{tr}$, so that <0|1>= 0, $<S_+|S_->= 0$.

These two states can be thought of arising from a surface given by a closed oriented band projecting on the xy plane. Thus to each side of the surface we can associate its normal unit vectors, (1 0) and (0 1). Suppose that we now cut this surface and introduce a twist on the band and we glue it to get thus a Möbius surface. We recall that this is what in cognitive psychology led to identify the genesis of memory, up to the topological identification, as discussed before. Now the surface has lost its orientability and we can identify one side with the other, so that we can generate the superpositions $|S_+>$ and $S_->$. Remarkably, these states are related by a change of phase, namely a rotation of 90 degrees. What the twisting and loss of orientability produced, can be equivalently understood by the fact that TRUE and FALSE are no longer as in Boolean logic and the Aristotelian propositional frame. Indeed, they are no longer dual and what matters is their difference, which in the case of scalar truth values does not exist. The other state also can be interpreted as a state that represents the fact that the states as represented by vectors, have components standing for truth and falsity values which are independent, so that the Aristotelian link that makes one the trivial reflexive value of the other one is no longer present: they each have a value of their own. The situation is radically different to Aristotelian-Boolean logic in that now (0 0) is another state, 'false and true'. This is the situation of 'Schroedinger's cat' and the Liar Paradox; they are topologically embodied. This state together with (1 1) , 'nor false nor true' state together with (0 1), true, and (1 0) false states we have a 4-state logic in which the logical connectives have been promoted to operators. We have thus returned to Nagarjuna's four-loci logic, the Hindu philosopher who for some reason -seemingly unclarified- called Aristotelian logic the 'student's logic' [299].

We have the expressions

$|0> = (|S_+> + |S_->)/2$ and $|1>= (|S_+> - |S_->)/2$, and

$|S_+> = 0> + |1>, |S_->= |0>- |1>$.

In other words, the classical states are the equally weighted signed superposition states, which themselves are the equal weighted signed Boolean states. While the classical states are ½ weighted the superposition states are not, though we can introduce the normalization factor $1/\sqrt{2}$. Describing the unbroken topological phase by the Hadamard 2x2 matrix H= $\begin{pmatrix} 1 & 1 \\ 1 & -1 \end{pmatrix}$ we obtain:

$H|0> = |S_+>$ and $H|1> = ||S_->$,

thus transforming the classical states into superposed states. In the state of true and false coordinates, the action of H on 0> rotated it by 45° , while the action of H on |1> flipped it. Reversing the previous, we still operate with H but halved, so that

$½ H|S_+> = |0>$ and $½ H|S_{-.}> = |1>$,

thus transforming the superposed states into the classical states. It is possible still to make the system unitary by normalizing by dividing by √2. To resume, through topological transformation tantamount to introduce nonorientability we can transform the classical eigenstates basis to the quantum states basis thus completing a selfreferential cycle upon doing this. Both complete basis of eigenstates are functionally complete and mutually spanning. We also have the decomposition of $M = |S_+> < S_-|$.

We recall that the Hadamard operator is the gate that allows transcending classical Boolean computations. The fact that it is also the interface between the quantum and classical states indicates the supradual primality on the brain-mind. In Matrix Logic quantum (for spin-up and spin-down states) and cognitive identities can be related interchangeably, through a spin logophysical operator,σ, which is a symmetric term of M. The decomposition is M = TIME + σ, TIME being symmetric and unitary, and $\sigma = \sigma_z$ the Pauli matrix. This link is a logophysical interface between the quantum and informational realms, in which topology, torsion, logic and the quantum world operate jointly. TIME is tantamount to the operator of phase conjugation, crucial to holography. Yet, we are no longer constrained to measurements appearing through Hermitean operators as their eigenvalues: Indeed, M is non-Hermitean. An important eigenvalue of the logophysical OR operator of the decision process, is the Golden Mean.

The exchange between the two domains can go in both ways, given that the Klein bottle has no independent Inside nor Outside, the quantum realm can be incorporated into the classical cognitive dynamics, while the logical elements can take part in the quantum evolution. Negative probabilities can be understood too in terms of non-orientability. This connection is what we may venture to be the support for the Principle of Complementarity in cognitive psychology, the ecological conception of perception as a system (J J Gibson [300], and to a nonorientable topological monism which we have elaborated in our works and related to Max Velmans' Reflexive Monism [301].

Pribram conceived that memory was 'imprinted' and processed holographically on the brain operating as a parallel computer [125], which we suggested operating in terms of Matrix Logic for which the matricial product is a form of relational connection with several error correcting operations [265]. In holography the brain operates as a phase conjugation logical gate (Mitchell [124]), which points to TIME for its realization. We have suggested that brain interhemispheric coordination and particularly visual binocularity operates as such TIME phase conjugacy (Rapoport, [222]). Brain holographics would operate producing an imaginary image of the quantum emissions, say photons emitted by the perceived object, which rather than being seen are participants of the process. The brain would use adaptive resonance of the input signals with the electromagnetic triplet of triplet already discussed, in its input path to generate the output image representation of the object and further project it as if Outside where we locate and experience the real object -even in hallucinatory states-, as proposed by Velman's Reflexive Monism [323].

As a final comment, the implicit and explicit orders introduced by Bohm are, in principle, incorporated into this linkage, but making this manifest requires elaboration. Also, the formulation of topological chemistry as related to cognition through the informational phase is a natural field to be pursued. With relations to knots as chemical configurations [118] it was examined by Stern [265]

24. On the superposition and interference of as quantum "macroscopic" systems. Penrose & Hameroff's theory revisited

Supraduality rather than duality is also the case of the "interference of minds" as superposition of states elicited in Quantum Cognition, which shows the "quantum-like" nature of the mind, as in decision processes, conceptualization, perceptual multistability, etc. (Aerts [2], Busemeyer [29], Khrennikov [100, 101]). Departing with a criticism of the physicalist reductionism for which Penrose & Hameroff have been mostly disapproved by neuroscientists, Khrennikov modeled the mind as if quantum-mechanical applying the so called Växjö interpretation of quantum mechanics, proposing a Hilbert space formalization of the mind in terms of "contextual probabilities". The Klein Bottle is contextual by nature. Khrennikov did not consider the neurological aspects in particular the supradual continuous subthreshold phenomenology of neurons which subtends the action potentials, nor the tubulin analogical code: The subthreshold phenomenology is indeed proper to quantum topological phases yet as the physical correlate of Quantum Cognition and the Klein Bottle logophysics of perception and cognition. That Boolean logic is not the case of the mind states already transpires in the ambiguity and uncertainty in our daily ponderings. It is the states of superposition which calls our attention, so that besides true and false states and operators, commutative in Boolean logic, a combination of both is also a mind-state as already discussed (Conte et.al., [37,38]). This is their take on the non-dual (viz. supradual) logic quantum-like structure of the mind understood that there is no quantum physical element committed to the mind but rather the interference and superposition of cognitive states. This runs counter the allegation that "Quantum behaviors are difficult to reconcile with our observed "classical" world. For example, we don't encounter superpositions in our perceived world, as the very act of measurement, or of conscious observation seems to cause superpositions to reduce to definite states - "collapse of the wave function." (Hameroff, p.7, [76]). These comments while appropriate to physical quantum behaviour in the atomic and particle scale, fails to acknowledge that in the mental world and its behavioural manifestations, say decision making, superposition, entanglement and disentanglement, or still gestaltic perception, quantum-like behaviour are the normal rather than unobservable or unencounterable events. They exist per se as if independent of actual microtubule behaviour though as claimed by Penrose & Hameroff they can be explained in those terms. In other terms, Penrose and Hameroff implicitly *assume the Cartesian Cut, ab initio*. Then naturally, in such a dualistic setting, it is hard, actually impossible to achieve surmounting the initial dualistic onto-epistemology. For the dual-logic logophysics, the mental domain is projected as an object to the presumably physical world of measurement free of selfreference [230], positivism as its implicit epistemology. According to McGilchrist this is the signature of the breaking of unity produced by the dominance of the left hemisphere over the right hemisphere of the brain [312], and very much responsible for the predicaments of Western culture after Descartes. For the latter see Rapoport [220,222,226,230,236]).

In other words, the 2-value logic is projected to a single value, all things other excluded., as logician and cybernetist Gotthard Günther claimed to be the inexorable case [302]. Günther criticized the nature of logic being related to propositional value, and considered it terms of relations displayed locationally, as appears to be the case of genomes in terms of the Klein Bottle logic [233]. Furthermore, in our previous discussion of the shape-based topological chemistry, or still in the plethora of physical phenomena where topological phases are omnipresent, say liquid crystals, acoustics, phonon modes propagating in lattices, shallow water waves and in scales of experience going down to nanomaterials, optics, they all manifest a supradual logophysics which is not beyond our pragmatics.

Remarkably, nonorientability and topological anholonomity as well as their bearing in the topological chemistry paradigm crucial to their unmentioned at all, as it arises in the superposition of molecular open and closed conformations of proteins, basic to their theory. Penrose & Hameroff do mention the relevance of π- electron delocalization in aromatics towards the formation of dipoles basic to ORC OR and further discussed them with regards to both consciousness blocking and enhancement. As already discussed, Hameroff in his study of computation in microtubules points to the central important of electron delocalizations at the very basis for hybrid conformations of molecules and the relevance to consciousness without mentioning their role in topological changes from nonorientability to orientability and the converse (Hameroff et.al., [74]), and this was carried out to the ORC OR theory entirely.

Most remarkably, Hameroff considered the 6-layered structure of cortices to suggest the so called giant pyramidal neurons which constitute layer V as the probable support for consciousness for being the location apex of the integration of Outside and Inside, the unity of action and perception which for the visual and somatosensory systems is embodied as a Klein Bottle, the latter as elicited by Werner & Whitsel [119,281] and proposed by Turvey as a Möbius strip unity [273] somewhat following the Pragmatist school of the 19th century after William James.

25. The integrated coherence of the brain-mind and the proposal of a non-orientable architecture associated to its synchronicity

Remarkably, in ORC OR theory, it is claimed that microtubules practice quantum computation in intervals in which they are isolated from the environment which they alternate with phases of classical interaction in which coherence is due to synchrony, the 40 Hz oscillation. It is this oscillation which is identified as providing the integration of the brain-mind, which has been identified as the neuronal function proper to consciousness (Buszáki [123])

This synchronization has been suggested to be based on an hexagonal lattice of somatosensory and visual primal cortical areas in which the hypercolumnar structure is unmanifested. In this model Möbius strip like synaptic patterns support the coherence across the cortex maximising synchronous oscillation (Wright et.al. [283, 284]). This lattice would arise by selforganisation during embryogenesis and development of neurons, consistently with the Free Energy Principle (Friston [55],[317]) and operating as a heterarchy of Markov Blankets -Klein Bottles (Rapoport, [234,236] optimizing adaptive learning in an analogical transition from vagueness to crispness of percepts. The Möbius strip would arise as the antenatal organisation of synaptic connection while the receptive field lattice architecture would be the later post-natal development of learned functional connections overlayed on the antenatal connection system. The authors suggest that this organisation could be general across the cortex and the possibility that the underlying principles could extend to the mutual self-organisation of synapses and interactions between cortical areas. Furthermore, it suggests a modular functional synaptic system, which, only in some parts of the cortex, and in some species, is manifest as anatomical ordering into columns. In other words, it would be more basic to the observed periodic pinwheel architecture supporting topological anholonomy, which would appear only locally across the cortex and for some species.

26. Further analysis of Penrose & Hameroff ORC OR theory

Upon Penrose & Hameroff presenting their ORC OR theory following the early findings of the Hindu group they claimed that they provide confirmation of their take that fundamental resonances that arise in microtubules such as anticipated by them [75,76,192]. This may be the case with regards to the microtubule dynamics of neurons, but to include this with regards to the theory of "quantum gravity" in completion of Quantum Mechanics due to the superposition of protein states in microtubules, might turn to be a step too far, which the authors seem to refrain from stating it explicitly. However, they also refrain from mentioning the vortical structures and the topological phases revealed by this group in their investigations of the action potentials arising from the subthreshold cytoskeleton phenomenology. They chose to posit the appearance of consciousness as arising from the decoherence of the purported quantum mechanical phenomenology which they refer to massive particles which somehow alter the coherence of the quantum fluctuations to produce the actual world of experience, consciousness the mediating agency. For them the collapse of the wave function is an agency-independent physical phenomenon, thus the "objective reduction" (OR) qualification, which further "provides the bridge between quantum and classical physics". They depart from extending the Schroedinger equation modified by random terms associated to quantum fluctuations, which somehow are resolved into decoherence by the microtubules.

Rather than consciousness producing the collapse of the quantum coherence of the superposition of protein multiconformations – which they attribute to a misinterpretation of their theory- they chose to posit the appearance of consciousness as arising from the said collapse. The latter, in their theory, is an agency independent physical process which produces the world of experience, thus the "objective reduction" (OR), a materialism of sorts. It is further produced by a superposition of curvature-gravitational fields which each lump is one of the conformations, which we shall discuss below.

As already discussed in the Klein Bottle logophysics topological superposition is the case of physics, chemistry, cognition, perception, semiosis, etc. Penrose conceives superposition differently. In his theory the superposition is a spacetime geometry property, entirely unrelated to torsion and nonorientability, nor the topological phases of the neuron's cytoskeleton but rather to the curved space of General Relativity. Hameroff: "…superpositions as tiny separated curvatures in spacetime geometry. Were such separations to continue… each spacetime curvature would evolve its own universe…" (p.2,[76]), Penrose deduced that spacetime separations would be unstable, and undergo objective reductions (OR) independent of a subject and based on the quantum uncertainty principle. In their theory, OR would stand for an extension of standard quantum mechanics, as the bridge between quantum- and classical-level physics as a 'quantum-gravitational' phenomenon. Whatever this bridge may originate from, be that environmental decoherence, the participation of an observer through a measurement of observation of any kind, multibranched universe from which a particular one arises from a "choice" or any interpretation of how the classical world of one actual alternative may be taken to arise out of fundamentally quantum-superposed ingredients". So what "quantum gravity" means for this theory, is in principle, as recognized by the authors, any theory in which the standard theory, viz. Schroedinger unitary evolution of quantum states in isolation, is superseded by a theory which considers the environment interacting with the system as a single system. This has been formalized in several proposals as the Dynamic Reduction Program or Collapse Theories (Diósi [44], Ghirardi, [62,63] , Pearle [186], Percival [188], Hughston [309], Adler and Horwitz [310], Rapoport [218]).

The metric-geometric proposal of ORC OR somewhat neglects the possibility of topological superposition embodied as nonorientability, and their omnipresent manifestation, say in liquid crystals pervasive to organisms and the brain itself. which does not presume neither the quantum nor the classical domains as primal. The Klein Bottle embodies and controls the cycle of transition from coherence to decoherence and its reconstitution, as a transition from supraduality to duality and the converse.

Furthermore, rather than consciousness producing the OR, Penrose conceives consciousness ("or an equivalent of it") as emerging from the OR. The OR process is considered to occur when quantum superpositions between slightly differing space-times take place, differing from one another by an integrated space-time measure which compares with the fundamental and extremely tiny Planck (4-volume) scale of space-time geometry. Since this is a 4-volume Planck measure, involving both time and space, we find that the time measure would be particularly tiny when the space-difference measure is relatively large (as with Schrödinger's cat)

To this effect Penrose conceived the selfenergy of the superposition acting as a threshold which upon trespassing it, specific curvature values are selected and the collapse to a single world is eventuated. This threshold would be connected to the gravitational constant, and though smaller than other constants arising from chemical or other processes, would be of crucial importance for the OR reduction to proceed, in a time range of 20 orders or magnitude smaller than usual quantum processes. Neuronal material would be involved in this reduction in order to produce the infinitesimal displacement and modification of curvature to allow the time threshold to operate in a process in which microtubules play a crucial role. However, rather than the environment playing a role for the reduction, this would be spontaneous and operate independently of the environment and in the cases of major masses, still a quantum entanglement of the environment and the quantum system would operate. Any reduction of either the environment or the system would be accompanied mutually by the other. This mutuality would introduce an uncontrollable stochastic element into the result of the reduction. Consequently any non-random element (which according to Orch OR is non-computable) which may influence the particular choice of state produced from the superposition would be completely masked by this randomness.

Further, according to Penrose a quantum-like unit of phenomenal experience, a "quale" occurs, as a "protoconscious" as a first step for the assessment or constitution of causality. To explain the reduction which Penrose & Hameroff to be alike a musical morphology of several chords and instruments -an "orchestra"- proposed that "microtubules inside brain neurons *orchestrate* quantum vibrational superpositions through resonance, entanglement and memory, guiding wavefunction evolution to threshold for "orchestrated OR" ("Orch OR"), events proposed to result in moments of full, rich conscious experience (more like music than random notes and tones). Sequences of such moments would give rise to our "stream of consciousness"" (Hameroff, p.3 [76])

Let us analyse the above. In their take consciousness rather than primal is somewhat derivative and emergent from a spacetime geometry related to the theory of general relativity (GR). Furthermore, on priming curvature rather than torsion from which in the theory of Cartan connections, i.e. nonnull torsion, they ignored that the covariant differential of torsion yields the curvature: the so-called Bianchi identities of the theory of connections, in differential geometry (Cartan [33] , Hehl [83], Sharpe [111], Sternberg [264] Rapoport and Sternberg [200]). This neglect is quite remarkable since these identities stand for the constitutive relational aspects of the geometries. Their theory does not consider that Minkowski or Euclidean space, with their flat null-curvature metric, can be the case with non-null torsion and still the curvature can be alternatively null or nontrivial, so their curvature is entirely a field which is derivative of a metric. There are several conceptual issues too. As much as our theoretizations are linked to the image-schemas we apply and the logic we chose, concepts and their consistence is a must. Given a metric, say in GR, the geometry is nonrelational -one object distant to the other but no field proper to the interaction. The metric geometry of GR is nonparticipative, and the principles of selfreference and hetero-reference- the two core principles of consciousness - are not its basis, torsion is null and thus nonorientability is not the case. Finally, this notion of spacetime is that of a container, for physical masses, which is the dual-logophysics purported by Descartes and Newton. But consciousness which for ORC OR is to emerge from this passive spacetime is not subjected to CONTAIN, as if a classical though relativistic mass would be the case. Consciousness is not characterized by containment, as the classical notion of matter is, and at its most basic level is relational and undissociable from the principles of selfreference and hetero-reference. Since the geometry of GR is based on metrics, it measures distances, separations between objects, a measure of objectivity which rules out interpenetration and thus setups a logophysics in which the principle of identity is dominant. Metamorphosis is outruled as well as setting up for derivatives of this distantiation pragmatics the hegemony of tertium-non-datur is emplaced: there is not third state but true and false ascriptions. Thus the mind is deleted. Already at its departure ORC OR vindicates a dual logophysics, Aristotelian-Boolean logic for its implicit onto-epistemology. Further, the metaphor of music ascribed to the flow of consciousness begs us to recall that the cognition of music reveals nonorientability (Mazzola [159], Tymoczko, [274] Shepard [250-253],Risset [237]). The stream of music and that of consciousness are strongly related [1].

We can state our case invoking the considerations that led to reveal the Klein Bottle as the metaform of pattern recognition as unappropiate to that effect, using instead the so-called persistent homology ([31,32,127]), which has come to be applied to study neural networks through their topological invariants. The underlying notion is that data has shape (Carlsson [31,32], Rapoport [232-234], be that the notion that the fundamental parameter to ponder is the discrete number of firings of a neuron. Accordingly the choice of the metric geometry in ORC OR appears to be inorganic, an abstraction from a world of unconnected entities signed by their masses, small for more, and entirely unrelatable, in principle and in concept, to the process they set themselves to explain, consciousness. With regards to the spacetime to emerge from the collapse of the quantum states in their theory, shape is not relevant at all, though it is implicitly the case of the morphology of tubulin in different conformations, particularly superposition states which they indicate their crucial role to their theory.

Their choice of geometry contravenes the very observation by Hameroff: "Atoms and small particles appear to exist as quantum superpositions of multiple possibilities, but yet we observe discrete, definite objects in our perceived "classical" world" " (p.14 [76]). Notwithstanding this observation in the "classical" world of forms, metamorphosis mediating, Aristotelian-Boolean logic is already superseded by supraduality. This is so particularly when instead of focusing in material structures considered synchronically we do so in considering processes, diachronically, processually, particularly those that arise in perception and cognition, say of the Necker cube, or Gestalt figures. Still to mention, also upon considering the mind, certainly is no object but for itself and for others exercising their theory of the mind making.

To explain the biophysics on which ORC is established and the relations with anaesthesia these authors considered for the main characteristics of the molecules, mainly liquid crystals and in particular water, their polarity and non-polarity properties. However, the nonorientable topologies of liquid crystals and water are entirely disconsidered, despite being crucial to cell physiology, namely to establish its overall coherence. The topological chemistry paradigm and in particular the two-version topology (Huckel and Huckel-Möbius) due to π- -electrons delocalization in aromatics is explicited by their theory, though no reference to topology but for tagging. After all the binding of molecules which is crucial to their model as the anaesthetic binds to tubulin is all about shaping accommodations, a topological issue. As early discussed by Sokolov they can be represented as modifications of the molecules electron distribution and the associated shape modifications and of topology. As Mezey puts it, the electron charge distribution of molecules is "fuzzy" ([118]). The three-dimensional shape of the molecule begs for topology, rather than valence skeleton representations: "molecules are not geometrical but topological objects" (p.8) and molecular shape is dynamical. The shape of a molecule is the molecule's electron distribution shape.

To put in terms of Hameroff & Penrose: "Non-polar anaesthetic binding sites in tubulin (and other proteins) are comprised largely of π- electron resonance clouds of benzene-like organic rings in aromatic amino acids tryptophan, phenylalanine and tyrosine. The basis for organic chemistry, π-electron resonance clouds are delocalized quantum objects, non-polar, but polarizable, and can induce and couple quantum dipoles, participate in quantum optical effects, and oscillate coherently in terahertz frequency ranges" [75]. This is it, in essence and concept. However, upon presenting the main argument for superposition in microtubules they do so in presenting the multiconformations of tubulin as shapes which are conceived either as unsuperposed or superposed interpreted as qubits. This is the basis for their "topological quantum computing" without ever mentioning the Hadamard-Klein Bottle information phase which is central to Quantum Information theory and supports transcending from classical to quantum computations, as already discussed.

These terahertz range frequencies are crucial to the whole range of oscillations of the cytoskeleton as a hierarchical *triplet of triplets* architecture which resonate in a range from hertz to terahertz eliciting topological phases (Bandyopadhyay [14], Gosh et.al. [64], Singh et.al. [134,254-256], Saxena et.al. [246]), and which make of the cytoskeleton a time-crystal. However, the topological anholonomities are still wanting for consideration to the genesis of spacetime consciousness. They claim a "quantum gravity" which would resolve the superposed states each producing a spacetime curvature into a single state, that of the manifest reality, a resolution which -they pledge- is tantamount to the emergence of conscious agency, through the so-called collapse of the wave function. Again, the left hemisphere dual over-rule is rather absolute, as McGilchrist puts it.

27 . Objective reductions and the dynamical reduction program

Basil Hiley showed that the quantum behaviour of particles can be described solely by using a hierarchy of Clifford algebras without recourse of wave functions nor Hilbert space. The essential aspect is the non-commutativity of Clifford algebras and the existence of idempotent elements which substitute the projectors of the standard theory [28]. We have just shown that in Matrix Logic the torsion produced by the commutator of FALSE and TRUE which are not dual in distinction with scalar Boolean logic, is nilpotent due to their status as projectors/idempotents. In the Klein Bottle logophysics non-commutativity is found in the chirality of taking the path around the singularity as in fig 3, either clockwise or anticlockwise produces a change of sign in the torsion closure. Covariantly represented in differential geometry terms, the torsion as a (2,1)-tensor $T^{a}_{b\,c}$ is anticommutative : $T^{a}_{bc} = -T^{a}_{cb}$. Would we change the order of the covariant indices, a change of sign is introduced [33,81-84,264,276].

Elio Conte upon noticing that the fundamental novelty of Quantum Mechanics is the non-commutatitivity of the product of operators of which Quantum Cognition sets to represent the decision process in terms of projectors in Hibert space, proposed instead a formulation using Clifford algebras, just alike Hiley (see Conte [37,38]). In this setting the e_3 idempotent, which he identified as the core element of the formalism, operates precisely as the previous decoherence. This topological decoherence contrasts not only with the quantum to classical physics transitions, but fundamentally with the resolution of quantum superpositions in Hameroff & Penrose ORC OR, where multiple states are reduced to one single state.

As we stated already a geometry with null torsion is such that selfreference and hetero-reference do not operate to its constitution if not banally, by their erasure. The issue is that there is no consciousness, whatever its mode, which is not supported by these two principles. For proof, the mirror-neuron phenomenology introduced by Rizzolatti et.al.. [238,50]. In particular we note that in the mind's resolution of superposition, as shown by Quantum Cognition, the mind entertains to consideration the different possible alternatives, consciously we can say at least of some of the possibilities and unconsciously too. It is the process of topological decoherence which produces the single unsuperposed classical state by action of the Hadamerd-Klein-Bottle-gate, and this appears to operate as if unconsciously. So consciousness appears to be primal rather than appearing as the process of singling out one mind-state alone from a superposition or multiple possible states.

The geometry of quantum states whose state vectors arise from the wave-like, quantum-level objects which evolve deterministically by the Schrodinger equation on state-space, also known as Quantum Geometry, is a projective Hilbert space [20] . Indeed a phase is a component in Quantum Mechanics which cannot be disconsidered, and very much so in logic with imaginary states too (Kauffman [112]). It is precisely this space on which Berry discovered the topological phase. A trajectory of a state-vector under the natural antipodal state-vectors identification produces a Möbius strip.

We already mentioned the Dynamical Reduction Program (DRP) or wave collapse theories as an attempt to complete Quantum Mechanics by accounting for the measuring or observing subject. On the one hand what we observe is discrete though seemingly continuous, while the cognitive process is analogical, continuous though framed in finite intervals, and mostly so when we are in a state of unconsciousness. We already discussed that despite the uncertainty whether perception is ultimately discrete or continuous, the latter can optimally be reconstituted from a discrete signal by sampling as if on a Möbius surface or a Klein Bottle. However, this reconstitution is always under noise, which is conceived as random. We also discussed that both the visual and somatosensory Klein Bottle maps embody these reconstitution under uncertainty.

The DRP approach consists in accepting that the dynamical equation of the standard theory should be modified by the addition of stochastic and nonlinear terms. After all, both the system, the observer and the instruments being applied to elicit the interrelations are all subjected to quantum fluctuations arising from the vacuum. The remarkable fact is that the resulting theory is capable, on the basis of a single dynamics which is assumed to govern all natural processes, to account at the same time for all well-established facts about microscopic systems as described by the standard theory, as well as for the so-called postulate of wave packet reduction (WPR), which accompanies the interaction of a microscopic system with a measuring device.

Given a Hilbert space of dimension n+1 , since the quantum state vectors are defined up to a phase, this leads to CP(n), the n-dimensional complex projective space of "rays" of quantum vector states whose evolution is deterministic as characterized by the Schroedinger equation for quantum state vectors. Upon measurement a state-vector reduction to a particular eigenstate is eventuated, the so called collapse of the wave function. This process can be described in "objective" agent-free mode by a nonlinear Schroedinger equation on CP(n) which assumes a non-isolation of the system vis-à-vis the environment.

This requires Stochastic Differential Geometry, yet not in configuration space or spacetime [180] but on CP(n) (see Hughston [309], Rapoport [218]). Remarkably, the Schroedinger equation on configuration space or spacetime manifold can be derived from Brownian motion with the drift vectorfield given by the logarithmic gradient of the logarithm of the wavefunction (Rapoport, [201-206,208,210-219,226,227]. They are crucial to Nelson's derivation of the Schroedinger equation in Stochastic Mechanics [172]) in terms of Brownian motions arising from quantum vacuum fluctuations and diffusion processes [174]). The logarithmic singularities of the eikonal- propagating torsion field may undergo quantum jumps [220,221], a primal distinction to the creation of visual experience [220,221,226].These Brownian motions are produced by dynamical geometries with torsion vectorfields or differential form 1-forms, Riemann-Cartan-Weyl connections: they are the basis for Stochastic Differential Geometry [201-204,2010]). The Laplacian operator has an additional to the metric Laplacian a first-order term given by the trace-torsion, the drift of the Brownian motions, so it fits exactly to the Fokker-Planck diffusion operator, or still, the differential generator of the generalized Brownian motion ([200-221,226]).Both the Fokker-Planck operator and the stochastic differential equation which extends the Schroedinger equation to the open case are intrinsic, they do not depend on any coordinate system ([218]). In the latter geometric formulation there is no separation of the quantum system and the apparatus/observer.

This extends to Brownian motion on CP(n) as a stochastic differential equation whose stochastic solution is a diffeomorphism of state-space, while the closed system case evolution produces a diffeomorphism of configuration space. With regards to the metric in the Quantum Geometry it is the Fubini-Study metric, while in configuration space metrics can be either positive-definite for the non-relativistic case or degenerate in the relativistic case (Oron and Hurwitz [180] , respectively (In the case of the Navier-Stokes of fluid-dynamics where the drift-torsion field is the velocity of the fluid, these diffeomorphisms stand for the solution flow of the Navier-Stokes equation [211-217;89]) .

In short, the standard Schroedinger flow on CP(n) is generalized to a random flow which has an additional term, a drift-torsion vector field, given by the gradient of the expectation value of the quantum Hamiltonian and the gradient of the variance of the Hamiltonian, the squared energy quantum uncertainty (Rapoport [218]). It is this torsion-drift term which drives the system to a fixed eigenstate. The noise term of this random flow also proportional to the gradient of the expectation value of the Hamiltonian can be written in terms of the Planck's constant, the gravitational constant and the velocity of light c, in a form that for the superposition of macroscopic systems which differ in little as a proton mass, the collapse to a single eigenstate is almost immediate (Hughston [309], Adler and Horwitz [310]) In this Stochastic Differential Geometry setting, the reduction of the quantum state to a single eigenstate does not require an observer for the system to collapse, so hardly we can attribute this as associated to an emerging consciousness, or a consciousness at all if not that probabilities and their expected values are involved. But then probabilities have a subjective-objective nature, rather than builtin into the quantum system. They have an epistemic function, a mode of knowledge through description, yet supported by a supradual ontology based on the principle of selfreference and hetero-reference.

28. Conclusions

We have presented a geometrical-topological theory which surmounts the paradigmatic divisions produced by the Cartesian Cut, and applied it to the study of consciousness, proposing a novel paradigm. Upon unfolding the presentation, it has been shown that the diverse manifestations of the physical -in the broad sense- and mental domains are derivatives of the supradual logophysics which is based on the principles of selfreference and hetero-reference, which operates even when an agent is absent. Having said and done, briefness is compelling, a long monograph forthcoming. It has transpired that the transdisciplinarian approach which the supradual logophysics supports, transcending the dual logic hegemonic in science and philosophy, also transcends the somewhat piecemeal knowledge that dualism can afford and manage through the recourse to interdisciplinarity, notwithstanding the courage, competence and brilliance that may be exercised in doing the job. The issue is that a choice of what logic is, means and does, is tantamount to the choice of an ontology fused with an epistemology: This is already the case of dual Aristotelian -Boolean logic which by default, and transparently so, produces fractures which cannot be salvaged if not transcending it, ab initio (Günther, [302], Rapoport [230]). Methodology ultimately is the choice of the logic, in the first place, rather than a decantation of what is left implicit. The usual consideration is that it comes without commitments: if so why the effects of dualism show so much on the contrary and we can still claim blindness?

Rather than a putative theory of quantum gravity for explaining the purported physical source for consciousness or a no less definable panpsychism that can profess little more than its argumentation, an omnipresent supradual logophysics which supports a world of signs and meaning as unseparable of the material-energetic-information domain, offers its cornucopia cum integral meaning (Rapoport,2022).

Acknowledgements My deep appreciation to Prof. Martin Land and Prof. Larry Horwitz for their kind invitation to present a correspondent contribution for the Proceedings of IARD 2022.Our gratitude as well to Prof. Lou Kauffman and Prof. Stein Johansen for their interest in my work.I dedicate this work to two persons. Firstly, my wife Sonia; without her support this would have ever been. This work was somewhat initiated in México, where I arrived in December 1985 to leave early 1987 following the earthquake and unfortunately, loss contact with ever after. Fortune and destiny, I soon met scientist-thinker Dr. Jacobo Grinberg-Zylberbaum (México City, 1946-?,?), and started to participate together with my wife in the seminar on consciousness, at the Universidad Nacional Autónoma de México, where Jacobo had his neuropsychology laboratory at the Psychology Faculty. The experience I had developed prior to my arrival in México, torsion geometry and supraduality in mathematical physics became a central topic of discussion and related to Jacobo's own work, particularly his syntergics. From these meetings we came to conceive the brain-mind as a torsion dislocated lattice extending beyond the body, and most fundamentally operating selfreferentially and hetero-referentially. At that time, the Klein Bottle was not considered at all. In parting away, Jacobo insisted that I should start putting into writing our understandings and enquiries, an invitation which I had repeatedly declined to do so. I took thirty years to start putting it into practice. This work is an homage to Jacobo, a man of integrity, extremely gifted, moved by an extraordinary curiosity, enthusiasm, and the will to service humankind. Wish he would have been here.

References

[1] Addis L 1999 *Of Mind and Music* (Ithaca and London: Cornell University Press)
[2] Aerts D and Sozzo S 2012 Quantum structures in cognition: Why and how are entangled. *Quantum Interaction 2011* D Song, M. Melucci and I Frommholz eds. LNCS 7052 (Berlin: Springer) pp. 116–127 ISBN 978-3-642-24970-9.
[3] Aharonov D 2003 A simple proof that Toffoli and Hadamard are Quantum Universal preprint arXiv:quantum- ph/0301040
[4] Aitchison I 2010 The Holiverse': holistic eversion of the 2-sphere in R^3 https://arxiv.org/abs/1008.0916
[5] Alberts B, Johnson A, Lewis J et.al. 2002 *Biology of the Cell* (New York: Garland Science)
[6] Alvarez G A and Cavanagh P 2004. The capacity of visual short-term memory is set both by information load and by number of objects. *Psychological Science*, 15, 106-111
[7] Andrews D L 2008 Structured Light and Its Applications: An Introduction to Phase-Structured Beam and Nanoscale Optical Forces (New York: Academic Press)

[8] Atasoy S, Donnelly I and J Pearson 2016 Human brain networks function in connectome specific harmonic waves *Nat Commun* 7, 10340 https://doi.org/10.1038/ncomms10340
[9] Bachelard G 1994 *The Poetics of Space.* (Boston: Beacon Press)
[10] Baiano C, Job X, Santangelo G, Auvray M and Kirsch L P 2021 Interactions between interoception and perspective-taking: Current state of research and future directions *Neuroscience & Biobehavioral Reviews* 130, 252-262,ISSN 0149-763
[11] Ballon, D. J. & Voss, H. U. Classical Möbius-ring resonators exhibit fermion–boson rotational symmetry. *Phys. Rev. Lett.* 101, 247701 (2008).
[12] Bartolo D and Carpentier D 2019 Topological elasticity of nonorientable ribbons *Phys Rev X* 9, 041058
[13] Thompson, D W., 1992. *On Growth and Form.* Dover reprint of 1942 2nd ed. (1st ed., 1917). ISBN 0-486-67135-6
[14] Bandyopadhyay A 2020 Nanobrain: *The Making of an Artificial Brain From a Time Crystal* (Boca Raton, Florida: CRC Press)
[15] Battaglia D, Witt A,Wolf F and Geisel T 2012 Dynamic Effective Connectivity of Inter-Areal Brain Circuits. *PLoS Comput Biol* 8(3): e100243 https://doi.org/10.1371/journal.pcbi.1002438
[16] Batterman R 2003 Falling cats,parallel parking,and polarized light, *Stud Hist.Phil. Mod .Phys.* 34, 527–557
[17] Bauer T, Banzer P, Karimi E, Orlov S, Rubano A, Marrucc L, Santamato E Boyd R and Leuchs G 2015 Observation of optical polarization Möbius strips *Science* 347, 6225
[18] Bell Pettigrew J 1907 *Design in Nature* (London: Longmans)
[19] Ben Jacob E 2003 Bacterial self-organisation: coenhancement of complexification and adaptability in a dynamic environment. *Phil Trans Royal Soc Lond A* 361, 1283-1312.
[20] Bengtsson I and Zyczkowski K 2006 *Geometry of Quantum States: An Introduction to Quantum Entanglement* (Cambridge University Press)
[21] Bliokh KY, Alonso M M, Sugic D, Perrin M, Nori F and Brasselet A 2017 Polarization Singularities and Möbius strips in sound and water-surface waves *Phys. Fluids* 33, 077122.
[22] Boeyens J and Levendis D 2008 *Number Theory and the Periodicity of Matter* (Berlin:Springer)
[23] Bóna A and Slawinski M 2011 Wavefronts and Rays as Characteristics and Asymptotics (Singapore: World Scientific)
[24] Bouligand Y 1978 Liquid Crystals and their Analogs in Biological Systems *Liquid Crystals* L. Liebert ed. (New York: Academic Press) pp 259-293
[25] Bouligand Y 1999 Defects and Textures *Physical Properties of Liquid Crystals* D. Demus, J. Goodby, G. Gray G , H.Spiess and V.Vill eds.(Weinheim,Germany:Wiley VCH)
[26] Brette R 2018 Is coding a relevant metaphor for the brain? *Behav Brain Sci.* 16;42:e215. doi: 10.1017/S0140525X19000049. PMID: 30714889.
[27] Brier S 2008 Bateson and Peirce on the Pattern that Connects and the Sacred *A Legacy for Living Systems* S. Brier ed Springer Series in Biosemiotics (Berlin: Springer) pp. 229-250
[28] Hiley B J and Callaghan 2012 Clifford Algebras and the Dirac-Bohm Quantum Hamilton-Jacobi Equation. *Foundations of Physics*. 42 (1): 192–208 **doi:10.1007/s10701-011-9558-z** ; ibid.2011Process, Distinction, Groupoids and Clifford Algebras: an Alternative View of the Quantum Formalism *New Structures for Physics*, ed B. Coecke, Lecture Notes in Physics, vol. 813, 705-750, Springer
[29] Busemeyer J and Bruza P 2012 *Quantum Models of Cognition and Decision* (Cambridge: Cambridge University Press)
[30] Cariani P 2000 Anaesthesia, neural information processing, and conscious awareness *Consciousness &Cognition* **9**:387-395 doi: 10.1006/ccog.1999.0420.
[31] Carlsson G 2009 Topology and data. *Bull AMS* 46, 2: 255-308
[32] Carlsson G, Ishkhanov T, De Silva V and Zomorodian A 2008 On the local behavior of spaces of natural images *Int. J. Comput. Vis.* **76**, 1,12, 1 https://doi.org/10.1007/s11263-007-0056-x

[33] Cartan E 1923 Sur les variétés à connexion affine et la théorie de la relativité généralisée (première partie, *Annales Scientifiques de l'École Normal Supériere* **41**, 1-25 doi:10.24033/asens.751;ibid.1925 Sur les varietes a conexión affine et la theorie de la generalisee (second partie) **42**, 17-88
[34] Charó G, Letelier C. and Schiaramella D 2022 Templex: a bridge between homology and templates for chaotic attractors *Chaos* **32,** 083108.
[35] Charó G, Artana G. and Schiaramella D 2021 Topological colouring of fluid particle unravels, finite-time chaotic sets, *Fluid Mechanics*, 929, doi.10.1017/jfm.2021.561
[36] Chow M H K, Yan T H, Bennett M J and J Wong J 2010 Liquid Crystal chromosomes, birefringence and DNA *Eukaryotic Cell* **9,** 10: 1577-1587
[37] Conte E 2013 A Clifford algebraic analysis gives mathematical explanationof quantization of quantum theory and delineates a model of quantum reality in which information, primitive cognition entities and a principle of existence are intrinsically represented ab initio *World Journal of Neuroscience, 3*, 157-170 WJNSdoi:10.4236/wjns.2013.33021
[38] Conte E, Khrennikov A, Todarello O, Federici A, De Robertis R and Zbilut J 2012 On the Possibility That We Think in a Quantum Mechanical Manner: An Experimental Verification of Existing Quantum Interference Effects In Cognitive Anomaly of Conjunction Fallacy *On the Borders of Complexity* (Hauppage,NY:Nova Science Publishers) Chap. 11 p. 149-163
[39] Cosic I and Cosic D 2021 Macromolecular Resonances *Rhythmic Oscillations in Proteins to Human Cognition* A. Bandyopadhyay and K. Ray eds. (Berlin: Springer), pp1-37
[40] Crawford J D.and Omohundro S 1984 On the global structure of period doubling flows. *Phys. D*,13:161–180 https://doi.org/10.1016/0167-2789(84)90275-6
[41] Damasio A 2022 *The Feeling of What Happens* (New York:Mariner Books)
[42] Damasio A 2018 *The Strange Order of Things* (New York : Pantheon)
[43] Deymier P and Runge K 2017 Sound Topology, Duality Coherence and Wave Mixing: An Introduction to the Emerging New Science of Sound (Berlin:Springer)
[44] Diósi L1988 Quantum Stochastic Processes as Models for State Vector Reduction *Journal of Physics A: Mathematical and General*, **21**(13): 2885. 2898. doi:10.1088/0305-4470/21/13/013; ibid.1989 Models for universal reduction of macroscopic quantum fluctuations *Phys. Rev. A* **40**, 1165
[45] Dolce D 2015 Unification of Relativistic and Quantum Mechanics from Elementary Cycles Theory *Electronic Journal Theoretical Physics* December 201 doi10.4399/97888548913193.
[46] Eganova I 2001The world of events reality: instantaneous action as a connection of the events through time *Instantaneous Action at a Distance in Modern Physics: Pros and Contras* A.E. Chubykalo, V. Pope and R. Smirnov-Rueda eds. (Hauppage, NY: Nova Science) pp.193-202 ISBN 1-56072-689-9
[47] Elworthy K 1982 *Stochastic Differential Equations on Manifolds* (London Mathematical Society 70 (London:Cambridge Univ. Press)
[48] Emerson D J, Weiser B P, Psonis J, Liao Z, Taratula O, Fiamengo A, Wang, X, Sugasawa K,Smith A B, Eckenhoff R G, et.al.. 2013 Direct modulation of microtubule stability contributes to anthracene general anaesthesia. *J. Am. Chem. Soc.* 135, 5389–5398
[49] Carlton F and Shepard R 1990 Psychologically simple motions as geodesics II: symmetric objects paths. *J of Math Psychology*, 34, 189-228
[50] Ferrari PF and Rizzolatti G 2014 Mirror neuron research: the past and the future. *Philos Trans R Soc Lond B Biol Sci*. **28**;369(1644):20130169. doi: 10.1098/rstb.2013.0169.
[51] Flapan E 2000 When Topology Meets Chemistry: A Topological Look at Molecular Chirality(Cambridge University Press); ibid. 2015 Knots, Molecules, and the Universe: An Introduction to Topology (Providence: AMS)
[52] Floridi L 2013 *The Philosophy of Information* (Oxford: Oxford UniversityPress)
[53] Fock V 1964 *The Theory of Space, Time and Gravitation* (London:Pergamon Press)
[54] Freund I 2010 Multitwist optical Möbius strips. *Optics Lett.*, 35(2):148– 19 150; ibid. 2000 Optical vortex trajectories *Optics Communications* 181, 19-33

[55] Friston K 2010 The free-energy principle: A unified brain theory? *Nat Rev. Neurosci.,* 11(2):127–138; ibid. 2012 A free energy principle for biological systems. *Entropy,* **29** 14(11):2100–2121

[56] Fröhlich H 1970 Long range coherence and the actions of enzymes *Nature*, 228, 1093.

[57] Gellhorn E 1967 *Principles of Autonomic-Somatic Integration* (Minneapolis:University of Minnesota Press.

[58] Gellhorn E and Kiely W 1972 Mystical states of consciousness: neurophysiological and clinical aspects *Journal of Nervous and Mental Diseases* **154**:399-405

[59] Georgiev D D 2021 Electric and Magnetic Fields Inside Neurons and Their Impact upon the Cytoskeletal Microtubules *Rhythmic Oscillations in Proteins to Human Cognition* A. Bandopadhyay and K Ray eds. (Berlin:Springer) pp.51-103

[60] Gibaud T et.al. 2017 Achiral symmetry breaking and positive Gaussian modulus lead to Scalloped colloidal membranes Achiral symmetry breaking and positive Gaussian *Proc National Acad Sciences PNAS* E3376–E3384 www.pnas.org/cgi/doi/10.1073/pnas.161704311

[61] Gilmore R and Lefranc M 2011 *The Topology of Chaos* (Weinheim, Germany: Wiley)

[62] Ghirardi, GC, Grassi R and Benatti F1995 Describing the Macroscopic World: Closing the Circle within the Dynamical Reduction Program *Foundations of Physics*, **25**(1): 5–38. doi:10.1007/BF02054655

[63] Ghirardi G 2020 Collapse Theories *Stanford Encyclopaedia of Philosophy* https://plato.stanford.edu/entries/qm-collapse/

[64] Ghosh S, Singh P, Manna J, Saxena J ,Pathik S,Sahoo, Krishnanda, S, Ray K, Hill J & Bandyopadhyay A 2022 The century-old picture of a nerve spike is wrong: filaments fire, before membrane, *Communicative & Integrative Biology*, 15:1, 115-120, DOI: 10.1080/19420889.2022.2071101

[65] Gidron Y, Barak T, Henik A, Gurman G, Stiener O 2002 Implicit learning of emotional information under anaesthesia. *Neuroreport.* 2002 Jan **21**;13(1):139-42. doi: 10.1097/00001756-200201210-00032

[66] Globus G 2003 Quantum Closures and Disclosures: Thinking-together postphenomenology and quantum brain dynamics (New York: Joh Benjamin Cummings Publs)

[67] Goenner H 2004 On the history of unified field theories *Living Rev. Relativity,* 7(2*)* https://doi.org/10.12942/lrr-2004-2

[68] Grinberg-Zylberbaum 1997 Ideas about a new psychophysiology of consciousness: The syntergic theory *Journal of Mind and Behavior* 18 (4):443-458

[69] Grinberg-Zylberbaum J 1994 The Einstein- Podolsky-Rosen Paradox in the Brain; The Transferred Potential *Physics Essays* 7,(4)

[70] Guckenheimer J and Labouriau I S 1993 Bifurcation of the Hodgkin and Huxley Equations: A New Twist *Bull. of Math. Biol.* No. **5**, pp. 937 952, 1993DOI: 10.1007/BF0246069

[71] Hagiwara T 2014 Derrida and Zen: dessert and swamp *Philosophy East and West*, **64**(1), 123–150. http://www.jstor.org/stable/43285883

[72] Hameroff S R and Watt R C 1982 Information processing in microtubules. *J. Theor. Biol.* **98**, 549–561. doi: 10.1016/0022-5193(82)90137-0

[73] Hameroff S R 1987 Ultimate Computing: Biomolecular Consciousness and NanoTechnology (Amsterdam, Elsevier)

[74] Hameroff S, Nip A, Porter M and Tuszynski, J 2002 Conduction pathways in microtubules,biological quantum computation and microtubules. Biosystems, **64**(13), 149-68.

[75] Hameroff S and Penrose R 2014 Consciousness in the universe: a review of the 'Orch OR' theory. *Phys. Life Rev.* **11**, 39–78.

[76] Hameroff S 2022 Consciousness, Cognition and the Neuronal Cytoskeleton – A New Paradigm Needed in Neuroscience. *Front. Mol. Neurosci.* 15:869935. doi: 10.3389/fnmol.2022.869935

[77] Hamilton J K , Hooper I R & Lawrence C R 2021 Absorption modes of Möbius strip resonators. *Sci Rep* **11**, 9045 (2021). https://doi.org/10.1038/s41598-021-88280-x

[78] Hayasaka H and Takeuchi S 1989 Anomalous weight reduction on a gyroscope's right rotation around the vertical axis of the Earth *Phys.Rev.Lett.* 63, p.2701 2701.

[79] Ho, Mae Won 1996 Bioenergetics and Biocommunication *Computation in Cellular and Molecular Biological Systems,* R. Cuthbertson et.al. eds (Singapore:World Scientific) pp.252-26,

[80] Ho, Mae-Won 1994) What is (Schrödinger's) Negentropy? *Modern Trends In Bio* ThermoKinetics **3**, 50-6

[81] Hehl F, McCrea J, Mielke E and. Ne'eman N 1995 Metric-Affine Gauge Theory of Gravity: Field Equations, Noether Identities, World Spinors, and Breaking of Dilation Invariance *Phys. Rep.* **258**, 1-171

[82] Hehl F, von der Heyde P, Kerlic D G and Nester N 1976 General Relativity with spin and torsion: Foundation and prospects, *Rev. Mod. Phys* **48**,393-416

[83] Hehl F and Obukhov Yu 2007 Elie Cartan's torsion in geometry and in field theory, an essay, Annales de La Fondation Louis de Broglie**, 32** no 2-3,

[84] Hehl F and Obukhov Yu 2003 *Foundations of Classical Electrodynamics* (Boston: Birkhäuser)

[85] Hewitson O 2015 Why Topology Matters in Psychoanalysis – Part II https://www.lacanonline.com/2015/03/why-topology-matters-in-psychoanalysis-part-ii/

[86] Hoffmeyer J 2010 Semiotic freedom: an emerging for Information and the Nature of Reality *From Physics to Metaphysics* P Davies ed. (Cambridge: Cambridge Univ. Press);ibid. 2007 Semiotic Scaffolding of Living Systems *Introduction to Biosemiotics: The New Biological Synthesis* ed M Barbieri ed. (Berlin: Springer)

[87] 't Hooft G 2018 Virtual black holes and space–time structure, *Found Pys*. **48**:1134–1149; https://doi.org/10.1007/s10701-017-0133-0.

[88] Hunt T and Schooler W 2019 The easy part of the Hard Problem: A resonance theory of consciousness,*Frontiers in Human Neurosciences* **13**:378, doi:10.3389/fnhum.2019.00378

[89] Ikeda N and Watanabe S 1989 *Stochastic Differential Equations and Diffusion Processes* (Amsterdam: Elsevier)

[90] Ingber D E 2003 Tensegrity I. Cell structure and hierarchical systems biology. *J Cell Sci.* Apr **1**;116(Pt 7):1157-73. doi: 10.1242/jcs.00359. PMID: 12615960.

[91] Izhikevich E M 2007 Dynamical Systems in Neuroscience: The Geometry of Excitability and Bursting, Appendix www.izhikevich.com (Cambridge, Mass: MIT Press)

[92] Jagacinski R, Lieke C K, Peper E and Beek P 2000 Dynamic, Stochastic and Topological Aspect of Polyrhythmic Performance. *J. Motor Behavior*, **32**(4):323–336

[93] Janke C and Magiera M M 2020 The tubulin code and its role in controlling microtubule properties and functions *Nature Reviews Molecular Cell Biology* doi:10.1038/s41580-020-0214-3

[94] Janowicz P 2007 Möbius aromacity: which side are you on? February 22, https://www.semanticscholar.org/Corpus ID: 19142209

[95] Jibu M and Yasue K 1995 *Quantum Brain Dynamics and Consciousness*: An *introduction* (New York: John Benjamins Publishing Company)

[96] Johnson M 1987 *The Body in the Mind: The Bodily Basis of Meaning, Imagination and Reason* (Chicago: University of Chicago Press)

[97] Karp G 2005 *Cell and Molecular Biology: Concepts and Experiments* (USA: John Wiley & Sons) p. 355.

[98] Kaufmann L1991 *Knots and Physics* (Singapore:WorldScientific)

[99] Kelz M and George A. Mashour 2019 The Biology of General Anaesthesia from Paramecium to Primate *Current Biology* **29**, R1199–R1210, November 18.

[100] Khrennikov A 2004 *Information Dynamics in Cognitive, Psychological, Social, and Anomalous Phenomena*. Fundamental Theories of Physics (Dordrecht: Kluwer); ibid. 2010 *Ubiquitous* Quantum Structure:from Psychology to Finances (Berlin: Springer)

[101] Khrennikov A 2006 Quantum-like brain: "Interference of minds" *Biosystems* **84**, 3, June 2006, Pages 225-241

[102] Kingsley P 2020 *Reality* (London: Catafalque Press, 2020);1999 *In the Dark Places of Wisdom* (Point Reyes, CA: Golden Sufi Center Publishing)

[103] Kingsley P 1995 Ancient Philosophy, Mystery and Magic. Empedocles and Pythagorean Tradition (Oxford, UK: Oxford University Press)
[104] Kocika M, Corno A, Lackovic V and Kanish V 2007 The helical ventricular myocardial band of Torrent-Guasp *Semin Thorac Cardiovac Surg* Annu **10**,52-60 DOI: 10.1053/j.pcsu.2007.01.006
[105] Koenderink J J 1990 The brain, a geometry engine. *Psychol. Res* **52**, 122–127 https://doi.org/10.1007/BF00877519
[106] Korotaev S M 1996 On the way to understanding the time phenomenon *The Construction of Time in the Natural Sciences, Part 2*, A. Levich, ed. (World Scientific, pp. 60–74 www.chronos.msu.edu
[107] Kosterlitz M and Thouless DJ 1998 *Topological Quantum Numbers in Non- relativistic Physics* (Singapore: World Scientific Publishing)
[108] Kozyrev N A 1971 On the possibility of experimental investigation of the properties of time *Time in Science and Philosophy* (Prague) pp. 111–132.
[109] Kozyrev N A, Nasonov V V 1980 O nekhotoryh svoistvah vremeni, obnaruzhennykh astronomicheskimi nablyudeniyami, Problemy issledovaniya Vselennoi, 9, p.76 (1980). (Russian), www.chronos.msu.edu.
[110] Kozyrev A N 1980 Astronomical proofs of the reality of 4D Minkowski Geometry, in Manifestations of Cosmic Factors on Earth and Moon. www.chronos.msu.edu.
[111] Sharpe R W and Chern S S 1997 Differential Geometry: Cartan's Generalization of Klein's Erlangen Program (Berlin: Springer)
[112] Kauffman L 1987 Imaginary values in Mathematical Logic *Proceedings of* The 17th International Symposium on Multiple Valued Logic, Boston May 26-28, 1987 IEE Computer Society Press pp 282-289; 2002 I am a Fibonacci Form *Cybernetics And Human Knowing* 11, 3: 101-106**;** Kauffman 1980 Complex numbers and Algebraic Logic, 10th Int, Symp. Multiple Valued Logic, IEEE Pub; 1987 L. H. Kauffman Self-reference and Recursive Forms, J. Social Bio. Struct. **10**, 53-72
[113] Perez JC 2009 *Codex Biogenesis: Les 13 Codes de l'ADN.* (Embour, Belgique:Marco Pietteur); 2010 Codon populations in single-stranded whole human genome DNA are fractal and finetuned by the golden ratio 1.618, *Inter Sci Comput Life Sci* 2: 1–13
[114] Caelli T, Hoffman W and Lindman H 1978 Subjective LorentzTransformations and the perception of motion. *J Opt Soc A* Mar;**68**(3):402-11. doi:10.1364/josa.68.000402
[115] Mausfeld R 2013 The Biological Function of Sensory Systems *Neurosciences: From Molecule to Behavior* C.G. Galizia, P.-M. Lledo ed (Berlin:Springer)
[116] Avrin J S 2012 Knots on a Torus: A Model of the Elementary Particle *Symmetry* **4**, 39-115; doi:10.3390/sym4010039
[117] Sheari A et.al. 2008 A tale of two symmetrical tails: structural and functional characteristic of palindromes *Bioinformatics* 9,274:doi:10.1186[/1471- 21059-27 9-27
[118] Mezey P 1993 Shape in Chemistry: An Introduction to Molecular Shape and Topology New York:VCH)
[119] Werner and Whitsel 1973 Functional Organisation of the Somatosensory Cortex *Somatosensory Systems Handbook of Sensory Physiology*, vol 2 A. Iggo eds (Berlin:Springer). https://doi.org/10.1007/978-3-642-65438-1_17
[120] Takahashi Y and Umezawa H 1996 Thermo field dynamics *Int J of Mod Phys B* 10, (13-14)
[121] Rapoport D and Sternberg S 1984 Classical mechanics without Lagrangians and Hamiltonians. *Nuov Cim A* **80**, 371–383 https://doi.org/10.1007/BF02785808
[122] Dembowski C,Gräf H_D, Harney H L, Heine A, Heiss, W D, Rehfeld H and Richter A 2001 Experimental observation of the topological structure of exceptional points *Phys. Rev. Lett*. **86**, 787–790 DOI:https://doi.org/10.1103/PhysRevLett.86.787
[123] Buszáki G 2021 *The Brain Inside Out* (Oxford: Oxford University Press); ibid 2011 *Rhythms of the Brain* (Oxford University Press)
[124] Mitchell E 2003 Quantum Holography: A Basis for the Interface Between Mind and Matter, Bioelectromagnetic Medicine (New York. Marcel Dekker); Mitchell E and Staretz R 2017 The

Hologram & the Nature of Consciousness Chap IV Consciousness in the Universe,Neuroscience, Quantum Physics, Evolution, Brain & Mind R.Penrose et.al. eds. (Cosmology Scs. Publs.)

[125] Pribram K 1999 Quantum Holography: Is it relevant to Brain Function? *Information Sciences* Vol. 115, pp. 97-102

[126] Solms M and Panksepp J 2012 The "Id" Knows More than the "Ego" Admits: Neuropsycho-analytic and Primal Consciousness Perspectives on the Interface Between Affective and Cognitive Neuroscience *Brain Sci.* **2**,(2), 147-175; https://doi.org/10.3390/brainsci2020147

[127] Aktas M E, Akbas E & Fatmaoui A E 2019 Persistence homology of networks: methods and applications. *Appl Netw Sci* **4**, 61 https://doi.org/10.1007/s41109-019-0179-3

[128] Miller K 2010 π = visual cortex. *Science* **330** Nov 1059-1061,10.1126/science.119885

[129] Myers T 2009 *Anatomy Trains* Amsterdam: Elsevier)

[130] Popov I A, Bozhenko K V & Boldyrev A I 2012 Is graphene aromatic?. *Nano Res.* **5**, 117–123 https://doi.org/10.1007/s12274-011-0192-z

[131] Kauffman L et.al. 2022 *Laws of Form: the 50th year* (Singapore: World Scientific)

[132] Gordon R 2002 The Hierarchical Genome and Differentiation Waves (Singapore:World Scientific): Gordon N K and Gordon R 2016 *Embryogenesis Explained*, (Singapore: World Scientific)

[133] Agrawal L *et.al.*. 2018 Fractal Information Theory (FIT)-Derived Geometric Musical Language (GML) for Brain-Inspired Hypercomputing *Soft Computing: Theories and Applications* Advances in Intelligent Systems and Computing **584** M.Pant et.al. eds (Singapore: Springer). https://doi.org/10.1007/978-981-10-5699-4_33

[134] Singh P *et.al.* 2018 Fractal and Periodical Biological Antennas: Hidden Topologies in DNA, Wasps and Retina in the Eye *Soft Computing Applications* Studies in Computational Intelligence vol **761**. Ray, K., Pant, M., Bandyopadhyay, A. (eds) Springer, Singapore. https://doi.org/10.1007/978-981-10-8049-4_6

[135] Ling G 1992 *A Revolution in the Physiology of the Living Cell* (Malabar,FL (US):Krieger Publs)

[136] Bonchev D 2010 and Rouvray RH. *Chemical Topology*. (London: Gordon & Breach)

[137] VanRullen R and Koch C 2003 Is perception discrete or continuous? *Trends Cogn. Sci.* 7, 207–213. doi: 10.1016/s1364-6613(03)00095-0

[138] Wilczek F and Shapere A eds 1989 *Geometric Phases in Physics* (Singapore:World Scientific)

[139] Lakoff G and Johnson M 1999 Philosophy in the Flesh: The Embodied Mind and Its Challenge to Western Thought (New York: Basic Books)

[140] Laughlin Jr C and Throop J 2009 Husserlian Meditations and Anthropological Reflections: Toward a Cultural Neurophenomenology of Experience and Reality *Anthropology of* Consciousness, **20** ,2 Pages 130-170

[141] Laughlin Jr C, McManus J and. d'Aquili E 1990 *Brain, Symbol and* Experience: Toward a *Neurophenomenology of Consciousness*. (New York: Columbia University Press)

[142] Laughlin Jr C, McManus J, Rubinstein R A and Shearer J 1986 The ritual transformation of experience *Studies in Symbolic Interaction* N Kenzin ed.Vol **7** Part A (London: Jai Press) pp107- 136

[143] Lavrentiev M M , Gusev V A, Yegonova Y A, Lutset M K and Fominykh S F 1990, O Registratsii istinnogo polozheniya solntsa (Russian) On the registration of a true position of the sun *Dokl. Akad. Nauk SSSR* **315**(2),368 370

[144] Lavrentiev M M, Yeganova I, Medvedev V G, Oleinik V K , Fominykh S F, Skanirovanii O 1992 Zvyeozdnogo neba Datchikom Kozyreva", *Dokl. Akad. Nauk* **323**(4),649-652 (On the Scanning of the Star Sky with Kozyrev's Detector)

[145] Lavrent' ev M M and Eganova I 2001 Kozyrev' method of astronomical observations ; information from true positions of stars, stellar system and planets *Instantaneous Action at a Distance in Modern Physics: Pros and Contras* A.E. Chubykalao, V. Pope and R. Smirnov-Rueda eds. (Hauppage, NY: Nova Science) pp.91-104 ISBN 1-56072-689-9

[146] Lebedev M A, Ossadtchi A, Mill N A 2019.Analysis of neuronal ensemble activity reveals the

pitfalls and shortcomings of rotation dynamics. *Sci Rep* **9**, 18978 https://doi.org/10.1038/s41598-019-54760-4
[147] Leyton M 1991 *Symmetry, Causality, Mind* (Cambridge, MA: MIT Press)
[148] Lima de Faria A 1988 *Evolution without Selection. Form and Function by Autoevolution* (Amsterdam:Elsevier).
[149] Lin Yi 1998 Mystery of nonlinearity and Lorenz's chaos *Int. J. Sys. & Cyber.* **27**, 6& 7,605-854
[150] Lopez U, Forster A, Annoni J M, Habre W, Iselin-Chaves I A 2006 Near-death experience in a boy undergoing uneventful elective surgery under general anaesthesia. *Paediatr Anaesth* **16**(1):85- 8. doi: 10.1111/j.1460- 9592.2005.01607.x.
[151] Lundholm I V ; Rodilla H, Weixiao Y, Wahlgren A D, Bourenkov G et.al.. 2015 Terahertz radiation induces non-thermal structural changes associated with Fröhlich condensation in a protein crystal. *Struct. Dyn.* **2**:054702. doi: 10.1063/1.4931825
[152] Lyre H 2014 Berry phase and quantum structure, *Stud. in Hist. and Phil. Mod.Phys.* **48**, 45-51
[153] Machon T and Alexander G 2013 Knots and nonorientable surfaces in chiral nematics. *PNAS* August 27,110;35:14174–14179. Doi 10.1073/pnas.1308225110
[154] Machon T 2016 Aspects of geometry and topology in liquid crystalline phases Ph.D. thesis, Univ. of Warwick, UK. http://wrap.warwick.ac.uk/79818/
[155] Malpas J 2007 *Heidegger's Topology: Being, Place, World* (Cambridge, MA: MIT Press)
[156] Marcer P and Rowlands P 2017 Nilpotent Quantum Mechanics: Analogs and *Front. Phys.* **5**:28. doi:10.3389/fphy.2017.00028
[157] Marks R J 1991 *Introduction to Shannon Sampling and Interpolation Theory* (Berlin:Springer)
[158] Mathologes 2015. Smale's Inside-out Paradox, https://www.youtube.com/watch?v=ixduANVe0gg
[159] Mazzola G 2004 The Topos of Music: Geometric Logic of Concepts, Theory, and Performance (Berlin:Birkhauser)
[160] McClare C F W 1971 Chemical machines, Maxwell's demon and living *Jour Theor Biol.***30**,1 Pages 1-34
[161] McNaughton B L et.al. 2006 The hippocampus as a path integration syste *J. Exp. Biol.,* **42** 199:173– 185
[162] McNaughton B, Battaglia F, Jensen O, Moser E , Moser M-B 2006, Path integration and the neural basis of the'cognitive map'. *Nature Revi., Neurosci.* **7**: *663. 10.1038/nrn1932*
[163] Meiske W and Schneider K 1987 Topological structure of integral manifolds and period-doubling bifurcation. *Z. Agnew. Math.Phys.***38**,302–314
[164] Melle M, Schlotthauer S, C Hall, Diaz-Herrera E and Schoen M 2014 Disclination lines at homogeneous and heterogeneous colloids immersed in a chiral liquid crystal *Soft Matter* **10**: 5489-5502.doi: 10.1039/C4SM00959B
[165] Miller J-A 1999 Mathemes Topology in the Teaching of Lacan, trans. by Mahlon Stoutz *Critical Essays on Jacques Lacan* E. Ragland ed (New York: Macmillan), pp. 28—48.
[166] Mindlin B and Solari H 1997 Tori and Klein Bottles in 4D Chaotic Flows *Phys. D* **102**,97,177-186 DOI:10.1016/S0167-2789(96)00189-3
[167] Moerman N, Bonke B and Oosting J 1993 Awareness and recall during general anaesthesia. Facts and feelings. *Anaesthesiology* Sep;**79**(3):454-64. PMID: 8363069
[168] Montagnier L, Jamal A, Ferris S, Montagnier JL and Lavallee 2009 Electromagnetic Signals Are Produced by Aqueous Nanostructures Derived from Bacterial DNA Sequences. *Interdiscip Sci Comput Life Scs.* **1** (2),1-10
[169] Montagnier L, Aissa,J, Del Giudice E, Lavallee C, Tedeschi A and Vitiello G. DNA waves and water 2011 *Journal of Physics: Conference Series 306* , 012007.doi:10.1088/17426596/306/1/012007.
[170] Montagnier L, Del Giudice E, Aïssa J, Lavallee C, Motschwiller S, Capolupo A, Polcari A, Romano P, Tedeschi A, Vitiello G 2015 Transduction of DNA

information through water and electromagnetic waves *Electromagn Biol Med.* **34**(2):106-12. doi: 10.3109/15368378.2015.1036072. PMID: 26098521
[171] Musevic I 2013Nematic colloids, topology and photonics *Phil Trans R Soc A* **371**:20120266. http://dx.doi.org/10.1098/rsta.2012.026
[172] Nagasawa M 1993 *Schrödinger Equations and Diffusion Theory* (Basel:Springer)
[173] Neuman Y 2008 *Reviving the Living: Meaning Making in Living Systems* (Amsterdam: Elsevier)
[174] Nielsen M and Chuang I 2000 *Quantum Computation and Quantum Information* (Cambridge: Cambridge University Press)
[175] Nelson E 1985 *Quantum Fluctuations* (Princeton, New Jersey: PrincetonUniversity Press)
[176] Ning Z and Haken H 1992 The geometric phase in nonlinear dissipative systems, *Mod. Phys. Let. B*, **6**,25,1141 DOI:https://doi.org/10.1103/PhysRevLett.68.2109
[177] Northoff G 2012 Psychoanalysis and the brain - why did Freud abandon neuroscience? *Front Psychol.* Apr **2**;3:71. doi:10.3389/fpsyg.2012.00071.
[178] Olivelle P 1998 *Upaniṣads* (Oxford: Oxford University Press). p. 77. ISBN
[179] Orlov Y 1994 The Logical Origins of Quantum Mechanics *Annals of Physics* **234**, 2, September Pages 245-254
[180] Oron O, Horwitz L P 2005 Relativistic Brownian Motion and Gravity as an Eikonal Approximation to a Quantum Evolution Equation *Found Phys* **35**, 1181–1203 https://doi.org/10.1007/s10701-005-6406-z
[181] Osinga H 2003 Nonorientable Manifolds in Three-DimensionalVector Fields *Int. J. Bifurcation Chaos.* **13**:553-570 https://doi.org/10.1142/S0218127403006777
[182] Panksepp J 2004 Affective neuroscience: The Foundations of Human and Animal Emotions (New York: Oxford University Press)
[183] Pang R, Lansdell B J, and Fairgall A L 2016 Dimensionality reduction in neuroscience *Curr Bio* Jul 25;**26**(14):R656-60.doi: 10.1016/j.cub.2016.05.029.
[184] Pascual-Leone J 1970 A mathematical model for the transition rule in Piaget`s developmental stages. *Acta Psychologica,* **32**, 301-345
[185] Payeron D 2011 On musical self-similarity *Acta Semiotica Fennica*n(Approch to Musical Semiotics Series) (Imatra: Univ. of Helsinki Press) 9,39
[186] Pearle P 1976 Reduction of the State Vector by a Nonlinear Schrödinger Equation *Physical Review D*, **13**(4): 857–868. doi:10.1103/PhysRevD.13.85
[187] Penner R C, Knudsen M, Wiuf C, Andersen J E 2011An Algebro-Topological Description of Protein Domain Structure. *PLoS ONE* **6**(5): e19670. https://doi.org/10.1371/journal.pone.0019670
[188] Percival I A 1995 Quantum spacetime fluctuations and primary state diffusion Quantum spacetime fluctuations and primary state diffusion *Proc. R. Soc Lond. A* **451** 503–513 https://doi.org/10.1098/rspa.1995.0139
[189] Petitmengin C 2017 Enaction as a Lived Experience Towards a Radical Neurophenomenology Constructivist Foundation **12**,2,319 http://constructivist.info/12/2/139
[190] Petitot J. 2003. The neurogeometry of pinwheels as a sub-Riemannian structure. *J. Physio., Paris*, **97**:265–300 DOI: 10.1016/j.jphysparis.2003.10.010
[191] Penrose R 1994 Shadows of the Mind: A Search for the Missing Science of Consciousness (Cambridge, UK, Oxford Univrsity Press)
[192] Penrose R and Hameroff S 2017 Consciousness in the Universe: Neuroscience:Quantum Space-Time Geometry and Orch OR Theory *Consciousness and the Universe Quantum Physics, Evolution, Brain & Mind, chap I* R.Penrose, S. Hameroff and S Kak eds. (Cambridge, MA : Cosmology Science Publishers)
[193] Pollack G 2013 *The Fourth Phase of Water: Beyond Solid, Liquid, and Vapor* (Seattle: Ebner & Sons)
[194] Pond J M 2000 Möbius dual-mode resonators and bandpass filters. *IEEE Trans. Microw.* Theory Tech. **48**, 2465–2471

[195] Pond J M, Liu S and Newman N 2001 Bandpass filters using dual-mode and quad-mode Möbius resonators. *IEEE Trans. Microw. Theory Tech.* **49**, 236

[196] Pribram K 2013 *The Form Within* (Westport (CA): Prospecta Press)

[197] Putz M and Ori O 2020 Topological Symmetry Transition between Toroidal and Klein Bottle Graphenic Systems *Symmetry* **12**, 1233;doi:10.3390/sym12081233

[198] Rainer H. Topology in chemistry: designing Möbius molecules 2006 *Chem Rev*, 106: 4820–4842

[199] Ragland E 2015 *Jacques Lacan and the Logic of Structure* (New York: Routledge

[200] Rapoport D L and Sternberg S 1984 On the interactions of spin with torsion *Annals Phys.* **158**,447-475

[201] Rapoport D L 1991 Stochastic processes in conformal Riemann-Cartan-Weyl gravitation *Int. J.Theor Phys.* **30** 11,1497https://doi.org/10.1007/BF0067561

[202] Rapoport D L and Tilli M 1987 Scale Fields as a Simplicity Principle *Hadronic Jour. Suppl.* 10(1), 25-90

[203] Rapoport D L 1995 The Geometry of Quantum Fluctuations, the Quantum Lyapounov Exponents and the Perron- Frobenius Stochastic Semigroups *Dynamical Systems and Chaos*, Proceedings International Conference on Dynamical Systems (Tokyo, 1994), ed Y. Aizawa (Singapore:World Scientific), 73-77

[204] Rapoport D L 1995 The Cartan Structure of Classical and Quantum Gravitation *Gravitation, The Space-Time Structure* Proc. of the 8thLatinoamerican Symposium in Relativity and Gravitation, Aguas De Lindoia, Brazil, July 25-29,1993, 220–229, eds W Rodrigues W and P Letelier (Singapore:WorldScientific)

[205] Rapoport D L 1996 The geometry of fluctuations I *New Frontiers in algebras, groups and geometries*,179-223, ed Tsagas et.al., Molise, Italy, 1995,August 9-12, The geometry of fluctuations II, Nonequilibrium statistical mechanics, 225-250 (Palm Harbor, Fl, US: Hadronic Press)

[206] Rapoport D L 1997 Riemann-Cartan-Weyl quantum geometry II : Cartan stochastic copying method Fokker-Planck operator and Maxwell-de Rham equations *Int J Theor Phys* **36**, 2115–2152 https://doi.org/10.1007/BF02435948

[207] Rapoport D L 1997 Torsion and non-linear quantum mechanics *Group XXI, Physical Applications and Mathematical Aspects of Algebras, Groups and Geometries, vol. I,* Proceedings (Clausthal, 1996), eds. H D Doebner, H.D. (Singapore:World Scientific), 446-ISBN-13: 978-9810231354

[208] Rapoport D L 1998 Riemann-Cartan-Weyl geometries, quantum diffusions, and the equivalence of the free Maxwell equations and Dirac-Hestenes equations. *Adv. Appl.Clifford Algebra* **8**,1, 129-146 DOI https://doi.org/10.1007/BF03041930

[209] Rapoport D L 1999 On the random geometry of fluid mechanics, general relativity and quantum mechanics *Fundamental open problems in science at the end of the millennium*, Bejing, China,August 1997,ed G *Tepper (Palm Harbor, Fl:* Hadronic Press) vols. I-III, 243-275

[210] Rapoport D 2000 Covariant Non-Linear Non-Equilibrium Thermodynamics and the Ergodic Theory of Stochastic and Quantum Flows *Instabilities and Nonequilibrium Structures VI.* Nonlinear Phenomena and Complex Systems vol 5 E Tirapegui J Martínez and R Tiemann eds Dordrecht. Springer). https://doi.org/10.1007/978-94-011-4247-2_19

[211] Rapoport D L 2002 Random diffeormorphism and integration of the classical Navier-Stokes equations, *Reports Math. Phys.* **49**, 1-47 https://doi.org/10.1016/S0034-4877(02)80002-7

[212] Rapoport D L 2002 On the geometry of the random representations for viscous fluids and a remarkable pure noise representation. *Reports Math. Phys.* **50**, 2, 211-250 DOI: 10.1016/S0034-4877(02)80055-6

[213] Rapoport D L 2003 Random symplectic geometry and the realizations of the Navier-Sokes equations by ordinary differential equations, *Random Operators and Stochastic Equations*

11, 4, 371-401 https://doi.org/10.1515/156939703771891888

[214] Rapoport D L 2003 Martingale problem approach to the representations of the Navier-Stokes equations on smooth-boundary manifolds and semispace *Random Operators* StochasticEquations,**11**(2), 359–380 DOI:10.1515/156939703322386887

[215] Rapoport D L 2004 Realizations of the Random Representations of the Navier-Stokes Equations by Ordinary Differential Equations. *Instabilities and Structures VII & VIII. Nonlinear Phenomena and Complex Systems, vol 8* O.Descalzi,J Martínez and E Tirapegui eds (Dordrecht:Springer) https://doi.org/10.1007/978-1-4020-2149- 7_23

[216] Rapoport D L 2004 Stochastic Geometry Approach to the Kinematic Dynamo Equation of Magnetohydrodynamics *Trends on Partial Differential Equations of Mathematical Physics* (Obidos, Portugal, June 2003), 225–] 241,ed J. Rodrigues (Birkhauser Progress in Nonlinear Differential Equations and Their Applications, vol. 61, Boston) DOI: 10.1007/3-7643-7317-2_17

[217] Rapoport D L 2005 Unification of Geometric and Random Structures through Torsion Fields: Brownian Motions, Viscous and Magneto-fluid-dynamics. *Found Phys* **35**, 1205–1244 DOI: 10.1007/s10701-005-6407-y; ibid. 2005 Cartan-Weyl Dirac and Laplacian Operators, Brownian Motions: The Quantum Potential and Scalar Curvature Maxwell's and Dirac—HestenesEquations, and Supersymmetric Systems *Found Phys* **35** (8) 1383-1431

[218] Rapoport D L 2007 On the spacetime and state-space geometries of random processes in geometric quantum mechanics *Foundations of Probability and Physics-4*, Proceedings, Vaxho, Sweden, 4-9 June, AIP Conference Proceedings, vol. **889**, G Adenier G et.al., p.225-229 (Berlin:Springer) DOI 10.1063/1.2713460

[219] Rapoport D L 2007 Torsion fields, Cartan-Weyl spacetime and Statespace Quantum Geometries and their Brownian motions ant the time variables *Foundations of* Physics **37** (4-5): 813-854

[220] Rapoport D L 2009Torsion Fields,the Extended Photon,Quantum Jumps, Eikonal Equations, the Twistor Geometry of Cognitive Space and the Laws of Thought *Ether, Spacetime and* Cosmology vol. 3, Physical Vacuum, Relativity and Quantum Mechanics ed M Duffy & J Levy; Physical Interpretations Relativity Theory (PIRT) Conference, Imperial College, June 2008, London, 389-457 (Apeiron Press) ISBN 13:978098649260

[221] Rapoport D L 2010 Torsion, propagating singularities, nilpotence, quantum jumps and the equations *Computing Anticipatory Systems Proceedings CASYS'09* , Liege, Belgium, ed D M Dubois,American Institute of Physics Conf. Series **1303** (Berlin:Springer) DOI:10.1063/1.3527144

[222] Rapoport D L 2010 Selfreference, the Möbius and Klein Bottle surfaces, Multivalued Logic and Cognition. *Inter. J. Comput. Anticip. Syst.* **23**: 103-113

[223] Rapoport,D L 2011 Surmounting the Cartesian Cut Through Philosophy, Physics, Logic, Cybernetics and Geometry: Selfreference, Torsion, the Klein Bottle, the Time Operator, Multivalued Logics and Quantum Mechanics. *Found Phys* ; **41**, 1: 33-76 https://doi.org/10.1007/s10701-009-9334-5

[224] Rapoport D L 2011 Surmounting the Cartesian Cut: Klein Bottle Logophysics, the Dirac Algebra and the Genetic Code, *NeuroQuantology* **9**,4, Special issue: Classical and "Quantlike" Views of the Genetic Code DOI: 10.14704/nq.2011.9.4.493

[225] Rapoport D L 2011 On the Fusion of Physics and Klein Bottle Logic in Biology, Embryogenesis and Evolution. *NeuroQuantology*; **9**, 4: 842-86

[226] Rapoport D L 2011 Surmounting the Cartesian Cut Further: Torsion Fields, the Extended Photon, Quantum Jumps,the Klein Bottle, Multivalued Logic, the Time Operator, Chronomes,Perception, Semiosis, Neurology & Cognition *Focus in Quantum Mechanics* eds. D Hathaway & E Randolph (NewYork:Nova Science) ISBN-13: 978-1621006800

[227] Rapoport D L 2012 Torsion Fields, Quantum Geometries, Brownian Motions and Statistical Thermodynamics. *Inter J Maths, Game Theory, and Algebra*, 21,no.6,465-543
[228] Rapoport D L 2013 Klein Bottle logophysics a unified principle for non-linear systems, cosmology, geophysics, biology, biomechanics ,perception. Jour of Phys (UK) : Conf. Ser 2013; **437**, 012024. doi:10.1088/17426596/437/1/01202.
[229] Rapoport D L 2014 Surmounting the Cartesian Cut: Torsion, Klein Bottle,Stereochemistry, the Biomechanics of the Cell Splitter in Embryogenesis & Bauplans. *Int. J- Comp. Anticip. Sys, (Liege, Proc. CASYS Conference)*; **29**: 225-246
[230] Rapoport D L 2014. Hyper Klein Bottle Logophysics Ontopoiesis of the Cosmos and Life *Phenomenology of Space and Time: The Forces of the Cosmos and the Ontopoietic Genesis of Life: Book Two* A Tyminieckia ed Volume **117** , Series Analecta Husserliana pp 275-350
[231] Rapoport D L 2016 Klein Bottle Logophysics, Selfreference, Heterarchies, Genomic Topologies, Harmonic and Evolution. Part I: Morphomechanics Space and Time in Biology & Physics, Cognition, Non-Linearity and the Structure of Uncertainty, *Quantum Biosystems*, Nov 2016, **7,** 1 ,Page 1-72
[232] Rapoport D L 2016 Klein Bottle Logophysics, Selfreference, Heterarchies, Genomic Topologies, Harmonics and Evolution. Part II: Nonorientability,Cognition, Chemical Topology and Eversions in Nature *Quantum Biosystems* **7**, 1, page 73-105
[233] Rapoport D L 2016 Klein Bottle Logophysics, Selfreference, Heterarchies, Logic of Genomics and its Dynamics, Quantum Information,Complexity and Palindromic Repeats in Evolution. *Quantum Biosystems* **7**, 1, 106-172
[234] Rapoport D L and Perez J-C 2018 Golden Ratio and Klein Bottle Logophysics: the Keys of the Codes of Life and Cognition, *Quantum Biosystems*, November **9** , 2 , Page 8-76
[235] Rapoport D L 2020 Supradual Critique of the Toroidal Model of Consciousness: A Simulacrum by Meijer, Jerman, Melkhik & Sbitnev, *Quantum Biosystems* **20**, no.11, 34- 43, 2020.
[236] Rapoport D L 2022 Klein Bottle Logophysics, the Primeval Distinction , Semiosis, Perception and the Topology of Consciousness *Laws of Form 50th Year* ed L Kauffman et.al. (Singapore: World Scientific 436-507, chap XIII
[237] Risset J C 1986 Pitch and rhythm paradoxes: comments on "Auditory paradox based on fractal waveform" *J Acoust Soc Ame* Sep **80**(3):961-2
[238] Rizzolatti G, Fadiga L, Fogassi L and Gallesse V 1999 Resonance behaviors and mirror neurons Arch Ital Biol May;**137**(2-3):85-100.
[239] Rodhe U, Poddar A and Sundararajan K 2013 Printed resonators, Möbius strip theory and applications *Microwaves Journal*, November issue, 24-54
[240] Rohrer T 2005 Image Schemata in the Brain *From Perception to Meaning: Image Schemas in Cognitive Linguistic*, eds B Hampe and J. Grady (Berlin: Mouton de Gruyter), pp. 165-196
[241] Rosen S 2004 Dimensions of Apeiron: A topological phenomenology of space, time and individuation (Amsterdam: Rodopi Publishers)
[242] Rosenfeld B 1997 *Geometry of Lie Groups* (Dordrecht: Kluwer)
[243] Rowlands P 2007 From Zero to Infinity: The Foundations of Physics (Series Knots and Everything) (Singapore: World Scientific)
[244] Ross D K 1989 Planck constant, torsion and spacetime defects, *Int. J. Theor. Phys* 28, 1333-1340.
[245] Rzepa HS. Möbius aromaticity and delocalization. *Chem Rev*, 2005, **105**: 3 697–3de 715
[246] Saxena K, Singh P, Sahoo P, et.al.. Fractal, scale free electromagnetic resonance of a single brain extracted microtubule nanowire, a single tubulin protein, and a single neuron. *Fractal Fract*. 2020;**4**(2):11.

[247] Scheinen S A et.al. 2018, Differentiating drug -related and state-related effects of Dexmedetomidine and propofol on the electroencelophalogram *Anaesthesiology* **129** (1),22-6 doi: 10.1097/ALN.0000000000002192
[248] Shanon B 2013 *The Antipodes of the Mind: Charting the Phenomenology of the Ayahuasca Experience* (Oxford: Oxford University Press)
[249] Shelton J 2016 Building a Möbius strip of good vibrations *Yale News* July 25.
[250] Shepard R 1981 Psychophysical Complementarity *Perceptual Organization*, eds M Kubovy and J Pomerantz (Mahwah, NJ, US: Lawrence Erlbaum Publishers); 2019, reprinted (London: Routledge)
[251] Shepard R 1964 Circularity in judgments of relative pitch. *J. Acous Soc Amer* **39**, 36:2346–2353.
[252] Shepard R 1982 Geometrical approximation of the structure of pitch space. *Psycholo. Rev.* **89**(4):305–333.
[253] Shepard R 2001 Perceptual-cognitive universals as reflections of the world *Behavioral and Brain Sciences* **24**, 581–601
[254] Singh P, Sahoo P, Saxena K, et.al. 2021 Cytoskeletal filaments deep inside a neuron are not silent: they regulate the precise timing of nerve spikes using a pair of vortices. *Symmetry* **13**(5):821
[255] Singh P, Saxena K, Sahoo P, et.al.2021 Electrophysiology using coaxial atom probe array: live imaging reveals hidden circuits of a hippocampal neuralnetwork. *J Neurophysiol.* **125**(6):2107–2116.
[256] Singh P, Sahoo P, Ghosh S, et.al.2021 Filaments and four ordered structures inside a neuron fire a thousand times faster than membrane: theory experiment. *J Integr Neurosci.* 2021c;20(4):777–790
[257] Skrbina D 2009 *Mind That Abides: Panpsychism in the New Millennium* (Amsterdam: Benjamin Cummins)
[258] Sokolov I V 1973 Topological Ideas in Stereochemistry *Russian Chem Rev* **42**: 6,452-463
[259] Solms J and Turnbull O 2017 The Brain and the Inner World: An Introduction to the Neurosciences of the Subjective Experience (London:Other/Karnac)
[260] Solms J 2019 The Hard Problem of Consciousness and the Free Energy Principle *Front. Psychol.,* 30 January Sec. Psychoanalysis and europsychoanalysis https://doi.org/10.3389/fpsyg.2018.02714
[261] Solms J 2017 What is “the unconscious,” and where is it located in the brain? A neuropsychoanalytic perspective *Ann. N.Y. Acad. Sci.* 1406 90–97
[262] Solms J 2021 *The Hidden Spring: A Journey to the Source of Consciousness* (New York; N N Norton Publishers)
[263] Stapien M, Latos-Grażyński L, Sprutta N,Chwalisz P and Szterenberg L 2007 Expanded Porphyrin with a Split Personality: A Hückel– Möbius Aromaticity Switch *Angewandte* Chemie **119**, 41: 7859-8047
[264] Sternberg S 1999 *Lectures on Differential Geometry* (Providence: AMS Pub)
[265] Stern A 2000 Quantum Theoretic Machines: What is Thought From the Point of View of Physics,(Amsterdam: Elsevier);ibid 1994 The Quantum BrainTheory and Implications (Amsterdam:North-Holland)
[266] Swindale N 1996. Visual cortex: Looking into a Klein Bottle. *Current Biol.,* **6** (7)776–779. DOI: 10.1016/s0960-9822(02)00592-4
[267] Taborsky E 1998 Architectonics of Semiosis (Semaphores and Signs) (London:Palmgrave Mac Millan)
[268] Tanaka S 1997.Topology of cortex visual maps, *Forma* **12**:101–108;
[269] Tanaka S 1995 Topological Analysis of Point Singularities in Stimulus Preference Maps of the Primary Visual Cortex *Proceedings Royal Soc: Biological Sciences* **261**, 1360, pp. 81-88
[270] Tanda S 2002 A Möbius strip of single *crystals Nature* 417(6887):379- DOI: 10.1038/417397
[271] Trewavas A 2017 The foundations of plant intelligence The foundations of plant intelligence *InterfaceFocus* 72016009820160098 http://doi.org/10.1098/rsfs.2016.0098

[272] Tripathi A and Bankaitis V A 2017 Molecular Docking: From Lock and Key to Combination Lock *J Mol Med Clin Appl.* **2**(1): 10.16966/2575-0305.106.
[273] Turvey M T 1992 Ecological foundations of cognition: Invariants of perception and action *Cognition: Conceptual and Methodological Issues,* Herbert et.al. eds (Washington, DC:Amer.Psychol. Assoc,), pp. 85-117
[274] Tymoczko D 2006 The Geometry of Musical Chords, *Science* **313**: 72-74; ibid. 2011, *A Geometry of Music* (Oxford University Press).
[275] Varela F J and Shear J 1999 The view from within: First-person approaches to the study of consciousness (London: Imprint Academic)
[276] Vargas J 2012 *Differential Forms for Cartan-Klein Geometry* (Suffolk, UK: Abramis)
[277] Vernooij E *et.al.* 2016 Listening to the Shepard–Risset Glissando: The relationship between emotional response, disruption of equilibrium and personality. *Front. Psychol.*, **7**:300.
[278] Vitiello G 2005 *My Double Unveiled: The dissipative quantum model of the brain* (New York, Benjamin Cummings)
[279] Voeikov V 2001 Reactive oxygen species, water, photons and life *Riv. Biol.* May-Aug;**94**(2):237-58. PMID: 11702650.
[280] Weiss H and Weiss V 2003 The golden mean as clock cycle of brain waves, *Chaos, Solitons and Fractals*, **18**(4):643-652 https://doi.org/10.1016/S0960-0779(03)00026-2
[281] Werner G and Whitsel B L 1968 Topology of the body representation in somatosensory area I of primates. *J. Neurophysiol.* **31**:856–86.
[282] Wittgenstein L 1922 *Tractatus Logico-Philosophicus* C.K. Ogden (trans.), (London: Routledge & Kegan Paul)
[283] Wright J J, Bourke P D, Favorov O V 2014Möbius-strip-like columnar functional connections are revealed in somato-sensory receptive field centroids *Front Neuroanat.* Oct 31;8:119. doi: 10.3389/fnana.2014.00119.
[284] Wright J J, Bourke P D. 2021The growth of cognition: Free energy minimization and embryogenesis of cortical computation *Phys Life Rev.* 2021 Mar;36:83-99. doi: 10.1016/j.plrev.2020.05.004. Epub 2020 Jun 1. PMID: 32527680.
[285] Woolf N J 2006 Microtubules in the cerebral cortex: Role in memory and consciousness *The* emerging physics of consciousness Tuszynski J A ed (Berlin:Springer) pp. 49–94.
[286] Wu Yong and Lin Yi 2002 Beyond Nonstructural Analysis: Blow-ups,Spinning Currents and the Modern Science (Singapore, World Scientific Publishers)
[287] Xu H, Mason D, Jiang J and Harris M 2016 Topological energy transfer in an optomechanical system with an exceptional point *Nat Com* **7**, 13551
[288] Berry M V 1984 Quantal Phase Factors Accompanying Adiabatic Changes *Proceedings of the Royal Society A*. **392** (1802): 45–57.
[289] Berry M V 1990 Anticipations of the geometric phase geometric phase *Physics Today* **43**(12): 340
[290] Pattee H 2007 The Necessity of Biosemiotics:Matter-Symbol Complementarity *Introduction to Biosemiotics*, M. Barbieri, Ed., Springer (Dordrecht; Springer) pp. 115-132
[291] Molinari S et.al. 2011 A 100-Parsec Elliptical and Twisted Ring of Cold and Dense Molecular Clouds Revealed by Herschel around theGalactic Centre *The Astrophysical Journal Letters* **735** 2), L33
[292] Bohm D 1986 Time, the implicate order, and pre-space Physics and the Ultimate Significance of Time (State University of New York Press)
[293] Bohm D and Hiley B 1993 The Undivided Universe: an Ontological Interpretation of Quantum Theory (London:Routledge)
[294] Johansen S E 1991 *Outline of Differential Epistemology*.Translation of the Norwegian edition, University of Trondheim, Trondheim, Norway; to appear
[295] Kauffman S 2000 *Investigations* (Oxford: Oxford University Press)
[296] *Heidegger* M 1972 *On Time and Being* Joan Stambaugh trans (New York: Harper and Row
[297] de Pablo P J, Schaap I, MacKintosh F and Schmidt C F 2003 Deformation and Collapse of Microtubules on the Nanometer Scale *Phys Rev Letters* **91**, 9, 098101 1

[298] Debever R 2015 *Elie Cartan and Albert Einstein: Letters on Absolut Parallelism, 1929-1932* (Princeton, NJ: Princeton University Press)
[299] Westerhoff J 2009 *Nagarjuna's Madhyamaka: A Philosophical Introduction* (Oxford University Press)
[300] Gibson J 1966 *The Senses Considered as Perceptual* Systems (Boston:Mifflin); ibid 1979 *The* Ecological Approach to Visual Perception (Boston: Houghton Mifflin)
[301] Velmans M 2009 *Understanding Consciousness* (London: Routledge)
[302] Günther G 1967 Time, Timeless Logic, Self-referential Systems, Ann. N.Y.Ac. Sciences **138**, 317-346; ibid 1962 Cybernetic Ontology and Transjunctional Operations *Selforganizing Systems*, M.C. Yovits, G.T. Jacobs, and G.D.Goldstein, (Spartan Books, Washington (D.C.)), (1962), 313-392
[303] Spencer Brown G 1969 *The Laws of Form* (London: Allen &Unwin)
[304] Hellerstein N S 2009 *Diamond: A Paradox* Logic (Singapore: World Scientific)
[305] Breault G, Hunter C and Mayers P 1999 Supramolecular topology *Tetrahedron* **55**, 5265-5293
[306] De Valois R and De Valois K 1990 *Spatial Vision* (Oxford University Press)
[307] Solms M and Friston 2018 M How and why consciousness arises: some considerations from Physics and Physiology Jour Consciousness Studies **25**, 202–238.
[308] Spinoza B *The Collected Writings of Spinoza*, 2 vols., Edwin Curley, translator, Princeton: Princeton University Press, vol. 1: 1985; vol. 2: 2016.
[309] Hughston L P 1996 Geometry of stochastic state vector reduction *Proceedings Royal Society A* 452, 1947 **https://doi.org/10.1098/rspa.1996.0048**
[310] Adler S and Horwitz L 2000 Structure and Properties of Hughston's Stochastic Extension of the Schroedinger Equation *Journal of Mathematical Physics* **41**, 248 DOI: 10.1063/1.533255
[311] Celliers P (2008) Complexity theory as a General Framework for sustainability science *Exploring* Sustainability Science: A Southern African Perspective M. Burns and A. Weaver (African Sun Media, SUN PreSS)
[312] Mc Gilchrist I 2021 *The Matter with Things*: *Our Brains, Our Delusions, and the Unmaking of the World*. (London: Perspectiva Press); ibid. 2009 *The Master and His Emissary*: *The* Divided Brain and the Making of Western World (USA:Yale Univ. Press)
[313] Bousso R 2002 The holographic principle *Reviews of Modern Physics*. **74** (3): 825–874.
[314] Bethea D 2009 *The Superstitious Muse* (Brighton, MASS: Academic Studies Press)
[315] Fitzhugh R 1955, Mathematical models of threshold phenomena in the nerve membrane, *Bull. Math. Biophysics*, **17**, 25
[316] Craig A D 2015 How do you feel? An interoceptive moment with your neurobiological self (Princeton, New Jersey, Oxford: Princeton University Press)
[317] Kirchhoff M, Parr T, Palacios E, Friston K and Kiverstein J 2018 The Markov blankets of life: autonomy, active inference and the free energy principle . *R. Soc. Interface* **15:** 2017079220170792 http://doi.org/10.1098/rsif.2017.0792
[318] Barbieri M 2007 *Introduction to Biosemiotics: The New Biological Synthesis* (Berlin:Springer Verlag)
[319] Piaget J and Inhelder B 1956 *The Child's Conception of Space* (London:Routledge)
[320] Merrick R 2011 *Interference: A Grand Scientific Musical Theory* (http://interferencetheory.com/files/Interference.pdf

AUTHOR INDEX

Institute of Physics
Dirac House, Temple Back
Bristol BS1 6BE UK

ISSN: 1742-6588
ISBN 978-1-7138-7453-9

9 781713 874539

CURRAN ASSOCIATES INC.
proceedings.com